MODERN CODING THEORY

Iterative techniques have revolutionized the theory and practice of coding and have been adopted in the majority of next-generation communications standards. *Modern Coding Theory* summarizes the state of the art in iterative coding, with particular emphasis on the underlying theory. Starting with Gallager's original ensemble of low-density parity-check codes as a representative example, focus is placed on the techniques to analyze and design practical iterative coding systems. The basic concepts are then extended for several general codes, including the practically important class of turbo codes. This book takes advantage of the simplicity of the binary erasure channel to develop analytical techniques and intuition, which are then applied to general channel models. A chapter on factor graphs helps to unify the important topics of information theory, coding, and communication theory.

Covering the most recent advances in the field, this book is a valuable resource for graduate students in electrical engineering and computer science, as well as practitioners who need to decide which coding scheme to employ, how to design a new scheme, or how to improve an existing system.

Additional resources, including instructor's solutions and figures, are available online: www.cambridge.org/9780521852296.

Tom Richardson received his Ph.D. in electrical engineering from Massachusetts Institute of Technology in 1990. For ten years, until 2000, he was a member of Bell Labs' Mathematical Sciences Research Center. In 2000, he joined Flarion Technologies, a wireless startup, which was acquired by Qualcomm, Inc. in 2006. He is currently an associate editor for the *IEEE Transactions on Information Theory.*

Rüdiger Urbanke is a professor in the School of Computer and Communication Sciences at EPFL, Switzerland. He was awarded his Ph.D. in electrical engineering in 1995 from Washington University, after which he worked at Bell Labs' Mathematical Sciences Research Center until joining the faculty at Ecole Polytechnique Fédérale de Lausanne in 1999.

Modern Coding Theory

TOM RICHARDSON
Qualcomm, Inc.

RÜDIGER URBANKE
Ecole Polytechnique Fédérale de Lausanne

CAMBRIDGE
UNIVERSITY PRESS

CAMBRIDGE
UNIVERSITY PRESS

University Printing House, Cambridge CB2 8BS, United Kingdom

Published in the United States of America by Cambridge University Press, New York

Cambridge University Press is part of the University of Cambridge.

It furthers the University's mission by disseminating knowledge in the pursuit of education, learning and research at the highest international levels of excellence.

www.cambridge.org
Information on this title: www.cambridge.org/9780521852296

First published 2008
Reprinted with corrections 2009

A catalogue record for this publication is available from the British Library

Library of Congress Cataloguing in Publication data

Richardson, Thomas J. (Thomas Joseph), 1961–
Modern coding theory / By T. Richardson and R. Urbanke.
p. cm.
Includes bibliographical references and index.
ISBN 978-0-521-85229-6 (hardback)
1. Coding theory. I. Urbanke, R. (Rüdiger), 1966– II. Title.
QA268.R53 2008
003'.54–dc22 2007039544

ISBN 978-0-521-85229-6 Hardback

To Karen,
Grammel, and Mathilde

CONTENTS

PREFACE

This book is all about *iterative* channel decoding. Two other names which are often used to identify the same area are *probabilistic* coding and *codes on graphs*. Iterative decoding was originally conceived by Gallager in his remarkable Ph.D. thesis of 1960. Gallager's work was, evidently, far ahead of its time. Limitations in computational resources in the 1960s were such that the power of his approach could not be fully demonstrated, let alone developed. Consequently, iterative decoding attracted only passing interest and slipped into a long dormancy. It was rediscovered by Berrou, Glavieux, and Thitimajshima in 1993 in the form of turbo codes, and then independently in the mid 1990s by MacKay and Neal, Sipser and Spielman, as well as Luby, Mitzenmacher, Shokrollahi, Spielman, and Stemann in a form much closer to Gallager's original construction. Iterative techniques have subsequently had a strong impact on coding theory and practice and, more generally, on the whole of communications.

The title *Modern Coding Theory* is clearly a hyperbole. There have been several other important recent developments in coding theory. To mention one prominent example: Sudan's algorithm and the Guruswami-Sudan improvement for list decoding of Reed-Solomon codes and their extension to soft-decision decoding have sparked new life into this otherwise mature subject. So what is our excuse? Iterative methods and their theory are strongly tied to advances in current computing technology and they are therefore inherently modern. They have also brought about a break with the past. Moreover, the techniques are influencing a wide range of applications within and beyond communications, connecting that area with many modern topics in, among others, statistical mechanics and complexity theory. Nevertheless, the font on the book cover expresses the irony that the roots of "modern" coding go back to a time when typewriters ruled the world.

The field of iterative decoding has not settled in the same way that classical coding has. There are nearly as many flavors of iterative decoding systems – and graphical models to represent them – as there are researchers in the field. We have therefore decided to focus more on techniques to analyze and design such systems rather than on specific instances. In order to present the theory, we have elected Gallager's original ensemble of low-density parity-check (LDPC) codes as a representative example. This ensemble is perhaps the most elegant example and it provides a framework within which the main results can be presented easily. Once the basic concepts are absorbed, their extensions to more general cases is typically routine and several such extensions (but not an exhaustive list) are discussed. In particular, we have included a thorough investigation of turbo codes.

A noticeable feature of this book is that we spend a considerable number of pages discussing iterative decoding over the binary erasure channel. Why spend so much time on a very specific and limited channel model? It is probably fair to say that what we now know about iterative decoding we learned first for the binary erasure channel. The basic analysis of iterative coding in the context of the binary erasure channel needs little more than pen and paper and some knowledge of calculus and probability. Nearly all important concepts developed during the study of the binary erasure channel carry over to general channels, although our current ability to extend the results is, in some cases, frustrated by technical challenges.

This book is written with several audiences in mind. First, we hope that it will be a useful text for a course in coding theory. If such a course is dedicated solely to iterative techniques, most necessary material should be contained in this book. If the course covers both classical algebraic coding and iterative topics, this book can be used in conjunction with one of the many excellent books on classical coding. We have intentionally excluded virtually all classical material, except for the most basic definitions. Second, we hope that this book will also be of use to the practitioner in the field who is trying to design or choose a coding scheme for a new communication system or to improve an existing system. Third, we hope that the book will serve as a useful reference for researchers in the field.

There are many possible paths through this book. Our own personal preference is to start with the chapter on factor graphs (Chapter 2). The material covered in this chapter has the special appeal that it unifies many themes of information theory, coding, and communication. Although all three areas trace their origin to Shannon's 1948 paper, they have subsequently diverged and specialized to a point where a typical textbook in one area treats each of the other two topics as distant cousins and gives them just passing reference. The factor graph approach is a nice way to glue them back together. The same technique allows for the computation of capacity, and deals with equalization, modulation, and coding on an equal footing. Following Chapter 2, we recommend covering the core of the material in Chapter 3 (binary erasure channel) and Chapter 4 (general binary memoryless symmetric channels) in a linear fashion.

The remaining material can be read in almost any order according to the preferences of the reader. One may choose to broaden the view and to go through some of the material on more general channels (Chapter 5). Alternatively, you might be more interested in general ensembles. Chapter 6 discusses turbo codes and Chapter 7 deals with various further ensembles and some issues of graph design.

Chapter 8 gives a brief look at a complementary way of analyzing iterative systems in terms of the expansion of the underlying bipartite graph. These techniques are usually aimed at proving guaranteed error-correcting capability and are usually not capable of predicting typical error-correcting performance.

The Appendices contain various chapters on topics which either describe tools for analysis or are simply too technical to fit into the main part. Appendix A takes a look at the encoding problem. Curiously, for iterative schemes the encoding task can be of equal (or even higher) complexity than the decoding task. Appendix B discusses efficient and accurate ways of implementing density evolution. In Appendix C we describe various techniques from probability which are useful in asserting that most elements of a properly chosen ensemble behave "close" to the ensemble average. We take a close look at generating functions in Appendix D. In particular we discuss how to accurately estimate the coefficients of powers of polynomials – a recurrent theme in this book. Finally, in Appendix E we collected a few proofs deemed too lengthy to include in the main text.

Although we have tried to make the material as accessible as possible, the prerequisites for different portions of the book vary considerably. Some seemingly simple issues require sophisticated tools for their resolution. A good example is the material related to the weight distribution of LDPC codes. When the density of equations increases to a painful level, the casual reader is advised not to get discouraged but rather to skip the proofs. Fortunately, in all these cases the subsequent material depends very little on the mathematical details of the proof.

If you are a lecturer and you are giving a beginning graduate-level course we recommend that you follow the basic course outlined above but skip some of the less accessible topics. For general binary memoryless symmetric channels one can first focus on Gallager's decoding algorithm A. The analysis for this case is very similar to the one for the binary erasure channel. A subsequent discussion of the belief propagation decoder can skip some of the proofs and so avoid a discussion of some of the technical difficulties. If your course is positioned as an advanced graduate-level course then most of the material should be accessible to the students.

We intended to write a thin book containing all there is to know about iterative decoding. We ended up with a rather thick one with a number of regrettable omissions: We do not cover the emerging theory of pseudo codewords and their connections to the error floor for general channels and we only scratched the surface of the rich area of interleaver design. The theory of rateless codes is deserving of a much more detailed look. We have not discussed the powerful techniques borrowed from statistical mechanics, which have been used successfully in the analysis of iterative systems. Finally, we mention, but do not discuss, source coding by iterative techniques.

Even within the topics we have covered many interesting extensions and details have been set aside and not been included. For these shortcomings, to paraphrase Descartes, "[We] hope that posterity will judge [us] kindly, not only as to the things which [we] have explained, but also as to those which [we] have intentionally omitted so as to leave to others the pleasure of discovery." ;-)

We have received much help over the years. We would like to thank A. Ahmed, A. Amraoui, B. Bauer, J. Boutros, S.-Y. Chung, H. Cronie, C. Di, N. Dütsch, J. Ezri, T. Filler, W. H. Fong, M. P. C. Fossorier, A. Guillen i Fabregas, M. Haenggi, T. Hehn, D. Huang, A. Karbasi, B. Konsbruck, S. Korada, S. Kudekar, L. Gong, N. Macris, A. Orlitsky, H. D. Pfister, D. Porrat, V. Rathi, K. S. Reddy, V. Skachek, A. Shokrollahi, D. A. Spielman, I. Tal, A. J. van Wijngaarden, L. Varshney, P. O. Vontobel, X. Wang, G. Wiechman, and L. Zhichu for providing us with feedback, and we apologize to all of those whom we missed.

Special thanks go to A. Barg, C. Berrou, M. Durvy, G. D. Forney, Jr., G. Kramer, D. J. C. MacKay, C. Méasson, S. S. Pietrobon, B. Rimoldi, D. Saad, E. Telatar, and Y. Yu for their extensive reviews and the large number of suggestions they provided. Probably nobody read the initial manuscript more carefully and provided us with more feedback than Igal Sason. We are very thankful to him for this invaluable help.

A. Chebira, J. Ezri, A. Gueye, T. Ktari, C. Méasson, C. Neuberg, and P. Reymond were of tremendous help in producing the figures and simulations. Thank you for all the work you did.

A considerable portion of this book is the direct result of our various collaborations. We enjoyed working with A. Amraoui, L. Bazzi, S.-Y. Chung, C. Di, S. Dusad, J. Ezri, G. D. Forney, Jr., H. Jin, N. Kahale, N. Macris, C. Méasson, A. Montanari, V. Novichkov, H. D. Pfister, D. Proietti, V. Rathi, A. Shokrollahi, I. Sason, and E. Telatar on many topics related to this book.

Nobody has contributed more to the realization of this book than E. Telatar (a.k.a. "Emre the Wise"). He hand-picked the font (MinionPro), designed the layout (using the "memoir" package by P. Wilson), showed us how to program figures in PostScript by hand, was our last resort for any LaTeX questions, and generously provided his expertise and advice in many other areas.

This book would not have been finished without the constant encouragement by P. Meyler at Cambridge. We thank A. Littlewood at Cambridge and P. Rote at Aptara for their expert handling.

Last but not least, M. Bardet has managed to keep the chaos at bay at EPFL despite RU's best efforts to the contrary. I would like to thank her for this miraculous accomplishment.

The idea for this book was born at the end of the last millennium when we were both happy members of the Mathematics of Communications group at Bell Labs headed by the late A. D. Wyner and then by J. Mazo. We would like to thank both of them for giving us the freedom to pursue our ideas. Since our departure from Bell Labs, EPFL and Flarion Technologies/Qualcomm have been our hospitable homes.

T. Richardson — R. Urbanke
South Orange, NJ — Lausanne, Switzerland
November, 2007

Chapter 1

INTRODUCTION

§1.1. Why You Should Read This Book

The technology of communication and computing advanced at a breathtaking pace in the 20th century, especially in the second half. A significant part of this advance in communication began some 60 years ago when Shannon published his seminal paper "A Mathematical Theory of Communication." In that paper Shannon framed and posed a fundamental question: how can we efficiently and reliably transmit information? Shannon also gave a basic answer: coding can do it. Since that time the problem of finding practical coding schemes that approach the fundamental limits established by Shannon has been at the heart of information theory and communications. Recently, significant advances have taken place that bring us close to answering this question. Perhaps, at least in a practical sense, the question has been answered. This book is about that answer.

The advance came with a fundamental paradigm shift in the area of coding that took place in the early 1990s. In *Modern Coding Theory*, codes are viewed as large complex systems described by *random sparse graphical models*, and encoding as well as decoding are accomplished by efficient *local* algorithms. The local interactions of the codebits are simple but the overall code is nevertheless complex (and so sufficiently powerful to allow reliable communication) because of the large number of interactions. The idea of random codes is in the spirit of Shannon's original formulation. What is new is the sparseness of the description and the local nature of the algorithms.

These are exciting times for coding theorists and practitioners. Despite all the progress made, many fundamental questions are still open. Even if you are not interested in coding itself, however, you might be motivated to read this book. Although the focus of this book is squarely on coding, the larger view holds a much bigger picture. Sparse graphical models and message-passing algorithms, to name just two of the notions that are fundamental to our treatment, play an increasingly important role in many other fields as well. This is not a coincidence. Many of the innovations were brought into the field of coding by physicists or computer scientists. Conversely, the success of modern coding has inspired work in several other fields.

Modern coding will not displace classical coding anytime soon. At any point in time hundreds of millions of Reed-Solomon codes work hard to make your life less error prone. This is unlikely to change substantially in the near future. But modern coding offers an alternative way of solving the communications problem. Most current wireless communications systems have already adopted modern coding.

Technically, our aim is focused on Shannon's classical problem: we want to transmit a *message* across a *noisy channel* so that the *receiver* can determine this message with *high probability* despite the imperfections of the channel. We are interested in *low-complexity* schemes that introduce *little delay* and allow *reliable* transmission close to the ultimate limit, the *Shannon capacity*.

We start with a review of the communications problem (Section 1.2), we cover some classical notions of codes (Sections 1.3, 1.4, 1.5, 1.7, and 1.8), and we review the channel coding theorem (Section 1.6). Section 1.9 gives an outline of the modern approach to coding. Finally, we close in Section 1.10 with a review of the notational conventions and some useful facts.

§1.2. Communications Problem

Consider the following communications scenario – the *point-to-point* communications problem depicted in Figure 1.1. A *source* transmits its information (speech, au-

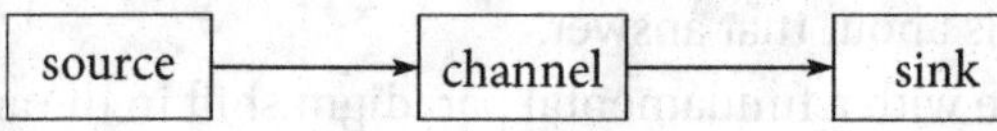

Figure 1.1: Basic point-to-point communications problem.

dio, data, etc.) via a noisy channel (phone line, optical link, wireless, storage medium, etc.) to a *sink*. We are interested in reliable transmission, i.e., we want to recreate the transmitted information with as little *distortion* (number of wrong bits, mean squared error distortion, etc.) as possible at the sink.

In his seminal paper in 1948, Shannon formalized the communications problem and showed that the point-to-point problem can be decomposed into two separate problems as shown in Figure 1.2. First, a *source encoder* transforms the source into a bit stream. Ideally, the source encoder removes all redundancy from the source so that the resulting bit stream has the smallest possible number of bits while still representing the source with enough accuracy. The *channel encoder* then processes the bit stream to add redundancy. This redundancy is carefully chosen to combat the noise that is introduced by the channel.

To be mathematically more precise: we model the output of the source as a stochastic process. For example, we might represent text as the output of a Markov chain, describing the local dependency structure of letter sequences. It is the task of the source encoder to represent this output as efficiently as possible (using as few bits as possible) given a desired distortion. The *distortion measure* reflects the "cost" of deviating from the original source output. If the source emits points in $\mathbb{R}^n$ it might

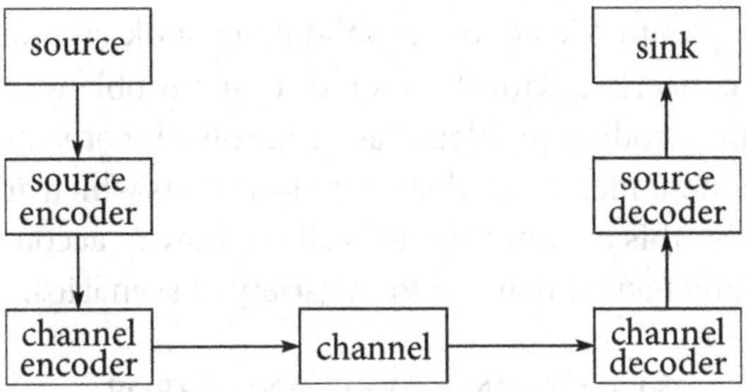

Figure 1.2: Basic point-to-point communications problem in view of the source-channel separation theorem.

be natural to consider the squared Euclidean distance, whereas if the source emits binary strings a more natural measure might be to count the number of positions in which the source output and the word that can be reconstructed from the encoded source differ. Shannon's *source coding theorem* asserts that, for a given source and distortion measure, there exists a minimum rate $R = R(d)$ (bits per emitted source symbol) which is necessary (and sufficient) to describe this source with distortion not exceeding d. The plot of this rate R as a function of the distortion d is usually called the *rate-distortion* curve. In the second stage an appropriate amount of *redundancy* is added to these source bits to protect them against the errors in the channel. This process is called *channel coding*. Throughout the book we model the channel as a probabilistic mapping and we are typically interested in the *average* performance, where the average is taken over all channel realizations. Shannon's *channel coding theorem* asserts the existence of a maximum rate (bits per channel use) at which information can be transmitted reliably, i.e., with vanishing probability of error, over a given channel. This maximum rate is called the *capacity* of the channel and is denoted by C. At the receiver we first decode the received bits to determine the transmitted information. We then use the decoded bits to reconstruct the source at the receiver. Shannon's *source-channel separation* theorem asserts that the source can be reconstructed with a distortion of at most d at the receiver if $R(d) < C$, i.e., if the rate required to represent the given source with the allowed distortion is smaller than the capacity of the channel. Conversely, no scheme can do better. One great benefit of the separation theorem is that a communications link can be used for a large variety of sources: one good channel coding solution can be used with any source. Virtually all systems in use today are based on this principle. It is important though to be aware of the limitations of the source-channel separation theorem. The optimality is only in terms of the achievable distortion when large blocks of data are encoded together. Joint schemes can be substantially better in terms of complexity or delay. Also, the separation is no longer valid if one looks at multi-user scenarios.

We will not be concerned with the source coding problem or, equivalently, we assume that the source coding problem has been solved. For us, the source emits a sequence of independent identically distributed (iid) bits which are equally likely to be zero or one. Under this assumption, we will see how to accomplish the channel coding problem in an efficient manner for a variety of scenarios.

§1.3. Coding: Trial and Error

How can we transmit information reliably over a noisy channel at a strictly positive rate? At some level we have already given the answer: add redundancy to the message that can be exploited to combat the distortion introduced by the channel. By starting with a special case we want to clarify the key concepts.

Example 1.3 (Binary Symmetric Channel). Consider the *binary symmetric channel* with *cross-over probability* ϵ depicted in Figure 1.4. We denote it by BSC(ϵ). Both

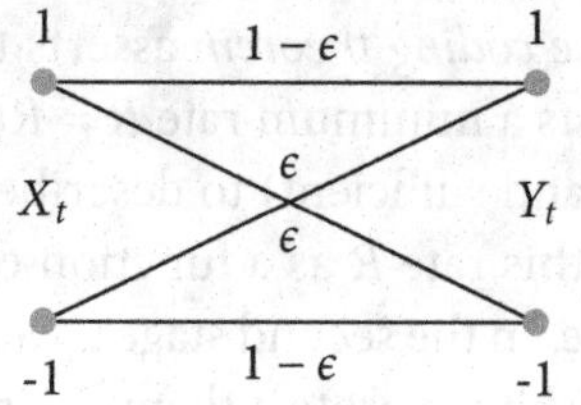

Figure 1.4: BSC(ϵ).

input X_t and output Y_t are elements of $\{\pm 1\}$. A transmitted bit is either received correctly or received *flipped*, the latter occurring with probability ϵ, and different bits are flipped or not flipped independently. We can assume that $0 < \epsilon < \frac{1}{2}$ without loss of generality.

The BSC is the generic model of a binary-input memoryless channel in which hard decisions are made at the front end of the receiver, i.e., where the received value is quantized to two values. ◇

First Trial: Suppose that the transmitted bits are independent and that $\mathbb{P}\{X_t = +1\} = \mathbb{P}\{X_t = -1\} = \frac{1}{2}$. We start by considering *uncoded* transmission over the BSC(ϵ). Thus, we send the source bits across the channel as is, without the insertion of redundant bits. At the receiver we estimate the transmitted bit X based on the observation Y. As we will learn in Section 1.5, the decision rule that minimizes the bit-error probability, call it $\hat{x}^{\text{MAP}}(y)$, is to choose that element of $\{\pm 1\}$ which maximizes $p_{X|Y}(x|y)$ for the given y. Since the prior on X is uniform, an application of Bayes's rule shows that this is equivalent to maximizing $p_{Y|X}(y|x)$ for the given

y. Since $\epsilon < \frac{1}{2}$ we conclude that the optimal estimator is $\hat{x}^{\text{MAP}}(y) = y$. The probability that the estimate differs from the true value, i.e., $P_b = \mathbb{P}\{\hat{x}^{\text{MAP}}(Y) \neq X\}$, is equal to ϵ. Since for every information bit we want to convey we send exactly one bit over the channel we say that this scheme has *rate* 1. We conclude that with uncoded transmission we can achieve a (rate, P_b)-pair of $(1, \epsilon)$.

Second Trial: If the error probability ϵ is too high for our application, what transmission strategy can we use to lower it? The simplest strategy is *repetition coding*. Assume we repeat each bit k times. To keep things simple, assume that k is odd. So if X, the bit to be transmitted, has value x then the input to the BSC(ϵ) is the k-tuple $x, \dots, x$. Denote the k associated observations by $Y_1, \dots, Y_k$. It is intuitive, and not hard to prove, that the estimator that minimizes the bit-error probability is given by the majority rule

$$\hat{x}^{\text{MAP}}(y_1, \dots, y_k) = \text{majority of } \{y_1, \dots, y_k\}.$$

Hence the probability of bit error is given by

$$P_b = \mathbb{P}\left\{\hat{x}^{\text{MAP}}(Y) \neq X\right\} \stackrel{k \text{ odd}}{=} \mathbb{P}\{\text{at least } \lceil k/2 \rceil \text{ errors occur}\} = \sum_{i > k/2} \binom{k}{i} \epsilon^i (1-\epsilon)^{k-i}.$$

Since for every information bit we want to convey we send k bits over the channel we say that such a scheme has rate $\frac{1}{k}$. So with repetition codes we can achieve the (rate, P_b)-pairs $(\frac{1}{k}, \sum_{i>k/2} \binom{k}{i} \epsilon^i (1-\epsilon)^{k-i})$. For P_b to approach zero we have to choose k larger and larger and as a consequence the rate approaches zero as well.

Can we keep the rate positive and make the error probability go to zero?

§1.4. Codes and Ensembles

Information is inherently discrete. It is natural and convenient to use *finite* fields to represent it. The most important instance for us is the *binary field* $\mathbb{F}_2$, consisting of $\{0, 1\}$ with mod-2 addition and mod-2 multiplication ($0 + 0 = 1 + 1 = 0$; $0 + 1 = 1$; $0 \cdot 0 = 1 \cdot 0 = 0$; $1 \cdot 1 = 1$). In words, if we use $\mathbb{F}_2$ then we represent information in terms of (sequences of) *bits*, a natural representation and convenient for the purpose of processing. If you are not familiar with finite fields, very little is lost if you replace any mention of a generic finite field $\mathbb{F}$ with $\mathbb{F}_2$. We write $|\mathbb{F}|$ to indicate the number of elements of the finite field $\mathbb{F}$, e.g., $|\mathbb{F}_2| = 2$. Why do we choose finite *fields*? As we will see, by using algebraic operations in both the encoding as well as the decoding we can significantly reduce the complexity.

Definition 1.5 (Code). A *code* C of *length* n and *cardinality* M over a field $\mathbb{F}$ is a collection of M elements from $\mathbb{F}^n$, i.e.,

$$C(n, M) = \{x^{[1]}, \dots, x^{[M]}\}, \; x^{[m]} \in \mathbb{F}^n, 1 \leq m \leq M.$$

The elements of the code are called *codewords*. The parameter n is called the *block-length*. ▽

EXAMPLE 1.6 (REPETITION CODE). Let $\mathbb{F} = \mathbb{F}_2$. The binary *repetition code* of length 3 is defined as $C(n = 3, M = 2) = \{000, 111\}$. ◇

In the preceding example we have introduced *binary* codes, i.e., codes whose components are elements of $\mathbb{F}_2 = \{0, 1\}$. Sometimes it is more convenient to think of the two field elements as $\{\pm 1\}$ instead (see, e.g., the definition of the BSC in Example 1.3). The standard mapping is $0 \leftrightarrow 1$ and $1 \leftrightarrow -1$. It is convenient to use both notations. We freely and frequently switch. With some abuse of notation, we make no distinction between these two cases and talk about binary codes and $\mathbb{F}_2$ even if the components take values in $\{\pm 1\}$.

DEFINITION 1.7 (RATE). The *rate* of a code $C(n, M)$ is $r = \frac{1}{n} \log_{|\mathbb{F}|} M$. It is measured in *information symbols per transmitted symbol.* ▽

EXAMPLE 1.8 (REPETITION CODE). Let $\mathbb{F} = \mathbb{F}_2$. We have $r(C(3, 2)) = \frac{1}{3} \log_2 2 = \frac{1}{3}$. It takes three channel symbols to transmit one information symbol. ◇

The following two definitions play a role only much later in the book, but it is convenient to collect them here for reference.

DEFINITION 1.9 (SUPPORT SET). The *support set* of a codeword $x \in C$ is the set of locations $i \in [n] = \{1, \ldots, n\}$ such that $x_i \neq 0$. ▽

DEFINITION 1.10 (MINIMAL CODEWORDS). Consider a *binary* code C, i.e., a code over $\mathbb{F}_2$. We say that a codeword $x \in C$ is *minimal* if its support set does not contain the support set of any other (non-zero) codeword. ▽

The Hamming distance introduced in the following definition and the derived minimum distance of a code (see Definition 1.12) are *the* central characters in all of classical coding. For us they only play a minor role. This is probably one of the most distinguishing factors between classical and modern coding.

DEFINITION 1.11 (HAMMING WEIGHT AND HAMMING DISTANCE). Let $u, v \in \mathbb{F}^n$. The *Hamming weight* of a word u, which we denote by $w(u)$, is equal to the number of non-zero symbols in u, i.e., the cardinality of the support set. The *Hamming distance* of a pair (u, v), which we denote by $d(u, v)$, is the number of positions in which u differs from v. We have $d(u, v) = d(u - v, 0) = w(u - v)$. Further, $d(u, v) = d(v, u)$ and $d(u, v) \geq 0$, with equality if and only if $u = v$. Also, $d(\cdot, \cdot)$ satisfies the *triangle inequality*

$$d(u, v) \leq d(u, t) + d(t, v),$$

for any triple $u, v, t \in \mathbb{F}^n$. In words, $d(\cdot, \cdot)$ is a true *distance* in the mathematical sense (see Problem 1.2). ▽

Definition 1.12 (Minimum Distance of a Code). Let C be a code. Its *minimum distance* $d(C)$ is defined as

$$d(C) = \min\{d(u,v) : u, v \in C, u \neq v\}. \qquad \triangledown$$

Let $x \in \mathbb{F}^n$ and $t \in \mathbb{N}$. A *sphere* of radius t centered at the point x is the set of all points in $\mathbb{F}^n$ that have distance at most t from x. If, for a code C of minimum distance d, we place spheres of radius $t = \lfloor \frac{d-1}{2} \rfloor$ around each codeword, then these spheres are disjoint. This follows from the triangle inequality: if u and v are codewords and x is any element in $\mathbb{F}^n$ then $d(C) \leq d(u,v) \leq d(u,x) + d(v,x)$. If x is in the sphere of radius t around u then this implies that $d(v,x) \geq d(C) - d(u,x) \geq \frac{d+1}{2} > t$. In words, x is not in the sphere of radius t around v. Further, by definition of d, t is the largest such radius.

The radius t has an important operational meaning that explains why much of classical coding is centered on the construction of codes with large minimum distance. To be concrete, consider the binary case. Assume we use a code $C(n, M, d)$ (i.e., a code with M codewords of length n and minimum distance d) for transmission over a BSC and assume that we employ a *bounded distance* decoder with decoding radius t, $t \leq \lfloor \frac{d-1}{2} \rfloor$. More precisely, given y the decoder chooses $\hat{x}^{\text{BD}}(y)$ defined by

$$\hat{x}^{\text{BD}}(y) = \begin{cases} x \in C, & \text{if } d(x,y) \leq t, \\ \text{error}, & \text{if no such } x \text{ exists,} \end{cases}$$

where by "error" the decoder declares that it is unable to decode. As we have just discussed, there can be *at most one* $x \in C$ so that $d(x,y) \leq t$. Therefore, if the weight of the error does not exceed t, then such a combination finds the correct transmitted word. A large t hence implies a large resilience against channel errors.

How large can d (and hence t) be made in the binary case? Let $\delta = d/n$ denote the *normalized* distance and consider for a fixed rate r, $0 < r < 1$,

$$\delta^*(r) = \limsup_{n\to\infty} \max\left\{\frac{d(C)}{n} : C \in \mathcal{C}\left(n, 2^{\lfloor nr \rfloor}\right)\right\},$$

where $\mathcal{C}\left(n, 2^{\lfloor nr \rfloor}\right)$ denotes the set of all binary block codes of length n containing at least $2^{\lfloor nr \rfloor}$ codewords. Problem 1.15 discusses the asymptotic *Gilbert-Varshamov* bound

$$h_2^{-1}(1-r) \leq \delta^*(r),$$

where $h_2(x) = -x\log_2 x - (1-x)\log_2(1-x)$ is the *binary entropy function* and where for $y \in [0,1]$, $h_2^{-1}(y)$ is the unique element $x \in [0, \frac{1}{2}]$ such that $h_2(x) = y$. *Elias* introduced the following upper bound,

$$\delta^*(r) \leq 2h_2^{-1}(1-r)(1-h_2^{-1}(1-r)). \tag{1.13}$$

Both bounds are illustrated in Figure 1.14. We can now answer the question posed

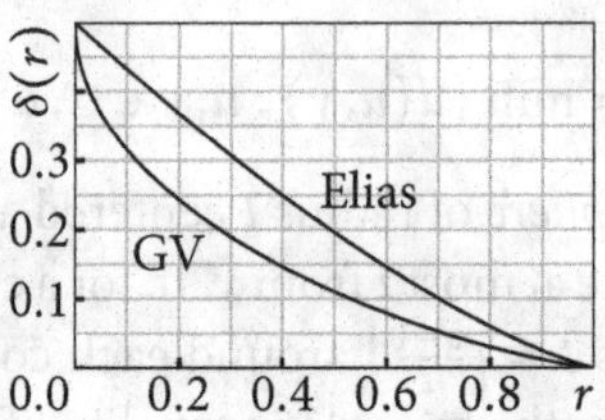

Figure 1.14: Upper and lower bound on $\delta^*(r)$.

at the end of the previous section. For a fixed channel BSC(ϵ) pick a rate r such that $\delta^*(r) > 2\epsilon + \omega$, where ω is some arbitrarily small but strictly positive quantity. We see from the Gilbert-Varshamov bound that such a strictly positive r and ω exist if $\epsilon < 1/4$. By the definition of δ^*, we can find a code of rate r of arbitrarily large blocklength n which has a relative minimum distance at least $\delta = 2\epsilon + \omega$. By Chebyshev's inequality (see Lemma C.3 on page 480), for every positive probability P bounded away from 1 there exists a positive constant c such that the number of channel flips in a block of length n is at most $n\epsilon + c\sqrt{n}$ with probability P. Assume that we employ a bounded distance decoder. If we choose n sufficiently large so that $n\epsilon + c\sqrt{n} < \delta n/2 = n\epsilon + n\omega/2$, then the bounded distance decoder succeeds with probability at least P. Since P can be chosen arbitrarily close to 1 we see that there exist codes that allow transmission at a positive rate with arbitrarily small positive probability of error. The above procedure is by no means optimal and does not allow us to determine up to what rates reliable transmission is possible. We will see in Section 1.6 how we can characterize the *largest* such rate.

Constructing provably good codes is difficult. A standard approach to show the *existence* of good codes is the probabilistic method: an *ensemble* $\mathcal{C}$ of codes is "constructed" using some random process and one proves that good codes occur with positive probability within this ensemble. Often the probability is close to 1 – almost all codes are good. This approach, used already by Shannon in his 1948 landmark paper, simplifies the code "construction" task enormously (at the cost of a less useful result).

Definition 1.15 (Shannon's Random Ensemble). Let the field $\mathbb{F}$ be fixed. Consider the following ensemble $\mathcal{C}(n, M)$ of codes of length n and cardinality M. There are nM degrees of freedom in choosing a code, one degree of freedom for each component of each codeword. The ensemble consists of all $|\mathbb{F}|^{nM}$ possible codes of length n and cardinality M. We endow this set with a uniform probability distribution. To sample from this ensemble proceed as follows. Pick the codewords

$x^{[1]}, \dots, x^{[M]}$ randomly by letting each component $x_i^{[m]}$ be an independently and uniformly chosen element of $\mathbb{F}$. ▽

We will see that such a code is likely to be "good" for many channels.

§1.5. MAP and ML Decoding and APP Processing

Assume we transmit over a channel with input $\mathbb{F}$ and output space $\mathcal{Y}$ using a code $C(n,M) = \{x^{[1]}, \dots, x^{[M]}\}$. Let the channel be specified by its transition probability $p_{Y\,|\,X}(y\,|\,x)$. The transmitter chooses the codeword $X \in C(n,M)$ with probability $p_X(x)$. (In communications the idea is that the transmitter wants to transmit one of M messages and uses one codeword for each possible message.) This codeword is then transmitted over the channel. Let Y denote the observation at the output of the channel. To what codeword should Y be decoded? If we decode Y to $\hat{x}(Y) \in C$, then the probability that we have made an error is $1 - p_{X\,|\,Y}(\hat{x}(Y)\,|\,y)$. Thus, to minimize the probability of block error we should choose $\hat{x}(Y)$ to maximize $p_{X\,|\,Y}(\hat{x}(Y)\,|\,y)$. The *maximum a posteriori* (MAP) decoding rule reads

$$\begin{aligned}
\hat{x}^{\text{MAP}}(y) &= \operatorname{argmax}_{x \in C} p_{X\,|\,Y}(x\,|\,y) \\
\text{by Bayes's rule} \qquad &= \operatorname{argmax}_{x \in C} p_{Y\,|\,X}(y\,|\,x) \frac{p_X(x)}{p_Y(y)} \\
&= \operatorname{argmax}_{x \in C} p_{Y\,|\,X}(y\,|\,x) p_X(x).
\end{aligned}$$

Ties can be broken in some arbitrary manner without affecting the error probability. As we indicated, this estimator minimizes the probability of (block) error $P_B = \mathbb{P}\{\hat{x}^{\text{MAP}}(Y) \neq X\}$. If all codewords are equally likely, i.e., if p_X is uniform, then

$$\hat{x}^{\text{MAP}}(y) = \operatorname{argmax}_{x \in C} p_{Y\,|\,X}(y\,|\,x) p_X(x) = \operatorname{argmax}_{x \in C} p_{Y\,|\,X}(y\,|\,x) = \hat{x}^{\text{ML}}(y),$$

where the right-hand side represents the decoding rule of the *maximum likelihood* (ML) decoder. In words, for a uniform prior p_X the MAP and the ML decoders are equivalent.

The key step in the MAP decoding process is to compute the *a posteriori probability* (APP) $p_{X\,|\,Y}(x\,|\,y)$, i.e., the distribution of X given the observation Y. So we call a MAP decoder also an APP decoder. Also, we will say that we perform *APP processing* to mean that we compute the a posteriori probabilities.

§1.6. Channel Coding Theorem

We have already seen that transmission at a strictly positive rate and an arbitrarily small positive probability of error is possible. What is the *largest* rate at which we can achieve a vanishing probability of error? Let us now investigate this question for transmission over the BSC.

We are interested in the scenario depicted in Figure 1.16. For a given binary code $C(n, M)$ the transmitter chooses with uniform probability a codeword $X \in C(n, M)$ and transmits this codeword over the channel $\text{BSC}(\epsilon)$. The output of the channel is denoted by Y. At the receiver the decoder estimates the transmitted codeword given the observation Y using the MAP rule $\hat{x}^{\text{MAP}}(y)$. How small can we make the incurred *block error probability* $\text{P}_{\text{B}}^{\text{MAP}}(C, \epsilon) = \mathbb{P}\{\hat{x}^{\text{MAP}}(Y) \neq X\}$ for given parameters n and M? Let $\hat{\text{P}}_{\text{B}}^{\text{MAP}}(n, M, \epsilon)$ be the minimum of $\text{P}_{\text{B}}^{\text{MAP}}(C, \epsilon)$ over all choices of $C \in \mathcal{C}(n, M)$.

Figure 1.16: Transmission over the $\text{BSC}(\epsilon)$.

Theorem 1.17 (Shannon's Channel Coding Theorem). If $0 < r < 1 - h_2(\epsilon)$ then $\hat{\text{P}}_{\text{B}}^{\text{MAP}}(n, 2^{\lfloor rn \rfloor}, \epsilon) \overset{n\to\infty}{\longrightarrow} 0$.

Proof. Pick a code C from Shannon's random ensemble $\mathcal{C}(n, 2^{\lfloor rn \rfloor})$ introduced in Definition 1.15. Since the MAP decoder is hard to analyze we use the following suboptimal decoder. For some fixed Δ, $\Delta > 0$, define $\rho = n\epsilon + \sqrt{2n\epsilon(1-\epsilon)/\Delta}$. If $x^{[m]}$ is the only codeword such that $d(y, x^{[m]}) \leq \rho$ then decode y as $x^{[m]}$ – otherwise declare an error.

For $u, v \in \{\pm 1\}^n$ let

$$f(u, v) = \begin{cases} 0, & \text{if } d(u, v) > \rho, \\ 1, & \text{if } d(u, v) \leq \rho, \end{cases}$$

and define

$$g^{[m]}(y) = 1 - f(x^{[m]}, y) + \sum_{m' \neq m} f(x^{[m']}, y).$$

Note that $g^{[m]}(y)$ equals zero if $x^{[m]}$ is the only codeword such that $d(y, x^{[m]}) \leq \rho$ and that it is at least one otherwise. Let $\text{P}_{\text{B}}^{[m]}$ denote the conditional block error probability assuming that $X = x^{[m]}$, i.e., $\text{P}_{\text{B}}^{[m]} = \mathbb{P}\{\hat{x}(Y) \neq X \mid X = x^{[m]}\}$. We have

$$\begin{aligned} \text{P}_{\text{B}}^{[m]}(C, \epsilon) &= \sum_{y: g^{[m]}(y) \geq 1} p_{Y \mid X^{[m]}}(y \mid x^{[m]}) \leq \sum_{y \in \{\pm 1\}^n} p_{Y \mid X^{[m]}}(y \mid x^{[m]}) g^{[m]}(y) \\ &= \sum_{y \in \{\pm 1\}^n} p_{Y \mid X^{[m]}}(y \mid x^{[m]}) [1 - f(x^{[m]}, y)] \end{aligned}$$

$$+ \sum_{y \in \{\pm 1\}^n} \sum_{m' \neq m} p_{Y \mid X^{[m]}}(y \mid x^{[m]}) f(x^{[m']}, y)$$
$$= \mathbb{P}\left\{d(Y, x^{[m]}) > \rho \mid X^{[m]} = x^{[m]}\right\}$$
$$+ \sum_{y \in \{\pm 1\}^n} \sum_{m' \neq m} p_{Y \mid X^{[m]}}(y \mid x^{[m]}) f(x^{[m']}, y).$$

Note that $d(y, x^{[m]}) = w(y + x^{[m]})$, where $y + x^{[m]}$ is the vector of channel errors (recall that addition and subtraction over $\mathbb{F}_2$ are the same). It follows that $d(Y, x^{[m]})$ is the sum of n independent Bernoulli random variables; call it Z. Then Z is a random variable with mean $n\epsilon$ and variance $n\epsilon(1-\epsilon)$. Recall from earlier that $\rho = n\epsilon + \sqrt{2n\epsilon(1-\epsilon)/\Delta}$. Therefore, from Chebyshev's inequality (see Lemma C.3 on page 480) we get

$$\mathbb{P}\left\{|Z - n\epsilon| \geq \sqrt{2n\epsilon(1-\epsilon)/\Delta}\right\} \leq \frac{n\epsilon(1-\epsilon)\Delta}{2n\epsilon(1-\epsilon)} = \frac{\Delta}{2}.$$

We can write $\mathrm{P_B}(C, \epsilon)$ as

$$\frac{1}{M} \sum_{m=1}^{M} \mathrm{P}_{\mathrm{B}}^{[m]}(C, \epsilon) \leq \frac{\Delta}{2} + \frac{1}{M} \sum_{m=1}^{M} \sum_{y \in \{\pm 1\}^n} \sum_{m' \neq m} p_{Y \mid X^{[m]}}(y \mid x^{[m]}) f(x^{[m']}, y).$$

Let $\mathbb{E}_{\mathcal{C}(n,M)}[\cdot]$ denote the expectation with respect to the ensemble $\mathcal{C}(n, M)$. We conclude that

$$\begin{aligned}
\hat{\mathrm{P}}_{\mathrm{B}}(n, M, \epsilon) &\leq \mathbb{E}_{\mathcal{C}(n,M)}\left[\mathrm{P_B}(C, \epsilon)\right] \\
&\leq \frac{\Delta}{2} + \frac{1}{M} \sum_{m=1}^{M} \sum_{y \in \{\pm 1\}^n} \sum_{m' \neq m} \mathbb{E}\left[p_{Y \mid X^{[m]}}(y \mid X^{[m]}) f(X^{[m']}, y)\right] \\
&\overset{(a)}{=} \frac{\Delta}{2} + \frac{1}{M} \sum_{m=1}^{M} \sum_{y \in \{\pm 1\}^n} \sum_{m' \neq m} \mathbb{E}\left[p_{Y \mid X^{[m]}}(y \mid X^{[m]})\right] \mathbb{E}\left[f(X^{[m']}, y)\right] \\
&= \frac{\Delta}{2} + \frac{1}{M} \sum_{m=1}^{M} \sum_{y \in \{\pm 1\}^n} \sum_{m' \neq m} \mathbb{E}\left[p_{Y \mid X^{[m]}}(y \mid X^{[m]})\right] \frac{\sum_{k=0}^{\lfloor \rho \rfloor} \binom{n}{k}}{2^n} \\
&= \frac{\Delta}{2} + (M-1) \frac{\sum_{k=0}^{\lfloor \rho \rfloor} \binom{n}{k}}{2^n},
\end{aligned}$$

where in step (a) we used the fact that if we consider for $m' \neq m$ the two associated codewords $X^{[m]}$ and $X^{[m']}$ as random variables then they are by construction *(pairwise) independent*. If we now use the bound $\sum_{k=0}^{m} \binom{n}{k} \leq 2^{n h_2(m/n)}$, which is valid for $m \leq n/2$ (see (1.59) and Problem 1.25), then as $\rho \leq n/2$ for sufficiently large n

$$\hat{\mathrm{P}}_{\mathrm{B}}(n, M, \epsilon) \leq \frac{\Delta}{2} + (M-1) 2^{-n\left(1 - h_2\left(\epsilon + \sqrt{\frac{2\epsilon(1-\epsilon)}{n\Delta}}\right)\right)}$$

$$\leq \frac{\Delta}{2} + 2^{nr} 2^{-n(1-h_2(\epsilon+\sqrt{\frac{2\epsilon(1-\epsilon)}{n\Delta}}))}$$
$$= \frac{\Delta}{2} + 2^{-n(1-h_2(\epsilon+\sqrt{\frac{2\epsilon(1-\epsilon)}{n\Delta}})-r)}$$
$$\leq \Delta \text{ for } n \text{ large enough if } r < 1 - h_2(\epsilon).$$

The proof is complete if we observe that this upper bound is valid for any $\Delta > 0$. □

The preceding proof shows that there exist codes in $\mathcal{C}(n, 2^{\lfloor nr \rfloor})$ that permit reliable transmission over the BSC(ϵ) up to a rate of $1 - h_2(\epsilon)$ bits per channel use. Actually, a much stronger statement is true, namely *almost any* code in the preceding ensemble can be used for transmission at vanishing probabilities of error.

Although we do not prove this here, the converse is true as well (see Problem 1.29): any attempt to transmit at a rate higher than $1 - h_2(\epsilon)$ must result in error probabilities bounded away from zero. Indeed, one can show that the block error probability P_B must tend to 1 for any sequence of codes of increasing blocklength and rate strictly above $1 - h_2(\epsilon)$. Therefore, $1 - h_2(\epsilon)$ is a *threshold* value, separating what is achievable from what is not. It is called the *Shannon capacity* of the BSC(ϵ) and we denote it by $C_{BSC}(\epsilon) = 1 - h_2(\epsilon)$.

As mentioned before, the minimum distance plays a central role in all of classical coding. The paradigm of classical coding can be summarized as follows: (i) find a code with a large minimum distance and a strong algebraic structure; (ii) devise a decoding algorithm which exploits the algebraic structure to accomplish bounded distance decoding efficiently (see page 7). This philosophy works well if we transmit at a rate that is bounded away from capacity. But, as the next example shows, we cannot hope to achieve capacity in this way.

EXAMPLE 1.18 (BOUNDED DISTANCE DECODER IS NOT SUFFICIENT). Consider a code of rate r, $r \in (0, 1)$. By the Elias bound (1.13) the normalized minimum distance $\delta(r)$ is upper bounded by $\delta(r) \leq 2h_2^{-1}(1-r)(1-h_2^{-1}(1-r))$, so that

$$h_2^{-1}(1-r) \geq \frac{1}{2}\left(1 - \sqrt{1 - 2\delta(r)}\right),$$

for $\delta(r) \in (0, 1/2)$. From this we deduce the weaker bound $h_2^{-1}(1-r) > \frac{1}{2}\delta(r) + \left(\frac{1}{2}\delta(r)\right)^2$ which is easier to handle. If we transmit over the BSC(ϵ) then the expected number of errors in a block of length n is $n\epsilon$. Further, for large n with high probability the actual number of errors is within $O(\sqrt{n})$ of this expected number (see the previous proof starting on page 10). Therefore, if we employ a bounded distance decoder we need $\frac{1}{2}\delta(r) \geq \epsilon$. If we combine this with the previous bound we get $h_2^{-1}(1-r) > \epsilon + \epsilon^2$. This is not possible if $\epsilon > \frac{\sqrt{3}-1}{2}$ since then the right-hand side exceeds $1/2$. It follows that such a bounded distance decoder cannot be used for reliable transmission over a BSC(ϵ) if $\epsilon \in (\frac{\sqrt{3}-1}{2}, \frac{1}{2})$. And for $\epsilon \in (0, \frac{\sqrt{3}-1}{2})$ we conclude that $r < 1 - h_2(\epsilon + \epsilon^2) < 1 - h_2(\epsilon) = C_{\text{BSC}}(\epsilon)$, i.e., capacity cannot be achieved either. ◇

The preceding example is not to say that bounded distance decoders are not useful. If we are willing to back off a little bit from the Shannon capacity and if ϵ is not too large then a bounded distance decoder might work well.

§1.7. Linear Codes and Complexity

By our preceding remarks, almost any code in $\mathcal{C}(n, 2^{\lfloor nr \rfloor})$ is suitable for reliable transmission at rates close to Shannon capacity at low error probability provided only that the length n is sufficiently large (we have limited the proof of the channel coding theorem to the BSC but this theorem applies in a much wider setting). So why not declare the coding problem solved and stop here? The answer is that Shannon's theorem does not take into account the *description*, the *encoding*, and the *decoding complexities.*

First consider the description complexity. Informally, it is the amount of memory required to define a code. Without further restriction on the structure, already the description of a particular code quickly becomes impractical as n grows, since it requires $n2^{\lfloor nr \rfloor}$ bits. Hence, as a first step toward a reduction in complexity we restrict our attention to *linear* codes.

§1.7.1. Linear Codes

We say that a code C over a field $\mathbb{F}$ is *linear* if it is closed under n-tuple addition and scalar multiplication:

$$\alpha x + \alpha' x' \in C, \quad \forall x, x' \in C \text{ and } \forall \alpha, \alpha' \in \mathbb{F}.$$

In fact, it suffices to check that

$$\alpha x - x' \in C, \quad \forall x, x' \in C \text{ and } \forall \alpha \in \mathbb{F}. \tag{1.19}$$

Choosing $\alpha = 0$ in (1.19) shows that if $x' \in C$ then so is $-x'$. Further, choosing $\alpha = 1$ and $x' = x$ shows that the all-zero word is a codeword of any linear code. Equivalently, since $\mathbb{F}^n$ is a *vector space*, condition (1.19) implies that a linear code is a *subspace* of $\mathbb{F}^n$.

For a linear code C the minimum distance $d(C)$ is equal to the *minimum of the weight* of all non-zero codewords:

$$d(C) = \min\left\{d(x, x') : x, x' \in C, x \neq x'\right\} = \min\left\{d(x - x', 0) : x, x' \in C, x \neq x'\right\}$$

$$= \min\{w(x-x') : x, x' \in C, x \neq x'\} = \min\{w(x) : x \in C, x \neq 0\}.$$

Since a linear code C of length n over $\mathbb{F}$ is a subspace of $\mathbb{F}^n$, there must exist an integer k, $0 \leq k \leq n$, so that C has a *dimension* k. This means that C contains $|\mathbb{F}|^k$ codewords. In the sequel we denote by $[n, k, d]$ the parameters of a linear code of length n, dimension k, and minimum distance d. It is customary to call a $k \times n$ matrix G, whose rows form a linearly independent basis for C, a *generator* matrix for C. Conversely, given a matrix $G \in \mathbb{F}^{k \times n}$ of rank k we can associate with it the code $C(G)$:

$$C(G) = \{x \in \mathbb{F}^n : x = uG, u \in \mathbb{F}^k\}. \tag{1.20}$$

In general many generator matrices G describe the same code (see Problem 1.6).

Definition 1.21 (Proper Codes). We say that the i-th position of a binary linear code C is *proper* if there exists a codeword $x \in C$ so that $x_i = 1$. If all positions of C are proper we say that C is proper. Equivalently, a linear code is proper if its generator matrix G contains no zero columns. ▽

Zero columns convey zero information – no pun intended – and so we can safely restrict our attention to proper codes in the sequel. Note also that if the i-th position of C is proper then the number of codewords that contain a 1 in the i-th position is equal to the number of codewords that contain a 0 in the i-th position (see Problem 1.5).

Definition 1.22 (Systematic Generator Matrix). A generator matrix G of a linear code $C[n, k, d]$ is said to be in *systematic form* if $G = (I_k\ P)$, where I_k is a $k \times k$ identity matrix and where P is a $k \times (n-k)$ matrix with entries in $\mathbb{F}$. If G is in systematic form and $u \in \mathbb{F}^k$ is a vector containing the information bits (i.e., an *information* word), then the corresponding codeword $x = uG$ has the form (u, uP), i.e., the first k components of x are equal to the information word u. ▽

To each linear code C we associate the *dual* code $C^\perp$:

$$C^\perp = \{v \in \mathbb{F}^n : xv^T = 0, \forall x \in C\} = \{v \in \mathbb{F}^n : Gv^T = 0^T\}. \tag{1.23}$$

Assume that v and v' are elements of $C^\perp$ and that x is an element of C. Since $xv^T = 0 = x(v')^T$ implies $x(\alpha v - v')^T = 0$ for any $\alpha \in \mathbb{F}$, it follows that $C^\perp$ is a linear code as well. Therefore, it has a basis. It is customary to denote such a basis by H. This basis H is a generator matrix of the code $C^\perp$. It is also said to be a *parity-check* matrix of the original code C. Let G be a generator matrix for a code C and let H be a corresponding parity-check matrix. By (1.23) the dual code is the set of solutions to the system of equations $Gv^T = 0^T$. Therefore, since G has k (linearly independent) rows we know, by linear algebra, that the dual code has dimension $n-k$, and therefore H has dimension $(n-k)\times n$: in fact, assume without loss of generality that G is in systematic form, $G = (I_k\ P)$. Represent v as $v = (v_s\ v_p)$, where v_s is of length k and where v_p is of length $n-k$. From $Gv^T = v_s^T + Pv_p^T$, we see that for each of the $|\mathbb{F}|^{n-k}$ distinct choices of v_p there is exactly one $v_s = -Pv_p^T$ so that $Gv^T = 0^T$.

The dual code is therefore characterized by

$$C^\perp = \{v \in \mathbb{F}^n : v = uH, u \in \mathbb{F}^{n-k}\} = \{v \in \mathbb{F}^n : Gv^T = 0^T\}.$$

In the same manner we have

$$C = \{x \in \mathbb{F}^n : x = uG, u \in \mathbb{F}^k\} = \{x \in \mathbb{F}^n : Hx^T = 0^T\}.$$

That the second description is true can be seen as follows. Clearly, for every $x \in C$, $Hx^T = 0^T$. This shows that $C \subseteq \{x \in \mathbb{F}^n : Hx^T = 0^T\}$. But by assumption $|C| = |\mathbb{F}|^k = |\{x \in \mathbb{F}^n : Hx^T = 0^T\}|$, since H has rank $n-k$.

As we will see, this latter description is particularly useful for our purpose. Given a generator matrix G, it is easy to find a corresponding parity-check matrix H and vice versa; see Problem 1.8.

Example 1.24 (Binary Hamming Codes). Let $\mathbb{F} = \mathbb{F}_2$ and let $m \in \mathbb{N}$. Let H be an $m \times (2^m - 1)$ binary matrix whose columns are formed by all the binary m-tuples except the all-zero m-tuple. We claim that H is the parity-check matrix of a binary linear code of length $n = 2^m - 1$, dimension $2^m - m - 1$, and minimum distance 3. To see that the minimum distance is 3, note that any two columns of H are linearly independent, but that there are triples of columns which are linearly dependent. Therefore, $Hx^T = 0^T$ has no solution for $x \in \mathbb{F}_2^n$ with $1 \le w(x) \le 2$ but it has solutions with $w(x) = 3$. Clearly, C has dimension at least $2^m - m - 1$ since H has m rows. Let us now show that C must have dimension *at most* $2^m - m - 1$, i.e., we have equality. Since C has distance 3, the spheres of radius 1 centered at each codeword are disjoint. Let us count the total number of words contained in all these spheres:

$$|C|\left(\binom{n}{0} + \binom{n}{1}\right) = |C|(1+n) = |C|2^m.$$

From this it follows that $|C|2^m \le 2^n = 2^{2^m-1}$, where the right-hand side represents the total number of points in the space $\mathbb{F}_2^n$. Turning this around we get $|C| \le 2^{2^m-m-1}$. This shows that C has dimension at most $2^m - m - 1$. Codes such that the spheres of radius $t = \lfloor \frac{d-1}{2} \rfloor$ centered around the codewords *cover* the whole space are called *perfect*. As a particular example consider the case $m = 3$. Then

$$H = \begin{pmatrix} 1 & 1 & 0 & 1 & 1 & 0 & 0 \\ 1 & 0 & 1 & 1 & 0 & 1 & 0 \\ 0 & 1 & 1 & 1 & 0 & 0 & 1 \end{pmatrix}, \tag{1.25}$$

and C_{Ham}, the code defined by the parity-check matrix H, is the $[7,4,3]$ binary Hamming code. ◇

In the preceding definitions we have assumed that the rows of G and H are linearly independent. For the sequel it is useful to relax this definition. We call a $k \times n$ matrix G a *generator* matrix *even* if G has rank strictly less than k. We say that k/n is the *design* rate of the code. The true rate is of course $\text{rank}(G)/n$. An equivalent statement is true for an $(n-k) \times n$ parity-check matrix H.

Definition 1.26 (Elias's Generator and Gallager's Parity-Check Ensemble). Fix the blocklength n and the *design* dimension k. To sample from Elias's *generator* ensemble, construct a $k \times n$ generator matrix by choosing each entry iid according to a Bernoulli random variable with parameter one-half. To sample from Gallager's *parity-check* ensemble, proceed in the same fashion to obtain a sample $(n-k) \times n$ parity-check matrix. Although both ensembles behave quite similarly, they are not identical. This is most easily seen by noting that every code in the generator ensemble has rate *at most* k/n and that some codes have a strictly smaller rate, whereas all codes in the parity-check ensemble have rate *at least* k/n. A closer investigation of the weight distribution of both ensembles is the topic of Problems 1.17 and 1.18. We denote these two ensembles by $\mathcal{G}(n,k)$ and $\mathcal{H}(n,k)$, respectively. ▽

For the most part we are only concerned with the binary case and therefore, unless explicitly stated otherwise, we assume in the sequel that $\mathbb{F} = \mathbb{F}_2$.

Are linear ensembles capable of achieving capacity? Consider, e.g., the generator ensemble $\mathcal{G}(n,k)$ and transmission over the BSC(ϵ). That the answer is in the affirmative can be seen as follows: consider a slight twist on the ensemble $\mathcal{G}(n,k)$. Pick a random element C from $\mathcal{G}(n,k)$ and a random *translation* vector c. The code is the set of codewords $C + c$. We translate so as to eliminate the special role that the all-zero word plays (since it is contained in any linear code). In this new ensemble the codewords are uniformly distributed and pairwise statistically independent (see Problem 1.16.) An inspection of the proof of Theorem 1.17 shows that these are the only two properties which are used in the proof. But a translation of the codewords leaves the error probability invariant so that also the ensemble $\mathcal{G}(n,k)$ itself is capable of achieving capacity. More generally, for any binary-input output-symmetric (see Definition 4.8) memoryless (see Definition 4.3) channel, linear codes achieve capacity.

§1.7.2. Description/Encoding Complexity of Linear Codes

From the generator and parity-check representation of a linear code we see that its description complexity is at most $\min\{rn^2, (1-r)n^2\}$ bits, where r is the rate of the code. Further, from (1.20) it is clear that the *encoding* task, i.e., the mapping of the information block onto the codeword, can be accomplished in $O(n^2)$ operations.

§1.7.3. MAP Decoding Complexity of Linear Codes

Let us now focus on the *decoding complexity*. To keep things simple, we restrict ourselves to the case of transmission over the BSC(ϵ). Assume we have a uniform prior on the set of codewords. The MAP or ML decoding rule then reads

$$
\begin{aligned}
\hat{x}^{\mathrm{ML}}(y) &= \operatorname{argmax}_{x:Hx^T=0^T} p_{Y\,|\,X}(y\,|\,x) \\
&= \operatorname{argmax}_{x:Hx^T=0^T} \epsilon^{d(x,y)}(1-\epsilon)^{n-d(x,y)} \\
\text{since } \epsilon \le \tfrac{1}{2} \qquad &= \operatorname{argmin}_{x:Hx^T=0^T} d(x,y) \\
&= \operatorname{argmin}_{x:Hx^T=0^T} w(x+y) \\
e = x + y \qquad &= \operatorname{argmin}_{e+y:He^T=Hy^T} w(e) \\
s^T = Hy^T \qquad &= \operatorname{argmin}_{e+y:He^T=s^T} w(e).
\end{aligned}
$$

The quantity $s^T = Hy^T$ is called the *syndrome* and it is known at the receiver. Consider the following related decision problem.

Problem Π: *ML Decision Problem*

Instance: A binary $(n-k)\times n$ matrix H, a vector $s \in \{0,1\}^{n-k}$, and an integer $w > 0$.

Question: Is there a vector $e \in \{0,1\}^n$ of weight at most w such that $He^T = s^T$?

Clearly, we can solve the ML decision problem once we have solved the associated ML decoding problem: For a given s find $\hat{x}^{\mathrm{ML}}$ and, therefore, the "error" vector e. By definition this is the lowest weight vector which "explains" the data and therefore the ML decision problem has an affirmative answer for $w \ge w(e)$ and a negative answer otherwise. We conclude that the ML decoding problem is at least as "difficult" as the ML decision problem.

In the theory of complexity, a problem Π is said to belong to the class P if it can be solved by a deterministic Turing machine in *polynomial* time in the length of the input. Instead of a Turing machine one may think of a program written in (let's say) C running on a standard computer, except that this computer has infinite memory. Simply speaking, problems in *P* are problems for which efficient algorithms are known: solving a system of linear equations and finding the minimum spanning tree or sorting are well-known examples in this class. Unfortunately, many problems which occur in practice appear not to belong to this class. A broader class is the class NP, which contains all problems that can be solved by a *non-deterministic* Turing machine in *polynomial* time. A non-deterministic algorithm is one that, when confronted with a choice between two alternatives, can create two copies of itself and simultaneously follow the consequences of both courses. For our discussion it suffices to know that all problems of the form "Does there exist a subset with a specific property?", assuming that this property is easily checked for any given subset, belong to the class NP. Of course, this may lead to an exponentially growing number of copies. The algorithm is said to solve the given problem if any one of these copies produces the correct answer. Clearly, we have $P \subseteq NP$ and whether this inclusion is proper is an important open problem. Not necessarily all problems in NP are equally hard. Assume that a specific problem Π in NP has the property that any problem in NP can be reduced to Π in polynomial time. Then, ignoring polynomial factors, it is reasonable to say that Π is as hard as any problem in NP. We say that such a problem is NP-*complete*. It is widely believed that $NP \neq P$. Suppose this is true. This implies that there are no efficient (polynomial-time in the size of the input) algorithms that solve *all* instances of an NP-*complete* problem. We now are faced with the following discouraging result.

THEOREM 1.27. The *ML Decision Problem* for the BSC is NP-complete.

This means that there are no efficient (polynomial in the blocklength) algorithms known to date to solve the *general* ML decoding problem and that it is highly likely that no such algorithm exist.

Does this mean that we should throw in the towel and declare that the efficient transmission of information at low error probability is a hopeless task? Not at all. First, the preceding result only says that no efficient algorithm is known which solves the ML decoding problem for *all* codes. But it leaves open the possibility that some subclasses of codes have an efficient ML decoding algorithm. We are interested in codes which allow transmission of information close to capacity at low probability of error and which are efficiently decodable. Furthermore, as we saw already in the proof of the channel coding theorem, there exist suboptimal decoders which are powerful enough for the task.

§1.8. RATE, PROBABILITY, COMPLEXITY, AND LENGTH

The most important parameters for the transmission problem are rate, probability of (block or bit) error, delay, and complexity (encoding and decoding). Delay is not an easy quantity to work with. It is therefore customary to consider instead the blocklength of the coding system. If you know convolutional codes (see Section 6.1) then you are aware that these two concepts are not necessarily the same. The situation gets even more complicated if we consider transmission over channels with feedback. But in the context of block coding schemes over channels without feedback (which is the focus of this book) we do not commit any fundamental error by equating the two.

Rate, probability of error, and blocklength are well-defined concepts. The notion of complexity on the other hand is fuzzy. As we have just discussed, we can easily distinguish polynomial and exponential complexity. However, we typically discuss algorithms that have linear complexity in the blocklength and we are concerned with the constants involved. These constants depend on the technology we use to implement the system. In a hardware realization we typically mean the number of gates necessary or the number of connections required. If the system is implemented as a program, then we are concerned about the number of operations and the amount of memory. It is therefore only of limited value to give a formal definition of complexity and we are content with a more engineering-oriented notion.

For a fixed channel, we want to transmit at large rates r with low probability of error P using simple encoding and decoding algorithms and short codes (small n). Clearly there are trade-offs: by their very nature longer codes allow more reliable transmission than shorter ones (or transmission at higher rates), and in a similar way more complex decoding algorithms (like, e.g., ML decoders) perform better than suboptimal but lower-complexity algorithms. We want to determine all achievable tuples

$$(r, \mathrm{P}, \chi_E, \chi_D, n)$$

and their associated coding schemes, where χ_E and χ_D denote the encoding and decoding complexity, respectively. This is a tall order and we are currently far from such a complete characterization. The task becomes much simpler if we ignore one of the quantities and investigate the trade-offs between the remaining ones. Clearly, the problem becomes trivial if we set no bounds on either the rate or the probability of error. On the other hand the problem stays non-trivial if we allow *unbounded complexity* or *unbounded blocklengths*.

A particularly useful transformation is to let $\delta = 1 - r/\mathrm{C}$, so that $r = (1-\delta)\mathrm{C}$. We call δ the *multiplicative gap*[1] to capacity and we are interested in the region of achievable tuples $(\delta, \mathrm{P}, \chi_E, \chi_D, n)$.

§1.8.1. Unbounded Complexity

If we ignore the issue of complexity, we enter the realm of classical information theory. We are interested in the behavior of the error probability as a function of the blocklength n and the gap δ.

Example 1.28 (Error Exponents for Block Codes). Consider transmission using a binary linear block code $C[n, nr]$ (a code of length n and dimension nr) via a discrete binary memoryless channel with capacity C, $\mathrm{C} > 0$, using a MAP decoder. Let $\mathrm{P}_{\mathrm{B}}^{\mathrm{MAP}}(n, r)$ be the resulting block error probability for the optimal choice of the

[1] Not to be confused with the relative minimum distance.

code (the minimum of the error probability over all choices of the code with a given rate). It is a celebrated result of information theory that $P_B^{MAP}(n, r)$ decreases exponentially fast in the blocklength, i.e., we have

$$e^{-n(E(r)+o_n(1))} \leq P_B^{MAP}(n, r) \leq e^{-nE(r)}. \tag{1.29}$$

$E(r)$ is called the *error exponent*. We have $E(r) > 0$ for all rates r in the range $r \in [0, C)$, where C is the (Shannon) *capacity* of the channel. For the BSC we determined its capacity in Section 1.6. (See Section 4.1.11 for the derivation of the capacity for more general channels.) Finally, $o_n(1)$ denotes a quantity which tends to zero as n tends to infinity. Figure 1.30 depicts the error exponent $E(r)$ (solid line) for the BSC($\epsilon \approx 0.11$). This error exponent is known exactly for $r \in [r_c, C]$, where

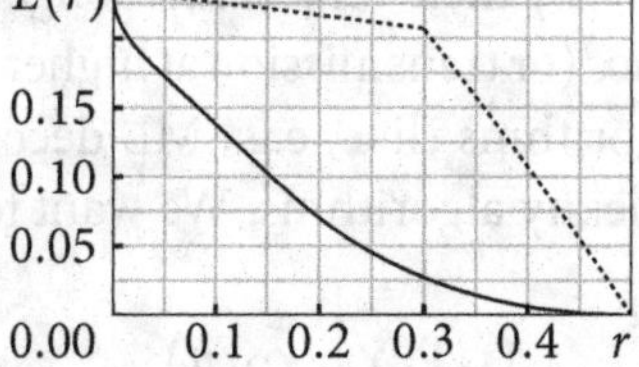

Figure 1.30: Error exponent of block codes (solid line) and of convolutional codes (dashed line) for the BSC($\epsilon \approx 0.11$).

r_c, $0 < r_c < C$, is called the *critical rate*. For rates $r \in [0, r_c)$, only a lower bound on $E(r)$ is known. At C the (left) derivative of $E(r)$ vanishes but the second (left) derivative is strictly positive, i.e., $E(r) = (C-r)^2\alpha + O((C-r)^3)$, where α is strictly positive. Therefore, if $r(\delta) = (1-\delta)C$, $\delta \in [0, 1]$, then $E(\delta) = \delta^2 C^2 \alpha + O(\delta^3)$. More generally, for a "typical" discrete binary memoryless channel, the error exponent as a function of δ, $\delta \in [0, 1]$, has the form $E(\delta) = \delta^2 C^2 \alpha + O(\delta^3)$, where α is strictly positive. We summarize: the error probability P_b^{MAP} behaves roughly like $e^{-n\delta^2 C^2 \alpha}$; i.e., it decreases exponentially in n with an exponent that is proportional to δ^2. ◇

Example 1.31 (Error Exponents for Convolutional Codes). Consider transmission using a binary convolutional code (see Section 6.1) of rate r and with *memory m* via a discrete binary-input memoryless channel with capacity C, $C > 0$, using a MAP decoder. Let $P_b^{MAP}(m, r)$ be the resulting bit error probability for an optimal choice of the code. In analogy to the case of block codes, $P_b^{MAP}(m, r)$ decreases exponentially fast in m, i.e.,

$$e^{-\frac{m}{r}(E(r)+o_m(1))} \leq P_b^{MAP}(m, r) \leq e^{-\frac{m}{r}E(r)}. \tag{1.32}$$

Similar to the block code case, this error exponent is known exactly for $r \in [C_c, C]$, where C_c, $0 < C_c < C$, is called the *critical rate*. Figure 1.30 depicts the error exponent $E(r)$ (dashed line) for the BSC($\epsilon \approx 0.11$). For rates $r \in [0, C_c)$ only a lower bound on $E(r)$ is known. As the main difference to the block code case, for convolutional codes $E(r) = (C - r)\alpha + O((C - r)^2)$, where α is strictly positive. Therefore, if $r(\delta) = (1 - \delta)C$, then $E(\delta) = \delta C\alpha + O(\delta^2)$. More generally, for any discrete binary-input memoryless channel the error exponent as a function of δ, $\delta \in [0, 1]$, is given by $E(\delta) = \delta C\alpha + O(\delta^2)$, where α is strictly positive. We summarize: the error probability P_B^{MAP} decreases exponentially in m but now the exponent is proportional to δ. ◇

§1.8.2. Unbounded Delay

For some applications fairly large blocklengths (delays) are perfectly acceptable. Therefore it is worthwhile investigating the behavior of coding schemes if we lift the restriction on blocklengths and only focus on rate, probability of error, and complexity. In the following we assume a naive computational model in which infinite precision arithmetic can be accomplished with unit cost.

Example 1.33 (Complexity of Block Codes). Consider again transmission with a binary-input linear code $C[n, nr]$ via a discrete binary memoryless channel with capacity C, $C > 0$, using a MAP decoder. Generically, the complexity of a MAP decoder, measured per information bit, is equal to $\frac{c}{nr}2^{n\min(r,1-r)}$, where c is a constant depending on the implementation. This can be seen as follows: there are 2^{nr} codewords. One straightforward approach is to determine the APP for each of them given the observation and to choose the argument which maximizes this measure. Normalized per information bit, this gives rise to a complexity of $\frac{c}{nr}2^{nr}$. On the other hand, as discussed in Problem 1.19, the MAP decoder can also be based on the parity-check matrix of the code, and this gives rise to a complexity of $\frac{c}{nr}2^{n(1-r)}$. An alternative generic MAP decoding scheme which also gives rise to a complexity $\frac{c}{nr}2^{n(1-r)}$ and which is based on the dual code is discussed in Problem 4.43. If in (1.29) we fix P_B and solve for n as a function of δ we get

$$n(\delta) = O(\frac{1}{\delta^2}).$$

It follows that the decoding complexity $\chi_D(\delta)$ is exponential in $1/\delta^2$. The encoding complexity normalized per information bit is equal to $\chi_E(n, r) = cn$, where c is again a small constant. We conclude that

$$\chi_E(\delta) = \frac{c}{\delta^2 C^2 \alpha}. \qquad \diamond$$

EXAMPLE 1.34 (COMPLEXITY OF CONVOLUTIONAL CODES). Consider once more transmission with a binary convolutional code via a discrete binary memoryless channel with capacity C, C > 0, using an ML decoder. ML decoding of these codes can be accomplished efficiently by means of the so-called *Viterbi algorithm* and the decoding complexity per information bit is equal to $\chi_D(r) = \frac{1}{r}2^m$ (see Section 6.1). By going through essentially the same steps as before, we see that the decoding complexity grows exponentially in $\frac{1}{\delta}$. This is a large improvement compared to block codes but still exponential in the inverse of the gap.

The preceding argument might give rise to confusion. How can convolutional codes be better than block codes? After all, we can always terminate a convolutional code with negligible loss in rate and so construct an equivalent block code. The answer is that in our discussion of block codes we assumed a *generic* decoder which does not exploit any structure that might be present in the block code. If a suitable structure is present (as is the case for convolutional codes) we can do better. ◇

There are many ways of combining given codes to arrive at a new code (see Problems 1.3 and 1.4). A basic and fundamental such construction is the one of *code concatenation*, an idea originally introduced by Forney. In this construction one uses an *inner code* to convert the very noisy bits arriving from the channel into fairly reliable ones and then an *outer code* to "clean up" any remaining errors in the block. Such code constructions can substantially decrease the decoding complexity viewed as a function of the blocklength n. But if we consider the decoding complexity as a function of the gap to capacity δ, it still increases exponentially in $1/\delta$.

Another avenue for exploration is the field of *suboptimal* decoding algorithms. Although up to this point we have tacitly assumed that we perform MAP decoding, this is not necessary as we can see from the proof of the channel coding theorem, which itself uses a suboptimal decoder.

EXAMPLE 1.35 (ITERATIVE CODING). Iterative coding is the focus of this book. Unfortunately the exact complexity versus performance trade-off for iterative decoding schemes is not known to date. In fact, it is not even known whether such schemes are capable of achieving the capacity for a wide range of channels. It is *conjectured* that

$$\chi_D(\delta, p) = \chi_E(\delta, p) = \frac{c}{\delta},$$

for some constant c which depends on the channel. To furnish a proof of the preceding conjecture is without doubt the biggest open challenge in the realm of iterative decoding. According to this conjecture the complexity (per information bit) grows *linearly* in the inverse to the gap as compared to exponentially. This is the main motivation for studying iterative decoding schemes. ◇

§1.9. Tour of Iterative Decoding

We have seen that iterative decoding holds the promise of approaching the capacity with unprecedentedly low complexity. Before delving into the details (and possibly getting lost therein) let us give a short overview of the most important components and aspects. We opt for the simplest non-trivial scenario. There is plenty of opportunity for generalizations later.

The first important ingredient is to represent codes in a *graphical* way. Consider again the Hamming code $C_{\text{Ham}}[7,4,3]$ given in terms of the parity-check matrix shown on page 15. By definition, $x = (x_1, \ldots, x_7)$, $x \in \mathbb{F}_2^7$, is a codeword of C_{Ham} if and only if

$$x_1 + x_2 + x_4 + x_5 = 0,$$
$$x_1 + x_3 + x_4 + x_6 = 0,$$
$$x_2 + x_3 + x_4 + x_7 = 0.$$

We associate $C_{\text{Ham}}[7,4,3]$ with the following graphical representation. We rewrite the three parity-check constraints in the form (recall that we work over $\mathbb{F}_2$ so that $-x = x$)

(1.36) $$x_1 + x_2 + x_4 = x_5,$$

(1.37) $$x_1 + x_3 + x_4 = x_6,$$

(1.38) $$x_2 + x_3 + x_4 = x_7.$$

Think of x_1, x_2, x_3, x_4 as the four independent *information* bits and x_5, x_6, x_7 as the derived *parity* bits. The following visualization of the situation was introduced by McEliece: represent the constraints via a *Venn* diagram. This is shown in Figure 1.39. Each circle corresponds to one parity-check constraint – the number of ones in each circle must be even. The topmost circle corresponds to the constraint expressed in (1.36). Consider first the *encoding* operation. Assume we are given $(x_1, x_2, x_3, x_4) = (0,1,0,1)$. It is then easy to fill in the missing values for (x_5, x_6, x_7) by applying one parity-check constraint at a time as shown in Figure 1.40: from (1.36) we deduce that $x_5 = 0$, from (1.37) it follows that $x_6 = 1$, and (1.38) shows that $x_7 = 0$. The complete codeword is therefore $(x_1, x_2, x_3, x_4, x_5, x_6, x_7) = (0,1,0,1,0,1,0)$. Next, consider the *decoding* problem. Assume that transmission takes place over the binary erasure channel with parameter ϵ (see page 71 for more details on this channel) and that the received message is $(0,?,?,1,0,?,0)$. We want to reconstruct the transmitted word. Figure 1.41 shows how this can be done. The decoding proceeds in a fashion similar to the encoding – we check each constraint (circle) to see if we can reconstruct a

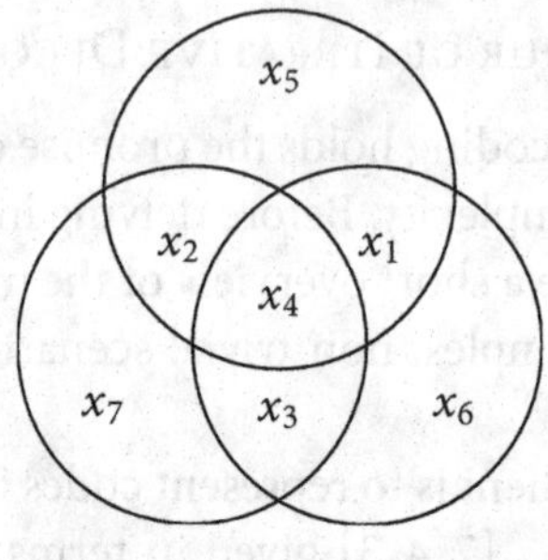

Figure 1.39: Venn diagram representation of C.

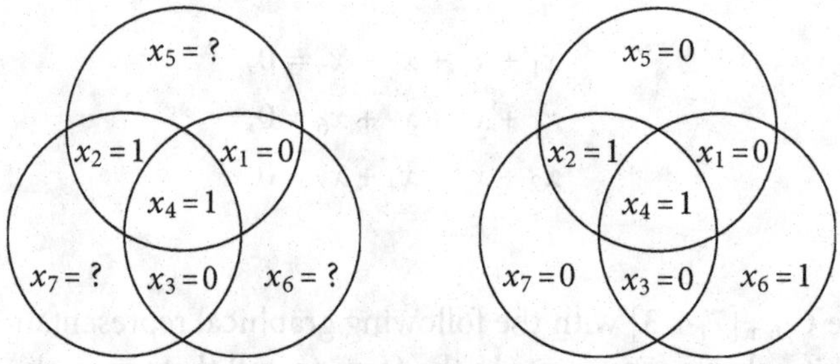

Figure 1.40: Encoding corresponding to $(x_1, x_2, x_3, x_4) = (0, 1, 0, 1)$. The number of ones contained in each circle must be even. By applying one such constraint at a time the initially unknown components x_5, x_6, and x_7 can be determined. The resulting codeword is $(x_1, x_2, x_3, x_4, x_5, x_6, x_7) = (0, 1, 0, 1, 0, 1, 0)$.

missing value from the values we already know. In the first step we recover $x_2 = 1$ using the constraint implied by the top circle. Next we determine $x_3 = 0$ by resolving the constraint given by the left circle. Finally, using the last constraint, we recover $x_6 = 1$. Unfortunately this "local" decoder – discovering one missing value at a time – does not always succeed. To see this, consider the case when the received message is $(?, ?, 0, ?, 0, 1, 0)$. A little thought (or an exhaustive check) show(s) that there is a unique codeword in C_{Ham}, namely $(0, 1, 0, 1, 0, 1, 0)$, which is compatible with this message. But the local decoding algorithm fails. As one can see in Figure 1.42, none of the three parity-check equations by themselves can resolve any ambiguity.

In practice we encounter codes that have hundreds, thousands, or even millions of nodes. In this case the Venn diagram representation is no longer convenient and we instead use the so-called *Tanner graph* description (see Sections 2.2 and 3.3). The basic principle however stays the same. Encoding and decoding are accomplished *locally*: specifically, we send messages along the edges of the Tanner graph and process these messages locally at each node.

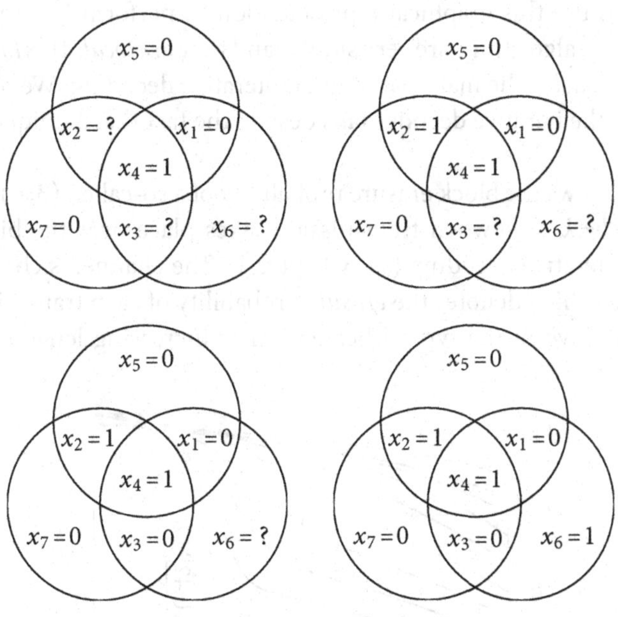

Figure 1.41: Decoding corresponding to the received message $(0, ?, ?, 1, 0, ?, 0)$. First we recover $x_2 = 1$ using the constraint implied by the top circle. Next we determine $x_3 = 0$ by resolving the constraint given by the left circle. Finally, using the last constraint, we recover $x_6 = 1$.

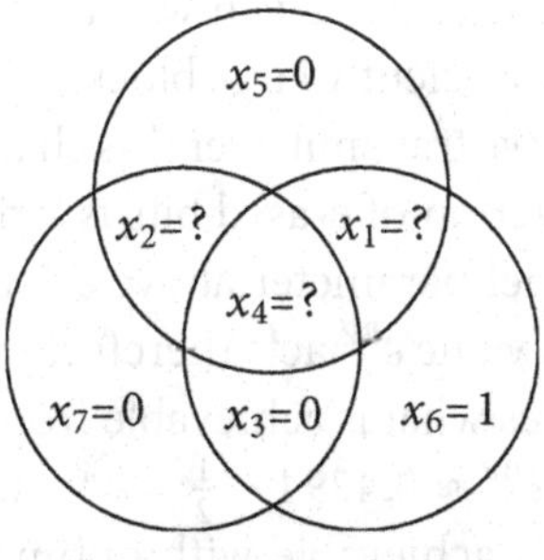

Figure 1.42: Decoding corresponding to the received message $(?, ?, 0, ?, 0, 1, 0)$. The local decoding fails since none of the three parity-check equations by themselves can resolve any ambiguity.

We summarize: we have seen that we can represent codes in a graphical manner and that we can use this graphical representation to perform both encoding and decoding. The two algorithms are very similar and they are *local*. This *local processing on a graphical model* is the main paradigm of iterative decoding. We have also seen that sometimes the iterative decoder fails despite the fact that a unique decoding is possible.

Figure 1.43 shows the block erasure probability of a so-called $(3,6)$-*regular low-density parity-check code* when transmission takes place over the binary erasure channel under iterative decoding (see Chapter 3). The channel is characterized by the parameter ϵ, which denotes the *erasure* probability of each transmitted bit. This plot is representative of the typical behavior. For increasing lengths the individ-

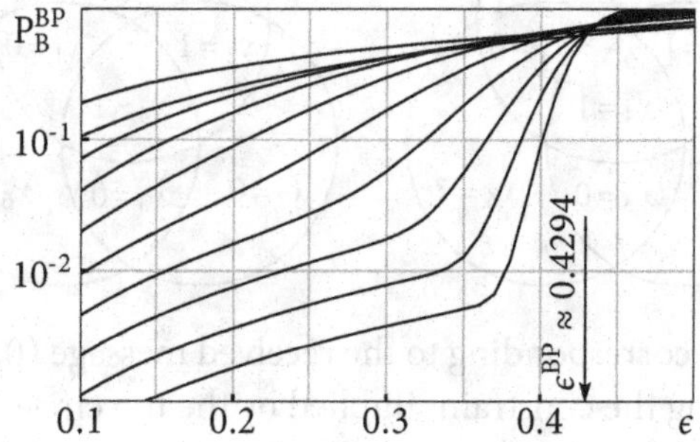

Figure 1.43: $\mathbb{E}_{\text{LDPC}(nx^3, \frac{n}{2}x^6)}\left[P_B^{BP}(G,\epsilon)\right]$ as a function of ϵ for $n = 2^i$, $i \in [10]$.

ual curves become steeper and steeper and they converge to a limiting asymptotic curve. Of particular importance is the value ϵ^{BP}, which marks the zero crossing of this asymptotic curve. For the example given we have $\epsilon^{BP} \approx 0.4294$. Its operational meaning is the following: for sufficiently large blocklengths we can achieve arbitrarily small probability of error if we transmit over this channel with channel parameter strictly less than ϵ^{BP} (if the fraction of erased bits is strictly less than ϵ^{BP}); but if we choose the value of the channel parameter above ϵ^{BP} then the error probability is bounded away from zero. The value ϵ^{BP} acts therefore like a *capacity* for this particular coding scheme: it separates what is achievable from what is not. The rate of the code is one-half and we have $\epsilon^{BP} \approx 0.4294 < \frac{1}{2} = \epsilon^{Sha}$, where ϵ^{Sha} denotes the Shannon threshold (the rate that is achievable with optimal codes and under optimal decoding). This indicates that even in the limit of infinite blocklengths we cannot achieve capacity with this particular coding scheme.

Is this failure to reach capacity due to the code or due to the suboptimal nature of the decoder? We will see that if our example code were decoded optimally the threshold would be $\epsilon^{MAP} \approx 0.48815$. This still falls short of the Shannon threshold, namely 0.5. We conclude that both code and decoding algorithm are to blame. Why don't we just use elements from Gallager's parity-check ensemble, which we know can achieve capacity? Unfortunately, iterative decoding does not work well on elements of this ensemble. Therefore, in constructing codes that are capable of approaching Shannon capacity under iterative decoding we have to worry both about constructing good codes (which under optimal decoding could achieve capacity) and about the performance of the suboptimal decoder compared to the optimal one.

Figure 1.43 also gives insight into the *finite-length* behavior. If we consider each finite-length curve, we see that it can be fairly cleanly separated into two regions – the so-called *waterfall* region in which the error probability falls off sharply and the *error floor* region in which the curves are much more shallow. As we discuss in Sections 3.23 and 4.13, the waterfall region is due to large decoding failures and in this region the decay of the error probability is exponential in the blocklength. On the other hand, the error floor is due to small failures and in this region the decay is only polynomial (see Sections 3.24 and 4.14).

In Figure 1.43 we did not plot the performance of a code but the *average* performance of an ensemble. It is this ensemble approach that makes an analysis possible. What can we say about the performance of individual instances? For individual instances the contributions stemming from large failures are sharply concentrated around this ensemble average (see Sections 3.6.2 and 4.3.2). But the contributions due to small weaknesses in the graph are no longer concentrated. In fact, in the limit of large blocklengths, the *distribution* of these small weaknesses converges to a well-defined limit (see Theorem D.32). The code design involves therefore two stages. First, we find ensembles that exhibit a large threshold (good behavior in the waterfall regime). Within this ensemble we then find elements that have few small weaknesses and, therefore, low error floors.

§1.10. Notation, Conventions, and Useful Facts

Large parts of this book should be accessible to a reader with a standard background in engineering mathematics but no prior background in coding. Some of the less standard techniques that are useful for our investigation are summarized in the appendices.

A familiarity with some basic notions of information theory is helpful in places. In fact not much is needed. We sometimes refer to the *entropy* of a random variable X, which we denote by $H(X)$. Entropy is a measure of the "unpredictability" of a random variable. The smaller the entropy the easier it is to predict the outcome of a random experiment. We use small letters for variables and capital letters for random variables: we say that the random variable X takes on the value

x. We write densities as $p_X(x)$, and sometimes we use the shorthand $p(x)$. In the case that X is discrete with probability distribution $p_X(x)$, the entropy is defined as $H(X) = -\sum_x p_X(x)\log p_X(x)$ where $0\log 0$ is taken to be 0. If X has a density then the equivalent quantity is called *differential entropy* and is defined in the natural way as $h(X) = -\int_x p_X(x)\log p_X(x)\mathrm{d}x$. Entropy and differential entropy share the basic properties mentioned next and so we will not make any further notational distinctions and simply refer to entropy. We sometimes invoke the so-called *chain rule*: if X and Y are random variables then

$$H(X,Y) = H(X) + H(Y\,|\,X), \tag{1.44}$$

where $H(Y\,|\,X) = \sum_x H(Y\,|\,X=x)p_X(x)$ and where $H(Y\,|\,X=x)$ is the entropy of the random variable with probability distribution $p_{Y\,|\,X}(y\,|\,x)$, where x is fixed. This chain rule is a direct consequence of $p_{X,Y}(x,y) = p_X(x)p_{Y\,|\,X}(y\,|\,x)$ (see Problem 1.26). The rule extends in the natural way to more than two random variables: e.g., $H(X,Y,Z) = H(X) + H(Y\,|\,X) + H(Z\,|\,X,Y)$.

For the relatively simple channels we are concerned with, the *Shannon capacity*, i.e., the *maximal rate* at which we can *reliably* transmit, is given by

$$C = \max_{p_X(x)} I(X;Y), \tag{1.45}$$

where X denotes the input of the channel and Y the output and where the *mutual information* $I(X;Y)$ is equal to $H(X) - H(X\,|\,Y) = H(Y) - H(Y\,|\,X)$. The mutual information $I(X;Y)$ is a measure of how much information Y contains about X (or vice versa). Problem 1.29 asks you to rederive the capacity of the BSC(ϵ) from this general formula.

A fundamental fact is that mutual information is non-negative, i.e., $I(X;Y) \geq 0$ (see Problem 1.27). Using the representation $I(X;Y) = H(X) - H(X\,|\,Y)$, we see that *conditioning does not increase entropy*, i.e., we have

$$H(X) \geq H(X\,|\,Y). \tag{1.46}$$

We write

$$X \to Y \to Z \tag{1.47}$$

to indicate that the triple X, Y, and Z forms a *Markov chain*, i.e., to indicate that $p_{X,Y,Z}(x,y,z) = p_X(x)p_{Y\,|\,X}(y\,|\,x)p_{Z\,|\,Y}(z\,|\,y)$. In this case we have

$$p_{X,Z\,|\,Y}(x,z\,|\,y) = p_{X\,|\,Y}(x\,|\,y)p_{Z\,|\,X,Y}(z\,|\,x,y) = p_{X\,|\,Y}(x\,|\,y)p_{Z\,|\,Y}(z\,|\,y).$$

In words, if $X \to Y \to Z$ then X and Z are independent given Y. Conversely, $p_{X,Z|Y}(x,z|y) = p_{X|Y}(x|y)p_{Z|Y}(z|y)$ implies that $X \to Y \to Z$. By symmetry of this condition we see that $X \to Y \to Z$ implies $Z \to Y \to X$. We get a Markov chain in a natural way if, for a pair of random variables X and Y, we let $Z = f(Y)$ for some function $f(\cdot)$. We can characterize Markov chains also in terms of mutual information: the random variables X, Y, and Z form the Markov chain $X \to Y \to Z$ if and only if $I(X;Z|Y) = 0$.

We make use of the *data processing inequality* which states that for any triple of random variables X, Y, and Z such that $X \to Y \to Z$, we have

$$H(X|Y) \le H(X|Z). \tag{1.48}$$

Equivalently, we have $I(X;Y) \ge I(X;Z)$. This is a natural statement: *processing can never increase the mutual information.*

Given a channel $p_{Y|X}(y|x)$ and a function $f(\cdot)$, we say that $Z = f(Y)$ is a *sufficient statistic* for X given Y if $X \to Z \to Y$, i.e., if X is independent of Y given Z (the relationship $X \to Y \to Z$ is always true if $Z = f(Y)$ as we have just discussed). A convenient necessary and sufficient condition that $Z = f(Y)$ constitutes a sufficient statistic is that $p_{Y|X}(y|x)$ can be written in the form $a(x,z)b(y)$ for some suitable functions $a(\cdot,\cdot)$ and $b(\cdot)$ (see Problem 1.22). For us the two most important consequences of knowing that $Z = f(Y)$ is a sufficient statistic for X given Y are that (i) the optimum decision on X can be based on Z alone and (ii) $H(X|Y) = H(X|Z)$. The first claim is a direct consequence of the fact that $p_{Y|X}(y|x)$ can be written in the form $a(x,z)b(y)$ and so the second term can be canceled in the MAP rule (see Section 1.5). The second claim follows from the data processing inequality since we know that in this case we have both $X \to Y \to Z$ (which proves $H(X|Y) \le H(X|Z)$) but also $X \to Z \to Y$ (which proves $H(X|Z) \le H(X|Y)$). A further relationship between the above quantities is discussed in Problem 1.23.

It is helpful to know the *Fano* inequality: assume that we want to estimate the random variable X taking values in the finite alphabet $\mathcal{X}$ knowing the random variable Y. Let $\hat{x}(y)$ denote any estimation rule and let $\mathbb{P}\{\hat{x}(Y) \ne X\}$ denote the resulting probability of error. Fano's inequality asserts that (see Problem 1.28)

$$h(\mathrm{P}) + \mathrm{P}\log(|\mathcal{X}| - 1) \ge H(X|Y). \tag{1.49}$$

For the special case where X is a *binary* random variable, Fano's inequality reads

$$h(\mathrm{P}) \ge H(X|Y). \tag{1.50}$$

Random variables X which have a Gaussian distribution appear frequently. In the Gaussian case we have $p_X(x) = \frac{1}{\sqrt{2\pi\sigma^2}}\exp(-(x-\mu)^2/(2\sigma^2))$, where μ is the *mean* of X, $\mu = \mathbb{E}[X]$, and σ^2 is the *variance* of X, $\sigma^2 = \mathbb{E}[(X-\mu)^2]$. We denote this density by $\mathcal{N}(\mu, \sigma^2)$.

If x is a real parameter taking values in $[0,1]$, e.g., if x is a probability, we write $\bar{x}$ for $1-x$. If $n \in \mathbb{N}$ then the set $\{1,\dots,n\}$ is often abbreviated as $[n]$.

Occasionally we want to bound the number of positive roots of a real-valued polynomial. Let $p(x) = \sum_{i=0}^{d} p_i x^i$ be a *real-valued* polynomial of degree d. Consider the associated piecewise linear function whose graph interpolates the points (i, p_i), $i \in \{0,\dots,d\}$. We say that $p(x)$ has v *sign changes* if this associated function crosses the x-axis v times.

EXAMPLE 1.51. Consider the polynomial $p(x) = 1 + x^4 - 3.4x^5 + x^{11}$. The associated piecewise linear graph passes the x-axis twice so that $v = 2$. ◇

THEOREM 1.52 (DESCARTE'S RULES OF SIGNS). Let $p(x)$ be a real-valued polynomial with v sign changes and r positive real roots. Then $r \le v$ and $v - r$ is even.

EXAMPLE 1.53. Consider again the polynomial $p(x) = 1 + x^4 - 3.4x^5 + x^{11}$. Since $v = 2$ there exists either no positive root or there are exactly two. In fact, since $p(0) = 1 > 0$, $p(1) = -0.4 < 0$, and $p(2) = 1956.2 > 0$, we know that there are exactly two. A numerical calculation shows that the two roots are at $x \approx 0.89144$ and $x \approx 1.11519$. ◇

We frequently need to refer to some basic relations and estimates of *factorials* and *binomials*. We summarize them here for reference:

$$n! = \int_0^\infty e^{-s} s^n ds, \tag{1.54}$$

$$n! = \sqrt{2\pi n}(n/e)^n(1 + O(1/n)), \tag{1.55}$$

$$(n/e)^n \le n!, \tag{1.56}$$

$$e^{-\frac{k^2}{n}} n^k/k! \le \binom{n}{k} \le n^k/k!, \tag{1.57}$$

$$\frac{1}{n+1} 2^{n h_2(k/n)} \le \binom{n}{k}, \tag{1.58}$$

$$\sum_{k=0}^{m} \binom{n}{k} \le 2^{n h_2(m/n)}, \quad m \le n/2, \tag{1.59}$$

$$\binom{2n}{n} = 4^n/\sqrt{n\pi}(1 - 1/(8n) + O(n^{-2})). \tag{1.60}$$

As you can see in (1.54), we write e for $\exp(1)$ to distinguish it from frequently occurring variables named e.

In general, we make no special distinction between *scalars* and *vectors*. In a few places, where confusion might otherwise arise, we denote a vector as $\underline{x}$. We assume that vectors are row vectors. We sometimes write x_i^j as a shorthand for the set of elements $x_i, x_{i+1}, \dots, x_j$. Given a matrix A and a suitable index set $\mathcal{I}$, we let $A_{\mathcal{I}}$ denote the submatrix of A, restricted to the columns indexed by $\mathcal{I}$. If we deal with a set of (random) variables $X_1, \dots, X_n$ it is often convenient to refer to *all but one* of them. In such a case we write $X_{\sim j}$ as a shorthand for $X_1, \dots, X_{j-1}, X_{j+1}, \dots, X_n$.

For a convex-∩ function $f(x)$ (we write "convex-∩" for a concave function like $f(x) = \log(x)$ and "convex-∪" for a convex function like x^2) and a density $p_X(x)$ *Jensen's* inequality asserts that

$$\int f(x) p_X(x) \mathrm{d}x \leq f\Big(\int x p_X(x) \mathrm{d}x\Big). \tag{1.61}$$

Given a polynomial $p(x)$ (or a function which is analytic around zero) we write $\text{coef}\{p(x), x^k\}$ to denote the coefficient of x^k in its Taylor series expansion: $\text{coef}\{(1+x)^n, x^k\} = \binom{n}{k}$.

For a function $a(t)$ let $\mathcal{G}_a(f) = \int a(t) e^{-2\pi j t f} dt$ denote its *Fourier* transform. The Parseval theorem asserts that

$$\int a(t) b^*(t) dt = \int \mathcal{G}_a(f) \mathcal{G}_b^*(f) df,$$

where $*$ denotes complex conjugation.

If $n \in \mathbb{N}$ we say that a function $f(n)$ is $O(g(n))$ if there exists a constant c so that $|f(n)| \leq c|g(n)|$ for all sufficiently large $n \in \mathbb{N}$. We say that $f(n)$ is $o(g(n))$ if $\lim_{n\to\infty} |f(n)|/|g(n)| = 0$. Finally, we write $f(n) = \Theta(g(n))$ if there exist two strictly positive constants c' and c'' so that $c'|g(n)| \leq |f(n)| \leq c''|g(n)|$ for n sufficiently large. If the underlying parameter n is not clear from context we explicitly denote it by writing $O_n(\cdot)$, $o_n(\cdot)$, and $\Theta_n(\cdot)$.

Unless said otherwise, all performance plots are for *ensemble averages* and not for specific instances. (Expectations are denoted as $\mathbb{E}[\cdot]$.) This makes the plots reproducible. Of course, typically better performance can be achieved by carefully selecting a specific instance from the ensemble.

We mark the end of a definition with ▽, the end of an example with ◇, and the end of a proof with □.

Theorems, lemmas, equations, figures, tables, etc. are labeled by a *common sequence of consecutive* numbers within a chapter. This facilitates locating a referenced item.

Maybe slightly unusually, the main text does not contain pointers to the literature. The typical lemma and theorem is a synthesis of either a sequence of papers or has been discovered independently by several authors. Rather than interrupting the flow of ideas in the main text we have opted to collect a history of the developments of the main results at the end of each chapter.

By now, the list of papers in the area of iterative decoding measures in the (tens of) thousands. We have therefore not even attempted to give an exhaustive account of references but to give a sufficient number of pointers to the existing literature so that you can locate the relevant material quickly. We apologize to all authors whose papers we have not included in our list. For the convenience of the reader we have placed a list of references at the end of each chapter. Each bibliographic entry is followed by a list of numbers within square brackets which indicate the pages on which this particular reference is cited.

The order of the exercises roughly follows the order of the topics discussed in the text. We have not ranked the exercises by difficulty. Those exercises that you can solve are easy, the other ones are hard.

Notes

Information theory and *coding* were born in Shannon's paper "A Mathematical Theory of Communication" [57]. It was in this paper that Shannon formulated the communications problem as we have presented it in Section 1.2 and where he showed that the transmission task can be broken down into a *source coding* and a *channel coding* problem. Our exposition of Shannon's *channel coding theorem* in Section 1.6 for the BSC closely follows van Lint [62].

At around the same time, and working at the same location (Bell Labs), Hamming constructed the first *class* of error-correcting codes [34]. Interestingly, Hamming's motivation was not Shannon's capacity but the practical concern of correctly operating a "computing" machine consisting of many relays. Such relays were inherently unreliable and every couple of million relay operations an error would occur. A large body of literature on coding was thereafter concerned with finding dense packings of "Hamming" spheres of radius half the minimum (Hamming) distance.

To date there exists an extensive literature on bounds on *weight distributions*. We refer the reader to the *Handbook of Coding Theory* [52], which contains several survey articles. The lower bound on the *minimum distance* we discuss in Problem 1.13 is due to Gilbert [32]. A similar bound was independently discovered by Varshamov [64]. Asymptotically these two bounds yield the same result, which is the statement discussed in Section 1.4 and in Problem 1.15. The upper bound on the minimum distance stated in Section 1.4 is due to Elias, who presented this result in class lectures in the early 1960s. The bound was also discovered by Bassalygo and first appeared in print in [5]. A substantially tighter upper bound was derived by McEliece, Rodemich, Rumsey, and Welch [48]. It is a tighter version of the linear *programming bound* based on Delsarte's *association schemes* [13]. For a discussion of the weight distribution of random codes (linear and non-linear) see Barg and Forney [3] and the references therein.

The *union bound* on the performance of a code based on its weight distribution is the simplest of a large class of such bounds (see Problem 1.21). A considerably more sophisticated approach was developed by Gallager in his thesis [28]. Good starting points are the article and the monograph by Sason and Shamai [54, 56], which discuss in depth the relationships between the numerous bounds proposed to date and which contain an exhaustive list of references.

In 1971 Cook proved that any problem in NP can be reduced in deterministic polynomial time to the *satisfiability* problem [11]. Later, Karp showed that the satisfiability problem can be reduced to many other problems in NP in deterministic polynomial time. The collection of all these problems is the class of NP-complete problems. The fact that the ML decision problem is NP-complete for binary linear codes was shown by Berlekamp, McEliece, and van Tilborg [7] by reducing the problem to *three-dimensional matching*. The intractability of computing the minimum distance was later shown by Vardy [63]. The classic reference relating to complexity is the book by Garey and Johnson [31]. For survey articles concerning complexity issues in coding theory see Barg [2], Sudan [60], Dumer, Micciancio, and Sudan [16], as well as Spielman [59].

Elias showed that linear codes achieve capacity on the BSC [17]. A proof that linear codes achieve capacity on *any* BMS channel (Theorem 4.68) is due to Dobrushin [15] and can also be found in Gallager [30].

BCH codes were discovered by *B*ose and Ray-*C*haudhuri [9] and independently by *H*ocquenghem [39]. *Reed-Solomon* codes (see Problem 1.9), which can be seen as a special case of BCH codes, were proposed around the same time by Reed and Solomon [53]. They are part of many standards and products. Fairly recently, there have been exciting new developments concerning the decoding of Reed-Solomon codes beyond the minimum distance. This development was initiated by Sudan who introduced his list decoding algorithm [61] for Reed-Solomon codes. This was quickly followed by an improved list decoding algorithm due to Guruswami and Sudan [33]. Further, it was shown by Kötter and Vardy how to turn the list decoding algorithm into a soft-decision decoding algorithm [41].

Convolutional codes (Example 1.31) are due to Elias [17]. The theory of convolutional codes was developed by Forney in a series of papers [22, 23, 24]. Further excellent references on the theory of convolutional codes are the books by Johannesson and Zigangirov [40], Viterbi and Omura [66], and the survey article by McEliece, which is contained in the collection [52]. *Sequential decoding* was invented by Wozencraft [69] with important contributions due to Fano [18]. The *Viterbi algorithm* was introduced by Viterbi [65] originally as a proof technique. It was pointed out by Forney [21] and later independently by Omura [49] that the Viterbi algorithm was optimal, i.e., that it accomplished MAP decoding. Heller, who at that time worked at the Jet Propulsion Laboratory, was the first to recognize that the Viterbi algorithm was not only of theoretical interest but also immensely practical [36, 37]. The classical reference is [25]. If you are looking for a historical account we recommend [27].

Forney developed the theory of *concatenated codes* (see page 22) in his thesis [19, 20].

A simple derivation of the random coding exponent was given by Gallager [29], extending Shannon's random coding technique [57]. Further bounds were developed by Shannon, Gallager, and Berlekamp [58]. The equivalent statements for convolutional codes were developed by Yudkin [70], Viterbi [65], and Forney [26]. For the state of the art in this area we refer the reader to Barg and McGregor [4] as well as Burnashev [10].

For a history of the development of iterative decoding see the Notes at the end of Chapter 4. The representation of Hamming codes via Venn diagrams which we used in Section 1.9 is due to McEliece [47].

The Descartes rule of signs can be found in [14] and [38, Chapter 6]. The idea of performing bit MAP decoding of a block code via a trellis, as discussed in Problem 1.19, is due to Bahl, Cocke, Jelinek, and Raviv [1]. The trellis is typically called the *Wolf trellis* [68], in reference to Wolf's work on the related block MAP decoding problem (another name is *syndrome trellis*). The MacWilliams identities, discussed in Problem 1.20, give a fundamental connection between the weight distribution of a code and its dual [44].

There is a large list of excellent books on classical coding theory. To mention just a few of the most popular ones in alphabetic order: Berlekamp [6], Blahut [8], Lin and Costello [42], MacWilliams and Sloane [45], McEliece [46], Pless [51], and van Lint [62]. A list of survey articles was collected by Pless and Huffman [52]. For books that focus exclusively on convolutional codes see Johannesson and Zigangirov [40], Piret [50], and Viterbi and Omura [66]. More recently, some books on iterative decoding also have become available: Heegard and Wicker [35], Schlegel and Perez [55], and Vucetic and Yuan [67]. Readers interested in information theory need look no further than to Gallager [30] or Cover and Thomas [12]. A further recommended source which spans a wide range of topics is the book by MacKay [43].

Codes also play an important role in many areas of science. To name just a few: in *mathematics* they give rise to dense point lattices and allow the construction of designs for use in *statistics*. In the area of *computer science*, public-key crypto systems may be based on codes, they are the crucial ingredient for some extractors – generating many weakly random bits from a few truly random ones – and they make efficient hashing functions. Codes are helpful in minimizing the communication complexity and they can be used to turn worst-case hardness into average-case hardness. In the area of *communications*, besides helping to combat the noise introduced by the communication channel, codes can help to represent sources efficiently, to avoid large peak-to-average power ratios, or to minimize interference in multi-user systems. Most of these areas are well outside the scope of this book.

PROBLEMS

1.1 (INNER PRODUCT). Let $\mathbb{F}$ be a field and consider the vector space $\mathbb{F}^n$, $n \in \mathbb{N}$. For $u, v \in \mathbb{F}^n$ define the *inner product* of u and v by $\langle u, v \rangle = \sum_{i=1}^n u_i v_i$, where all operations are performed in $\mathbb{F}$. Show that this inner product has the following properties:

1. $\langle t + u, v \rangle = \langle t, v \rangle + \langle u, v \rangle, \quad t, u, v \in \mathbb{F}^n,$
2. $\langle \alpha u, v \rangle = \alpha \langle u, v \rangle, \quad \alpha \in \mathbb{F}, u, v \in \mathbb{F}^n,$
3. $\langle u, v \rangle = \langle v, u \rangle, \quad u, v \in \mathbb{F}^n.$

Unfortunately, $\langle \cdot, \cdot \rangle$ is not necessarily an inner product in the mathematical sense: show that, for $\mathbb{F} = \mathbb{F}_2$, $\langle u, u \rangle = 0$ does not imply $u = 0$ by exhibiting a counterexample. Therefore, $\mathbb{F}_2^n$, equipped with $\langle \cdot, \cdot \rangle$, is not an inner-product space. As a consequence, if G is a generator matrix over $\mathbb{F}_2$ and H is a corresponding parity-check matrix, then

$$\begin{pmatrix} G \\ H \end{pmatrix}$$

does not necessarily span the whole space $\mathbb{F}_2^n$ – the row-space spanned by H is not the orthogonal complement of the row-space spanned by G. Consider, e.g., the generator matrix

$$G = \begin{pmatrix} 1 & 1 & 0 & 0 \\ 0 & 0 & 1 & 1 \end{pmatrix}$$

and the associated code $C(G)$. Determine $C^\perp$. A code C such that $C = C^\perp$ is called *self-dual.*

1.2 (HAMMING DISTANCE). Show that the Hamming distance introduced in Definition 1.11 fulfills the triangle inequality.

1.3 (EXTENDING, PUNCTURING, AND SHORTENING). In this exercise we are concerned with certain simple procedures that allow us to construct new codes from given ones. Assume we have a code $C(n, M, d)$ (not necessarily linear) over a field $\mathbb{F}$. If d is odd then we can *extend* the code by appending to each codeword $x = (x_1, \dots, x_n)$ the extra symbol $x_{n+1} = -\sum_{i=1}^n x_i$. The inverse operation, namely to delete a symbol, is called *puncturing*. Finally, consider any subset of codewords that have the same last symbol. Keep only this subset and delete this common symbol. This procedure is called *shortening*. How do these procedures change the parameters of the code? Under which of these operations do linear codes stay linear?

1.4 ($u + v$ Construction). Assume we are given two binary codes $C_1(n, M_1, d_1)$ and $C_2(n, M_2, d_2)$. Define a new code C by

$$C = \{(u, u + v) \mid u \in C_1, v \in C_2\}.$$

Show that C is binary and has parameters $C(2n, M_1M_2, d = \min(2d_1, d_2))$.

1.5 (Proper Codes). Let C be a proper (see Definition 1.21 on page 14) linear code of length n over the field $\mathbb{F}$. Prove that, for every position i, $1 \le i \le n$, and every field element α, $\alpha \in \mathbb{F}$, the number of codewords that have an α at position i is equal to $|C|/|\mathbb{F}|$. What happens if the code is not proper? Is the equivalent statement true if you consider a *set* of positions?

1.6 (One C, many G). Show that a binary linear code C of length n and dimension k has $2^{\binom{k}{2}} \prod_{i=1}^{k} (2^i - 1)$ distinct $k \times n$ generator matrices.

1.7 (Conversion of G into Systematic Form). Let G be a $k \times n$ generator matrix of rank k. Show that G can be brought into systematic form by elementary row/column operations (i.e., permutations of rows/columns, multiplication of a row by a non-zero element of $\mathbb{F}$, and addition of one row to another.) Prove that these operations do not change the minimum distance of the code.

Hint: Show that G must contain a $k \times k$ rank-k submatrix, call it A, formed by k columns of G. Multiply G from the left by A^{-1} and perform column permutations if necessary to bring the result into systematic form.

1.8 (Conversion $G \leftrightarrow H$). Show that if G is a generator matrix in systematic form, $G = (I_k\ P)$, then $H = \left(-P^T\ I_{n-k}\right)$ is a corresponding parity-check matrix. The parity-check matrix of the $[7, 4, 3]$ binary Hamming code given in (1.25) has the form $\left(-P^T\ I_{n-k}\right)$. Use this fact to write down a corresponding generator matrix G.

1.9 (Reed-Solomon Codes). Reed-Solomon (RS) codes are one of the most widely used codes today. In addition they possess several fundamental properties (see, e.g., Problem 1.11). Here is how they are defined.

Given a finite field $\mathbb{F}$, choose n and k such that $n \le |\mathbb{F}|$ and $1 \le k \le n$. To construct a RS code with parameters n and k over the field $\mathbb{F}$ choose n *distinct* elements from $\mathbb{F}$, call them $x_0, \dots, x_{n-1}$. Let $\mathbb{F}[x]$ denote the ring of polynomials with coefficients in $\mathbb{F}$, i.e., the set of polynomials in the indeterminate x with coefficients in $\mathbb{F}$ and the standard addition and multiplication of polynomials. Then C is defined as

$$C = \{(A(x_0), \dots, A(x_{n-1})) : A(x) \in \mathbb{F}[x] \text{ s.t. } \deg(A(x)) < k\}.$$

In words, we consider the set of polynomials with coefficients in $\mathbb{F}$ and degree at most $k - 1$. Each such polynomial we evaluate at the n distinct points x_i, $0 \le i < n$, and the result of these evaluations form the n components of a codeword.

Show that an RS code as defined earlier with parameters n and k over the field $\mathbb{F}$ has dimension k and minimum distance $n-k+1$. We summarize the parameters of such a code in the compact form $[n, k, n-k+1]$.

Hint: Over any field $\mathbb{F}$, a polynomial of degree d has at most d roots in $\mathbb{F}$. This is called the *fundamental theorem of algebra.*

1.10 (Singleton Bound). Show that for any code $C(n, M, d)$ over a field $\mathbb{F}$, $d \leq n - \log_{|\mathbb{F}|}(M) + 1$. This is called the *Singleton* bound. Codes that achieve this bound are called *maximum-distance separable* (MDS).

Hint: Arrange the M codewords of length n in the form of an $M \times n$ matrix and delete all but the first $\lfloor \log_{|\mathbb{F}|}(M) \rfloor$ columns. Argue that the minimum distance between the resulting rows is at most one.

1.11 (Maximum Distance Separable Codes). As discussed in Problem 1.10, a code C which fulfills the Singleton bound is called MDS. Examples of such codes are repetition codes with parameters $[n, 1, n]$ and RS codes with parameters $[n, k, n-k+1]$. Let C be an MDS code with parameters $[n, k, d = n-k+1]$. Show that C has the following property. Any subset of $[n]$ of cardinality k is an *information set.* More precisely, for any such subset and any choice of elements of $\mathbb{F}$ at these k positions there is exactly one codeword which agrees with this choice.

Next let C be a binary linear code with parameters $[n, k]$ and generator matrix G. Assume that its dual, $C^{\perp}$, is an MDS code (hence this dual has parameters $[n, n-k, k+1]$). Show that also in this case C has the property that any subset of $[n]$ of cardinality k is an information set.

1.12 (Hamming Bound). Consider a code $C(n, M, d)$ over a field $\mathbb{F}$. Prove that $M \sum_{i=0}^{\lfloor \frac{d-1}{2} \rfloor} \binom{n}{i} (|\mathbb{F}|-1)^i \leq |\mathbb{F}|^n$.

Hint: Consider spheres of radius $t = \lfloor \frac{d-1}{2} \rfloor$ centered around each codeword.

Note: A code C which fulfills the Hamming bound with equality is said to be *perfect.* It is easy to check that binary repetition codes of odd length are perfect codes. Further all Hamming codes are perfect. Without proof we state that the only other perfect multiple-error correcting codes are the $[23, 12, 7]$ binary Golay code and the $[11, 6, 5]$ ternary Golay code.

1.13 (Gilbert-Varshamov Bound). Show that for any $d \leq n$, $n \in \mathbb{N}$, and any field $\mathbb{F}$ there exist codes $C(n, M, d)$ over $\mathbb{F}$ with $M \geq \frac{|\mathbb{F}|^n}{\sum_{i=0}^{d-1} \binom{n}{i} (|\mathbb{F}|-1)^i}$.

Hint: Consider spheres of radius $d-1$ centered around each codeword.

1.14 (Greedy Code Search Algorithm). The Gilbert-Varshamov bound we discussed in Problem 1.13 gives rise to the following greedy code search algorithm. Start by picking a point in $\mathbb{F}^n$ where the choice is made with uniform probability. The initial code C consists of this chosen codeword. At any following step delete from $\mathbb{F}^n$ all codewords currently in C as well as all points contained in spheres of radius $d-1$ around these codewords, where d is the design distance. If the resulting set is nonempty pick another point and add it to C, otherwise the algorithm terminates.

Show that this algorithm results in a code $C(n, M, d)$ over $\mathbb{F}$ with minimum distance at least d and cardinality at least

$$\left\lceil |\mathbb{F}|^n / \left(\sum_{i=0}^{d-1} \binom{n}{i} (|\mathbb{F}|-1)^i\right) \right\rceil.$$

Note: Such a greedy algorithm does in general not result in a linear code. Surprisingly, the bound is still true if we restrict C to be linear, and a slightly modified greedy algorithm generates a linear code with the desired properties. Define $V(\mathbb{F}, n, t) = \sum_{i=0}^{t} \binom{n}{i} (|\mathbb{F}|-1)^i$. Let C_0 be the trivial $[n, 0, n]$ code consisting of the zero word only. We proceed by induction. Suppose that we constructed the linear $[n, k, \geq d]$ code C_k. Such a code contains $M = |\mathbb{F}|^k$ codewords. If $|\mathbb{F}|^k V(\mathbb{F}, n, d) \geq |\mathbb{F}|^n$ then we stop and output C. Otherwise proceed as follows. Delete from $\mathbb{F}^n$ all codewords of C_k as well as all points contained in the spheres of radius $d-1$ around these codewords. Since $|\mathbb{F}|^k V(\mathbb{F}, n, d-1) < |\mathbb{F}|^n$, the resulting set is not empty. Pick any element from this set and call it g_{k+1}. Let C_{k+1} be the linear code resulting from joining g_{k+1} to a set of generators $\{g_1, \dots, g_k\}$ for C_k. We claim that C_{k+1} has minimum distance at least d. Every element c' in C_{k+1} has a representation of the form $c' = a g_{k+1} + c$ where $a \in \mathbb{F}$ and $c \in C_k$. If $a = 0$ then $w(c') = w(c) \geq d$ by the assumption that C_k is a $[n, k, \geq d]$ code. If on the other hand $a \neq 0$ then $w(c') = w(a g_{k+1} + c) = w(g_{k+1} + a^{-1} c) = d(g_{k+1}, -a^{-1} c) \geq d$ by our choice of g_{k+1} and since $-a^{-1} c \in C_k$ by linearity.

1.15 (Asymptotic Gilbert-Varshamov Bound). Fix $\mathbb{F} = \mathbb{F}_2$. Consider codes of increasing blocklength n with $2^{\lfloor nr \rfloor}$ codewords, where r, $r \in (0, 1)$, is the rate. Let $d(C)$ denote the minimum distance of a code C and define the normalized distance $\delta = d/n$. Starting with the Gilbert-Varshamov bound discussed in Problem 1.13, show that $\delta^*(r)$ as defined on page 7 fulfills $\delta^*(r) \geq h_2^{-1}(1-r)$.

1.16 (Pairwise Independence for Generator Ensemble). Let $\mathbb{F} = \mathbb{F}_2$ and consider the following slight generalization of Elias's generator ensemble $\mathcal{G}(n, k)$; call it $\tilde{\mathcal{G}}(n, k)$. To sample from $\tilde{\mathcal{G}}(n, k)$ choose a random element C from $\mathcal{G}(n, k)$ and a random translation vector c from $\mathbb{F}_2^n$. The random sample $\tilde{C}$ is $\tilde{C} = C + c$, i.e., a translated version of the code C. Prove that in $\tilde{\mathcal{G}}(n, k)$ codewords have a uniform distribution and that pairs of codewords are independent. More precisely, let $\{u^{[i]}\}_i$, where i ranges from 0 to $2^k - 1$, denote the set of binary k-tuples, ordered in some fixed but arbitrary way. For $G \in \mathcal{G}(n,k)$ and $c \in \mathbb{F}_2^n$, let $x^{[i]}(G,c) = c + u^{[i]}G$ be the i-th codeword. Prove that for $j \neq i$, $\mathbb{P}\{X^{[j]} \mid X^{[i]} = x\}$ is uniform, i.e., for any $v \in \mathbb{F}_2^n$, $\mathbb{P}\{X^{[j]} = v \mid X^{[i]} = x\} = \frac{1}{2^n}$.

1.17 (Mean and Second Moment for $\mathcal{G}(n,k)$). Consider Elias's generator ensemble $\mathcal{G}(n,k)$, $0 < k \leq n$, as described in Definition 1.26. Recall that its design rate is $r = k/n$. For $C \in \mathcal{G}(n,k)$, let $A(C,w)$ denote the codewords in C of Hamming weight w. Show that

$$\mathbb{E}_C[A(C,w=0)] = 1 + \frac{2^{nr}-1}{2^n} = \bar{A}(w=0),$$
$$\mathbb{E}_C[A(C,w)] = \binom{n}{w}\frac{2^{nr}-1}{2^n} = \bar{A}(w), w \geq 1,$$
$$\mathbb{E}_C[A^2(C,w)] = \bar{A}(w) + \frac{(2^{nr}-1)(2^{nr}-2)}{2^{2n}}\binom{n}{w}^2, w \geq 1,$$
$$\mathbb{E}_C[(A(C,w)-\bar{A}(w))^2] = \bar{A}(w) - \frac{2^{nr}-1}{2^{2n}}\binom{n}{w}^2, w \geq 1.$$

What is the expected number of distinct codewords in an element chosen uniformly at random from $\mathcal{G}(n,k)$?

1.18 (Mean and Second Moment for $\mathcal{H}(n,k)$). Consider Gallager's parity-check ensemble $\mathcal{H}(n,k)$, $0 < k \leq n$, as described in Definition 1.26. Recall that its design rate is equal to $r = k/n$. For $C \in \mathcal{H}(n,k)$, let $A(C,w)$ denote the codewords in C of Hamming weight w. Show that

$$A(C,w=0)] = 1,$$
$$\mathbb{E}_C[A(C,w)] = \binom{n}{w}2^{-n(1-r)} = \bar{A}(w), \quad w \geq 1,$$
$$\mathbb{E}_C[A^2(C,w)] = \bar{A}(w) + \binom{n}{w}\left(\binom{n}{w}-1\right)2^{-2n(1-r)}, \quad w \geq 1.$$
$$\mathbb{E}_C[(A(C,w)-\bar{A}(w))^2] = \bar{A}(w) - \binom{n}{w}2^{-2n(1-r)}, \quad w \geq 1.$$

What is the expected number of distinct codewords in an element chosen uniformly at random from $\mathcal{H}(n,k)$?

1.19 (Wolf Trellis – Bahl, Cocke, Jelinek, and Raviv [1], Wolf [68]). There is an important graphical representation of codes, called a *trellis*. A trellis $T = (V,E)$ of rank n is a finite directed graph with vertex set V and edge set E. Each vertex is assigned a "depth" in the range $\{0,\ldots,n\}$, and each edge connects a vertex at depth i to one at depth $i+1$, for some $i = 0,1,\ldots,n-1$. The Wolf trellis for a binary linear code C with $(n-k) \times n$ parity-check matrix $H = (h_1^T, \dots, h_n^T)$, where h_j^T is the j-th column of H, is defined as follows. At depth i, $i = 0, \dots, n$, the vertex set consists of 2^{n-k} vertices, which we identify with the set of binary $(n-k)$-tuples. A vertex v at depth i is connected to a vertex u at depth $(i+1)$ with label $\lambda \in \{0, 1\}$ if there exists a codeword $c = (c_1, \dots, c_n) \in C$ with $c_{i+1} = \lambda$ such that

$$v = \sum_{j=1}^{i} c_j h_j \quad \text{and} \quad u = \sum_{j=1}^{i+1} c_j h_j,$$

where by convention $\sum_{j=1}^{0} c_j h_j = (0, \dots, 0)$. This is the key to this representation: there is a one-to-one correspondence between the labeled paths in the Wolf trellis starting and ending in the zero state (the fact that the final state is zero follows from $Hc^T = 0^T$) and the codewords of C. Note that codewords "share" edges and so the total number of edges is, in general, much smaller than the number of codewords times the length of the code. Hence, the Wolf trellis is a compact representation of a code. Many important algorithms can be performed on the trellis of a code (e.g., decoding, determination of minimum distance, etc.) We revisit this topic in Section 6.1 when discussing convolutional codes.

Draw the Wolf trellis for the $[7, 4, 3]$ Hamming code with parity-check matrix given in (1.25).

1.20 (MacWilliams Identities – MacWilliams [44] – presentation that follows due to Barg). For linear codes there is a fundamental relationship between the weight distribution of a code and the weight distribution of its dual code. Let us explore this relationship for the binary case. Let C be a binary linear $[n, k]$ code. Let A_i, $i = 0, \dots, n$, be the number of codewords in C of weight i (to lighten the notation we write A_i instead of $A(C, i)$). Note that $A_0 = 1$ and that $|C| = \sum_{i=0}^{n} A_i$. We call the collection $\{A_i\}_{0 \le i \le n}$ the weight distribution of the code C. In the same manner, we call $\{A_i^\perp\}_{0 \le i \le n}$ the weight distribution of the dual code $C^\perp$. The MacWilliams identity states that

$$A_j^\perp = \frac{1}{|C|} \sum_{i=0}^{n} A_i P_j(i), \; 0 \le j \le n,$$

where $P_j(i) = \sum_{l=0}^{n} (-1)^l \binom{i}{l} \binom{n-i}{j-l}$.

1. Let E be a subset of $\{1, 2, \dots, n\}$ and $|E|$ denote its size. Let $C(E)$ be the subcode of C with zeros outside E. More precisely, $C(E)$ is the collection of those elements of C which have non-zero elements only within the coordinates indexed by E. Note that the zero codeword is always in $C(E)$ so that $C(E)$ is not empty. Prove that $\sum_{i=0}^{n} A_i \binom{n-i}{n-w} = \sum_{|E|=w} |C(E)|$ for every $0 \le w \le n$, where the sum is over all E of size w.

2. Let H_E denote the submatrix formed by the columns of H that are indexed by E. Assume that $|E| = w$. Prove that $C(E)$ is a binary linear code with dimension $\dim(C(E)) = w - \text{rank}(H_E)$.

3. Prove that $w - \text{rank}(H_E) = k - \text{rank}(G_{\bar{E}})$ where $\bar{E}$ is the complement of E within $\{1, \ldots, n\}$. (This property plays a central role in Lemma 3.74 and Theorem 3.77.)

4. Justify each of the following steps:

$$\begin{aligned}\sum_{i=0}^{n-u} A_i^{\perp} \binom{n-i}{u} &\overset{\text{(i)}}{=} \sum_{|E|=n-u} |C^{\perp}(E)| \\ &\overset{\text{(ii)}}{=} \sum_{|E|=n-u} 2^{n-u-\text{rank}(G_E)} \\ &\overset{\text{(iii)}}{=} 2^{n-k-u} \sum_{|E|=n-u} 2^{u-\text{rank}(H_{\bar{E}})} \\ &\overset{\text{(iv)}}{=} 2^{n-k-u} \sum_{i=0}^{u} A_i \binom{n-i}{n-u}.\end{aligned}$$

5. Use $\sum_i (-1)^{j-i} \binom{n-j}{n-i} \binom{n-i}{n-w} = \delta_{j,w}$ to prove $A_j^{\perp} = \frac{1}{|C|} \sum_{i=0}^{n} A_i P_j(i)$, $0 \leq j \leq n$, where $P_j(i) = \sum_{l=0}^{n} (-1)^l \binom{i}{l} \binom{n-i}{j-l}$.

Hint: Consider the generating function $\sum_{i=0}^{n} P_j(i) x^i = (1+x)^{n-i}(1-x)^i$ and expand it using the form $(1-x)^n (1 + \frac{2x}{1-x})^{n-i}$.

6. Use your result to calculate the weight distribution of the dual of the $[7, 4, 3]$ Hamming code.

1.21 (Upper Bound on Error Probability via Weight Distribution). Consider a linear binary code C of length n and dimension k. Let A_w denote the number of words of weight w and let

$$A(D) = \sum_{w=0}^{n} A_w D^w.$$

$A(D)$ is called the *weight enumerator* of the code. Consider the parity-check ensemble $\mathcal{H}(n, k)$.

Let x be a *fixed* binary word of length n. Argue that for a randomly chosen element $C \in \mathcal{H}(n, k)$

$$\mathbb{P}\{x \in C\} = \begin{cases} 1, & x = 0, \\ 2^{-(n-k)}, & \text{otherwise.} \end{cases}$$

Hint: What is the probability that x fulfills *one* parity-check equation, i.e., that $\sum_j H_{i,j} x_j = 0$ for some fixed i?

Since H is random, A_w is a random variable as well and we denote it by $A_w(H)$. Use the previous result and argue that *every* code in this ensemble contains the all-zero word, in particular

$$\bar{A}_0 = \mathbb{E}[A_0(H)] = 1,$$

and that the *expected* number of codewords of weight w, $w > 0$, is given by

$$\bar{A}_w = \mathbb{E}[A_w(H)] = \binom{n}{w} 2^{-(n-k)}.$$

Define the *average* weight enumerator as

$$\bar{A}(D) = \sum_{w=0}^{n} \bar{A}_w D^w.$$

Using the preceding result on $\bar{A}_w$ and your mastery of formal power sums, write $\bar{A}(D)$ in a compact form.

For a fixed code C, we will now derive a simple upper bound on its block error probability under ML decoding when transmitting over an additive white Gaussian noise channel with noise variance σ^2. More precisely, suppose the input takes values in $\{\pm 1\}$ and that we add to each component of the transmitted codeword an iid Gaussian random variable with variance σ^2. Let Y denote the received word and let $\hat{x}^{\mathrm{ML}}(y)$ denote the ML decoded vector. Finally, let $\mathrm{P_B}$ denote the resulting probability of block error. Justify each step in the following sequence:

$$\begin{aligned}
\mathrm{P_B} &= \mathbb{P}\{\hat{x}^{\mathrm{ML}}(Y) \neq 0 \mid X = 0\} = \mathbb{P}\{\hat{x}^{\mathrm{ML}}(Y) \in C \setminus \{0\} \mid X = 0\} \\
&= \mathbb{P}\{\max_{x \in C \setminus \{0\}} p(Y \mid x) \geq p(Y \mid 0) \mid X = 0\} \leq \sum_{x \in C \setminus \{0\}} \mathbb{P}\{p(Y \mid x) \geq p(Y \mid 0) \mid X = 0\} \\
&= \sum_{x \in C \setminus \{0\}} \mathbb{P}\{|x - Y|^2 \leq |Y|^2 \mid X = 0\} = \sum_{x \in C \setminus \{0\}} Q\left(\frac{\sqrt{w(x)}}{\sigma}\right) \\
&= \sum_{w=1}^{n} A_w Q\left(\frac{\sqrt{w}}{\sigma}\right) \leq \frac{1}{2} \sum_{w=1}^{n} A_w \mathrm{e}^{-\frac{w}{2\sigma^2}} = \frac{1}{2}\left[A(\mathrm{e}^{-\frac{1}{2\sigma^2}}) - A_0\right].
\end{aligned}$$

Collect all the previous results to conclude that the *average* block error probability for our ensemble can be bounded by

$$\mathbb{E}[\mathrm{P_B}(H)] \leq \frac{\left(1 + \mathrm{e}^{-\frac{1}{2\sigma^2}}\right)^n - 1}{2^{n-k+1}} \leq 2^{n(\log_2(1 + \mathrm{e}^{-\frac{1}{2\sigma^2}}) - (1-r))}.$$

We conclude that as long as $r < 1 - \log_2(1 + \mathrm{e}^{-\frac{1}{2\sigma^2}})$, we can drive the expected expected error probability to 0 by increasing the length. Better bounds can improve this rate function.

1.22 (Sufficient Statistic). Consider transmission of X chosen with probability $p_X(x)$ from some code C and let Y denote the received observation. Further, let $Z = f(Y)$, where $f(\cdot)$ is a given function. Prove that Z constitutes a sufficient statistic for X given Y if and only if $p_{Y|X}(y|x)$ can be brought into the form $a(x,z)b(y)$ for some suitable functions $a(\cdot,\cdot)$ and $b(\cdot)$.

1.23 (More on Sufficient Statistic). Consider three random variables X, Y, and Z. Suppose that $X \to Y \to Z$, where $Z = f(Y)$. We discussed on page 29 that if $X \to Z \to Y$ then $H(X|Y) = H(X|Z)$.

Show that the converse is true as well. More precisely, show that if $H(X|Y) = H(X|Z)$ then $X \to Z \to Y$.

Hint: Show that $I(X;Y|Z) = 0$.

1.24 (Bound on Binomials). Let $0 \le k \le m$ and $k, m \in \mathbb{N}$. Justify the following steps.

$$\frac{1}{\binom{m}{k}} = \frac{k!}{m(m-1)\dots(m-k+1)} = \frac{k!}{m^k}\mathrm{e}^{-\sum_{i=1}^{k-1}\ln(1-i/m)} \le \frac{k!}{m^k}\mathrm{e}^{k^2/m}.$$

Hint: Bound the sum by an integral.

1.25 (Bound on Sum of Binomials). Prove the upper bound stated in (1.59).

Hint: Consider the binomial identity $\sum_{k=0}^{n}\binom{n}{k}x^k = (1+x)^n$ with $x = m/(n-m)$.

1.26 (Chain Rule). Give a proof of the chain rule $H(X,Y) = H(X) + H(Y|X)$.

Hint: Write $p_{X,Y}(x,y)$ as $p_X(x)p_{Y|X}(y|x)$.

1.27 (Non-negativity of Mutual Information). Prove that $I(X;Y) \ge 0$.

Hint: Write $-I(X;Y)$ as $\sum_{x,y} p_{X,Y}(x,y)\log\frac{p_X(x)p_Y(y)}{p_{X,Y}(x,y)}$ and apply Jensen's inequality (1.61).

1.28 (Fano's Inequality). Prove Fano's inequality (1.49).

Hint: Define the $\{0,1\}$-valued random variable E, which takes on the value 1 if $\hat{x}(Y) \ne X$. Expand $H(E,X|Y)$ as $H(X|Y) + H(E|X,Y)$ and $H(E|Y) + H(X|E,Y)$, equate the two terms, and bound $H(E|X,Y)$, $H(E|Y)$, as well as $H(X|E,Y)$.

1.29 (The Capacity of the BSC Rederived). Start with (1.45) and show that the capacity of the BSC(ϵ) is equal to $C_{\text{BSC}}(\epsilon) = 1 - h_2(\epsilon)$ bits per channel use.

1.30 (Descarte's Rule of Signs). Prove Theorem 1.52.

Hint: Use induction on r. First prove the claim for $r = 0$ using the fundamental theorem of algebra.

References

[1] L. Bahl, J. Cocke, F. Jelinek, and J. Raviv, *Optimal decoding of linear codes for minimizing symbol error rate*, IEEE Trans. Inform. Theory, 20 (1974), pp. 284–287. [34, 39, 66]

[2] A. Barg, *Complexity issues in coding theory*, in Handbook of Coding Theory, V. S. Pless and W. C. Huffman, eds., Elsevier Science, Amsterdam, Holland, 1998, pp. 649–754. [33]

[3] A. Barg and G. D. Forney, Jr., *Random codes: Minimum distances and error exponents*, IEEE Trans. Inform. Theory, 48 (2002), pp. 2568–2573. [32]

[4] A. Barg and A. McGregor, *Distance distribution of binary codes and the error probability of decoding*, IEEE Trans. Inform. Theory, 51 (2005), pp. 4237–4246. [34]

[5] L. A. Bassalygo, *New upper bounds for error correcting codes*, Problemy Peredachi Informatsii, 1 (1965), pp. 41–44. [32]

[6] E. R. Berlekamp, *Algebraic Coding Theory*, Aegean Park Press, Walnut Creek, CA, USA, revised ed., 1984. [34]

[7] E. R. Berlekamp, R. J. McEliece, and H. C. A. van Tilborg, *On the inherent intractability of certain coding problems*, IEEE Trans. Inform. Theory, 24 (1978), pp. 384–386. [33]

[8] R. E. Blahut, *Algebraic Codes for Data Transmission*, Cambridge Univ. Press, 2003. [34]

[9] R. C. Bose and D. K. Ray-Chaudhuri, *On a class of error correcting binary group codes*, Inform. Control, 3 (1960), pp. 68–79. [33]

[10] M. Burnashev, *Code spectrum and reliability function: Binary symmetric channel.* E-print: cs./0612032, 2007. [34]

[11] S. A. Cook, *The complexity of theorem proving procedures*, in Proc. of STOC, 1971, pp. 151–158. [33]

[12] T. M. Cover and J. A. Thomas, *Elements of Information Theory*, Wiley, New York, NY, USA, 1991. [34]

[13] P. Delsarte, *An algebraic approach to the association schemes of coding theory*. Philips Research Reports Supplements no. 10, 1973. [32]

[14] R. Descartes, Geometrie *1636*, *A Source Book in Mathematics*, Harvard Univ. Press, Cambridge, MA, USA, 1969. [34]

[15] R. L. Dobrushin, *Asymptotic optimality of group and systematic codes for certain channels*, Teor. Veroyat. i Primenen., 8 (1963), pp. 52–66. [33]

[16] I. Dumer, D. Micciancio, and M. Sudan, *Hardness of approximating the minimum distance of a linear code*, IEEE Trans. Inform. Theory, 49 (2003), pp. 22–37. [33]

[17] P. Elias, *Coding for noisy channels*, in IRE International Convention Record, Mar. 1955, pp. 37–46. [33]

[18] R. M. Fano, *A heuristic discussion of probabilistic decoding*, IEEE Trans. Inform. Theory, 9 (1963), pp. 64–74. [33]

[19] G. D. Forney, Jr., *Concatenated Codes*, PhD thesis, MIT, 1966. [34]

[20] ——, *Concatenated Codes*, MIT Press, 1966. [34]

[21] ——, *Review of random tree codes*. Appendix A, Final Report, Contract NAS2-3637, NASA CR73176, NASA Ames Res. Ctr., Dec. 1967. [33]

[22] ——, *Convolutional codes I: Algebraic structure*, IEEE Trans. Inform. Theory, 16 (1970), pp. 720–738. [33, 366]

[23] ——, *Correction to 'Convolutional codes I: Algebraic structure'*, IEEE Trans. Inform. Theory, 17 (1971), p. 360. [33, 366]

[24] ——, *Structural analysis of convolutional codes via dual codes*, IEEE Trans. Inform. Theory, 19 (1973), pp. 512–518. [33]

[25] ——, *The Viterbi algorithm*, Proc. IEEE, 61 (1973), pp. 268–278. [34]

[26] ——, *Convolutional codes II: Maximum-likelihood decoding*, Inform. Contr., 25 (1974), pp. 222–266. [34, 366]

[27] ——, *The Viterbi algorithm: A personal history*. E-print: cond-mat/0104079, 2005. [34]

[28] R. G. Gallager, *Low-density parity-check codes*, IRE Trans. Inform. Theory, 8 (1962), pp. 21–28. [33, 66, 159, 264, 419, 434]

[29] ——, *A simple derivation of the coding theorem and some applications*, IEEE Trans. Inform. Theory, 11 (1965), pp. 3–18. [34]

[30] ——, *Information Theory and Reliable Communication*, Wiley, New York, NY, USA, 1968. [33, 34]

[31] M. R. Garey and D. S. Johnson, *Computers and Intractability: A Guide to the Theory of NP-Completeness*, W. H. Freeman, San Francisco, CA, USA, 1979. [33]

[32] E. N. Gilbert, *A comparison of signaling alphabets*, Bell System Tech. J., 31 (1952), pp. 504–522. [32]

[33] V. Guruswami and M. Sudan, *Improved decoding of Reed-Solomon and algebraic-geometry codes*, IEEE Trans. Inform. Theory, 45 (1999), pp. 1757–1767. [33]

[34] R. W. Hamming, *Error detecting and error correcting codes*, Bell System Tech. J., 26 (1950), pp. 147–160. [32]

[35] C. Heegard and S. B. Wicker, *Turbo Coding*, Kluwer Academic Publ., New York, NY, USA, 1999. [34, 66]

[36] J. A. Heller, *Short constraint length convolutional codes.* Jet Prop. Lab., Space Prog. Summary 37-54, 1968. [34]

[37] ——, *Improved performance of short constraint length convolutional codes.* Jet Prop. Lab., Space Prog. Summary 37-56, 1969. [34]

[38] P. Henrici, *Applied and Computational Complex Analysis, Vol. 1*, Wiley, New York, NY, USA, 1974. [34]

[39] A. Hocquenghem, *Codes correcteurs d'erreurs*, Chiffres, 2 (1959), pp. 147–156. [33]

[40] R. Johannesson and K. S. Zigangirov, *Fundamentals of Convolutional Coding*, IEEE Press, Piscataway, NJ, USA, 1999. [33, 34, 366]

[41] R. Kötter and A. Vardy, *Algebraic soft-decision decoding of Reed-Solomon codes*, IEEE Trans. Inform. Theory, 49 (2003), pp. 2809–2825. [33]

[42] S. Lin and D. J. Costello, Jr., *Error Control Coding*, Prentice Hall, Englewood Cliffs, NJ, USA, 2nd ed., 2004. [34]

[43] D. J. C. MacKay, *Information Theory, Inference, and Learning Algorithms*, Cambridge Univ. Press, 2003. [34]

[44] F. J. MacWilliams, *A theorem on the distribution of weights in a systematic code*, Bell System Tech. J., 42 (1963), pp. 79–94. [34, 40]

[45] F. J. MacWilliams and N. J. Sloane, *The Theory of Error-Correcting Codes*, North-Holland, 1977. [34]

[46] R. J. McEliece, *The Theory of Information and Coding: A Mathematical Framework for Communication*, Addison-Wesley, Reading, MA, USA, 1977. [34]

[47] ——, *The 2004 Shannon lecture*, in Proc. of the IEEE Int. Symposium on Inform. Theory, Chicago, IL, USA, June 27–July 2, 2004, IEEE. [34]

[48] R. J. McEliece, E. R. Rodemich, H. Rumsey, Jr., and L. R. Welch, *New upper bounds on the rate of a code via the Delsarte-MacWilliams inequalities*, IEEE Trans. Inform. Theory, 23 (1977), pp. 157–166. [32]

[49] J. K. Omura, *On the Viterbi decoding algorithm*, IEEE Trans. Inform. Theory, 15 (1969), pp. 177–179. [33]

[50] P. M. Piret, *Convolutional Codes: An Algebraic Approach*, MIT Press, 1988. [34, 366]

[51] V. S. Pless, *Introduction to the Theory of Error-Correcting Codes*, Wiley, 1989. [34]

[52] V. S. Pless and W. C. Huffman, eds., *Handbook of Coding Theory*, Elsevier Science, Amsterdam, Holland, 1998. [32–34]

[53] I. S. REED AND G. SOLOMON, *Polynomial codes over certain finite fields*, SIAM J., 8 (1960), pp. 300–304. [33]

[54] I. SASON AND S. SHAMAI, *Performance Analysis of Linear Codes under Maximum-Likelihood Decoding: A Tutorial*, vol. 3 of Foundations and Trends in Communications and Information Theory, NOW, Delft, Holland, July 2006. [33]

[55] C. SCHLEGEL AND L. C. PEREZ, *Trellis and Turbo Coding*, Wiley-IEEE Press, 2004. [34]

[56] S. SHAMAI AND I. SASON, *Variations on the Gallager bounds, connections, and applications*, IEEE Trans. Inform. Theory, 48 (2002), pp. 3029–3051. [33]

[57] C. E. SHANNON, *A mathematical theory of communication*, Bell System Tech. J., 27 (1948), pp. 379–423, 623–656. [32, 34]

[58] C. E. SHANNON, R. G. GALLAGER, AND E. R. BERLEKAMP, *Lower bounds to error probability for coding on discrete memoryless channels II*, Inform. Control, 10 (1967), pp. 522–552. [34]

[59] D. A. SPIELMAN, *The complexity of error-correcting codes*, Lect. Notes Comput. Sci., 1279 (1997), pp. 67–84. [33]

[60] M. SUDAN, *Algorithmic issues in coding theory*, in 17th Conf. on Foundations of Software Technology and Theoretical Computer Science, Kharagpur, India, 1997. Invited paper. [33]

[61] ———, *Decoding Reed-Solomon codes beyond the error-correction diameter*, in Proc. of the Allerton Conf. on Commun., Control, and Computing, Monticello, IL, USA, 1997. [33]

[62] J. H. VAN LINT, *Introduction to Coding Theory*, Springer Verlag, New York, NY, USA, 2nd ed., 1992. [32, 34]

[63] A. VARDY, *The intractability of computing the minimum distance of a code*, IEEE Trans. Inform. Theory, 43 (1997), pp. 1757–1766. [33]

[64] R. R. VARSHAMOV, *Estimate of the number of signals in error correcting codes*, Doklady Akad. Nauk SSSR, 117 (1957), pp. 739–741. [32]

[65] A. J. VITERBI, *Error bounds of convolutional codes and an asymptotically optimum decoding algorithm*, IEEE Trans. Inform. Theory, 13 (1967), pp. 260–269. [33, 34, 66]

[66] A. J. VITERBI AND J. K. OMURA, *Principles of Digital Communication and Coding*, McGraw-Hill, New York, NY, USA, 1979. [33, 34]

[67] B. VUCETIC AND J. YUAN, *Turbo codes: Principles and applications*, Kluwer Academic Publ., New York, NY, USA, 2000. [34]

[68] J. K. WOLF, *Efficient maximum likelihood decoding of linear block codes using a trellis*, IEEE Trans. Inform. Theory, 24 (1978), pp. 76–80. [34, 39]

[69] J. M. Wozencraft, *Sequential decoding for reliable communication*, Research Lab. of Electron. Tech. Rept. 325, MIT, Cambridge, MA, USA, 1957. [33]

[70] H. L. Yudkin, *Channel state testing in information decoding*, PhD thesis, MIT, Cambridge, MA, USA, 1964. [34]

Chapter 2

FACTOR GRAPHS

This chapter is largely about the following question: how can we efficiently compute marginals of multivariate functions. A surprisingly large number of computational problems can be phrased in this way. The decoding problem, which is the focus of this book, is an important particular case.

§2.1. Distributive Law

Let $\mathbb{F}$ be a field (think of $\mathbb{F} = \mathbb{R}$) and let $a, b, c \in \mathbb{F}$. By the *distributive law*

$$ab + ac = a(b + c). \tag{2.1}$$

This simple law, properly applied, can significantly reduce computational complexity: consider, e.g., the evaluation of $\sum_{i,j} a_i b_j$ as $(\sum_i a_i)(\sum_j b_j)$. Factor graphs provide an appropriate framework to systematically take advantage of the distributive law.

Example 2.2 (Simple Example). Let's start with an example. Consider a function f with factorization

$$f(x_1, x_2, x_3, x_4, x_5, x_6) = f_1(x_1, x_2, x_3) f_2(x_1, x_4, x_6) f_3(x_4) f_4(x_4, x_5). \tag{2.3}$$

We are interested in computing the *marginal* of f with respect to x_1. With some abuse of notation, we denote this marginal by $f(x_1)$:

$$f(x_1) = \sum_{x_2, x_3, x_4, x_5, x_6} f(x_1, x_2, x_3, x_4, x_5, x_6) = \sum_{\sim x_1} f(x_1, x_2, x_3, x_4, x_5, x_6).$$

In the previous line we introduced the notation $\sum_{\sim \ldots}$ to denote a summation over all variables contained in the expression *except* the ones listed. This convention will save us from a flood of notation. Assume that all variables take values in a finite alphabet, call it $\mathcal{X}$. Determining $f(x_1)$ for all values of x_1 by brute force requires $\Theta\left(|\mathcal{X}|^6\right)$ operations, where we assume a naive computational model in which all operations (addition, multiplication, function evaluations, etc.) have the same cost. But we can do better: taking advantage of the factorization, we can rewrite $f(x_1)$ as

$$f(x_1) = \Big[\sum_{x_2, x_3} f_1(x_1, x_2, x_3)\Big]\Big[\sum_{x_4} f_3(x_4)\Big(\sum_{x_6} f_2(x_1, x_4, x_6)\Big)\Big(\sum_{x_5} f_4(x_4, x_5)\Big)\Big].$$

Fix x_1. The evaluation of the first factor can be accomplished with $\Theta\left(|\mathcal{X}|^2\right)$ operations. The second factor depends only on x_4, x_5, and x_6. It can be evaluated efficiently in the following manner. For each value of x_4 (and x_1 fixed), determine $\sum_{x_5} f_4(x_4, x_5)$ and $\sum_{x_6} f_2(x_1, x_4, x_6)$. Multiply by $f_3(x_4)$ and sum over x_4. Therefore, the evaluation of the second factor requires $\Theta\left(|\mathcal{X}|^2\right)$ operations as well. Since there are $|\mathcal{X}|$ values for x_1, the overall task has complexity $\Theta\left(|\mathcal{X}|^3\right)$. This compares favorably to the complexity $\Theta\left(|\mathcal{X}|^6\right)$ of the brute force approach. ◇

§2.2. Graphical Representation of Factorizations

Consider a function and its factorization. Associate with this factorization a *factor graph* as follows. For each variable draw a *variable node* (circle) and for each factor draw a *factor node* (square). Connect a variable node to a factor node by an *edge* if and only if the corresponding variable appears in this factor. The resulting graph for the function of Example 2.2 is shown on the left of Figure 2.4. The factor graph is *bipartite*. This means that the set of vertices is partitioned into two groups (the set of nodes corresponding to variables and the set of nodes corresponding to factors) and that an edge always connects a variable node to a factor node. For our particular example the factor graph is a (bipartite) *tree*. This means that there are no *cycles* in the graph; i.e., there is one and only one path between each pair of nodes. As

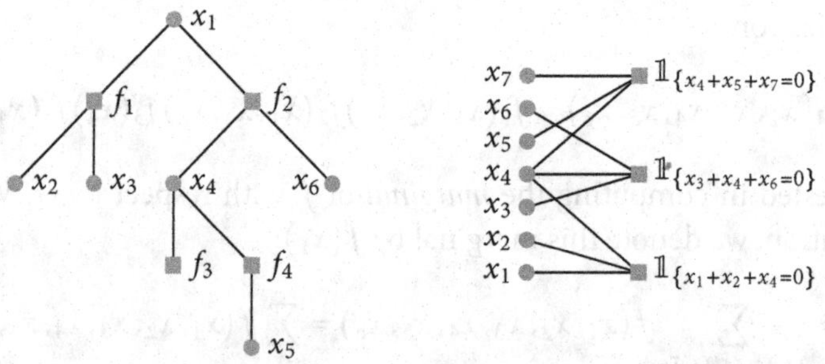

Figure 2.4: Left: Factor graph of f given in Example 2.2. Right: Factor graph for the code membership function defined in Example 2.5.

we will show in the next section, for factor graphs that are trees marginals can be computed efficiently by *message-passing* algorithms. This remains true in the slightly more general scenario where the factor graph forms a *forest*; i.e., the factor graph is disconnected and it is composed of a collection of trees. In order to keep things simple we will assume a single tree and ignore this straightforward generalization.

EXAMPLE 2.5 (SPECIAL CASE: TANNER GRAPH). Consider the binary linear code $C(H)$ defined[1] by the parity-check matrix

$$H = \begin{array}{c} \begin{matrix} x_1 & x_2 & x_3 & x_4 & x_5 & x_6 & x_7 \end{matrix} \\ \begin{pmatrix} 1 & 1 & 0 & 1 & 0 & 0 & 0 \\ 0 & 0 & 1 & 1 & 0 & 1 & 0 \\ 0 & 0 & 0 & 1 & 1 & 0 & 1 \end{pmatrix} \end{array}.$$

Let $\mathbb{F}_2$ denote the binary field with elements $\{0,1\}$ and let $x = (x_1, \dots, x_7)$. Consider the function $f(x_1, \dots, x_7)$ from $\mathbb{F}_2^7$ to $\{0,1\} \subset \mathbb{R}$ that is defined by

$$f(x_1, \dots, x_7) = \mathbb{1}_{\{x \in C(H)\}} = \begin{cases} 1, & \text{if } Hx^T = 0^T, \\ 0, & \text{otherwise.} \end{cases}$$

We can factor f as

$$f(x_1, \dots, x_7) = \mathbb{1}_{\{x_1+x_2+x_4=0\}} \mathbb{1}_{\{x_3+x_4+x_6=0\}} \mathbb{1}_{\{x_4+x_5+x_7=0\}}.$$

Each term $\mathbb{1}_{\{\cdot\}}$ is an *indicator function*: it is 1 if the condition inside the braces is fulfilled and 0 otherwise. The function f is sometimes also called the *code membership* function since it tests whether a particular word is a member of the code or not. The factor graph of f is shown on the right in Figure 2.4. It is called the *Tanner* graph of H. We will have much more to say about it in Section 3.3. ◇

It is hopefully clear at this point that *any* (binary) linear block code has a Tanner graph representation. But more is true: e.g., if you are familiar with convolutional codes take a peek at Figure 6.5 on page 327. It represents a convolutional code in terms of a factor graph. Throughout the rest of this book we will encounter factor graphs for a wide range of codes and a wide range of applications.

§2.3. RECURSIVE DETERMINATION OF MARGINALS

Consider the factorization of a generic function g and suppose that the associated factor graph is a tree (by definition it is always bipartite). Suppose that we are interested in marginalizing g with respect to the variable z; i.e., we are interested in computing $g(z) = \sum_{\sim z} g(z, \dots)$. Since the factor graph of g is a bipartite tree, g has a generic factorization of the form

$$g(z, \dots) = \prod_{k=1}^{K} [g_k(z, \dots)]$$

[1] We mean here the code C whose parity-check matrix is H and not the dual code.

for some integer K with the following crucial property: z appears in each of the factors g_k, but all other variables appear in *only one* factor. To see this assume to the contrary that another variable is contained in two of the factors. This implies that besides the path that connects these two factors via variable z another path exists. But this contradicts the assumption that the factor graph is a tree.

For the function f of Example 2.2 this factorization is

$$f(x_1,\dots) = [f_1(x_1,x_2,x_3)]\,[f_2(x_1,x_4,x_6)f_3(x_4)f_4(x_4,x_5)],$$

so that $K = 2$. The generic factorization and the particular instance for our running example f are shown in Figure 2.6. Taking into account that the individual factors

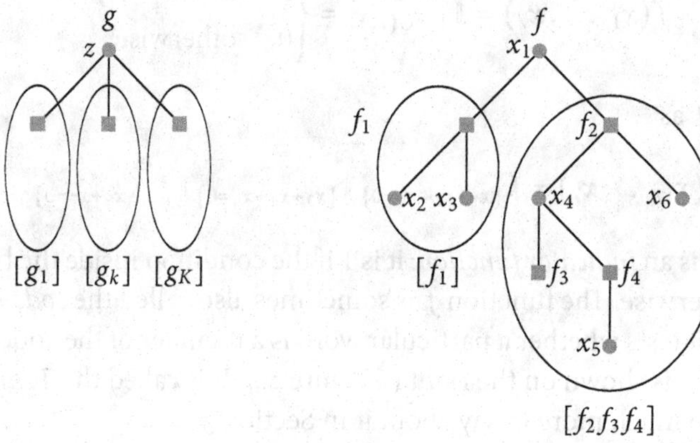

Figure 2.6: Generic factorization and the particular instance.

$g_k(z,\dots)$ only share the variable z, an application of the distributive law leads to

$$\sum_{\sim z} g(z,\dots) = \underbrace{\sum_{\sim z}\prod_{k=1}^{K}[g_k(z,\dots)]}_{\text{marginal of product}} = \underbrace{\prod_{k=1}^{K}\Big[\sum_{\sim z} g_k(z,\dots)\Big]}_{\text{product of marginals}}. \tag{2.7}$$

In words, the marginal $\sum_{\sim z} g(z,\dots)$ is the product of the individual marginals $\sum_{\sim z} g_k(z,\dots)$. In terms of our running example we have

$$f(x_1) = \Big[\sum_{\sim x_1} f_1(x_1,x_2,x_3)\Big]\Big[\sum_{\sim x_1} f_2(x_1,x_4,x_6)f_3(x_4)f_4(x_4,x_5)\Big].$$

This single application of the distributive law leads, in general, to a non-negligible reduction in complexity. But we can go further and apply the same idea recursively to each of the terms $g_k(z,\dots)$.

In general, each g_k is itself a product of factors. In Figure 2.6 these are the factors of g that are grouped together in one of the ellipsoids. Since the factor graph is a bipartite tree, g_k must in turn have a generic factorization of the form

$$g_k(z,\dots) = \underbrace{h(z, z_1, \dots, z_J)}_{\text{kernel}} \prod_{j=1}^{J} \underbrace{[h_j(z_j, \dots)]}_{\text{factors}},$$

where z appears only in the "kernel" $h(z, z_1, \dots, z_J)$ and each of the z_j appears *at most twice*, possibly in the kernel and in at most one of the factors $h_j(z_j, \dots)$. All other variables are again unique to a single factor. For our running example we have

$$f_2(x_1, x_4, x_6) f_3(x_4) f_4(x_4, x_5) = \underbrace{f_2(x_1, x_4, x_6)}_{\text{kernel}} \underbrace{[f_3(x_4) f_4(x_4, x_5)]}_{x_4} \underbrace{[1]}_{x_6}.$$

The generic factorization and the particular instance for our running example f are shown in Figure 2.8. Another application of the distributive law gives

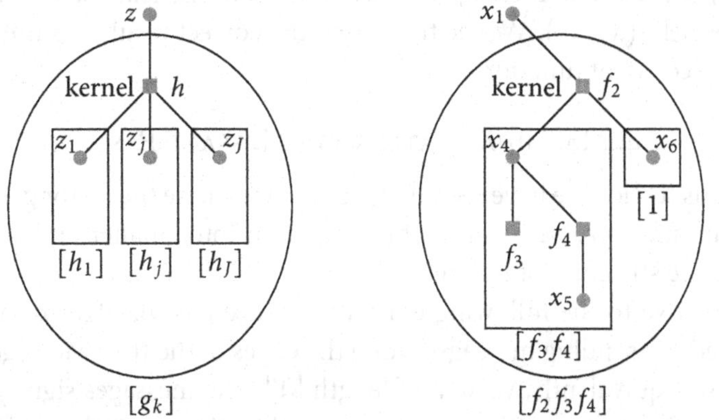

Figure 2.8: Generic factorization of g_k and the particular instance.

$$\sum_{\sim z} g_k(z,\dots) = \sum_{\sim z} h(z, z_1, \dots, z_J) \prod_{j=1}^{J} [h_j(z_j, \dots)]$$

(2.9)
$$= \sum_{\sim z} h(z, z_1, \dots, z_J) \underbrace{\prod_{j=1}^{J} \Big[\sum_{\sim z_j} h_j(z_j, \dots)\Big]}_{\text{product of marginals}}.$$

In words, the desired marginal $\sum_{\sim z} g_k(z, \dots)$ can be computed by multiplying the kernel $h(z, z_1, \dots, z_J)$ with the individual marginals $\sum_{\sim z_j} h_j(z_j, \dots)$ and summing out all remaining variables other than z.

We are back to where we started. Each factor $h_j(z_j, \dots)$ has the same generic form as the original function $g(z, \dots)$, so that we can continue to break down the marginalization task into smaller pieces. This recursive process continues until we have reached the leaves of the tree. The calculation of the marginal then follows the recursive splitting in reverse. In general, nodes in the graph compute marginals, which are functions over $\mathcal{X}$, and pass these on to the next level. In the next section we will elaborate on this method of computation, known as message passing: the marginal functions are messages. The message combining rules at function nodes is explicit in (2.9). And at a variable node we simply perform pointwise multiplication.

Let us consider the initialization of the process. At the leaf nodes the task is simple. A function leaf node has the generic form $g_k(z)$, so that $\sum_{\sim z} g_k(z) = g_k(z)$: this means that the initial message sent by a function leaf node is the function itself. To find out the correct initialization at a variable leaf node consider the simple example of computing $f(x_1) = \sum_{\sim x_1} f(x_1, x_2)$. Here, x_2 is the variable leaf node. By the message-passing rule (2.9) the marginal $f(x_1)$ is equal to $\sum_{\sim x_1} f(x_1, x_2) \cdot \mu(x_2)$, where $\mu(x_2)$ is the initial message that we send from the leaf variable node x_2 towards the kernel $f(x_1, x_2)$. We see that to get the correct result this initial message should be the constant function 1.

§2.4. Marginalization via Message Passing

In the previous section we have seen that, in the case where the factor graph is a tree, the marginalization problem can be broken down into smaller and smaller tasks according to the structure of the tree.

This gives rise to the following efficient *message-passing* algorithm. The algorithm proceeds by sending messages along the edges of the tree. Messages are *functions* on $\mathcal{X}$, or, equivalently, vectors of length $|\mathcal{X}|$. The messages signify marginals of parts of the function and these parts are combined to form the marginal of the whole function. Message passing originates at the leaf nodes. Messages are passed up the tree and as soon as a node has received messages from all its children, the incoming messages are processed and the result is passed up to the parent node.

Example 2.10 (Message-Passing Algorithm for f of Example 2.2). Consider this procedure in detail for the case of our running example as shown in Figure 2.11. The top leftmost graph is the factor graph. Message passing starts at the leaf nodes as shown in the middle graph on the top. The variable leaf nodes x_2, x_3, x_5, and x_6 send the constant function 1 as discussed at the end of the previous section. The factor leaf node f_3 sends the function f_3 up to its parent node. In the next time step the factor node f_1 has received messages from both its children and can therefore proceed. According to (2.9), the message it sends up to its parent node x_1 is the product of the incoming messages times the "kernel" f_1, after summing out all variable nodes except x_1; i.e., the message is $\sum_{\sim x_1} f_1(x_1, x_2, x_3)$. In the same manner factor node f_4 forwards to its parent node x_4 the message $\sum_{\sim x_4} f_4(x_4, x_5)$. This is shown in the rightmost figure in the top row. Now, variable node x_4 has received messages from all its children. It forwards to its parent node f_2 the product of its incoming messages, in agreement with (2.7), which says that the marginal of a product is the product of the marginals. This message, which is a function of x_4, is $f_3(x_4) \sum_{\sim x_4} f(x_4, x_5) = \sum_{\sim x_4} f_3(x_4) f_4(x_4, x_5)$. Next, function node f_2 can forward its message, and, finally, the marginalization is achieved by multiplying all incoming messages at the root node x_1. ◇

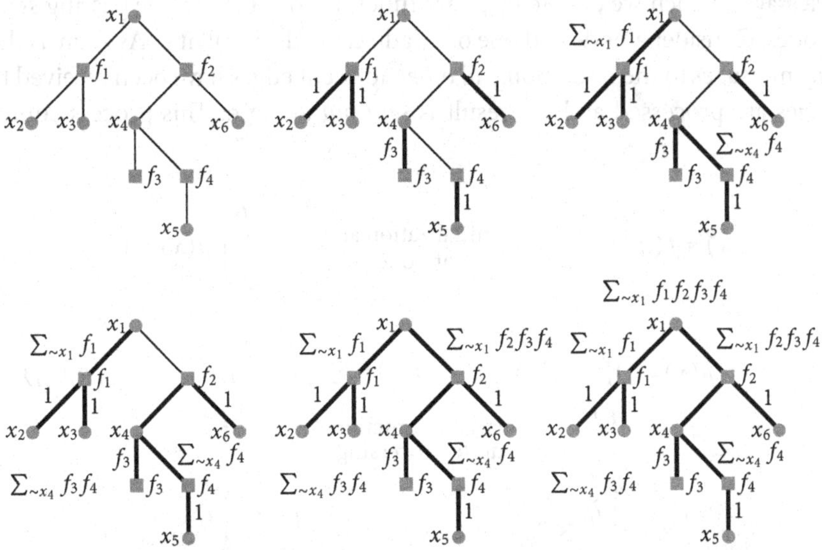

Figure 2.11: Marginalization of function f from Example 2.2 via message passing. Message passing starts at the leaf nodes. A node that has received messages from all its children processes the messages and forwards the result to its parent node. Bold edges indicate edges along which messages have already been sent.

Before stating the message-passing rules formally, consider the following important generalization. Whereas so far we have considered the marginalization of a function f with respect to a *single* variable x_1 we are actually interested in marginalizing for *all* variables. We have seen that a single marginalization can be performed efficiently if the factor graph of f is a *tree*, and that the complexity of the computation essentially depends on the largest degree of the factor graph and the size of the underlying alphabet. Consider now the problem of computing *all* marginals. We can draw for each variable a tree rooted in this variable and execute the single marginal message-passing algorithm on each rooted tree. It is easy to see, however, that the algorithm does not depend on which node is the root of the tree and that in fact all the computations can be performed simultaneously on a single tree. Simply start at all leaf nodes and for every edge compute the outgoing message along this edge as soon as you have received the incoming messages along all *other* edges that connect to the given node. Continue in this fashion until a message has been sent in both directions along every edge. This computes *all* marginals so it is more complex than computing a single marginal but only by a factor roughly equal to the average degree of the nodes. We now summarize the set of message-passing rules.

Messages, which we denote by μ, are functions on $\mathcal{X}$. Message passing starts at leaf nodes. Consider a node and one of its adjacent edges, call it e. As soon as the *incoming* messages to the node along all *other* adjacent edges have been received these messages are processed and the result is *sent out* along e. This process continues

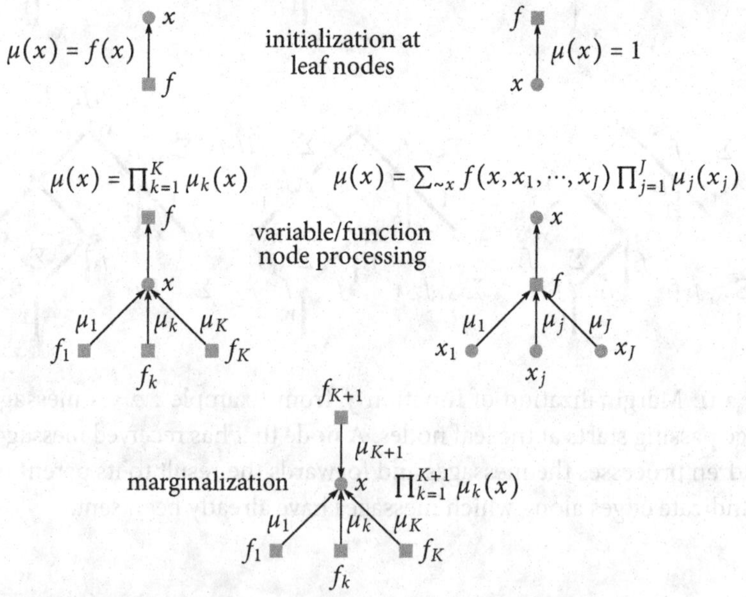

Figure 2.12: Message-passing rules. The top row shows the initialization of the messages at the leaf nodes. The middle row corresponds to the processing rules at the variable and function nodes, respectively. The bottom row explains the final marginalization step.

until messages along all edges in the tree have been processed. In the final step the marginals are computed by combining *all* messages which enter a particular variable node. The initial conditions and processing rules are summarized in Figure 2.12. Since the messages represent probabilities or *beliefs*, the algorithm is also known as the *belief propagation* (BP) algorithm. From now on we will mostly refer to it under this name.

§2.5. Decoding via Message Passing

§2.5.1. Bit-wise MAP Decoding

Assume we transmit over a binary-input ($X_i \in \{\pm 1\}$) memoryless ($p_{Y\,|\,X}(y\,|\,x) = \prod_{i=1}^{n} p_{Y_i\,|\,X_i}(y_i\,|\,x_i)$) channel using a linear code $C(H)$ defined by its parity-check matrix H and assume that codewords are chosen uniformly at random. The rule for the *bit-wise* maximum a posteriori (MAP) decoder reads:

$$
\begin{aligned}
& \hat{x}_i^{\text{MAP}}(y) = \text{argmax}_{x_i \in \{\pm 1\}} p_{X_i\,|\,Y}(x_i\,|\,y) \\
\text{(law of total probability)} \quad & = \text{argmax}_{x_i \in \{\pm 1\}} \sum_{\sim x_i} p_{X\,|\,Y}(x\,|\,y) \\
\text{(Bayes's)} \quad & = \text{argmax}_{x_i \in \{\pm 1\}} \sum_{\sim x_i} p_{Y\,|\,X}(y\,|\,x) p_X(x) \\
\text{(2.13)} \quad & = \text{argmax}_{x_i \in \{\pm 1\}} \sum_{\sim x_i} \Big(\prod_j p_{Y_j\,|\,X_j}(y_j\,|\,x_j) \Big) \mathbb{1}_{\{x \in C\}},
\end{aligned}
$$

where in the last step we have used the fact that the channel is memoryless and that codewords have uniform prior. This is important: in the preceding formulation we consider y a *constant* (since it is given to the decoding algorithm as an input). Therefore, we write $\sum_{\sim x_i}$ to indicate a summation over all components of x (except x_i) *and not the components of* y.

Assume that the code membership function $\mathbb{1}_{\{x \in C\}}$ has a factorized form. From (2.13) it is then clear that the bit-wise decoding problem is equivalent to calculating the marginal of a factorized function and choosing the value that maximizes this marginal.

Example 2.14 (Bit-wise MAP Decoding). Consider the parity-check matrix given in Example 2.5. In this case $\text{argmax}_{x_i \in \{\pm 1\}} p_{X_i\,|\,Y}(x_i\,|\,y)$ can be factorized as

$$
\text{argmax}_{x_i \in \{\pm 1\}} \sum_{\sim x_i} \Big(\prod_{j=1}^{7} p_{Y_i\,|\,X_j}(y_j\,|\,x_j) \Big) \mathbb{1}_{\{x_1+x_2+x_4=0\}} \mathbb{1}_{\{x_3+x_4+x_6=0\}} \mathbb{1}_{\{x_4+x_5+x_7=0\}}.
$$

The corresponding factor graph is shown in Figure 2.15. This graph includes the Tanner graph of H but additionally contains the factor nodes which represent the

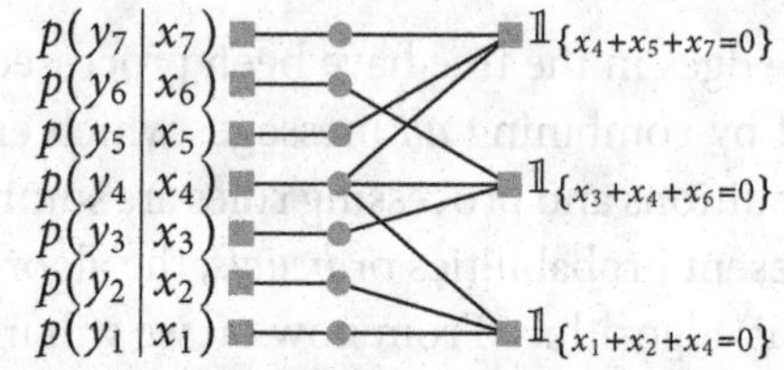

Figure 2.15: Factor graph for the MAP decoding of our running example.

effect of the channel. For this particular case the resulting graph is a tree. We can therefore apply the message-passing algorithm to this example to perform bit-wise MAP decoding. ◇

§2.5.2. Simplification of Message-Passing Rules for Bit-wise MAP Decoding

In the binary case a message $\mu(x)$ can be thought of as a real-valued vector of length 2, $(\mu(1), \mu(-1))$ (here we think of the bit values as $\{\pm 1\}$). The initial such message sent from the factor leaf node representing the i-th channel realization to the variable node i is $(p_{Y_i \mid X_i}(y_i \mid 1), p_{Y_i \mid X_i}(y_i \mid -1))$ (see Figure 2.15). Recall that at a variable node of degree $K+1$ the message-passing rule calls for a pointwise multiplication:

$$\mu(1) = \prod_{k=1}^{K} \mu_k(1), \qquad \mu(-1) = \prod_{k=1}^{K} \mu_k(-1).$$

Introduce the *ratio* $r_k = \mu_k(1)/\mu_k(-1)$. The initial such ratios are the likelihood ratios associated with the channel observations. We have

$$r = \frac{\mu(1)}{\mu(-1)} = \frac{\prod_{k=1}^{K} \mu_k(1)}{\prod_{k=1}^{K} \mu_k(-1)} = \prod_{k=1}^{K} r_k;$$

i.e., the ratio of the outgoing message at a variable node is the product of the incoming ratios. If we define the log-likelihood ratios $l_k = \ln(r_k)$, then the processing rule reads $l = \sum_{k=1}^{K} l_k$.

Consider now the ratio of an outgoing message at a check node which has degree $J+1$. For a check node the associated "kernel" is

$$f(x, x_1, \dots, x_J) = \mathbb{1}_{\{\prod_{j=1}^{J} x_j = x\}}.$$

Since in the current context we assume that the x_i take values in $\{\pm 1\}$ (and not $\mathbb{F}_2$) we write $\prod_{j=1}^{J} x_j = x$ (instead of $\sum_{j=1}^{J} x_j = x$). We therefore have

$$r = \frac{\mu(1)}{\mu(-1)} = \frac{\sum_{\sim x} f(1, x_1, \dots, x_J) \prod_{j=1}^{J} \mu_j(x_j)}{\sum_{\sim x} f(-1, x_1, \dots, x_J) \prod_{j=1}^{J} \mu_j(x_j)}$$

$$
\begin{aligned}
&= \frac{\sum_{x_1,\dots,x_J:\prod_{j=1}^J x_j=1} \prod_{j=1}^J \mu_j(x_j)}{\sum_{x_1,\dots,x_J:\prod_{j=1}^J x_j=-1} \prod_{j=1}^J \mu_j(x_j)} = \frac{\sum_{x_1,\dots,x_J:\prod_{j=1}^J x_j=1} \prod_{j=1}^J \frac{\mu_j(x_j)}{\mu_j(-1)}}{\sum_{x_1,\dots,x_J:\prod_{j=1}^J x_j=-1} \prod_{j=1}^J \frac{\mu_j(x_j)}{\mu_j(-1)}} \\
&= \frac{\sum_{x_1,\dots,x_J:\prod_{j=1}^J x_j=1} \prod_{j=1}^J r_j^{(1+x_j)/2}}{\sum_{x_1,\dots,x_J:\prod_{j=1}^J x_j=-1} \prod_{j=1}^J r_j^{(1+x_j)/2}} = \frac{\prod_{j=1}^J (r_j+1) + \prod_{j=1}^J (r_j-1)}{\prod_{j=1}^J (r_j+1) - \prod_{j=1}^J (r_j-1)}. \qquad (2.16)
\end{aligned}
$$

The last step warrants some remarks. If we expand out $\prod_{j=1}^J (r_j+1)$, then we get the sum of all products of the individual terms r_j, $j = 1, \dots, J$ (e.g., $\prod_{j=1}^3 (r_j+1) = 1 + r_1 + r_2 + r_3 + r_1r_2 + r_1r_3 + r_2r_3 + r_1r_2r_3$). Similarly, $\prod_{j=1}^J (r_j-1)$ is the sum of all products of the individual terms r_j, where all products consisting of d terms such that $J-d$ is odd have a negative sign (e.g., we have $\prod_{j=1}^3 (r_j-1) = -1 + r_1 + r_2 + r_3 - r_1r_2 - r_1r_3 - r_2r_3 + r_1r_2r_3$). From this it follows that

$$
\prod_{j=1}^J (r_j+1) + \prod_{j=1}^J (r_j-1) = 2 \sum_{x_1,\dots,x_J:\prod_{j=1}^J x_j=1} \prod_{j=1}^J r_j^{(1+x_j)/2}.
$$

Applying the analogous reasoning to the denominator, the equality follows. If we divide both numerator and denominator by $\prod_{j=1}^J (r_j+1)$, we see that (2.16) is equivalent to the statement

$$
r = \frac{1 + \prod_j \frac{r_j-1}{r_j+1}}{1 - \prod_j \frac{r_j-1}{r_j+1}},
$$

which in turn implies $\frac{r-1}{r+1} = \prod_j \frac{r_j-1}{r_j+1}$. From $r = e^l$ we see that $\frac{r-1}{r+1} = \tanh(l/2)$. Combining theses two statements we have

$$
\tanh(l/2) = \frac{r-1}{r+1} = \prod_{j=1}^J \frac{r_j-1}{r_j+1} = \prod_{j=1}^J \tanh(l_j/2), \quad \text{so that}
$$

$$
l = 2\tanh^{-1}\Big(\prod_{j=1}^J \tanh(l_j/2)\Big). \qquad (2.17)
$$

To summarize, in the case of transmission over a binary channel the messages can be compressed to a single real quantity. In particular, if we choose this quantity to be the log-likelihood ratio (log of the ratio of the two likelihoods) then the processing rules take on a particularly simple form: at variables nodes messages add, and at check nodes the processing rule is stated in (2.17).

§2.5.3. Forney-Style Factor Graphs

Factor graphs (FGs) represent one particular language to formulate the relationship between a function and its local components. One popular alternative is the representation in terms of *Forney-style factor graphs* (FSFGs). These graphs are sometimes also called *normal graphs.*

Figure 2.18: Left: Standard FG in which each variable node has degree at most 2. Right: Equivalent FSFG. The variables in the FSFG are associated with the edges in the FG.

Consider the FG shown in the left-hand side of Figure 2.18. Note that each variable node has degree 1 or 2. We can convert the *bipartite* graph into a *regular* (in the sense of not bipartite) graph by representing variable nodes as (half-)edges. The result is shown on the right-hand side of Figure 2.18. This is the FSFG representation. In general, a variable node might have degree larger than 2. In this case we can *replicate* such a variable node a sufficient number of times by an *equality factor* as shown in Figure 2.19. The left-hand side shows a variable node of degree $K + 1$. The right-hand side shows the representation as an FSFG. The K additional variables $x_1, \dots, x_K$ are enforced to be equal to the original variable x by an "equality factor," i.e., $x = x_1 = \cdots = x_K$. Figure 2.20 compares the standard FG for the MAP decoding

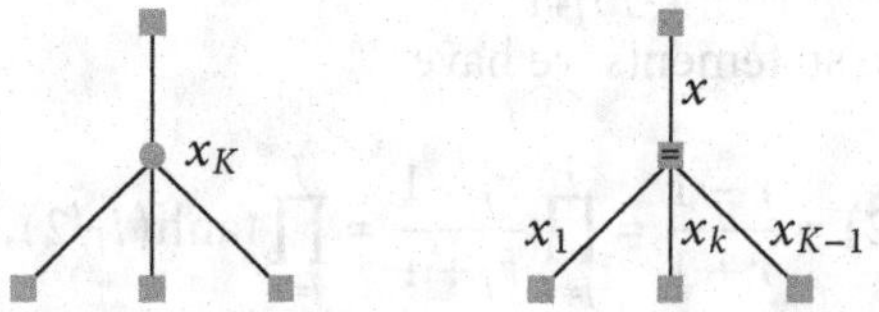

Figure 2.19: Representation of a variable node of degree K as an FG (left) and the equivalent representation as an FSFG (right).

problem of our running example with the corresponding FSFG.

The relationship between the standard FG and the FSFG is straightforward and little effort is required to move from one representation to the other. The message-passing rules carry over verbatim. In fact, in the setting of FSFGs we only need the factor node processing rule: for a generic node of degree $J+1$, the outgoing message

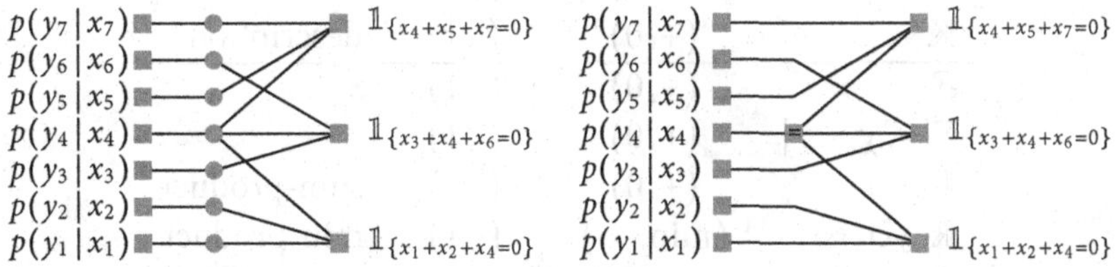

Figure 2.20: Standard FG and the corresponding FSFG for the MAP decoding problem of our running example.

along edge x is

$$\mu(x) = \sum_{\sim x} f(x, x_1, \dots, x_J) \prod_{j=1}^{J} \mu_j(x_j). \tag{2.21}$$

Recall that variables in the FSFG are represented by edges. Some thought shows that to compute the final marginalization with respect to a certain variable, we need to multiply the two messages that flow along the corresponding edge.

FSFGs have fewer nodes and are therefore typically simpler. Note that "internal" edges represent internal or "state" variables, whereas "half-edges" (like the bottom edge on the right of Figure 2.18) represent external variables which can be connected to other systems on graphs.

§2.5.4. Generalization to Commutative Semirings

We started with a discussion of the distributive law assuming that the underlying algebraic structure was a field $\mathbb{F}$ and used it to derive efficient marginalization algorithms. A closer look at our derivation shows that actually all that was needed was that we were working in a *commutative semiring*. In a commutative semiring $\mathbb{K}$ the two operations, which we call "+" and "·," satisfy the following axioms: (i) the operations "+" and "·" are commutative ($x + y = y + x$; $x \cdot y = y \cdot x$) and associative ($x + (y + z) = (x + y) + z$; $x \cdot (y \cdot z) = (x \cdot y) \cdot z$) with identity elements denoted by "0" and "1," respectively; (ii) the *distributive law* $(x + y) \cdot z = (x \cdot z) + (y \cdot z)$ holds for any triple $x, y, z \in \mathbb{K}$. In comparison to a field $\mathbb{F}$ we do not require the existence of an inverse with respect to either operation.

Table 2.22 lists several commutative semirings which are useful in the context of coding. The task of verifying that each of these examples indeed fulfills the stated axioms is relegated to Problem 2.1. The most important example in the context of iterative coding is the so-called *sum-product* semiring, in which the two operations are standard addition and multiplication. This is the example which we have used so far. As we have seen, the sum-product semiring is the relevant semiring if we want to minimize the *bit* error probability.

$\mathbb{K}$	"$(+,0)$"	"$(\cdot,1)$"	description
$\mathbb{F}$	$(+,0)$	$(\cdot,1)$	
$\mathbb{F}[x,y,\dots]$	$(+,0)$	$(\cdot,1)$	
$\mathbb{R}_{\geq 0}$	$(+,0)$	$(\cdot,1)$	sum-product
$\mathbb{R}_{>0}\cup\{\infty\}$	$(\min,\infty)$	$(\cdot,1)$	min-product
$\mathbb{R}_{\geq 0}$	$(\max,0)$	$(\cdot,1)$	max-product
$\mathbb{R}\cup\{\infty\}$	$(\min,\infty)$	$(+,0)$	min-sum
$\mathbb{R}\cup\{-\infty\}$	$(\max,-\infty)$	$(+,0)$	max-sum
$\{0,1\}$	$(\texttt{OR},0)$	$(\texttt{AND},1)$	Boolean

Table 2.22: List of commutative semirings which are relevant for iterative decoding. The entry $\mathbb{F}[x,y,\dots]$ denotes the set of polynomials in the variables $x,y,\dots$, with coefficients in the field $\mathbb{F}$, and the usual polynomial arithmetic.

The second most important example is the *max-sum* semiring. As before we operate over the reals but now addition is replaced by maximization and multiplication is replaced by addition. All previous statements and algorithms stay valid if we perform this simple substitution of operations. We will soon see that the max-sum semiring is the proper setting for performing *block* decoding. To be concrete, consider the distributive law for the max-sum semiring. If in (2.1) we replace addition with maximization and multiplication with addition then we get

$$\max\{x+y, x+z\} = x + \max\{y,z\},$$

and, more generally,

$$\max_{i,j}\{x_i + y_j\} = \max_i\{x_i\} + \max_j\{y_j\}.$$

What is the *marginalization* of a function $f(x_1,\dots,x_n)$ of n real variables in the context of the max-sum semiring? By replacing the operations we see that it is *maximization*, i.e.,

$$f(x_1) = \max_{x_2,\dots,x_n} f(x_1,\dots,x_n) = \max_{\sim x_1} f(x_1,\dots,x_n). \tag{2.23}$$

As before, if the FG of the function $f(x_1,\dots,x_n)$ is a tree, this maximization can be accomplished efficiently by a message-passing algorithm operating over the max-sum semiring. The message-passing rules are formally identical. More precisely, the original variable node processing rule $\mu(z) = \prod_{k=1}^{K}\mu_k(z)$ is transformed into the rule $\mu(z) = \sum_{k=1}^{K}\mu_k(z)$, and the function node processing rule, which previously was $\mu(z) = \sum_{\sim z} f(z, z_1, \dots, z_J) \prod_{j=1}^{J} \mu_j(z_j)$, now reads

$$\mu(z) = \max_{\sim z}\{f(z, z_1, \dots, z_J) + \sum_{j=1}^{J} \mu_j(z_j)\}.$$

The final marginalization step, which used to consist of computing the product $\prod_{k=1}^{K+1} \mu_k(z)$, now requires us to evaluate the sum $\sum_{k=1}^{K+1} \mu_k(z)$.

§2.5.5. Block-wise MAP Decoding

Assume we are transmitting over a binary memoryless channel using a linear code $C(H)$ defined by its parity-check matrix H and assume that codewords are chosen uniformly at random from $C(H)$. The processing rule for the block-wise MAP decoder is

$$\begin{aligned} &\text{(Bayes's)} \qquad \hat{x}^{\text{MAP}}(y) = \text{argmax}_x p_{X \mid Y}(x \mid y) = \text{argmax}_x p_{Y \mid X}(y \mid x) p_X(x) \\ &\text{(memoryless)} \qquad = \text{argmax}_x \Big(\prod_j p_{Y_j \mid X_j}(y_j \mid x_j)\Big) \mathbb{1}_{\{x \in C\}}. \end{aligned}$$

To see the similarity to bit-wise MAP decoding, consider the i-th bit of $\hat{x}^{\text{MAP}}(y)$; write it as $(\hat{x}^{\text{MAP}}(y))_i$. We have

$$\begin{aligned} \big(\hat{x}^{\text{MAP}}(y)\big)_i &= \text{argmax}_{x_i \in \{\pm 1\}} \max_{\sim x_i} \Big(\prod_j p_{Y_j \mid X_j}(y_j \mid x_j)\Big) \mathbb{1}_{\{x \in C\}} \\ &= \text{argmax}_{x_i \in \{\pm 1\}} \max_{\sim x_i} \sum_j \log p_{Y_j \mid X_j}(y_j \mid x_j) + \log\big(\mathbb{1}_{\{x \in C\}}\big). \end{aligned}$$

If we compare this with (2.13) we see that the two criteria only differ by a substitution of the two basic operations: addition goes into maximization and multiplication goes into addition (the initial messages are of course also different – we use likelihoods for bit-wise decoding and log-likelihoods for block-wise decoding). Therefore, block-wise decoding can be accomplished if we employ the max-sum algebra instead of the sum-product algebra.

It is common to write the block-wise decoder in the equivalent form

$$\big(\hat{x}^{\text{MAP}}(y)\big)_i = \text{argmin}_{x_i \in \{\pm 1\}} \min_{\sim x_i} \sum_{j=1}^{n} - \log p_{Y_j \mid X_j}(y_j \mid x_j) - \log\big(\mathbb{1}_{\{x \in C\}}\big).$$

If the channel output is discrete, so that we deal with probability mass functions, then this form is more convenient since the involved metric $-\log p_{Y_j \mid X_j}(y_j \mid x_j)$ is positive. Formally this means that we use the min-sum algebra instead of the max-sum algebra. In the sequel we adhere to this custom and use the min-sum algebra for block-wise MAP decoding.

§2.6. Limitations of Cycle-Free Codes

The previous sections have shown a way of performing MAP decoding efficiently, assuming that the corresponding Tanner graph is a tree. Unfortunately, the class of codes that admit a tree-like (binary) Tanner graph is not powerful enough to perform well.

Lemma 2.24 (Bad News about Cycle-Free Codes). Let C be a binary linear code of rate r that admits a binary Tanner graph that is a forest. Then C contains at least $\frac{2r-1}{2}n$ codewords of weight 2.

Proof. Without loss of generality we can assume that the Tanner graph is connected. Otherwise the code $C = C[n,k]$ is of the form $C = C_1 \times C_2$, where $C_1 = C_1[n_1,k_1]$, $C_2 = C_2[n_2,k_2]$, $n = n_1 + n_2$, $n_1, n_2 \geq 1$, and $k = k_1 + k_2$, i.e., each codeword is the concatenation of a codeword from C_1 with a codeword from C_2. Applying the bound to each component (to keep things simple we assume there are only two such components),

$$\frac{2r_1-1}{2}n_1 + \frac{2r_2-1}{2}n_2 = \frac{2\frac{k_1}{n_1}-1}{2}n_1 + \frac{2\frac{k_2}{n_2}-1}{2}n_2$$
$$= \frac{2k_1-n_1}{2} + \frac{2k_2-n_2}{2}n_2 = \frac{2k-n}{2} = \frac{2r-1}{2}n.$$

Let us therefore assume that the Tanner graph of the code consists of a single tree. The graph has n variable nodes and $(1-r)n$ check nodes since by the tree property all check nodes (i.e., the respective equations) are linearly independent. The total number of nodes in the tree is therefore $(2-r)n$. Again by the tree property, there are $(2-r)n-1 < (2-r)n$ edges in this graph. Since each such edge connects to exactly one variable node, the average variable node degree is upper bounded by $2-r$. It follows that there must be at least nr variable nodes that are leaf nodes, since each internal variable node has degree at least 2. Since there are in total $(1-r)n$ check nodes and since every leaf variable node is connected to exactly one check node, it follows that at least $rn-(1-r)n = (2r-1)n$ leaf variable nodes are connected to check nodes that are adjacent to multiple leaf variable nodes. Each such variable node can be paired up with one of the other such leaf nodes to give rise to a codeword of weight 2. □

We see that cycle-free codes (of rate above one-half) necessarily contain many low-weight codewords and, hence, have a large probability of error. This is bad news indeed. As discussed in more detail in Problems 4.52 and 4.53, codes of rate below one-half also contain low-weight codewords, and the problem persists even if we allow a small number of cycles.

§2.7. Message Passing on Codes with Cycles

We started with an efficient algorithm to compute the marginals of functions whose factor graph is a tree. Next we saw that the decoding task can be phrased as such a marginalization problem, for minimizing either bit or block error probability. But we now know that codes with a cycle-free (binary) Tanner graph are not powerful enough for transmission at low error rates. Tanner graphs of good codes necessarily have many cycles. So how shall we proceed?

First, one can resort to more powerful graphical models. We discuss in Chapter 6 (terminated) convolutional codes. Although terminated convolutional codes are linear block codes (with a particular structure) and therefore they have a standard binary Tanner graph representation, we will see that convolutional codes possess a cycle-free representation (and therefore the BP algorithm can be used to perform MAP decoding) if we allow *state* nodes. By increasing the size of the allowed state space one can approach capacity. However, these state nodes come at the price of increased decoding complexity and, as discussed in the introduction, the complexity-gap trade-off is not very favorable. Another possibility is to consider non-binary codes (see Section 7.4). Unfortunately, complexity is again increased considerably by allowing non-binary alphabets. Finally, one can define the message-passing algorithm also in the case where cycles are present. Except for some degenerate cases, message passing in the presence of cycles is strictly suboptimal; see Problem 3.11. But as we will see in Chapters 3 and 4, excellent performance can be achieved. For codes with cycles, message passing no longer performs MAP decoding. We will therefore spend a considerable effort on learning tools that allow us to determine the performance of such a combination.

Notes

Tanner [35] proposed to represent codes as bipartite graphs and to visualize iterative decoding as a message-passing algorithm on such a graph. The framework of FGs discussed in this chapter is the result of a collaborative effort by Wiberg [37], Wiberg, Loeliger, and Kötter [37, 38], as well as Kschischang, Frey, and Loeliger [23]. It is not the only graphical model suitable for iterative decoding. Indeed, we have discussed the notion of FSFGs in Section 2.5.3. These were introduced by Forney [14], who called them *normal* graphs. As shown in [14], normal graphs allow for an elegant local *dualization* of the graph. Extensions of this idea were discussed by Mao and Kschischang [29]. A further equivalent graphical language was put forth around the same time by Aji and McEliece [1] (see also the article by Shafer and Shenoy [34]). The message-passing algorithm which we derived via the FG approach is known under many different names (iterative decoding, belief-propagation, message passing, probabilistic decoding, etc.). It was gradually realized that what might appear as different algorithms (invented in many different communities) are in fact special cases of the same basic principle. Let us trace here just a few of those instances. Probably the first such instance is the transfer-matrix method of statistical mechanics. It is explored in detail in [9] in conjunction with the so-called *Bethe Ansatz* [10, 11] and it goes back at least to the 1930s. In the setting of communications, it was Gallager [16] who introduced low-density parity-check (LDPC) codes and the related message-passing algorithm in 1960. Viterbi introduced his so-called Viterbi algorithm for the decoding of convolutional codes [36]. For a historical perspective see the Notes at the end of Chapter 1. The connection between the Viterbi algorithm and message-passing decoding is discussed in detail in Section 6.1 and Problem 6.2.

In the mid-1960s, Baum and Welch developed an algorithm to estimate the parameters of hidden Markov models. This algorithm is known as the *Baum-Welch* algorithm. For a list of publications we refer the reader to the papers by Baum and Petrie [6], Baum and Sell [8], and Baum, Petrie, Soules, and Weiss [7]. Closely related is the *BCJR* algorithm which was used by Bahl, Cocke, Jelinek, and Raviv to perform bit MAP decoding for a convolutional code, see [2]. In 1977, Dempster, Laird, and Rubin investigated the *expectation-maximization* algorithm [12], which in turn includes the Baum-Welch algorithm as a special case (see [31]).

In 1983 Kim and Pearl introduced the *belief propagation* algorithm [21, 33] to solve statistical inference problems. That the turbo decoding algorithm is in fact an instance of belief propagation was realized by MacKay, McEliece, and Cheng [30] as well as Frey and Kschischang [15]. An in-depth discussion of all these connections can be found in the article of McEliece, MacKay, and Cheng [30], the article of Kschischang and Frey [22], as well as the book of Heegard and Wicker [20].

Our exposition of the FG approach follows closely the one in [23]. It was Wiberg who realized that the sum-product and the min-sum algorithm are formally equivalent [37]. The formalization and further generalization of this equivalence in terms of semirings is due to Aji and McEliece [1]. As we have seen, this generalization makes it possible to view a large class of algorithms simply as special instances of the same principle.

A set of applications for the FG framework is discussed in the paper by Worthen and Stark [39]. If you are looking for tutorial articles concerning FGs we recommend the paper by Loeliger [25].

It was shown by Etzion, Trachtenberg, and Vardy [13] that binary codes which possess a cycle-free Tanner graph (without state nodes) necessarily have small minimum distance. We discussed in Lemma 2.24 only the simple case of codes with rate r of at least one-half. In this case we saw that the minimum distance is at most 2. If $r < \frac{1}{2}$, then the aforementioned authors showed that the minimum distance is at most $2/r$.

Battail, Decouvelaere, and Godlewski were early pioneers in the area of combining "soft information" stemming from various partial descriptions of a code into one final estimate [4]. However, they did not discuss the notion of feedback, i.e., iterative decoding. Battail, Decouvelaere, and Godlewski termed their coding method *replica* coding (see also [3, 5]). Hagenauer, Offer, and Papke [18] introduced the "log-likelihood algebra," which contains the message-passing rule at variable and check nodes.

The FG approach has also inspired an implementation of message-passing decoding by analog computation. This has been pioneered by two research groups: Hagenauer, Winklhofer, Offer, Méasson, Mörz, Gabara, and Yan [17, 19, 32], as well as Loeliger, Lustenberger, Helfenstein, and Tarköy [26, 27, 28].

Problems

2.1 (Factor Graphs for Semirings). Consider the examples listed in Table 2.22. Show in each case that it indeed forms a commutative semiring.

2.2 (Message-Passing Algorithm for BEC). Starting from the message-passing rules summarized in Figure 2.12, derive the decoding algorithm for the binary erasure channel (BEC) (see Section 3.1 for a discussion of this channel model). What is the message alphabet and what are the computation rules? Simplify the rules as far as possible.

2.3 (Min-Sum Algorithm for BEC). Apply the min-sum algebra to the decoding of LDPC ensembles over the BEC. What are the initial messages and what are the processing rules? Show that the messages that are a priori two-tuples can be compressed into a single number. Finally, show that the resulting message-passing rules are identical to the ones using the sum-product semiring. In words, over the BEC (locally optimal) iterative bit- and block-wise decoding are identical.

2.4 (Hansel and Gretel Take a Field Trip in the Dark Forest). Hansel and Gretel, together with all their classmates, take a field trip. The forest in which they are walking is so dark that each kid can only see its immediate neighbors. Assume that communication is limited to these nearest neighbors as well and that the whole group of schoolchildren forms a tree (in the graph sense) with respect to this neighborhood structure.

Construct a message-passing algorithm which allows them to count to ensure that none of the children was eaten by the wolf. What is the initialization and what are the message-passing rules? Can you modify the algorithm to only count a prescribed subset, e.g., the set of girls?

2.5 (Message Passing for Mappers – Loeliger [24]). Assume that the two binary symbols x and y are mapped by a function m into one 4-AM symbol, call it z, as

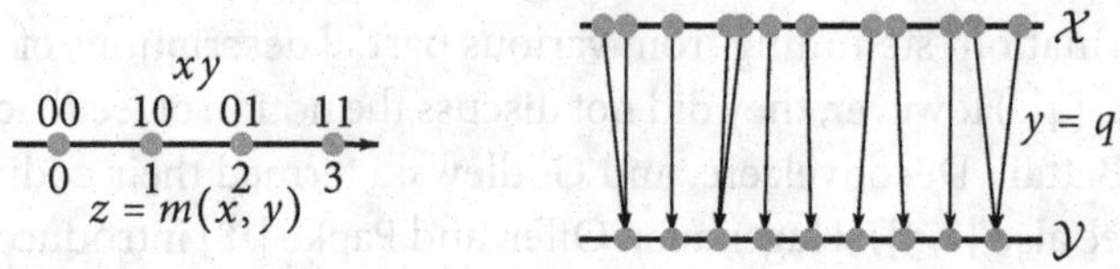

Figure 2.25: Left: Mapping $z = m(x, y)$. Right: Quantizer $y = q(x)$.

shown on the left of Figure 2.25. In more detail, $m : \mathcal{X} \times \mathcal{Y} \to \mathcal{Z}$. Such a mapper is, e.g., useful as part of a multilevel transmission scheme. Draw the corresponding FSFG. Starting from the general message-passing rule stated in (2.21) and assuming that the incoming messages are $\mu_{x,m}(x)$, $\mu_{y,m}(y)$, and $\mu_{z,m}(z)$, respectively, what are the outgoing messages $\mu_{m,x}(x)$, $\mu_{m,y}(y)$, and $\mu_{m,z}(z)$?

2.6 (Message Passing for Quantizers – Loeliger [24]). Consider a quantizer as shown on the right in Figure 2.25. More precisely, let $\mathcal{X}$ be a finite input alphabet and $\mathcal{Y}$ be a finite output alphabet at let q be the quantization function, $q : \mathcal{X} \to \mathcal{Y}$. Draw the corresponding FSFG. Starting from the general message-passing rule stated in (2.21) and assuming that the incoming messages are $\mu_{x,q}(x)$ and $\mu_{y,q}(y)$, respectively, what are the outgoing messages $\mu_{q,x}(x)$ and $\mu_{q,y}(y)$?

References

[1] S. M. Aji and R. J. McEliece, *The generalized distributive law*, IEEE Trans. Inform. Theory, 46 (2000), pp. 325–343. [65, 66]

[2] L. Bahl, J. Cocke, F. Jelinek, and J. Raviv, *Optimal decoding of linear codes for minimizing symbol error rate*, IEEE Trans. Inform. Theory, 20 (1974), pp. 284–287. [34, 39, 66]

[3] G. Battail, *Building long codes by combination of simple ones, thanks to weighted-output decoding*, in Proc. URSI ISSSE, Erlangen, Germany, Sept. 1989, pp. 634–637. [67]

[4] G. Battail, M. Decouvelaere, and P. Godlewski, *Replication decoding*, IEEE Trans. Inform. Theory, 25 (1979), pp. 332–345. [67, 277]

[5] G. Battail and M. S. El-Sherbini, *Coding for radio channels*, Ann. Télécommun., 37 (1982), pp. 75–96. [67]

[6] L. E. Baum and T. Petrie, *Statistical inference for probabilistic functions of finite state Markov chains*, Ann. Math. Stat., 37 (1966), pp. 1554–1536. [66]

[7] L. E. Baum, T. Petrie, G. Soules, and N. Weiss, *A maximization technique occurring in the statistical analysis of probabilistic functions of Markov chains*, Ann. Math. Stat., 41 (1970), pp. 164–171. [66]

[8] L. E. BAUM AND G. R. SELL, *Growth transformations for functions on manifolds*, Pac. J. Math., 27 (1968), pp. 211–227. [66]

[9] R. J. BAXTER, *Exactly Solved Models in Statistical Mechanics*, Academic Press, London, 1982. [66]

[10] H. A. BETHE, *On the theory of metals, I. Eigenvalues and eigenfunctions of a linear chain of atoms*, Zeits. Physik, 71 (1931), pp. 205–226. [66]

[11] ——, *Statistical theory of superlattices*, Proc. Roy. Soc., A150 (1935), pp. 552–75. [66]

[12] A. P. DEMPSTER, N. M. LAIRD, AND D. B. RUBIN, *Maximum likelihood from incomplete data via the EM algorithm*, J. Roy. Stat. Soc. B, 39 (1977), pp. 1–38. [66]

[13] T. ETZION, A. TRACHTENBERG, AND A. VARDY, *Which codes have cycle-free Tanner graphs?*, IEEE Trans. Inform. Theory, 45 (1999), pp. 2173–2181. [66]

[14] G. D. FORNEY, JR., *Codes on graphs: Normal realizations*, IEEE Trans. Inform. Theory, 47 (2001), pp. 520–548. [65]

[15] B. J. FREY AND F. R. KSCHISCHANG, *Probability propagation and iterative decoding*, in Proc. of the Allerton Conf. on Commun., Control, and Computing, Monticello, IL, USA, Sept. 1996. [66]

[16] R. G. GALLAGER, *Low-density parity-check codes*, IRE Trans. Inform. Theory, 8 (1962), pp. 21–28. [33, 66, 159, 264, 419, 434]

[17] J. HAGENAUER, E. OFFER, C. MÉASSON, AND M. MÖRZ, *Decoding and equalization with analog non-linear networks*, Eur. Trans. Telecomm. (ETT), (1999), pp. 107–128. [67]

[18] J. HAGENAUER, E. OFFER, AND L. PAPKE, *Iterative decoding of binary block and convolutional codes*, IEEE Trans. Inform. Theory, 42 (1996), pp. 429–445. [67]

[19] J. HAGENAUER AND M. WINKLHOFER, *The analog decoder*, in Proc. of the IEEE Int. Symposium on Inform. Theory, Cambridge, MA, USA, Aug. 1998, p. 145. [67]

[20] C. HEEGARD AND S. B. WICKER, *Turbo Coding*, Kluwer Academic Publ., New York, NY, USA, 1999. [34, 66]

[21] J. H. KIM AND J. PEARL, *A computational model for causal and diagnostic reasoning in inference systems*, in IJCAI, 1983, pp. 190–193. [66]

[22] F. R. KSCHISCHANG AND B. J. FREY, *Iterative decoding of compound codes by probability propagation in graphical models*, IEEE J. Sel. Area. in Comm., (1998), pp. 219–230. [66]

[23] F. R. KSCHISCHANG, B. J. FREY, AND H.-A. LOELIGER, *Factor graphs and the sum-product algorithm*, IEEE Trans. Inform. Theory, 47 (2001), pp. 498–519. [65, 66]

[24] H.-A. LOELIGER, *Some remarks on factor graphs*, in Proc. of the Int. Conf. on Turbo Codes and Related Topics, Brest, France, Sept. 2003, pp. 111–115. [67, 68]

[25] ——, *An introduction to factor graphs*, Signal Process., 21 (2004), pp. 28–41. [66]

[26] H.-A. Loeliger, F. Lustenberger, M. Helfenstein, and F. Tarköy, *Probability propagation and decoding in analog VLSI*, in Proc. of the IEEE Int. Symposium on Inform. Theory, Cambridge, MA, USA, Aug. 1998, p. 146. [67]

[27] ——, *Decoding in analog VLSI*, IEEE Commun. Mag., (1999), pp. 99–101. [67]

[28] ——, *Probability propagation and decoding in analog VLSI*, IEEE Trans. Inform. Theory, 47 (2001), pp. 837–843. [67]

[29] Y. Mao and F. R. Kschischang, *On factor graphs and the Fourier transform*, IEEE Trans. Inform. Theory, 51 (2005), pp. 1635–1649. [65]

[30] R. J. McEliece, D. J. C. MacKay, and J.-F. Cheng, *Turbo decoding as an instance of Pearl's 'belief propagation' algorithm*, IEEE J. Sel. Area. Commun., 16 (1998), pp. 140–152. [66, 262]

[31] G. J. McLachlan and T. Krishnan, *The EM Algorithm and Extensions*, Wiley, New York, NY, USA, 1997. [66]

[32] M. Mörz, T. Gabara, R. Yan, and J. Hagenauer, *An analog 0.25 μm BiCMOS tailbiting MA, USAP decoder*, in Proc. of the IEEE Int. Solid-State Circuits Conf., San Francisco, CA, Feb. 2000, pp. 356–357. [67]

[33] J. Pearl, *Probabilistic Reasoning in Intelligent Systems: Networks of Plausible Inference*, Morgan Kaufmann Publ., San Mateo, CA, USA, 1988. [66]

[34] G. R. Shafer and P. P. Shenoy, *Probability propagation*, Ann. Math. Art. Intel., 2 (1990), pp. 327–352. [65]

[35] R. M. Tanner, *A recursive approach to low complexity codes*, IEEE Trans. Inform. Theory, 27 (1981), pp. 533–547. [65, 161, 261, 263]

[36] A. J. Viterbi, *Error bounds of convolutional codes and an asymptotically optimum decoding algorithm*, IEEE Trans. Inform. Theory, 13 (1967), pp. 260–269. [33, 34, 66]

[37] N. Wiberg, *Codes and Decoding on General Graphs*, PhD thesis, Linköping University, S-581 83, Linköping, Sweden, 1996. [65, 66, 263, 266, 366]

[38] N. Wiberg, H.-A. Loeliger, and R. Kötter, *Codes and iterative decoding on general graphs*, Eur. Trans. Telecomm. (ETT), 6 (1995), pp. 513–526. [65, 263]

[39] A. P. Worthen and W. E. Stark, *Unified design of iterative receivers using factor graphs*, IEEE Trans. Inform. Theory, 47 (2001), pp. 843–850. [66]

Chapter 3
BINARY ERASURE CHANNEL

The *binary erasure channel* (BEC) is perhaps the simplest non-trivial channel model imaginable. It was introduced by Elias as a toy example in 1954. The emergence of the Internet promoted the erasure channel into the class of "real-world" channels. Indeed, erasure channels can be used to model data networks, where packets either arrive correctly or are lost due to buffer overflows or excessive delays.

A priori, one might well doubt that studying the BEC will significantly advance our understanding of the general case. Quite surprisingly, however, most properties and statements that we encounter in our investigation of the BEC hold in much greater generality. Thus, the effort invested in fully understanding the BEC case will reap substantial dividends later on.

You do not need to read the whole chapter to know what iterative decoding for the BEC is about. The core of the material is contained in Sections 3.1–3.14 as well as 3.24. The remaining sections concern either more specialized or less accessible topics. They can be read in almost any order.

§3.1. Channel Model

Erasure channels model situations where information may be lost but is never corrupted. The BEC captures erasure in the simplest form: single *bits* are transmitted and either received correctly or known to be lost. The decoding problem is to find the values of the bits given the locations of the erasures and the non-erased part of the codeword. Figure 3.1 depicts the BEC(ϵ). Time, indexed by t, is discrete and

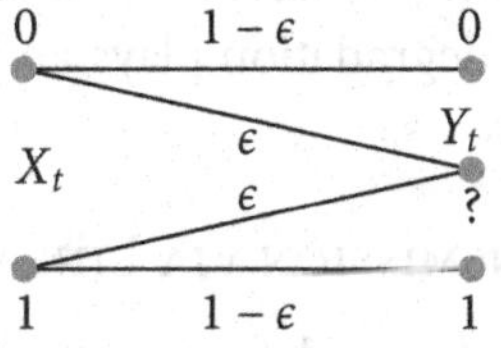

Figure 3.1: Binary erasure channel with parameter ϵ.

the transmitter and receiver are synchronized (they both know t). The channel input at time t, denoted by X_t, is binary: $X_t \in \{0, 1\}$. The corresponding output Y_t takes on values in the alphabet $\{0, 1, ?\}$, where ? indicates an *erasure*. Each transmitted bit is either erased with probability ϵ, or received correctly: $Y_t \in \{X_t, ?\}$ and $\mathbb{P}\{Y_t = ?\} = \epsilon$. Erasure occurs for each t independently. For this reason we say that the channel is *memoryless*. The capacity of the BEC(ϵ) is $C_{BEC}(\epsilon) = 1 - \epsilon$ bits per channel use. It is easy to see that $C_{BEC}(\epsilon) \leq 1 - \epsilon$: assume that the transmitted is told in advance which bits will be erased. Clearly, this additional information can only increase the achievable rate. Since on average only $(1 - \epsilon)n$ of the n transmitted positions are usable (are not erased) the best that the transmitter can do is to fill these $(1 - \epsilon)n$ slots with information. Since the transmitter sees which positions are erased, he can simply read out the non-erased positions to retrieve the information. Thus, even for this case where the transmitter knows in advance which bits will be erased, information can be transmitted reliably at a rate of at most $1 - \epsilon$ bits per channel use. Perhaps surprisingly, reliable transmission at a rate arbitrarily close to $1 - \epsilon$ is possible even without knowledge of the erased positions in advance. This is confirmed in Example 3.6.

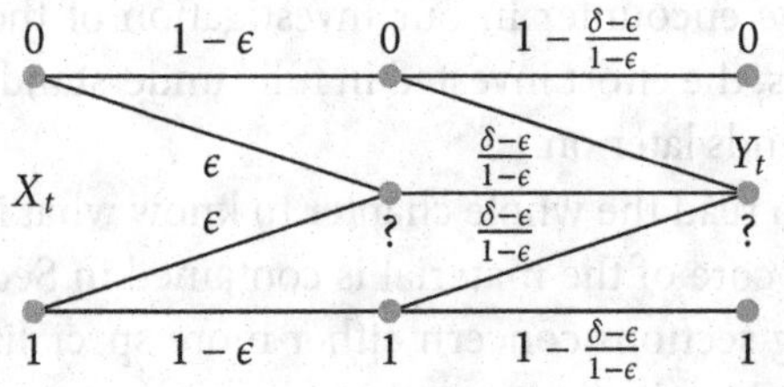

Figure 3.2: For $\epsilon \leq \delta$, the BEC(δ) is degraded with respect to the BEC(ϵ).

Consider the channel *family* $\{\text{BEC}(\epsilon)\}_{\epsilon=0}^{1}$. This family is *ordered* in the following sense: given two BECs, lets say with parameter ϵ and δ, $\epsilon < \delta$, we can represent the BEC(δ) as the *concatenation* of the BEC(ϵ) with a memoryless ternary-input channel as shown in Figure 3.2. Hence, the output of the BEC(δ) can be interpreted as a *degraded* version, i.e., a further perturbed version, of the output of the BEC(ϵ). We say, because this interpretation is possible, that the family $\{\text{BEC}(\epsilon)\}_{\epsilon=0}^{1}$ is ordered by *degradation*. This notion of degradation plays an important role in the analysis of message-passing coding systems.

§3.2. Transmission via Linear Codes

Consider a binary linear code $C[n, k]$ defined in terms of a parity-check matrix H. Assume that the transmitter chooses the codeword X uniformly at random from C and that transmission takes place over the BEC(ϵ). Let Y be the received word with elements in the extended alphabet $\{0, 1, ?\}$, where ? indicates an erasure. Let $\mathcal{E}$, $\mathcal{E} \subseteq [n] = \{1, \ldots, n\}$, denote the *index set of erasures* and let $\bar{\mathcal{E}} = [n] \setminus \mathcal{E}$. More precisely, $i \in \mathcal{E}$ if and only if $Y_i =?$, i.e., if the channel erased the i-th bit. Recall from page 31 that $H_{\mathcal{E}}$ denotes the submatrix of H indexed by the elements of $\mathcal{E}$ and that $X_{\mathcal{E}}$ denotes the corresponding subvector.

§3.2.1. Block-wise MAP Decoding

Consider *block-wise* maximum a posteriori (MAP) decoding; i.e., the decoding rule is

$$\hat{x}^{\text{MAP}}(y) = \text{argmax}_{x \in C} p_{X \mid Y}(x \mid y). \tag{3.3}$$

Write the defining equation $Hx^T = 0^T$ in the form $H_{\mathcal{E}} x_{\mathcal{E}}^T + H_{\bar{\mathcal{E}}} x_{\bar{\mathcal{E}}}^T = 0^T$, which, since we are working over $\mathbb{F}_2$, is equivalent to

$$H_{\mathcal{E}} x_{\mathcal{E}}^T = H_{\bar{\mathcal{E}}} x_{\bar{\mathcal{E}}}^T. \tag{3.4}$$

Note that $s^T = H_{\bar{\mathcal{E}}} x_{\bar{\mathcal{E}}}^T$, the right-hand side of (3.4), is *known* to the receiver since $x_{\bar{\mathcal{E}}} = y_{\bar{\mathcal{E}}}$. Consider the equation $H_{\mathcal{E}} x_{\mathcal{E}}^T = s^T$. Since, by assumption, the transmitted word is a valid codeword, we know that this equation has *at least one* solution. In particular, $\text{rank}(H_{\mathcal{E}}) \leq |\mathcal{E}|$. If $\text{rank}(H_{\mathcal{E}}) = |\mathcal{E}|$, then block-wise MAP decoding can be accomplished by solving $H_{\mathcal{E}} x_{\mathcal{E}}^T = s^T$. On the other hand, there are *multiple* solutions (i.e., the MAP decoder is not able to recover the codeword uniquely) if and only if $\text{rank}(H_{\mathcal{E}}) < |\mathcal{E}|$. More formally, let

$$\mathcal{X}^{\text{MAP}}(y) = \{x \in C : H_{\mathcal{E}} x_{\mathcal{E}}^T = H_{\bar{\mathcal{E}}} y_{\bar{\mathcal{E}}}^T ; x_{\bar{\mathcal{E}}} = y_{\bar{\mathcal{E}}}\},$$

i.e., $\mathcal{X}^{\text{MAP}}(y)$ is the set of all codewords *compatible* with the received word y. Since the prior is uniform, (3.3) becomes

$$\hat{x}^{\text{MAP}}(y) = \text{argmax}_{x \in C} p_{Y \mid X}(y \mid x).$$

Now, for any codeword x, if $x_{\bar{\mathcal{E}}} \neq y_{\bar{\mathcal{E}}}$, then $p_{Y \mid X}(y \mid x) = 0$ and if $x_{\bar{\mathcal{E}}} = y_{\bar{\mathcal{E}}}$, then $p_{Y \mid X}(y \mid x) = (1-\epsilon)^{n-|\mathcal{E}|}\epsilon^{|\mathcal{E}|}$. Thus, all elements of $\mathcal{X}^{\text{MAP}}(y)$ are equally likely and the transmitted vector x is either uniquely determined by y or there are multiple solutions. Therefore, we say

$$\hat{x}^{\text{MAP}}(y) = \begin{cases} x \in \mathcal{X}^{\text{MAP}}(y), & \text{if } |\mathcal{X}^{\text{MAP}}(y)| = 1, \\ ?, & \text{otherwise.} \end{cases}$$

We remark that a ? (erasure) is not the same as an *error*. The correct solution x is an element of $\mathcal{X}^{\text{MAP}}(y)$.

§3.2.2. Bit-wise MAP Decoding

Consider now the *bit-wise* MAP decoder that uses the decoding rule

$$\hat{x}_i^{\text{MAP}}(y) = \text{argmax}_{\alpha \in \{0,1\}} p_{X_i \mid Y}(\alpha \mid y). \tag{3.5}$$

If $i \in \mathcal{E}$, when can x_i be recovered? Intuitively, we expect that x_i can be recovered if and only if all elements of $\mathcal{X}^{\text{MAP}}(y)$ have the same value for the i-th bit. This is in fact correct. More specifically, we claim that x_i can *not* be recovered if and only if $H_{\{i\}}$ is an element of the space spanned by the columns of $H_{\mathcal{E}\setminus\{i\}}$. This is equivalent to the statement that $Hw^T = 0^T$ has a solution with $w_i = 1$ and $w_{\bar{\mathcal{E}}} = 0$. Now, if there is such a solution then for every element x of $\mathcal{X}^{\text{MAP}}(y)$ we also have $x+w \in \mathcal{X}^{\text{MAP}}(y)$. It follows that exactly half the elements $x \in \mathcal{X}^{\text{MAP}}(y)$ have $x_i = 0$ and half have $x_i = 1$. Conversely, if we can find two elements x and x' of $\mathcal{X}^{\text{MAP}}(y)$ with $x_i \neq x'_i$, then $w = x + x'$ solves $Hw^T = 0^T$ and has $w_i = 1$ and $w_{\bar{\mathcal{E}}} = 0$. Proceeding formally, we get

$$\hat{x}_i^{\text{MAP}}(y) = \text{argmax}_{\alpha\in\{0,1\}} p_{X_i\,|\,Y}(\alpha\,|\,y) = \text{argmax}_{\alpha\in\{0,1\}} \sum_{x\in\{0,1\}^n : x_i=\alpha} p_{X\,|\,Y}(x\,|\,y)$$

$$= \text{argmax}_{\alpha\in\{0,1\}} \sum_{x\in C: x_i=\alpha} p_{X\,|\,Y}(x\,|\,y) = \begin{cases} \alpha, & \text{if } \forall x \in \mathcal{X}^{\text{MAP}}(y),\ x_i = \alpha, \\ ?, & \text{otherwise.} \end{cases}$$

We conclude that optimal (block or bit) decoding for the BEC can be accomplished in complexity at most $O(n^3)$ by solving a linear system of equations (e.g., by Gaussian elimination). Further, we have a characterization of decoding failures of a MAP decoder for both the block and the bit erasure case in terms of rank conditions.

Example 3.6 (Performance of $\mathcal{H}(n,k)$). In Problem 3.22 you are asked to show that the average block erasure probability of Gallager's parity-check ensemble (see Definition 1.26) satisfies

$$\mathbb{E}_{\mathcal{H}(n,k)}[\text{P}_{\text{B}}^{\text{MAP}}(H,\epsilon)] \leq \sum_{e=0}^{n-k} \binom{n}{e} \epsilon^e (\bar{\epsilon})^{n-e} 2^{e-n+k} + \sum_{e=n-k+1}^{n} \binom{n}{e} \epsilon^e (\bar{\epsilon})^{n-e}$$

$$= 2^{k-n} (\bar{\epsilon})^n \sum_{e=0}^{n-k} \binom{n}{e} \left(\frac{2\epsilon}{\bar{\epsilon}}\right)^e + (\bar{\epsilon})^n \sum_{e=n-k+1}^{n} \binom{n}{e} \left(\frac{\epsilon}{\bar{\epsilon}}\right)^e,$$

where the bound is loose by a factor of at most 2. Let the blocklength n tend to infinity. Suppose that $r = k/n = (1-\delta)\text{C}_{\text{BEC}}(\epsilon) = \bar{\delta}\bar{\epsilon}$, where $\frac{1}{1+\epsilon} < \bar{\delta} < 1$. The elements of both sums are *unimodal* sequences (see Section D.5); i.e., they are first increasing up to their maximum value, after which they decrease. Consider the terms $\{\binom{n}{e}\left(\frac{2\epsilon}{\bar{\epsilon}}\right)^e\}_{e=0}^{n-k}$ of the first sum. Because of our assumption $\frac{1}{1+\epsilon} < \bar{\delta}$, the upper summation index $n-k$ is to the left of the maximum, which occurs at $e \approx \frac{2\epsilon}{1+\epsilon} n$. The first sum can therefore be upper bounded by the last term times the number of summands. Similarly, the maximum term of $\{\binom{n}{e}\left(\frac{\epsilon}{\bar{\epsilon}}\right)^e\}_{e=n-k+1}^{n}$ occurs at $e \approx \epsilon n$. Since $\bar{\delta} < 1$, this is to the left of the lower summation index $e = n-k+1$. In fact, it is to the left of the index $e = n-k$. This second sum can therefore be upper bounded by the term corresponding to $e = n - k$ times the number of summands. This leads to the bound

$$\mathbb{E}_{\mathcal{H}(n,rn)}[\mathrm{P}_{\mathrm{B}}^{\mathrm{MAP}}(H,\epsilon)] \leq (n+1)\binom{n}{n-k}(\bar{\epsilon})^n\left(\frac{\epsilon}{\bar{\epsilon}}\right)^{n-k}$$
$$\overset{(1.59)}{\leq} (n+1)2^{nh_2(\bar{\delta}\bar{\epsilon})+n\log_2(\bar{\epsilon})+n(1-\bar{\delta}\bar{\epsilon})\log_2(\frac{\epsilon}{\bar{\epsilon}})}$$
$$= (n+1)2^{-nD_2(1-\bar{\delta}\bar{\epsilon}\,\|\,\epsilon)}, \tag{3.7}$$

where we defined $D_2(\alpha\,\|\,\beta) = -\alpha\log_2\frac{\beta}{\alpha} - \bar{\alpha}\log_2\frac{\bar{\beta}}{\bar{\alpha}}$. The quantity $D_2(\cdot,\cdot)$ is known as the *Kullback-Leibler* distance (between two Bernoulli distributions with parameters α and β, respectively). Let $\alpha, \beta \in (0,1)$. Using Jensen's inequality (1.61), we see that

$$D_2(\alpha\,\|\,\beta) = -\alpha\log_2\frac{\beta}{\alpha} - \bar{\alpha}\log_2\frac{\bar{\beta}}{\bar{\alpha}} \geq -\log_2\left(\alpha\frac{\beta}{\alpha} + \bar{\alpha}\frac{\bar{\beta}}{\bar{\alpha}}\right) = 0,$$

with strict inequality if $\alpha \neq \beta$. In our case $1-\bar{\delta}\bar{\epsilon} \geq \epsilon$ with strict inequality unless $\delta = 0$ or $\epsilon = 1$. Therefore, for $\delta \in (0,1]$ and $\epsilon \in [0,1)$ the right-hand side of (3.7) tends to zero exponentially fast in the blocklength n. We conclude that reliable transmission at any rate $r = \bar{\delta}\bar{\epsilon} = (1-\delta)\mathrm{C}_{\mathrm{BEC}}(\epsilon)$, $\delta > 0$, is possible. In words, reliable transmission up to $\mathrm{C}_{\mathrm{BEC}}(\epsilon)$ is possible, as promised. ◇

§3.3. Tanner Graphs

Let C be binary linear code and let H be a parity-check matrix of C, i.e., $C = C(H)$. Recall that, by our convention, we do not require the rows of H to be linearly independent. Assume that H has dimensions $m \times n$. In Example 2.5 on page 50 we introduced the Tanner graph associated with a code C. This is the graph which visualizes the factorization of the code membership function. Since this graph plays a central role let us repeat its definition.

The Tanner graph associated with H is a *bipartite* graph. It has n *variable* nodes, corresponding to the components of the codeword, and m *check* nodes, corresponding to the set of parity-check constraints (rows of H). Check node j is connected to variable node i if $H_{ji} = 1$, i.e., if variable i participates in the j-th parity-check constraint. Since there are many parity-check matrices representing the same code, there are many Tanner graphs corresponding to a given C. Although all of these Tanner graphs describe the same code, they are not equivalent from the point of view of the message-passing decoder (see Problem 3.15).

EXAMPLE 3.8 ((3, 6)-REGULAR CODE). Consider the parity-check matrix

$$
\begin{array}{c} \\ 1 \\ 2 \\ 3 \\ 4 \\ 5 \\ 6 \\ 7 \\ 8 \\ 9 \\ 10 \end{array}
\begin{array}{c}
\begin{array}{cccccccccccccccccccc} 1 & 2 & 3 & 4 & 5 & 6 & 7 & 8 & 9 & 10 & 11 & 12 & 13 & 14 & 15 & 16 & 17 & 18 & 19 & 20 \end{array} \\
\left(\begin{array}{cccccccccccccccccccc}
0&0&0&0&1&0&0&0&1&1&1&0&0&0&0&1&0&0&0&1 \\
0&0&0&0&0&0&1&1&0&0&1&1&0&1&0&1&0&0&0&0 \\
0&1&1&0&0&0&1&0&0&0&0&0&0&0&0&1&0&1&0&1 \\
0&0&0&0&0&1&0&1&0&1&0&0&0&0&0&0&1&1&1&0 \\
1&1&0&0&1&0&0&0&0&0&0&0&1&0&0&0&1&0&1&0 \\
0&0&0&0&0&0&1&0&0&0&1&1&0&1&1&0&0&0&0&1 \\
0&0&0&1&1&1&0&1&0&0&0&0&1&0&1&0&0&0&0&0 \\
1&0&1&0&0&0&0&0&1&0&0&0&1&1&1&0&0&0&0&0 \\
1&1&1&1&0&0&0&0&0&1&0&0&0&0&0&0&1&0&0&0 \\
0&0&0&1&0&1&0&0&1&0&0&1&0&0&0&0&0&1&1&0
\end{array}\right)
\end{array} = H. \tag{3.9}
$$

The bipartite graph representing $C(H)$ is shown on the left of Figure 3.10. Each check

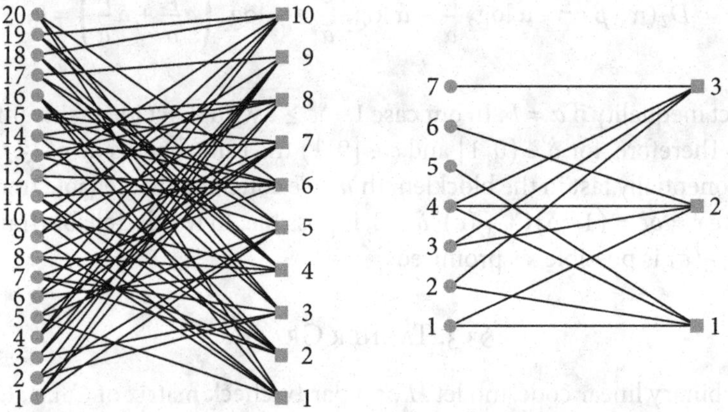

Figure 3.10: Left: Tanner graph of H given in (3.9). Right: Tanner graph of $[7, 4, 3]$ Hamming code corresponding to the parity-check matrix on page 15. This graph is discussed in Example 3.11.

node represents one linear constraint (one row of H). For the particular example we start with 20 degrees of freedom (20 variable nodes). The 10 constraints reduce the number of degrees of freedom by at most 10 (and exactly by 10 if all these constraints are linearly independent as in this specific example). Therefore at least 10 degrees of freedom remain. It follows that the shown code has rate (at least) one-half. ◇

§3.4. LOW-DENSITY PARITY-CHECK CODES

In a nutshell, *low-density parity-check* (LDPC) codes are linear codes that have at least one *sparse* Tanner graph. The primary reason for focusing on such codes is that, as we will see shortly, they exhibit good performance under message-passing decoding.

Consider again the Tanner graph on the left of Figure 3.10. Each variable node has degree 3 and every check node has degree 6. We call such a code a $(3,6)$-regular LDPC code. More generally, an $(\mathtt{l},\mathtt{r})$-regular LDPC code is a binary linear code such that every variable node has degree $\mathtt{l}$ and every check node has degree $\mathtt{r}$. Why *low-density*? The number of edges in the Tanner graph of a $(\mathtt{l},\mathtt{r})$-regular LDPC code is $\mathtt{l}n$, where n is the length of the code. As n increases, the number of edges in the Tanner graph grows *linearly* in n. This is in contrast to codes in the parity-check ensemble $\mathcal{H}(n,nr)$ where the number of edges in their Tanner graph grows like the *square* of the code length n.

The behavior of LDPC codes can be significantly improved by allowing nodes of different degrees as well as other structural improvements. We define an *irregular* LDPC code as an LDPC code for which the degrees of nodes are chosen according to some distribution.

Example 3.11 (Tanner Graph of $[7,4,3]$ Hamming Code). The right-hand side of Figure 3.10 shows the Tanner graph of the $[7,4,3]$ Hamming code corresponding to the parity-check matrix on page 15. The three check nodes have degree 4. There are three variable nodes of degree 1, three variable nodes of degree 2, and one variable node of degree 3. ◇

Assume that the LDPC code has length n and that the number of variable nodes of degree i is Λ_i, so that $\sum_i \Lambda_i = n$. In the same fashion, denote the number of check nodes of degree i by P_i, so that $\sum_i P_i = n\bar{r}$, where r is the *design* rate (ratio of length minus number of constraints and the length) of the code and $\bar{r}$ is a shorthand for $1-r$. Further, since the edge counts must match up, $\sum_i i\Lambda_i = \sum_i iP_i$. It is convenient to introduce the following compact notation:

$$\Lambda(x) = \sum_{i=1}^{\mathtt{l}_{\max}} \Lambda_i x^i, \qquad P(x) = \sum_{i=1}^{\mathtt{r}_{\max}} P_i x^i, \tag{3.12}$$

i.e., $\Lambda(x)$ and $P(x)$ are polynomials with non-negative expansions around zero whose integral coefficients are equal to the number of nodes of various degrees. From these definitions we see immediately the following relationships:

$$\Lambda(1) = n, \quad P(1) = n\bar{r}, \quad r(\Lambda,P) = 1 - \frac{P(1)}{\Lambda(1)}, \quad \Lambda'(1) = P'(1). \tag{3.13}$$

We call Λ and P the variable and check *degree distributions from a node perspective*. Sometimes it is more convenient to use the *normalized* degree distributions

$$L(x) = \frac{\Lambda(x)}{\Lambda(1)}, \qquad R(x) = \frac{P(x)}{P(1)}.$$

EXAMPLE 3.14 (DEGREE DISTRIBUTION OF [7, 4, 3] HAMMING CODE). We have

$$\Lambda(x) = 3x + 3x^2 + x^3, \qquad P(x) = 3x^4,$$
$$L(x) = \frac{3}{7}x + \frac{3}{7}x^2 + \frac{1}{7}x^3, \qquad R(x) = x^4. \qquad \diamond$$

DEFINITION 3.15 (THE STANDARD ENSEMBLE LDPC (Λ, P)). Given a degree distribution pair (Λ, P), define an *ensemble* of bipartite graphs LDPC (Λ, P) in the following way. Each graph in LDPC (Λ, P) has $\Lambda(1)$ variable nodes and $P(1)$ check nodes: Λ_i variable nodes and P_i check nodes have degree i. A node of degree i has i *sockets* from which the i edges emanate, so that in total there are $\Lambda'(1) = P'(1)$ sockets on each side. Label the sockets on each side with the set $[\Lambda'(1)] = \{1, \ldots, \Lambda'(1)\}$ in some arbitrary but fixed way. Let σ be a permutation on $[\Lambda'(1)]$. Associate to σ a bipartite graph by connecting the i-th socket on the variable side to the $\sigma(i)$-th socket on the check side. Letting σ run over the set of permutations on $[\Lambda'(1)]$ generates a set of bipartite graphs. Finally, we define a probability distribution over the set of graphs by placing the uniform probability distribution on the set of permutations. This is the ensemble of bipartite graphs LDPC (Λ, P). In the random graph literature this is what is called the *configuration model.*

It remains to associate a code with every element of LDPC (Λ, P). We will do so by associating a parity-check matrix to each graph. Because of possible multiple edges and since the encoding is done over the field $\mathbb{F}_2$, we define the parity-check matrix H as the $\{0, 1\}$-matrix that has a non-zero entry at row i and column j if the i-th check node is connected to the j-th variable node an *odd* number of times.

Since to every graph we can associate a code, we use these two terms interchangeably and we refer, e.g., to codes as elements of LDPC (Λ, P).

This is a subtle point: graphs are *labeled* (they have labeled sockets) and have a uniform probability distribution; the induced codes are unlabeled and the probability distribution on the set of codes is not necessarily the uniform one. Therefore, if in the sequel we say that we pick a code uniformly at random we really mean that we pick a graph at random from the ensemble of graphs and consider the induced code. This convention should not cause any confusion and simplifies our notation considerably. ∇

As discussed in Problem 3.6, ensembles with a positive fraction of degree 1 variable nodes have non-zero bit error probability for *all* non-zero channel parameters even in the limit of infinite blocklengths: by our definition of the ensemble (where variable nodes are matched randomly to check nodes) there is a positive probability that two degree 1 variable nodes connect to the same check node and such a code contains codewords of weight 2. Therefore, we only consider ensembles without degree 1 nodes. But, as we will discuss in Chapter 7, it is possible to introduce degree 1 variable nodes if their edges are placed with care.

For the asymptotic analysis it is more convenient to take on an *edge perspective*. Define

$$(3.16)\quad \lambda(x) = \sum_i \lambda_i x^{i-1} = \frac{\Lambda'(x)}{\Lambda'(1)} = \frac{L'(x)}{L'(1)}, \quad \rho(x) = \sum_i \rho_i x^{i-1} = \frac{P'(x)}{P'(1)} = \frac{R'(x)}{R'(1)}.$$

Note that λ and ρ are polynomials with non-negative expansions around zero. Some thought shows that λ_i (ρ_i) is equal to the *fraction of edges* that connect to variable (check) nodes of degree i; see Problem 3.2. In other words, λ_i (ρ_i) is the probability that an edge chosen uniformly at random from the graph is connected to a variable (check) node of degree i. We call λ and ρ the variable and check *degree distributions from an edge perspective*. The inverse relationships read

$$(3.17)\quad \frac{\Lambda(x)}{n} = L(x) = \frac{\int_0^x \lambda(z)\mathrm{d}z}{\int_0^1 \lambda(z)\mathrm{d}z}, \qquad \frac{P(x)}{n\bar{r}} = R(x) = \frac{\int_0^x \rho(z)\mathrm{d}z}{\int_0^1 \rho(z)\mathrm{d}z}.$$

As discussed in Problems 3.3 and 3.4, the average variable and check degrees, call them $\mathtt{l}_{\text{avg}}$ and $\mathtt{r}_{\text{avg}}$, can be expressed as

$$(3.18)\quad \mathtt{l}_{\text{avg}} = L'(1) = \frac{1}{\int_0^1 \lambda(x)\mathrm{d}x}, \qquad \mathtt{r}_{\text{avg}} = R'(1) = \frac{1}{\int_0^1 \rho(x)\mathrm{d}x},$$

respectively, and the *design* rate is given by

$$(3.19)\quad r(\lambda,\rho) = 1 - \frac{\mathtt{l}_{\text{avg}}}{\mathtt{r}_{\text{avg}}} = 1 - \frac{L'(1)}{R'(1)} = 1 - \frac{\int_0^1 \rho(x)\mathrm{d}x}{\int_0^1 \lambda(x)\mathrm{d}x}.$$

The design rate is the rate of the code assuming that all constraints are linearly independent.

Example 3.20 (Conversion of Degree Distributions: Hamming Code). For the $[7,4,3]$ Hamming code we have

$$\lambda(x) = \frac{1}{4} + \frac{1}{2}x + \frac{1}{4}x^2, \qquad \rho(x) = x^3. \qquad \diamond$$

Example 3.21 (Conversion of Degree Distributions: Second Example). Consider the pair (Λ, P):

$$\Lambda(x) = 613x^2 + 202x^3 + 57x^4 + 84x^7 + 44x^8, \qquad P(x) = 500x^6,$$

with

$$\Lambda(1) = 1000, \qquad P(1) = 500, \qquad \Lambda'(1) = P'(1) = 3000.$$

This pair represents an ensemble of codes of length 1000 and of (design) rate one-half. Converting into edge perspective we get

$$\lambda(x) = \frac{1226}{3000}x + \frac{606}{3000}x^2 + \frac{228}{3000}x^3 + \frac{588}{3000}x^6 + \frac{352}{3000}x^7, \quad \rho(x) = x^5. \quad \diamond$$

Since (Λ, P), (n, L, R), and (n, λ, ρ) contain equivalent information, we frequently and freely switch between these various perspectives. We write $(\Lambda, P) \triangleq (n, L, R) \triangleq (n, \lambda, \rho)$ if we want to express the fact that degree distributions in different formats are equivalent. We therefore often refer to the standard ensemble as LDPC (n, λ, ρ). For the asymptotic analysis it is convenient to fix (λ, ρ) and to investigate the performance of the ensemble LDPC (n, λ, ρ) as the blocklength n tends to infinity. For some n the corresponding (Λ, P) is not integral. We assume in such a scenario that the individual node distributions are rounded to the closest integer (while observing the edge equality constraint). In any case, sublinear (in n) deviations of degree distributions have no effect on the asymptotic performance or rate of the ensemble. In the sequel we therefore ignore this issue.

The design rate as defined in (3.19) is in general only a lower bound on the actual rate because the parity-check matrix can have linearly dependent rows. The following lemma asserts that, under some technical conditions, the actual rate of a random element of an ensemble is close to the design rate with high probability as the blocklength increases.

Lemma 3.22 (Rate versus Design Rate). Consider the ensemble LDPC $(n, \lambda, \rho) \triangleq$ LDPC (n, L, R). Let $r(\lambda, \rho)$ denote the design rate of the ensemble and let $r(\mathtt{G})$ denote the actual rate of a code $\mathtt{G}$, $\mathtt{G} \in$ LDPC (n, λ, ρ). Consider the function $\Psi(y)$,

$$\Psi(y) = -L'(1)\log_2\left[\frac{(1+yz)}{(1+z)}\right] + \sum_i L_i \log_2\left[\frac{1+y^i}{2}\right]$$

$$+ \frac{L'(1)}{R'(1)}\sum_j R_j \log_2\left[1 + \left(\frac{1-z}{1+z}\right)^j\right], \tag{3.23}$$

$$z = \left(\sum_i \frac{\lambda_i y^{i-1}}{1+y^i}\right) \Big/ \left(\sum_i \frac{\lambda_i}{1+y^i}\right). \tag{3.24}$$

Assume that, for $y \geq 0$, $\Psi(y) \leq 0$ with equality only at $y = 1$. Then for $\xi > 0$ and $n \geq n(\xi)$, sufficiently large,

$$\mathbb{P}\{r(\mathtt{G}) - r(\lambda, \rho) > \xi\} \leq e^{-n\xi \ln(2)/2}.$$

Example 3.25 (Degree Distribution $(\lambda(x) = \frac{3x+3x^2+4x^{13}}{10}, \rho(x) = x^6)$). The function $\Psi(y)$ is shown in Figure 3.26. According to this plot, $\Psi(y) \leq 0$ with equality

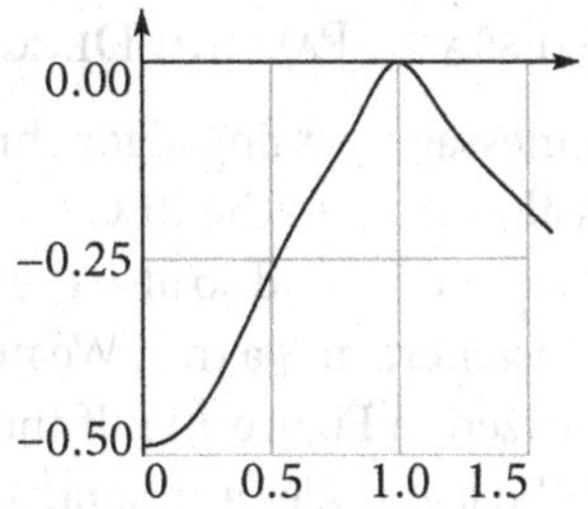

Figure 3.26: Function $\Psi(y)$.

only at $y = 1$. Lemma 3.22 therefore applies: the rate of most codes in this ensemble is not much larger than the design rate $r(\lambda, \rho) = \frac{19}{39}$. ◇

An example where the technical condition is *not* fulfilled is discussed in Problem 3.38.

Discussion: The proof of Lemma 3.22 is technical and so we relegate it to page 517. As we discuss there in more detail, the function $\Psi(y)$ represents the log of the number of codewords of a given weight divided by the length of the code minus the design rate. The parameterization is chosen so that $y = 1$ corresponds to the codewords of relative weight one-half. If the maximum of $\Psi(y)$ is taken on at 1 then this means that most codewords have relative weight one-half (as one would expect). In this case one can show that the maximum is 0 which implies that the log of the expected number of codewords divided by n is equal to the design rate. This bound is crude and in fact a much stronger statement is valid. We demonstrate this by means of *regular* ensembles. The proof of the following lemma can be found on page 519.

LEMMA 3.27 (DESIGN RATE EQUALS RATE FOR REGULAR ENSEMBLES). Consider the regular ensemble LDPC $\left(nx^{\mathtt{l}}, n\frac{\mathtt{l}}{\mathtt{r}}x^{\mathtt{r}}\right)$ with $2 \leq \mathtt{l} < \mathtt{r}$. Let $r(\mathtt{l}, \mathtt{r}) = 1 - \frac{\mathtt{l}}{\mathtt{r}}$ denote the design rate of the ensemble and let $r(\mathtt{G})$ denote the actual rate of a code $\mathtt{G}$, $\mathtt{G} \in$ LDPC $\left(nx^{\mathtt{l}}, n\frac{\mathtt{l}}{\mathtt{r}}x^{\mathtt{r}}\right)$. Then

$$\mathbb{P}\{r(\mathtt{G})n = r(\mathtt{l}, \mathtt{r})n + \nu\} = 1 - o_n(1),$$

where $\nu = 1$ if $\mathtt{l}$ is even, and $\nu = 0$ otherwise.

Discussion: The extra constant ν is easily explained. If the variable nodes have an even degree then each variable appears in an even number of constraints and so the sum of all constraints is zero. So at least one linearly dependent equation exists in this case. The lemma asserts that all other equations are linearly independent with high probability.

§3.5. Message-Passing Decoder

In Chapter 2 we introduced a message-passing algorithm to accomplish the decoding task. Let us specialize this algorithm to the BEC.

The Tanner graph of an LDPC code (and so also the factor graph corresponding to bit-wise MAP decoding) is in general not a tree. We nevertheless use the standard message-passing rules summarized in Figure 2.12. If the factor graph is a tree there is a natural *schedule* – start at the leaf nodes and send a message once all incoming messages required for the computation have arrived. But to completely define the algorithm for a code with cycles we need to *specify* a schedule. In general, different schedules can lead to different performance. This is our convention: we proceed in *rounds* or *iterations*; a round starts by processing incoming messages at check nodes and then sending the resulting outgoing messages to variable nodes along all edges. These messages are subsequently processed at the variable nodes and the outgoing messages are sent back along all edges to the check nodes. This constitutes one round of message passing. In general, decoding consists of several such rounds. In iteration zero, no problem-specific information exists that we can send from the check nodes to the variable nodes. Therefore, in this initial round, the variable nodes simply send the messages received from the channel to their neighboring check nodes.

By the standard message-passing rules the initial messages are $(\mu_j(0), \mu_j(1)) = (p_{Y_j \mid X_j}(y_j \mid 0), p_{Y_j \mid X_j}(y_j \mid 1))$. Specializing this to our case, we see that the initial messages are either $(1-\epsilon, 0)$, (ϵ, ϵ), or $(0, 1-\epsilon)$. This corresponds to the three possibilities, namely that the received value is 0, ? (erasure), or 1, respectively.[1] Recall that the *normalization* of the messages plays no role. We saw in Section 2.5.2 that we only need to know the ratio and this conclusion stays valid if the graph contains cycles. Therefore, equivalently we can work with the set of messages $(1,0)$, $(1,1)$, and $(0,1)$. In the sequel we will call these also the "0" (zero), the "?" (erasure), and the "1" (one), message. Therefore, e.g., 0, $(1,0)$, and "zero," all refer to the same message.

We now get to the processing rules. We claim that for the BEC the general message-passing rules summarized in Figure 2.12 simplify to the following: at a variable node the outgoing message is an erasure if *all* incoming messages are erasures. Otherwise, since the channel *never* introduces errors, all non-erasure messages must agree and either be 0 or 1. In this case the outgoing message is equal to this common value. At a check node the outgoing message is an erasure if *any* of the incoming messages is an erasure. Otherwise, if all of the incoming messages are either 0 or 1 then the outgoing message is the mod-2 sum of the incoming messages.

Consider the first claim: if all messages entering a variable node are from the set $\{(1,0), (1,1)\}$, then the outgoing message (which is equal to the component-

[1] We assume in this chapter that the bits take on the values 0 and 1.

wise product of the incoming messages according to the general message-passing rules) is also from this set. Further, it is equal to $(1,1)$ (i.e., an erasure) only if *all* incoming messages are of the form $(1,1)$ (i.e., erasures). The equivalent statement is true if all incoming messages are from the set $\{(0,1),(1,1)\}$. (Since the channel never introduces errors we only need to consider these two cases.)

Next consider the claim concerning the message-passing rule at a check node: it suffices to consider a check node of degree 3 with two incoming messages since check nodes of higher degree can be modeled as the cascade of several check nodes, each of which has two inputs and one output (e.g., $x_1+x_2+x_3=(x_1+x_2)+x_3$). Let $(\mu_1(0),\mu_1(1))$ and $(\mu_2(0),\mu_2(1))$ denote the incoming messages. By the standard message-passing rules the outgoing message is

$$\begin{aligned}(\mu(0),\mu(1)) &= \Big(\sum_{x_1,x_2} \mathbb{1}_{\{x_1+x_2=0\}}\mu_1(x_1)\mu_2(x_2), \sum_{x_1,x_2} \mathbb{1}_{\{x_1+x_2=1\}}\mu_1(x_1)\mu_2(x_2)\Big)\\ &= (\mu_1(0)\mu_2(0)+\mu_1(1)\mu_2(1), \mu_1(0)\mu_2(1)+\mu_1(1)\mu_2(0)).\end{aligned}$$

If $(\mu_2(0),\mu_2(1))=(1,1)$ then, up to normalization, $(\mu(0),\mu(1))=(1,1)$. This shows that if any of the inputs is an erasure then the output is an erasure. Further, if both messages are known then an explicit check shows that the message-passing rules correspond to the mod-2 sum.

Figure 3.28 shows the application of the message-passing decoder to the $[7,4,3]$ Hamming code assuming that the received word is $(0,?,?,1,0,?,0)$. In iteration 0 the variable-to-check messages correspond to the received values. Consider the check-to-variable message sent in iteration 1 from check node 1 to variable node 2 (this is shown on the left of Figure 3.28, second from the top). This message is 1 (the mod-2 sum of incoming messages) according to the message-passing rule. This is intuitive: this message reflects the fact that through the parity-check constraint $x_1+x_2+x_4+x_5=0$ we can find x_2 given x_1, x_4, and x_5. Although this might not be completely obvious at this point, this message-passing algorithm is entirely equivalent to the greedy algorithm based on the Venn diagram description which we discussed in Section 1.9. In other words: for the BEC, message-passing is equivalent to greedily checking whether any of the parity-check constraints allows us to find a yet unknown value from already known ones. After three iterations the transmitted word is found to be $(0,1,0,1,0,1,0)$.

§3.6. Two Basic Simplifications

In the previous sections we have introduced code ensembles and a low-complexity message-passing decoder. We start our investigation of how well this combination performs.

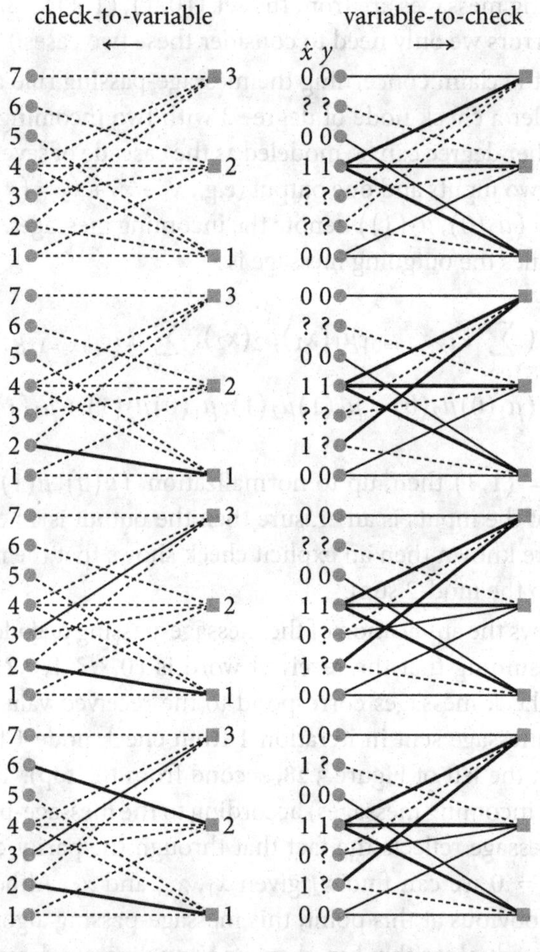

Figure 3.28: Message-passing decoding of the [7, 4, 3] Hamming code with the received word $y = (0, ?, ?, 1, 0, ?, 0)$. The vector $\hat{x}$ denotes the current estimate of the transmitted word x. A 0 message is indicated as thin line, a 1 message is indicated as thick line, and a ? message is drawn as dashed line. The four rows correspond to iterations 0 to 3. After the first iteration we recover $x_2 = 1$, after the second $x_3 = 0$, and after the third we know that $x_6 = 1$. The recovered codeword is $x = (0, 1, 0, 1, 0, 1, 0)$.

§3.6.1. Restriction to the All-Zero Codeword

The first big simplification stems from the realization that the performance is *independent* of the transmitted codeword and is only a function of the erasure pattern: at any iteration the set of known variable nodes is only a function of the *set* of known messages but independent of their values. The equivalent statement is true for the set of known check nodes.

Fact 3.29 (Conditional Independence of Erasure Probability). Let $\mathtt{G}$ be the Tanner graph representing a binary linear code C. Assume that C is used to transmit over the BEC(ϵ) and assume that the decoder performs message-passing decoding on $\mathtt{G}$. Let $\mathrm{P}^{\mathrm{BP}}(\mathtt{G}, \epsilon, \ell, x)$ denote the conditional (bit or block) probability of erasure after the ℓ-th decoding iteration, assuming that x was sent, $x \in C$. Then $\mathrm{P}^{\mathrm{BP}}(\mathtt{G}, \epsilon, \ell, x) = \frac{1}{|C|} \sum_{x' \in C} \mathrm{P}^{\mathrm{BP}}(\mathtt{G}, \epsilon, \ell, x') = \mathrm{P}^{\mathrm{BP}}(\mathtt{G}, \epsilon, \ell)$, i.e., $\mathrm{P}^{\mathrm{BP}}(\mathtt{G}, \epsilon, \ell, x)$ is independent of the transmitted codeword.

As a consequence, we are free to choose a particular codeword and to analyze the performance of the system assuming that this codeword was sent. It is natural to assume that the all-zero word, which is contained in every linear code, was sent. We refer to this assumption as the "all-zero codeword" assumption.

A word about notation: *iterative* decoding is a generic term referring to decoding algorithms which proceed in iterations. A subclass of iterative algorithms are *message-passing* algorithms (like the algorithm which we introduced in the previous section). Message-passing algorithms are iterative algorithms which obey the *message-passing paradigm*: this means that an outgoing message along an edge only depends on the incoming messages along all edges *other* than this edge itself. The message-passing algorithm which we introduced in Chapter 2 and which we adopted in this chapter is a special case in which the messages represent probabilities or "beliefs." The algorithm is therefore also known as *belief propagation* (BP) algorithm. For the BEC essentially any meaningful message-passing algorithm is equivalent to the BP algorithm. But for the general case message-passing algorithms other than the BP algorithm play an important role. For the remainder of this chapter we use the shorthand BP to refer to the decoder.

§3.6.2. Concentration

Rather than analyzing individual codes it suffices to assess the *ensemble average performance*. This is true, since, as the next theorem asserts, the individual elements of an ensemble behave with high probability close to the ensemble average.

Theorem 3.30 (Concentration around Ensemble Average). Let $\mathtt{G}$, chosen uniformly at random from LDPC (n, λ, ρ), be used for transmission over the BEC(ϵ). Assume that the decoder performs ℓ rounds of message-passing decoding and let

$P_b^{BP}(G, \epsilon, \ell)$ denote the resulting bit erasure probability. Then, for ℓ fixed and for any given $\delta > 0$, there exists an $\alpha > 0$, $\alpha = \alpha(\lambda, \rho, \epsilon, \delta, \ell)$, such that

$$\mathbb{P}\{|P_b^{BP}(G, \epsilon, \ell) - \mathbb{E}_{G' \in \text{LDPC}(n,\lambda,\rho)}[P_b^{BP}(G', \epsilon, \ell)]| > \delta\} \leq e^{-\alpha n}.$$

In words, the theorem asserts that all except an exponentially (in the blocklength) small fraction of codes behave within an arbitrarily small δ from the ensemble average. Assuming sufficiently large blocklengths, the ensemble average is a good indicator for the individual behavior. We therefore focus our effort on the design and construction of ensembles whose average performance approaches the Shannon theoretic limit. In Theorem 3.30 we assume a fixed number of iterations and the theorem leaves open the possibility that the constant α approaches zero when the number of iterations increases. Fortunately, as we discuss in Section 3.19, this does not happen.

Example 3.31 (Concentration for LDPC $(\Lambda(x) = 512x^3, P(x) = 256x^6)$). Figure 3.32 shows the erasure probability curves under BP decoding for 10 randomly chosen elements. We see that for this example the plotted curves are within a ver-

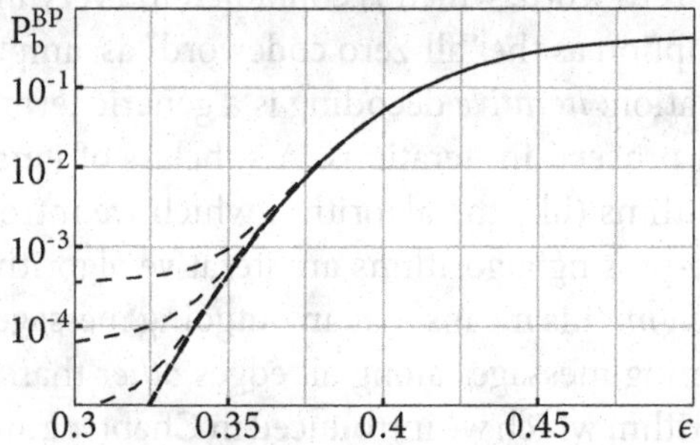

Figure 3.32: Bit erasure probability of 10 random samples from LDPC $(512x^3, 256x^6) \triangleq$ LDPC $(512, x^2, x^5)$.

tical distance of $\delta \approx 10^{-3}$: all samples follow the "main" curve up to some point. At this point, which depends on the sample, the curve of the individual sample flattens out. We will see in Section 3.23 that the main curve is due to large-sized decoding failures (i.e., errors whose support size is a linear fraction of the blocklength) and we will give an analytic characterization of this curve. On the other hand, as we will discuss in Section 3.24, the error floor is due to certain "weaknesses" in the graph which typically can be *expurgated*. We derive the limiting distribution of the error floor in Lemma 3.166. ◇

We do not prove Theorem 3.30 here. Rather this is done in a much broader context in Appendix C, where a variety of probabilistic tools and theorems are discussed that are useful in the context of message-passing coding. The main idea behind Theorem 3.30 is easy to explain: a message-passing decoder is a *local* algorithm. This means that local changes in the graph only affect local decisions. Consider a code G and some small modification, e.g., switch the endpoints of two randomly chosen edges. Because of the local nature of the message-passing algorithm this switch has (in the limit of large blocklengths) only a negligible effect on the resulting performance. Since, finally, LDPC codes have only a linear number of edges, any two codes can be converted into each other by a linear number of such elementary steps, each of which has only a small effect on its performance.

In Theorem 3.30 we have not given any explicit constants. Such constants can be furnished, and indeed they are given in Appendix C. Unfortunately though, even the best constants that have been proved to date cannot explain the actual empirically observed tight concentration. Theorem 3.30 should therefore be thought more as a moral support for the approach taken, rather than a relevant engineering tool by which to judge the performance.

§3.7. Computation Graph and Tree Ensemble

In the previous section we have reduced the analysis already in two essential ways. First, we can assume that the all-zero word was transmitted, and second, we only need to find the ensemble-average performance. Assuming these simplifications, how can we determine the performance of LDPC codes under BP decoding?

§3.7.1. Computation Graph

Message passing takes place on the local neighborhood of a node/edge. As a first step we characterize this neighborhood.

Example 3.33 (Computation Graph – Node and Edge Perspective). Consider again the parity-check matrix H given in (3.9). Its Tanner graph is shown in Figure 3.10. Let us focus on the decoding process for bit x_1 assuming that two rounds of message passing are performed. By convention, since no real processing is done in iteration 0, we do not count it. Recall from the description of the decoder that the decision for bit x_1 is based on the messages received from its adjoined check nodes c_5, c_8, and c_9. These in turn process the information received from their *other* neighbors to form their opinion. As an example, the outgoing message of c_5 is a function of the messages arriving from x_2, x_5, x_{13}, x_{17}, and x_{19}. If we unroll this dependency structure for bit x_1 we arrive at the *computation graph* depicted in Figure 3.34. The figure depicts the computation graph for two iterations. With some abuse of notation we say that the computation graph has *height* 2. It is rooted in the variable node x_1; it is bipartite, i.e., each edge connects a variable node to a check node; and all leaves are variable nodes. This computation graph is depicted as a tree, but in fact *it*

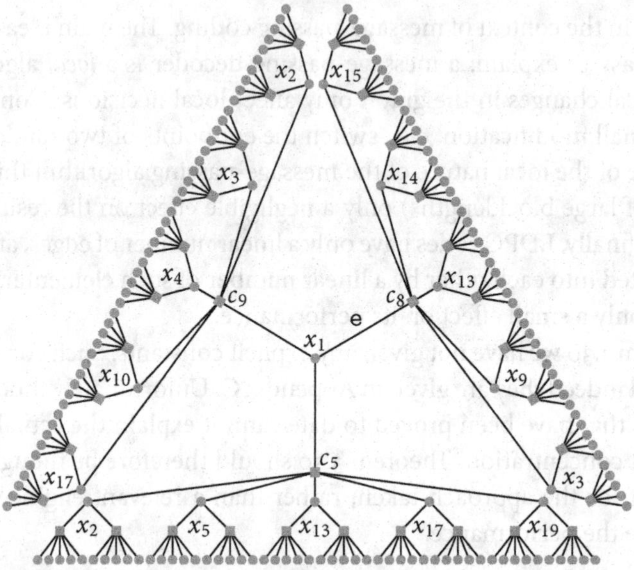

Figure 3.34: Computation graph of height 2 (two iterations) for bit x_1 and the code $C(H)$ for H given in (3.9). The computation graph of height 2 (two iterations) for edge e is the subtree consisting of edge e, variable node x_1, and the two subtrees rooted in check nodes c_5 and c_9.

is not: several of the variable and check nodes appear *repeatedly*. For example, x_3 appears as a child of both c_8 and c_9. Therefore, more properly, this computation graph should be drawn as a rooted graph in which each distinct node appears only once.

The preceding graph is a computation graph from a *node perspective* since we start from a variable node. We can also start from an *edge* and unravel the dependencies of the message sent along this edge. We call the result the computation graph from an *edge perspective*. In Figure 3.34 the resulting computation graph of height 2 for edge e is shown as well. It is the subtree consisting of variable node x_1 and the two subtrees rooted in check nodes c_5 and c_9. ◇

Definition 3.35 (Computation Graph Ensemble – Node and Edge Perspective). Consider the ensemble LDPC (n, λ, ρ). The associated ensemble of computation graphs of *height* ℓ from a node perspective, denoted by $\mathring{\mathcal{C}}_\ell(n, \lambda, \rho)$, is defined as follows. To sample from this ensemble, pick a graph G from LDPC (n, λ, ρ) uniformly at random and draw the computation graph of height ℓ of a randomly chosen variable node of G. Each such computation graph, call it T, is an *unlabeled rooted graph* in which each distinct node is drawn exactly once. The ensemble $\mathring{\mathcal{C}}_\ell(n, \lambda, \rho)$ consists of the set of such computation graphs together with the probabilities $\mathbb{P}\{\mathtt{T} \in$

$\mathring{C}_\ell(n, \lambda, \rho)\}$, which are induced by the preceding sampling procedure.

In the same way, to sample from the ensemble of computation graphs from an edge perspective, denote it by $\vec{C}_\ell(n, \lambda, \rho)$, pick randomly an edge e, and draw the computation graph of e of height ℓ in G. Since $\mathring{C}_\ell(n, \lambda, \rho)$ and $\vec{C}_\ell(n, \lambda, \rho)$ share many properties it is convenient to be able to refer to both of them together. In this case we write $C_\ell(n, \lambda, \rho)$. ▽

Example 3.36 ($\mathring{C}_1(n, \lambda(x) = x, \rho(x) = x^2)$). In this simple example every variable node has two outgoing edges and every check node has three attached edges. Figure 3.37 shows the six elements of this ensemble together with their associated probabilities $\mathbb{P}\{\mathrm{T} \in \mathring{C}_1(n, x, x^2)\}$. All these probabilities behave like $O(1/n)$, except for the tree in the top row, which asymptotically has probability 1. Also shown is the conditional probability of error, $\mathrm{P}_\mathrm{b}^{\mathrm{BP}}(\mathrm{T}, \epsilon)$. This is the probability of error which we incur if we decode the root node of the graph T assuming that transmission takes place over the BEC(ϵ) and assuming that we perform BP decoding for one iteration. ◇

T	$\mathbb{P}\{\mathrm{T} \in \mathring{C}_1(n, x, x^2)\}$	$\mathrm{P}_\mathrm{b}^{\mathrm{BP}}(\mathrm{T}, \epsilon)$
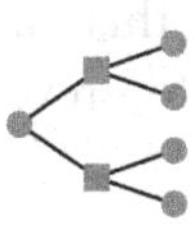	$\frac{(2n-6)(2n-8)}{(2n-1)(2n-5)}$	$\epsilon(1-(1-\epsilon)^2)^2$
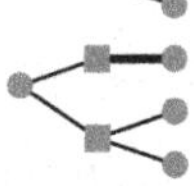	$\frac{2(2n-6)}{(2n-1)(2n-5)}$	$\epsilon^2(1-(1-\epsilon)^2)$
	$\frac{1}{(2n-1)(2n-5)}$	ϵ^3
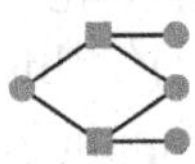	$\frac{4(2n-6)}{(2n-1)(2n-5)}$	$\epsilon^2 + \epsilon^3(1-\epsilon)$
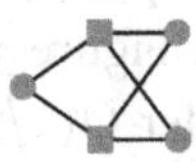	$\frac{2}{(2n-1)(2n-5)}$	$\epsilon(1-(1-\epsilon)^2)$
	$\frac{2}{2n-1}$	ϵ^2

Figure 3.37: Elements of $\mathring{C}_1(n, \lambda(x) = x, \rho(x) = x^2)$ together with their probabilities $\mathbb{P}\{\mathrm{T} \in \mathring{C}_1(n, x, x^2)\}$ and the conditional probability of error, $\mathrm{P}_\mathrm{b}^{\mathrm{BP}}(\mathrm{T}, \epsilon)$. Thick lines indicate double edges.

The operational meaning of the ensembles $\mathring{C}_\ell(n, \lambda, \rho)$ and $\vec{C}_\ell(n, \lambda, \rho)$ is clear: $\mathring{C}_\ell(n, \lambda, \rho)$ represents the ensemble of computation graphs that the BP decoder encounters when making a decision on a randomly chosen bit from a random sample of LDPC (n, λ, ρ), assuming the decoder performs ℓ iterations and $\vec{C}_\ell(n, \lambda, \rho)$ represents the ensemble of computation graphs which the BP decoder encounters when determining the variable-to-check message sent out along a randomly chosen edge in the ℓ-th iteration.

For $\mathrm{T} \in \mathring{\vec{C}}_\ell(n, \lambda, \rho)$, let $\mathrm{P}_\mathrm{b}^{\mathrm{BP}}(\mathrm{T}, \epsilon)$ denote the conditional probability of error incurred by the BP decoder, assuming that the computation graph is T. With this notation we have

$$\mathbb{E}_{\mathrm{LDPC}(n,\lambda,\rho)}[\mathrm{P}_\mathrm{b}^{\mathrm{BP}}(\mathrm{G}, \epsilon, \ell)] = \sum_{\mathrm{T}} \mathbb{P}\{\mathrm{T} \in \mathring{\vec{C}}_\ell(n, \lambda, \rho)\} \, \mathrm{P}_\mathrm{b}^{\mathrm{BP}}(\mathrm{T}, \epsilon). \tag{3.38}$$

In principle the right-hand side of (3.38) can be computed exactly as shown in Figure 3.37. For a fixed ℓ, there are only a finite number of elements T in $\mathring{\vec{C}}_\ell(n, \lambda, \rho)$. The probability $\mathbb{P}\{\mathrm{T} \in \mathring{\vec{C}}_\ell(n, \lambda, \rho)\}$ is a combinatorial quantity, independent of the channel, and it can be determined by counting. To determine $\mathrm{P}_\mathrm{b}^{\mathrm{BP}}(\mathrm{T}, \epsilon)$ proceed as follows: recall that we can assume that the all-zero word was transmitted. Therefore, assume that all variable nodes of T are initially labeled with zero. Each such label is now erased with probability ϵ, where the choice is independent for each node. For each resulting constellation of erasures the root node can either be determined (by the BP decoder) or not. We get $\mathrm{P}_\mathrm{b}^{\mathrm{BP}}(\mathrm{T}, \epsilon)$ if we sum over all possible erasure constellations with their respective probabilities. If we perform this calculation for the example shown in Figure 3.37, the exact expression is not too revealing, but if we expand the result in powers of $1/n$ we get

$$\begin{aligned} \mathbb{E}_{\mathrm{LDPC}(n,x,x^2)}[\mathrm{P}_\mathrm{b}^{\mathrm{BP}}(\mathrm{G}, \epsilon, \ell = 1)] =& \epsilon(1 - (1-\epsilon)^2)^2 + \\ & \epsilon^2(3 - 12\epsilon + 13\epsilon^2 - 4\epsilon^3)/n + O(1/n^2). \end{aligned} \tag{3.39}$$

We see that for increasing blocklengths the expected bit erasure probability after one iteration converges to the constant value $\epsilon(1-(1-\epsilon)^2)^2$. This is equal to the conditional erasure probability $\mathrm{P}_\mathrm{b}^{\mathrm{BP}}(\mathrm{T}, \epsilon)$ of the tree-like computation graph shown in the top of Figure 3.37. This result is not surprising: from Figure 3.37 we see that this computation graph has essentially (up to factors of order $O(1/n)$) probability 1 and that the convergence speed to this asymptotic value is therefore of order $1/n$.

Although this approach poses no conceptual problems, it quickly becomes computationally infeasible since the number of computation graphs grows exponentially in the number of iterations. Faced with these difficulties, we start with a simpler task – the determination of the limiting (in the blocklength) performance; we will come back to the finite-length analysis in Section 3.22. We will see that in this limit the repetitions in the computation graphs vanish and that the limiting performance can be characterized in terms of a recursion. In order to give a clean setting for this recursion, and also to introduce concepts which are important for the general case, we start by giving a formal definition of the limiting objects.

§3.7.2. Tree Ensemble

For a fixed number of iterations and increasing blocklengths, it is intuitive that fewer and fewer cycles occur in the corresponding computation graphs. In fact, in the limit of infinitely long blocklengths the computation graph becomes a tree with probability 1 and each subtree of a computation graph tends to an independent sample whose distribution is determined only by the degree distribution pair (λ, ρ).

Definition 3.40 (Tree Ensembles – Node and Edge Perspective). The tree ensembles $\mathring{\mathcal{T}}_\ell(\lambda, \rho)$ and $\vec{\mathcal{T}}_\ell(\lambda, \rho)$ are the *asymptotic versions* of the computation graph ensembles $\mathring{\mathcal{C}}_\ell(n, \lambda, \rho)$ and $\vec{\mathcal{C}}_\ell(n, \lambda, \rho)$. We start by describing $\vec{\mathcal{T}}_\ell(\lambda, \rho)$. Each element of $\vec{\mathcal{T}}_\ell(\lambda, \rho)$ is a bipartite tree rooted in a variable node. The ensemble $\vec{\mathcal{T}}_0(\lambda, \rho)$ contains a single element – the trivial tree consisting only of the root variable node – and it will serve as our anchor. Let $\mathtt{L}(i)$ denote a bipartite tree rooted in a variable node which has i (check-node) children and, in the same manner, let $\mathtt{R}(i)$ denote a bipartite tree rooted in a check node which has i (variable-node) children as shown in Figure 3.41. To sample from $\vec{\mathcal{T}}_\ell(\lambda, \rho)$, $\ell \geq 1$, first sample an element from $\vec{\mathcal{T}}_{\ell-1}(\lambda, \rho)$. Next substitute each of its leaf variable nodes with a random element from $\{\mathtt{L}(i)\}_{i \geq 1}$, where $\mathtt{L}(i)$ is chosen with probability λ_{i+1}. Finally, substitute each of its leaf check nodes with a random element from $\{\mathtt{R}(i)\}_{i \geq 1}$, where $\mathtt{R}(i)$ is chosen with probability ρ_{i+1}. The preceding definition implies the following recursive decomposition. In order to sample from $\vec{\mathcal{T}}_\ell(\lambda, \rho)$, sample from $\vec{\mathcal{T}}_i(\lambda, \rho)$, $0 \leq i \leq \ell$, and replace each of its (variable) leaf nodes by independent samples from $\vec{\mathcal{T}}_{\ell-i}(\lambda, \rho)$. This recursive structure is the key to the analysis of BP decoding. The description

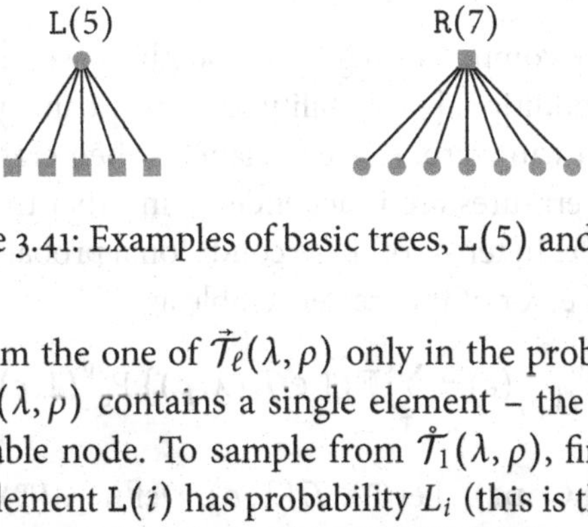

Figure 3.41: Examples of basic trees, $\mathtt{L}(5)$ and $\mathtt{R}(7)$.

of $\mathring{\mathcal{T}}_\ell(\lambda, \rho)$ differs from the one of $\vec{\mathcal{T}}_\ell(\lambda, \rho)$ only in the probabilistic choice of the root node. Again, $\mathring{\mathcal{T}}_0(\lambda, \rho)$ contains a single element – the trivial tree consisting only of the root variable node. To sample from $\mathring{\mathcal{T}}_1(\lambda, \rho)$, first choose an element of $\{\mathtt{L}(i)\}_{i \geq 1}$, where element $\mathtt{L}(i)$ has probability L_i (this is the difference from the previous definition). Next, substitute each of its leaf check nodes with a random element from $\{\mathtt{R}(i)\}_{i \geq 1}$, where $\mathtt{R}(i)$ is chosen with probability ρ_{i+1}. For all further steps we proceed as in the case of the ensemble $\vec{\mathcal{T}}$. It follows that we have again the following recursive decomposition. In order to sample from $\mathring{\mathcal{T}}_\ell(\lambda, \rho)$, sample from $\mathring{\mathcal{T}}_i(\lambda, \rho)$, $1 \leq i \leq \ell$, and replace each of its (variable) leaf nodes by independent samples from $\vec{\mathcal{T}}_{\ell-i}(\lambda, \rho)$. As for the computation graph ensembles, we apply the same convention and write $\mathcal{T}_\ell(\lambda, \rho)$ if the statement refers equally to the node or the edge tree ensemble. ▽

Example 3.42 ($\mathcal{T}_\ell(\lambda(x) = x^{\mathtt{l}-1}, \rho(x) = x^{\mathtt{r}-1})$ Ensemble). The tree ensemble is particularly simple for the regular case. Then each $\vec{\mathcal{T}}_\ell(\lambda(x) = x^{\mathtt{l}-1}, \rho(x) = x^{\mathtt{r}-1})$ consists of a single element, a bipartite graph of height ℓ, rooted in a variable node, where each variable node has $\mathtt{l}-1$ check-node children and each check node has $\mathtt{r}-1$ variable-node children. The same is true for $\mathring{\mathcal{T}}_\ell(\lambda(x) = x^{\mathtt{l}-1}, \rho(x) = x^{\mathtt{r}-1})$, except that the root variable node has $\mathtt{l}$ check-node children. ◇

Example 3.43 ($\mathring{\mathcal{T}}_\ell(\lambda(x) = \frac{1}{2}x + \frac{1}{2}x^2, \rho(x) = \frac{1}{5}x^3 + \frac{4}{5}x^4)$ Ensemble). As discussed before, $\mathring{\mathcal{T}}_0(\lambda, \rho)$ consists of a single element – a root variable node. The 12 elements of $\mathring{\mathcal{T}}_1(\lambda, \rho)$ are shown in Figure 3.44, together with their probabilities. (Note that $\lambda(x) = 1/2x + 1/2x^2$ implies that $L(x) = 3/5x^2 + 2/5x^3$.) ◇

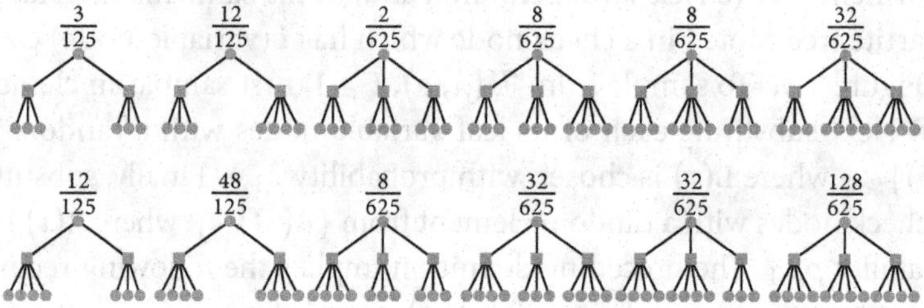

Figure 3.44: Twelve elements of $\mathring{\mathcal{T}}_1(\lambda, \rho)$ together with their probabilities.

As in the case of the computation graph ensembles, we can associate with each element of $\mathring{\mathcal{T}}_\ell(\lambda, \rho)$ a conditional probability of error: we imagine that each variable node is initially labeled with zero, that each label is then erased with probability ϵ by the channel, where erasures are independent, and that the BP decoder tries to determine the root node. In terms of these conditional probabilities of error we can write the probability of error of the tree ensemble as

$$\mathrm{P}^{\mathrm{BP}}_{\mathring{\mathcal{T}}_\ell(\lambda,\rho)}(\epsilon) = \sum_{\mathrm{T}} \mathbb{P}\{\mathrm{T} \in \mathring{\mathcal{T}}_\ell(\lambda, \rho)\}\, \mathrm{P}^{\mathrm{BP}}_{\mathrm{b}}(\mathrm{T}, \epsilon).$$

Definition 3.45 (Tree Code). Let $\mathrm{T} \in \mathcal{T}_\ell(\lambda, \rho)$. Define $C(\mathrm{T})$ to be the set of valid codewords on T. More precisely, $C(\mathrm{T})$ is the set of $0/1$ assignments on the variables contained in T that fulfill the constraints on the tree. Further, let $C^{0/1}(\mathrm{T})$ denote the valid codewords on T such that the root variable node is $0/1$. Clearly, $C^0(\mathrm{T})$ and $C^1(\mathrm{T})$ are disjoint and their union equals $C(\mathrm{T})$. Finally, we need the subset of $C^1(\mathrm{T})$ consisting only of the *minimal codewords*. Denote this set by $C^1_{\min}(\mathrm{T})$: a codeword in $C^1_{\min}(\mathrm{T})$ has a 1 at the root node and then for each of its connected check nodes *exactly one* of its child variable nodes is also 1. This continues until we have reached the boundary. Figure 3.46 shows a tree with a non-minimal (left) and a minimal (right) assignment. ▽

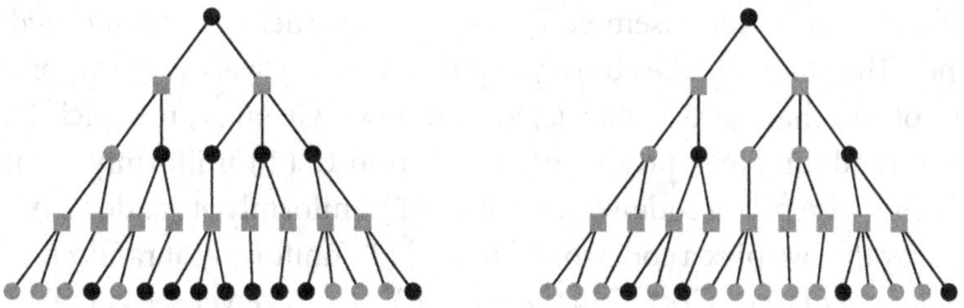

Figure 3.46: Left: Element of $C^1(\mathtt{T})$ for a given tree $\mathtt{T} \in \mathcal{T}_2$. Right: Minimal such element, i.e., an element of $C^1_{\min}(\mathtt{T})$. Every check node has either no connected variables of value 1 or exactly two such neighbors. Black and gray circles indicate variables with associated values of 1 or 0, respectively.

Discussion: Consider a code C and the computation graph $\mathtt{T}$ for a fixed number of iterations. This computation graph may or may not be a tree. Project the global codewords of C onto the set of variables contained in $\mathtt{T}$. It can happen that this set of projections is a strict subset of $C(\mathtt{T})$. Here is a simple example: a position indexed by $\mathtt{T}$ may not be proper and so it is permanently fixed to 0 by global constraints which do not appear in the set of local constraints that define $\mathtt{T}$ (e.g., consider a variable node on the boundary of $\mathtt{T}$ which has 3 edges connected to a check node of degree 3 outside $\mathtt{T}$). As the next lemma shows, for large blocklengths this rarely happens. We will not need this fact for our analysis for the BEC but it will come in handy when we talk about GEXIT functions for more general channels.

Consider a position i of the code and let $\ell \in \mathbb{N}$. Consider the node-perspective computation graph of bit i of depth ℓ, call it $\mathtt{T}$. Let $C(\mathtt{T})$ denote the tree code. Let $P(C, \mathtt{T})$ denote the projection of the set of global codewords onto $\mathtt{T}$. We say that $\mathtt{T}$ is *proper* if $\mathtt{T}$ is a tree and $P(C, \mathtt{T}) = C(\mathtt{T})$, i.e., if the set of projections of global codewords onto $\mathtt{T}$ is equal to the set of all local codewords. Note that for $\ell = 0$ this notion coincides with the notion that the code $\mathtt{G}$ is proper at position i. You can find the proof of the following lemma on page 523.

Lemma 3.47 (Most Finite Projections Are Proper). Let $\ell \in \mathbb{N}$. Let $\mathtt{G}$ be chosen uniformly at random from LDPC $(n, \mathtt{l}, \mathtt{r})$, $3 \leq \mathtt{l} < \mathtt{r}$. Let $i \in [n]$ and let $\mathtt{T}$ be the node-perspective computation graph of depth ℓ rooted in position i. Then

$$\mathbb{P}\{\mathtt{T} \text{ is proper}\} = 1 + o(1).$$

§3.8. Tree Channel and Convergence to Tree Channel

§3.8.1. Tree Channel

Definition 3.48 ($(\mathcal{T}_\ell, \epsilon)$-Tree Channel). Given the BEC characterized by its erasure probability ϵ and a tree ensemble $\mathcal{T}_\ell = \mathcal{T}_\ell(\lambda, \rho)$, we define the associated $(\mathcal{T}_\ell, \epsilon)$-tree channel. The channel takes binary input $X \in \{0, 1\}$ with uniform probability. The output of the channel is constructed as follows. Given X, first pick $\mathtt{T}$ from $\mathcal{T}_\ell$ uniformly at random. Next, pick a codeword from $C^0(\mathtt{T})$ uniformly at random if $X = 0$ and otherwise pick a codeword from $C^1(\mathtt{T})$ uniformly at random. As a shorthand, let us say that we pick a codeword from $C^X(\mathtt{T})$ uniformly at random. Transmit this codeword over the BEC(ϵ). Call the output Y. The receiver sees $(\mathtt{T}, Y)$ and estimates X. Let $\mathrm{P}^{\mathrm{BP}}_{\mathcal{T}_\ell}(\epsilon)$ denote the resulting bit error probability, assuming that $(\mathtt{T}, Y)$ is processed by a BP decoder. ▽

Discussion: We know already that the error probability depends on the tree but not on the codeword that is sent. The distribution of the codeword is therefore irrelevant for the subsequent discussion. For the sake of definiteness we have chosen this distribution to be the uniform one. This is also consistent with our previous discussion. We know that most projections onto computation graphs are proper. And for a proper projection the induced probability distribution is the uniform one.

§3.8.2. Convergence to Tree Channel

Theorem 3.49 (Convergence to Tree Channel). For a given degree distribution pair (λ, ρ) consider the sequence of associated ensembles LDPC (n, λ, ρ) for increasing blocklength n under ℓ rounds of BP decoding. Then

$$\lim_{n\to\infty} \mathbb{E}_{\mathrm{LDPC}(n,\lambda,\rho)}[\mathrm{P}^{\mathrm{BP}}_{\mathrm{b}}(\mathtt{G}, \epsilon, \ell)] = \mathrm{P}^{\mathrm{BP}}_{\mathring{\mathcal{T}}_\ell(\lambda,\rho)}(\epsilon).$$

Proof. From characterization (3.38) we have

$$\lim_{n\to\infty} \mathbb{E}_{\mathrm{LDPC}(n,\lambda,\rho)}[\mathrm{P}^{\mathrm{BP}}_{\mathrm{b}}(\mathtt{G}, \epsilon, \ell)] = \lim_{n\to\infty} \sum_{\mathtt{T}} \mathbb{P}\{\mathtt{T} \in \mathring{\mathcal{C}}_\ell(n, \lambda, \rho)\}\, \mathrm{P}^{\mathrm{BP}}_{\mathrm{b}}(\mathtt{T}, \epsilon).$$

Consider

$$\sum_{\mathtt{T}} \mathbb{P}\{\mathtt{T} \in \mathring{\mathcal{C}}_\ell(n, \lambda, \rho)\}\, \mathrm{P}^{\mathrm{BP}}_{\mathrm{b}}(\mathtt{T}, \epsilon) - \sum_{\mathtt{T}} \mathbb{P}\{\mathtt{T} \in \mathring{\mathcal{T}}_\ell(\lambda, \rho)\}\, \mathrm{P}^{\mathrm{BP}}_{\mathrm{b}}(\mathtt{T}, \epsilon).$$

Since the conditional probability of error is identical, the preceding difference can be written as

$$\sum_{\mathtt{T}} \left(\mathbb{P}\{\mathtt{T} \in \mathring{\mathcal{C}}_\ell(n, \lambda, \rho)\} - \mathbb{P}\{\mathtt{T} \in \mathring{\mathcal{T}}_\ell(\lambda, \rho)\}\right) \mathrm{P}^{\mathrm{BP}}_{\mathrm{b}}(\mathtt{T}, \epsilon).$$

The proof now follows by observing that, for each T,

$$\lim_{n\to\infty} \mathbb{P}\{\mathrm{T} \in \mathring{\mathcal{C}}_\ell(n,\lambda,\rho)\} = \mathbb{P}\{\mathrm{T} \in \mathring{\mathcal{T}}_\ell(\lambda,\rho)\}.$$

If you go back to Figure 3.37 you see that, in the limit of infinite blocklengths, the only computation graph which we encounter is the tree shown in the top row. In the general case, any computation graph that contains at least one repetition (edge or node) has a probability that tends to zero at a speed of at least $1/n$. Finally, this convergence is uniform since there are only a finite number of possible graphs of a fixed height. □

§3.9. Density Evolution

Theorem 3.49 asserts that in the limit of large blocklengths the average performance of an ensemble LDPC (n,λ,ρ) converges to the performance of the corresponding tree channel.

Theorem 3.50 (Performance of Tree Channel). Consider a degree distribution pair (λ,ρ) with associated normalized variable degree distribution from a node perspective $L(x)$. Let ϵ be the channel parameter, $\epsilon \in [0,1]$. Define $x_{-1} = 1$ and for $\ell \geq 0$ let

$$x_\ell = \epsilon\lambda\left(1-\rho\left(1-x_{\ell-1}\right)\right). \tag{3.51}$$

Then for $\ell \geq 0$

$$\mathrm{P}^{\mathrm{BP}}_{\vec{\mathcal{T}}_\ell}(\epsilon) = x_\ell, \qquad \mathrm{P}^{\mathrm{BP}}_{\mathcal{T}_\ell}(\epsilon) = \epsilon L(1-\rho(1-x_{\ell-1})).$$

Proof. Consider first $\mathrm{P}^{\mathrm{BP}}_{\vec{\mathcal{T}}_\ell}(\epsilon)$. By definition, the initial variable-to-check message is equal to the received message, which is an erasure message with probability ϵ. It follows that $\mathrm{P}^{\mathrm{BP}}_{\vec{\mathcal{T}}_0}(\epsilon) = \epsilon$, as claimed. We use induction. Assume that $\mathrm{P}^{\mathrm{BP}}_{\vec{\mathcal{T}}_\ell}(\epsilon) = x_\ell$. Consider $\mathrm{P}^{\mathrm{BP}}_{\vec{\mathcal{T}}_{\ell+1}}(\epsilon)$. We start with the check-to-variable messages in the $(\ell+1)$-th iteration. Recall that by definition of the algorithm a check-to-variable message emitted by a check node of degree i along a particular edge is the erasure message if any of the $i-1$ incoming messages is an erasure. By assumption, each such message is an erasure with probability x_ℓ and all messages are independent, so that the probability that the outgoing message is an erasure is equal to $1-(1-x_\ell)^{i-1}$. Since the edge has probability ρ_i to be connected to a check node of degree i it follows that the expected erasure probability of a check-to-variable message in the $(\ell+1)$-th iteration is equal to $\sum_i \rho_i(1-(1-x_\ell)^{i-1}) = 1-\rho(1-x_\ell)$. Now consider the erasure probability of the variable-to-check messages in the $(\ell+1)$-th iteration. Consider an edge e that is connected to a variable node of degree i. The outgoing variable-to-check message along this edge in the $(\ell + 1)$-th iteration is an erasure if the received value of the associated variable node is an erasure and all $i - 1$ incoming messages are erasures. This happens with probability $\epsilon(1 - \rho(1 - x_\ell))^{i-1}$. Averaging again over the edge degree distribution λ, we get that $\text{P}^{\text{BP}}_{\vec{\mathcal{T}}_{\ell+1}}(\epsilon) = \epsilon\lambda(1 - \rho(1 - x_\ell)) = x_{\ell+1}$, as claimed.

The proof for $\text{P}^{\text{BP}}_{\mathring{\mathcal{T}}_\ell}(\epsilon)$ is very similar. Recall that $\vec{\mathcal{T}}_\ell$ and $\mathring{\mathcal{T}}_\ell$ are identical except for the choice of the root node, which is chosen according to the normalized node degree distribution $L(x)$ instead of the edge degree distribution $\lambda(x)$. □

Example 3.52 (Density Evolution for $(\lambda(x) = x^2, \rho(x) = x^5)$). For the degree distribution pair $(\lambda(x) = x^2, \rho(x) = x^5)$ we have $x_0 = \epsilon$ and for $\ell \geq 1$, $x_\ell = \epsilon(1 - (1 - x_{\ell-1})^5)^2$. For example, for $\epsilon = 0.4$ the sequence of values of x_ℓ is 0.4, 0.34, 0.306, 0.2818, 0.2617, 0.2438, and so forth. ◇

Theorem 3.50 gives a precise characterization of the asymptotic performance in terms of the recursions stated in (3.51). Those recursions are termed *density evolution* equations since they describe how the erasure probability evolves as a function of the iteration number.[2]

§3.10. Monotonicity

Monotonicity either with respect to the channel parameter or with respect to the number of iterations ℓ plays a fundamental role in the analysis of density evolution. The first lemma is a direct consequence of the non-negativity of the coefficients of the polynomials $\lambda(x)$ and $\rho(x)$ and the fact that $\rho(1) = 1$. We skip the proof.

Lemma 3.53 (Monotonicity of $f(\cdot, \cdot)$). For a given degree distribution pair (λ, ρ) define $f(\epsilon, x) = \epsilon\lambda(1 - \rho(1 - x))$. Then $f(\epsilon, x)$ is increasing in both its arguments for $x, \epsilon \in [0, 1]$.

Lemma 3.54 (Monotonicity with Respect to Channel). Let (λ, ρ) be a degree distribution pair and $\epsilon \in [0, 1]$. If $\text{P}^{\text{BP}}_{\vec{\mathcal{T}}_\ell}(\epsilon) \xrightarrow{\ell \to \infty} 0$ then $\text{P}^{\text{BP}}_{\vec{\mathcal{T}}_\ell}(\epsilon') \xrightarrow{\ell \to \infty} 0$ for all $0 \leq \epsilon' \leq \epsilon$.

Proof. We prove the claim for $\text{P}^{\text{BP}}_{\vec{\mathcal{T}}_\ell}(\epsilon)$. The corresponding claim for $\text{P}^{\text{BP}}_{\mathring{\mathcal{T}}_\ell}(\epsilon)$ can be treated in a nearly identical manner and we skip the details. Recall from Theorem 3.50 that $\text{P}^{\text{BP}}_{\vec{\mathcal{T}}_\ell}(\epsilon) = x_\ell(\epsilon)$, where $x_0(\epsilon) = \epsilon$, $x_\ell(\epsilon) = f(\epsilon, x_{\ell-1}(\epsilon))$, and $f(\epsilon, x) = \epsilon\lambda(1 - \rho(1 - x))$. Assume that for some $\ell \geq 0$, $x_\ell(\epsilon') \leq x_\ell(\epsilon)$. Then

$$x_{\ell+1}(\epsilon') = f(\epsilon', x_\ell(\epsilon')) \overset{\text{Lem. 3.53}}{\leq} f(\epsilon, x_\ell(\epsilon)) = x_{\ell+1}(\epsilon).$$

But if $\epsilon' \leq \epsilon$, then $x_0(\epsilon') = \epsilon' \leq \epsilon = x_0(\epsilon)$ and we conclude by induction that $x_\ell(\epsilon') \leq x_\ell(\epsilon)$. So if $x_\ell(\epsilon) \xrightarrow{\ell \to \infty} 0$, then $x_\ell(\epsilon') \xrightarrow{\ell \to \infty} 0$. □

[2]For the BEC this "density" simplifies to a probability (of erasure) but for the general case discussed in Chapter 4, density evolution really describes an evolution of densities.

LEMMA 3.55 (MONOTONICITY WITH RESPECT TO ITERATION). Let $\epsilon, x_0 \in [0,1]$. For $\ell = 1, 2, \ldots$, define $x_\ell(x_0) = f(\epsilon, x_{\ell-1}(x_0))$. Then $x_\ell(x_0)$ is a monotone sequence converging to the *nearest* (in the direction of monotonicity) solution of the equation $x = f(\epsilon, x)$.

Proof. If $x_0 = 0$ or $\epsilon = 0$ then $x_\ell = 0$ for $\ell \geq 1$ and the fixed point is $x = 0$. If for some $\ell \geq 1$, $x_\ell \geq x_{\ell-1}$ then $x_{\ell+1} = f(\epsilon, x_\ell) \overset{\text{Lem. 3.53}}{\geq} f(\epsilon, x_{\ell-1}) = x_\ell$, and the corresponding conclusion holds if $x_\ell \leq x_{\ell-1}$. This proves the monotonicity of the sequence $\{x_\ell\}_{\ell \geq 0}$.

Since for $\epsilon \geq 0$ we have $0 \leq f(\epsilon, x) \leq \epsilon$ for all $x \in [0,1]$, it follows that x_ℓ converges to an element of $[0, \epsilon]$ – call it x_∞. By the continuity of f we have $x_\infty = f(\epsilon, x_\infty)$. It remains to show that x_∞ is the nearest (in the sense of monotonicity) fixed point. Consider a fixed point z such that $x_\ell(x_0) \leq z$ for some $\ell \geq 0$. Then $x_{\ell+1}(x_0) = f(\epsilon, x_\ell(x_0)) \overset{\text{Lem. 3.53}}{\leq} f(\epsilon, z) = z$, which shows that $x_\infty \leq z$. Similarly, if $x_\ell(x_0) \geq z$ for some $\ell \geq 0$ then $x_\infty \geq z$. This shows that x_ℓ cannot "jump" over any fixed point and must therefore converge to the nearest one. □

§3.11. THRESHOLD

From the density evolution equations (3.51) we see that for every non-negative integer ℓ

$$\mathrm{P}^{\mathrm{BP}}_{\vec{\mathcal{T}}_\ell}(\epsilon = 0) = 0, \quad \text{but} \quad \mathrm{P}^{\mathrm{BP}}_{\vec{\mathcal{T}}_\ell}(\epsilon = 1) = 1,$$

and in particular these equalities are satisfied if $\ell \to \infty$. Combined with the preceding monotonicity property this shows the existence of a well-defined supremum of ϵ for which $\mathrm{P}^{\mathrm{BP}}_{\vec{\mathcal{T}}_\ell}(\epsilon) \overset{\ell\to\infty}{\longrightarrow} 0$. This supremum is called the *threshold*. Further, in Problem 3.16 you will show that $\mathrm{P}^{\mathrm{BP}}_{\vec{\mathcal{T}}_\ell}(\epsilon) \overset{\ell\to\infty}{\longrightarrow} 0$ implies $\mathrm{P}^{\mathrm{BP}}_{\vec{\mathcal{T}}_\ell}(\epsilon) \overset{\ell\to\infty}{\longrightarrow} 0$, and vice versa. We can therefore generically consider $\mathrm{P}^{\mathrm{BP}}_{\mathcal{T}_\ell}(\epsilon)$.

DEFINITION 3.56 (THRESHOLD OF DEGREE DISTRIBUTION PAIR). The *threshold* associated with the degree distribution pair (λ, ρ), call it $\epsilon^{\mathrm{BP}}(\lambda, \rho)$, is defined as

$$\epsilon^{\mathrm{BP}}(\lambda, \rho) = \sup\{\epsilon \in [0,1] : \mathrm{P}^{\mathrm{BP}}_{\mathcal{T}_\ell(\lambda,\rho)}(\epsilon) \overset{\ell\to\infty}{\longrightarrow} 0\}. \qquad \triangledown$$

EXAMPLE 3.57 (THRESHOLD OF $(\lambda(x) = x^2, \rho = x^5)$). Numerical experiments show that $\epsilon^{\mathrm{BP}}(3,6) \approx 0.42944$. ◇

What is the operational meaning of $\epsilon^{\mathrm{BP}}(\lambda, \rho)$? Using an ensemble LDPC (n, λ, ρ) of sufficient length we can transmit reliably over the channel BEC(ϵ) if $\epsilon < \epsilon^{\mathrm{BP}}(\lambda, \rho)$ but we cannot hope to do so for channel parameters exceeding this threshold. More

l	r	$r(\mathtt{l},\mathtt{r})$	$\epsilon^{\text{Sha}}(\mathtt{l},\mathtt{r})$	$\epsilon^{\text{BP}}(\mathtt{l},\mathtt{r})$
3	6	$\frac{1}{2}$	$\frac{1}{2} = 0.5$	≈ 0.4294
4	8	$\frac{1}{2}$	$\frac{1}{2} = 0.5$	≈ 0.3834
3	5	$\frac{2}{5}$	$\frac{3}{5} = 0.6$	≈ 0.5176
4	6	$\frac{1}{3}$	$\frac{2}{3} \approx 0.667$	≈ 0.5061
3	4	$\frac{1}{4}$	$\frac{3}{4} = 0.75$	≈ 0.6474

Table 3.58: Thresholds $\epsilon^{\text{BP}}(\mathtt{l},\mathtt{r})$ under BP decoding and the corresponding Shannon thresholds $\epsilon^{\text{Sha}}(\mathtt{l},\mathtt{r})$ for some regular degree distribution pairs.

precisely, given $\epsilon < \epsilon^{\text{BP}}(\lambda,\rho)$ there exists an iteration number ℓ so that $\text{P}^{\text{BP}}_{\mathring{\mathcal{T}}_\ell}(\epsilon)$ is below the desired bit erasure probability. Therefore, by Theorems 3.30 and 3.49, elements of the ensemble LDPC (n,λ,ρ), if decoded with ℓ rounds of BP, will show a performance approaching $\text{P}^{\text{BP}}_{\mathring{\mathcal{T}}_\ell}(\epsilon)$ as the blocklength increases. Table 3.58 lists thresholds for some regular degree distribution pairs.

§3.12. Fixed Point Characterization of Threshold

The preceding definition of the threshold is not very convenient for the purpose of analysis. We therefore state a second equivalent definition based on the fixed points of density evolution.

Theorem 3.59 (Fixed Point Characterization of the Threshold). For a given degree distribution pair (λ,ρ) and $\epsilon \in [0,1]$ let $f(\epsilon,x) = \epsilon\lambda(1-\rho(1-x))$.

(i) $\epsilon^{\text{BP}}(\lambda,\rho) = \sup\{\epsilon \in [0,1] : \ x = f(\epsilon,x) \text{ has no solution } x \text{ in } (0,1]\}$.

(ii) $\epsilon^{\text{BP}}(\lambda,\rho) = \inf\{\epsilon \in [0,1] : \ x = f(\epsilon,x) \text{ has a solution } x \text{ in } (0,1]\}$.

Proof. Let $x(\epsilon)$ be the largest solution in $[0,1]$ to $x = f(\epsilon,x)$. Note that for any $x \in [0,1]$ we have $0 \le f(\epsilon,x) \le \epsilon$. We conclude that $x(\epsilon) \in [0,\epsilon]$.

By Lemma 3.55 we have $x_\ell(\epsilon) \overset{\ell\to\infty}{\longrightarrow} x(\epsilon)$. We conclude that if $x(\epsilon) > 0$ then ϵ is above the threshold, whereas if $x(\epsilon) = 0$ then ϵ is below the threshold. □

Definition 3.60 (Critical Point). Given a degree distribution pair (λ,ρ) that has threshold ϵ^{BP} we say that x^{BP} is a *critical point* if

$$f(\epsilon^{\text{BP}},x^{\text{BP}}) = x^{\text{BP}} \quad \text{and} \quad \frac{\partial f(\epsilon^{\text{BP}},x)}{\partial x}\Big|_{x=x^{\text{BP}}} = 1.$$

In words, x^{BP} is (one of) the point(s) at which $f(\epsilon^{\text{BP}},x)-x$ tangentially touches the horizontal axis. ▽

The fixed point characterization gives rise to the following convenient graphical method for determining the threshold. Draw $f(\epsilon, x) - x$ as a function of x, $x \in (0, 1]$. The threshold ϵ^{BP} is the largest ϵ such that the graph of $f(\epsilon, x) - x$ is negative.

EXAMPLE 3.61 (GRAPHICAL DETERMINATION OF THRESHOLD). The graphical determination of the threshold of $(\lambda, \rho) = (x^2, x^5)$ is shown in Figure 3.62. The graphs of $f(\epsilon, x) - x = \epsilon(1 - (1 - x)^5)^2 - x$ for the values $\epsilon = 0.4, 0.42944$, and 0.45 are depicted. We see that the supremum of all ϵ such that this plot is strictly negative for $x \in (0, 1]$ is achieved at $\epsilon^{\text{BP}} \approx 0.42944$. For this ϵ^{BP} there is one *critical* value of x, $x^{\text{BP}} \approx 0.2606$. At this point the expected decrease in the erasure fraction per iteration reaches zero so that the decoder is expected to slow down critically and come to a halt. ◇

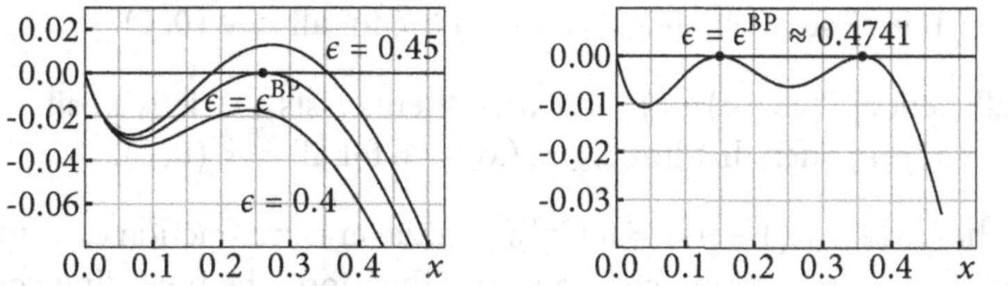

Figure 3.62: Left: Graphical determination of the threshold for $(\lambda, \rho) = (x^2, x^5)$. There is one critical point, $x^{\text{BP}} \approx 0.2606$ (black dot). Right: Graphical determination of the threshold for optimized degree distribution described in Example 3.63. There are two critical points, $x^{\text{BP}}_{1,2} \approx 0.1493, 0.3571$ (two black dots).

EXAMPLE 3.63 (OPTIMIZED ENSEMBLE). In the general case there can be more than one critical point. This happens in particular with optimized degree distributions. The more degrees we allow and the more highly optimized ensembles we consider the more simultaneous critical points we are likely to find. Consider the degree distribution pair

$$\lambda(x) = 0.106257x + 0.486659x^2 + 0.010390x^{10} + 0.396694x^{19},$$
$$\rho(x) = 0.5x^7 + 0.5x^8.$$

It has a design rate of one-half, a threshold of $\epsilon^{\text{BP}} \approx 0.4741$, and two critical points $x^{\text{BP}} \approx 0.35713565$ and $x^{\text{BP}} \approx 0.14932401$. Why do we get multiple critical points if we optimize degree distributions? In Section 3.14.4 we will see that to achieve capacity we need $f(\epsilon^{\text{BP}}, x) = x$ for the whole range $x \in [0, \epsilon^{\text{BP}}]$, i.e., we need equality everywhere so that all points in this range are critical points. This is called the *matching condition*. Therefore, the closer we get to capacity the more critical points we expect to see. ◇

§3.13. Stability

Expanding the right-hand side of (3.51) into a Taylor series around zero we get

$$x_\ell = \epsilon\lambda'(0)\rho'(1)x_{\ell-1} + O(x_{\ell-1}^2). \tag{3.64}$$

For sufficiently small x_ℓ the convergence behavior is determined by the term linear in x_ℓ. More precisely, the convergence depends on whether $\epsilon\lambda'(0)\rho'(1)$ is smaller or larger than 1.

Theorem 3.65 (Stability Condition). Assume that we are given a degree distribution pair (λ, ρ) and $\epsilon, x_0 \in [0, 1]$. Let $x_\ell(x_0)$ be defined as in Lemma 3.55.

> [Necessity] If $\epsilon\lambda'(0)\rho'(1) > 1$ then there exists a strictly positive constant $\xi = \xi(\lambda, \rho, \epsilon)$ such that $\lim_{\ell\to\infty} x_\ell(x_0) \geq \xi$ for all $x_0 \in (0, 1)$.
>
> [Sufficiency] If $\epsilon\lambda'(0)\rho'(1) < 1$ then there exists a strictly positive constant $\xi = \xi(\lambda, \rho, \epsilon)$ such that $\lim_{\ell\to\infty} x_\ell(x_0) = 0$ for all $x_0 \in (0, \xi)$.

Note that $\epsilon\lambda(1 - \rho(1 - 0)) = 0$ for any initial erasure fraction ϵ, so that zero is a *fixed point* of the recursion given in (3.51). Therefore, the preceding condition is the *stability* condition of the fixed point at zero. The most important consequence of the stability condition is the implied upper bound on the threshold:

$$\epsilon^{\mathrm{BP}}(\lambda, \rho) \leq \frac{1}{\lambda'(0)\rho'(1)}. \tag{3.66}$$

In fact, this bound also applies to MAP decoding.

Lemma 3.67 (Stability Condition and MAP Threshold). Assume that we are given a degree distribution pair (λ, ρ) and a real number ϵ, $\epsilon \in [0, 1]$.

> [Necessity] If $\epsilon\lambda'(0)\rho'(1) > 1$ then there exists a strictly positive constant $\xi = \xi(\lambda, \rho, \epsilon)$ such that $\lim_{n\to\infty} \mathrm{P}_{\mathrm{b}}^{\mathrm{MAP}}(n, \lambda, \rho, \epsilon) \geq \xi$.

Proof. To each element $\mathtt{G} \in \mathrm{LDPC}(n, \lambda, \rho)$ and each channel realization associate a "normal" graph (not bipartite); call it Γ. The nodes of Γ are the check nodes of $\mathtt{G}$. The edges of Γ correspond to the variable nodes of degree 2 in $\mathtt{G}$ whose values were erased by the channel: each such variable node of degree 2 has exactly two outgoing edges in $\mathtt{G}$ and so naturally forms an edge in Γ. This connection between the bipartite graph $\mathtt{G}$ and the graph Γ is discussed in more detail in Section C.5.

Let $\mu = \epsilon\lambda'(0)\rho'(1)$. From Lemma C.37 and Lemma C.38 we know that if $\mu > 1$ then a positive fraction of nodes in Γ lie on cycles.

A cycle in Γ corresponds to a cycle in $\mathtt{G}$ so that all involved nodes are of degree 2 and have been erased by the channel. Such a cycle constitutes a codeword, all of whose components have been erased. No decoder (not even a MAP decoder) can recover the value associated with these bits. Since the fraction of concerned bits is positive, the bit erasure probability is positive. In other words, we are transmitting above the MAP bit threshold. □

§3.14. EXIT Charts

An EXIT chart is a helpful visualization of the asymptotic performance under BP decoding. For the BEC it is equivalent to density evolution.

§3.14.1. Graphical Representation of Density Evolution

Consider a degree distribution pair (λ, ρ). Recall from Section 3.12 that the asymptotic behavior of such a degree distribution pair is characterized by $f(\epsilon, x) = \epsilon\lambda(1 - \rho(1 - x))$, which represents the evolution of the fraction of erased messages emitted by variable nodes, assuming that the system is in *state* (has a current such fraction of) x and that the channel parameter is ϵ. It is helpful to represent $f(\epsilon, x)$ as the composition of two functions, one which represents the "action" of the variable nodes (which we can think of as a repetition code) and the second which describes the "action" of check nodes (a simple parity-check code). We define $v_\epsilon(x) = \epsilon\lambda(x)$ and $c(x) = 1 - \rho(1 - x)$, so that $f(\epsilon, x) = v_\epsilon(c(x))$. Recall that the condition for convergence reads $f(\epsilon, x) < x$, $\forall x \in (0, 1)$. Observe that $v_\epsilon(x)$ has an inverse for $x \geq 0$ since $\lambda(x)$ is a polynomial with non-negative coefficients. The condition for convergence can hence be written as

$$c(x) < v_\epsilon^{-1}(x), \quad x \in (0, 1).$$

This has a pleasing graphical interpretation: $c(x)$ has to lie strictly below $v_\epsilon^{-1}(x)$ over the whole range $x \in (0, 1)$. The threshold ϵ^{BP} is the supremum of all numbers ϵ for which this condition is fulfilled. The local such condition around $x = 0$, i.e., the condition for small positive values x, reads $\rho'(1) = c'(0) < \frac{dv_\epsilon^{-1}(x)}{dx}\big|_{x=0} = \frac{1}{\epsilon\lambda'(0)}$. This is the stability condition.

Example 3.68 (Graphical Representation for $(\lambda(x) = x^2, \rho(x) = x^5)$). We have $v_\epsilon^{-1}(x) = (x/\epsilon)^{1/2}$ and $c(x) = 1-(1-x)^5$. The curves corresponding to $v_\epsilon^{-1}(x)$ for $\epsilon = 0.35, 0.42944$, and 0.50 as well as the curve corresponding to $c(x)$ are plotted in the left-hand graph of Figure 3.69. For $\epsilon \approx 0.42944$, $v_\epsilon^{-1}(x)$ just touches $c(x)$ for some $x \in (0, 1)$, i.e., $\epsilon^{\text{BP}} \approx 0.42944$. The graph on the right-hand side of Figure 3.69 shows the evolution of the decoding process for $\epsilon = 0.35$. ◇

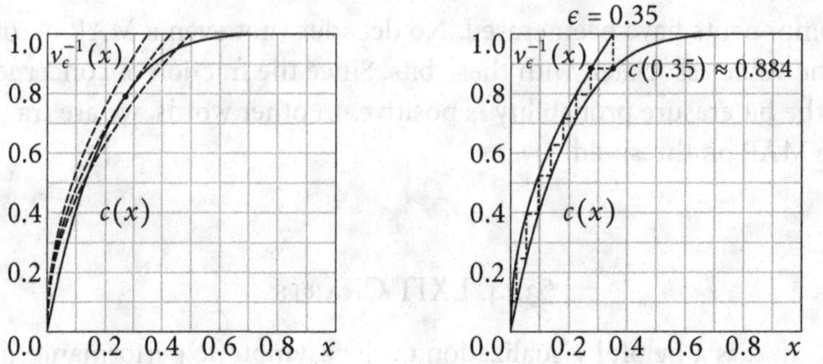

Figure 3.69: Left: Graphical determination of the threshold for $(\lambda(x) = x^2, \rho(x) = x^5)$. The function $v_\epsilon^{-1}(x) = (x/\epsilon)^{1/2}$ is shown as a dashed line for $\epsilon = 0.35$, $\epsilon = \epsilon^{\text{BP}} \approx 0.42944$, and $\epsilon = 0.5$. The function $c(x) = 1 - (1-x)^5$ is shown as a solid line. Right: Evolution of the decoding process for $\epsilon = 0.35$. The initial fraction of erasure messages emitted by the variable nodes is $x = 0.35$. After half an iteration (at the output of the check nodes) this fraction has evolved to $c(x = 0.35) \approx 0.88397$. After one full iteration, i.e., at the output of the variable nodes, we see an erasure fraction of $x = v_\epsilon(0.88397)$, i.e., x is the solution to the equation $0.883971 = v_\epsilon^{-1}(x)$. This process continues in the same fashion for each subsequent iteration, corresponding graphically to a *staircase* function which is bounded below by $c(x)$ and bounded above by $v_\epsilon^{-1}(x)$.

§3.14.2. EXIT Function

We will now see that both $c(x)$ and $v_\epsilon(x)$ have an interpretation in terms of entropy.

Definition 3.70 (EXIT Function). Let C be a binary code. Let X be chosen with probability $p_X(x)$ from C and let Y denote the result of letting X be transmitted over a BEC(ϵ). The *extrinsic information transfer* (EXIT) function associated with the i-th bit of C, call it $h_i(\epsilon)$, is defined as

$$h_i(\epsilon) = H(X_i \mid Y_{\sim i}).$$

The *average* EXIT function is

$$h(\epsilon) = \frac{1}{n} \sum_{i=1}^{n} h_i(\epsilon). \qquad \triangledown$$

Discussion: The input parameter ϵ represents an entropy: if we assume that X_i is chosen uniformly at random from $\{0,1\}$ and that Y_i is the result of sending X_i over a BEC then $H(X_i \mid Y_i) = \epsilon H(X_i \mid Y_i = ?) = \epsilon$. The EXIT function therefore characterizes how entropy is transferred from input to output. The word "extrinsic" in EXIT refers to the fact that we consider $H(X_i \mid Y_{\sim i})$ instead of $H(X_i \mid Y)$, i.e., we do not include the observation Y_i itself. You are warned that in the literature mutual information is often considered instead of entropy. Assuming that we impose a uniform prior on X_i, this differs from our definition only in a trivial way ($H(X_i \mid Y_{\sim i}) = 1 - I(X_i; Y_{\sim i})$).

Example 3.71 (Parity-Check Code). Consider the binary parity-check code with parameters $C[n, n-1, 2]$ and a uniform prior on the set of codewords. By symmetry, $h_i(\epsilon) = h(\epsilon)$ for all $i \in [n]$ and

$$h(\epsilon) = 1 - (1 - \epsilon)^{n-1}.$$

A check node of degree i in the Tanner graph represents a parity-check code $C[i, i-1, 2]$. Let the channel parameter of the BEC be x. From earlier discussion the associated EXIT function is $1 - (1 - x)^{i-1}$. The function $c(x)$ depicted in Figure 3.69 is the *average* over the set of EXIT functions corresponding to check nodes of degree i, where the average is with respect to the edge degree distribution: indeed, $\sum_i \rho_i(1 - (1 - x)^{i-1}) = 1 - \rho(1 - x) = c(x)$. ◇

Example 3.72 (Repetition Code). Consider the binary repetition code $C[n, 1, n]$ with a uniform prior on the set of codewords. By symmetry, $h_i(\epsilon) = h(\epsilon)$ for all $i \in [n]$ and

$$h(\epsilon) = \epsilon^{n-1}.$$ ◇

Example 3.73 ([7, 4, 3] Hamming Code). The parity-check matrix H of the $[7, 4, 3]$ binary Hamming code is stated explicitly in Example 1.24. Assuming a uniform prior on the set of codewords, a tedious calculation reveals (see Problem 3.30) that

$$h_{\text{Ham}}(\epsilon) = 3\epsilon^2 + 4\epsilon^3 - 15\epsilon^4 + 12\epsilon^5 - 3\epsilon^6.$$ ◇

Figure 3.75 depicts the EXIT functions of Examples 3.71, 3.72, and 3.73.

There are many alternative characterizations of EXIT functions, each with its own merit. Let us list the most useful ones.

Lemma 3.74 (Various Characterizations of EXIT Functions). Let $C[n, k]$ be a binary linear code and let X be chosen with uniform probability from $C[n, k]$. Further, let H and G be a parity and a generator matrix representing C, respectively. Let Y denote the result of letting X pass over a BEC(ϵ). Let $\hat{x}_i^{\text{MAP}}(y_{\sim i})$ denote the MAP estimator function of the i-th bit given the observation $y_{\sim i}$. Then the following are equivalent:

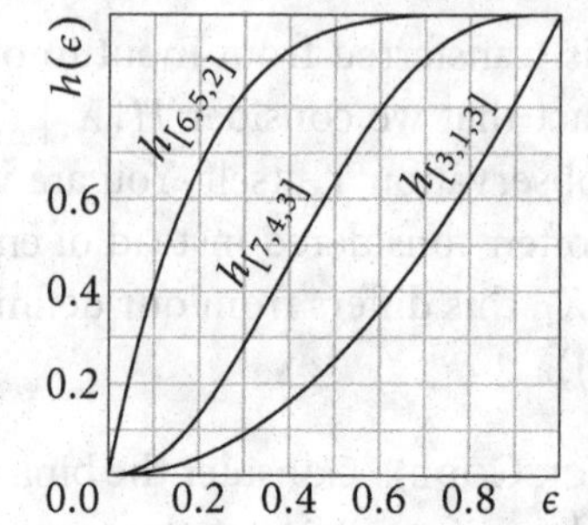

Figure 3.75: EXIT function of the $[3,1,3]$ repetition code, the $[6,5,2]$ parity-check code, and the $[7,4,3]$ Hamming code.

(i) $h_i(\epsilon) = H(X_i \mid Y_{\sim i})$

(ii) $h_i(\epsilon) = \mathbb{P}\{\hat{x}_i^{\text{MAP}}(Y_{\sim i}) = ?\}$

(iii) $h(\epsilon) = \frac{\mathrm{d}H(X \mid Y)}{n\mathrm{d}\epsilon}$

(iv) $h_i(\epsilon) = \sum_{\mathcal{E} \subseteq [n] \smallsetminus \{i\}} \epsilon^{|\mathcal{E}|}(1-\epsilon)^{n-1-|\mathcal{E}|}(1 + \text{rank}(H_{\mathcal{E}}) - \text{rank}(H_{\mathcal{E} \cup \{i\}}))$

(v) $h_i(\epsilon) = \sum_{\mathcal{E} \subseteq [n] \smallsetminus \{i\}} \epsilon^{n-1-|\mathcal{E}|}(1-\epsilon)^{|\mathcal{E}|}(\text{rank}(G_{\mathcal{E} \cup \{i\}}) - \text{rank}(G_{\mathcal{E}}))$.

Discussion: The second characterization states that the EXIT value equals the erasure probability of an extrinsic MAP decoder. This is useful when doing actual computations. The most fundamental characterization is the third one: it states that the (average) EXIT function equals the derivative of the conditional entropy with respect to the channel parameter divided by the blocklength n.

Proof. The proof of the equivalence of characterizations (i), (ii), (iv), and (v) is left as Problem 3.28.

In what follows we concentrate on the third characterization, which is the most fundamental one (if you would like to define an EXIT function for a non-uniform prior or a non-linear code, this is the proper starting point). Although all bits are sent through the same channel BEC(ϵ), it is convenient to imagine that bit i is sent through a BEC with parameter ϵ_i, where (by chance) $\epsilon_i = \epsilon$ for all $i \in [n]$. We claim that

$$\frac{\mathrm{d}H(X \mid Y(\epsilon_1, \ldots, \epsilon_n))}{\mathrm{d}\epsilon} = \sum_{i=1}^{n} \frac{\partial H(X \mid Y(\epsilon_1, \ldots, \epsilon_n))}{\partial \epsilon_i}\Big|_{\epsilon_j = \epsilon, \forall j \in [n]}$$
$$= \sum_{i=1}^{n} \frac{\partial H(X_i \mid Y(\epsilon_1, \ldots, \epsilon_n))}{\partial \epsilon_i}\Big|_{\epsilon_j = \epsilon, \forall j \in [n]}.$$

To see the second transition, use the chain rule to write $H(X \mid Y) = H(X_i \mid Y) + H(X_{\sim i} \mid X_i, Y)$. Due to the memoryless property of the channel, we further have $H(X_{\sim i} \mid X_i, Y) = H(X_{\sim i} \mid X_i, Y_{\sim i})$. But the right-hand side is no longer a function of ϵ_i. Therefore its contribution vanishes when we take the (partial) derivative with respect to ϵ_i. Finally, note that

$$H(X_i \mid Y) = \mathbb{P}\{Y_i = ?\}\,\mathbb{P}\{\hat{x}_i^{\text{MAP}}(Y_{\sim i}) = ?\} = \epsilon_i h_i(\epsilon),$$

so that $\frac{\partial H(X_i \mid Y)}{\partial \epsilon_i} = h_i(\epsilon)$ and $\frac{\mathrm{d}H}{n\mathrm{d}\epsilon} = h(\epsilon)$. □

As we have seen in the preceding proof, it is useful to generalize to a situation where the i-th bit is sent through the channel BEC(ϵ_i). Let us assume that the individual channel parameters ϵ_i are parameterized in a differentiable way by a common parameter ϵ, i.e., $\epsilon_i = \epsilon_i(\epsilon)$. Now ϵ is no longer the channel parameter itself but if we change ϵ in a smooth manner then we can imagine that the set of individual channels $\{\text{BEC}(\epsilon_i)\}_{i=1}^n$ describes some smooth curve in "channel space." Depending on the parameterization, this description includes, e.g., the case where all channels stay fixed except for one, or the case where all channels are the same and are changed in lock-step. Characterization (iii) of Lemma 3.74 is still meaningful; i.e., we can define the (average) EXIT function as $h(\epsilon) = \frac{\mathrm{d}H(X \mid Y)}{n\mathrm{d}\epsilon}$ and the individual EXIT functions as $h_i(\epsilon) = \frac{\partial H(X_i \mid Y)}{\partial \epsilon_i}\frac{\mathrm{d}\epsilon_i}{\mathrm{d}\epsilon}$.

Example 3.76 (EXIT Function of Variable Nodes). Consider the EXIT function of an $[i+1, 1, i]$ repetition code, where the last bit is passed through a BEC with parameter ϵ and bit 1 to bit i are passed through a BEC with parameter x. This is the case for a variable node of degree i in the Tanner graph of an LDPC code. Consider the EXIT function corresponding to one of the first i positions. We have $H(X \mid Y) = \epsilon \prod_{k=1}^{i} x_k$, where all the x_k have value equal to x. Taking the derivative with respect to, e.g., x_1, and setting $x_k = x$ we get ϵx^{i-1}. If we average this over the edge degree distribution λ we get $\sum_i \lambda_i \epsilon x^{i-1} = \epsilon\lambda(x) = v_\epsilon(x)$. We conclude that $v_\epsilon(x)$ is the average of a collection of EXIT functions. ◇

As we have seen in Examples 3.71 and 3.76, $v_\epsilon(x)$ and $c(x)$ are (averaged) EXIT functions. For this reason we call Figure 3.69 an *EXIT chart*.

§3.14.3. Basic Properties of EXIT Functions

Theorem 3.77 (Duality Theorem). Let C be a binary linear code, let $C^\perp$ be its dual, and, assuming uniform priors on the codewords, let $h_i(\epsilon)$ and $h_i^\perp(\epsilon)$ denote the corresponding EXIT functions. Then

$$h_i(\epsilon) + h_i^\perp(1-\epsilon) = 1.$$

Proof. Recall that if H is a parity-check matrix of C then it is also a generator matrix of $C^{\perp}$. The claim now follows from characterizations (iv) and (v) of Lemma 3.74. □

Example 3.78 (Repetition Code and Single Parity-Check Code). The codes $[n, 1, n]$ and $[n, n-1, 2]$ are duals and from Examples 3.72 and 3.71 we know that for all $i \in [n]$ their EXIT functions are ϵ^{n-1} and $1-(1-\epsilon)^{n-1}$, respectively. Therefore, in agreement with Theorem 3.77,

$$\epsilon^{n-1} + 1 - (1-(1-\epsilon))^{n-1} = 1. \quad \diamond$$

Theorem 3.79 (Minimum Distance Theorem). Let C be a binary linear code of length n with minimum distance d and with EXIT functions $h_i(\epsilon)$, $i \in [n]$. Then $h_i(\epsilon)$, $i \in [n]$, is a polynomial of minimum degree *at least* $d-1$ and the average EXIT function has minimum degree *exactly* $d-1$.

Proof. Consider characterization (iv) of Lemma 3.74. Let $\mathcal{E}$ be a subset of cardinality strictly lower than $d-1$. Since any $d-1$ or fewer columns of H are linearly independent, it follows that $(1+\text{rank}(H_{\mathcal{E}}) - \text{rank}(H_{\mathcal{E}\cup\{i\}}))$ is zero for any such subset $\mathcal{E}$. Therefore, $h_i(\epsilon)$ does not contain monomials of a degree less than $d-1$. On the other hand, if $\mathcal{E}\cup\{i\}$ is chosen to correspond to the support of a minimum distance codeword, then $(1+\text{rank}(H_{\mathcal{E}})-\text{rank}(H_{\mathcal{E}\cup\{i\}}))$ is one and this will contribute a monomial of degree $d-1$. Since these minimum degree terms cannot be canceled by any other terms it follows that $h(\epsilon)$ has minimum degree exactly $d-1$. □

Example 3.80 (Repetition Code, Parity-Check Code, Hamming Code). We see from our previous examples that the average EXIT functions of the $[n, 1, n]$ repetition code, the $[n, n-1, 2]$ single parity-check code, and the $[7, 4, 3]$ Hamming code have minimum degree $n-1$, 1, and 2, respectively, as predicted by Theorem 3.79. ◇

The most fundamental property of EXIT functions is a direct consequence of the third characterization in Lemma 3.74. For historical reasons we label it as a theorem, even though there is really nothing to be proved.

Theorem 3.81 (Area Theorem). Let C be a binary code. Let X be chosen with uniform probability from C and let Y denote the received word after transmission over a BEC(ϵ). To emphasize that Y depends on the channel parameter ϵ write $Y(\epsilon)$. If $h(\epsilon)$ denotes the corresponding EXIT function then

$$H(X \mid Y(\delta)) = n\int_0^{\delta} h(\epsilon)\,\mathrm{d}\epsilon. \tag{3.82}$$

Proof. Using the characterization $h(\epsilon) = \frac{\mathrm{d}H(X \mid Y(\epsilon))}{n\mathrm{d}\epsilon}$ of Lemma 3.74, we have

$$\int_0^\delta \frac{\mathrm{d}H(X \mid Y(\epsilon))}{\mathrm{d}\epsilon} \mathrm{d}\epsilon = H(X \mid Y(\delta)) - H(X \mid Y(0)) = H(X \mid Y(\delta)). \qquad \square$$

Problem 3.31 discusses a combinatorial proof of this theorem which starts with the characterization of $h(\epsilon)$ given in Definition 3.70.

Example 3.83 (Area Theorem Applied to [7, 4, 3] Hamming Code).

$$\int_0^1 h_{\text{Ham}}(\epsilon)\mathrm{d}\epsilon = \int_0^1 \left(3\epsilon^2 + 4\epsilon^3 - 15\epsilon^4 + 12\epsilon^5 - 3\epsilon^6\right) \mathrm{d}\epsilon = \frac{4}{7}. \qquad \diamond$$

§3.14.4. Matching Condition

Let us go back to the EXIT chart method as depicted on the right-hand side of Figure 3.69. The area under the curve $c(x)$ equals $1 - \int\rho$ and the area to the left of the curve $v_\epsilon^{-1}(x)$ is equal to $\epsilon \int\lambda$. This is easily verified by a direct calculation, integrating the respective functions, but it also follows from (the Area) Theorem 3.81. Assume that we integrate from 0 to 1. The normalized integral is then equal to $H(X \mid Y(1))/n = H(X)/n$. If we assume a uniform prior on the set of codewords this equals r, the rate of the code.

Consider the check-node side. The rate of a parity-check code of length i is $\frac{i-1}{i} = 1 - \frac{1}{i}$, so that the area under an individual EXIT function of such a code is $1 - \frac{1}{i}$. If we average over the edge degree distribution we get $1 - \sum_i \rho_i \frac{1}{i} = 1 - \int\rho$.

The argument at the variable node side is more interesting. First note that the area "to the left" of the curve $v_\epsilon^{-1}(x)$ is equal to the area "under" $v_\epsilon(x)$. By definition, if we integrate the average EXIT function then we get the difference of entropies at the two endpoints of the "path." For the EXIT function corresponding to a variable node one position is fixed to the channel BEC(ϵ) and the other ones go from "no information" to "perfect channel." If all the other positions have "no information" then the uncertainty is ϵ, if all the other positions see a "perfect channel" then the uncertainty is 0. The difference is therefore ϵ. The EXIT function is the average of the individual EXIT functions: the EXIT function associated with the fixed position is zero (since the input does not change) and the remaining i individual EXIT functions are equal by symmetry. We conclude that the integral under one of these EXIT functions is $\frac{\epsilon}{i}$. If we average with respect to the edge degree distribution λ this gives us $\epsilon \int\lambda$.

We know that a necessary condition for successful BP decoding is that these two areas do not overlap (the two curves $v_\epsilon^{-1}(x)$ and $c(x)$ must not cross). Since the total area equals 1, we get the necessary condition (for successful BP decoding) $(\epsilon \int\lambda) + (1 - \int\rho) \leq 1$. Rearranging terms, this is equivalent to the condition

$$1 - C_{\text{Sha}} = \epsilon \leq \frac{\int\rho}{\int\lambda} = 1 - r(\lambda, \rho).$$

In words, the design rate $r(\lambda, \rho)$ of any LDPC ensemble which, for increasing blocklengths, allows successful decoding over the BEC(ϵ) cannot surpass the Shannon limit $1 - \epsilon$.

The final result (namely that transmission above capacity is not possible) is trivial, but the method of proof shows how capacity enters in the calculation of the performance under BP decoding and it shows that for the design rate to achieve capacity the component curves have to be perfectly matched. In the next section we construct capacity-achieving degree distributions. We use the matching condition $\epsilon\lambda(1 - \rho(1 - x)) = x$ as our starting point. This matching condition also explains why optimized ensembles tend to have many critical points as we discussed in Example 3.63.

§3.15. Capacity-Achieving Degree Distributions

Consider a pair of degree distributions (λ, ρ) of design rate $r = r(\lambda, \rho)$ and with threshold $\epsilon^{\text{BP}} = \epsilon^{\text{BP}}(\lambda, \rho)$. By Shannon, we must have $r \leq 1 - \epsilon^{\text{BP}}$, and it is natural to use the following definition of the (multiplicative) gap as a measure of performance.

Definition 3.84 (Multiplicative Gap). Let (λ, ρ) be a degree distribution pair with rate $r = r(\lambda, \rho)$ and threshold $\epsilon^{\text{BP}} = \epsilon^{\text{BP}}(\lambda, \rho)$. Further, let $\delta = \delta(\lambda, \rho)$ be the unique non-negative number such that $r = (1 - \delta)(1 - \epsilon^{\text{BP}})$, i.e., $\delta = \frac{1-\epsilon^{\text{BP}}-r}{1-\epsilon^{\text{BP}}}$. We then say that (λ, ρ) *achieves a fraction* $(1 - \delta)$ *of capacity*. Equivalently, we say that (λ, ρ) has a *multiplicative gap* δ. ▽

Unfortunately, as the next theorem shows, no fixed pair (λ, ρ) can have zero gap. The proof of this theorem is the topic of Problem 3.18.

Theorem 3.85 (Lower Bound on Gap). Let (λ, ρ) be a degree distribution pair of design rate r, $r \in (0, 1)$, and with average check-node degree $\mathtt{r}_{\text{avg}}$. Then

$$\delta(\lambda, \rho) \geq \frac{r^{\mathtt{r}_{\text{avg}}-1}(1 - r)}{1 + r^{\mathtt{r}_{\text{avg}}-1}(1 - r)}.$$

The best we can therefore hope for is to construct a *sequence* of degree distribution pairs that achieve capacity.

Definition 3.86 (Capacity-Achieving Sequence of Degree Distributions). We say that a sequence $\{(\lambda^{(N)}, \rho^{(N)})\}_{N\geq 1}$ *achieves capacity* on the BEC(ϵ) if

$$\lim_{N\to\infty} r(\lambda^{(N)}, \rho^{(N)}) = 1 - \epsilon, \text{ and}$$

$$\lim_{N\to\infty} \delta(\lambda^{(N)}, \rho^{(N)}) = 0. \tag{3.87}$$

Note that (3.87) implies that $\epsilon^{\text{BP}}(\lambda^{(N)}, \rho^{(N)})$ converges to ϵ. ▽

If you are mainly interested in how in *practice* we can find degree distributions that allow reliable transmission close to capacity, fast forward to Section 3.18. Our current aim is to *prove* that capacity-achieving degree distributions exist.

EXAMPLE 3.88 (CHECK-CONCENTRATED DEGREE DISTRIBUTION). For $\alpha \in (0,1)$ so that $1/\alpha \in \mathbb{N}$, choose

$$\hat{\lambda}_\alpha(x) = 1-(1-x)^\alpha = \sum_{i=1}^{\infty}\binom{\alpha}{i}(-1)^{i-1}x^i, \qquad \rho_\alpha(x) = x^{\frac{1}{\alpha}}.$$

Recall that for $\alpha \in \mathbb{R}$ and $i \in \mathbb{N}$, $\binom{\alpha}{i}(-1)^{i-1}$ is defined as

$$\binom{\alpha}{i}(-1)^{i-1} = \frac{\alpha(\alpha-1)\cdots(\alpha-i+1)}{i!}(-1)^{i-1} = \frac{\alpha}{i}(1-\frac{\alpha}{i-1})\cdots(1-\alpha).$$

Since $\alpha \in (0,1)$, this shows that all coefficients of the expansion of $\hat{\lambda}_\alpha(x)$ are non-negative. Note that $\hat{\lambda}_\alpha(x)$ and $\rho_\alpha(x)$ are matched, i.e.,

$$\hat{\lambda}_\alpha(1-\rho_\alpha(1-x)) = x, \quad \forall x \in [0,1). \tag{3.89}$$

Unfortunately we cannot use $(\hat{\lambda}_\alpha(x), \rho_\alpha(x))$ directly since this pair has an associated rate of 0. We will therefore have to modify it. Let $\hat{\lambda}_\alpha^{(N)}(x)$ denote the function consisting of the first N terms of the Taylor series expansion of $\hat{\lambda}_\alpha(x)$ (up to and including the term x^{N-1}) and define the *normalized* function $\lambda_\alpha^{(N)}(x) = \frac{\hat{\lambda}_\alpha^{(N)}(x)}{\hat{\lambda}_\alpha^{(N)}(1)}$. We then have

$$\int_0^1 \hat{\lambda}_\alpha^{(N)}(x)\mathrm{d}x = \sum_{i=1}^{N-1}\binom{\alpha}{i}\frac{(-1)^{i-1}}{i+1} = \frac{\alpha-\binom{\alpha}{N}(-1)^{N-1}}{\alpha+1},$$

$$\hat{\lambda}_\alpha^{(N)}(1) = \sum_{i=1}^{N-1}\binom{\alpha}{i}(-1)^{i-1} = 1-\frac{N}{\alpha}\binom{\alpha}{N}(-1)^{N-1},$$

$$\int_0^1 \rho_\alpha(x)\mathrm{d}x = \frac{\alpha}{1+\alpha}.$$

To verify the summations first check the correctness for $N = 2$ and then apply induction. Let us start by computing the rate $r(\alpha, N)$. Using (3.19) and the above summations we have

$$r(\alpha,N) = 1-\frac{\int_0^1\rho_\alpha(x)\mathrm{d}x}{\int_0^1\lambda_\alpha^{(N)}(x)\mathrm{d}x} = 1-\hat{\lambda}_\alpha^{(N)}(1)\frac{\int_0^1\rho_\alpha(x)\mathrm{d}x}{\int_0^1\hat{\lambda}_\alpha^{(N)}(x)\mathrm{d}x}$$
$$= \frac{\frac{N}{\alpha}\binom{\alpha}{N}(-1)^{N-1}(1-1/N)}{1-\frac{1}{N}\frac{N}{\alpha}\binom{\alpha}{N}(-1)^{N-1}}.$$

Next, consider the gap to capacity. For any $x \in [0,1]$,

$$\hat{\lambda}_\alpha(x) \geq \hat{\lambda}_\alpha^{(N)}(x),$$

since the coefficients of $\hat{\lambda}_\alpha(x)$ are non-negative, so that from (3.89)

$$x = \hat{\lambda}_\alpha(1-\rho_\alpha(1-x)) \geq \hat{\lambda}_\alpha^{(N)}(1-\rho_\alpha(1-x)) = \hat{\lambda}_\alpha^{(N)}(1)\lambda_\alpha^{(N)}(1-\rho_\alpha(1-x)).$$

It follows that $\epsilon^{\text{BP}}(\alpha,N) \geq \hat{\lambda}_\alpha^{(N)}(1)$. Therefore,

$$\delta(\alpha,N) = \frac{1-\epsilon^{\text{BP}}(\alpha,N)-r(\alpha,N)}{1-\epsilon^{\text{BP}}(\alpha,N)} \leq \frac{\hat{\lambda}_\alpha^{(N)}(1)}{1-\hat{\lambda}_\alpha^{(N)}(1)}\left[\frac{\int_0^1 \rho_\alpha(x)\text{d}x}{\int_0^1 \hat{\lambda}_\alpha^{(N)}(x)\text{d}x}-1\right]$$
$$\leq \frac{1-\frac{N}{\alpha}\binom{\alpha}{N}(-1)^{N-1}}{N-\frac{N}{\alpha}\binom{\alpha}{N}(-1)^{N-1}}.$$

Suppose that we want to construct a capacity-achieving sequence of degree distributions for the channel parameter ϵ. This means that we want to choose $\alpha = \alpha(N)$ in such a way that the rate converges to the design rate, $r = 1-\epsilon$, and that the gap converges to zero. In order to proceed we require an estimate on $\frac{N}{\alpha}\binom{\alpha}{N}(-1)^{N-1}$. In Problem 3.19 you are asked to show that

$$\ln\left(\frac{N}{\alpha}\binom{\alpha}{N}(-1)^{N-1}\right) = -\alpha H(N-1) - c(N,\alpha)\alpha^2,$$

where $H(N)$ is the N-th Harmonic number and where, for $\alpha < \frac{1}{2}$, $0 \leq c(N,\alpha) \leq 5$. Here is our choice: pick $\alpha = \alpha(N)$ so that $\frac{N}{\alpha}\binom{\alpha}{N}(-1)^{N-1} = 1-\epsilon$. Since $H(N) \approx \ln(N)$ this is possible and $\alpha \approx \frac{\ln\frac{1}{1-\epsilon}}{H(N-1)} \approx \frac{\ln\frac{1}{1-\epsilon}}{\ln N}$ (from the last expression we see that α is indeed in $[0,\frac{1}{2})$ for N sufficiently large as required in the estimate). It follows that

$$r(N) = 1-\epsilon + O(1/N), \qquad \delta(N) \leq \frac{1-(1-\epsilon)}{N-(1-\epsilon)} = O(1/N). \qquad \Diamond$$

Example 3.90 (Heavy-Tail Poisson Distribution). There are infinitely many other capacity-achieving degree distributions. We get a second example if we start with

$$\hat{\lambda}_\alpha(x) = -\frac{1}{\alpha}\ln(1-x) = \frac{1}{\alpha}\sum_{i=1}^{\infty}\frac{x^i}{i}, \text{ and}$$

$$\rho_\alpha(x) = \text{e}^{\alpha(x-1)} = \text{e}^{-\alpha}\sum_{i=0}^{\infty}\frac{\alpha^i x^i}{i!},$$

and proceed along the same lines. The details of the calculations for this case are left to Problem 3.20. $\Diamond$

§3.16. Gallager's Lower Bound on Density

We have seen in the previous section two sequences of capacity-achieving degree distribution pairs and infinitely more examples exist. Faced with such ample choice, which one is "best"? There are many possible and reasonable criteria upon which we could decide. As an example, we could investigate how quickly the finite-length behavior approaches the asymptotic limit for the various choices. This is done in Sections 3.22 and 3.23. Another important point of view is to introduce the notion of *complexity*. As we discussed in Section 3.5, on the BEC the decoding complexity is in one-to-one correspondence with the number of edges in the graph, since we use each edge at most once. How sparse can a graph be to achieve a fraction $(1-\delta)$ of capacity? In the setting of LDPC ensembles under BP decoding Theorem 3.85 gives the answer. We will now see a more general information-theoretic bound which applies to any (sequence of) code(s) and any decoding algorithm.

Definition 3.91 (Density of Parity-Check Matrices). Let C be a binary linear code of length n and rate r and let H be a parity-check matrix of C. The *density* of H, denoted by $\Delta(H)$, is defined as

$$\Delta(H) = \frac{1}{nr}|\{(i,j) : H_{i,j} \neq 0\}|. \qquad \triangledown$$

Example 3.92 (Density of Standard LDPC Ensemble). The density of a degree distribution pair $(\Lambda, P) \triangleq (n, L, R) \triangleq (n, \lambda, \rho)$ is equal to

$$\Lambda'(1)\frac{1}{nr} = P'(1)\frac{1}{nr} = L'(1)\frac{1}{r} = R'(1)\frac{1-r}{r} = \frac{1}{\int\lambda}\frac{1}{r} = \frac{1}{\int\rho}\frac{1-r}{r}.$$

This density is equal to the decoding complexity of the BP decoder, when measured *per information bit.* $\diamond$

Theorem 3.93 (Lower Bound on Density). Let $\{C_N\}$ be a sequence of binary linear codes in which codewords are used with uniform probability and which achieve a fraction $(1-\delta)$ of the capacity of the BEC(ϵ) with vanishing bit erasure probability. Let $\{\Delta_N\}$ be the corresponding sequence of densities of their parity-check matrices. Then

$$\liminf_{N\to\infty} \Delta_N > \frac{K_1 + K_2 \ln\frac{1}{\delta}}{1-\delta}, \tag{3.94}$$

where

$$K_1 = \frac{\epsilon \ln\left(\frac{\epsilon}{1-\epsilon}\right)}{(1-\epsilon)\ln\left(\frac{1}{1-\epsilon}\right)}, \quad K_2 = \frac{\epsilon}{(1-\epsilon)\ln\left(\frac{1}{1-\epsilon}\right)}.$$

Proof. Consider a binary linear code $C[n, nr]$. Let X be chosen uniformly at random from $C[n, nr]$ and let Y be the received word after transmission over a BEC(ϵ). Let $\mathcal{E}$ denote the random index set of erasure positions. There is a one-to-one correspondence between Y and $(Y_{\bar{\mathcal{E}}}, \mathcal{E})$ as well as between $(X, \mathcal{E})$ and $(X_{\mathcal{E}}, X_{\bar{\mathcal{E}}}, \mathcal{E})$. Further, $X_{\bar{\mathcal{E}}} = Y_{\bar{\mathcal{E}}}$. We have

$$\begin{aligned}
H(X \mid Y) &= H(X \mid Y_{\bar{\mathcal{E}}}, \mathcal{E}) = H(X, \mathcal{E} \mid Y_{\bar{\mathcal{E}}}, \mathcal{E}) \\
&= H(X_{\mathcal{E}}, X_{\bar{\mathcal{E}}}, \mathcal{E} \mid Y_{\bar{\mathcal{E}}}, \mathcal{E}) = H(X_{\mathcal{E}}, X_{\bar{\mathcal{E}}} \mid Y_{\bar{\mathcal{E}}}, \mathcal{E}) = H(X_{\mathcal{E}} \mid Y_{\bar{\mathcal{E}}}, \mathcal{E}) \\
&= \sum_{y_{\bar{\mathcal{E}}}, E} H(X_{\mathcal{E}} \mid Y_{\bar{\mathcal{E}}} = y_{\bar{\mathcal{E}}}, \mathcal{E} = E) p(Y_{\bar{\mathcal{E}}} = y_{\bar{\mathcal{E}}}, \mathcal{E} = E) \\
&= \sum_{y_{\bar{\mathcal{E}}}, E} (|E| - \text{rank}(H_E)) p(Y_{\bar{\mathcal{E}}} = y_{\bar{\mathcal{E}}}, \mathcal{E} = E) \\
&= \sum_{E} (|E| - \text{rank}(H_E)) p(\mathcal{E} = E) = n\epsilon - \sum_{E} \text{rank}(H_E) p(\mathcal{E} = E).
\end{aligned}$$

The rank of H_E is upper bounded by the number of non-zero rows of H_E. This number in turn is equal to the number of parity-check nodes which involve erased bits. Therefore, $\sum_E \text{rank}(H_E) p(\mathcal{E} = E)$ is upper bounded by the average (where the average is over the channel realization) number of parity checks which involve erased bits. Let $P(x)$ denote the check-node degree distribution from a node perspective (such a degree distribution is defined regardless whether the code is low-density or not) and let $R(x)$ denote the normalized such quantity, $R(x)n(1-r) = P(x)$. If a parity-check node is of degree i then the probability that it involves at least one erased bit is $1 - (1-\epsilon)^i$, and therefore the average number of parity-check nodes which involve at least one erased bit is

$$\sum_i P_i(1 - (1-\epsilon)^i) = n(1-r) - P(1-\epsilon) \leq n(1-r)\big(1 - (1-\epsilon)^{R'(1)}\big).$$

In the last step we have used Jensen's inequality (1.61). We therefore have

$$\frac{H(X \mid Y)}{n} \geq \epsilon - (1-r)\big(1 - (1-\epsilon)^{R'(1)}\big). \tag{3.95}$$

Assume now that we have a sequence of codes $\{C_N\}$, where C_N has length n_N, and that this sequence has asymptotic gap δ. This means that $\delta_N \overset{N\to\infty}{\longrightarrow} \delta$, and therefore $r_N \overset{N\to\infty}{\longrightarrow} (1-\delta)(1-\epsilon)$. Apply the lim inf to both sides of (3.95). From Fano's inequality (1.49) we know that $\frac{1}{n_N} H(X \mid Y)$ must converge to zero for the bit erasure probability to tend to zero. We therefore have

$$\liminf_{N\to\infty} (1 - (1-\delta_N)(1-\epsilon))(1 - (1-\epsilon)^{R'_N(1)}) \geq \epsilon.$$

Solving this equation for $R'_N(1)$ we get

$$\liminf_{N\to\infty} R'_N(1) \geq \frac{\ln\left(1+\frac{\epsilon}{\delta_N(1-\epsilon)}\right)}{\ln\left(\frac{1}{1-\epsilon}\right)} > \frac{\ln\left(\frac{\epsilon}{\delta_N(1-\epsilon)}\right)}{\ln\left(\frac{1}{1-\epsilon}\right)}.$$

To finish the proof observe that

$$\begin{aligned}\liminf_{N\to\infty} \Delta_N &= \liminf_{N\to\infty} R'_N(1)\frac{1-r_N}{r_N} = \liminf_{N\to\infty} R'_N(1)\frac{1-(1-\delta_N)(1-\epsilon)}{(1-\delta_N)(1-\epsilon)} \\ &\geq \liminf_{N\to\infty} R'_N(1)\frac{\epsilon}{(1-\delta_N)(1-\epsilon)} > \frac{\ln\left(\frac{\epsilon}{\delta(1-\epsilon)}\right)}{\ln\left(\frac{1}{1-\epsilon}\right)}\frac{\epsilon}{(1-\delta)(1-\epsilon)} \\ &= \frac{K_1+K_2\ln\frac{1}{\delta}}{1-\delta}.\end{aligned}$$

□

§3.17. Optimally Sparse Degree Distribution Pairs

Surprisingly, check-concentrated degree distribution pairs are essentially optimal with respect to a complexity versus performance trade-off. More precisely, the next theorem states that if we pick N and α for the check-concentrated ensemble carefully then the trade-off between gap δ and complexity Δ is, up to a small additive constant, the best possible. We skip the proof.

Theorem 3.96 (Optimality of Check-Concentrated Distribution). Consider the check-concentrated ensembles introduced in Example 3.88 and transmission over the BEC(ϵ). Choose

$$N = \max\left(\left\lceil\frac{1-c(\epsilon)(1-\epsilon)(1-\delta)}{\delta}\right\rceil, \lceil(1-\epsilon)^{-\frac{1}{\epsilon}}\rceil\right), \qquad \alpha = \frac{\ln\frac{1}{1-\epsilon}}{\ln N},$$

where $c(\epsilon) = (1-\epsilon)^{\frac{\pi^2}{6}}e^{(\frac{\pi^2}{6}-\gamma)\epsilon}$ and where γ is the Euler-Mascheroni constant, $\gamma \approx 0.5772$. This degree distribution pair achieves at least a fraction $1-\delta$ of the channel capacity with vanishing bit erasure probability under BP decoding. Further, the density Δ is upper bounded by

$$\Delta \leq \frac{K_1+K_2\log\frac{1}{\delta}+K(\epsilon,\delta)}{1-\delta}\left(1+\frac{1-\epsilon}{\epsilon}\delta\right) \leq \frac{K_1+K_2\log\frac{1}{\delta}}{1-\delta}+\kappa+O(\delta), \tag{3.97}$$

where K_1 and K_2 are the constants of Theorem 3.93,

$$K(\epsilon,\delta) = \frac{\epsilon\ln\left(1+\frac{1-\epsilon}{\epsilon}(\delta c(\epsilon)+(1-c(\epsilon)))\right)}{(1-\epsilon)\ln\frac{1}{1-\epsilon}},$$

and where $\kappa = \max_{\epsilon \in [0,1]} K(\epsilon, 0) \approx 0.5407$. Comparing (3.97) with (3.94), we see that, up to a small constant κ, right-concentrated degree distribution pairs are optimally sparse.

From the preceding discussion one might get the impression that essentially no further improvement is possible in terms of the performance-complexity trade-off. This is true within the current framework. But we will see in Chapter 7 that, by allowing more complicated graphical models (in particular by introducing so-called *state* nodes), better trade-offs can be achieved.

§3.18. Degree Distributions with Given Maximum Degree

In practice we are concerned not only with complexity (average degree) but also with the maximum degree since large degrees imply slow convergence of the finite-length performance to the asymptotic threshold (see Section 3.22). A glance at Theorem 3.96 shows that, although the average degree only grows like $\ln \frac{1}{\delta}$, the maximum grows exponentially faster, namely like $\frac{1}{\delta}$. Assume therefore that we have a given bound on the maximum variable-node degree; call it $\mathtt{l}_{\max}$. Recall that a degree distribution pair (λ, ρ) has a threshold of at least ϵ if

$$\epsilon\lambda(1 - \rho(1 - x)) - x \leq 0, \quad x \in [0, 1].$$

Assume that we fix the check-node degree distribution ρ and define the function

$$f(x, \lambda_2, \ldots, \lambda_{\mathtt{l}_{\max}}) = \epsilon\lambda(1 - \rho(1 - x)) - x = \epsilon \sum_{i \geq 2} \lambda_i (1 - \rho(1 - x))^{i-1} - x.$$

The function f is *linear* in the variables λ_i, $i = 2, \ldots, \mathtt{l}_{\max}$ (note that $\lambda_1 = 0$), and from $r(\lambda, \rho) = 1 - \int \rho / \int \lambda$, we see that the rate is an increasing function in $\sum \lambda_i / i$ (for fixed ρ). Therefore, we can use the continuous linear program

$$\max_{\lambda} \Big\{ \sum_{i \geq 2} \lambda_i / i \mid \lambda_i \geq 0; \sum_{i \geq 2} \lambda_i = 1; f \leq 0; x \in [0, 1] \Big\} \tag{3.98}$$

to find the degree distribution λ that maximizes the rate for a given right-hand-side ρ and a given threshold ϵ. In practice, we use a finite program. We can avoid numerical problems for x around zero caused by this discretization by incorporating the stability condition $\lambda_2 \leq 1/(\epsilon\rho'(1))$ explicitly into the linear program.

If we exchange the roles of the variable and check nodes and let y denote the fraction of the erased messages emitted at the check nodes, then an equivalent condition is

$$1 - y - \rho(1 - \epsilon\lambda(y)) \leq 0, \quad y \in [0, 1].$$

Proceeding in the same fashion as before, we see that we can optimize the check degree distribution for a fixed variable degree distribution. In order to find a good degree distribution pair (λ, ρ) we can proceed as follows. Start with a given degree distribution pair and iterate between these two optimization problems. In practice an even simpler procedure suffices. It is conjectured that *check-concentrated* degree distributions achieve optimal performance. In this case $\rho(x)$ is completely specified by fixing the average check-node degree $\mathtt{r}_{\text{avg}}$:

$$\rho(x) = \frac{\mathtt{r}(\mathtt{r}+1-\mathtt{r}_{\text{avg}})}{\mathtt{r}_{\text{avg}}} x^{\mathtt{r}-1} + \frac{\mathtt{r}_{\text{avg}} - \mathtt{r}(\mathtt{r}+1-\mathtt{r}_{\text{avg}})}{\mathtt{r}_{\text{avg}}} x^{\mathtt{r}},$$

where $\mathtt{r} = \lfloor \mathtt{r}_{\text{avg}} \rfloor$. Now run the linear program (3.98) for various values of $\mathtt{r}_{\text{avg}}$ until the optimum solution has been found.

EXAMPLE 3.99 (COMPARISON: OPTIMUM VERSUS CHECK-CONCENTRATED). Assume that we choose $\mathtt{l}_{\max} = 8$ and $\mathtt{r} = 6$, and that we want to obtain a design rate of one-half. Following the preceding procedure we find the pair

$$\lambda(x) = 0.409x + 0.202x^2 + 0.0768x^3 + 0.1971x^6 + 0.1151x^7, \quad \rho(x) = x^5,$$

which yields a rate of $r(\lambda, \rho) \approx 0.5004$. This degree distribution pair has a threshold of $\epsilon^{\text{BP}}(\lambda, \rho) \approx 0.4810$, which corresponds to a gap of $\delta \approx 0.0359$.

In comparison, a quick check shows that the check-concentrated degree distribution pair with equal complexity and comparable rate has parameters $\alpha = \frac{1}{5}$ and $N = 13$. We get

$$\begin{aligned} \lambda(x) &= 0.416x + 0.166x^2 + 0.1x^3 + 0.07x^4 + 0.053x^5 + 0.042x^6 + \\ &\quad 0.035x^7 + 0.03x^8 + 0.026x^9 + 0.023x^{10} + 0.02x^{11} + 0.0183x^{12}, \\ \rho(x) &= x^5. \end{aligned}$$

The design rate is $r(\lambda, \rho) \approx 0.499103$ and the threshold is $\epsilon^{\text{BP}}(\lambda, \rho) \approx 0.480896$ (it is determined by the stability condition), so that $\delta \approx 0.0385306$. The second degree distribution pair is slightly worse in all respects (higher maximum degree, smaller rate, smaller threshold, higher fraction of edges of degree 2). ◇

§3.19. PEELING DECODER AND ORDER OF LIMITS

Density evolution computes the limit

$$\lim_{\ell\to\infty} \lim_{n\to\infty} \mathbb{E}_{\text{LDPC}(n,\lambda,\rho)}[\mathrm{P}_{\mathrm{b}}^{\text{BP}}(\mathtt{G}, \epsilon, \ell)]; \tag{3.100}$$

i.e., we determined the limiting performance of an ensemble under a *fixed* number of iterations as the blocklength tends to infinity and then let the number of iterations tend to infinity. What happens if the order of limits is exchanged, i.e., how does the limit

$$\lim_{n\to\infty}\lim_{\ell\to\infty}\mathbb{E}_{\text{LDPC}(n,\lambda,\rho)}[\text{P}_\text{b}^{\text{BP}}(\text{G},\epsilon,\ell)] \tag{3.101}$$

behave? This limit corresponds to the more typical operation in practice: for each fixed length the BP decoder continues until no further progress is achieved. We are interested in the performance as the blocklength tends to infinity. In Section 3.22 we discuss the finite-length analysis of LDPC ensembles under BP decoding. We will see how we can compute the performance for a particular length (and an infinite number of iterations). The following example shows the typical behavior.

EXAMPLE 3.102 (CONVERGENCE TO THRESHOLD FOR LDPC (n, x^2, x^5)). Consider the ensemble LDPC (n, x^2, x^5). Figure 3.103 shows $\mathbb{E}_{\text{LDPC}(n,x^2,x^5)}[\text{P}_\text{b}^{\text{BP}}(\text{G},\epsilon,\ell=\infty)]$ as a function of ϵ for $n = 2^i$, $i = 6,\dots,20$. More precisely, we consider an *expurgated* ensemble as discussed in Section 3.24. In the limit of large blocklengths, the bit erasure probability converges to zero for $\epsilon < \epsilon^{\text{BP}} \approx 0.4294$ and to a non-zero constant for values above the threshold. In particular, the threshold for the limit (3.101) is the same as the threshold for the limit (3.100). ◇

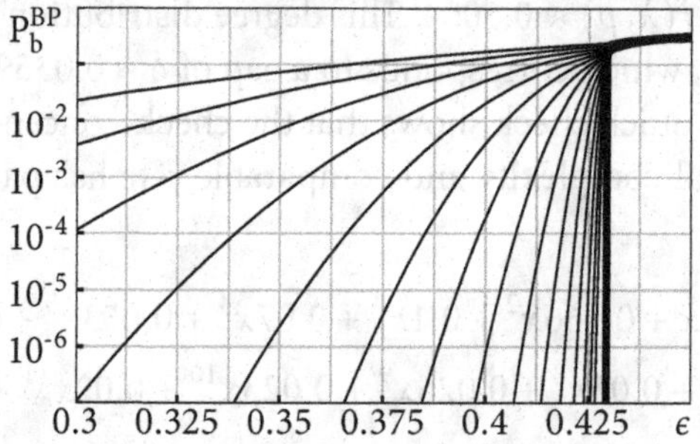

Figure 3.103: $\mathbb{E}_{\text{LDPC}(n,x^2,x^5)}[\text{P}_\text{b}^{\text{BP}}(\text{G},\epsilon,\ell=\infty)]$ as a function of ϵ for $n = 2^i$, $i = 6,\dots,20$. Also shown is the limit $\mathbb{E}_{\text{LDPC}(\infty,x^2,x^5)}[\text{P}_\text{b}^{\text{BP}}(\text{G},\epsilon,\ell\to\infty)]$, which is discussed in Problem 3.17 (thick curve).

Motivated by this example, we will show that the thresholds corresponding to the two limits are always the same. Assume this for the moment and consider its consequence: it follows that the threshold is the same regardless of how the limit is taken (sequentially or jointly) as long as both n and ℓ tend to infinity. To see this, let $\tilde{\ell}(n)$ be any increasing function in n so that $\lim_{n\to\infty}\tilde{\ell}(n) = \infty$. Then, for any fixed ℓ,

$$\lim_{n\to\infty}\mathbb{E}_{\text{LDPC}(n,\lambda,\rho)}[\text{P}_\text{b}^{\text{BP}}(\text{G},\epsilon,\tilde{\ell}(n))] \le \lim_{n\to\infty}\mathbb{E}_{\text{LDPC}(n,\lambda,\rho)}[\text{P}_\text{b}^{\text{BP}}(\text{G},\epsilon,\ell)],$$

because the error probability is a decreasing function in ℓ. Since this is true for any ℓ we can take the limit on the right-hand side to get

$$\lim_{n\to\infty} \mathbb{E}_{\mathrm{LDPC}(n,\lambda,\rho)}[\mathrm{P}_{\mathrm{b}}^{\mathrm{BP}}(\mathtt{G},\epsilon,\tilde{\ell}(n))] \leq \lim_{\ell\to\infty}\lim_{n\to\infty} \mathbb{E}_{\mathrm{LDPC}(n,\lambda,\rho)}[\mathrm{P}_{\mathrm{b}}^{\mathrm{BP}}(\mathtt{G},\epsilon,\ell)].$$

On the other hand, using again the monotonicity with respect to ℓ, we have for a fixed n

$$\lim_{\ell\to\infty} \mathbb{E}_{\mathrm{LDPC}(n,\lambda,\rho)}[\mathrm{P}_{\mathrm{b}}^{\mathrm{BP}}(\mathtt{G},\epsilon,\ell)] \leq \mathbb{E}_{\mathrm{LDPC}(n,\lambda,\rho)}[\mathrm{P}_{\mathrm{b}}^{\mathrm{BP}}(\mathtt{G},\epsilon,\tilde{\ell}(n))].$$

Taking the limit with respect to n we get

$$\lim_{n\to\infty}\lim_{\ell\to\infty} \mathbb{E}_{\mathrm{LDPC}(n,\lambda,\rho)}[\mathrm{P}_{\mathrm{b}}^{\mathrm{BP}}(\mathtt{G},\epsilon,\ell)] \leq \lim_{n\to\infty} \mathbb{E}_{\mathrm{LDPC}(n,\lambda,\rho)}[\mathrm{P}_{\mathrm{b}}^{\mathrm{BP}}(\mathtt{G},\epsilon,\tilde{\ell}(n))].$$

If we assume that

$$\lim_{n\to\infty}\lim_{\ell\to\infty} \mathbb{E}_{\mathrm{LDPC}(n,\lambda,\rho)}[\mathrm{P}_{\mathrm{b}}^{\mathrm{BP}}(\mathtt{G},\epsilon,\ell)] = \lim_{\ell\to\infty}\lim_{n\to\infty} \mathbb{E}_{\mathrm{LDPC}(n,\lambda,\rho)}[\mathrm{P}_{\mathrm{b}}^{\mathrm{BP}}(\mathtt{G},\epsilon,\ell)],$$

then it follows from the preceding two inequalities that

$$\lim_{n\to\infty} \mathbb{E}_{\mathrm{LDPC}(n,\lambda,\rho)}[\mathrm{P}_{\mathrm{b}}^{\mathrm{BP}}(\mathtt{G},\epsilon,\tilde{\ell}(n))]$$

also has the same limit.

The key to the analysis is the so-called *peeling* decoder. This decoder has identical performance to that of the message-passing decoder: if you take a sneak preview of Section 3.22 you will see that the message-passing decoder gets stuck in the largest "stopping set" that is contained in the set of erased bits. Such a stopping set is a subset of the variable nodes together with its outgoing edges so that no check node has induced degree 1. From the following description of the peeling decoder you will see that the peeling decoder gets stuck in exactly the same structure. The performance of the two decoders is therefore identical (assuming an infinite number of iterations).

Although the two decoders have identical performance, the computation rules of the peeling decoder differ from those of the message-passing one in two aspects: (i) at the variable nodes we do not obey the message-passing principle but we replace the received value with the current estimate of the bit based on *all* incoming messages and the received message; (ii) rather than updating all messages in parallel we pick in each step one check node and update its outgoing messages as well as the messages of its neighboring variable nodes.

In addition we can apply the following simplifications without changing the behavior of the algorithm: once a non-erasure message has been sent out along a check node this check node has served its purpose and it no longer plays a role in the future of the decoding. This is true since a check node sends out a non-erased message only if all but possibly one of its neighbors are known. Therefore, after processing this check node *all* its neighbors are known. It follows from this observation that we can safely delete from the graph any such check node and all its attached edges. In the same manner, each known variable node can send to its neighboring check node its value and these values are accumulated at the check node. After that we can remove the known variable node and its outgoing edges from the graph. This procedure gives rise to a sequence of *residual* graphs. Successful decoding is equivalent to the condition that the sequence of residual graphs reaches the empty graph.

Example 3.104 (Peeling Decoder Applied to $[7,4,3]$ Hamming Code). Rather than giving a formal definition of the decoder, let us apply the peeling decoder to the $[7,4,3]$ Hamming code with received word $(0,?,?,1,0,?,0)$. This is shown in Figure 3.105. The top leftmost picture shows the initial graph and the received word. The following two pictures are part of the initialization: (i) known variable nodes send their values along all outgoing edges; (ii) these values are accumulated at the check nodes (black indicates an accumulated value of 1 and white an accumulated value of 0) and all known variable nodes and their connected edges are removed. After the initialization each decoding step consists of the following: (i) choose a check node of residual degree 1 uniformly at random and forward the accumulated value to the connected variable node whose value is now determined, (ii) delete the chosen check node and its connected edge and forward the value of the newly determined variable node to all its remaining neighbors (check nodes), and (iii) accumulate the forwarded values at the check nodes and delete the variable node and its connected edges.

In our example at each step only a single check node of residual degree 1 exists. After three decoding steps the residual graph is the empty graph – the decoder has succeeded in determining the codeword. ◇

We will now see that if we apply the peeling decoder to elements of an ensemble and if we increase the blocklength, then the sequence of residual graphs closely follows a "typical path." We will describe this path, characterize the typical deviation from it, and relate this path to standard density evolution. As we discussed earlier, running the peeling decoder until it is stuck is equivalent to running the message-passing decoder for an infinite number of iterations. Therefore, if we can show that the peeling decoder has a threshold equal to the threshold ϵ^{BP} computed via density evolution, then we have in effect shown that the two limits (3.100) and (3.101) agree.

The proof of the following theorem is relegated to Section C.4.

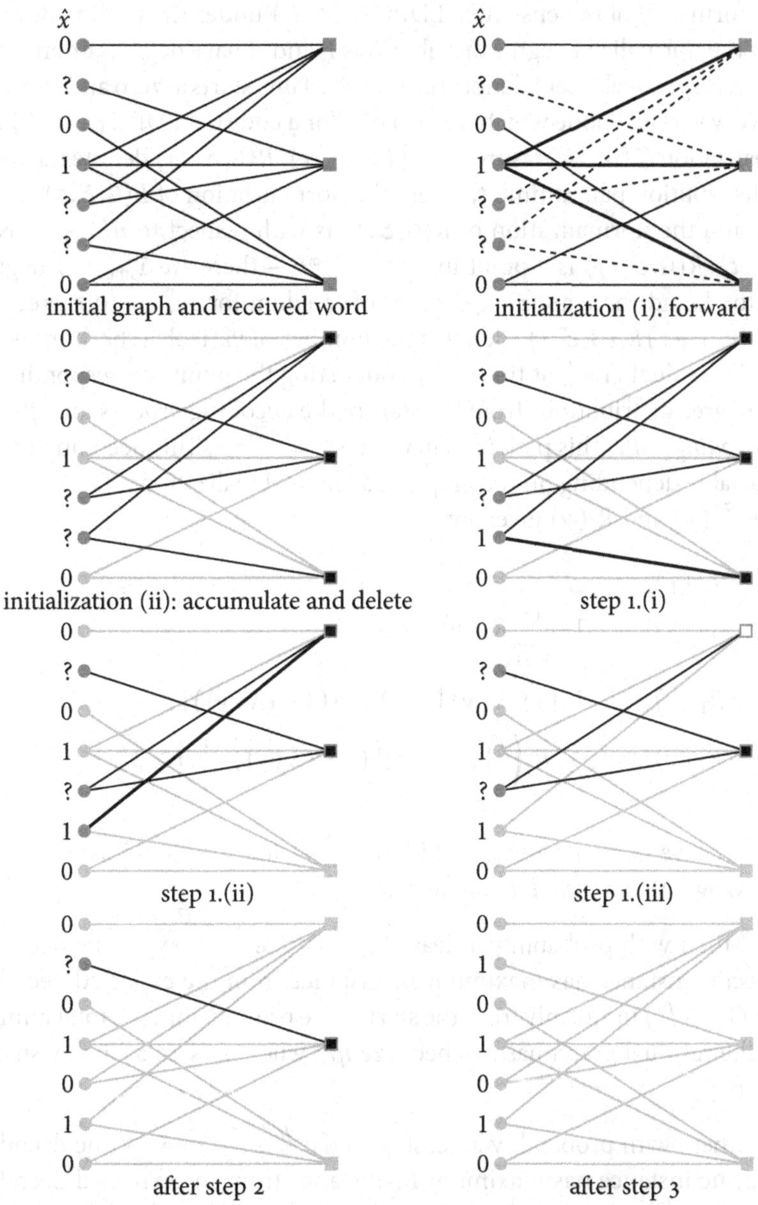

Figure 3.105: Peeling decoder applied to the $[7, 4, 3]$ Hamming code with the received word $y = (0, ?, ?, 1, 0, ?, 0)$. The vector $\hat{x}$ indicates the current estimate of the decoder of the transmitted codeword x. After three decoding steps the peeling decoder has successfully recovered the codeword.

Theorem 3.106 (Evolution of Residual Graph for Peeling Decoder). Consider the performance of the ensemble LDPC (n, L, R) under the peeling decoder. Assume that the normalized degree distributions L and R have degrees bounded by $\mathtt{l}_{\max}$ and $\mathtt{r}_{\max}$, respectively. Let t denote time, $t \in \mathbb{N}$. Time starts at zero and increases by one for every variable node which we peel off. For a code $\mathtt{G} \in$ LDPC (n, L, R) and a channel realization $\mathcal{E}$ (set of erasures), let $(L(\mathtt{G}, \mathcal{E}, t), R(\mathtt{G}, \mathcal{E}, t))$ denote the residual degree distribution pair at time t, where the normalization of $L(\mathtt{G}, \mathcal{E}, t)$ is with respect to n and the normalization of $R(\mathtt{G}, \mathcal{E}, t)$ is with respect to $n(1-r)$. Note that $(L(\mathtt{G}, \mathcal{E}, t), R(\mathtt{G}, \mathcal{E}, t))$ is a point in $\mathbb{R}^{\mathtt{l}_{\max}-1+\mathtt{r}_{\max}}$ – there are $\mathtt{l}_{\max} - 1$ degrees of freedom for L and there are $\mathtt{r}_{\max}$ degrees of freedom for $\mathtt{r}_{\max}$. More precisely, $nL_i(\mathtt{G}, \mathcal{E}, t)$ $(n(1-r)R_i(\mathtt{G}, \mathcal{E}, t))$ denotes the number of variable (check) nodes of degree i in the residual graph at time t. By connecting the points corresponding to the residual degree distributions from the start of the decoding process until its end we get the *decoding path*. This path is a curve in $\mathbb{R}^{\mathtt{l}_{\max}-1+\mathtt{r}_{\max}}$. This decoding path is a random variable depending on the graph and channel realization.

Consider $\tilde{L}_i(y)$ and $\tilde{R}_i(y)$ given by

$$\begin{aligned}
(3.107) \qquad \tilde{L}_i(y) &= \epsilon L_i y^i, \quad i \geq 2, \\
\tilde{R}_0(y) &= 1 - \sum_{j \geq 1} \tilde{R}_j(y), \\
(3.108) \qquad \tilde{R}_1(y) &= R'(1)\epsilon\lambda(y)\left[y - 1 + \rho(1 - \epsilon\lambda(y))\right], \\
(3.109) \qquad \tilde{R}_i(y) &= \sum_{j \geq 2} R_j \binom{j}{i} (\epsilon\lambda(y))^i (1 - \epsilon\lambda(y))^{j-i}, i \geq 2.
\end{aligned}$$

If we plot the curve corresponding to $(\tilde{L}(y), \tilde{R}(y))$ for $y \in [0, 1]$ as a curve in $\mathbb{R}^{\mathtt{l}_{\max}-1+\mathtt{r}_{\max}}$ we get the *expected decoding path*.

If $\epsilon < \epsilon^{\mathrm{BP}}$ then with probability at least $1 - O(n^{1/6} e^{-\frac{\sqrt{nL'(1)}}{(\mathtt{l}_{\max}\mathtt{r}_{\max})^3}})$ the decoding path of a specific instance has maximum L_1 distance from the expected decoding path at most $O(n^{-1/6})$ uniformly from the start of the process until the total number of nodes in the residual graph has reached size ηn, where η is an arbitrary strictly positive constant.

If $\epsilon > \epsilon^{\mathrm{BP}}$ then, with probability at least $1 - O(n^{1/6} e^{-\frac{\sqrt{nL'(1)}}{(\mathtt{l}_{\max}\mathtt{r}_{\max})^3}})$, the decoding path of a specific instance has maximum L_1-distance from the expected decoding path at most $O(n^{-1/6})$ uniformly from the start of the process until the decoder gets stuck.

Discussion: The actual decoding path is parameterized by t and the expected decoding path is parameterized by y. But the plots are parametric (neither t nor y appear) and both paths are curves in $\mathbb{R}^{\mathtt{l}_{\max}-1+\mathtt{r}_{\max}}$. It is sometimes convenient to plot the curves in a non-parametric form, as is done in Figure 3.111. In this case it is good to know the relationship between t and y: after t steps the expected number of remaining variable nodes is $n\epsilon - t$, since there are $n\epsilon$ variable nodes remaining at the start of the decoding process and we remove one at each step. From (3.107) we know that the expected number of remaining variables is also equal to $n\sum_i \tilde{L}_i(y) = n\epsilon L(y)$. It follows that $n\epsilon - t = n\epsilon L(y)$, which is equivalent to $t = n\epsilon(1 - L(y))$ or $y = L^{-1}(1 - \frac{t}{n\epsilon})$.

Example 3.110 (Evolution for (3,6)-Regular Ensemble). Assume that we apply the peeling decoder to the $(3,6)$-regular ensemble. What is the evolution of the degree distribution of the residual graph? From the description of the algorithm, a variable is either removed or it retains all of its edges. Therefore, all remaining variable nodes are of degree 3 for our example. According to Theorem 3.106, the fraction of remaining variables (with respect to n) as a function of the parameter y is ϵy^3. Here, $y = 1$ corresponds to the beginning of the process just after the initialization step.

The degree distribution on the check-node side is more interesting. Consider this degree distribution just after the initial step. Each edge of a check node is contained in the initial residual graph with probability ϵ. This means that, e.g., the number of check nodes of degree 1, should be $\sum_j R_j \binom{j}{1}\epsilon^1(1-\epsilon)^{j-1}$. We can rewrite this as $R'(1)\epsilon\rho(1-\epsilon)$, which agrees with the stated formula if we specialize it to $y = 1$. Figure 3.111 shows the evolution of the residual degrees $R_j(y)$, $j = 0, \ldots, 6$. ◇

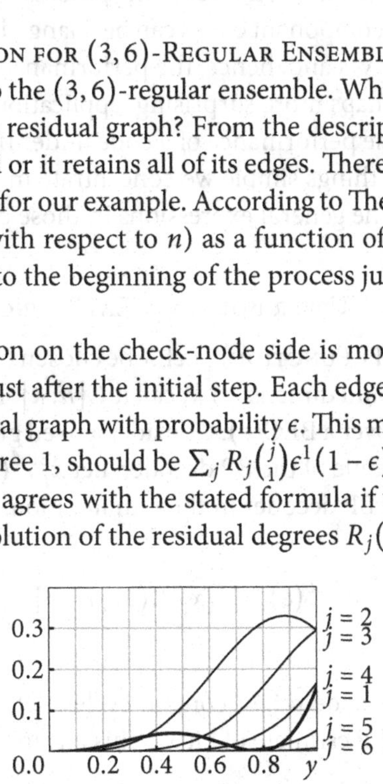

Figure 3.111: Evolution of the residual degrees $R_j(y)$, $j = 0, \ldots, 6$, as a function of the parameter y for the $(3,6)$-regular degree distribution. The channel parameter is $\epsilon = \epsilon^{\text{BP}} \approx 0.4294$. The curve corresponding to nodes of degree 1 is shown as a thick line.

From the stated expressions we can verify that the threshold under the peeling decoder is identical to the threshold under BP decoding. Consider the term $y - 1 + \rho(1 - \epsilon\lambda(y))$, which appears within the square brackets in the expression for $\tilde{R}_1(y)$. We recognize this as the density evolution equation in disguise: if we choose $\epsilon < \epsilon^{\text{BP}}$, then this expression is strictly positive over the whole range, implying that the fraction of nodes of degree 1 along the expected path stays strictly positive over the whole decoding range. But if we choose $\epsilon > \epsilon^{\text{BP}}$, then this fraction hits zero at some critical point. At this point the algorithm stops.

§3.20. EXIT Function and MAP Performance

In Section 3.14, EXIT functions appear as a handy tool to visualize the decoding process: from EXIT curves, such as the ones depicted in Figure 3.69, one can immediately read off the "bottlenecks" in the decoding process. Once these critical regions have been identified, the component codes can be changed appropriately to improve the "matching" of the curves and, hence, the performance of the system.

There is another, perhaps more surprising, application of EXIT functions: they can be used to connect the performance of a code under BP decoding to that under MAP decoding. To keep things simple we concentrate in the sequel on the regular case but we write down the general expressions in those cases where it requires no additional effort.

§3.20.1. Upper Bound on EXIT Function

Definition 3.112 (BP EXIT Function). Let X be chosen uniformly at random from a linear code of length n and dimension k, call it $C[n,k]$. Let Y denote the received word after transmission over a BEC(ϵ). Consider a fixed graphical representation of the code and a fixed schedule of the BP decoder. Let $\hat{x}_i^{\text{BP},\ell}(Y_{\sim i})$ denote the extrinsic estimate delivered by the BP decoder in the ℓ-th iteration. The BP EXIT function is defined as

$$h_i^{\text{BP},\ell}(\epsilon) = \mathbb{P}\{\hat{x}_i^{\text{BP},\ell}(Y_{\sim i}) = ?\}.$$

▽

Discussion: Since the code might contain cycles, set Y_i to be an erasure when computing the extrinsic BP estimate of bit x_i. This ensures that this estimate is only a function of $Y_{\sim i}$.

Lemma 3.113 (EXIT versus BP EXIT Function).

$$h_i(\epsilon) \leq h_i^{\text{BP},\ell}(\epsilon).$$

Proof. We have

$$h_i(\epsilon) = H(X_i \mid Y_{\sim i}) \overset{\text{(i)}}{\leq} H(X_i \mid \hat{x}_i^{\text{BP},\ell}(Y_{\sim i})) \overset{\text{(ii)}}{\leq} \mathbb{P}\{\hat{x}_i^{\text{BP},\ell}(Y_{\sim i}) = ?\} = h_i^{\text{BP},\ell}(\epsilon).$$

Step (i) is a direct consequence of (the data processing inequality) (1.48). Consider step (ii). We are given the BP estimate $\hat{x}_i^{\text{BP},\ell}(Y_{\sim i})$ and the structure of the code and we want to compute the entropy of X_i. If $\hat{x}_i^{\text{BP},\ell}(Y_{\sim i}) \neq ?$ then the entropy is zero. On the other hand, if $\hat{x}_i^{\text{BP},\ell}(Y_{\sim i}) = ?$, which happens with probability $\mathbb{P}\{\hat{x}_i^{\text{BP},\ell}(Y_{\sim i}) = ?\}$, then the entropy is at most 1. □

It would seem that not much more can be said about the relationship of the EXIT function and the BP EXIT function. But in the limit of large blocklengths, a fundamental connection between these two quantities appears. Therefore, we turn our attention to the MAP performance of long codes.

§3.20.2. Asymptotic BP and MAP EXIT Function

Definition 3.114 (EXIT Function of (λ, ρ)). The EXIT function associated with the degree distribution (λ, ρ) is defined as

$$h(\epsilon) = \limsup_{n \to \infty} \mathbb{E}_{\text{LDPC}(n,\lambda,\rho)}\Big[\frac{1}{n}\sum_{i=1}^{n} h_{\text{G},i}(\epsilon)\Big],$$

where G denotes an element taken uniformly at random from LDPC (n, λ, ρ). ▽

Discussion: We used the lim sup instead of the ordinary limit. Toward the end of this section we will have proved that, at least for the regular case, the ordinary limit exists. If you prefer, simply ignore this technicality and think of the standard limit.

Definition 3.115 (BP EXIT Function of (λ, ρ) for the BEC(ϵ)). The BP EXIT function associated with the degree distribution pair (λ, ρ) is defined as

$$h^{\text{BP}}(\epsilon) = \lim_{\ell \to \infty} \lim_{n \to \infty} \mathbb{E}_{\text{LDPC}(n,\lambda,\rho)}\Big[\frac{1}{n}\sum_{i=1}^{n} h_{\text{G},i}^{\text{BP},\ell}(\epsilon)\Big].$$ ▽

Contrary to $h(\epsilon)$, $h^{\text{BP}}(\epsilon)$ can be computed easily.

Lemma 3.116 (BP EXIT Function for Regular Ensembles). Consider the regular degree distribution pair $(\lambda(x) = x^{\mathtt{l}-1}, \rho(x) = x^{\mathtt{r}-1})$. Then the BP EXIT function is given in parametric form by

$$h^{\text{BP}}(\epsilon) = \begin{cases} (\epsilon, 0), & \epsilon \in [0, \epsilon^{\text{BP}}), \\ (\epsilon(x), L(1-\rho(1-x))), & \epsilon \in (\epsilon^{\text{BP}}, 1] \leftrightarrow x \in (x^{\text{BP}}, 1], \end{cases}$$

where $\epsilon(x) = \frac{x}{\lambda(1-\rho(1-x))}$, and where x^{BP} denotes the location of the unique minimum of $\epsilon(x)$ in the range $(0, 1]$ and $\epsilon^{\text{BP}} = \epsilon(x^{\text{BP}})$.

Proof. Suppose that we first let the blocklength tend to infinity and then let the number of iterations grow. We know from Section 3.12 that in this case the erasure probability emitted by the variable nodes, call it x, tends to a limit and that this limit is a fixed point of the density evolution equation $\epsilon\lambda(1-\rho(1-x)) = x$. Solving this fixed point equation for ϵ, we get $\epsilon(x) = \frac{x}{\lambda(1-\rho(1-x))}$, $x \in (0, 1]$. In words, for each non-zero fixed point x of density evolution, there is a unique channel parameter ϵ. If at the fixed point the erasure probability emitted by the variable nodes is x, then the extrinsic erasure probability of the decision equals $L(1-\rho(1-x))$. This is also the value of the BP EXIT function at this point. That $\epsilon(x)$ has indeed a unique minimum as claimed in the lemma and that this minimum determines the threshold ϵ^{BP} is the topic of Problem 3.14. □

Example 3.117 (BP EXIT Function). The BP EXIT function $h^{\text{BP}}(\epsilon)$ for the $(\mathtt{l}=3, \mathtt{r}=6)$-regular case is shown on the left-hand side of Figure 3.118. By explicit com-

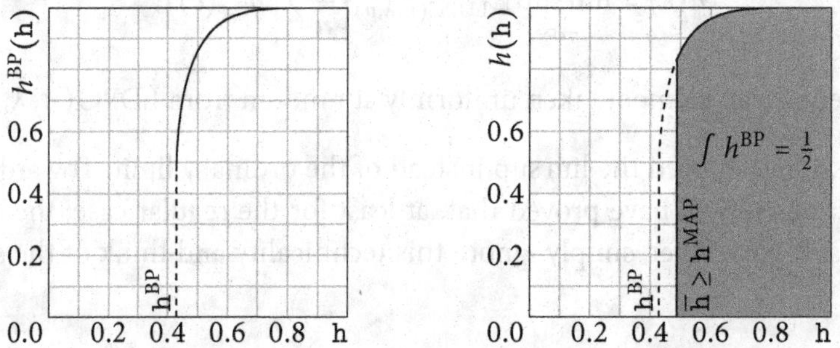

Figure 3.118: Left: BP EXIT function $h^{\text{BP}}(\epsilon)$; Right: Corresponding EXIT function $h(\epsilon)$ constructed according to Theorem 3.120.

putation we see that the unique minimum of $\epsilon(x)$ appears at $x^{\text{BP}} \approx 0.2605710$ and, therefore, $\epsilon^{\text{BP}} = \epsilon(x^{\text{BP}}) \approx 0.429439814$. As predicted by Lemma 3.116, $h^{\text{BP}}(\epsilon)$ is zero for $\epsilon \in [0, \epsilon^{\text{BP}})$. At $\epsilon = \epsilon^{\text{BP}}$ it jumps to the value $L(1-\rho(1-x^{\text{BP}})) \approx 0.472646$. Finally, to the right of ϵ^{BP}, both $\epsilon(x)$ and $L(1-\rho(1-x))$ are increasing and continuous, and so $h^{\text{BP}}(\epsilon)$ increases smoothly until it reaches 1 at $\epsilon = 1$. ◇

The integral under the curve $(\epsilon(x), L(1-\rho(1-x)))$ appears frequently in what follows. It is therefore handy to compute it once and for all and to give it a name. We call this integral the *trial entropy*, a choice that will hopefully become clear after we have stated and proved Theorem 3.120.

Definition 3.119 (Trial Entropy). Consider a degree distribution pair (λ, ρ) and define $\epsilon(x) = \frac{x}{\lambda(1-\rho(1-x))}$. The associated *trial entropy* $P(x)$ is defined as

$$P(x) = \int_0^x L(1-\rho(1-z))\epsilon'(z)\text{d}z$$
$$= \epsilon(x)L(1-\rho(1-x)) + L'(1)x\rho(1-x) - \frac{L'(1)}{R'(1)}(1-R(1-x)),$$

where the second line follows from the first one by applying integration by parts twice. In the case of $(\mathtt{l}, \mathtt{r})$-regular degree distributions we get

$$P(x) = x + \frac{1}{\mathtt{r}}(1-x)^{\mathtt{r}-1}(1 + \mathtt{l}(\mathtt{r}-1)x - \mathtt{r}x) - \frac{\mathtt{l}}{\mathtt{r}}. \qquad \triangledown$$

Here is the punch line: the EXIT function can be constructed from the BP EXIT function as shown on the right-hand side of Figure 3.118. This explains why we called $P(x)$ the trial entropy.

Theorem 3.120 (EXIT Function for Regular Degree Distributions). Consider the $(\mathtt{l}, \mathtt{r})$-regular degree distribution, let $P(x)$ denote the associated trial entropy, and define $\epsilon(x) = x/\lambda(1-\rho(1-x))$. Let x^{MAP} be the unique positive solution of $P(x) = 0$ and define $\epsilon^{\text{MAP}} = \epsilon(x^{\text{MAP}})$. Then

$$h(\epsilon) = \begin{cases} 0, & \epsilon \in [0, \epsilon^{\text{MAP}}), \\ h^{\text{BP}}(\epsilon), & \epsilon \in (\epsilon^{\text{MAP}}, 1], \end{cases}$$

and for $\epsilon^{\text{MAP}} \leq \epsilon \leq 1$

$$\lim_{n\to\infty} \mathbb{E}_{\text{G}}[H_{\text{G}}(X \mid Y(\epsilon))/n] = \int_0^{\epsilon} h(\epsilon')\text{d}\epsilon' = P(x(\epsilon)),$$

where $x(\epsilon)$ is the largest solution of the equation $\epsilon(x) = \epsilon$.

Discussion: In words, the theorem states that $\lim_{n\to\infty} \mathbb{E}_{\text{G}}[H_{\text{G}}(X \mid Y(\epsilon))/n]$ exists and that the quantity ϵ^{MAP} defined earlier is the MAP threshold. In the limit of infinite blocklengths, the average conditional entropy converges to zero for $\epsilon < \epsilon^{\text{MAP}}$ and it is non-zero for $\epsilon > \epsilon^{\text{MAP}}$. Further, for $\epsilon > \epsilon^{\text{MAP}}$, the EXIT function $h(\epsilon)$ coincides with the BP EXIT function $h^{\text{BP}}(\epsilon)$ and $\lim_{n\to\infty} \mathbb{E}_{\text{G}}[H_{\text{G}}(X \mid Y(\epsilon))/n]$ is equal to the integral of $h(\epsilon)$, which in turn is equal to $P(x(\epsilon))$. This is the reason we called P the trial *entropy*.

Example 3.121 (EXIT Curve for $(\mathtt{l} = 3, \mathtt{r} = 6)$-Regular Ensemble). The right-hand side of Figure 3.118 shows the application of Theorem 3.120 to the $(3, 6)$-regular degree distribution. The result is $x^{\text{MAP}} \approx 0.432263$ and $\epsilon^{\text{MAP}} = \epsilon(x^{\text{MAP}}) \approx 0.488151$. $\diamond$

Proof of Theorem 3.120. We prove the theorem by establishing the following four claims:

$$\text{(i)}\ h(\epsilon) = 0, \epsilon \in [0, \epsilon^{\text{MAP}}), \qquad \text{(ii)}\ h(\epsilon) \leq h^{\text{BP}}(\epsilon), \epsilon \in [0, 1],$$

$$\text{(iii)} \int_0^1 h(\epsilon)\text{d}\epsilon \geq r(\lambda, \rho), \qquad \text{(iv)} \int_{\epsilon^{\text{MAP}}}^1 h^{\text{BP}}(\epsilon)\text{d}\epsilon = r(\lambda, \rho).$$

Let us first see how the theorem follows from these claims. From (i) we see that $h(\epsilon)$ is zero for $\epsilon \in [0, \epsilon^{\text{MAP}})$. Consider $h(\epsilon)$ for $\epsilon > \epsilon^{\text{MAP}}$. We have

$$\int_{\epsilon^{\text{MAP}}}^{1} h(\epsilon)\text{d}\epsilon \overset{\text{(i)}}{=} \int_{0}^{1} h(\epsilon)\text{d}\epsilon \overset{\text{(iii)}}{\geq} r(\lambda,\rho) \overset{\text{(iv)}}{=} \int_{\epsilon^{\text{MAP}}}^{1} h^{\text{BP}}(\epsilon)\text{d}\epsilon.$$

But since on the other hand $h(\epsilon) \overset{\text{(ii)}}{\leq} h^{\text{BP}}(\epsilon)$ for $\epsilon \in (\epsilon^{\text{MAP}}, 1]$, it must be true that $h(\epsilon) = h^{\text{BP}}(\epsilon)$ for almost all $\epsilon \in (\epsilon^{\text{MAP}}, 1]$. Since further $h(\epsilon)$ is a monotone function and $h^{\text{BP}}(\epsilon)$ is continuous we have equality everywhere. From this argument we know that $r(\lambda,\rho) = \int_0^1 h(\epsilon)\text{d}\epsilon$. Using Lemma 3.27, which asserts that for regular ensembles the actual rate converges to the design rate, we can rewrite this as

$$\limsup_{n\to\infty} \int_0^1 \mathbb{E}_{\text{G}}[h_{\text{G}}(\epsilon)]\text{d}\epsilon = \int_0^1 \limsup_{n\to\infty} \mathbb{E}_{\text{G}}[h_{\text{G}}(\epsilon)]\text{d}\epsilon \overset{\text{Def. 3.114}}{=} \int_0^1 h(\epsilon)\text{d}\epsilon.$$

In fact, we can conclude the more general equality

$$\limsup_{n\to\infty} \int_E \mathbb{E}_{\text{G}}[h_{\text{G}}(\epsilon)]\text{d}\epsilon = \int_E h(\epsilon)\text{d}\epsilon,$$

where E denotes any subset of $[0,1]$. This is true since the left-hand side is always upper bounded by the right-hand side (Fatou-Lebesgue). Since we have equality on $[0,1]$, we must have equality for all subsets E. Therefore,

$$\limsup_{n\to\infty} \int_0^{\epsilon} \mathbb{E}_{\text{G}}[h_{\text{G}}(\epsilon')]\text{d}\epsilon' = \int_0^{\epsilon} h(\epsilon')\text{d}\epsilon' = r(\lambda,\rho) - \int_{\epsilon}^{1} h(\epsilon')\text{d}\epsilon'.$$

But conversely, we have

$$\begin{aligned}\liminf_{n\to\infty} \int_0^{\epsilon} \mathbb{E}_{\text{G}}[h_{\text{G}}(\epsilon')]\text{d}\epsilon' &= \liminf_{n\to\infty} \int_0^{1} \mathbb{E}_{\text{G}}[h_{\text{G}}(\epsilon')]\text{d}\epsilon' - \limsup_{n\to\infty} \int_{\epsilon}^{1} \mathbb{E}_{\text{G}}[h_{\text{G}}(\epsilon')]\text{d}\epsilon' \\ &= r(\lambda,\rho) - \limsup_{n\to\infty} \int_{\epsilon}^{1} \mathbb{E}_{\text{G}}[h_{\text{G}}(\epsilon')]\text{d}\epsilon' \\ &= r(\lambda,\rho) - \int_{\epsilon}^{1} h(\epsilon')\text{d}\epsilon'.\end{aligned}$$

Comparing the last two expressions we conclude that the limit exists and is given in terms of the integral of $h(\epsilon)$. That the latter integral is equal to $P(x(\epsilon))$ follows from the definition of the trial entropy, the fact that $P(x^{\text{MAP}}) = 0$, and that, for $x > x^{\text{MAP}}$, $h = h^{\text{BP}}$.

We prove the four claims in order and start with (i), which is the most difficult one. Let ϵ denote the channel parameter and let x denote the corresponding fixed point of density evolution, i.e., the largest solution of the equation $\epsilon\lambda(1-\rho(1-x)) =$

x. Further, define $y = 1 - \rho(1-x)$. Assume we use the peeling decoder discussed in Section 3.19. At the fixed point the expected degree distribution of the residual graph, call it $(\tilde{L}(z), \tilde{R}(z))$, has the form

$$\tilde{L}(z) = \frac{L(zy)}{L(y)}, \tag{3.122}$$

$$\tilde{R}(z) = \frac{R(1-x+zx) - R(1-x) - zxR'(1-x)}{1 - R(1-x) - xR'(1-x)}. \tag{3.123}$$

First, let us show that, at $\epsilon = \epsilon^{\text{MAP}}$, $r(\tilde{L}, \tilde{R}) = 0$, i.e., the design rate of the residual ensemble is zero. This means that, for this parameter, the residual graph has in expectation the same number of variable nodes as check nodes. Write $r(\tilde{L}, \tilde{R}) = 1 - \frac{\tilde{L}'(1)}{\tilde{R}'(1)} = 0$ in the form $\tilde{R}'(1) = \tilde{L}'(1)$. If we express the latter condition explicitly using the definitions of $\tilde{L}$ and $\tilde{R}$ shown as follows as well as the mentioned relationships between ϵ, x, and y, we find after a few steps of calculus the equivalent condition $P(x) = 0$ (where $P(x)$ is the trial entropy of Definition 3.119). Now note that $q(x) = P(1-x) = (\mathtt{rl} - \mathtt{r} - 1)x^{\mathtt{r}} - \mathtt{r}(\mathtt{l}-1)x^{\mathtt{r}-1} + \mathtt{r}x - (\mathtt{r}-1)$ has three sign changes and therefore by Descarte's rule of signs (see Theorem 1.52) the polynomial equation $q(x) = 0$ has either one or three non-negative real solutions. If $\mathtt{l} > 2$, it is easy to check that $q(1) = q'(1) = 0$, i.e., $q(x)$ has a double root at $x = 0$, and since $q(0) = \mathtt{r} - 1 > 0$, $q''(1) = -(\mathtt{l}-2)(\mathtt{r}-1)\mathtt{r} < 0$, and $\lim_{x\to\infty} q(x) = -\infty$, there must be exactly one root of $q(x)$ for $x \in (0,1)$ and so exactly one root of $P(x)$ for $x \in (0,1)$. If $\mathtt{l} = 2$, then $q(x)$ has a triple root at $x = 1$ and $\epsilon^{\text{MAP}} \le \frac{1}{\mathtt{r}-1}$.

It remains to be shown that the actual rate indeed is zero, i.e., that (up to a sublinear number) all the check-node equations are linearly independent with high probability. This will settle the claim since this implies that in the limit of infinite blocklengths the normalized (by n) conditional entropy is zero. We use the technique introduced in Lemma 3.22 and show that

$$\Psi(u) = \log\Big(\frac{1}{2}(1+u^{\mathtt{l}})^{\mathtt{l}-1}\prod_{i=2}^{\mathtt{r}}((1+u^{\mathtt{l}-1})^i + (1-u^{\mathtt{l}-1})^i)^{\tilde{R}_i}\Big) \le 0, \tag{3.124}$$

for $u \in [0,1]$, where we have equality only at $u = 0$ and $u = 1$. To show that this inequality indeed holds does not require any sophisticated math but the proof is lengthy. We therefore relegate it to Problem 3.33.

Let us prove (ii). Using the upper bound discussed in Lemma 3.113, we know that for any $\mathtt{G} \in \text{LDPC}(n, \lambda, \rho)$ and $\ell \in \mathbb{N}$

$$\frac{1}{n}\sum_{i=1}^{n} h_{\mathtt{G},i}(\epsilon) \le \frac{1}{n}\sum_{i=1}^{n} h_{\mathtt{G},i}^{\text{BP},\ell}(\epsilon).$$

If we take first the expectation over the elements of the ensemble, then the lim sup on both sides with respect to n, and finally the limit $\ell \to \infty$, we get the desired result.

Consider (iii). By (the Area) Theorem 3.81 we have for any $\mathrm{G} \in \mathrm{LDPC}(n, \lambda, \rho)$

$$r(\mathrm{G}) = \int_0^1 \frac{1}{n} \sum_{i=1}^n h_{\mathrm{G},i}(\epsilon) \mathrm{d}\epsilon.$$

If we take the expectation over the elements of the ensemble and the limit as n tends to infinity, then by Lemma 3.27 the left-hand side converges to the design rate $r(\lambda, \rho)$. Therefore, taking into account that the integrand is non-negative and bounded (by the value 1), we have

$$
\begin{aligned}
&\text{(Lemma 3.27)} && r(\lambda, \rho) = \lim_{n \to \infty} \mathbb{E}\Big[\int_0^1 \frac{1}{n} \sum_{i=1}^n h_{\mathrm{G},i}(\epsilon) \mathrm{d}\epsilon\Big] \\
&\text{(Fubini)} && = \lim_{n \to \infty} \int_0^1 \mathbb{E}\Big[\frac{1}{n} \sum_{i=1}^n h_{\mathrm{G},i}(\epsilon)\Big] \mathrm{d}\epsilon \\
&\text{(Fatou-Lebesgue)} && \leq \int_0^1 \limsup_{n \to \infty} \mathbb{E}\Big[\frac{1}{n} \sum_{i=1}^n h_{\mathrm{G},i}(\epsilon)\Big] \mathrm{d}\epsilon \\
&\text{(Definition 3.114)} && = \int_0^1 h(\epsilon) \mathrm{d}\epsilon.
\end{aligned}
$$

To see (iv) let us determine that number ϵ^{MAP} so that $r(\lambda, \rho) = \int_{\epsilon^{\mathrm{MAP}}}^1 h^{\mathrm{BP}}(\epsilon) \mathrm{d}\epsilon$. We see from Lemma 3.116 and Definition 3.119 that

$$\int_{\epsilon^{\mathrm{MAP}}}^1 h^{\mathrm{BP}}(\epsilon) \mathrm{d}\epsilon = P(1) - P(x^{\mathrm{MAP}}) = P(1),$$

where the last step follows since we have already seen that $P(x^{\mathrm{MAP}}) = 0$. If we specialize the expression $P(x)$ for the regular case given in Definition 3.119 we readily see that $P(1) = r(\lambda, \rho)$. □

§3.20.3. Maxwell Construction

It is surprising that (at least in the regular case) the MAP performance can be derived directly from the performance of the BP decoder. So far this connection is seen through a sequence of lemmas. Let us give a more direct operational interpretation of this connection. The central character in this section is the *extended* BP (EBP) EXIT curve.

Definition 3.125 (Extended BP EXIT Curve). For a given a degree distribution pair (λ, ρ) define $\epsilon(x) = x / \lambda(1 - \rho(1 - x))$. Then the EBP EXIT curve is defined as

$$h^{\mathrm{EBP}} = (\epsilon(x), L(1 - \rho(1 - x))), \quad x \in [0, 1]. \qquad \triangledown$$

We know from Theorem 3.120 that, for regular ensembles and $x > x^{\text{BP}}$, $h^{\text{BP}} = h^{\text{EBP}}$. But h^{EBP} also contains a second "spurious" branch ($x < x^{\text{BP}}$). This branch corresponds to *unstable* fixed points of density evolution.

Example 3.126 (EBP EXIT Curve for $(\mathtt{l} = 3, \mathtt{r} = 6)$-Regular Ensemble). The EBP EXIT curve is shown in Figure 3.127. For small values of x, the EBP curve goes

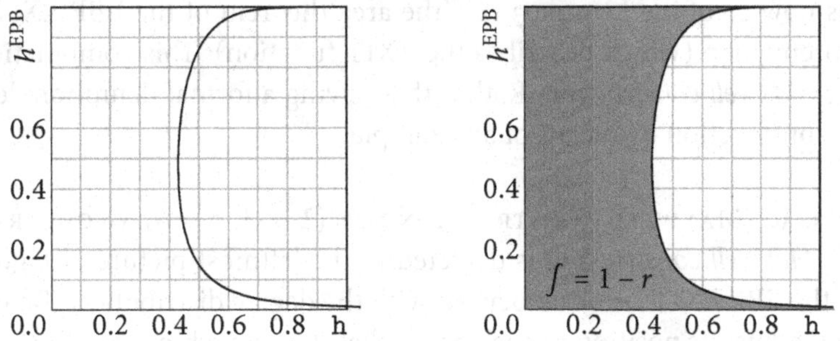

Figure 3.127: Left: EBP EXIT curve of the $(\mathtt{l} = 3, \mathtt{r} = 6)$-regular ensemble. Note that the curve goes "outside the box" and tends to infinity. Right: According to Lemma 3.128 the gray area is equal to $1 - r(\mathtt{l}, \mathtt{r}) = \frac{\mathtt{l}}{\mathtt{r}} = \frac{1}{2}$.

"outside the box." This is a consequence of $\lambda'(0)\rho'(1) = 0 < 1$: for small values of x we have $\epsilon\lambda(1 - \rho(1 - x)) = \epsilon\lambda'(0)\rho'(1)x + O(x^2) = O(x^2)$. Therefore, $\epsilon(x) \overset{x\to 0}{\to} 1/(\lambda'(0)\rho'(1)) = \infty$. But in general, even for ensembles for which $\lambda'(0)\rho'(1) > 1$, part of the EBP curve might have "ϵ" coordinates larger than 1. One such example is discussed in Problem 3.39. ◇

The EBP EXIT curve has its own area theorem. A priori, this area theorem has no connection to (the Area) Theorem 3.81 – after all, the EBP EXIT curve is defined in terms of the (in general) suboptimal BP decoder, whereas the EXIT curve to which Theorem 3.81 applies concerns optimal (MAP) decoding. That there is nevertheless a connection between the two is discussed in Problem 3.81.

Lemma 3.128 (Area Theorem for EBP EXIT Curve). Assume that we are given a degree distribution pair (λ, ρ) of design rate r. Then the area under the EBP EXIT curve satisfies

$$\int_0^1 h^{\text{EBP}}(x)\text{d}\epsilon(x) = r(\lambda, \rho).$$

Proof. According to Definitions 3.119 and 3.125 we have $\int_0^1 h^{\text{EBP}}(x)\text{d}\epsilon(x) = P(1) = r(\lambda, \rho)$, where the last step follows from a direct computation, which we have already discussed in the proof of Theorem 3.120. □

Example 3.129 (Area of EBP EXIT Curve for ($\mathtt{l} = 3, \mathtt{r} = 6$)-Regular Ensemble). The right-hand side picture of Figure 3.127 shows the area of the EBP EXIT curve for this case. Since part of the EPB EXIT curve lies outside the unit box it is slightly more convenient to regard the complement of this area which is shown in gray. As predicted by Lemma 3.128, the gray area is equal to $1 - r(\mathtt{l}, \mathtt{r}) = \frac{1}{\mathtt{r}} = \frac{1}{2}$. ◇

Let us now combine Lemma 3.128 (the area theorem of the EBP EXIT curve) with Theorem 3.120 (which describes the EXIT function). This combination gives rise to the *Maxwell construction*. Rather than giving a formal definition, let us explain this construction by means of an example.

Example 3.130 (Maxwell Construction for ($\mathtt{l} = 3, \mathtt{r} = 6$)-Regular Ensemble). The *Maxwell construction* is depicted in the leftmost picture of Figure 3.131. Consider the EBP EXIT curve associated with the degree distribution. Take a vertical line and adjust its position in such a way that the area which is to the left of this line and bounded to the left by the EBP EXIT curve is equal to the area which is to the right of this line and bounded above by the EBP EXIT curve. These two areas are shown in dark gray in the leftmost picture. The claim is that the unique such location of the vertical line is at $\epsilon = \epsilon^{\text{MAP}}$. In fact, some thought shows that this is a straightforward consequence of Theorem 3.120 and Lemma 3.128.

Instead of looking at the balance of the two dark gray areas shown in the leftmost picture we can consider the balance of the two dark gray areas shown in the middle and the rightmost picture. These two areas differ only by a constant from the previous such areas. In the next section we give an *operational* interpretation of the latter two areas in terms of the so-called *Maxwell* decoder. ◇

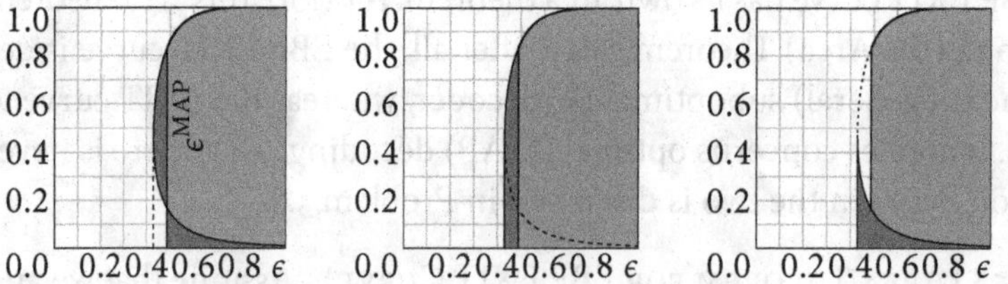

Figure 3.131: Left: Because of Theorem 3.120 and Lemma 3.128, at the MAP threshold ϵ^{MAP} the two dark gray areas are in balance. Middle: The dark gray area is proportional to the total number of variables which the M decoder introduces. Right: The dark gray area is proportional to the total number of equations which are produced during the decoding process and which are used to resolve variables.

§3.21. Maxwell Decoder

Let us define the *Maxwell* (M) *decoder*: Given the received word, which was transmitted over the BEC(ϵ), the M decoder proceeds like the standard peeling decoder described in Section 3.19. At each time step a parity-check equation involving a single undetermined variable is chosen and used to determine the value of the variable. This value is substituted in any parity-check equation involving the same variable. If at any time the peeling decoder gets stuck in a non-empty stopping set, a position $i \in [n]$ is chosen uniformly at random from the set of yet undetermined bits and a binary (symbolic) variable v_i representing the value of bit i is associated with this position. In what follows, the decoder proceeds as if position i was known and whenever the value of bit i is called for it is referred to as v_i. This means that messages consist not only of numbers 0 or 1 but in general contain (combinations of) symbolic variables v_i. In other words, the messages are really equations that state how some quantities can be expressed in terms of other quantities. It can happen that during the decoding process of the peeling decoder a yet undetermined variable is connected to several nodes of degree 1. It will then receive a message describing its value from each of these connected check nodes of degree 1. Of course, all these messages describe the *same* value (recall that, over the BEC, no errors occur). Therefore, if and only if at least one of these messages contains a symbolic variable, then the condition that all these messages describe the same value gives rise to linear equations which have to be fulfilled. Whenever this happens, the decoder resolves this set of equations with respect to some of the previously introduced variables v_i and eliminates those resolved variables in the whole system. The decoding process finishes once the residual graph is empty. By definition of the process, the decoder always terminates.

At this point there are two possibilities. The first is that all introduced variables $\{v_i\}_{i \in \mathcal{I}}, \mathcal{I} \subseteq [n]$, were resolved at some later stage of the decoding process (a special case of this being that no such variables ever had to be introduced). In this case each bit has an associated value (either 0 or 1) and this is the only solution compatible with the received information). In other words, the decoded word is the MAP estimate. The other possibility is that there are some undetermined variables $\{v_i\}_{i \in \mathcal{I}}$ remaining. In this case each variable node either already has a specific value (0 or 1) or by definition of the decoder can be expressed as a linear combination of the variables $\{v_i\}_{i \in \mathcal{I}}$. In such a case each realization (choice) of $\{v_i\}_{i \in \mathcal{I}} \in \{0,1\}^{|\mathcal{I}|}$ gives rise to a valid codeword and all codewords compatible with the received information are the result of a particular such choice. In other words, we have accomplished a complete list decoding, so that $|\mathcal{I}|$ equals the conditional entropy $H(X \mid Y)$. All this is probably best understood by an example.

Example 3.132 (M Decoder Applied to $(\mathtt{l} = 3, \mathtt{r} = 6)$ Code). Figure 3.133 shows

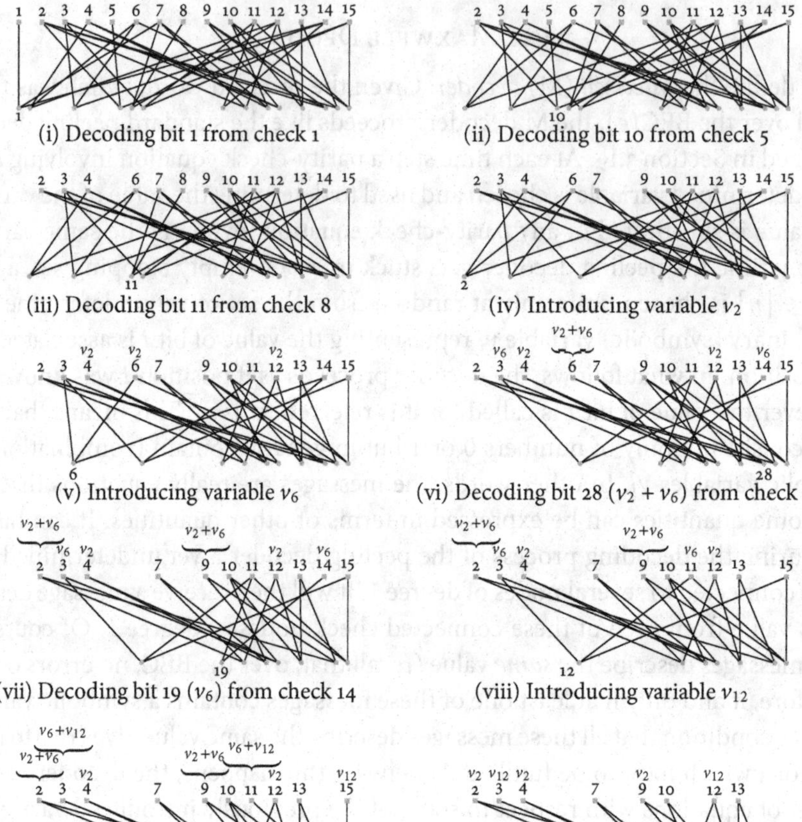

(i) Decoding bit 1 from check 1

(ii) Decoding bit 10 from check 5

(iii) Decoding bit 11 from check 8

(iv) Introducing variable v_2

(v) Introducing variable v_6

(vi) Decoding bit 28 ($v_2 + v_6$) from check 6

(vii) Decoding bit 19 (v_6) from check 14

(viii) Introducing variable v_{12}

(ix) Decoding bit 30 ($v_6 + v_{12} = v_{12}$) from checks 11 and 15 $\longrightarrow v_6 = 0$

(x) Decoding bit 24 (v_{12}) from check 3

(xi) Decoding bit 23 ($v_2 + v_{12}$) from check 4

(xii) Decoding bit 21 ($v_2 + v_{12}$) from check 7

(xiii) Decoding bit 29 ($v_{12} = v_2 + v_{12}$) from checks 2 and 9 $\longrightarrow v_2 = 0$

(xiv) Decoding bit 26 ($v_{12} = 0$) from checks 10, 12, and 13 $\longrightarrow v_{12} = 0$

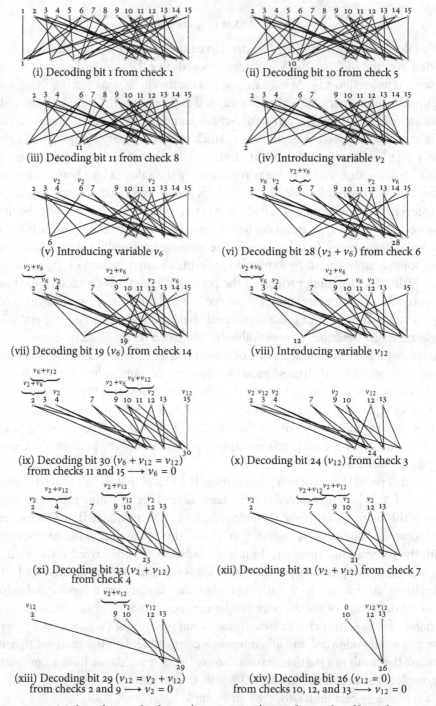

Figure 3.133: M decoder applied to a $(\mathtt{l} = 3, \mathtt{r} = 6)$-regular code of length $n = 30$.

the workings of the M decoder applied to a simple code of length $n = 30$. Assume that the all-zero codeword has been transmitted. In the initial decoding step the received (i.e., known and equal to 0) bits are removed from the bipartite graph. The remaining graph is shown in (i). The first phase is equivalent to the standard peeling algorithm: in the first three steps, the decoder determines the bits 1 (from check node 1), 10 (from check node 5), and 11 (from check node 8). At this point the peeling decoder is stuck in the constellation shown in (iv). The second phase is distinct to the M decoder: the decoder assigns the variable v_2 to the (randomly chosen) bit 2, which is now considered to be *known*. The decoder then proceeds again as the standard peeling algorithm. Any time it gets stuck, it assigns a new variable v_i to a yet undetermined and randomly chosen position i. This process continues until some of the previously introduced variables can be eliminated. For example, consider step (ix): the variable node at position 30 receives the messages $v_6 + v_{12}$ as well as the message v_{12}. This means that the decoder has deduced from the received word that the only compatible codewords are the ones for which the value of bit 30 is equal to the value of bit 12 and also equal to the sum of the values of bit 6 and bit 12. The decoder can now deduce from this that $v_6 = 0$, i.e., the previously introduced variable v_6 is eliminated from the system. Phases in which new variables are introduced, phases during which some previously introduced variables are resolved, and regular BP decoding phases might alternate. Decoding is successful (in the sense that a MAP decoder would have succeeded) if at the end of the decoding process, all introduced variables have been resolved. This is the case for the shown example. ◇

The following lemma, whose proof we skip, explains why at the MAP threshold the two dark gray areas are in balance. In short, the area on the left is proportional to the total number of variables which the decoder introduces and the area on the right is proportional to the total number of equations which are generated and which are used to resolve those variables. Further, as long as the number of generated equations is no larger than the number of introduced variables then these equations are linearly independent with high probability. Therefore, when these two areas are equal then the number of unresolved variables at the end of the decoding process is (essentially) zero, which means that MAP decoding is possible.

Lemma 3.134 (Asymptotic Number of Unresolved Variables). Consider the $(\mathtt{l}, \mathtt{r})$-regular degree distribution pair. Let $\epsilon(x) = x/\lambda(1-\rho(1-x))$ and let $P(x)$ denote the trial entropy of Definition 3.119.

Let $\mathtt{G}$ be chosen uniformly at random from LDPC $\left(n, \lambda(x) = x^{\mathtt{l}-1}, \rho(x) = x^{\mathtt{r}-1}\right)$. Assume that transmission takes place over the BEC(ϵ) where $\epsilon \geq \epsilon^{\text{MAP}}$ and that we apply the M decoder. Let $S(\mathtt{G}, \ell)$ denote the number of variable nodes in the residual graph after the ℓ-th decoding step and let $V(\mathtt{G}, \ell)$ denote the number of unresolved variables $\{v_i\}_{i\in\mathcal{I}}$ at this point, i.e., $V(\mathtt{G},\ell) = |\mathcal{I}|$. Then, as n tends to infinity, $(s(x), v(x)) = \lim_{n\to\infty}(\mathbb{E}_{\mathtt{G}}[S(\mathtt{G},\lfloor xn \rfloor)/n], \mathbb{E}_{\mathtt{G}}[V(\mathtt{G},\lfloor nx \rfloor)/n])$ exists and is given by

$$s(x) = \epsilon(x)h^{\text{EBP}}(x), \quad v(x) = P(\bar{x}) - P(x) + (\epsilon^{\text{BP}} - \epsilon(x))h^{\text{EBP}}(x)\mathbb{1}_{\{x \leq x^{\text{BP}}\}},$$

where $\bar{x}$ is the largest real solution of $\epsilon = \epsilon(\bar{x})$. Further, the individual instances $(S(\mathtt{G},\lfloor zn \rfloor)/n, V(\mathtt{G},\lfloor nz \rfloor)/n)$ concentrate around this asymptotic limit.

Example 3.135 (Unresolved Variables for $(\mathtt{l} = 3, \mathtt{r} = 6)$-Ensemble). Figure 3.136 compares the evolution of the number of unresolved variables as a function of the size as predicted by Lemma 3.134 with empirical samples. We see a good agreement of the predicted curves with the empirical samples, even for the case $\epsilon = 0.46$ for which the lemma is not guaranteed to apply (since $0.46 < \epsilon^{\text{MAP}}$). ◇

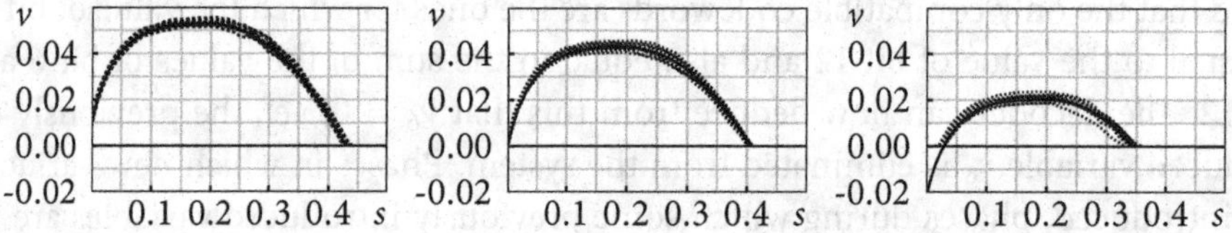

Figure 3.136: Comparison of the number of unresolved variables for the Maxwell decoder applied to the LDPC $(n, x^{\mathtt{l}-1}, x^{\mathtt{r}-1})$ ensembles as predicted by Lemma 3.134 with samples for $n = 10,000$. The asymptotic curves are shown as solid lines, whereas the sample values are printed as dashed lines. The parameters are $\epsilon = 0.50$ (left), $\epsilon = \epsilon^{\text{MAP}} \approx 0.48815$ (middle), and $\epsilon = 0.46$ (right). The parameter $\epsilon = 0.46$ is not covered by Lemma 3.134. Nevertheless, up to the point where the predicted curve dips below zero the experimental data agrees well.

§3.22. Exact Finite-Length Analysis

The density evolution equations give a precise and simple characterization of the asymptotic (in the blocklength) performance under BP decoding. Nevertheless, this approach has several shortcomings. The concentration of the individual performance around the ensemble average is exponential in the blocklength but the convergence of the ensemble performance to the asymptotic limit can be as slow as $O(1/n)$. Therefore, for moderate blocklengths the performance predicted by the density evolution equations might deviate significantly from the actual behavior. This problem is particularly pronounced when we are interested in the construction of moderately long ensembles and impose stringent conditions on the required error probabilities. In these cases the finite-length effects cannot be ignored.

§3.22.1. Stopping Sets

We see how a finite-length analysis for the BEC can be accomplished by studying how the BP decoder fails. The key objects for the analysis are the so-called *stopping sets* (ss).

Definition 3.137 (Stopping Sets). A *stopping set* (ss) $\mathcal{S}$ is a subset of $\mathcal{V}$, the set of variable nodes, such that all neighbors of $\mathcal{S}$, i.e., all check nodes which are connected to $\mathcal{S}$, are connected to $\mathcal{S}$ at *least twice*. The support set of any codeword is an ss and, in particular, so is the empty set. ▽

Example 3.138. Consider the Tanner graph shown in Figure 3.139. The subset $\mathcal{S}$ =

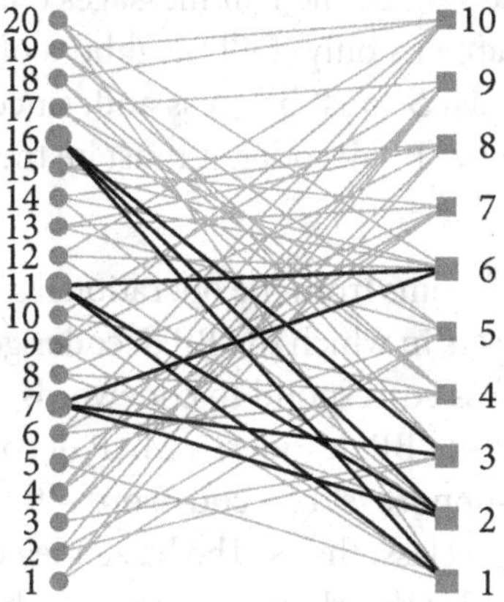

Figure 3.139: The subset of variable nodes $\mathcal{S} = \{7, 11, 16\}$ is a stopping set.

$\{7, 11, 16\}$ is an ss – the set of check nodes connected to $\mathcal{S}$ is $\{1, 2, 3, 6\}$, and each of these nodes is connected to $\mathcal{S}$ at least twice. Note that this subset does *not* form the support set of a codeword – check node 2 has *three* incoming connections. ◇

The basic structural properties of ss and their operational significance is summarized in the following lemma.

Lemma 3.140 (Basic Properties of Stopping Sets).

1. Let $\mathcal{S}_1$ and $\mathcal{S}_2$ be two ss. Then $\mathcal{S}_1 \cup \mathcal{S}_2$ is an ss.
2. Each subset of $\mathcal{V}$ contains a maximum ss (which might be the empty set).
3. Let G be a binary linear code. Assume that we use G to transmit over the BEC and that we employ a BP decoder until either the codeword has been recovered or until the decoder fails to progress further. Let $\mathcal{E}$ denote the subset of $\mathcal{V}$ that is erased by the channel. Then the set of erasures that remains when the decoder stops is equal to the maximum ss contained in $\mathcal{E}$.

Proof. To see the first claim note that if $\mathtt{c}$ is a neighbor of $\mathcal{S}_1 \cup \mathcal{S}_2$ then it must be a neighbor of at least one of $\mathcal{S}_1$ or $\mathcal{S}_2$. Without loss of generality assume that $\mathtt{c}$ is a neighbor of $\mathcal{S}_1$. Since $\mathcal{S}_1$ is an ss, $\mathtt{c}$ has at least two connections to $\mathcal{S}_1$ and therefore at least two connections to $\mathcal{S}_1 \cup \mathcal{S}_2$. Each subset of $\mathcal{V}$ thus contains a maximum ss – the union of all ss contained in $\mathcal{V}$ (which might be the empty set). It remains to verify the third claim. Let $\mathcal{S}$ be an ss contained in $\mathcal{E}$. We claim that the BP decoder cannot determine the variable nodes contained in $\mathcal{S}$. This is true since, even if all other bits were known, every neighbor of $\mathcal{S}$ has at least two connections to the set $\mathcal{S}$ and so all messages to $\mathcal{S}$ are erasure messages. It follows that the decoder cannot determine the variables contained in the maximum ss contained in $\mathcal{E}$. Conversely, if the decoder terminates at a set $\mathcal{S}$, then all messages entering this subset must be erasure messages, which happens only if all neighbors of $\mathcal{S}$ have at least two connections to $\mathcal{S}$. In other words, $\mathcal{S}$ must be an ss and, since no erasure contained in an ss can be determined by the BP decoder, it must be the maximum such ss. □

The name "stopping set" stems from the operational significance outlined in the last property: these are the sets in which the BP decoder gets stuck. Although codewords give rise to ss, not all ss correspond to codewords. This gives a characterization of the suboptimality of the BP decoder: a MAP decoder fails if and only if the set of erasures include the support set of a codeword – the BP decoder on the other hand gets trapped by a larger class, the ss. The basic idea of the finite-length analysis is straightforward. Consider the block erasure probability. Fix a certain erasure pattern. Over the ensemble of all graphs with a certain degree distribution determine the number of constellations which contain stopping sets within the support set of this erasure pattern. Since by definition of the ensemble all constellations have equal probability, the probability (averaged over the ensemble and conditioned on the specific erasure pattern) that this erasure pattern cannot be corrected is then the ratio of the number of such "bad" constellations to the total number. To arrive at the unconditional erasure probability, average over all erasure patterns. In the following we describe this procedure and its extension in some more detail. The combinatorial problem at the heart of this procedure, namely the determination of "bad" constellations, is involved. As an example we treat one simple case in Section 3.22.4. To keep the discussion simple, we restrict our attention to the case of $(\mathtt{l}, \mathtt{r})$-regular ensembles.

§3.22.2. Infinite Number of Iterations

Consider the regular ensemble LDPC $\left(nx^{\mathtt{l}}, n\frac{\mathtt{l}}{\mathtt{r}}x^{\mathtt{r}}\right)$. Fix v variable nodes and number the corresponding $v\mathtt{l}$ variable-node sockets as well as all $n\mathtt{l}$ check-node sockets in some fixed but arbitrary way. A *constellation* is a particular choice of how to attach the $v\mathtt{l}$ variable-node sockets to the $n\mathtt{l}$ check-node sockets. Let $T(v)$ denote the cardinality of all such choices. We have $T(v) = \binom{n\mathtt{l}}{v\mathtt{l}}(v\mathtt{l})!$.

Let $\mathcal{B}(v)$ denote the set of all constellations on v fixed variable nodes which *contain* (non-trivial) ss and let $B(v)$ denote the cardinality of $\mathcal{B}(v)$. Clearly, $0 \leq B(v) \leq T(v)$. Let $\mathcal{B}(s,v)$, $\mathcal{B}(s,v) \subseteq \mathcal{B}(v)$, denote the set of constellations on v fixed variable nodes whose maximum ss is of size s. As before, let $B(s,v)$ denote the corresponding cardinality. For $\chi \in \mathbb{C}$, and with some abuse of notation, define

$$B(\chi, v) = \sum_s B(s,v)\left(\frac{s}{n}\right)^{\chi}, \qquad P(\chi,\epsilon) = \sum_{v=0}^{n}\binom{n}{v}\epsilon^{v}(1-\epsilon)^{n-v}\frac{B(\chi,v)}{T(v)}.$$

Consider first the case $\chi = 0$, so that $B(\chi = 0, v) = B(v)$. Note that $B(\chi = 0, v)/T(v)$ denotes the probability that v chosen variable nodes contain a non-empty ss. Since $\binom{n}{v}\epsilon^{v}(1-\epsilon)^{n-v}$ is the probability that there are v erased bits,

$$P(\chi = 0, \epsilon) = \mathbb{E}[\mathrm{P}_{\mathrm{B}}^{\mathrm{BP}}(\mathtt{G}, \epsilon)],$$

where the expectation is over the ensemble LDPC $\left(nx^{\mathtt{l}}, n\frac{\mathtt{l}}{\mathtt{r}}x^{\mathtt{r}}\right)$. In words, $P(\chi = 0, \epsilon)$ denotes the average *block* erasure probability. If, on the other hand, we pick $\chi = 1$, then $B(\chi = 1, v)/T(v)$ equals the expected fraction (normalized to n) of bits participating in an ss contained in the v fixed variable nodes. It follows that

$$P(\chi = 1, \epsilon) = \mathbb{E}[\mathrm{P}_{\mathrm{b}}^{\mathrm{BP}}(\mathtt{G}, \epsilon)],$$

the average *bit* erasure probability. It is shown in Section 3.22.4 how the quantity B can be determined.

Example 3.141 (Block Erasure Probability of LDPC $\left(nx^3, \frac{n}{2}x^6\right)$). Figure 3.142 shows the resulting curves for $\mathbb{E}_{\mathrm{LDPC}(nx^3, \frac{n}{2}x^6)}[\mathrm{P}_{\mathrm{B}}^{\mathrm{BP}}(\mathtt{G}, \epsilon)]$ as a function of ϵ for $n = 2^i$, $i \in [10]$. For increasing lengths the curves take on the characteristic shape which is composed of a "waterfall" region and an "error floor" region. ◇

As we have seen, in many aspects practically all elements of an ensemble behave the same. This is in particular true for the behavior in the initial phase of the BP decoder where the number of erasures in the system is still large. Toward the very end, however, the various elements of the ensembles show significant differences and this difference does not tend to zero as we increase the length. Some elements of the ensemble exhibit small "weaknesses." In practice it is therefore important to eliminate these weaknesses. A simple generic method of accomplishing this is to "expurgate" the ensemble. If we were using a MAP decoder the expurgation procedure should eliminate codes which contain low-weight *codewords*. Since we are using the BP decoder, the appropriate expurgation is to eliminate codes which contain low-weight *stopping sets*.

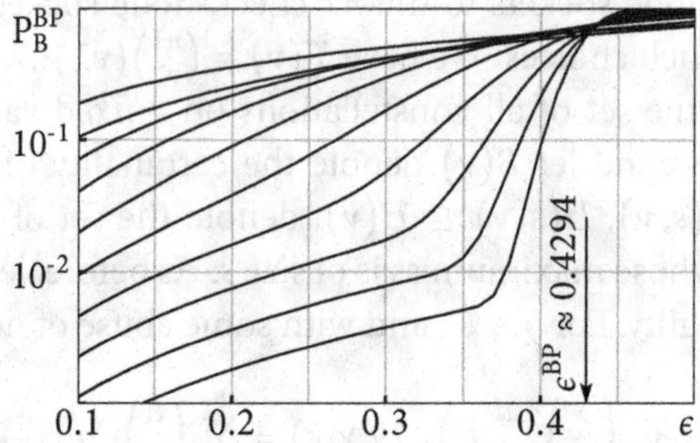

Figure 3.142: $\mathbb{E}_{\text{LDPC}(nx^3, \frac{n}{2}x^6)}[\text{P}_\text{B}^{\text{BP}}(\text{G}, \epsilon)]$ as a function of ϵ for $n = 2^i$, $i \in [10]$.

Definition 3.143 (Expurgated Ensembles). Assume we are given the ensemble LDPC(n, λ, ρ) and an expurgation parameter s, $s \in \mathbb{N}$. The expurgated ensemble ELDPC(n, s, λ, ρ) consists of the subset of codes in LDPC(n, λ, ρ) that contain no stopping sets of size in the range $[1, \dots, s-1]$. ▽

All our previous expressions can be generalized. In particular let $\mathcal{B}(s_{\min}, s, v)$, $\mathcal{B}(s_{\min}, s, v) \subseteq \mathcal{B}(v)$, denote the set of constellations on v fixed variable nodes whose maximum ss is of size s and which contain ss only of size at least $s_{\min}$. Define $B(\chi, s_{\min}, v) = \sum_s B(s_{\min}, s, v) \left(\frac{s}{n}\right)^\chi$, and let $P(\chi, s_{\min}, \epsilon) = \sum_{v=0}^{n} \binom{n}{v} \epsilon^v (1-\epsilon)^{n-v} \frac{B(\chi, s_{\min}, v)}{T(v)}$.

Example 3.144 (Expurgation of LDPC$\left(nx^3, \frac{n}{2}x^6\right)$ Ensemble). Figure 3.145 depicts $P(\chi = 0, s_{\min}, \epsilon)$ for the ensemble LDPC$\left(nx^3, \frac{n}{2}x^6\right)$, where $n = 500$, 1000, 2000. The dashed curves correspond to the case $s_{\min} = 1$, whereas the solid curves correspond to the case where $s_{\min}$ was chosen to be 12, 22, and 40, respectively. How can we determine realistic values for $s_{\min}$? In Section 3.24 we see how to determine the *expected* number of stopping sets of a given size. Applying these techniques we can determine that, for the given choices, the expected number of ss of size less than $s_{\min}$ is 0.787263, 0.902989, and 0.60757, respectively. Since in each case the expected number is less than 1 it follows that there exist elements of the ensemble which do not contain ss of size less than the chosen $s_{\min}$. This bound is crude since it is only based on the *expected* value and we can get better estimates by investigating the *distribution* of the number of low-weight ss. Indeed, this is done in the asymptotic setting in Theorem D.32. The plots in Figure 3.145 demonstrate the importance of considering expurgated ensembles. ◇

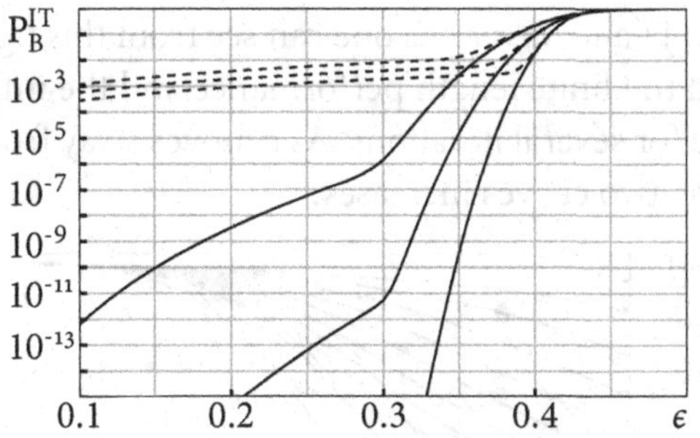

Figure 3.145: $P(\chi = 0, s_{\min}, \epsilon)$ for the ensemble LDPC $\left(nx^3, \frac{n}{2}x^6\right)$, where $n = 500, 1000, 2000$. The dashed curves correspond to the case $s_{\min} = 1$, whereas the solid curves correspond to the case where $s_{\min}$ was chosen to be 12, 22, and 40, respectively. In each of these cases the expected number of ss of size smaller than $s_{\min}$ is less than 1.

§3.22.3. Fixed Number of Iterations

As a further generalization, consider the quantity

$$P(\chi, s_{\min}, \ell, \epsilon) = \sum_{v=0}^{n} \binom{n}{v} \epsilon^v (1-\epsilon)^{n-v} \frac{B(\chi, s_{\min}, \ell, v)}{T(v)}.$$

We define the *depth* of a constellation as the smallest number of iterations that is required to decode this constellation. Stopping sets have infinite depth. Here, $\mathcal{B}(s_{\min}, s, \ell, v)$, $\mathcal{B}(s_{\min}, s, \ell, v) \subseteq \mathcal{B}(v)$, denotes the set of constellations on v fixed variable nodes whose depth is *greater* than ℓ, which contain ss only of size at least $s_{\min}$, and whose size after ℓ iterations is equal to s. Further define

$$B(\chi, s_{\min}, \ell, v) = \sum_{s} B(s_{\min}, s, \ell, v) \left(\frac{s}{n}\right)^{\chi}.$$

As a particular instance, $P(\chi = 1, s_{\min} = 1, \ell, \epsilon)$ is the expected bit erasure probability at the ℓ-th iteration and it is therefore the natural generalization of the asymptotic density evolution equations to finite-length ensembles.

Example 3.146 (Equations for LDPC $\left(500x^3, 250x^6\right)$ Ensemble). Figure 3.147 depicts $P(\chi = 1, s_{\min} = 12, \ell, \epsilon)$ for the ensemble LDPC $\left(500x^3, 250x^6\right)$ and the first 10 iterations (solid curves). Also shown are the corresponding curves of the asymptotic density evolution for the first 10 iterations (dashed curves). More precisely, from the density evolution equations (3.51) we know that asymptotically the bit erasure probability in the ℓ-th iteration is equal to $\epsilon L(1 - \rho(1 - x_{\ell-1}))$, where

$x_\ell = \epsilon\lambda(1-\rho(1-x_{\ell-1}))$ and $x_0 = \epsilon$. As one can see from this figure, for parameters close to the threshold the finite length performance and the infinite-length performance are fairly close for several iterations. As ϵ moves away from the threshold the difference between the two curves increases. ◇

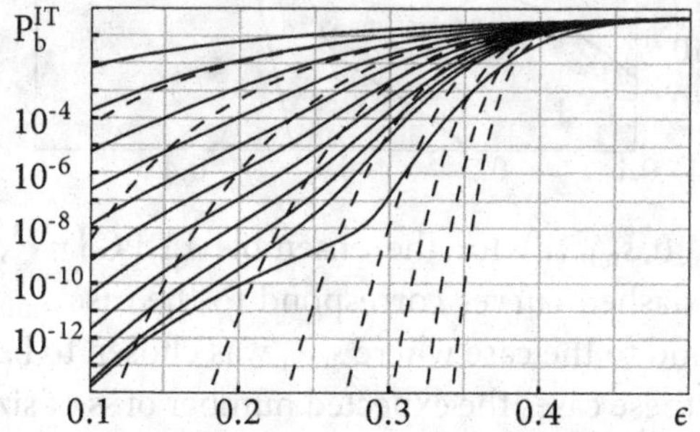

Figure 3.147: $P(\chi = 1, s_{\min} = 12, \ell, \epsilon)$ for the ensemble LDPC $(500x^3, 250x^6)$ and the first 10 iterations (solid curves). Also shown are the corresponding curves of the asymptotic density evolution for the first 10 iterations (dashed curves).

Using the quantity $P(\chi = 0, s_{\min} = 1, \ell, \epsilon)$ we can make statements about the *distribution of the required number of iterations*. More precisely, for $\ell \geq 1$,

$$P(\chi = 0, s_{\min} = 1, \ell-1, \epsilon) - P(\chi = 0, s_{\min} = 1, \ell, \epsilon)$$

is the probability that the required number of iterations is ℓ.

Example 3.148 (Distribution of Iteration Number for LDPC $(nx^3, \frac{n}{2}x^6)$ Ensemble). Figure 3.149 depicts this probability distribution of the number of iterations. Over a large interval the curves are approximately straight lines (in a log-log plot), which indicates that over this range the probability distribution follows a power law, i.e., has the form $\ell^\alpha \beta$ for some suitable non-negative constants α and β. In other words, the probability distribution has large tails. This observation cannot be ignored in the design of a message-passing system. If low erasure probabilities are desired then the decoder has to be designed in a flexible enough manner so that for some rare instances the number of iterations is significantly larger than the average. Otherwise the additional erasure probability incurred by a premature termination of the decoding process can be significant. ◇

§3.22.4. Recursions

So far we discussed how to accomplish the finite-length analysis for the BEC(ϵ), assuming that we are given the probability of occurrence of ss of all sizes. We now show how this latter task can be accomplished via a recursive approach.

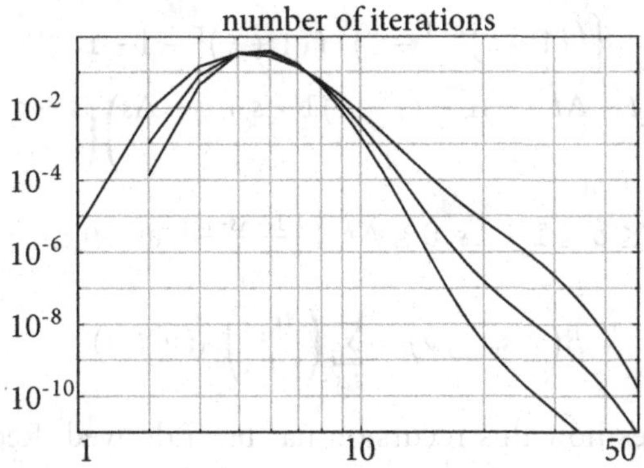

Figure 3.149: Probability distribution of the iteration number for the LDPC $\left(nx^3, \frac{n}{2}x^6\right)$ ensemble, lengths $n = 400$ (top curve), 600 (middle curve), and 800 (bottom curve) and $\epsilon = 0.3$. The typical number of iterations is around 5, but, e.g., for $n = 400$, 50 iterations are required with a probability of roughly 10^{-10}.

In the sequel we assume that the parameters $s_{\min}$ and χ have been fixed and we suppress the dependency of the recursions on these parameters in our notation. Although we give the expressions for general χ, the combinatorial interpretation of the quantities is simplest if we assume that $\chi = 0$. Our language therefore reflects this case.

Consider the set of all constellations which *contain* non-trivial ss. Let $A(v, t, s)$ be the number of such constellations on v *fixed* variable nodes which have t *fixed* check nodes of degree at least 2 and s *fixed* check nodes of degree 1. Note that for $s = 0$ these are the ss of size v, and we have

$$A(v, t, s = 0) = \text{coef}\left\{\left((1+x)^{\mathtt{r}} - 1 - \mathtt{r}x\right)^t, x^{v\mathtt{l}}\right\}(v\mathtt{l})!\,.$$

The first term describes in how many ways we can choose the $v\mathtt{l}$ sockets from t fixed check nodes so that no check node has degree 1 and the term $(v\mathtt{l})!$ takes into account all possible permutations. Therefore, initialize $A(v, t, s)$ by

$$A(v, t, s) = \text{coef}\left\{\left((1+x)^{\mathtt{r}} - 1 - \mathtt{r}x\right)^t, x^{v\mathtt{l}}\right\}(v\mathtt{l})!\left(\frac{v}{n}\right)^{\chi} \mathbb{1}_{\{s_{\min} \leq v\}} \mathbb{1}_{\{s=0\}},$$

where the extra factor $\left(\frac{v}{n}\right)^{\chi}$ distinguishes between the block and the bit erasure case, $\mathbb{1}_{\{s_{\min} \leq v\}}$ enables us to look at expurgated ensembles, and $\mathbb{1}_{\{s=0\}}$ indicates that this expression is only valid for $s = 0$ and that $A(v, t, s)$ is initialized to zero otherwise.

For v running from 1 to n, recursively determine $A(v, t, s)$ for all $0 \leq t, s \leq n$ by means of the update rule

$$A(v, t, s) \mathrel{+}= A(v-1, t-\Delta t-\sigma, s+\sigma-\Delta s)v(\mathtt{l})!\binom{t+s}{\Delta t+\Delta s}\binom{\Delta t+\Delta s}{\Delta t}$$

$$\operatorname{coef}\left\{\left((1+x)^{\mathtt{r}-1}-1\right)^{\sigma}\left((1+x)^{\mathtt{r}}-1-\mathtt{r}x\right)^{\Delta t}, x^{\mathtt{l}-\Delta s-\tau}\right\}$$
$$\binom{(t-\Delta t-\sigma)\mathtt{r}-(v-1)\mathtt{l}+s+\sigma-\Delta s}{\tau}\binom{s+\sigma-\Delta s}{\sigma}\mathtt{r}^{\Delta s}\frac{\Delta s}{s},$$

where $1 \leq \Delta s \leq \mathtt{l}$, $0 \leq \sigma \leq \mathtt{l} - \Delta s$, $0 \leq \Delta t \leq \lfloor \frac{\mathtt{l}-\Delta s-\sigma}{2} \rfloor$, and $0 \leq \tau \leq \mathtt{l} - \Delta s - \sigma - 2\Delta t$. We then have

$$B(\chi, s_{\min}, v) = \sum_{t,s}\binom{n\frac{\mathtt{l}}{\mathtt{r}}}{t+s}A(v,t,s).$$

Figure 3.150 indicates how this recursion has been derived. Recall that in the BP

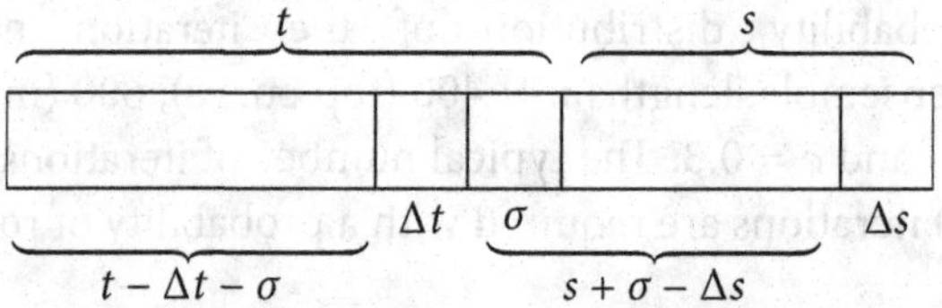

Figure 3.150: Derivation of the recursion for $A(v,t,s)$ for LDPC $\left(\Lambda(x) = nx^{\mathtt{l}}, P(x) = n\frac{\mathtt{l}}{\mathtt{r}}x^{\mathtt{r}}\right)$ and an unbounded number of iterations.

decoding process one variable node which is connected to a check node of degree 1 is removed together with all involved edges. This process is continued until we have reached an ss. For the recursion we proceed in the opposite direction – we start with an ss and *add* one variable node at a time, each time ensuring that this variable node has at least one check node of degree 1 attached to it. The recursion describes in how many ways this can be done, where some care has to be taken to ensure that each constellation is counted only once.

More precisely, assume we add one variable node of degree $\mathtt{l}$. Assume that Δs of its attached check nodes have degree 1, where by assumption $\Delta s \geq 1$. Assume further that σ of the check nodes that previously had degree 1 are being "covered" by edges emanating from the new variable node, so that each of these check nodes now has degree at least 2. Finally, assume that Δt new check nodes of degree at least 2 are added (through multiple edges). All remaining edges go to check nodes that already had degree at least 2. If the new constellation has parameters (v,t,s) then the original constellation must have had parameters $(v-1, s+\sigma-\Delta s, t-\Delta t-\sigma)$.

Let's look in detail at all the factors in the recursion: The term $v(\mathtt{l})!$ accounts for the fact that we can choose the position of the new variable node within the total v variable nodes and that we can permute its edges. The factors $\binom{t+s}{\Delta t+\Delta s}\binom{\Delta t+\Delta s}{\Delta t}$ count in how many ways the $\Delta t + \Delta s$ new check nodes can be placed among the total $t+s$ check nodes. For each new check node of degree 1 we can choose the socket to which we want to connect, resulting in a factor $\mathtt{r}^{\Delta s}$. Next, the σ check nodes which are being covered are chosen from $s + \sigma - \Delta s$ check nodes which had degree 1, accounting for the factor $\binom{s+\sigma-\Delta s}{\sigma}$. If we assume that τ edges are connected to check nodes that already had degree at least 2, then there are $\binom{(t-\Delta t-\sigma)\mathtt{r}-(\nu-1)\mathtt{1}+s+\sigma-\Delta s}{\tau}$ many ways of choosing these τ sockets. This is true since $(t - \Delta t - \sigma)\mathtt{r} - (\nu - 1)\mathtt{1} + s + \sigma - \Delta s$ is the total number of open such sockets. By assumption, $\mathtt{1} - \Delta s - \tau$ edges are used to add Δt new check nodes of degree at least 2 and to cover σ check nodes that used to have degree 1. This accounts for the factor

$$\operatorname{coef}\left\{\left((1+x)^{\mathtt{r}-1} - 1\right)^{\sigma} \left((1+x)^{\mathtt{r}} - 1 - \mathtt{r}x\right)^{\Delta t}, x^{\mathtt{1}-\Delta s-\tau}\right\}.$$

Only the last factor $\frac{\Delta s}{s}$, which ensures that each constellation is counted exactly once, merits a more detailed discussion. To see its function, consider again the peeling decoder. If a constellation has, e.g., u variable nodes with at least one check node of degree 1 attached to it then the peeling decoder has u distinct possibilities in choosing which node to remove next. In the same way, such a constellation can be "reconstructed" in u distinct ways. Therefore, in the aforementioned recursion we should weigh each such reconstruction with the corresponding fraction $\frac{1}{u}$. Unfortunately, u does not appear explicitly in the recursion. Rather, we are given s, the number of check nodes of degree 1. Since one variable node can be attached to multiple check nodes of degree 1, u and s are not in one-to-one correspondence. Fortunately, we are also given Δs, the number of check nodes of degree 1 that the new variable node is connected to. Let Δs be called the *multiplicity* of a variable node. We then see that if we add the multiplicities of the u variable nodes then this multiplicity adds to s. Therefore, summed over all possible reconstruction paths the factor $\frac{\Delta s}{s}$ ensures that each constellation is counted exactly once.

The complexity of the preceding recursion is $\Theta(n^3)$ in time and $\Theta(n^2)$ in space.

§3.23. Finite-Length Scaling

We have seen in the previous section how an *exact* (ensemble-average) finite-length analysis can be accomplished. This solves the problem in principle. In practice this approach becomes computationally challenging as we increase the degree of irregularity or the blocklength. Further, although the analysis is exact, it reveals little about the nature of the system.

Let us therefore consider an alternative approach. First, we separate the contributions to the error probability into those that stem from "large" and those that are due to "small" failures. These two contributions are fundamentally different and are therefore best dealt with separately.

In Section 3.24 we discuss methods that allow us to assess the contribution due to small failures. In the present section we discuss a scaling law which applies to large failures. The left-hand graph in Figure 3.62 shows the situation for the $(3,6)$-regular ensembles that has one critical point. The right-hand graph in Figure 3.62 shows an optimized degree distribution which has two critical points. For a critical point x^{BP} let $\nu^{BP} = \epsilon^{BP} L(1-\rho(1-x^{BP}))$. The operational meaning of ν^{BP} is the following: assume that a degree distribution pair has a single non-zero critical point. Suppose that $\epsilon \approx \epsilon^{BP}$ and that for a large blocklength the BP decoder fails. If the residual graph (of the peeling decoder) is still large (i.e., the failure is not just due to a small weakness in the graph) then with probability approaching 1 (as n tends to infinity) the fraction of undetermined bits is close to ν^{BP}.

Theorem 3.151 (Basic Scaling Law). Consider transmission over the BEC using random elements from an ensemble LDPC (n, λ, ρ) which has a single non-zero critical point x^{BP}. Let $\epsilon^{BP} = \epsilon^{BP}(\lambda, \rho)$ denote the threshold and let $\nu^{BP} = \epsilon^{BP} L(1-\rho(1-x^{BP}))$. Set $z = \sqrt{n}(\epsilon^{BP} - \epsilon)$. Let $P_{b,\gamma}(n, \lambda, \rho, \epsilon)$ and $P_{B,\gamma}(n, \lambda, \rho, \epsilon)$ denote the expected bit/block erasure probability *due to errors of size at least* $\gamma\nu^{BP}$, where $\gamma \in (0,1)$. Then as n tends to infinity (with z held fixed),

$$P_{B,\gamma}(n, \lambda, \rho, \epsilon) = Q(z/\alpha)(1 + o(1)),$$
$$P_{b,\gamma}(n, \lambda, \rho, \epsilon) = \nu^{BP} Q(z/\alpha)(1 + o(1)),$$

where $\alpha = \alpha(\lambda, \rho)$ is a constant which depends on the ensemble.

Figure 3.153 shows the application of the basic scaling law (dotted curves) to the $(3,6)$-regular ensemble. As one can see, although the scaling curves follow the actual exact curves reasonably well, and indeed approach the true curves for increasing lengths, the curves are shifted.

Conjecture 3.152 (Refined Scaling Law). Consider transmission over the BEC using random elements from an ensemble LDPC (n, λ, ρ) which has a single non-zero critical point x^{BP}. Let $\epsilon^{BP} = \epsilon^{BP}(\lambda, \rho)$ denote the threshold and let $\nu^{BP} = \epsilon^{BP} L(1-\rho(1-x^{BP}))$. Set $z = \sqrt{n}(\epsilon^{BP} - \beta n^{-\frac{2}{3}} - \epsilon)$. Let $P_{b,\gamma}(n, \lambda, \rho, \epsilon)$ and $P_{B,\gamma}(n, \lambda, \rho, \epsilon)$ denote the expected bit/block erasure probability *due to errors of size at least* $\gamma\nu^{BP}$, where $\gamma \in (0,1)$. Then as n tends to infinity (with z held fixed),

$$P_{B,\gamma}(n, \lambda, \rho, \epsilon) = Q(z/\alpha)\left(1 + O(n^{-1/3})\right),$$
$$P_{b,\gamma}(n, \lambda, \rho, \epsilon) = \nu^{BP} Q(z/\alpha)\left(1 + O(n^{-1/3})\right),$$

where $\alpha = \alpha(\lambda, \rho)$ and $\beta = \beta(\lambda, \rho)$ are given by

$$\alpha = \left(\frac{\rho(\bar{x}^{BP})^2 - \rho((\bar{x}^{BP})^2) + \rho'(\bar{x}^{BP})(1 - 2x^{BP}\rho(\bar{x}^{BP})) - (\bar{x}^{BP})^2\rho'((\bar{x}^{BP})^2)}{L'(1)\lambda(y^{BP})^2\rho'(\bar{x}^{BP})^2} + \right.$$

$$\left.\frac{(\epsilon^{\text{BP}})^2\lambda(y^{\text{BP}})^2 - (\epsilon^{\text{BP}})^2\lambda((y^{\text{BP}})^2) - (y^{\text{BP}})^2(\epsilon^{\text{BP}})^2\lambda'((y^{\text{BP}})^2)}{L'(1)\lambda(y^{\text{BP}})^2}\right)^{1/2},$$

$$\beta/\Omega = \left(\frac{(\epsilon^{\text{BP}})^4 r_2^2 (\epsilon^{\text{BP}}\lambda'(y^{\text{BP}})^2 r_2 - x^{\text{BP}}(\lambda''(y^{\text{BP}})r_2 + \lambda'(y^{\text{BP}})x^{\text{BP}}))^2}{L'(1)^2\rho'(\bar{x}^{\text{BP}})^3(x^{\text{BP}})^{10}(2\epsilon^{\text{BP}}\lambda'(y^{\text{BP}})^2 r_3 - \lambda''(y^{\text{BP}})r_2 x^{\text{BP}})}\right)^{1/3},$$

$$r_i = \sum_{m\geq j\geq i}(-1)^{i+j}\binom{j-1}{i-1}\binom{m-1}{j-1}\rho_m(\epsilon^{\text{BP}}\lambda(y^{\text{BP}}))^j,$$

where x^{BP} is the unique critical point, $\bar{x}^{\text{BP}} = 1 - x^{\text{BP}}$, $y^{\text{BP}} = 1 - \rho(1 - x^{\text{BP}})$, and Ω is a universal constant whose value can be taken equal to 1 for all practical purposes.[3]

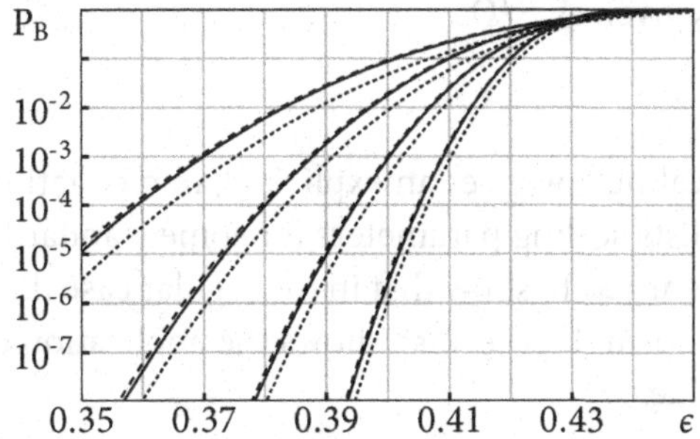

Figure 3.153: Scaling of $\mathbb{E}_{\text{LDPC}(n,x^2,x^5)}[P_B(G,\epsilon)]$ for transmission over the BEC(ϵ) and BP decoding. The threshold for this combination is $\epsilon^{\text{BP}} \approx 0.4294$. The blocklengths/expurgation parameters are $n/s = 1024/24$, $2048/43$, $4096/82$, and $8192/147$, respectively. The solid curves represent the exact ensemble averages. The dotted curves are computed according to the basic scaling law stated in Theorem 3.151. The dashed curves are computed according to the refined scaling law stated in Conjecture 3.152. The scaling parameters are $\alpha = 0.56036$ and $\beta/\Omega = 0.6169$; see Table 3.154.

Discussion: The conjecture improves upon the basic scaling law in two aspects. First, the refined scaling law predicts a *finite-length shift* of the BP threshold of the form $\beta n^{-\frac{2}{3}}$. From Figure 3.153 we see that the refined scaling law indeed corrects for the shift (dashed curves) and that it gives a much better match to the true curves.

[3] Define

$$K(z) = \frac{1}{2}\int \frac{\text{Ai}(iy)\text{Bi}(2^{1/3}z + iy) - \text{Ai}(2^{1/3}z + iy)\text{Bi}(iy)}{\text{Ai}(iy)}\,dy,$$

where Ai(·) and Bi(·) are the so-called Airy functions. Then $\Omega = \int_0^\infty [1 - K(z)^2]\,dz$. Numerically, we get $\Omega = 0.996 \pm 0.005$.

l	r	ϵ^{Sha}	ϵ^{BP}	α	β/Ω
3	4	$\frac{3}{4} = 0.75$	0.6473	0.5425	0.5936
3	5	$\frac{3}{5} = 0.6$	0.5176	0.5657	0.6162
3	6	$\frac{1}{2} = 0.5$	0.4294	0.5604	0.6170
4	5	$\frac{4}{5} = 0.8$	0.6001	0.5533	0.5716
4	6	$\frac{2}{3} \approx 0.6666$	0.5061	0.5605	0.5744
5	6	$\frac{5}{6} \approx 0.8333$	0.5510	0.5606	0.5597
6	7	$\frac{6}{7} \approx 0.8571$	0.5079	0.5623	0.5478
6	12	$\frac{1}{2} = 0.5$	0.3075	0.5116	0.5063

Table 3.154: Thresholds and scaling parameters for some regular standard ensembles. The shift parameter is given as β/Ω.

Second, the refined scaling law gives an explicit characterization of the scaling parameters. Table 3.154 lists scaling parameters for some standard regular ensembles. It is the topic of Problem 3.41 to show that in the regular case the scaling parameters take on a very simple form. Figure 3.155 shows the application of the scaling result to an irregular ensemble.

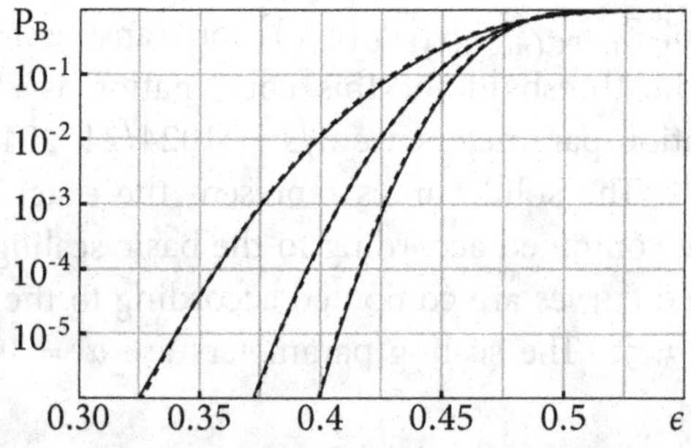

Figure 3.155: Scaling of $\mathbb{E}_{\text{LDPC}(n,\lambda=\frac{1}{6}x+\frac{5}{6}x^3,\rho=x^5)}[P_B(G,\epsilon)]$ for transmission over BEC(ϵ) and BP decoding. The threshold for this combination is $\epsilon^{\text{BP}} \approx 0.482803$. The blocklengths/expurgation parameters are $n/s = 350/14$, $700/23$, and $1225/35$. The solid curves represent the simulated ensemble averages. The dashed curves are computed according to the refined scaling law stated in Conjecture 3.152 with scaling parameters $\alpha = 0.5791$ and $\beta/\Omega = 0.6887$. The two curves are almost on top of each other and are hard to distinguish.

We will not give a proof of the scaling law but we provide in the remainder of this section an informal justification of the aforementioned scaling forms. It is convenient to visualize the decoder in the form of peeling off one variable node at a time as was discussed in Section 3.19. Keep track of the evolution of the residual graph as a function of time. We know from Theorem 3.106 that the trajectories corresponding to individual instances of decoding closely follow the expected value (given by the density evolution equations). For the $(3,6)$-regular ensemble the evolution of the expected degree distribution is shown in Figure 3.111. Focus on the evolution of the residual nodes of degree 1, denoted R_1 (the corresponding curve is the one shown in bold in Figure 3.111). Several individual trajectories are shown for $\epsilon = 0.415$ and $n = 2048$ and $n = 8192$ in Figure 3.156. From Theorem 3.106 we know that the deviations of individual curves are no larger than $O(n^{5/6})$. In fact the deviations are even smaller and, not surprisingly, they are of order $O(\sqrt{n})$. Assume that the channel parameter ϵ is close to ϵ^{BP}. If $\epsilon = \epsilon^{\text{BP}}$ then at the critical point the expected number of check nodes of degree 1 is zero. Assume now that we vary ϵ slightly. If we vary ϵ so that $\Delta\epsilon$ is of order $\Theta(1)$, then the expected number of check nodes of degree 1 at the critical point is of order $\Theta(n)$. Since the standard deviation is of order $\Theta(\sqrt{n})$, then with high probability the decoding process will either succeed (if $(\epsilon-\epsilon^{\text{BP}}) < 0$) or die (if $(\epsilon-\epsilon^{\text{BP}}) > 0$). The interesting scaling happens if we choose our variation of ϵ in such a way that $\Delta\epsilon = z/\sqrt{n}$, where z is a constant. In this case the expected gap at the critical point scales in the same way as the standard deviation and one would expect that the probability of error stays constant. Varying the constant z gives rise to the scaling function.

What is the distribution of the states around the mean? A closer look shows that this distribution (of the nodes of degree 1) is initially Gaussian and remains Gaussian throughout the decoding process. The mean of this Gaussian is the solution of the density evolution equations. Therefore, all that is required to specify the full distribution is the evolution of the covariance matrix. In a similar manner as there are differential equations that describe the evolution of the mean, one can write down a system of differential equations that describe the evolution of the covariance matrix.

Consider Figure 3.156. Those trajectories that hit the $R_1 = 0$ plane die. One can quantify the probability for the process to hit the $R_1 = 0$ plane as follows. Stop density and covariance evolution when the number of variables reaches the critical value ν^{BP}. At this point the probability distribution of the state approaches a Gaussian (as the length increases) with a given mean and covariance for $R_1 \geq 0$ (while it is obviously 0 for $R_1 < 0$). Estimate the survival probability (i.e., the probability of not hitting the $R_1 = 0$ plane at any time) by summing the Gaussian distribution over $R_1 \geq 0$. Obviously this integral can be expressed in terms of a Q-function.

The preceding description leads to the basic scaling law. Where does the shift in Conjecture 3.152 come from? It is easy to understand that we underestimated the error probability in the preceding calculation: we correctly excluded from the sum

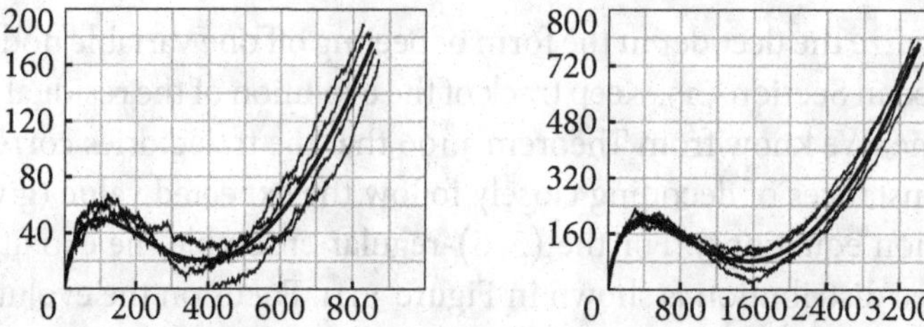

Figure 3.156: Evolution of $n(1-r)R_1$ as a function of the size of the residual graph for several instances for the ensemble LDPC $\left(n, \lambda(x)=x^2, \rho(x)=x^5\right)$ for $n=2048$ (left) and $n=8192$ (right). The transmission is over the BEC($\epsilon=0.415$).

the part of the Gaussian distribution lying in the $R_1 < 0$ half-space – trajectories contributing to this part must have hit the $R_1 = 0$ plane at some point in the past. On the other hand, we cannot be certain that trajectories such that $R_1 > 0$ when ν crosses ν^{BP} didn't hit the $R_1 = 0$ plane at some time in the past and bounce back (or will not hit it at some later point). Taking this effect into account gives us the shift.

§3.24. Weight Distribution and Error Floor

In the previous section we have seen how we can model large-sized failures of the decoder. We focus now on those errors that are caused by small weaknesses in the graph. We start by determining the average weight distribution. The regular case is discussed in more detail in Lemma D.17 on page 513. If you are unfamiliar with manipulations of generating functions you might want to review the material in Appendix D before proceeding. If patience is not your strongest virtue, fast forward to Lemma 3.166 to see how the following results help in finding expressions for the error floor.

Lemma 3.157 (Weight Distribution). Let $A_{\text{cw}}(\mathtt{G}, w)/A_{\text{ss}}(\mathtt{G}, w)$ denote the number of codewords/stopping sets of weight w of a code $\mathtt{G}$, $\mathtt{G} \in \text{LDPC}(\Lambda, P)$. Then

$$\mathbb{E}[A_{\text{cw/ss}}(\mathtt{G}, w)] = \sum_e \alpha(w, e)\beta_{\text{cw/ss}}(e),$$

where

$$\alpha(w, e) = \operatorname{coef}\Big\{\prod_i (1 + xy^i)^{\Lambda_i}, x^w y^e\Big\},$$

$$\beta_{\text{cw}}(e) = \frac{\operatorname{coef}\left\{\prod_i \left((1+z)^i + (1-z)^i\right)^{P_i}, z^e\right\}}{2^{n\bar{r}}\binom{\Lambda'(1)}{e}},$$

$$\beta_{\text{ss}}(e) = \frac{\text{coef}\{\prod_i((1+z)^i - iz)^{P_i}, z^e\}}{\binom{\Lambda'(1)}{e}}.$$

Proof. Consider the $\Lambda(1)$ variable nodes, out of which there are Λ_i nodes of degree i. In how many ways can we select w of those nodes so that the total number of attached edges is e? Denote this number by $\alpha(w,e)$ and consider its generating function $\sum_{w,e}\alpha(w,e)x^w y^e$. This generating function depends on $\Lambda(x)$. To start, consider a single variable node of degree i. Then this generating function is equal to $1 + xy^i$; we can either skip this node (corresponding to 1) or, if we include this node, then we have one node and i edges (corresponding to the term xy^i). More generally, if we have Λ_i variable nodes of degree i then the generating function is $\prod_i(1 + xy^i)^{\Lambda_i}$, since for each such node we can decide to include it in our set or not. The expression of $\alpha(w,e)$ follows from this observation by letting the product range over all node degrees. It remains to determine the probability that we get a codeword/ss if we connect e edges uniformly at random to the $P'(1) = \Lambda'(1)$ check-node sockets. Denote this probability by $\beta_{\text{cw/ss}}(e)$. In total there are $\binom{\Lambda'(1)}{e}$ choices. How many of those give rise to codewords/ss? A constellation gives rise to a codeword if each check-node socket is connected to an even number of edges. For a check node of degree i the corresponding generating function is $\sum_j\binom{i}{2j}z^{2j} = \frac{1}{2}\left((1+z)^i + (1-z)^i\right)$ (see Section D.3). On the other hand, we get an ss if no check node contains exactly one connection. This is encoded by the generating function $\sum_{j\neq 1}\binom{i}{j}z^j = (1+z)^i - iz$. Taking into account that there are P_i check nodes of degree i, the expressions for $\beta_{\text{cw/ss}}(e)$ follow. □

For large blocklengths and weights, $\mathbb{E}[A_{\text{cw/ss}}(\mathtt{G}, w)]$ becomes expensive to compute. Fortunately, we can derive a very accurate asymptotic expansion using the Hayman method discussed in Appendix D. This expansion is substantially easier to evaluate.

Lemma 3.158 (Hayman Approximation of Weight Distribution). Let $\mathtt{G}$ be an element of LDPC (n, L, R) and let $A_{\text{cw}}(\mathtt{G}, w)/A_{\text{ss}}(\mathtt{G}, w)$ denote the number of codewords/ss of $\mathtt{G}$ of weight w of $\mathtt{G}$. Define

$$p_{\text{cw}}(x,y,z) = \left(\prod_j(1+xy^j)^{L_j}\right)\prod_i\left(\frac{(1+z)^i + (1-z)^i}{2}\right)^{R_i\bar{r}},$$

$$p_{\text{ss}}(x,y,z) = \left(\prod_j(1+xy^j)^{L_j}\right)\prod_i\left((1+z)^i - iz\right)^{R_i\bar{r}}.$$

Let $\underline{a} = \left(\frac{x\partial p}{p\partial x}, \frac{y\partial p}{p\partial y}, \frac{z\partial p}{p\partial z}\right)$ and define the 3×3 symmetric matrix $\underline{B}(x,y,z)$: its first row consists of the vector $x\partial\underline{a}/\partial x$ and its second and third rows of the corresponding derivative with respect to y and z. Let $(x_\omega, y_\epsilon, z_\epsilon)$ be a positive solution of the system of equations $\underline{a}(x,y,z) = (\omega, \epsilon, \epsilon)$ and define

$$\frac{1}{\sigma^2} = (0,1,1)\underline{B}^{-1}(x_\omega, y_\epsilon, z_\epsilon)(0,1,1)^T + \frac{L'(1)}{\epsilon(L'(1)-\epsilon)}.$$

Then for $\omega \in (0,1)$ and $n \in \mathbb{N}$ so that $n\omega \in \mathbb{N}$,

$$\mathbb{E}[A(\mathtt{G}, n\omega)] = \sqrt{\frac{\sigma^2\epsilon(1-\epsilon/L'(1))}{2\pi n|B(x_\epsilon, y_\epsilon, z_\epsilon)|}} \frac{\mu p(x_\omega, y_\epsilon, z_\epsilon)^n}{x_\omega^{n\omega} y_\epsilon^{n\epsilon} z_\epsilon^{n\epsilon}} \mathrm{e}^{-nL'(1)h(\epsilon/L'(1))}(1+o(1)),$$

where $\mu = 1$ except when all variable nodes either have even degree or all have odd degree, in which case $\mu = 2$. The *growth rate* $G(\omega)$ of the weight distribution, defined as

$$G(\omega) = \lim_{n\to\infty} \frac{1}{n} \log \mathbb{E}[A(\mathtt{G}, n\omega)],$$

is given by

$$G(\omega) = \log\left(\frac{p(x_\omega, y_\epsilon, z_\epsilon)}{x_\omega^\omega y_\epsilon^\epsilon z_\epsilon^\epsilon} \mathrm{e}^{-L'(1)h(\epsilon/L'(1))}\right).$$

Note: The matrix $\underline{B}$ becomes singular for left regular ensembles. The special case of regular ensembles is discussed in detail in Lemma D.17.

Proof. From Lemma 3.157 we know that

$$\mathbb{E}[A(\mathtt{G}, n\omega)] = \sum_e \mathrm{coef}\left\{p(x,y,z)^n, x^w y^e z^e\right\} / \binom{nL'(1)}{e}.$$

Fix a value ω, $0 < \omega < 1$, and set $w = n\omega$. Further fix ϵ, $0 < \epsilon < L'(1)$, and look at values e of the form $e = n\epsilon + \Delta e$, where $\Delta e = o(\sqrt{n \log n})$. What is the local expansion of the term coef$\{\cdots\}$, i.e., what is the behavior of coef$\{\cdots\}$ for small deviations of e around $n\epsilon$? The answer is furnished by the Hayman method stated in Lemma D.14. This local expansion reads

$$\frac{\mu p\left(x_\omega, y_\epsilon, z_\epsilon\right)^n}{\sqrt{(2\pi n)^3|\underline{B}|}\left(x_\omega^\omega y_\epsilon^\epsilon z_\epsilon^\epsilon\right)^n} \left(y_\epsilon z_\epsilon\right)^{-\Delta e} \mathrm{e}^{-\frac{(\Delta e)^2}{2n}(0,1,1)\underline{B}^{-1}(0,1,1)^T}(1+o(1)),$$

where $(x_\omega, y_\epsilon, z_\epsilon)$ is the unique positive solution to the set of equations $\underline{a}(x,y,z) = (\omega, \epsilon, \epsilon)$ and where μ is equal to 1 unless all powers of $L(x)$ are even or all powers of $L(x)$ are odd, in which case μ takes on the value 2.

For the term $\binom{nL'(1)}{e}$ we have from Example D.16 the local expansion

$$\binom{nL'(1)}{n\epsilon} = \frac{e^{nL'(1)h(\epsilon/L'(1))}}{\sqrt{2\pi n\epsilon(1-\epsilon/L'(1))}}(\epsilon/(L'(1)-\epsilon))^{-\Delta e} e^{-\frac{(\Delta e)^2}{2n\epsilon(1-\epsilon/L'(1))}}(1+o(1)).$$

Let us combine these two expressions to get a local expansion of the number of words with a given number of emanating edges. To simplify the notation define

$$A = \sqrt{\frac{\epsilon(1-\epsilon/L'(1))}{(2\pi n)^2|\underline{B}|}} \frac{\mu p\,(x_\omega, y_\epsilon, z_\epsilon)^n e^{-nL'(1)h(\epsilon/L'(1))}}{(x_\omega^\omega y_\epsilon^\epsilon z_\epsilon^\epsilon)^n},$$

$$B = y_\epsilon^{-\Delta e} z_\epsilon^{-\Delta e}(\epsilon/(L'(1)-\epsilon))^{\Delta e},$$

$$C = e^{-\frac{(\Delta e)^2}{2n}\left((0,1,1)\,\underline{B}^{-1}(0,1,1)^T + \frac{L'(1)}{\epsilon(L'(1)-\epsilon)}\right)} = e^{-\frac{(\Delta e)^2}{2n\sigma^2}},$$

where in the last step we have made use of the definition of σ^2 in the statement of the theorem. We see that the number of codewords of weight $n\omega$ which have $e = n\epsilon + \Delta e$ attached edges has a local expansion ABC. Here, A represents the number of such codewords with $e = n\epsilon$ edges, B represents the first-order change if we alter the number of edges, and C represents the second-order change. In order to determine the total number of codewords of weight $n\omega$ (regardless of the number of attached edges) we need to sum this expression over e. Look for the value of ϵ where the summands over e take on their maximum value. Since B represents the change of the summands with respect to a change in e, this value of ϵ is characterized by the condition that $B = 1$. Explicitly, the condition reads

$$y_\epsilon z_\epsilon = (\epsilon/(L'(1)-\epsilon)).$$

Around this spot the expression has the expansion $Ae^{-\frac{(\Delta e)^2}{2n\sigma^2}}$; i.e., it behaves like a Gaussian with variance $\sigma^2 n$. Therefore, the sum is equal to $A\sqrt{2\pi n\sigma^2}$. This is the result stated in the lemma. The growth rate is a special case of this general result. □

How can the preceding expression be evaluated efficiently?

Corollary 3.159 (Efficient Evaluation of Weight Distribution). Define

$$a_{\text{cw/ss}}(x,y) = \Big(\prod_j (1+xy^j)^{L_j}\Big), \qquad b_{\text{cw}}(z) = \prod_i \Big(\frac{(1+z)^i + (1-z)^i}{2}\Big)^{R_i \bar{r}},$$

$$b_{\text{ss}}(z) = \prod_i \big((1+z)^i - iz\big)^{R_i \bar{r}}, \quad c_{\text{cw/ss}}(x,y) = \frac{x\partial a_{\text{cw/ss}}(x,y)}{a_{\text{cw/ss}}(x,y)\partial x},$$

$$d_{\text{cw/ss}}(x,y) = \frac{y\partial a_{\text{cw/ss}}(x,y)}{a_{\text{cw/ss}}(x,y)\partial y}, \qquad e_{\text{cw/ss}}(z) = \frac{z\partial b_{\text{cw/ss}}(z)}{b_{\text{cw/ss}}(z)\partial z}.$$

Let z be the parameter, $z > 0$. Calculate in the given order

$$\epsilon(z) = e(z), \qquad y(z) = \epsilon(z)/(z(L'(1) - \epsilon(z))),$$
$$x(z) = \text{solution of } d(x, y(z)) = \epsilon(z), \qquad \omega(z) = c(x(z), y(z)).$$

Insert the calculated set of parameters $(x(z), y(z), z)$, and $\epsilon(z)$ into the expressions of Lemma 3.158. The result is the Hayman approximation of the number of words of weight $n\omega(z)$.

Discussion: The evaluation of the parameters ϵ and y as well as ω is immediate. The only non-trivial computation is the determination of x as a function of z.

Example 3.160 (Weight Distribution). To be specific, let us consider the weight distribution of codewords. The weight distribution of ss behaves in a similar way. Figure 3.161 shows the growth rate $G(\omega)$ for the $(3,6)$ ensemble as well as the $(2,4)$ ensemble. Note that those two cases behave differently. While $G_{(3,6)}(\omega)$, the growth

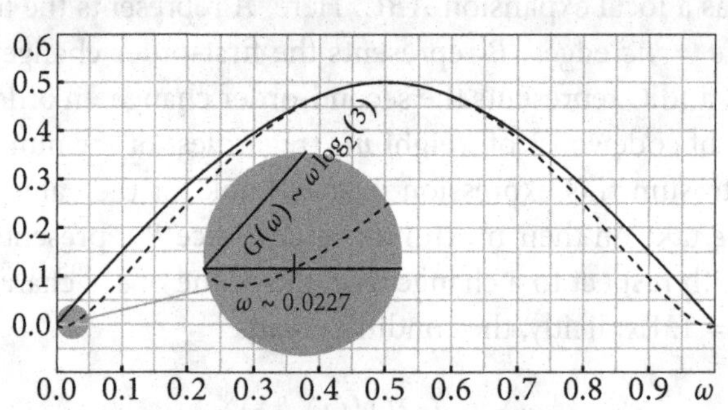

Figure 3.161: Growth rate $G(\omega)$ for the $(3,6)$ (dashed line) as well as the $(2,4)$ (solid line) ensemble.

rate of the $(3,6)$ ensemble, is initially (for small ω) negative and only turns positive around $\omega \approx 0.0227$, $G_{(2,4)}(\omega)$ is positive over the whole range. ◇

What causes this different behavior in the two ensembles? The following lemma, whose proof we skip, gives the answer.

Lemma 3.162 (Growth Rate). Consider an ensemble LDPC (n, L, R) and let $\mathtt{l}_{\min}$ be the minimal variable-node degree. Let $L_{\mathtt{l}_{\min}}$ denote the fraction of variable nodes having degree $\mathtt{l}_{\min}$. Then

$$G_{\text{cw/ss}}(\omega) = \omega\frac{\mathtt{l}_{\min} - 2}{2}(\log(\omega) - 1) + \omega \log\Big(L_{\mathtt{l}_{\min}}\Big(\frac{\mathtt{l}_{\min}\rho'(1)}{L'(1)}\Big)^{\mathtt{l}_{\min}/2}\Big) + O(\omega^2).$$

For $\mathtt{l}_{\min} = 2$ this specializes to

$$G_{\text{cw/ss}}(\omega) = \omega \ln(\lambda'(0)\,\rho'(1)) + O(\omega^2).$$

Note: If $\mathtt{l}_{\min} \geq 3$, then $G(\omega)$ is always negative for sufficiently small ω (due to the term $\omega \log(\omega)$). The situation is more interesting if $\mathtt{l}_{\min} = 2$. In this case $G(\omega)$ has a derivative at zero and this derivative is negative/positive if $\lambda'(0)\rho'(1)$ is smaller/bigger than 1. For example, for the growth rate of the $(2,4)$ ensemble shown in Figure 3.161 we have $\lambda'(0)\rho'(1) = 3$ so that $G_{(2,4)}(\omega) = \omega \log_2(3) + O(\omega^2)$.

At this point it is tempting to conjecture that the minimum distance grows linearly if and only if $\lambda'(0)\rho'(1) < 1$. Unfortunately the situation is more involved. First, note that the growth rate, as defined earlier, corresponds to

$$\lim_{n\to\infty} \frac{1}{n} \log \mathbb{E}[A(\mathtt{G}, \omega n)].$$

This quantity has the advantage of being relatively easy to compute. But we have not shown (and there is no reason to believe that this is in fact correct) that the growth rate of "typical" elements of the ensemble behaves like this average. To answer this question one should determine instead $\lim_{n\to\infty} \mathbb{E}[\frac{1}{n} \log A(\mathtt{G}, \omega n)]$. Unfortunately this quantity is much harder to compute. Using Jensen's inequality we know that

$$\lim_{n\to\infty} \mathbb{E}[\frac{1}{n} \log A(\mathtt{G}, \omega n)] \leq \lim_{n\to\infty} \frac{1}{n} \log \mathbb{E}[A(\mathtt{G}, \omega n)] = G(\omega). \tag{3.163}$$

This implies that the computed growth rate $G(\omega)$ is in general an upper bound on the "typical" growth rate. Fortunately, this issue does not play a role if we are interested in the question whether typical codes exhibit a linear minimum distance or not.

The second point *is* important in our current context. The growth rate $G(\omega)$ only captures the behavior of codewords/ss of weight linear in n. It remains to investigate the behavior of sublinear-sized constellations. For what follows it is convenient to recall from Definition 1.10 the notion of a minimal codeword.

Lemma 3.164 (Expected Number of Codewords of Fixed Weight in Asymptotic Limit). Consider the ensemble LDPC (n, λ, ρ) and define $\mu = \lambda'(0)\rho'(1)$. Let $\mathbb{E}[A_{\text{cw/ss}}(\mathtt{G}, w)]$ ($\mathbb{E}[\hat{A}_{\text{cw/ss}}(\mathtt{G}, w)]$) denote the expected number of (minimal) codewords/ss. Let $\bar{P}_{\text{cw/ss}}(x) = \sum_{w\geq 0} \bar{p}_{\text{cw/ss}}(w) x^w$ ($\hat{P}_{\text{cw/ss}}(x) = \sum_{w\geq 1} \hat{p}_{\text{cw/ss}}(w) x^w$) denote the asymptotic generating function counting the number of (minimal) cw/ss, i.e.,

$$\bar{p}_{\text{cw/ss}}(w) = \lim_{n\to\infty} \mathbb{E}[A_{\text{cw/ss}}(\mathtt{G}, w)], \qquad \hat{p}_{\text{cw/ss}}(w) = \lim_{n\to\infty} \mathbb{E}[\hat{A}_{\text{cw/ss}}(\mathtt{G}, w)].$$

Then

$$(3.165) \qquad \bar{P}_{\text{cw/ss}}(x) = \frac{1}{\sqrt{1-\mu x}}, \quad \hat{P}_{\text{cw/ss}}(x) = -\frac{1}{2}\ln(1-\mu x).$$

Proof. From (D.38) we know that $\bar{p}_{\text{cw/ss}}(w) = \mu^w \binom{2w}{w} 4^{-w}$, so that

$$\bar{P}_{\text{cw/ss}}(x) = \sum_{w \geq 0} \mu^w \binom{2w}{w} 4^{-w} x^w = \frac{1}{\sqrt{1-\mu x}},$$

and we know that $\hat{p}_{\text{cw/ss}}(w) = \frac{\mu^w}{2w}$, so that

$$\hat{P}_{\text{cw/ss}}(x) = \frac{1}{2} \sum_{w \geq 1} \frac{\mu^w x^w}{w} = -\frac{1}{2}\ln(1-\mu x). \qquad \square$$

LEMMA 3.166 (ASYMPTOTIC ERROR FLOOR). Consider the ensemble LDPC (n, λ, ρ) and define $\mu = \lambda'(0)\rho'(1)$. Let s denote the expurgation parameter introduced in Definition 3.143. Then for $\epsilon < \epsilon^{\text{BP}}$

$$\lim_{n\to\infty} \mathbb{E}_{\text{ELDPC}(n,s,\lambda,\rho)}[\text{P}_{\text{B}}^{\text{MAP/BP}}(\text{G}, \epsilon)] = 1 - e^{-\sum_{w=s}^{\infty} \frac{(\mu\epsilon)^w}{2w}},$$

$$\lim_{n\to\infty} n\,\mathbb{E}_{\text{ELDPC}(n,s,\lambda,\rho)}[\text{P}_{\text{b}}^{\text{MAP/BP}}(\text{G}, \epsilon)] = \frac{1}{2}\sum_{w\geq s}(\mu\epsilon)^w.$$

For $s = 1$ these expressions specialize to

$$\lim_{n\to\infty} \mathbb{E}_{\text{LDPC}(n,\lambda,\rho)}[\text{P}_{\text{B}}^{\text{MAP/BP}}(\text{G}, \epsilon)] = 1 - \sqrt{1-\mu\epsilon},$$

$$\lim_{n\to\infty} n\,\mathbb{E}_{\text{LDPC}(n,\lambda,\rho)}[\text{P}_{\text{b}}^{\text{MAP/BP}}(\text{G}, \epsilon)] = \frac{1}{2}\frac{\mu\epsilon}{1-\mu\epsilon}.$$

Discussion: For both the BP decoder and the MAP decoder we only consider the erasure floor below ϵ^{BP}. We conjecture that under MAP decoding the expression for the erasure floor stays valid in the regime $\epsilon^{\text{BP}} \leq \epsilon < \epsilon^{\text{MAP}}$.

Proof. We start with the unexpurgated case.

Consider BP decoding. We claim that for $\epsilon < \epsilon^{\text{BP}}$ the erasure probability due to large-weight stopping sets is negligible. Let α be any strictly positive constant. Since $\epsilon < \epsilon^{\text{BP}}$, there exists an ℓ so that x_ℓ, the expected fraction of erasures left in the ℓ-th iteration, is strictly less than α. By (the Concentration) Theorem 3.30 this implies that the contribution to the block erasure probability due to erasures of size exceeding αn decays exponentially in the length n. Since this is true under BP decoding this is also true under MAP decoding.

If $\mu = 0$, i.e., if $\mathtt{l}_{\min} \geq 3$, then we need to show that for $\epsilon < \epsilon^{\text{BP}}$

$$\lim_{n\to\infty} \mathbb{E}_{\text{LDPC}(n,\lambda,\rho)}[\text{P}_{\text{B}}^{\text{MAP/BP}}(\text{G},\epsilon)] = \lim_{n\to\infty} n\,\mathbb{E}_{\text{LDPC}(n,\lambda,\rho)}[\text{P}_{\text{b}}^{\text{MAP/BP}}(\text{G},\epsilon)] = 0.$$

We use the union bound. From (D.33), (D.34), and (D.36) we see that the block (bit) erasure probability due to cw/ss of size up to $n^{\frac{1}{3}}$ decays like $1/n$ $(1/n^2)$. This contribution therefore vanishes. Further, we know from Lemma 3.162 that for $\mathtt{l}_{\min} \geq 3$ the growth rate $G(\omega)$ is negative for ω sufficiently small. Redefine α if necessary so that $G(\omega) < 0$ for $\omega \in (0, \alpha]$. Therefore, also the contributions due to cw/ss of size between $n^{\frac{1}{3}}$ and αn vanish. This proves the claim since we know already from our discussion above that error events of size larger than αn do not contribute in the asymptotic limit.

Let us now turn to the case $\mu > 0$. We start with the bit-erasure probability. Fix a maximum weight W, $W \in \mathbb{N}$. By (D.39), if we fix a finite W and let n tend to infinity, then the minimal codewords/stopping sets of weight at most W become asymptotically non-overlapping with high probability (with high probability there is only a small number of such words; they have bounded weight and they are distributed over a length n which tends to infinity). The standard union bound is therefore tight. By (D.38) the expected number of minimal words of size w is equal to $\mu_w = \frac{\mu^w}{2w}$. Each such word is erased with probability ϵ^w and causes a bit erasure probability of w/n. If we sum up the individual contributions we see that the bit erasure probability caused by words of size at most W multiplied by n tends to $\frac{1}{2}\frac{\mu\epsilon - \mu^W \epsilon^W}{1-\mu\epsilon}$.

If $\mu\epsilon < 1$, then the contribution to the bit erasure probability stemming from larger words can be made arbitrarily small by choosing W sufficiently large. We do this again in two stages. For cw/ss of size up to $n^{\frac{1}{3}}$ we know from (D.37) that their number grows like μ^w. The union bound (which is an upper bound on the contribution) therefore shows that cw/ss of size between $W+1$ and $n^{\frac{1}{3}}$ contribute to the erasure probability only in a negligible way. By the same argument that we used above we also know that cw/ss of size between $n^{\frac{1}{3}}$ and αn as well as cw/ss of size larger than αn do not contribute to the error probability in the asymptotic limit. The claim now follows by letting W tend to infinity.

Consider now the block erasure probability. What is the contribution to the block erasure probability due to cw/ss that are entirely composed of minimal cw/ss of individual weight at most W, where W is a fixed integer? The number of such cw/ss and their composition is given by a Poisson distribution (see (D.39)). We make a mistake if at least one minimal cw/ss is contained in the set of erasures. Therefore, the block erasure probability due to such cw/ss is given by

$$1 - \sum_{a_1,\dots,a_W} \prod_{w=1}^{W} \frac{\mu_w^{a_w} e^{-\mu_w}}{a_w!} (1-\epsilon^w)^{a_w}.$$

In more detail: we make an error (represented by 1) unless for a given constellation (represented by the numbers $a_1, \dots, a_W$) each of the minimal cw/ss has at least one component which is not contained in the erasure set (represented by the factor $1 - \epsilon^w$). Now note that

$$\sum_i \frac{\lambda^i e^{-\lambda}}{i!}(1-\epsilon^w)^i \overset{\tilde{\lambda}=\lambda(1-\epsilon^w)}{=} \sum_i \frac{\tilde{\lambda}^i e^{-\tilde{\lambda}-\lambda\epsilon^w}}{i!} = e^{-\lambda\epsilon^w} \sum_i \frac{\tilde{\lambda}^i e^{-\tilde{\lambda}}}{i!} = e^{-\lambda\epsilon^w}.$$

If we apply the same sequence of steps to our problem we get

$$1 - \sum_{a_1,\dots,a_W} \prod_{w=1}^{W} \frac{\mu_w^{a_w} e^{-\mu_w}}{a_w!}(1-\epsilon^w)^{a_w} = 1 - e^{-\sum_{w=1}^{W} \mu_w \epsilon^w} = 1 - e^{-\sum_{w=1}^{W} \frac{(\mu\epsilon)^w}{2w}}.$$

By using a similar argument as for the bit erasure probability one can show that words of large weight (larger than a constant) have a negligible influence on the block erasure probability. We get the promised expression by letting W tend to infinity and observing that $e^{-\sum_{w=1}^{\infty} \frac{(\mu\epsilon)^w}{2w}} = \sqrt{1-\mu\epsilon}$.

The expurgated case follows in essentially the same manner by taking into account the following key fact: the distribution of the number of cw/ss of size w, $w \geq s$, in the expurgated ensemble is the same as the corresponding distribution in the unexpurgated ensemble. Therefore the only change in the above derivations is the lower summation index. □

Notes

Regular LDPC codes were defined by Gallager in his ground breaking thesis [19]. Rather than looking at *ensembles*, Gallager showed that one can construct specific instances so that the corresponding computation graph is tree-like for $c\log(n)$ iterations, where c is a suitable positive constant. He then argued that for that number of iterations the message-passing decoder can be applied and analyzed in terms of what would now be called density evolution. Although the limited computational resources available at that time did not allow him to compute actual thresholds, explicit computations on some examples for a few iterations gave an idea of the large potential of this scheme. Historically it is interesting to note that LDPC codes were never patented (neither by Gallager himself, nor by Codex for whom Gallager consulted, nor by Motorola which later acquired Codex). According to Forney, Codex did not foresee a market for LDPC codes except the government level. For the latter, patents are of no use.

In 1954 Elias introduced the *erasure channel* as a toy example [16]. With the advent of the Internet this channel was thrown into the limelight. For a general historical outline of message-passing decoding see the Notes at the end of Chapter 4 starting on page 261. We limit our present discussion to the BEC.

Although Gallager did not explicitly consider transmission over the BEC, the general (class of) message-passing decoder(s) that he introduced specializes to the BP decoder discussed in this chapter.

Zyablov and Pinsker were the first authors to consider the decoding problem for sparse graph codes on the erasure channel [55]. They introduced a particular type of regular code ensemble and a decoding algorithm that is equivalent to the peeling decoder (and therefore also equivalent to the message-passing decoder): at each decoding round the decoder recovers all erasures that are connected to checks which are otherwise connected only to known bits. They showed that in the asymptotic case and for a suitable choice of parameters such a decoder has a positive erasure correcting radius and that the number of decoding rounds is of the order $\log(n)$. Translating their results into modern language, they accomplished this by bounding the probability that a submatrix does not contain checks of degree 1 (i.e., by bounding the probability of containing a stopping set). They also computed the average weight distribution of the ensemble. Unfortunately, the paper was hidden in the Russian literature and was essentially forgotten until recently. It therefore had little impact on the ensuing development.

Without doubt one of most important post-Gallager developments in the realm of the analysis of message-passing systems was the series of papers [27, 24, 25, 26] by Luby, Mitzenmacher, Shokrollahi, Spielman, and Stemann. A good portion of the material presented in this chapter is taken from these papers. In particular, they introduced the notion of *irregular* ensembles together with the elegant and compact description of these ensembles in terms of degree distributions discussed in Section 3.3. It is only through this added degree of freedom introduced by degree distributions that capacity can be approached arbitrarily closely. These papers also contain the complete asymptotic analysis, which we have presented. In particular, the fact that the decoding performance is independent of the transmitted codeword (the *all-zero codeword assumption* discussed in Section 3.6.1), the notion of *concentration* and the proof technique to show concentration (Theorem 3.30), the *convergence to the tree ensemble* (Theorem 3.49), the density evolution equations (Theorem 3.50), the *stability condition* (Theorem 3.65), the proof that the heavy-tail Poisson distribution (which the authors called Tornado sequence) gives rise to capacity-achieving degree distributions (Example 3.90), and, finally, the idea of using *linear programming* to find good degree distributions (Section 3.18) are contained in these papers. The original analysis was based on the peeling decoder discussed in Section 3.19. It is fair to say that almost all we know about message-passing coding we first learned in the context of message-passing coding for the BEC, and a good portion of the fundamental ideas was developed in this sequence of papers.

The fact that there can be global constraints in addition to the local constraints described by the computation tree (see discussion on page 93) was pointed out independently by Macris, Montanari, and Xu as well.

The *lower bound on the gap to capacity* expressed in Theorem 3.85 is due to Shokrollahi [37]. It was the first bound in which the Shannon capacity was derived explicitly from the density evolution equations. One can therefore think of it as a precursor to the (asymptotic) area theorem. A systematic study of capacity-achieving degree distributions was undertaken by Oswald and Shokrollahi [37]. They showed that the check-concentrated (which was called right-concentrated in their paper) degree distribution has a better complexity-versus-performance trade-off than the Tornado sequence. Further properties of the class of capacity-achieving degree distribution pairs were discussed by Orlitsky, Viswanathan, and Zhang [35].

Sason and Urbanke [44] presented the so-called *Gallager* lower *bound* on the density of codes (Section 3.16 as well as Theorem 3.93). It is a variant of an argument originally put forth by Gallager in his thesis [19]. The same paper also contained the material of Section 3.17, which proves that check-concentrated ensembles are essentially optimal. Upper bounds on achievable rates of LDPC ensembles under message-passing decoding were derived by Barak, Burshtein, and Feder [7].

EXIT charts were introduced by ten Brink [49] as an efficient visualization of BP decoding. For the BEC, EXIT charts are equivalent to the density evolution analysis and they are exact. The basic properties of EXIT charts for transmission over the BEC, in particular the *duality result* (Theorem 3.77) as well as the *area theorem* (Theorem 3.81) are due to Ashikhmin, Kramer, and ten Brink [4, 5]. Rather than following one of the original proofs of the area theorem as given by these authors (one of which is discussed in Problem 3.31), we have taken the point of view of Méasson, Montanari, Richardson, and Urbanke [28], which uses characterization (iii) in Lemma 3.74 as a starting point. From this vantage point, the area theorem is just an application of the fundamental theorem of calculus (see Theorem 3.81).

The *MAP decoding threshold* for transmission over the BEC was first determined by Montanari [34] essentially by the method on page 126 which we use to prove claim (i) of Theorem 3.120. The realization that the EXIT curve can be derived from the BP curve via the area theorem is due to Méasson and Urbanke [31]. The material in Section 3.20 is taken from the papers of Méasson, Montanari, and Urbanke [29, 30].

Several authors have considered ways of improving the BP decoder. Although one can phrase this in many ways, the essential idea is to use the BP decoder to reduce the originally large linear system to a (hopefully) small one (think of the Maxwell decoder that introduces some symbolic variables and expresses the remaining bits in terms of these unknowns). The latter system can then be decoded directly by Gaussian elimination with manageable complexity. A patent application that contains this idea was filed by Shokrollahi, Lassen, and Karp [45]. A variation on this theme was independently suggested by Pishro-Nik and Fekri [38]. The idea just described is reminiscent of the efficient encoding method described in Appendix A.

The *Maxwell decoder* (see Section 3.21) as the hidden bridge between MAP and BP decoding is due to Méasson, Montanari, and Urbanke [29, 30]. Further connections relating to the upper bound on the MAP threshold can be found at the end of Chapter 4.

The concept of *stopping sets* (Section 3.22.1) was introduced by Richardson and Urbanke in the context of efficient encoding algorithms for LDPC ensembles [42]. Stopping sets play the same role for BP decoding over the BEC that codewords play under MAP decoding. The *exact finite-length analysis* (Section 3.22) for regular ensembles was initiated by Di, Proietti, Richardson, Telatar, and Urbanke in [13]. This paper contained recursions to compute the exact ensemble average block erasure probability for the regular case and more efficient recursions for the special case of left degree 2 and 3. This was quickly followed by extensions to irregular ensembles and the development of more efficient recursions by Zhang and Orlitsky [54]. Efficient expressions for the general case which allowed the determination of the bit as well as block erasure probability (with equal complexity) as well as the determination of the finite-length performance for a fixed number of iterations and expurgated cases were given by Richardson and Urbanke [43]. Our exposition follows closely this paper. An alternative approach to the finite-length analysis was put forth by Yedidia, Sudderth, and Bouchaud [53] (see also Wang, Kulkarni, and Poor [52]).

Scaling laws have a long and successful history in statistical physics. We refer the reader to the books by Fisher [17] and Privman [39]. The idea of scaling was introduced into the coding literature by Montanari [33]. The scaling law presented in Section 3.23 is due to Amraoui, Montanari, Richardson, and Urbanke [2]. The explicit determination of the scaling parameters for the irregular case as well as the optimization of finite-length ensembles is due to Amraoui, Montanari, and Urbanke [3] (for a description of the Airy functions see [1]). The refined scaling law stated in Conjecture 3.152 was shown to be correct for left-regular right-Poisson ensembles by Dembo and Montanari [11].

Miller and Cohen [32] proved that the rate of a regular LDPC code converges to the design rate (Lemma 3.27). Lemma 3.22, which gives a general condition of convergence of the code rate to the design rate, is due to Méasson, Montanari, and Urbanke [30].

The first investigations into the *weight distribution* of regular LDPC ensembles were already done by Gallager [18]. The combinatorial expressions and expressions for the growth rate were extended to the irregular case simultaneously for various flavors of LDPC codes by Burshtein and Miller [9], Litsyn and Shevelev [22, 23], as well as Di, Richardson, and Urbanke [14, 15]. The related weight distribution of stopping sets was investigated by Orlitsky, Viswanathan, and Zhang [36]. *Asymptotic expansions* for the weight distribution (not only its growth rate) using the Hayman method were first given by Di [12]. Rathi [40] used the Hayman method to prove concentration results for the weight distribution of regular codes (see also Barak and Burshtein [6]).

The error floor under BP decoding was investigated by Shokrollahi, Richardson, and Urbanke [41].

The weight distribution problem also received considerable attention in the statistical physics literature. Sourlas pointed out [46, 47] that a code can be considered a spin-glass system, opening the way for applying the powerful methods of statistical physics to the investigation of the performance of codes. The weight distribution of regular LDPC code ensembles was considered by Kabashima, Sazuka, Nakamura, and Saad [20] and it was shown both by Condamin [10] and by van Mourik, Saad, and Kabashima [50] that in this case the limit of $\frac{1}{n}$ log of the expected number of codewords equals the expected value of $\frac{1}{n}$ log of the number of codewords, i.e., that inequality (3.163) is in fact an equality. The general weight distribution was investigated by Condamin [10] and by van Mourik, Saad, and Kabashima [50, 51].

Problems

3.1 (Poisson Ensemble). Consider the infinite degree distribution $R_i = c\nu^i/i!$, $i \geq 0$, where ν is strictly positive. Find the constant c so that $\sum_i R_i = 1$. Express the generating function $R(x) = \sum_i R_i x^i$ in a compact form. What is the average degree $R'(1)$? Find the corresponding degree distribution from an edge perspective $\rho(x)$. Consider the ensemble LDPC (n, λ, ρ). How should we choose ν so that the ensemble has rate r?

3.2 (Edge Perspective). Let $(L(x), R(x))$ denote a degree distribution pair from the node perspective. Let $(\lambda(x), \rho(x))$ be the corresponding degree distribution pair from an edge perspective as defined in (3.16). Prove that λ_i (ρ_i) is the probability that a randomly chosen edge is connected to a variable (check) node of degree i.

3.3 (Average Variable and Check Node Degree). Prove that $(\int \lambda)^{-1}$ is the average variable-node degree and, similarly, that $(\int \rho)^{-1}$ is the average check-node degree.

3.4 (Design Rate). Prove that $1 - \int \rho / \int \lambda$ is equal to the design rate $r(\lambda, \rho)$ of the ensemble LDPC (n, λ, ρ).

3.5 (Your Friends Have More Friends Than You). The following standard exercise in graph theory has a priori no connection to coding. But it demonstrates very well the difference between node and edge perspective.

Consider a graph (not bipartite) with node degree distribution $\Lambda(x) = \sum_i \Lambda_i x^i$. This means that there are Λ_i nodes of degree i. The total number of nodes is $\Lambda(1)$

and, since every edge has two ends, in order that such a graph exists we need $\Lambda'(1)$ to be even. Think of the nodes as people and assume that each edge represents the relationship of "friendship," i.e, two nodes are connected if and only if the two respective people are friends.

Express the average number of friends, call it a, in terms of $\Lambda(x)$. Now express the average number of friends *of a friend* in terms of $\Lambda(x)$. Denote this quantity by b. Show that in average *a friend* has more friends than the average person; i.e., show that $b - a \geq 0$. What is the condition on $\Lambda(x)$ so that this inequality becomes an equality? Don't take it personally. This applies to everyone.

3.6 (Random Variable Nodes of Degree 1 Are Bad). Consider an ensemble $\text{LDPC}(n, \lambda, \rho) \triangleq \text{LDPC}(n, L, R)$ with $\lambda(0) = \lambda_1 > 0$. Prove that in the limit of large blocklengths, the fraction of variable nodes of degree 1 which are connected to check nodes which in turn are connected to at least two variable nodes of degree 1 converges to $\gamma = 1 - \rho(1 - \lambda_1)$. Use this result to conclude that if transmission takes place over the $\text{BEC}(\epsilon)$, then under any decoder the resulting bit erasure probability is lower bounded by $L_1 \epsilon^2 \gamma$, which is strictly positive for any $\epsilon > 0$.

Discussion: This is the reason why we did not include variable nodes of degree 1 in our definition of LDPC ensembles. The picture changes if edges are placed in a structured way as discussed in more detail in Chapter 7.

3.7 (Tanner Codes – Tanner [48]). The definition of the ensemble can be generalized in the following way. Consider a set of n variable nodes and m *generalized* check nodes. The check nodes are defined in terms of a code C, where C has length $\mathtt{r}$ and dimension k. Assume that a check node is satisfied if and only if the connected variable nodes form a codeword in C. What is the design rate of the code as a function of the parameters?

3.8 (Proper Projections). Prove Lemma 3.47 for general ℓ. Also, what happens if you consider a projection onto a set of k randomly chosen components?

Hint: Take $\mathtt{G}$ and remove from it the computation tree $\mathtt{T}$ except for the leaf variable nodes. Compute the expected number of codewords in such a graph assuming that the leaf nodes take on the value 0. Lemma D.25 might come in handy for this purpose. Then proceed as in the case $\ell = 0$.

3.9 (Elements of $\mathcal{H}(n, k)$ Have High Density). The purpose of this exercise is to show that almost no binary linear code has a low-density representation. Consider binary linear codes $C[n, k]$, where k is the *design* dimension.

(i) Write down the total number of binary $(n - k) \times n$ parity-check matrices H.

(ii) Prove that each code $C[n, k]$ is represented by at most $2^{(n-k)^2}$ distinct $(n - k) \times n$ parity-check matrices H.

(iii) Determine an upper bound on the number of binary $(n-k) \times n$ parity-check matrices H with at most na non-zero entries, where $a \in \mathbb{R}$.

(iv) Conclude from (i), (ii), and (iii) and using Problem 1.25 that if we pick a binary linear code $C[n, k]$ uniformly at random from Gallager's parity-check ensemble then the probability that one of its parity-check matrices has at most na ones tends to zero. More precisely, show that if $k = rn$, where $r \in (0, 1)$, then the probability is upper bounded by an expression that asymptotically (as n grows) reads $2^{-\alpha n^2}$ for some constant α that is positive (calculate α).

3.10 (Equivalence of Decoders). Show that the message-passing decoder of Section 3.5 and the iterative decoder introduced in Section 1.9 (more precisely, the generalization of this decoder working on the Tanner graph) lead to identical results. For any graph and any erasure pattern, the remaining set of erasures at the end of the decoding process is identical.

3.11 (Suboptimality of Iterative Decoder). Find a simple graph and an erasure pattern so that message-passing decoding fails on the BEC, but on the other hand the ML decoder recovers the transmitted codeword. This shows that in general the message-passing decoder is suboptimal. What is the smallest example you can find?

3.12 (Regular Example). Consider the regular degree distribution pair $(\lambda(x) = x^2, \rho(x) = x^6)$. Determine the rate, the threshold ϵ^{BP} (both graphically and analytically), the gap to capacity, and the stability condition. Further, determine the gap to capacity that is predicted by Theorem 3.85 and Theorem 3.93.

3.13 (Density Evolution for Regular Example). Consider the regular degree distribution pair $(\lambda(x) = x^2, \rho(x) = x^5)$. For $\epsilon = 4/10$, determine $\mathrm{P}^{\text{BP}}_{\vec{\mathcal{T}}_\ell}(\epsilon)$ and $\mathrm{P}^{\text{BP}}_{\mathcal{T}_\ell}(\epsilon)$ for $\ell \in [10]$.

3.14 (Analytic Determination of Threshold – Bazzi, Richardson, and Urbanke [8]). Although one can use the graphical method discussed in Section 3.12 to determine the threshold to any degree of accuracy, it is nevertheless pleasing (and not too hard) to derive analytic expressions. Solving (3.51) for ϵ we get

$$\epsilon(x) = \frac{x}{\lambda(1-\rho(1-x))}. \tag{3.167}$$

In words, for a given positive real number x there is a unique value of ϵ such that x fulfills the fixed point equation $f(\epsilon, x) = x$. Therefore, if $x \leq \epsilon$ then the threshold is upper bounded by ϵ.

Consider the regular degree distribution pair $(\mathtt{l}, \mathtt{r})$. Let $\bar{x}^{\text{BP}}$ denote the unique positive real root of the polynomial $p(x) = ((\mathtt{l}-1)(\mathtt{r}-1)-1)x^{\mathtt{r}-2} - \sum_{i=0}^{\mathtt{r}-3} x^i$.

Prove that the threshold $\epsilon^{\text{BP}}(\mathtt{l},\mathtt{r})$ is equal to $\epsilon^{\text{BP}}(\mathtt{l},\mathtt{r}) = \epsilon(1-\bar{x}^{\text{BP}})$, where $\epsilon(x)$ is the function defined in (3.167).

Show that $\epsilon^{\text{BP}}(3,4) = \frac{3125}{3672+252\sqrt{21}} \approx 0.647426$.

3.15 (Overcomplete Tanner Graphs – Vardy). Let C be a binary linear code with dual $C^{\perp}$. Let H be the parity-check matrix of C whose rows consist of *all* elements of $C^{\perp}$ and let $\mathtt{G}$ be the corresponding Tanner graph. Assume that transmission takes place over the BEC(ϵ) and that the BP decoder continues until it no longer makes progress. Show that in this case the BP decoder performs bit MAP decoding. What is the drawback of this scheme?

3.16 (Equivalence of Convergence). Show that $\text{P}^{\text{BP}}_{\vec{\mathcal{T}}_\ell}(\epsilon) \overset{\ell\to\infty}{\longrightarrow} 0$ if and only if $\text{P}^{\text{BP}}_{\mathring{\mathcal{T}}_\ell}(\epsilon) \overset{\ell\to\infty}{\longrightarrow} 0$.

3.17 (Asymptotic Bit Erasure Probability Curve). Regard the asymptotic bit erasure probability curve $\mathbb{E}_{\text{LDPC}(\infty,\lambda,\rho)}[\text{P}_{\text{b}}^{\text{BP}}(\mathtt{G},\epsilon,\ell\to\infty)]$. Prove that it has the parametric form

$$(\epsilon, \epsilon L(1-\rho(1-x(\epsilon)))),\quad \epsilon\in[\epsilon^{\text{BP}},1],$$

where $x(\epsilon)$ is the largest root in $[0,1]$ of the equation $\epsilon\lambda(1-\rho(1-x)) = x$. It is more convenient to parameterize the curve by x (instead of ϵ). Show that an equivalent parameterization is

$$\Big(\frac{x}{\lambda(1-\rho(1-x))}, \frac{xL(1-\rho(1-x))}{\lambda(1-\rho(1-x))}\Big),\quad x\geq x(\epsilon^{\text{BP}}).$$

3.18 (Lower Bound on Gap). Prove Theorem 3.85.

Hint: Start with (3.51) and argue that you can rewrite it in the form $1-\rho(1-x)\leq \lambda^{-1}(x/\epsilon^{\text{BP}})$, $0\leq x\leq 1$. Now integrate the two sides from zero to ϵ^{BP} and simplify.

3.19 (Bound on Fractional Binomial). Show that

$$\ln\left(\frac{N}{\alpha}\binom{\alpha}{N}(-1)^{N-1}\right) = -\alpha H(N-1) - c(N,\alpha)\alpha^2,$$

where, for $\alpha<\frac{1}{2}$ we have $0\leq c(N,\alpha)\leq 5$.

3.20 (Heavy-Tail Poisson Distribution Achieves Capacity). Show that the heavy-tail Poisson distribution introduced in Example 3.90 gives rise to capacity-achieving degree distribution pairs.

3.21 (Rank of Random Binary Matrix). Let $R(l,m,k)$ denote the number of binary matrices of dimension $l \times m$ and rank k, so that by symmetry $R(l,m,k) = R(m,l,k)$. Show that for $l \leq m$

$$R(l,m,k) = \begin{cases} 1, & 0 = k < l, \\ 2^{ml}\prod_{i=0}^{l-1}(1-2^{i-m}), & 0 < k = l, \\ R(l-1,m,k)2^k + R(l-1,m,k-1)(2^m - 2^{k-1}), & 0 < k < l, \\ 0, & \text{otherwise.} \end{cases} \tag{3.168}$$

3.22 (ML Performance of $\mathcal{H}(n,k)$). In this exercise we are concerned with the maximum likelihood performance of the ensemble $\mathcal{H}(n,k)$. Let $\mathrm{P}_{\mathrm{b}}^{\mathrm{ML}}(H,\epsilon)$ denote the *bit erasure probability* of a particular code defined by the parity-check matrix H when used to transmit over a BEC(ϵ) and when decoded by an ML decoder. Let $\mathrm{P}_{\mathrm{B}}^{\mathrm{ML}}(H,\epsilon)$ denote the corresponding *block erasure probability*. Show that

$$\mathbb{E}_{\mathcal{H}(n,k)}[\mathrm{P}_{\mathrm{b}}^{\mathrm{ML}}(H,\epsilon)] = \sum_{e=0}^{n}\binom{n}{e}\epsilon^e\bar{\epsilon}^{n-e}\frac{e}{n}\frac{\sum_j R(e-1,n-k,j)2^j}{2^{(n-k)e}}, \tag{3.169}$$

$$\mathbb{E}_{\mathcal{H}(n,k)}[\mathrm{P}_{\mathrm{B}}^{\mathrm{ML}}(H,\epsilon)] = \sum_{e=0}^{n-k}\binom{n}{e}\epsilon^e\bar{\epsilon}^{n-e}\left[1-\prod_{i=0}^{e-1}\left(1-2^{i-n+k}\right)\right] + \sum_{e=n-k+1}^{n}\binom{n}{e}\epsilon^e\bar{\epsilon}^{n-e}, \tag{3.170}$$

where $R(l,m,k)$ is the number of binary matrices of dimension $l \times m$ and of rank k (see Problem 3.21).

3.23 (Cycle Codes and the Stability Condition). Consider the degree distribution pair $(\lambda(x) = x, \rho(x))$. Codes from such an ensemble (with all variable nodes of degree 2) are called *cycle codes*. Show that $\epsilon^{\mathrm{BP}}(\lambda(x) = x, \rho(x)) = \frac{1}{\rho'(1)}$, i.e., the threshold of such a degree distribution pair is given by the stability condition.

3.24 (EXIT Function of Dual of Hamming Code). Find the dual of the $[7,4,3]$ Hamming code and determine its EXIT function by direct computation. Check that your result agrees with (the duality) Theorem 3.77.

3.25 (Bounds on the Girth of a Tanner Graph – Gallager [19]). We are given an undirected graph G with node set $\mathcal{V}$ and edge set $\mathcal{E}$, where each edge e is a an unordered pair of (not necessarily distinct) vertices, e.g., $\mathsf{e} = \{\mathsf{v}_i, \mathsf{v}_j\}$. In this case we say that e *connects* vertices v_i and v_j. A *path* is an ordered alternating sequence of nodes and edges $\mathsf{v}_{i_1}, \mathsf{e}_{i_1}, \mathsf{v}_{i_2}, \mathsf{e}_{i_2}, \dots$, so that $\mathsf{e}_{i_k} = \{\mathsf{v}_{i_k}, \mathsf{v}_{i_{k+1}}\}$. A *cycle* in a graph is a closed path. The *length* of the cycle is the number of nodes (or edges) on this path. The *girth* g of a graph is the length of the smallest cycle in the graph. Consider the Tanner graph of an $(\mathtt{l},\mathtt{r})$-regular LDPC code of length n. Show that the girth g of this graph grows at most logarithmically in n. More precisely, show that

$$g/4 - 1 < \frac{\log n}{\log(\mathtt{l}-1)(\mathtt{r}-1)}.$$

Conversely, can you show that it is always possible to construct such a graph with girth fulfilling

$$g/4 + 1 > \frac{\log n + \log \frac{\mathtt{lr}-\mathtt{l}-\mathtt{r}}{2\mathtt{r}}}{2\log(\mathtt{l}-1)(\mathtt{r}-1)} \geq g/4 - 1?$$

For large n the construction promises a girth which is at least half of the given upper bound.

3.26 (Connection Girth and Minimum Stopping Set Size – Orlitsky, Urbanke, Viswanathan, and Zhang). Let $\sigma(\mathtt{l},g)$ denote the size of the smallest possible stopping set in a bipartite graph with variable node degree $\mathtt{l}$ and girth g. Show that

(i) $\forall \mathtt{l} \geq 2, \sigma(\mathtt{l},2) = 1$.

(ii) $\forall \mathtt{l} \geq 2, \sigma(\mathtt{l},4) = 2$.

(iii) $\sigma(\mathtt{l},6) = \mathtt{l}+1$.

(iv) $\sigma(\mathtt{l},8) = 2\mathtt{l}$.

Can you show that, more generally,

$$\sigma(\mathtt{l},g) \geq \begin{cases} 1+\mathtt{l}+\mathtt{l}(\mathtt{l}-1)+\cdots+\mathtt{l}(\mathtt{l}-1)^{\frac{g-6}{4}}, & g/2 \text{ is odd}, \\ 1+\mathtt{l}+\mathtt{l}(\mathtt{l}-1)+\cdots+\mathtt{l}(\mathtt{l}-1)^{\frac{g-8}{4}}, & g/2 \text{ is even?} \end{cases}$$

Conversely, can you show that

$$\sigma(\mathtt{l},g) \leq \frac{4}{\mathtt{l}-2}(\mathtt{l}-1)^{\frac{g+4}{2}}$$

and $\sigma(2,g) \leq g/2$?

Note: If $\mathtt{l} = 2$ the bounds on g grow linearly in n and, therefore, the size of the smallest stopping set is at most $O(\log n)$. This suggests that if in a graph a significant fraction of variable nodes has degree 2 then small stopping sets are likely to exist.

3.27 (Finite Geometry Construction – Kou, Lin, and Fossorier [21]). Consider the following construction of an $(\mathtt{l}, \mathtt{r})$-regular LDPC code of length n. Pick a prime p. Let $n = p^2$ and arrange the n variable nodes in a $p \times p$ grid. Choose a slope and a shift and consider a line, i.e., the set of all points on this line. Each such line contains exactly p points. Associate to each such line a check node, which is connected to all the variable nodes on this line. If we pick all p shifts for a given slope then we get p check nodes and every variable node participates in exactly one check. Show that if we pick $\mathtt{l}$ distinct slopes then we get an LDPC code (i) of length p^2, (ii) with $\mathtt{l}p$ check nodes, (iii) with variable node degree $\mathtt{l}$, and (iv) with check-node degree p. Show further that the resulting graph has girth exactly 6.

3.28 (Various Characterizations of EXIT Function). Prove the equivalence of the various characterizations in Lemma 3.74.

3.29 (Cycle Codes). Consider a cycle code ensemble LDPC $(n, x, \rho(x))$. Show that the BP threshold and the MAP threshold are identical.

3.30 (EXIT Function of Hamming Code). Derive the EXIT function given in Example 3.73 for the $[7, 4]$ Hamming code.

3.31 (Alternative Proof of Area Theorem – Ashikhmin, Kramer, and ten Brink [4, 5]). For the BEC there are many alternative proofs of the area theorem. Let us consider one such alternative here. Let C be a binary linear code of rate r and length n. We want to show that $\int_0^1 h(\epsilon)\mathrm{d}\epsilon = r$.

Let $\Pi(n)$ denote the set of permutations on n letters. Assume that the 2^{nr} codewords of the code C are equally likely. Let $\mathcal{K}$ denote the index set of *known* bits. Justify each of the following steps:

$$
\begin{aligned}
n\int_0^1 h(\epsilon)\mathrm{d}\epsilon &= \int_0^1 \sum_{i=1}^{n} h_i(\epsilon)\mathrm{d}\epsilon = \sum_{i=1}^{n}\int_0^1 H(X_i \mid Y_{[n]\setminus\{i\}})\mathrm{d}\epsilon \\
&= \sum_{i=1}^{n}\int_0^1 \sum_{\mathcal{K}\in[n]\setminus\{i\}} (1-\epsilon)^{|\mathcal{K}|}\epsilon^{n-1-|\mathcal{K}|} H(X_i \mid X_{\mathcal{K}})\mathrm{d}\epsilon \\
&\overset{\text{(i)}}{=} \sum_{i=1}^{n}\sum_{\mathcal{K}\in[n]\setminus\{i\}} \frac{(n-1-|\mathcal{K}|)!(|\mathcal{K}|)!}{n!} H(X_i \mid X_{\mathcal{K}}) \\
&\overset{\text{(ii)}}{=} \frac{1}{n!}\sum_{j=1}^{n}\sum_{\sigma\in\Pi[n]} H(X_{\sigma(j)} \mid X_{\sigma([j-1])}) \\
&= \frac{1}{n!}\sum_{\sigma\in\Pi[n]}\sum_{j=1}^{n} H(X_{\sigma(j)} \mid X_{\sigma([j-1])}) \\
&\overset{\text{(iii)}}{=} \frac{1}{n!}\sum_{\sigma\in\Pi[n]} H(X_1, \dots, X_n) \overset{\text{(iv)}}{=} H(X) \overset{\text{(v)}}{=} nr.
\end{aligned}
$$

3.32 (Interpretation of Area Theorem). In the general case where bits are sent over different channels (the Area) Theorem 3.81 has the following interpretation: if we change the set of all channels from some starting state to some final state (change the individual channel parameters) then by doing so we change the conditional entropy $H(X \mid Y)$. Assume that we connect the initial and the final state by some smooth path; i.e., $\epsilon_i = \epsilon_i(\epsilon)$ is a piecewise differentiable function of ϵ for $i \in [n]$. Then the (average) EXIT function $h(\epsilon)$ measures the change per $d\epsilon$ of $H(X \mid Y)$. More interestingly, since $h(\epsilon) = \frac{1}{n}\sum_{i=1}^{n} h_i(\epsilon)$, every bit position i contributes locally to this change according to $h_i(\epsilon)$. For different curves that connect the same final and initial state the total change of $H(X \mid Y)$ along the path is the same but the individual contributions according to $h_i(\epsilon)$ are in general different. This is best seen by a simple example.

Consider the $[2,1,2]$ repetition code. Assume first that the two channels are parameterized by $\epsilon_1 = \epsilon = \epsilon_2$, where ϵ goes from 0 to 1.

Consider next the alternative parameterization $\epsilon_1(\epsilon) = \min\{1, \epsilon\}$ and $\epsilon_2(\epsilon) = \max\{0, \epsilon - 1\}$, where $\epsilon \in [0, 2]$.

For both cases compute the individual contributions of the two EXIT functions.

3.33 (EXIT for Regular Ensembles). Prove inequality (3.124).

Hint: The proof is conceptually simple but on the lengthy side. Either pick a particular $\mathtt{l}$ and prove the assertion for this $\mathtt{l}$ or prove the statement for some "sufficiently large" $\mathtt{l}$. Bon courage!

3.34 (Scaling of P_B for $\mathcal{H}$). Consider Gallager's parity-check ensemble $\mathcal{H}(n, r)$ and transmission over the BEC(ϵ). Fix $z = \sqrt{n}(\epsilon^* - \epsilon)$ and let n tend to infinity. Show that in this limit the average block error probability under MAP decoding behaves like

$$\mathbb{E}_{\mathcal{H}(n,r)}[\mathrm{P}_\mathrm{B}(H, z)] = Q\left(\frac{z}{\sqrt{\epsilon^*(1-\epsilon^*)}}\right)(1 + O(1/n)).$$

Hint: Make use of the results of Problem 3.21. More precisely, let A be a $k \times m$ random binary matrix where each entry is chosen independently uniformly at random from $\{0, 1\}$. Use the fact that

$$\mathbb{P}\{\operatorname{rank}(A) = k\} = \begin{cases} 0, & k > m, \\ \prod_{i=0}^{k-1}\left(1 - 2^{i-m}\right), & 0 \le k \le m. \end{cases}$$

3.35 (Minimal Codewords). Consider an ensemble LDPC (n, λ, ρ) with $\lambda'(0) > 0$. Let n tend to infinity and consider the ratio of the expected number of codewords of weight w and the expected number of minimal codewords of weight w. Show that for increasing w this ratio converges to $\sqrt{4w/\pi}$.

Hint: Expand out the two generating functions in Lemma 3.164.

3.36 (MAP versus BP Threshold). Consider the sequence of $(i, 2i)$-regular LDPC ensembles in the limit of large blocklengths. This exercise will show that the performance of BP and MAP decoding can be arbitrarily different but nevertheless the fixed points of BP completely specify the MAP performance.

(i) What is the design rate of these ensembles? What is the Shannon threshold?

(ii) Give the parametric expression $(\epsilon(x), h^{\text{EBP}}(x))$ for the EBP EXIT function.

(iii) Show that $\lim_{i\to\infty} \epsilon(x) = x$ for any $0 < x \leq 1$.

(iv) Use (iii) to show that the sequence of BP thresholds $\epsilon^{\text{BP}}(i)$ converges to 0 when $i \to \infty$.

(v) Show that the sequence of MAP thresholds $\epsilon^{\text{MAP}}(i)$ converges to $\epsilon^{\text{Sha}} = \frac{1}{2}$ if $i \to \infty$.

(vi) Plot the asymptotic EBP EXIT curves for $i = 2, 3, 4, 6, 12, 24$, and 72.

3.37 (Alternative Derivation of Stability Condition). Consider the asymptotic generating function as given in Lemma 3.164 and the resulting erasure floor as stated in Lemma 3.166. At what value of ϵ does this erasure floor expression diverge? Does this value look familiar to you?

3.38 (Dominant Codeword Type). Consider the degree distribution pair $(\lambda(x) = 0.15x + 0.15x^2 + 0.7x^{60}, \rho(x) = 0.1x^4 + 0.2x^5 + 0.2x^6 + 0.3x^7 + 0.2x^{20})$. Plot the function $\Psi(y)$ defined in Lemma 3.22 for this case and show that its maximum is not taken on at $y = 1$. Does this necessarily mean that the relative weight of the "dominant" codeword type is *not* one-half?

3.39 (EXIT Out of Box). We have seen in Figure 3.127 that the EBP EXIT curve can extend outside the "unit box." This happens, e.g., if $\lambda'(0)\rho'(1) < 1$. But this is not the only possibility.

Construct a degree distribution with $\lambda'(0)\rho'(1) > 1$ whose EBP EXIT curve goes "out of the unit box."

3.40 (Dual Stability Condition at the BP Threshold). Consider a degree distribution pair (λ, ρ). Let ϵ^{BP} denote its threshold and let x^{BP} denote a critical point. Assume that $x^{\text{BP}} > 0$. Let $(\tilde{\lambda}, \tilde{\rho})$ denote the degree distribution of the residual graph at the threshold. Show that $\tilde{\lambda}'(1)\tilde{\rho}'(0) = 1$.

Hint: Start with expressions (3.122) and (3.123).

3.41 (Scaling Parameters for Regular Case). Starting with the general expressions given in Conjecture 3.152, show that for the $(\mathtt{l},\mathtt{r})$-regular case the scaling parameters take on the simple form

$$\alpha = \epsilon^{\mathrm{BP}} \sqrt{\frac{\mathtt{l}-1}{\mathtt{l}}\Big(\frac{1}{x^{\mathrm{BP}}} - \frac{1}{y^{\mathrm{BP}}}\Big)},$$

and

$$\beta/\Omega = \epsilon^{\mathrm{BP}} \left(\frac{\mathtt{l}-2}{\mathtt{l}x^{\mathrm{BP}}y^{\mathrm{BP}}}\right)^{2/3} \left(\frac{1}{(\mathtt{l}-1)} + \frac{(\mathtt{r}-2)x^{\mathrm{BP}}}{1-x^{\mathrm{BP}}} - 2\right)^{-1/3}.$$

Hint: Use the relationships between ϵ^{BP}, x^{BP}, and y^{BP} obtained from the fixed point equations and from the fact that the derivative of $\epsilon(x) = \frac{x}{\lambda(1-\rho(1-x))}$ vanishes at the fixed point.

References

[1] M. Abramowitz and I. A. Stegun, *Handbook of Mathematical Functions*, Nat. Bur. Stand. 55, Washington, DC, USA, 1964. [159]

[2] A. Amraoui, A. Montanari, T. Richardson, and R. Urbanke, *Finite-length scaling for iteratively decoded LDPC ensembles*, in Proc. of the Allerton Conf. on Commun., Control, and Computing, Monticello, IL, USA, Oct. 2003. [159, 266, 500]

[3] A. Amraoui, A. Montanari, and R. Urbanke, *How to find good finite-length codes: From art towards science*, Euro. Trans. Telecomm., 18 (2007), pp. 491–508. [159]

[4] A. Ashikhmin, G. Kramer, and S. ten Brink, *Code rate and the area under extrinsic information transfer curves*, in Proc. of the IEEE Int. Symposium on Inform. Theory, Lausanne, Switzerland, June 30–July 5, 2002, p. 115. [158, 166]

[5] ———, *Extrinsic information transfer functions: Model and erasure channel property*, IEEE Trans. Inform. Theory, 50 (2004), pp. 2657–2673. [158, 166]

[6] O. Barak and D. Burshtein, *Lower bounds on the spectrum and error rate of LDPC code ensembles*, in Proc. of the IEEE Int. Symposium on Inform. Theory, Adelaide, Australia, Sept. 4–9, 2005, pp. 42–46. [160]

[7] O. Barak, D. Burshtein, and M. Feder, *Bounds on the achievable rates of LDPC codes over the binary erasure channel*, IEEE Trans. Inform. Theory, 50 (2004), pp. 2483–2492. [158]

[8] L. Bazzi, T. Richardson, and R. Urbanke, *Exact thresholds and optimal codes for the binary-symmetric channel and Gallager's decoding algorithm A*, IEEE Trans. Inform. Theory, 50 (2004), pp. 2010–2021. [162, 264]

[9] D. Burshtein and G. Miller, *Asymptotic enumeration methods for analyzing LDPC codes*, IEEE Trans. Inform. Theory, 50 (2004), pp. 1115–1131. [159, 529]

[10] S. Condamin, *Study of the weight enumerator function for a Gallager code*, project report, Cavendish Laboratory, University of Cambridge, July 2002. [160]

[11] A. Dembo and A. Montanari, *Finite size scaling for the core of large random hypergraphs*. E-print: math.PR/0702007, 2007. [159]

[12] C. Di, *Asymptotic and Finite-Length Analysis of Low-Density Parity-Check Codes*, PhD thesis, EPFL, Lausanne, Switzerland, 2004. Number 3072. [160, 530]

[13] C. Di, D. Proietti, T. Richardson, E. Telatar, and R. Urbanke, *Finite length analysis of low-density parity-check codes on the binary erasure channel*, IEEE Trans. Inform. Theory, 48 (2002), pp. 1570–1579. [159]

[14] C. Di, T. Richardson, and R. Urbanke, *Weight distribution of iterative coding systems: How deviant can you be?*, in Proc. of the IEEE Int. Symposium on Inform. Theory, Washington, DC, USA, June 2001. [159]

[15] ——, *Weight distribution of low-density parity-check codes*, IEEE Trans. Inform. Theory, 52 (2006), pp. 4839–4855. [159, 500]

[16] P. Elias, *Error-free coding*, IEEE Trans. Inform. Theory, 4 (1954), pp. 29–37. [156]

[17] M. E. Fisher, *Proc. of the Enrico Fermi school, Varenna, Italy, 1970, course N. 51*, in Critical Phenomena, International School of Physics Enrico Fermi, Course LI, edited by M. S. Green, Academic, New York, 1971. [159]

[18] R. G. Gallager, *Low-density parity-check codes*, IRE Trans. Inform. Theory, 8 (1962), pp. 21–28. [33, 66, 159, 264, 419, 434]

[19] ——, *Low-Density Parity-Check Codes*, MIT Press, Cambridge, MA, USA, 1963. [156, 158, 164, 261, 264, 420]

[20] Y. Kabashima, N. Sazuka, K. Nakamura, and D. Saad, *Evaluating zero error noise thresholds by the replica method for Gallager code ensembles*, in Proc. of the IEEE Int. Symposium on Inform. Theory, Lausanne, Switzerland, June 2002, IEEE, p. 255. [160, 266]

[21] Y. Kou, S. Lin, and M. P. C. Fossorier, *Low-density parity-check codes based on finite geometries: A rediscovery and new results*, IEEE Trans. Inform. Theory, 47 (2001), pp. 2711–2736. [166, 419]

[22] S. L. Litsyn and V. S. Shevelev, *On ensembles of low-density parity-check codes: Asymptotic distance distributions*, IEEE Trans. Inform. Theory, IT–48 (2002), pp. 887–908. [159]

[23] ——, *Distance distribution in ensembles of irregular low-density parity-check codes*, IEEE Trans. Inform. Theory, IT–49 (2003), pp. 3140–3159. [159]

[24] M. Luby, M. Mitzenmacher, A. Shokrollahi, and D. A. Spielman, *Analysis of low density codes and improved designs using irregular graphs*, in Proc. of the 30th Annual ACM Symposium on Theory of Computing, 1998, pp. 249–258. [157, 262, 500]

[25] ——, *Efficient erasure correcting codes*, IEEE Trans. Inform. Theory, 47 (2001), pp. 569–584. [157, 262, 500]

[26] ——, *Improved low-density parity-check codes using irregular graphs*, IEEE Trans. Inform. Theory, 47 (2001), pp. 585–598. [157, 262, 500]

[27] M. Luby, M. Mitzenmacher, A. Shokrollahi, D. A. Spielman, and V. Stemann, *Practical loss-resilient codes*, in Proc. of the 29th annual ACM Symposium on Theory of Computing, 1997, pp. 150–159. [157, 262, 434, 456]

[28] C. Méasson, A. Montanari, T. Richardson, and R. Urbanke, *Life above threshold: From list decoding to area theorem and MSE*, in Proc. of the IEEE Inform. Theory Workshop, San Antonio, TX, USA, Oct. 2004. E-print: cs.IT/0410028. [158, 265]

[29] C. Méasson, A. Montanari, and R. Urbanke, *Maxwell's construction: The hidden bridge between maximum-likelihood and iterative decoding*, in Proc. of the IEEE Int. Symposium on Inform. Theory, Chicago, IL, USA, 2004, p. 225. [158, 159]

[30] ——, *Maxwell's construction: The hidden bridge between maximum-likelihood and iterative decoding*. submitted to IEEE Transactions on Inform. Theory, 2005. [158, 159]

[31] C. Méasson and R. Urbanke, *An upper-bound on the ML thresholds of LDPC ensembles over the BEC*, in Proc. of the Allerton Conf. on Commun., Control, and Computing, Monticello, IL, USA, Oct. 2003, pp. 478–487. [158]

[32] G. Miller and G. Cohen, *The rate of regular LDPC codes*, IEEE Trans. Inform. Theory, 49 (2003), pp. 2989–2992. [159]

[33] A. Montanari, *The glassy phase of Gallager codes*, Eur. Phys. J. B, 23 (2001), pp. 121–136. E-print: cond-mat/0104079. [159, 265, 266]

[34] ——, *Why "practical" decoding algorithms are not as good as "ideal" ones?*, in Proc. DIMA, USACS Workshop on Codes and Complexity, Rutgers University, Piscataway, NJ, USA, Dec. 4–7, 2001, pp. 63–66. [158, 266]

[35] A. Orlitsky, K. Viswanathan, and J. Zhang, *On capacity-achieving sequences of degree distributions*, in Proc. of the IEEE Int. Symposium on Inform. Theory, Yokohama, Japan, June 29–July 4, 2003, IEEE, p. 269. [158]

[36] ——, *Stopping set distribution of LDPC code ensembles*, in Proc. of the IEEE Int. Symposium on Inform. Theory, Yokohama, Japan, June 29–July 4, 2003, IEEE, p. 123. [159]

[37] P. Oswald and A. Shokrollahi, *Capacity achieving sequences for the erasure channel*, in Proc. of the IEEE Int. Symposium on Inform. Theory, Washington, DC, USA, June 24–29, 2001, IEEE, p. 48. [158]

[38] H. Pishro-Nik and F. Fekri, *On decoding of low-density parity-check codes over the binary erasure channel*, IEEE Trans. Inform. Theory, 50 (2004), pp. 439–454. [158]

[39] V. Privman, *Finite-size scaling theory*, in Finite Size Scaling and Numerical Simulation of Statistical Systems, V. Privman, ed., World Scientific Publ., Singapore, 1990, pp. 1–98. [159]

[40] V. Rathi, *On the asymptotic weight and stopping set distribution of regular LDPC ensembles*, IEEE Trans. Inform. Theory, 52 (2006), pp. 4212–4218. [160, 530]

[41] T. Richardson, A. Shokrollahi, and R. Urbanke, *Finite-length analysis of various low-density parity-check ensembles for the binary erasure channel*, in Proc. of the IEEE Int. Symposium on Inform. Theory, Lausanne, Switzerland, June 2002, p. 1. [160]

[42] T. Richardson and R. Urbanke, *Efficient encoding of low-density parity-check codes*, IEEE Trans. Inform. Theory, 47 (2001), pp. 638–656. [159, 456]

[43] ——, *Finite-length density evolution and the distribution of the number of iterations for the binary erasure channel.* in preparation, 2003. [159]

[44] I. Sason and R. Urbanke, *Parity-check density versus performance of binary linear block codes over memoryless symmetric channels*, IEEE Trans. Inform. Theory, 49 (2003), pp. 1611–1635. [158, 264]

[45] A. Shokrollahi, S. Lassen, and R. M. Karp, *Systems and processes for decoding chain reaction codes through inactivation.* US Patent application, number 6,856,263, June 2003. [158]

[46] N. Sourlas, *Spin-glass models as error-correcting codes*, Nature, 339 (1989), pp. 693–695. [160, 261, 266, 367]

[47] ——, *Spin glasses, error-correcting codes and finite-temperature decoding*, Europhys. Lett., 25 (1994), pp. 159–164. [160]

[48] R. M. Tanner, *A recursive approach to low complexity codes*, IEEE Trans. Inform. Theory, 27 (1981), pp. 533–547. [65, 161, 261, 263]

[49] S. ten Brink, *Convergence of iterative decoding*, Electron. Lett., 35 (1999), pp. 806–808. [158, 264]

[50] J. van Mourik, D. Saad, and Y. Kabashima, *Critical noise levels for LDPC decoding*, Phys. Rev. E, 66 (2002). [160, 266]

[51] ——, *Magnetization enumerator for LDPC codes – A statistical physics approach*, in Proc. of the IEEE Int. Symposium on Inform. Theory, Lausanne, Switzerland, June 2002, IEEE, p. 256. [160]

[52] C.-C. Wang, S. R. Kulkarni, and H. V. Poor, *Upper bounding the performance of arbitrary finite LDPC codes on binary erasure channels*, in Proc. of the IEEE Int. Symposium on Inform. Theory, Seattle, WA, USA, July 2006, pp. 411–415. [159]

[53] J. S. Yedidia, E. Sudderth, and J.-P. Bouchaud, *Projection algebra analysis of error-correcting codes*, in Proc. of the Allerton Conf. on Commun., Control, and Computing, Monticello, IL, USA, Oct. 2001, p. 662. [159]

[54] J. Zhang and A. Orlitsky, *Finite-length analysis of LDPC codes with large left degrees*, in Proc. of the IEEE Int. Symposium on Inform. Theory, Lausanne, Switzerland, June 30–July 5, 2002, p. 3. [159]

[55] V. Zyablov and M. Pinsker, *Decoding complexity of low-density codes for transmission in a channel with erasures*, Problemy Peredachi Informatsii, 10 (1974), pp. 15–28. [157, 261]

Chapter 4

BINARY MEMORYLESS SYMMETRIC CHANNELS

We now broaden our scope of channels from the binary erasure channel (BEC) to the class of binary memoryless symmetric (BMS) channels. Many concepts encountered during our study of the BEC still apply and reappear suitably generalized. It might at first seem that this expanded class of channels is still restricted and special and that we are only covering a small portion of the large volume of channels encountered in practice. Actually, however, a wide range of situations can be dealt with rather straightforwardly once we have mastered BMS channels. One should therefore view the following as part of the foundation upon which much of communications rests.

Sections 4.2–4.10 recapitulate Sections 3.5–3.6, and 3.8–3.14 in this more general setting and these sections form the core material. The remaining sections can be read in essentially any order. They contain either more advanced topics or less accessible material.

General BMS channels are more mathematically challenging than the BEC. Section 4.1 summarizes the necessary prerequisites. Our advice: quickly skim it so you know what material it contains but do not study it in detail. At any point later, when the need arises, you can return to fill in gaps.

§4.1. Basic Definitions and Examples

§4.1.1. Log-Likelihood Ratios

It is convenient to let the input alphabet be $\{\pm 1\}$ instead of $\{0, 1\}$, which we used for the BEC. We use the standard mapping

$$\mathbb{F}_2 : \begin{array}{ccc} 0 & \longleftrightarrow & +1 \\ 1 & \longleftrightarrow & -1 \end{array} : \{\pm 1\}.$$

Under this mapping the additive group of $\mathbb{F}_2$, modulo-2 addition over $\{0, 1\}$, is faithfully represented as multiplication (of integers) over $\{\pm 1\}$.

We denote the input to the channel by X, $X \in \{\pm 1\}$, and the output by Y, $Y \in \mathcal{Y}$, where $\mathcal{Y}$ can be either discrete or continuous, finite or infinite. To avoid distracting and gratuitous abstraction we nearly always assume that $\mathcal{Y}$ is a subset of the extended reals $\bar{\mathbb{R}} = [-\infty, +\infty]$. We consider only *discrete time* channels and we indicate the channel input and output at time t by X_t and Y_t, respectively.

There are a few BMS channels of particular interest and we frequently illustrate concepts and properties by means of them. One is the BEC, which is discussed in detail in Chapter 3. But we also refer to the binary symmetric channel (BSC), which is discussed in Example 1.3, and the binary additive white Gaussian noise channel (BAWGNC), which we introduce now.

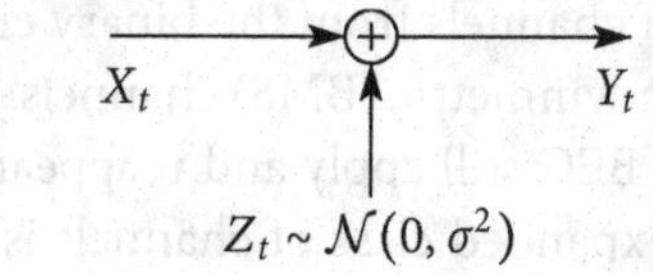

Figure 4.2: BAWGNC(σ).

Example 4.1 (BAWGNC(σ)). Figure 4.2 depicts the *binary additive white Gaussian noise channel* with noise variance σ^2. We denote it by BAWGNC(σ). The input X_t is an element of $\{\pm 1\}$ and the output Y_t is real-valued. More precisely, $Y_t = X_t + Z_t$, where Z_t is a Gaussian random variable with zero mean and variance σ^2. The random variables $\{Z_t\}$ are independent. In the engineering literature several alternative parameterizations are common. We can characterize the channel by E_N/σ^2, the ratio of the energy per transmitted bit E_N to the noise energy σ^2 (this is called the *signal-to-noise ratio*). For our setting we have $E_N = 1$ since $X_t \in \{\pm 1\}$. Alternatively, the measure E_b/N_0 is often quoted. Here, E_b is the energy per transmitted *information bit*, $E_b = E_N/r$, where r is the rate, and $N_0 = 2\sigma^2$ is the so-called double-sided power spectral density. We therefore have $E_b/N_0 = E_N/(2r\sigma^2)$. The signal-to-noise ratio is sometimes quoted in *dB*, this means as $10\log_{10}(E_N/\sigma^2)$, and the same is true for E_b/N_0. You can find a discussion on why these various characterizations are useful on page 195. ◇

Definition 4.3 (Memoryless Channels). A channel, characterized by its *transition probability* $p_{Y|X}(y|x)$, is said to be *memoryless*[1] if

$$p_{Y|X}(y|x) = \prod_t p_{Y_t|X_t}(y_t|x_t). \tag{4.4}$$

The BEC(ϵ), the BSC(ϵ), and the BAWGNC(σ) are all memoryless. Except for some extensions discussed in Section 5.3, all channels considered in this book are memoryless.

---

[1]The above definition is adequate if we restrict ourselves to channels without feedback, as is always the case in this book. More generally, a channel is memoryless if and only if $p_{Y_t|X_t,X_{t-1},\dots,Y_{t-1},\dots}(y_t|x_t, x_{t-1}, \dots, y_{t-1}, \dots) = p_{Y_t|X_t}(y_t|x_t)$.

DEFINITION 4.5 (LOG-LIKELIHOOD RATIO). Consider a binary memoryless channel defined by its transition probability $p_{Y|X}(y|x)$. The associated *log-likelihood ratio function* $l(y)$ is defined as

$$l(y) = \ln \frac{p_{Y|X}(y|1)}{p_{Y|X}(y|-1)}. \tag{4.6}$$

The *log-likelihood ratio* associated with the random variable Y is defined as $L = l(Y)$. It is a random variable itself. Sometimes, slightly abusing notation, we denote it by $L(Y)$. ▽

§4.1.2. L IS A SUFFICIENT STATISTIC

For a binary memoryless channel, L constitutes a *sufficient statistic* with respect to decoding. This means that an optimal decoder can be based on the log-likelihood ratio $l(y)$ instead of on y itself.

LEMMA 4.7 (SUFFICIENT STATISTIC). Assume that $X = (X_1, \dots, X_n)$ is chosen with probability $p_X(x)$ from a binary code C and that transmission takes place over a binary memoryless channel. Let $Y = (Y_1, \dots, Y_n)$ denote the observation at the output of the channel and let $L = (L_1, \dots, L_n)$ denote the corresponding vector of (bit-wise) log-likelihood ratios, i.e., $L_i = l(Y_i)$. Then L constitutes a sufficient statistic for estimating X given Y; L also constitutes a sufficient statistic for estimating X_i given Y.

Proof. To prove that L constitutes a sufficient statistic for estimating X given Y it suffices to show that $p_{Y|X}(y|x)$ can be factored as $a(x,l)b(y)$ for some suitable functions $a(\cdot,\cdot)$ and $b(\cdot)$ (see page 29). Start with $p_{Y|X}(y|x)$ and express it as $\prod_i p_{Y_i|X_i}(y_i|x_i)$ using the fact that the channel is memoryless. Now divide by $\prod_i p_{Y_i|X_i}(y_i|-1)$ and write each ratio

$$p_{Y_i|X_i}(y_i|x_i)/p_{Y_i|X_i}(y_i|-1)$$

as $\exp(l_i \frac{1}{2}(x_i+1))$. We have

$$p_{Y|X}(y|x) = \left(e^{\frac{1}{2}\sum_i x_i l_i}\right)\left(e^{\frac{1}{2}\sum_i l_i}\prod_i p_{Y_i|X_i}(y_i|-1)\right) = a(x,l)b(y),$$

where the last step is valid since l_i is a function of y_i. In a similar manner,

$$\begin{aligned} p_{Y|X_i}(y|x_i) &= \left(\sum_{c\in C: c_i = x_i} e^{\frac{1}{2}\sum_j x_j l_j}\frac{p_X(x)}{p_{X_i}(x_i)}\right)\left(e^{\frac{1}{2}\sum_j l_j}\prod_j p_{Y_j|X_i}(y_j|-1)\right) \\ &= a(x_i,l)b(y), \end{aligned}$$

showing that L constitutes a sufficient statistic for estimating X_i given Y. □

Without essential loss of generality we can therefore assume that the first step of a receiver is to apply the function $l(\cdot)$. In fact, this preprocessing can be interpreted as part of the channel.

§4.1.3. Symmetric Channels

Definition 4.8 (Channel Symmetry). Assume $\mathcal{Y} \subset \bar{\mathbb{R}}$. We say that a binary memoryless channel is *symmetric* (more precisely, output-symmetric) if

$$p_{Y|X}(y|1) = p_{Y|X}(-y|-1). \tag{4.9}$$

Example 4.10 (Symmetry of Standard Channels). Our three standard channels, the BEC, the BSC, as well as the BAWGNC, are all symmetric. ◇

§4.1.4. Distributions

Distributions of log-likelihood ratios $L = l(Y)$ associated with BMS channels play an important role in the study of iterative decoding. To clarify, assume that $X \in \{\pm 1\}$ is transmitted over a BMS channel. Let Y denote the observation and let $L(Y)$ denote the corresponding log-likelihood ratio. Let a denote the density of L assuming that $X = 1$ and let A denote the associated cumulative distribution. We say that a is an L-density and that A is an L-distribution.

Definition 4.11 (Symmetry of L-Distribution). We say that an L-density a is *symmetric* if

$$\mathsf{a}(x) = \mathsf{e}^{x}\mathsf{a}(-x)$$

for all $x \in \mathbb{R}$. Equivalently (see Problem 4.3), we call an L-distribution A symmetric if

$$\int f(x)\mathsf{d}\mathsf{A}(x) = \int \mathsf{e}^{-x} f(-x)\mathsf{d}\mathsf{A}(x)$$

for all bounded continuous functions $f(x)$ so that $f(-x)\mathsf{e}^{-x}$ is bounded. ▽

In Section 4.1.8 we show that L-distributions for BMS channels are always symmetric. Sometimes it is more convenient to work with alternative quantities, i.e., to make a change of variables. One example is the absolute value of the log-likelihood ratio $|L|$. If the distribution of L conditioned on $X = 1$ is symmetric, then it is determined by the distribution of $|L|$, which does not depend on the value of X. We have

$$\mathsf{a}(x) = \mathbb{1}_{\{x \geq 0\}} \frac{1}{1 + \mathsf{e}^{-x}} |\mathsf{a}|(x) + \mathbb{1}_{\{x \leq 0\}} \frac{\mathsf{e}^{x}}{1 + \mathsf{e}^{x}} |\mathsf{a}|(-x),$$

where $|\mathsf{a}|$ denotes the density of $|L|$. This can have important practical consequences; e.g., if you want to estimate $\mathsf{a}(-x)$ for large values of x (where $\mathsf{a}(-x)$ is typically very small since $\mathsf{a}(x) = \mathsf{a}(-x)\mathrm{e}^x$ and since the integral is normalized to 1) then it suffices to estimate the much larger quantity $|\mathsf{a}|(x)$ – an easier task.

Assume that we have *two* observations of X, call them Y_1 and Y_2, resulting from transmitting X over two *independent* BMS channels. Let L_1 and L_2 be the two associated log-likelihood ratios. Then the log-likelihood ratio associated with (Y_1, Y_2) is $L_1 + L_2$ since, due to the conditional independence of Y_1 and Y_2 given X,

$$l(y_1, y_2) = \ln \frac{p_{Y_1,Y_2\,|\,X}(y_1, y_2 \,|\, +1)}{p_{Y_1,Y_2\,|\,X}(y_1, y_2 \,|\, -1)}$$
$$= \ln\Big(\frac{p_{Y_1\,|\,X}(y_1 \,|\, +1)}{p_{Y_1\,|\,X}(y_1 \,|\, -1)} \frac{p_{Y_2\,|\,X}(y_2 \,|\, +1)}{p_{Y_2\,|\,X}(y_2 \,|\, -1)}\Big) = l(y_1) + l(y_2).$$

Recall that addition of independent random variables implies convolution of their densities. Therefore, if a_1 and a_2 are the L-densities associated with Y_1 and Y_2, conditioned on $X = 1$, then it follows that $\mathsf{a}_1 \circledast \mathsf{a}_2$ is the L-density associated with (Y_1, Y_2) conditioned on $X = 1$, where $\circledast$ denotes the standard convolution over $\mathbb{R}$.

A function which is closely related to $l(y)$ is

$$d(y) = \tanh(l(y)/2) = \frac{1 - \mathrm{e}^{-l(y)}}{1 + \mathrm{e}^{-l(y)}} = p_{X\,|\,Y}(1 \,|\, y) - p_{X\,|\,Y}(-1 \,|\, y), \tag{4.12}$$

where in the last step we have used Bayes's rule and assumed that $p_X(1) = p_X(-1) = \frac{1}{2}$. We see that $d(y)$ takes values in the interval $[-1, 1]$. By definition, from $l(y)$ one can compute $d(y)$, but the reverse is true as well. When we conceive of $d(y)$ as a random variable we write $D = d(Y)$. Conditioned on $X = 1$, we have $d(Y) \in (-1, 1]$. A distribution of D conditioned on $X = 1$ is termed a D-distribution and the associated density a D-density.

If $\mathfrak{a}$ is a D-density then symmetry takes the form

$$\frac{\mathfrak{a}(y)}{\mathfrak{a}(-y)} = \frac{1 + y}{1 - y}. \tag{4.13}$$

Equivalently, we say that the D-distribution $\mathfrak{A}$ is symmetric if

$$\int_{-1}^{1} f(y)\mathrm{d}\mathfrak{A}(y) = \int_{-1}^{1} f(-y)\frac{1 - y}{1 + y}\mathrm{d}\mathfrak{A}(y) \tag{4.14}$$

for all bounded continuous functions $f(y)$ on $[-1, 1]$ so that $f(-y)\frac{1-y}{1+y}$ is bounded. Thus, again, when symmetry holds, the D-distribution is completely determined by the distribution of $|D|$. This random variable is distributed on $[0, 1]$. Its distribution is termed a $|D|$-distribution and the associated density is termed a $|D|$-density.

Let us agree on the following slightly unconventional *probabilistic* definition of the *hard-decision* function. This function takes as input a log-likelihood ratio associated with a bit and outputs the hard decision of this bit:

$$\mathfrak{H}(x) = \begin{cases} +1, & \text{if } x > 0, \\ +1, & \text{with probability } \frac{1}{2} \text{ if } x = 0, \\ -1, & \text{with probability } \frac{1}{2} \text{ if } x = 0, \\ -1, & \text{if } x < 0. \end{cases} \tag{4.15}$$

The reason for flipping a fair coin to make the decision if the random variable takes on the value zero is that in some cases (e.g., in the case of transmission over the BEC) the observed random variable has a point mass at zero.

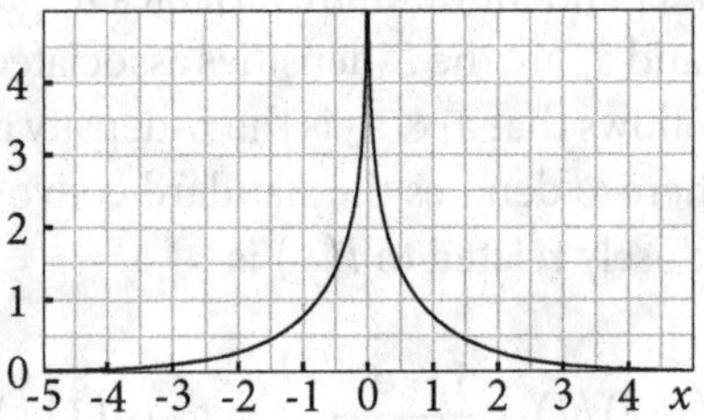

Figure 4.16: $\ln\coth\frac{|x|}{2}$.

Another important quantity is

$$g(y) = (\mathfrak{H}(l(y)), \ln\coth(|l(y)|/2)) = (\mathfrak{H}(d(y)), -\ln|d(y)|). \tag{4.17}$$

A plot of $\ln\coth\frac{|x|}{2}$ is shown in Figure 4.16. We write $G = g(Y)$, or simply $G(Y)$, if we refer to the corresponding random variable. Distributions of G conditioned on $X = 1$ are termed G-distributions and the associated densities G-densities. Note that $g(y)$ takes values in $\{\pm 1\} \times [0, +\infty]$. A G-density $a(s, x)$ therefore has the form

$$a(s, x) = \mathbb{1}_{\{s=1\}} a(1, x) + \mathbb{1}_{\{s=-1\}} a(-1, x).$$

A symmetric G-density exhibits the form

$$a(1, x) = a(-1, x)\coth(x/2). \tag{4.18}$$

As before, when symmetry holds, a G-distribution is completely specified by its corresponding $|G|$-distribution.

We have seen that the convolution of L-densities has an important operational interpretation. Is there a natural way to define convolutions of G-densities and what is its operational significance? G-densities are defined over the product of $\mathbb{R}^+$ and $\{\pm 1\}$. The space $\mathbb{R}^+$ has the well-defined standard convolution. Instead of $\{\pm 1\}$ think of $\mathbb{F}_2$, the set $\{0,1\}$ with modulo-2 addition. The associated convolution is the cyclic convolution of sequences of length two. Therefore, the convolution of G-densities is just the two-dimensional convolution which consists of the familiar convolution over $\mathbb{R}^+$ in one dimension and the convolution over $\mathbb{F}_2$ in the other dimension. In other words, the new convolution is a convolution over the group $\mathbb{F}_2 \times [0,+\infty]$. Explicitly, the convolution of

$$\mathbb{1}_{\{s=1\}} a_1(1,x) + \mathbb{1}_{\{s=-1\}} a_1(-1,x) \quad \text{and} \quad \mathbb{1}_{\{s=1\}} a_2(1,x) + \mathbb{1}_{\{s=-1\}} a_2(-1,x)$$

is

$$\begin{aligned} &\mathbb{1}_{\{s=1\}}\big(a_1(1,\cdot) \star a_2(1,\cdot) + a_1(-1,\cdot) \star a_2(-1,\cdot)\big) + \\ &\mathbb{1}_{\{s=-1\}}\big(a_1(-1,\cdot) \star a_2(1,\cdot) + a_1(1,\cdot) \star a_2(-1,\cdot)\big), \end{aligned}$$

where $\star$ denotes the (one-sided) convolution of standard distributions. We denote this convolution by the symbol $\circledast$. We will have much more to say about G-densities and their significance. Here is a quick preview. We are given X_1 and $X_2 \in \{\pm 1\}$. Define $X = X_1 \cdot X_2$. Let Y_i be the result of transmitting X_i over a BMS channel, $i = 1,2$, where the two channels are independent. If a_i is the G-density of Y_i conditioned on $X_i = 1$, then $a = a_1 \boxplus a_2$ is the G density of (Y_1, Y_2) conditioned on $X = 1$.

We have just seen that the variables of interest to us can be represented in the L-, D-, or G-domain and that the L- and the G-domain have associated convolutions. The following convention will limit our notational burden. We write $\circledast$ if we mean the convolution in the L-domain and $\boxast$ if we refer to the convolution in the G-domain, regardless of the representation of the density which is used: $\mathsf{a} \boxast \mathsf{b}$ denotes the density which is the result of transforming both a and b into the G-domain, then performing the $\boxast$ convolution, and finally transforming the result back into the L-domain. We will see shortly that $\circledast$ describes how the distribution of the messages changes at a *variable* node under the so-called *belief-propagation* decoder, while $\boxast$ describes the change of this distribution at the *check* node side.

Since channels can be represented by their distributions, very often we use a distribution to indicate the associated channel. It is sometimes convenient to represent this in the following graphical way:

L-densities	$x \xrightarrow{\mathsf{a}} Y,$	$Z \sim \mathsf{a},$	$Y = xZ,$
$\lvert L\rvert$-densities	$x \xrightarrow{\lvert\mathsf{a}\rvert} (S,Y),$	$Y \sim \lvert\mathsf{a}\rvert,$	$x \xrightarrow{\text{BSC}\left(\frac{e^{-Y}}{1+e^{-Y}}\right)} S,$
D-densities	$x \xrightarrow{\mathfrak{a}} Y,$	$Z \sim \mathfrak{a},$	$Y = xZ,$

$|D|$-densities $\qquad x \xrightarrow{|\mathfrak{a}|} (S, Y), \quad Y \sim |\mathfrak{a}|, \quad x \xrightarrow{\mathrm{BSC}(\frac{1-Y}{2})} S.$

To be concrete, consider the first line. It gives an operational interpretation of the L-representation of a symmetric density a: the transmitted bit $x \in \{\pm 1\}$ is multiplied by a random variable Z, which is distributed according to a. If, on the other hand, we characterize the channel by its $|L|$-density, then it is natural to represent the observation by the tuple (S, Y). Here, Y is distributed according to $|\mathrm{a}|$ and gives the *reliability* of the observation, while S is the result of sending x through a BSC with cross-over parameter $\frac{e^{-Y}}{1+e^{-Y}}$. We can think of S as the *sign*. The interpretations using D-densities is similar.

§4.1.5. Distributions - Examples

Let us determine the distributions associated with our three standard examples and and demonstrate symmetry in each case. For $z \in \mathbb{R}$, let H_z denote the (Heavyside) distribution defined by

$$\mathsf{H}_z(x) = \begin{cases} 0, & x < z, \\ 1, & x \geq z. \end{cases}$$

The density associated with H_z is Δ_z, defined by $\Delta_z(x) = \Delta_0(x - z)$, where $\Delta_0(x)$ is the (Dirac) delta of unit mass centered at zero.

Example 4.19 (Distributions for the BEC(ϵ)). Consider the BEC(ϵ) as in Chapter 3 but assume that the input takes values in $\{\pm 1\}$. Let $\mathsf{A}_{\mathrm{BEC}(\epsilon)}(y)$ denote the L-distribution, assuming that $X = 1$. Note that Y can only take on the values 1 or ?. We have $l(1) = \ln((\bar{\epsilon})/0) = +\infty$ and, since $p_{Y|X}(1 \mid 1) = \bar{\epsilon}$, this occurs with probability $\bar{\epsilon}$. In the same manner $l(?) = \ln(\epsilon/\epsilon) = 0$, which occurs with probability ϵ. It follows that

$$\mathsf{A}_{\mathrm{BEC}(\epsilon)}(y) = |\mathsf{A}_{\mathrm{BEC}(\epsilon)}|(y) = \epsilon \mathsf{H}_0(y),$$
$$\mathsf{a}_{\mathrm{BEC}(\epsilon)}(y) = |\mathsf{a}_{\mathrm{BEC}(\epsilon)}|(y) = \epsilon \Delta_0(y) + \bar{\epsilon}\Delta_{+\infty}(y).$$

Since both Δ_0 and $\Delta_{+\infty}$ are symmetric L-distributions, it follows that $\mathsf{a}_{\mathrm{BEC}(\epsilon)}$ is symmetric. The D-distribution and associated D-density are given by

$$\mathfrak{A}_{\mathrm{BEC}(\epsilon)}(y) = |\mathfrak{A}_{\mathrm{BEC}(\epsilon)}|(y) = \epsilon \mathsf{H}_0(y) + \bar{\epsilon}\mathsf{H}_1(y),$$
$$\mathfrak{a}_{\mathrm{BEC}(\epsilon)}(y) = |\mathfrak{a}_{\mathrm{BEC}(\epsilon)}|(y) = \epsilon \Delta_0(y) + \bar{\epsilon}\Delta_1(y).$$

In both cases it is understood that the functions are only defined on $[-1, 1]$. Since both Δ_0 and Δ_1 are symmetric D-distributions, it follows that $\mathfrak{a}_{\mathrm{BEC}(\epsilon)}$ is symmetric. Finally, the G-distribution and associated G-density are

$$A_{\mathrm{BEC}(\epsilon)}(1, y) = \bar{\epsilon}\mathsf{H}_0(y), \qquad A_{\mathrm{BEC}(\epsilon)}(-1, y) = 0,$$

$$a_{\mathrm{BEC}(\epsilon)}(1,y) = \bar{\epsilon}\Delta_0(y) + \frac{\epsilon}{2}\Delta_{+\infty}(y), \quad a_{\mathrm{BEC}(\epsilon)}(-1,y) = \frac{\epsilon}{2}\Delta_{+\infty}(y).$$

Symmetry is checked easily using (4.13). Of course, symmetry of a distribution in any representation (L, D, G) implies symmetry in all other representations. ◇

Example 4.20 (Distributions for the BSC(ϵ)). The L-distribution assuming $X = 1$ is

$$\mathsf{A}_{\mathrm{BSC}(\epsilon)}(y) = \epsilon \mathsf{H}_{-\ln\frac{\bar{\epsilon}}{\epsilon}}(y) + \bar{\epsilon}\mathsf{H}_{\ln\frac{\bar{\epsilon}}{\epsilon}}(y), \quad |\mathsf{A}_{\mathrm{BSC}(\epsilon)}|(y) = \mathsf{H}_{\ln\frac{\bar{\epsilon}}{\epsilon}}(y),$$
$$\mathsf{a}_{\mathrm{BSC}(\epsilon)}(y) = \epsilon \Delta_{-\ln\frac{\bar{\epsilon}}{\epsilon}}(y) + \bar{\epsilon}\Delta_{\ln\frac{\bar{\epsilon}}{\epsilon}}(y), \quad |\mathsf{a}_{\mathrm{BSC}(\epsilon)}|(y) = \Delta_{\ln\frac{\bar{\epsilon}}{\epsilon}}(y).$$

For $y \neq \pm y_0 = \ln\frac{\bar{\epsilon}}{\epsilon}$, we have $\mathsf{a}_{\mathrm{BSC}(\epsilon)}(\pm y) = 0$, so symmetry follows from

$$\mathsf{a}_{\mathrm{BSC}(\epsilon)}(y_0) = \bar{\epsilon} = \mathrm{e}^{\ln\frac{\bar{\epsilon}}{\epsilon}}\epsilon = \mathrm{e}^{y_0}\mathsf{a}_{\mathrm{BSC}(\epsilon)}(-y_0).$$

Similarly, the D-distribution and the corresponding D-density are

$$\mathfrak{A}_{\mathrm{BSC}(\epsilon)}(y) = \epsilon \mathsf{H}_{-(1-2\epsilon)}(y) + \bar{\epsilon}\mathsf{H}_{1-2\epsilon}(y), \quad |\mathfrak{A}_{\mathrm{BSC}(\epsilon)}|(y) = \mathsf{H}_{1-2\epsilon}(y),$$
$$\mathfrak{a}_{\mathrm{BSC}(\epsilon)}(y) = \epsilon \Delta_{-(1-2\epsilon)}(y) + \bar{\epsilon}\Delta_{1-2\epsilon}(y), \quad |\mathfrak{a}_{\mathrm{BSC}(\epsilon)}|(y) = \Delta_{1-2\epsilon}(y).$$

To verify the symmetry in this representation directly using (4.13) note that

$$\frac{\bar{\epsilon}}{\epsilon} = \frac{1 + (1-2\epsilon)}{1-(1-2\epsilon)}.$$

Finally, the G-distribution and associated G-density are

$$A_{\mathrm{BSC}(\epsilon)}(1,y) = \bar{\epsilon}\mathsf{H}_{-\ln(1-2\epsilon)}(y), \quad A_{\mathrm{BSC}(\epsilon)}(-1,y) = \epsilon \mathsf{H}_{-\ln(1-2\epsilon)}(y),$$
$$a_{\mathrm{BSC}(\epsilon)}(1,y) = \bar{\epsilon}\Delta_{-\ln(1-2\epsilon)}(y), \quad a_{\mathrm{BSC}(\epsilon)}(-1,y) = \epsilon \Delta_{-\ln(1-2\epsilon)}(y).$$

We can verify the symmetry in this representation directly using (4.18). We have

$$\bar{\epsilon} = \epsilon \coth\left(\frac{-\ln(1-2\epsilon)}{2}\right) = \epsilon\frac{\bar{\epsilon}}{\epsilon}.$$ ◇

Example 4.21 (Distributions for BAWGNC(σ)). For the BAWGNC(σ) it is easier to specify the L-density directly (instead of the L-distribution):

$$\mathsf{a}_{\mathrm{BAWGNC}(\sigma)}(y) = \sqrt{\frac{\sigma^2}{8\pi}}\mathrm{e}^{-\frac{(y-\frac{2}{\sigma^2})^2\sigma^2}{8}}.$$

This is a Gaussian with mean $\frac{2}{\sigma^2}$ and variance $\frac{4}{\sigma^2}$. Symmetry is verified by

$$\mathsf{a}_{\text{BAWGNC}(\sigma)}(y) = \sqrt{\frac{\sigma^2}{8\pi}}\mathrm{e}^{-\frac{(y-\frac{2}{\sigma^2})^2\sigma^2}{8}} = \sqrt{\frac{\sigma^2}{8\pi}}\mathrm{e}^{y}\mathrm{e}^{-\frac{(-y-\frac{2}{\sigma^2})^2\sigma^2}{8}} = \mathrm{e}^{y}\mathsf{a}_{\text{BAWGNC}(\sigma)}(-y).$$

The D-density is given by

$$\mathfrak{a}_{\text{BAWGNC}(\sigma)}(y) = \frac{\sigma}{\sqrt{2\pi}(1-y^2)}\mathrm{e}^{-\frac{(1-\sigma^2\tanh^{-1}(y))^2}{2\sigma^2}} = \frac{\mathrm{e}^{-\frac{4+\sigma^4\ln^2((1+y)/(1-y))}{8\sigma^2}}}{\sqrt{2\pi}(1-y^2)}\sqrt{\frac{1+y}{1-y}}.$$

Symmetry is easily checked using the last formulation. Finally, the G-density is

$$\begin{aligned} a_{\text{BAWGNC}(\sigma)}(\pm 1, y) &= \mathrm{e}^{-\frac{(2\mp\sigma^2\ln\coth(y/2))^2}{8\sigma^2}}\frac{\sigma}{\sqrt{8\pi}\sinh(y)} \\ &= \mathrm{e}^{-\frac{4+\sigma^4(\ln\coth(y/2))^2}{8\sigma^2}}\frac{\sigma}{\sqrt{8\pi}\sinh(y)}(\coth(y/2))^{\pm\frac{1}{2}}. \end{aligned}$$

The second representation allows us to check the symmetry condition in a straightforward fashion. Figure 4.22 shows the L-density, the D-density, as well as the corresponding G-density for $\sigma = 5/4$. ◇

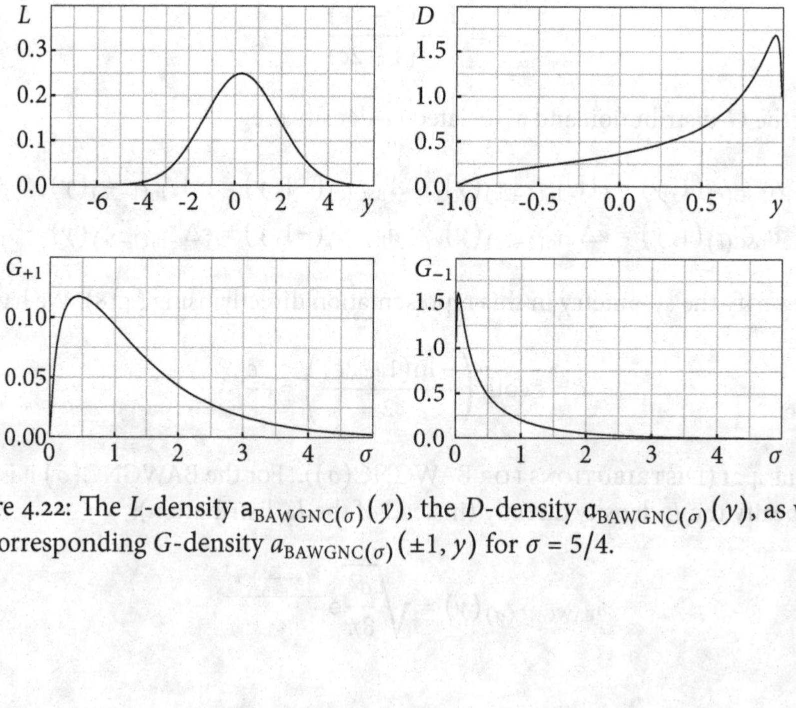

Figure 4.22: The L-density $\mathsf{a}_{\text{BAWGNC}(\sigma)}(y)$, the D-density $\mathfrak{a}_{\text{BAWGNC}(\sigma)}(y)$, as well as the corresponding G-density $a_{\text{BAWGNC}(\sigma)}(\pm 1, y)$ for $\sigma = 5/4$.

§4.1.6. Distributions - The Fine Print

As we have just seen, for some (standard) channels the L-, D-, and G-distributions show esoteric mathematical features, including the occurrence of (Dirac) delta functions at infinity. Some of the results in this book depend on technical properties of the space of densities. Thus, a relatively detailed description of fundamental definitions and their consequences is unavoidable.

We start with L-distributions. Let $\mathcal{A}_L$ denote the space of right-continuous, non-decreasing functions A defined over $\mathbb{R}$ satisfying

$$\lim_{x\to-\infty} \mathsf{A}(x) = 0, \qquad \lim_{x\to+\infty} \mathsf{A}(x) \le 1.$$

To each $\mathsf{A} \in \mathcal{A}_L$ we associate a random variable X over $(-\infty, +\infty]$. The random variable X has *law* or *distribution* A, i.e., $\mathbb{P}\{X \in (-\infty, x]\} = \mathsf{A}(x)$. The reason we allow $\lim_{x\to+\infty} \mathsf{A}(x) \le 1$ (i.e., we allow an inequality) is to permit X to have some probability mass at $+\infty$. Indeed, $\mathbb{P}\{X = +\infty\} = 1-\lim_{x\to+\infty} \mathsf{A}(x)$. Given an element $\mathsf{A} \in \mathcal{A}_L$ we define $\mathsf{A}^-(x)$ to be the left limit of A at x, i.e., $\mathsf{A}^-(x) = \lim_{y\uparrow x} \mathsf{A}(y)$. Note that $\mathsf{A}^-(x)$ is left continuous.

Generally, we work with "densities" over $(-\infty, +\infty]$. Formally, these densities are (Radon-Nikodyn) derivatives of elements of $\mathcal{A}_L$. The derivative, when it exists, is the density of the associated random variable X over $(-\infty, +\infty)$, although there may be an additional point mass at $+\infty$. We use densities primarily in the following way: given $\mathsf{A} \in \mathcal{A}_L$ the Lebesgue-Stieltjes integral $\int g(x)\mathrm{d}\mathsf{A}(x)$ is well defined for, e.g., non-negative continuous functions g. If a is the density corresponding to the distribution A we write $\int g(x)\mathsf{a}(x)\mathrm{d}x$ as a proxy for $\int g(x)\mathrm{d}\mathsf{A}(x)$. If $\mathsf{A}^-(+\infty) < 1$ and if the limit $\lim_{x\to+\infty} g(x)$ exists, then one has to include the term

$$(1 - \mathsf{A}^-(+\infty))(\lim_{x\to+\infty} g(x))$$

in the definition of $\int g(x)\mathsf{a}(x)\mathrm{d}x$.

Given $\mathsf{A}, \mathsf{B} \in \mathcal{A}_L$, their *convolution* $\mathsf{A} \circledast \mathsf{B}$ is defined by $(\mathsf{A} \circledast \mathsf{B})(x) = \int \mathsf{A}(x - y)\mathrm{d}\mathsf{B}(y) = \int \mathsf{B}(x - y)\mathrm{d}\mathsf{A}(y)$. The integral is defined for almost all x and right continuity determines the rest. Note that $\mathsf{A} \circledast \mathsf{B} \in \mathcal{A}_L$. This generalizes the notion of convolution of densities, for if A and B have corresponding densities a and b, respectively, then $(\mathsf{A} \circledast \mathsf{B})$ is the distribution corresponding to the density $\mathsf{a} \circledast \mathsf{b}$. We write $\mathsf{a} \circledast \mathsf{b}$ to indicate the density associated with the distribution $\mathsf{A} \circledast \mathsf{B} \in \mathcal{A}_L$. One can check that if Z_1 and Z_2 are independent random variables over $(-\infty, +\infty]$ with distributions A_{Z_1} and A_{Z_2}, respectively, then the distribution of $Z_1 + Z_2$ is $\mathsf{A}_{Z_1} \circledast \mathsf{A}_{Z_2}$ (as is the case for independent random variables defined over $(-\infty, +\infty)$). This provides the formal framework for L-distributions.

There is not much to add for D-distributions. We denote the space of right-continuous distributions on $(-1, 1]$ by $\mathcal{A}_D$. For each $\mathfrak{A} \in \mathcal{A}_D$ we have $\mathfrak{A}(1) = 1$ and $\lim_{y \to -1} \mathfrak{A}(y) = 0$.

Finally, the following is the fine print for G-distributions. Let $A(s, x)$ be a G-distribution. Write $A(s, x)$ in the form

$$A(s, x) = \mathbb{1}_{\{s=1\}} A(1, x) + \mathbb{1}_{\{s=-1\}} A(-1, x),$$

where $A(1, x)$ and $A(-1, x)$ are non-decreasing and right continuous,

$$\lim_{x \to +\infty} A(1, x) \geq \lim_{x \to +\infty} A(-1, x),$$

and where $A(1, 0) \geq 0$ and $A(-1, 0) = 0$ (the last two conditions correspond to the conditions $\lim_{x \to +\infty} \mathsf{A}(x) \leq 1$ and $\lim_{x \to -\infty} \mathsf{A}(x) = 0$ for functions in $\mathcal{A}_L$). We speak of densities over $\{\pm 1\} \times [0, +\infty]$,

$$a(s, x) = \mathbb{1}_{\{s=1\}} a(1, x) + \mathbb{1}_{\{s=-1\}} a(-1, x),$$

by substituting for $A(1, x)$ and $A(-1, x)$ their associated densities. The definition is analogous to that used for $\mathcal{A}_L$ except that, here, $a(1, x)$ has a point mass at $x = 0$ of magnitude $A(1, 0)$, and both $a(1, x)$ and $a(-1, x)$ have point masses at $x = +\infty$ of magnitude $\frac{1}{2}(1 - \lim_{x \to +\infty} A(1, x) - \lim_{x \to +\infty} A(-1, x))$. We split this point mass (which corresponds to the probability of erasure) evenly so that the symmetry condition (4.18) stays fulfilled also for $x = +\infty$. We are only interested in densities over $\{\pm 1\} \times [0, +\infty]$ that satisfy these conditions. We denote the space of such distributions by $\mathcal{A}_G$.

Let Γ be the map which maps L-distributions into G-distributions and let Γ^{-1} be its inverse. Let A be an L-distribution and let A be the corresponding G-distribution. We then have

$$\begin{aligned} A(-1, x \geq 0) &= \mathsf{A}(-\ln \coth(x/2)), \\ \mathsf{A}(x \geq 0) &= 1 - A^{-}(1, \ln \coth(x/2)), \ \ \mathsf{A}(x < 0) = A(-1, \ln \coth(-x/2)), \end{aligned}$$

and $\mathsf{A}(0) = \frac{1}{2}(1 - \lim_{x \to +\infty} A(1, x) - \lim_{x \to +\infty} A(-1, x))$. One can check that $\Gamma^{-1}(\Gamma(\mathsf{A})) = \mathsf{A}$ for all $\mathsf{A} \in \mathcal{A}_L$. Further, Γ and Γ^{-1} are linear operators on the spaces $\mathcal{A}_L$ and $\mathcal{A}_G$, respectively. For convenience, although it constitutes an abuse of notation, we apply Γ and Γ^{-1} to densities as well. It is implicitly understood that the notation is a representation of the appropriate operation applied to distributions.

The space $\mathcal{A}_G$ has a well-defined convolution: the convolution of

$$\mathbb{1}_{\{s=1\}} A(1, x) + \mathbb{1}_{\{s=-1\}} A(-1, x) \quad \text{and} \quad \mathbb{1}_{\{s=1\}} B(1, x) + \mathbb{1}_{\{s=-1\}} B(-1, x)$$

is the distribution

$$\mathbb{1}_{\{s=1\}}\big(A(1,\cdot) \star B(1,\cdot) + A(-1,\cdot) \star B(-1,\cdot)\big)+$$
$$\mathbb{1}_{\{s=-1\}}\big(A(-1,\cdot) \star B(1,\cdot) + A(1,\cdot) \star B(-1,\cdot)\big),$$

where $\star$ denotes the (one-sided) convolution of standard distributions. In other words, the new convolution is a convolution over the group $\mathbb{F}_2 \times [0,+\infty]$. We denote this convolution by the symbol $\circledast$. Again, we shall allow the convolution operator to act on the densities associated with elements of $\mathcal{A}_G$ with the implicit understanding that the preceding discussion provides the rigorous definition.

§4.1.7. Distributions - Convergence

In the analysis of iterative decoding we are often concerned with the issue of *convergence* of distributions. Let us make this precise.

Definition 4.23 (Convergence in Distribution/Weak Convergence). We say that a sequence of D-densities $\{\mathfrak{a}_i\}$ converges *in distribution* to a density $\mathfrak{a}$ if their cumulative distributions $\{\mathfrak{A}_i\}$ converge pointwise at all points of continuity of $\mathfrak{A}$. Rather than looking at D-distributions, it is equivalent to consider the pointwise convergence of L-distributions or G-distributions.

Convergence in distribution is equivalent to weak convergence. We say that a sequence of D-densities $\{\mathfrak{a}_i\}$ *converges weakly* to the density $\mathfrak{a}$ if for all bounded and continuous functions $f(x)$ on $[-1,1]$

$$\lim_{i\to\infty}\int_{-1}^{1} f(x)\mathrm{d}\mathfrak{A}_i(x) = \int_{-1}^{1} f(x)\mathrm{d}\mathfrak{A}(x).$$

Naturally, we can invoke the equivalent condition for densities in any of the standard representations. ▽

Lemma 4.24 (The Limit of Symmetric Densities is Symmetric). If a sequence of symmetric densities converges in distribution/weakly then the limit density is symmetric.

Proof. Assume that the sequence of symmetric D-densities with cumulative distributions $\{\mathfrak{A}_i\}$ converges weakly to the D-density with cumulative distribution $\mathfrak{A}$. By assumption each $\mathfrak{A}_i$ is symmetric. Using the characterization of symmetry (4.14), this means that for all bounded continuous $f(x)$ on $[-1,1]$ so that $f(-x)\frac{1-x}{1+x}$ is bounded we have

$$\int_{-1}^{1} f(x)\mathrm{d}\mathfrak{A}_i(x) = \int_{-1}^{1} f(-x)\frac{1-x}{1+x}\mathrm{d}\mathfrak{A}_i(x).$$

By assumption the sequence $\{\mathfrak{A}_i\}$ converges weakly to $\mathfrak{A}$. Since both $f(x)$ as well as $f(-x)\frac{1-x}{1+x}$ are continuous and bounded we can take the limit on both sides to conclude that $\mathfrak{A}_\infty$ is symmetric. □

A technical but important point is the following.

LEMMA 4.25 (SEQUENTIAL COMPACTNESS). The space of symmetric densities is *sequentially compact.*

Discussion: If you are not familiar with sequential compactness you need not worry. We will use this result sparingly and primarily as a technical convenience. It means that given a sequence of symmetric densities there exists a subsequence that converges to a limit density and that this limit density is symmetric (which we know from Lemma 4.24).

Proof. To be concrete, consider a sequence of D-densities $\{\mathfrak{a}_i\}$ and their associated D-distributions $\{\mathfrak{A}_i\}$. Let $\{y_i\}_{i\in\mathbb{N}}$ be an enumeration of the rational points in $[-1,1]$. (This is a countable set.)

We use a standard diagonalization procedure. First, find a subsequence $\{1i\}_{i\in\mathbb{N}}$ of $\mathbb{N}$ so that $\{\mathfrak{A}_{1i}(y_1)\}$ converges. This can be done since each distribution takes values in $[0,1]$, which is compact (and so also sequentially compact). Next find $\{2i\}_{i\in\mathbb{N}}$, a subsequence of $\{1i\}_{i\in\mathbb{N}}$, so that $\{\mathfrak{A}_{2i}(y_2)\}$ converges. Continue in this fashion. By construction, the sequence of distributions $\{\mathfrak{A}_{ii}\}$ (this is the sequence of "diagonal" elements) converges at all rational points of $[-1,1]$.

Since each $\mathfrak{A}_{ii}$ is non-decreasing it follows that the limit (on the rationals) is non-decreasing. Using right continuity we uniquely define the limit distribution $\mathfrak{A}_\infty$. If x is any point of continuity of $\mathfrak{A}_\infty$ then by looking at rational points arbitrarily close to x we see that $\{\mathfrak{A}_{ii}(x)\}$ converges to $\mathfrak{A}_\infty(x)$. It follows that $\{\mathfrak{A}_{ii}(x)\}$ converges in distribution/weakly to $\mathfrak{A}_\infty(x)$.

From Lemma 4.24 we know that the weak limit of symmetric densities is symmetric. □

§4.1.8. BMS CHANNELS HAVE SYMMETRIC DISTRIBUTIONS

The symmetry of the BEC, the BSC, and the BAWGNC is no coincidence. As the following theorem asserts, the L-distribution for any BMS channel is symmetric (and hence so are the D-distribution and the G-distribution).

THEOREM 4.26 (SYMMETRY OF L-DISTRIBUTION FOR BMS CHANNELS). Consider a BMS channel with transition probability $p_{Y\mid X}(y\mid x)$ and let A denote the associated L-distribution. Then A is symmetric.

Proof. Recall that the L-density is the density of $l(Y)$ conditioned on $X = 1$. In order to avoid technicalities, let us assume that a is continuous.

According to Definition 4.11 we need to show that $\mathsf{a}(z) = \mathrm{e}^z \mathsf{a}(-z)$ for all $z \in \mathbb{R}$. Consider a small interval $[z, z + \Delta z]$. Then we have

$$\mathsf{a}(z)\Delta z \approx \mathbb{P}\{Y \in l^{-1}([z, z + \Delta z])\} = \int_{y \in l^{-1}([z,z+\Delta z])} p_{Y|X}(y|1)\mathrm{d}y,$$
$$\mathsf{a}(-z)\Delta z \approx \mathbb{P}\{Y \in l^{-1}([-z - \Delta z, -z])\} = \int_{y \in l^{-1}([-z-\Delta z,-z])} p_{Y|X}(y|1)\mathrm{d}y.$$

Using the channel symmetry condition (4.9) we have

$$l(y) = \ln \frac{p_{Y|X}(y|1)}{p_{Y|X}(y|-1)} = \ln \frac{p_{Y|X}(-y|-1)}{p_{Y|X}(-y|1)} = -l(-y). \tag{4.27}$$

This implies that $l^{-1}(z) = -l^{-1}(-z)$. We therefore have

$$\begin{aligned}\mathsf{a}(z)\Delta z &\approx \int_{y \in l^{-1}([z,z+\Delta z])} p_{Y|X}(y|1)\mathrm{d}y \overset{\text{(i)}}{\approx} \int_{y \in l^{-1}([z,z+\Delta z])} \mathrm{e}^z p_{Y|X}(y|-1)\mathrm{d}y \\ &\overset{(4.9)}{=} \int_{y \in l^{-1}([z,z+\Delta z])} \mathrm{e}^z p_{Y|X}(-y|1)\mathrm{d}y \\ &\overset{(4.27)}{=} \mathrm{e}^z \int_{y \in l^{-1}([-z-\Delta z,-z])} p_{Y|X}(y|1)\mathrm{d}y \approx \mathrm{e}^z \mathsf{a}(-z)\Delta z.\end{aligned}$$

In step (i) we used the fact that for all $y \in \mathcal{Y}$ we have $p_{Y|X}(y|1) = \mathrm{e}^{l(y)} p_{Y|X}(y|-1)$. The proof concludes by letting Δz tend to zero. □

Motivated by Lemma 4.7, we say that two BMS channels are *equivalent* if they have the same L-distribution. It is often convenient to pick one representative from each equivalence class. This can be done as shown in the following lemma.

Lemma 4.28 (Channel Equivalence Lemma). Let $\mathsf{a}(y)$ be a symmetric L-density. The binary symmetric channel $p_{Y|X}(\cdot|\cdot)$ with $p_{Y|X}(y|1) = \mathsf{a}(y)$ (and, hence, by symmetry $p_{Y|X}(y|-1) = \mathsf{a}(-y)$) has an associated L-density equal to $\mathsf{a}(y)$.

Proof.

$$l(y) = \ln \frac{p_{Y|X}(y|1)}{p_{Y|X}(y|-1)} = \ln \frac{\mathsf{a}(y)}{\mathsf{a}(-y)} = \ln \frac{\mathsf{a}(y)}{\mathsf{a}(y)\mathrm{e}^{-y}} = y.$$ □

§4.1.9. APP Processing and Symmetry

We have just seen that for a BMS channel the distribution of L, conditioned on $X = 1$, is symmetric. This symmetry is preserved under *a posteriori probability* (APP) processing.

Theorem 4.29 (Linear Codes and APP Processing). Let X be a vector of length n chosen with probability $p_X(x)$ from a binary linear code C. Assume that transmission takes place over a binary and memoryless channel and let Y denote the output of the channel. Define

$$\phi_i(y_{\sim i}) = \ln \frac{p_{X_i \mid Y_{\sim i}}(+1 \mid y_{\sim i})}{p_{X_i \mid Y_{\sim i}}(-1 \mid y_{\sim i})}, \tag{4.30}$$

and $\Phi_i = \phi_i(Y_{\sim i})$. Further, let $l_i(y_i) = \ln \frac{p_{Y_i \mid X_i}(y_i \mid +1)}{p_{Y_i \mid X_i}(y_i \mid -1)}$ and $L_i = l_i(Y_i)$. Then Φ_i constitutes a sufficient statistic for estimating X_i given $Y_{\sim i}$ and $L_i + \Phi_i$ constitutes a sufficient statistic for estimating X_i given Y.

Suppose that $p_X(x)$ is uniform, that the i-th component of C is proper, and that the channel is symmetric. In this case the distribution of Φ_i, assuming that $X_i = 1$, is equal to the distribution of Φ_i, assuming that the all-one codeword was transmitted. Further, the "channel" $p_{\Phi_i \mid X_i}(\phi_i \mid x_i)$ is symmetric and ϕ_i is a log-likelihood ratio.

Discussion: The quantity $\Phi_i = \phi_i(Y_{\sim i})$ is called the *extrinsic* MAP estimate (of X_i). The word "extrinsic" refers to the fact that we base our estimate on $Y_{\sim i}$ and do not include the direct observation Y_i.

Proof. Start with $p_{Y_{\sim i} \mid X_i}(y_{\sim i} \mid x_i)$. Use Bayes's rule, divide the resulting expression by $p_{X_i \mid Y_{\sim i}}(-1 \mid y_{\sim i})$, and rewrite $p_{X_i \mid Y_{\sim i}}(x_i \mid y_{\sim i}) / p_{X_i \mid Y_{\sim i}}(-1 \mid y_{\sim i})$ as $e^{\Phi_i \frac{(x_i+1)}{2}}$. From this we see that

$$p_{Y_{\sim i} \mid X_i}(y_{\sim i} \mid x_i) = \frac{e^{\Phi_i \frac{x_i}{2}}}{p_{X_i}(x_i)} \Big(e^{\frac{\Phi_i}{2}} p_{X_i \mid Y_{\sim i}}(-1 \mid y_{\sim i}) p_{Y_{\sim i}}(y_{\sim i}) \Big) = \frac{e^{x_i \frac{\Phi_i}{2}}}{p_{X_i}(x_i)} b(y_{\sim i}).$$

According to our discussion on page 29, this shows that Φ_i constitutes a sufficient statistic for estimating X_i given $Y_{\sim i}$. In a similar manner,

$$p_{Y \mid X_i}(y \mid x_i) = \frac{e^{x_i \frac{\Phi_i + L_i}{2}}}{p_{X_i}(x_i)} b(y).$$

This shows that $L_i + \Phi_i$ constitutes a sufficient statistic for estimating X_i given Y.

If we assume that X is chosen uniformly at random from a binary linear code whose i-th component is proper then it follows that X_i has a uniform prior. Therefore, applying Bayes's rule to both denominator and numerator of the ratio (4.30), we see that Φ_i is in fact a log-likelihood ratio. We now show that the distribution of Φ_i conditioned on $X = c$, $c \in C$, is only a function of c_i. Let $\phi_i^{-1}(z)$ denote the set of all $y_{\sim i} \in \mathbb{R}^{n-1}$ such that $\phi_i(y_{\sim i}) = z$. Now, for any codeword $w \in C$, we have

$$\begin{aligned}
\frac{p_{Y_{\sim i}|X_i}(y_{\sim i}\,|\,1)}{p_{Y_{\sim i}|X_i}(y_{\sim i}\,|\,-1)} &= \frac{\sum_{u\in C:u_i=1} p_{Y_{\sim i}|X}(y_{\sim i}\,|\,u)}{\sum_{u\in C:u_i=-1} p_{Y_{\sim i}|X}(y_{\sim i}\,|\,u)} \\
\text{BMS channel} \qquad &= \frac{\sum_{u\in C:u_i=1} p_{Y_{\sim i}|X}(w_{\sim i}y_{\sim i}\,|\,uw)}{\sum_{u\in C:u_i=-1} p_{Y_{\sim i}|X}(w_{\sim i}y_{\sim i}\,|\,uw)} \\
\text{by linearity of } C \qquad &= \frac{\sum_{u\in C:u_i=w_i} p_{Y_{\sim i}|X}(w_{\sim i}y_{\sim i}\,|\,u)}{\sum_{u\in C:u_i=-w_i} p_{Y_{\sim i}|X}(w_{\sim i}y_{\sim i}\,|\,u)} \\
&= \frac{p_{Y|X_i}(w_{\sim i}y_{\sim i}\,|\,w_i)}{p_{Y|X_i}(w_{\sim i}y_{\sim i}\,|\,-w_i)},
\end{aligned}$$

so that for all $w \in C$

$$y_{\sim i} \in \phi_i^{-1}(z) \Leftrightarrow w_{\sim i}y_{\sim i} \in \phi_i^{-1}(w_i z). \tag{4.31}$$

Let $\mathsf{A}_i^u(z)$ denote the cumulative distribution of Φ_i conditioned that u was transmitted, $u \in C$. If $u, w \in C$ then

$$\begin{aligned}
\mathsf{A}_i^u(z) &= \int_{y_{\sim i}\in\phi_i^{-1}((-\infty,z])} p_{Y|X}(y_{\sim i}\,|\,u)\mathrm{d}y_{\sim i} \\
\text{BMS channel} \qquad &= \int_{y_{\sim i}\in\phi_i^{-1}((-\infty,z])} p_{Y|X}(u_{\sim i}w_{\sim i}y_{\sim i}\,|\,w)\mathrm{d}y_{\sim i} \\
&= \int_{u_{\sim i}w_{\sim i}y_{\sim i}\in\phi_i^{-1}((-\infty,z])} p_{Y|X}(y_{\sim i}\,|\,w)\mathrm{d}y_{\sim i}.
\end{aligned}$$

If $u_i w_i = 1$ then by (4.31) the last expression is equal to $\mathsf{A}_i^w(z)$. On the other hand if $u_i w_i = -1$ then by (4.31) the last expression is equal to $1 - \mathsf{A}_i^w(-z)$ (assuming that there is no point mass at z).

This shows that the distribution of Φ_i conditioned on $X_i = 1$ equals the distribution of Φ_i conditioned that the all-one codeword was transmitted, since the former is just the average of all choices where $c_i = 1$ and all these choices have the same distribution. Further, the distribution of Φ_i conditioned on $X_i = 1$ is the reverse (flip around vertical axis) of the distribution of Φ_i conditioned on $X_i = -1$. In other words, the channel $p_{\Phi_i|X_i}(\phi_i\,|\,x_i)$ is symmetric. □

Problem 4.6 discusses a simple generalization.

§4.1.10. Smooth Channel Families

Definition 4.32 (Smooth Channel Families). Consider a family of BMS channels characterized by their L-densities $\{\mathsf{a}_\epsilon\}$ and parameterized by ϵ, where ϵ takes values in some interval $I \subseteq \mathbb{R}$. The channel family is said to be *smooth* with respect to the parameter ϵ if for all continuously differentiable functions $f(y)$ so that $e^{y/2} f(y)$ is bounded, the integral $\int f(y)\mathsf{a}_\epsilon(y)\mathrm{d}y$ exists and is a continuously differentiable function with respect to ϵ, $\epsilon \in I$. ▽

Discussion: Why do we ask that $e^{y/2} f(y)$ is bounded rather than $f(y)$ itself? This is due to the symmetry of a: it is then "natural" to write

$$\int f(y)\mathsf{a}_\epsilon(y)\mathrm{d}y = \int \left(e^{y/2} f(y)\right) \underbrace{\left(e^{-y/2}\mathsf{a}_\epsilon(y)\right)}_{\text{even in } y} \mathrm{d}y.$$

In the sequel we often say as a shorthand that BMSC(ϵ) is smooth to mean that we are transmitting over BMSC(ϵ) and that the *family* {BMSC(ϵ)} is smooth at the point ϵ. Under the stated conditions, the derivate $\frac{\mathrm{d}}{\mathrm{d}\epsilon}\int f(y)\mathsf{a}_\epsilon(y)\mathrm{d}y$ exists and it is a linear functional of f. It is therefore consistent to formally *define* the derivative of $\mathsf{a}_\epsilon(y)$ with respect to ϵ by setting

$$\int f(y)\frac{\mathrm{d}\mathsf{a}_\epsilon(y)}{\mathrm{d}\epsilon}\mathrm{d}y = \frac{\mathrm{d}}{\mathrm{d}\epsilon}\int f(y)\mathsf{a}_\epsilon(y)\,\mathrm{d}y. \tag{4.33}$$

For a large class of channel families it is straightforward to check that they are smooth. This is, e.g., the case if the output alphabet $\mathcal{Y}$ is finite and the transition probabilities are differentiable functions of ϵ, or if it admits a density with respect to the Lebesgue measure, and the density is differentiable for each y. In these cases, the formal derivative (4.33) coincides with the ordinary derivative.

Example 4.34 (Smoothness of Standard Families). The families $\{\mathrm{BEC}(\epsilon)\}_{\epsilon=0}^{1}$, $\{\mathrm{BSC}(\epsilon)\}_{\epsilon=0}^{\frac{1}{2}}$, as well as $\{\mathrm{BAWGNC}(\sigma)\}_{\sigma=0}^{+\infty}$ are all smooth. ◇

§4.1.11. Capacity Functional for BMS Channels

The capacity of a BMS channel is a *linear* functional of its L-density. This simplifies the determination of the capacity itself. This fact can also be exploited when searching for good degree distribution pairs and it plays a vital role in the definition of the EXIT as well as the GEXIT function.

Lemma 4.35 (Capacity Functional). Let a be the L-density and $\mathfrak{a}$ be the D-density associated with a BMS channel. Then the capacity of this channel in bits per channel use, call it C(a), is

$$C(\mathsf{a}) = \int \mathsf{a}(y)\left(1 - \log_2(1 + e^{-y})\right)\mathrm{d}y = \int_0^{+\infty} |\mathsf{a}|(y)\left(1 - h_2\left(\frac{e^{-y}}{1+e^{-y}}\right)\right)\mathrm{d}y$$

$$= \int_{-1}^{1} \mathsf{a}(y)\log_2(1+y)\,dy \qquad = \int_{0}^{1} |\mathsf{a}|(y)\Big(1 - h_2((1-y)/2)\Big)dy,$$

where $h_2(x) = -x\log_2(x) - (1-x)\log_2(1-x)$.

Proof. Since the channel is symmetric, the optimal input distribution is the uniform one; see Problem 4.8. Further, by (the Channel Equivalence) Lemma 4.28 we can assume without loss of generality that $p_{Y|X}(y\,|\,x) = \mathsf{a}(xy)$.

$$\begin{aligned}
C(\mathsf{a}) &= I(X;Y) = H(Y) - H(Y\,|\,X)\\
&= \int \Big(-p_Y(y)\log_2 p_Y(y) + \frac{1}{2}\sum_{x=\pm1} p_{Y|X}(y\,|\,x)\log_2 p_{Y|X}(y\,|\,x)\Big)dy\\
&= \int \frac{1}{2}\sum_{x=\pm1} p_{Y|X}(y\,|\,x)\log_2 \frac{p_{Y|X}(y\,|\,x)}{\frac{1}{2}(p_{Y|X}(y\,|\,1) + p_{Y|X}(y\,|\,-1))}dy,\\
&= \int p_{Y|X}(y\,|\,1)\log_2 \frac{p_{Y|X}(y\,|\,1)}{\frac{1}{2}(p_{Y|X}(y\,|\,1) + p_{Y|X}(y\,|\,-1))}dy,\\
&= \int \mathsf{a}(y)\log_2 \frac{\mathsf{a}(y)}{\frac{1}{2}\mathsf{a}(y)(1+\mathsf{e}^{-y})}dy = \int \mathsf{a}(y)(1 - \log_2(1+\mathsf{e}^{-y}))dy,
\end{aligned}$$

where in the transition to the fourth line we have used the fact that $p_{Y|X}(-y\,|\,-x) = p_{Y|X}(y\,|\,x)$. The three further representations can be proved in a similar manner but it is more insightful to consider the following alternative proof.

Consider the operational characterization of the channel according to its $|L|$-density $|\mathsf{a}|$ as discussed on page 181: $x \xrightarrow{|\mathsf{a}|} (S, Y)$, where $Y \sim |\mathsf{a}|$ and $x \xrightarrow{\text{BSC}\left(\frac{\mathsf{e}^{-Y}}{1+\mathsf{e}^{-Y}}\right)} S$. In words, for each given *reliability* value y the channel acts like a BSC with cross-over probability $\frac{\mathsf{e}^{-y}}{1+\mathsf{e}^{-y}}$. The second representation of capacity follows since such a BSC has associated capacity $1 - h_2\left(\frac{\mathsf{e}^{-y}}{1+\mathsf{e}^{-y}}\right)$. The remaining two representations have a similar interpretation in terms of D and $|D|$-densities, and we skip the details. □

Example 4.36 (Capacity of BEC(ϵ)). Inserting $|\mathsf{a}_{\text{BEC}(\epsilon)}|(y) = \epsilon\Delta_0(y) + \bar{\epsilon}\Delta_1(y)$ into the previous formula gives us back the familiar

$$\begin{aligned}
C(\mathsf{a}_{\text{BEC}(\epsilon)}) &= \int_0^1 [\epsilon\Delta_0(y) + \bar{\epsilon}\Delta_1(y)]\,(1 - h_2((1-y)/2))dy\\
&= 1 - \epsilon \text{ bits per channel use.}
\end{aligned}$$
◇

Example 4.37 (Capacity of BSC(ϵ)). We recover from $|\mathsf{a}_{\text{BSC}(\epsilon)}|(y) = \Delta_{1-2\epsilon}(y)$ the familiar $C(\mathsf{a}_{\text{BSC}(\epsilon)}) = 1 - h_2(\epsilon)$ bits per channel use. ◇

EXAMPLE 4.38 (CAPACITY OF BAWGNC(σ)). Unfortunately the capacity for the BAWGNC(σ) cannot be expressed in an elementary form. To compute it numerically it is best to use the representation of capacity in terms of the D-density as stated in Lemma 4.35 with the D-density as given in Example 4.21:

$$\mathrm{C}(\mathsf{a}_{\mathrm{BAWGNC}(\sigma)}) = \int_{-1}^{+1} \frac{\sigma}{\sqrt{2\pi}(1-y^2)} \mathrm{e}^{-\frac{(1-\sigma^2\tanh^{-1}(y))^2}{2\sigma^2}} \log_2(1+y)\mathrm{d}y.$$

Alternatively, $\mathrm{C}(\mathsf{a}_{\mathrm{BAWGNC}(\sigma)})$ can be computed via the series (see Problem 4.11)

$$1+\frac{1}{\ln(2)}\Big(\Big(\frac{2}{\sigma^2}-1\Big)Q\Big(\frac{1}{\sigma}\Big)-\sqrt{\frac{2}{\pi\sigma^2}}\mathrm{e}^{-\frac{1}{2\sigma^2}}+\sum_{i=1}^{\infty}\frac{(-1)^i}{i(i+1)}\mathrm{e}^{\frac{2i(i+1)}{\sigma^2}}Q\Big(\frac{1+2i}{\sigma}\Big)\Big).$$

This series converges quickly: the error which we incur by considering only the first i terms in the series is of order $O(i^{-3})$.

Problem 4.12 discusses the limiting cases of $\mathrm{C}(\mathsf{a}_{\mathrm{BAWGNC}(\sigma)})$ for both very small and very large σ.

The capacity is a function of $1/\sigma^2$ alone. More generally, if we allow a scaling of the inputs then the capacity is a function of E_N/σ^2, where E_N is the energy expended per channel use (dimension). (This means that we use inputs from the set $\{\pm\sqrt{E_N}\}$ instead of $\{\pm 1\}$.) A plot of $\mathrm{C}_{\mathrm{BAWGNC}}$ as a function of E_N/σ^2 is shown in Figure 4.39. Also shown is the capacity of the additive white Gaussian noise channel (AWGNC)

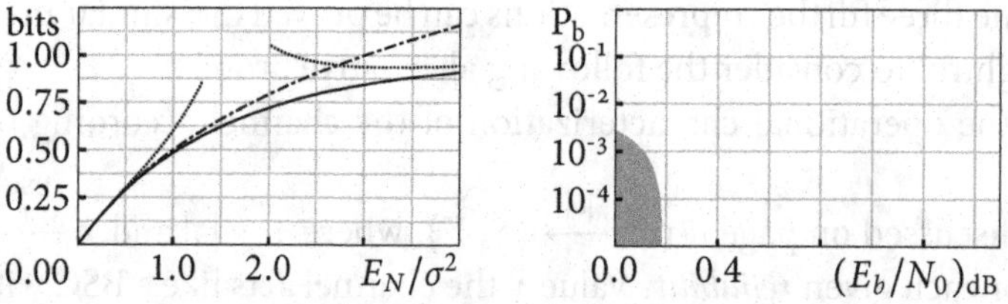

Figure 4.39: Left: Capacity of the BAWGNC (solid line) and the AWGNC (dashed line) in bits per channel use as a function of E_N/σ^2. Also shown are the asymptotic expansions (dotted) for large and small values of $\frac{E_N}{\sigma^2}$ discussed in Problem 4.12. Right: The achievable (white) region for the BAWGNC and $r = \frac{1}{2}$ as a function of $(E_b/N_0)_{\mathrm{dB}}$.

with real-valued inputs, which is equal to

$$\mathrm{C}_{\mathrm{AWGNC}} = \frac{1}{2}\log_2(1+E_N/\sigma^2) \text{ bits per channel use.}$$

As we can see from this figure, for low rates (small values of E_N/σ^2) we pay only a small penalty for restricting the input to be binary. In order to assess the performance of a code over the BAWGNC it is natural to plot the bit error probability

P_b as a function of E_N/σ^2. For low rates r, it is even more useful to plot the bit error probability as a function of E_b/N_0, where $E_b = E_N/r$ is the *energy* expended *per information bit* and $N_0 = 2\sigma^2$ is the *one-sided power spectral density*, so that $E_b/N_0 = \frac{1}{2r}E_N/\sigma^2$. Why is it convenient to use E_b/N_0 for low rates? In this case

$$C_{BAWGNC} \approx C_{AWGNC} = \frac{1}{2}\log_2(1 + 2rE_b/N_0) = \frac{r}{\log_2(e)}E_b/N_0 + O((E_b/N_0)^2).$$

Suppose that a given coding scheme achieves a fraction $(1 - \delta)$ of capacity; i.e., using this coding scheme we can transmit at rate $r = (1 - \delta)C_{AWGNC}$ (and achieve the desired error probability). If the rate is sufficiently small then we see from the preceding approximation that

$$C_{AWGNC} \sim \frac{r}{\log_2(e)}E_b/N_0 = \frac{(1-\delta)}{\log_2(e)}C_{AWGNC}E_b/N_0.$$

Comparing the left-hand and right-hand sides we see that $E_b/N_0 \approx \log_2(e)/(1-\delta)$, which is independent of r. In other words, measuring the performance with respect to E_b/N_0 allows us to compare codes of different (low) rates on an (almost) equal footing. Let C denote the Shannon capacity of a given channel. Then any rate below C can be achieved with vanishing probability of error and, vice versa, to achieve a vanishing probability of error we have to transmit below C. What if we allow a non-vanishing probability of error, let's say p? What is then the maximal rate at which we can transmit? Call this rate $C^{(p)}$. In this case we can proceed as follows: first compress the information such that the original bits can be reconstructed from the compressed version with a Hamming distortion of (at most) p. From rate-distortion theory we know that this requires a source code of rate $1 - h_2(p)$, where $h_2(\cdot)$ denotes the binary entropy function. These compressed bits can be transmitted over the channel at vanishing probability of error, so that the condition for successful transmission reads $r(1-h_2(p)) < C$. Further, by the source-channel separation theorem for point-to-point channels this is the best we can do. It follows that $C^{(p)} = \frac{C}{1-h_2(p)}$. To be concrete, consider the channel family $\{BAWGNC(E_b/N_0)\}$. The associated capacity $C(E_b/N_0)$ is a strictly increasing function of E_b/N_0. From our earlier remarks we see that to transmit over this channel at rate r with a bit error probability of at most P_b requires that $r < \frac{C(E_b/N_0)}{1-h_2(P_b)}$ or, vice versa, that $E_b/N_0 > C^{-1}(r(1-h_2(p)))$. Figure 4.39 shows the resulting achievable $(P_b, E_b/N_0)$-region. ◇

§4.1.12. Further Functionals for BMS Channels

Capacity plays an important role in the analysis and practice of iterative coding, but there are also other linear functionals that are of interest. The first is trivially related to capacity.

Definition 4.40 (Entropy Functional). The *entropy* $\mathrm{H}(\mathsf{a})$ associated with a symmetric L-density a is $\mathrm{H}(\mathsf{a}) = 1 - \mathrm{C}(\mathsf{a})$. Therefore, from Lemma 4.35,

$$\mathrm{H}(\mathsf{a}) = \int_{-\infty}^{+\infty} \mathsf{a}(y)\log_2(1+\mathrm{e}^{-y})\mathrm{d}y \quad = \int_{0}^{+\infty} |\mathsf{a}|(y)h_2\Big(\frac{\mathrm{e}^{-y}}{1+\mathrm{e}^{-y}}\Big)\mathrm{d}y$$

$$= \int_{-1}^{1} \mathfrak{a}(y)\log_2\frac{2}{1+y}\mathrm{d}y \quad = \int_{0}^{1} |\mathfrak{a}|(y)h_2((1-y)/2)\mathrm{d}y,$$

where $h_2(x) = -x\log_2(x) - (1-x)\log_2(1-x)$. ▽

Lemma 4.41 (Duality Rule for Entropy). Let a and b denote two symmetric L-densities. Then

$$\mathrm{H}(\mathsf{a} \circledast \mathsf{b}) + \mathrm{H}(\mathsf{a} \boxast \mathsf{b}) = \mathrm{H}(\mathsf{a}) + \mathrm{H}(\mathsf{b}).$$

Discussion: We call this the "duality" rule since it relates the output entropy at a repetition code to the output entropy at a parity-check code.

Proof. Let X, Y denote two independent random variables having associated densities a and b, respectively. Define $Z = 2\tanh^{-1}(\tanh(X/2)\tanh(Y/2))$. From the definition of $\boxast$ on page 181 and the discussion in Example 4.85, we know that Z has density $\mathsf{c} = \mathsf{a} \boxast \mathsf{b}$. Therefore,

$$\begin{aligned}\mathrm{H}(\mathsf{c}) &= \int \mathsf{c}(z)\log_2(1+\mathrm{e}^{-z})\mathrm{d}z \\ &= \int\int \mathsf{a}(x)\mathsf{b}(y)\log_2\big(1+\mathrm{e}^{-2\tanh^{-1}(\tanh(x/2)\tanh(y/2))}\big)\mathrm{d}x\,\mathrm{d}y \\ &= \int\int \mathsf{a}(x)\mathsf{b}(y)\log_2\Big(\frac{(1+\mathrm{e}^{-x})(1+\mathrm{e}^{-y})}{1+\mathrm{e}^{-x-y}}\Big)\mathrm{d}x\,\mathrm{d}y \\ &= \mathrm{H}(\mathsf{a}) + \mathrm{H}(\mathsf{b}) - \mathrm{H}(\mathsf{a} \circledast \mathsf{b}).\end{aligned}$$

To see the last step note that

$$\begin{aligned}\mathrm{H}(\mathsf{a} \circledast \mathsf{b}) &= \int\Big(\int \mathsf{a}(x)\mathsf{b}(y-x)\mathrm{d}x\Big)\log_2(1+\mathrm{e}^{-y})\mathrm{d}y \\ &= \int\int \mathsf{a}(x)\mathsf{b}(y)\log_2(1+\mathrm{e}^{-x-y})\mathrm{d}x\,\mathrm{d}y.\end{aligned}$$ □

In Section 3.14 we discussed the use of EXIT functions for the case of transmission over the BEC. These EXIT functions are defined in terms of entropies, a concept that, as we have just seen, carries over naturally to the general case. The notion of EXIT involves the computation of entropy only of the *extrinsic* part of the information (the information regarding a bit that we get via the code constraints by observing *other* bits). Although it constitutes an abuse of notation, it is convenient to introduce the notion of an EXIT functional in a more general way.

Definition 4.42 (EXIT Functional). The entropy functional $\mathrm{H}(\cdot)$ is also called the *EXIT* functional and the associated kernel the *EXIT* kernel. Assuming that we represent densities in the L-domain, this kernel[2] is $l(y) = \log_2(1 + \mathrm{e}^{-y})$. ▽

If you revisit Chapter 3 you will see that EXIT functions for the BEC have two interpretations. From Definition 3.70 we see that the EXIT value equals the uncertainty of a bit X_i given the extrinsic observation $Y_{\sim i}$. We generalize this notion in Definition 4.129 using the aforementioned EXIT functional. On the other, from characterization (iii) in Lemma 3.74 and its discussion in the proof, we see that the i-th EXIT function for the BEC is also equal to the rate of change of the mutual information of the overall system due to a small change in the capacity of the i-th channel. This interpretation formed the basis for (the Area) Theorem 3.81. In the general case these two notions are no longer equivalent and so we next introduce the functional which allows us to extend the second notion. Currently we are only interested in the functional itself and some of its properties. The operational significance of this functional and how it gives rise to a general area theorem is discussed in Section 4.12.

In what follows, we write $\{\mathrm{BMSC}(\mathtt{h})\}$ to denote a family of BMSC channels parameterized by entropy. Therefore, with some abuse of notation, we write in the sequel $\{\mathrm{BEC}(\mathtt{h})\}$, $\{\mathrm{BSC}(\mathtt{h})\}$, and $\{\mathrm{BAWGNC}(\mathtt{h})\}$. We have $\mathtt{h} = \epsilon$ for the BEC, $\mathtt{h} = h_2(\epsilon)$ for the BSC, and $\mathtt{h} = \mathrm{H}(\mathsf{a}_{\mathrm{BAWGNC}(\sigma)})$ for the BAWGNC.

Definition 4.43 (GEXIT Functional). Consider a family $\{\mathrm{BMSC}(\mathtt{h})\}$ of smooth BMS channels. Let $\{\mathsf{a}_{\mathrm{BMSC}(\mathtt{h})}\}$ denote the corresponding family of L-densities. The *generalized EXIT* (GEXIT) functional with respect to $\{\mathrm{BMSC}(\mathtt{h})\}$ applied to the symmetric L-density b is

$$G(\mathsf{a}_{\mathrm{BMSC}(\mathtt{h})}, \mathsf{b}) = \frac{\mathrm{d}}{\mathrm{d}\mathtt{h}} \mathrm{H}(\mathsf{a}_{\mathrm{BMSC}(\mathtt{h})} \circledast \mathsf{b}).$$

If we define the GEXIT kernel

$$l^{\mathsf{a}_{\mathrm{BMSC}(\mathtt{h})}}(y) = \int \frac{\mathrm{d}\mathsf{a}_{\mathrm{BMSC}(\mathtt{h})}(z)}{\mathrm{d}\mathtt{h}} \log_2(1 + \mathrm{e}^{-z-y})\mathrm{d}z, \tag{4.44}$$

then $G(\mathsf{a}_{\mathrm{BMSC}(\mathtt{h})}, \mathsf{b})$ can be written as

$$G(\mathsf{a}_{\mathrm{BMSC}(\mathtt{h})}, \mathsf{b}) = \int \mathsf{b}(y) l^{\mathsf{a}_{\mathrm{BMSC}(\mathtt{h})}}(y)\mathrm{d}y.$$ ▽

[2] Not to be confused with the log-likelihood ratio function, which is unfortunately also denoted by $l(y)$.

Discussion: Expression (4.44) has to be interpreted in the sense of Definition 4.32: we write $\int \frac{\mathrm{d}a_{\mathrm{BMSC(h)}}(z)}{\mathrm{dh}} \log_2(1+\mathrm{e}^{-z-y})\,\mathrm{d}z$ as a proxy for

$$\frac{\mathrm{d}}{\mathrm{dh}}\left\{\int a_{\mathrm{BMSC(h)}}(z)\log_2(1+\mathrm{e}^{-z-y})\,\mathrm{d}z\right\}.$$

The latter expression exists according to Definition 4.32: for a fixed y and as a function of z, $\log_2(1+\mathrm{e}^{-z-y})$ is continuously differentiable and $\log_2(1+\mathrm{e}^{-z-y})\mathrm{e}^{z/2}$ is bounded. Further, by assumption the channel family is smooth. Note further that $l^{a_{\mathrm{BMSC(h)}}}(y)$ is continuous and non-negative so that $G(a_{\mathrm{BMSC(h)}}, \mathrm{b})$ exists as well.

EXAMPLE 4.45 (GEXIT KERNEL FOR $\{\mathrm{BEC(h)}\}$). We have

$$\int a_{\mathrm{BEC(h)}}(z)\log_2(1+\mathrm{e}^{-z-y})\mathrm{d}z = \mathrm{h}\log_2(1+\mathrm{e}^{-y}),$$

so that $l^{a_{\mathrm{BEC(h)}}}(y) = \log_2(1+\mathrm{e}^{-y})$, which is the regular EXIT kernel. This agrees with our previous observation that for the BEC the notions of EXIT and GEXIT coincide. ◇

EXAMPLE 4.46 (GEXIT KERNEL FOR $\{\mathrm{BSC(h)}\}$). We have

$$\int a_{\mathrm{BSC(h)}}(z)\log_2(1+\mathrm{e}^{-z-y})\mathrm{d}z = \bar{\epsilon}\log_2(1+\frac{\epsilon}{\bar{\epsilon}}\mathrm{e}^{-y}) + \epsilon\log_2(1+\frac{\bar{\epsilon}}{\epsilon}\mathrm{e}^{-y}),$$

where $\epsilon = h_2^{-1}(\mathrm{h})$. Differentiating with respect to h gives

$$l^{a_{\mathrm{BSC(h)}}}(y) = \left(\log\left(\frac{1+\frac{\bar{\epsilon}}{\epsilon}\mathrm{e}^{-y}}{1+\frac{\epsilon}{\bar{\epsilon}}\mathrm{e}^{-y}}\right) + \frac{1}{\epsilon+\bar{\epsilon}\mathrm{e}^{y}} - \frac{1}{\bar{\epsilon}+\epsilon\mathrm{e}^{y}}\right)/\log\left(\frac{\bar{\epsilon}}{\epsilon}\right).$$

As we discuss in Example 4.48, the kernel is not unique. An equivalent kernel is

$$l^{a_{\mathrm{BSC(h)}}}(y) = \log\left(\frac{1+\frac{\bar{\epsilon}}{\epsilon}\mathrm{e}^{-y}}{1+\frac{\epsilon}{\bar{\epsilon}}\mathrm{e}^{-y}}\right)/\log\left(\frac{\bar{\epsilon}}{\epsilon}\right).$$

For a fixed $y \in \mathbb{R}$ and $\mathrm{h} \to 0$, the kernel converges to 1 as $1 + y/\ln(\epsilon)$, whereas the limit when $\mathrm{h} \to 1$ is equal to $\frac{2}{1+\mathrm{e}^{y}}$. ◇

EXAMPLE 4.47 (GEXIT KERNEL FOR $\{\mathrm{BAWGNC(h)}\}$). We get

$$l^{a_{\mathrm{BAWGNC(h)}}}(y) = \left(\int \frac{\mathrm{e}^{-\frac{(z-2/\sigma^2)^2\sigma^2}{8}}}{1+\mathrm{e}^{z+y}}\mathrm{d}z\right)/\left(\int \frac{\mathrm{e}^{-\frac{(z-2/\sigma^2)^2\sigma^2}{8}}}{1+\mathrm{e}^{z}}\mathrm{d}z\right),$$

where σ is the unique positive number so that the BAWGNC with noise variance σ has entropy h. ◇

EXAMPLE 4.48 (ALTERNATIVE KERNEL REPRESENTATIONS). Because of the symmetry property of L-densities we can write for *any* kernel $l(y)$

$$\int \mathsf{a}(y)l(y)\,\mathrm{d}y = \int_0^{+\infty} |\mathsf{a}|(y)\frac{l(y)+\mathrm{e}^{-y}l(-y)}{1+\mathrm{e}^{-y}}\mathrm{d}y.$$

In words, the kernel is uniquely specified on the absolute value domain $[0,+\infty]$, but for each $y \in [0,+\infty]$ we can split the weight of the kernel between $+y$ and $-y$ in an arbitrary fashion restricted only by the constraint that $l(y)+\mathrm{e}^{-y}l(-y)$ equals the desired value. We can use this degree of freedom to bring some kernels into a more convenient form. For example, the second kernel given in Example 4.46 is equivalent to the first one but it is simpler. To see this, consider the function $f(y) = \frac{1}{\epsilon+\bar{\epsilon}\mathrm{e}^y} - \frac{1}{\bar{\epsilon}+\epsilon\mathrm{e}^y}$ and note that $f(y)+\mathrm{e}^{-y}f(-y) = 0$. Problem 4.19 discusses some particularly insightful and convenient kernel representations for the Gaussian case. ◇

In general, for any kernel of a functional in the L-domain there is an associated kernel in the $|L|$, D, $|D|$, G, and $|G|$-domain and it is often useful to consider a particular domain if we want to exhibit a particular property. The differences (and similarities) of various kernels are best seen in the $|D|$-domain. The L-domain kernel and the associated $|D|$-domain kernel are linked by a change of variables as follows:

$$|d|(w) = \frac{1-w}{2}l(\ln\frac{1-w}{1+w}) + \frac{1+w}{2}l(\ln\frac{1+w}{1-w}). \tag{4.49}$$

In Figure 4.50 we compare the GEXIT kernel for the $\{\mathrm{BEC}(\mathtt{h})\}$ with the GEXIT kernels for the $\{\mathrm{BSC}(\mathtt{h})\}$ and the $\{\mathrm{BAWGNC}(\mathtt{h})\}$ in the $|D|$-domain for several channel parameters (see Problems 4.16, 4.17, and 4.18). These kernels are distinct but similar. In particular, for $\mathtt{h} = 0.5$ the GEXIT kernel with respect to $\{\mathrm{BAWGNC}(\mathtt{h})\}$ is hardly distinguishable from the regular EXIT kernel. The GEXIT kernel with respect to the family $\{\mathrm{BSC}(\mathtt{h})\}$ shows more variation.

Entropy plays a fundamental role in information theory, which helps motivate our interest in the EXIT and GEXIT functionals. In iterative decoding, especially for LDPC ensembles, convolutions of densities appear frequently. Consequently, Fourier transforms of densities in their various representations also play important roles. In fact, many information-theoretic functionals of interest are the evaluation of a Fourier transform at a point.

DEFINITION 4.51 (VARIABLE-DOMAIN FOURIER TRANSFORM). Let a be an L-density and let X denote a random variable distributed according to a. The Fourier transform of a is

$$\mathcal{F}_{\mathsf{a}}(s) = \mathbb{E}[\mathrm{e}^{-sX}],$$

defined for all $s \in \mathbb{C}$ where the expectation exists. ▽

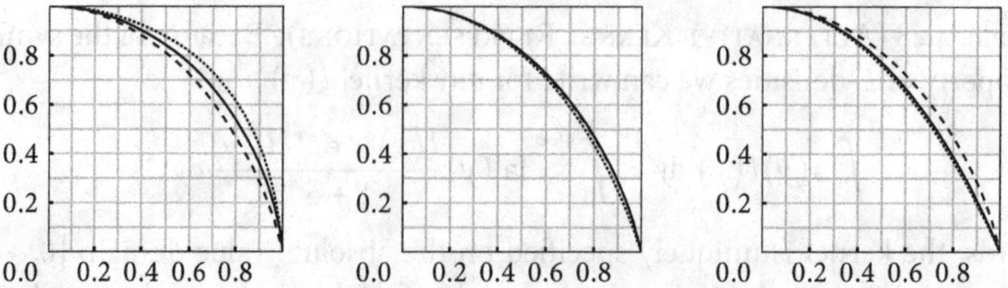

Figure 4.50: Comparison of the kernels $|d|^{\mathtt{aBEC(h)}}(\cdot)$ (dashed line) with $|d|^{\mathtt{aBSC(h)}}(\cdot)$ (dotted line) and $|d|^{\mathtt{aBAWGNC(h)}}(\cdot)$ (solid line) at channel entropy $\mathtt{h} = 0.1$ (left), $\mathtt{h} = 0.5$ (middle), and $\mathtt{h} = 0.9$ (right).

From standard properties of the Fourier transform it follows that for any L-densities a and b

$$\mathcal{F}_{\mathtt{a} \circledast \mathtt{b}} = \mathcal{F}_{\mathtt{a}} \mathcal{F}_{\mathtt{b}}.$$

Symmetry of the L-density a is equivalent to $\mathcal{F}_{\mathtt{a}}(s) = \mathcal{F}_{\mathtt{a}}(1-s)$:

$$\mathbb{E}[\mathrm{e}^{-sX}] = \int \mathtt{a}(x)\mathrm{e}^{-sx}\mathrm{d}x \overset{\text{Def. 4.11}}{=} \int \mathtt{a}(-x)\mathrm{e}^{x}\mathrm{e}^{-sx}\mathrm{d}x = \mathbb{E}[\mathrm{e}^{-(1-s)X}].$$

For symmetric L-densities a, $\mathcal{F}_{\mathtt{a}}$ exists and is analytic in the strip where the real part of s is contained in $[0, 1]$.

Definition 4.52 (Check-Domain Fourier Transform). Let a be a G-density and let (S, Y) denote a random variable distributed according to a. For this definition we use the convention $S \in \{0, 1\}$ and we identify $\{0, 1\}$ with the additive group of $\mathbb{F}_2$. The Fourier transform of this density is

$$\mathcal{G}_a(\mu, \nu) = \mathbb{E}[\mathrm{e}^{-\mu S - \nu Y}]$$

for all $\mu \in \{0, i\pi\}$, $\nu \in \mathbb{C}$ where the expectation exists. ▽

From standard properties of the Fourier transform it follows that for any G-densities a and b

$$\mathcal{G}_{a \boxast b} = \mathcal{G}_a \mathcal{G}_b.$$

Symmetry of the G-density a is equivalent to $\mathcal{G}_a(i\pi, \nu) = \mathcal{G}_a(0, \nu + 1)$:

$$\begin{aligned} \mathcal{G}_a(i\pi, \nu) &= \int_y (a(0, y) - a(1, y))\mathrm{e}^{-\nu y}\mathrm{d}y \\ &\overset{(4.18)}{=} \int_y (a(1, y)\coth(y/2) - a(1, y))\mathrm{e}^{-\nu y}\mathrm{d}y \end{aligned}$$

$$\begin{aligned}
&= \int_y a(1,y)\frac{e^{-y/2}}{\sinh(y/2)}e^{-\nu y}dy \\
&= \int_y a(1,y)\frac{e^{y/2}}{\sinh(y/2)}e^{-(\nu+1)y}dy \\
&= \int_y (a(1,y)\coth(y/2) + a(1,y))e^{-(\nu+1)y}dy \\
&\overset{(4.18)}{=} \int_y (a(0,y) + a(1,y))e^{-(\nu+1)y}dy \\
&= \mathcal{G}_a(0,\nu+1).
\end{aligned}$$

Sometimes we write $\mathcal{G}$ and $\mathcal{F}$ for densities which are not G-densities and L-densities, respectively. We mean by this that the functional is applied after the appropriate change of variables.

Definition 4.53 (Error Probability Functional). The *error probability* associated with the symmetric L-density a is

$$\mathfrak{E}(\mathsf{a}) = \frac{1}{2}\int \mathsf{a}(x)e^{-(|x/2|+x/2)}dx. \tag{4.54}$$

Note that $\mathfrak{E}(\mathsf{a})$ is the incurred probability of error if we (optimally) estimate the value of a bit based upon the channel output. ▽

Discussion: You might wonder why instead of (4.54) we did not just define $\mathfrak{E}(\mathsf{a}) = \int_{-\infty}^{0} \mathsf{a}(x)dx$. Using symmetry, the two definitions are equal provided a does not have a point mass at 0. When there is a point mass at 0, however, we need to include half of it in the error probability – formulation (4.54) is correct also in this case without modification. (In the case that the density is not symmetric we need to use the more awkward definition $\mathfrak{E}(\mathsf{a}) = \int_{-\infty}^{0} \mathsf{a}(x)dx$ and include half of the mass at 0.)

Example 4.55 (Error Probability). We have $\mathfrak{E}(\mathsf{a}_{\mathrm{BEC}(\epsilon)}) = \epsilon/2$, $\mathfrak{E}(\mathsf{a}_{\mathrm{BSC}(\epsilon)}) = \epsilon$, and $\mathfrak{E}(\mathsf{a}_{\mathrm{BAWGNC}(\sigma)}) = Q(\frac{1}{\sigma})$. ◇

Definition 4.56 (D Mean Functionals). The *D-mean* associated with the symmetric L-density a is

$$\mathfrak{D}(\mathsf{a}) = \int \mathsf{a}(x)\tanh(x/2)dx = \int_{-1}^{1} \mathfrak{a}(y)y\,dy. \tag{4.57}$$

More generally, the *D-k-moment* associated with the symmetric L-density a is

$$\mathfrak{D}_k(\mathsf{a}) = \int \mathsf{a}(x)\tanh^k(x/2)dx = \int_{-1}^{1} \mathfrak{a}(y)y^k dy, \tag{4.58}$$

where k is a non-negative integer. ▽

EXAMPLE 4.59 (D-MEAN). We have $\mathfrak{D}(\mathsf{a}_{\text{BEC}(\epsilon)}) = 1 - \epsilon$ and $\mathfrak{D}(\mathsf{a}_{\text{BSC}(\epsilon)}) = (1 - 2\epsilon)^2$. There does not seem to be an elementary expression for $\mathfrak{D}(\mathsf{a}_{\text{BAWGNC}(\sigma)})$, but it can be computed numerically. ◇

LEMMA 4.60 (MULTIPLICATIVITY OF $\mathfrak{D}_k$ UNDER $\circledast$-CONVOLUTION). Let a and b denote two symmetric L-densities. Then $\mathfrak{D}_k(\mathsf{a} \circledast \mathsf{b}) = \mathfrak{D}_k(\mathsf{a})\,\mathfrak{D}_k(\mathsf{b})$.

Proof. For even k we have $\mathfrak{D}_k(\mathsf{a}) = \mathcal{G}_{\mathsf{a}}(0, k)$, while for odd k we have $\mathfrak{D}_k(\mathsf{a}) = \mathcal{G}_{\mathsf{a}}(i\pi, k)$. □

Consider the following situation: we have a symmetric L-density a and we are interested in the behavior of $\mathfrak{E}(\mathsf{a}^{\circledast n})$ as a function of n, where $\mathsf{a}^{\circledast n}$ denotes the n-fold L-convolution of a with itself. This probability decays in general to zero exponentially fast in n. What is its exponent, i.e., what is $\lim_{n\to\infty} \frac{1}{n} \log \mathfrak{E}(\mathsf{a}^{\circledast n})$? The answer is given by the log of the so-called Bhattacharyya constant.

DEFINITION 4.61 (BHATTACHARYYA FUNCTIONAL). The *Bhattacharyya* constant associated with the symmetric L-density a is

$$\mathfrak{B}(\mathsf{a}) = \int \mathsf{a}(x) e^{-x/2} dx. \tag{4.62}$$

The Bhattacharyya constants for our standard channels are worked out in Examples 4.126, 4.127, and 4.128 in the context of the so-called stability analysis of BP decoding. Problems 4.24, 4.25, and 4.26 discuss the computation of $\mathfrak{B}(\cdot)$ for these three channels by directly computing $\mathfrak{E}(\mathsf{a}^{\circledast n})$ and Problem 4.23 shows yet another alternative using Bernstein's inequality (see Section C.2). Finally, Problem 4.20 gives a way of computing the Bhattacharyya constant directly from the transition probability $p_{Y|X}(y|x)$ without first computing the L-density a. ▽

The key to finding the exponent describing the decay of the error probability is to show that the Bhattacharyya constant behaves multiplicatively under convolution and to establish a link between the Bhattacharyya constant and the error probability. These two steps are achieved in the following two lemmas.

LEMMA 4.63 (MULTIPLICATIVITY OF $\mathfrak{B}$ UNDER $\circledast$-CONVOLUTION). Let a and b denote two symmetric L-densities. Then $\mathfrak{B}(\mathsf{a} \circledast \mathsf{b}) = \mathfrak{B}(\mathsf{a})\,\mathfrak{B}(\mathsf{b})$.

Proof. This follows from $\mathfrak{B}(\mathsf{a}) = \mathcal{F}_{\mathsf{a}}(\frac{1}{2})$. □

LEMMA 4.64 ($\mathfrak{B}$ VERSUS $\mathfrak{E}$). Let a be a symmetric L-density. Then

$$2\,\mathfrak{E}(\mathsf{a}) \le \mathfrak{B}(\mathsf{a}) \le 2\sqrt{\mathfrak{E}(\mathsf{a})(1 - \mathfrak{E}(\mathsf{a}))}, \tag{4.65}$$

where the left inequality is tight for the BEC and the right is tight for the BSC.

Proof. The tightness of the left and the right bound for the case of the BEC and the BSC is best verified by direct calculation (see Problem 4.21). The left bound follows from characterizations (4.54) and (4.62) since $e^{-(|x/2|+x/2)} \le e^{-x/2}$. To see the right bound, consider the BSC(ϵ). We have $\mathfrak{E}(a_{BSC(\epsilon)}) = \epsilon$ and $\mathfrak{B}(a_{BSC(\epsilon)}) = 2\sqrt{\epsilon\bar{\epsilon}}$. But $2\sqrt{\epsilon\bar{\epsilon}}$ is convex-∩ on $[0, \frac{1}{2}]$ (recall that we write "convex-∩" to indicate a concave function). Therefore, if $\epsilon = \alpha\epsilon_1 + (1-\alpha)\epsilon_2$, where $\alpha \in [0,1]$ and $\epsilon_1, \epsilon_2 \in [0, \frac{1}{2}]$, then

$$\alpha\,\mathfrak{B}(a_{BSC(\epsilon_1)}) + (1-\alpha)\,\mathfrak{B}(a_{BSC(\epsilon_2)}) \le \mathfrak{B}(a_{BSC(\epsilon)}).$$

The result now follows since any symmetric channel a_{BMSC} can be represented as a convex combination of elements from $\{a_{BSC(\epsilon)}\}$. □

If we combine the preceding two lemmas we see that $\mathfrak{E}(a^{\circledast n}) \le \frac{1}{2}\mathfrak{B}(a)^n$, so that $\lim_{n\to\infty} \frac{1}{n}\log\mathfrak{E}(a^{\circledast n}) \le \log\mathfrak{B}(a)$. A generalization and the reverse inequality are stated in the next lemma. The proof can be found on page 543 in Appendix E.

Lemma 4.66 (Large Deviation). Let the L-densities a_i, $i = 1, \dots, k$, be symmetric. Then for any set of natural numbers d_i, $i = 1, \dots, k$,

$$\frac{2\,\mathfrak{B}_{\min}^k}{3\pi}\,\frac{\left(e^2\frac{\mathfrak{B}_{\min}}{4D}\right)^{\frac{1}{2}}}{1+e^2\frac{\mathfrak{B}_{\min}}{4D}}\prod_{i=1}^{k}\mathfrak{B}^{d_i}(a_i) \le \mathfrak{E}(a) \le \frac{1}{2}\prod_{i=1}^{k}\mathfrak{B}^{d_i}(a_i), \tag{4.67}$$

where $a = a_1^{\circledast d_1} \circledast a_2^{\circledast d_2} \circledast \cdots \circledast a_k^{\circledast d_k}$, $D = \sum_i d_i$, and $\mathfrak{B}_{\min} = \min_i \mathfrak{B}(a_i)$.

Problems 4.59, 4.60, 4.61, and 4.62 explore further relationships between $H(\cdot)$, $\mathfrak{E}(\cdot)$, and $\mathfrak{B}(\cdot)$.

§4.1.13. Linear Codes and BMS Channels

The main theme of this book is the investigation of low-complexity coding schemes that are capable of approaching the capacity of a wide range of channels. Since all these schemes are based on *linear* codes, before venturing any further, we should first ensure that linear codes by themselves are powerful enough for our purpose. In this respect it is comforting to know the following theorem.

Theorem 4.68 (Linear Codes Achieve Capacity). Linear codes achieve the capacity of BMS channels.

Proof. We will not give a full proof here but rather just point out how it can be deduced from standard coding theorems. As already mentioned in Section 4.1.11, because of the symmetry of the channel the optimal input distribution is the uniform one. This is good news since for proper binary linear codes the induced marginal for each bit (assuming a uniform distribution on the codewords) is the uniform one (see Problem 1.5).

From Problem 1.16 we know that if we add a shift (chosen uniformly at random) to a random sample of Elias's generator ensemble $\mathcal{G}$, then pairs of codewords are independent. We have seen in the proof of the coding theorem for the BSC (Theorem 1.17) that we only need the *pairwise* independence of codewords. This is true also in the general case. We conclude that (shifted) random *linear* codes do not incur a penalty (at least in terms of rate) over truly random codes. Finally, all shifted versions of the same code have exactly the same performance over a BMS channel. Therefore, linear codes themselves can achieve the capacity of any BMS channel. □

§4.1.14. Degradation

In traditional information theory one frequently considers a sequence of codes, usually of increasing length, whose *rates* approach the capacity of a given *fixed* channel. For us it is more natural to take an alternate route where we *fix the rate* and consider an ordered *family of channels*, parameterized by a real-valued parameter. The parameter orders the channels within the family – with an increase in the parameter indicating a "worsening" of the channel. We are then interested in the largest channel parameter (the worst channel) for which reliable transmission at the given rate is possible.

A natural and very useful way of ordering the individual elements of the family of channels is by means of *degradation*, a notion that we encountered already for the BEC. In general, degradation induces a partial order on the space of all channels. This partial order plays an important role in our understanding of the asymptotic behavior of iterative systems.

Definition 4.69 (Stochastic and Physical Degradation). Consider two memoryless channels specified by means of their transition probabilities $p_{Y|X}$ and $p_{Z|X}$, respectively. Let the output alphabets be denoted by $\mathcal{Y}$ and $\mathcal{Z}$, respectively. We say that $p_{Z|X}$ is *stochastically degraded* with respect to $p_{Y|X}$ if there exists a memoryless channel with transition probability $p_{Z|Y}(z\,|\,y)$, $y \in \mathcal{Y}$ and $z \in \mathcal{Z}$, such that for all x in the input alphabet of the channel $p_{Z|X}(\cdot|\cdot)$ and all $z \in \mathcal{Z}$

$$p_{Z|X}(z\,|\,x) = \sum_{y} p_{Y|X}(y\,|\,x) p_{Z|Y}(z\,|\,y). \tag{4.70}$$

We speak of *physical degradation* if (see the discussion on Markov chains on page 28) $X \to Y \to Z$, i.e., if

$$p_{Y,Z|X}(y,z\,|\,x) = p_{Y|X}(y\,|\,x) p_{Z|Y}(z\,|\,y).$$

Physical degradation implies stochastic degradation since

$$p_{Z|X}(z|x) = \sum_y p_{Y,Z|X}(y,z|x) = \sum_y p_{Y|X}(y|x)p_{Z|Y}(z|y). \qquad \triangledown$$

Discussion: Stochastic degradation concerns the *marginal* distribution $p_{Z|X}$, whereas physical degradation concerns the *joint* distribution $p_{Y,Z|X}(y,z|x)$. For our purpose, however, there is essentially no difference between the two concepts. The quantities we are interested in (e.g., probability of error or capacity) are functions of the marginal distribution only. We are therefore free to choose the joint distribution as we see fit. Therefore, whenever we have a channel $p_{Z|X}$ that is stochastically degraded with respect to another channel $p_{Y|X}$ we will assume that the channel $p_{Z|X}$ is "realized" as a physically degraded version of $p_{Y|X}$. Since in this case $X \to Y \to Z$, we know already (see again page 28) that the error probability of the channel $p_{Z|X}$ cannot be smaller than that of the channel $p_{Y|X}$. We will soon discuss many other functionals that preserve ordering by degradation. These functionals are all functionals of the marginal distribution only. For this reason we make no further distinction in the future and simply speak of *degradation*. We denote the relationship of degradation by $p_{Y|X} \twoheadrightarrow p_{Z|X}$.

Example 4.71 (Degradation of the Family $\{\text{BSC}(\epsilon)\}$). Consider the serial concatenation of the BSC(ϵ) and the BSC($\Delta\epsilon$); i.e., the output of the BSC(ϵ) is used as the input to the BSC($\Delta\epsilon$). This serial concatenation (in any order) yields the BSC(ϵ'), with

$$\epsilon' = \bar{\epsilon}\Delta\epsilon + \bar{\Delta\epsilon}\epsilon. \tag{4.72}$$

Moreover, for any $\epsilon' \in [0, \frac{1}{2}]$ and any $0 \leq \epsilon < \epsilon'$ there exists a positive $\Delta\epsilon$, namely $\Delta\epsilon = (\epsilon' - \epsilon)/(1 - 2\epsilon)$, such that (4.72) is fulfilled. This shows that the family $\{\text{BSC}(\epsilon)\}$ is ordered by degradation. Pictorially,

$$\{\pm1\} \xrightarrow{\text{BSC}(\epsilon)} \{\pm1\} \xrightarrow{\text{BSC}(\Delta\epsilon)} \{+1\} \equiv \{\pm1\} \xrightarrow{\text{BSC}(\epsilon')} \{\pm1\}. \qquad \diamond$$

Equivalent statements are true for the family of BAWGNCs, the family of binary-input Cauchy channels (BCCs), and the family of binary-input Laplace channels (BLCs) except that the latter two examples are not self-degrading (see Problems 4.4, 4.27, and 4.28). In the preceding example, the degrading channel is symmetric. A natural question is whether it is sufficient to consider symmetric degrading channels when $p_{Y|X}$ and $p_{Z|X}$ are symmetric. This is answered in the affirmative in the next lemma, the proof of which is left as Problem 4.29.

Lemma 4.73 (Symmetric Degrading Channels). Let $p_{Y|X}$ and $p_{Z|X}$ be two memoryless symmetric channels and assume that $p_{Y|X} \multimap p_{Z|X}$. Then there exists a memoryless *symmetric* channel $p_{Z|Y}$ such that

$$p_{Z|X}(z|x) = \sum_{y \in \mathcal{Y}} p_{Y|X}(y|x) p_{Z|Y}(z|y).$$

§4.1.15. Functionals That Preserve Degradation

Degradation for BMS channels is intimately connected with convexity. In the sequel, we use the notation $\cdot \multimap \cdot$ to indicate the relationship of degradation also in conjunction with distributions that characterize the channels: e.g., we write in the next theorem $\mathfrak{A} \multimap \mathfrak{B}$ to indicate that the BMS channel characterized by its D-distribution $\mathfrak{B}$ is degraded with respect to the BMS channel characterized by its D-distribution $\mathfrak{A}$. You can find the proof of the next theorem on page 537 in Appendix E.

Theorem 4.74. Let $\mathfrak{A}$ and $\mathfrak{B}$ denote two D-distributions, and let $|\mathfrak{A}|$ and $|\mathfrak{B}|$ denote the two corresponding $|D|$-distributions, i.e., distributions on $[0,1]$. Then the following are equivalent:

(i) $\mathfrak{A} \multimap \mathfrak{B}$,

(ii) $\int_0^1 f(x)\, \mathrm{d}|\mathfrak{A}|(x) \leq \int_0^1 f(x)\, \mathrm{d}|\mathfrak{B}|(x)$, for all f that are non-increasing and convex-$\cap$ on $[0,1]$,

(iii) $\int_z^1 |\mathfrak{A}|(x) \mathrm{d}x \leq \int_z^1 |\mathfrak{B}|(x) \mathrm{d}x$, for all $z \in [0,1]$.

There is a rich class of functionals that preserve ordering by degradation. In fact, the class is rich enough to imply the following (see page 542 in Appendix E for the proof.)

Lemma 4.75 (Convergence under Degradation). Any sequence of symmetric densities ordered by degradation converges to a symmetric limit density.

Theorem 4.74 characterizes degradation of densities as an ordering of a certain class of linear functionals. It is also valuable to know the class of *all* functionals that respect the ordering of degradation. It turns out to be the same class. You can find the proof of the following theorem on page 543 in Appendix E.

Theorem 4.76 (Functionals That Preserve Degradation). A function f satisfies

$$\int_0^1 f(x)\, \mathrm{d}|\mathfrak{A}|(x) \leq \int_0^1 f(x)\, \mathrm{d}|\mathfrak{B}|(x)$$

for every degraded pair of densities $\mathfrak{A} \multimap \mathfrak{B}$ if and only if f is non-increasing and convex-$\cap$ on $[0,1]$.

Many functionals of interest either preserve (or reverse) the partial order induced by degradation. In Problem 4.59 the $|D|$-domain kernels of $\mathfrak{E}(\cdot)$, $\mathrm{H}(\cdot)$, and $\mathfrak{B}(\cdot)$ are discussed. Consider first the probability of error: the associated kernel in the $|D|$-domain is $\frac{1}{2}(1-w)$, which is decreasing and convex-$\cap$ on $[0,1]$. The EXIT kernel in the $|D|$-domain is $h_2((1-w)/2)$, which is decreasing and convex-$\cap$ on $[0,1]$ as well. Finally, the Bhattacharyya kernel in the $|D|$-domain is $\sqrt{1-w^2}$. This function has first derivative $-w/\sqrt{1-w^2}$ and second derivative $-1/(1-w^2)^{3/2}$. It is therefore also decreasing and convex-$\cap$ on $[0,1]$: hence degraded channels have a larger probability of error, higher entropy, and larger Bhattacharyya constant than the original one.

The proof of the following lemma is the topic of Problem 4.32.

Lemma 4.77 (GEXIT Kernels Preserve Ordering). Let $\{\mathsf{a}_{\mathrm{BMSC}(\mathtt{h})}\}$ represent a smooth family of symmetric L-densities and let $l^{\mathsf{a}_{\mathrm{BMSC}(\mathtt{h})}}(z)$ denote the associated GEXIT kernel. Then $l^{\mathsf{a}_{\mathrm{BMSC}(\mathtt{h})}}(z)$ is non-increasing and convex-$\cap$ on $[0,1]$.

Lemma 4.78 (Erasure Decomposition Lemma). A BMS channel characterized by its L-density a is degraded with respect to $\mathsf{a}_{\mathrm{BEC}(2\,\mathfrak{E}(\mathsf{a}))}$.

Proof. In Problem 4.33 you are asked to provide a proof by directly constructing the degrading channel. Here we give an alternative, more conceptual, proof.

Define $\epsilon = \mathfrak{E}(\mathsf{a})$. Let $|\mathfrak{A}|$ denote the $|D|$-distribution associated with a and let $|\mathfrak{A}_{\mathrm{BEC}(2\epsilon)}|$ denote the $|D|$-distribution associated with the BEC(2ϵ). From Example 4.19 we know that $|\mathfrak{A}_{\mathrm{BEC}(2\epsilon)}|(x) = 2\epsilon$ for $x \in [0,1)$ so that $\int_z^1 |\mathfrak{A}_{\mathrm{BEC}(2\epsilon)}|(x)dx = 2\epsilon(1-z)$. Consider the function $f(z) = \int_z^1 |\mathfrak{A}|(x)dx$. Clearly, $f(1) = 0$. We claim that $f(0) = 2\epsilon$. To see this, note that the kernel of the error probability functional in the $|D|$-domain is $\frac{1}{2}(1-w)$ as discussed in Problem 4.59. Therefore $\mathfrak{E}(\mathfrak{a}) = \frac{1}{2}\int_0^1 (1-w)|\mathfrak{a}|(w)dw$. If we use integration by parts we see that $\epsilon = \mathfrak{E}(\mathfrak{a}) = \frac{1}{2}\big((1-w)|\mathfrak{A}|(w)\big|_{w=0}^{1} + \int_0^1 |\mathfrak{A}|(w)dw\big) = \frac{1}{2}\int_0^1 |\mathfrak{A}|(w)dw$. Since $|\mathfrak{A}|(x)$ is increasing in x, it follows that f is a convex-$\cap$ function. We conclude that

$$\int_z^1 |\mathfrak{A}|(x)dx = f(z) \geq 2\epsilon(1-z) = \int_z^1 |\mathfrak{A}_{\mathrm{BEC}(2\epsilon)}|(x)dx.$$

By Theorem 4.74 we have $\mathfrak{A}_{\mathrm{BEC}(2\epsilon)} \twoheadrightarrow \mathfrak{A}$. □

Discussion: This lemma shows that one can have two channels, one degraded with respect to the other, which have the *same* error probability. Therefore, degradation does not imply a *strict* increase in the value of a *given* decreasing and convex-$\cap$ functional. But for any pair of degraded channels *some* such functional is strictly increasing.

§4.1.16. APP Processing and Degradation

Lemma 4.79 (Sufficient Statistic Preserves Degradation). Let X be a binary vector chosen with probability $p_X(x)$ from a code C. Consider transmission over the BMS channels $p_{Y|X}$ and $p_{Z|X}$ where $X \to Y \to Z$, i.e., $p_{Z|X}$ is degraded with respect to $p_{Y|X}$. If $\tilde{Y}_{\sim i} = f(Y_{\sim i})$ constitutes a sufficient statistic for X_i given $Y_{\sim i}$ and $\tilde{Z}_{\sim i} = g(Z_{\sim i})$ constitutes a sufficient statistic for X_i given $Z_{\sim i}$ then $X_i \to \tilde{Y}_{\sim i} \to \tilde{Z}_{\sim i}$, i.e., $p_{\tilde{Z}_{\sim i}|X_i}$ is degraded with respect to $p_{\tilde{Y}_{\sim i}|X_i}$.

Proof. As discussed on page 28, where we introduced Markov chains, we need to show that $p_{\tilde{Z}_{\sim i}|X_i,\tilde{Y}_{\sim i}}(\tilde{z}_{\sim i}\,|\,x_i,\tilde{y}_{\sim i}) = p_{\tilde{Z}_{\sim i}|\tilde{Y}_{\sim i}}(\tilde{z}_{\sim i}\,|\,\tilde{y}_{\sim i})$. We have

$$
\begin{aligned}
& p_{\tilde{Z}_{\sim i}|X_i,\tilde{Y}_{\sim i}}(\tilde{z}_{\sim i}\,|\,x_i,\tilde{y}_{\sim i}) \\
&= \sum_{y_{\sim i},z_{\sim i}} p_{Y_{\sim i},Z_{\sim i},\tilde{Z}_{\sim i}|X_i,\tilde{Y}_{\sim i}}(y_{\sim i},z_{\sim i},\tilde{z}_{\sim i}\,|\,x_i,\tilde{y}_{\sim i}) \\
&= \sum_{y_{\sim i},z_{\sim i}} p_{Y_{\sim i}|X_i,\tilde{Y}_{\sim i}}(y_{\sim i}\,|\,x_i,\tilde{y}_{\sim i})\, p_{Z_{\sim i},\tilde{Z}_{\sim i}|X_i,Y_{\sim i},\tilde{Y}_{\sim i}}(z_{\sim i},\tilde{z}_{\sim i}\,|\,x_i,y_{\sim i},\tilde{y}_{\sim i}) \\
&= \sum_{y_{\sim i},z_{\sim i}} p_{Y_{\sim i}|\tilde{Y}_{\sim i}}(y_{\sim i}\,|\,\tilde{y}_{\sim i})\, p_{Z_{\sim i},\tilde{Z}_{\sim i}|Y_{\sim i},\tilde{Y}_{\sim i}}(z_{\sim i},\tilde{z}_{\sim i}\,|\,y_{\sim i},\tilde{y}_{\sim i}) \\
&= \sum_{y_{\sim i},z_{\sim i}} p_{Y_{\sim i},Z_{\sim i},\tilde{Z}_{\sim i}|\tilde{Y}_{\sim i}}(y_{\sim i},z_{\sim i},\tilde{z}_{\sim i}\,|\,\tilde{y}_{\sim i}) = p_{\tilde{Z}_{\sim i}|\tilde{Y}_{\sim i}}(\tilde{z}_{\sim i}\,|\,\tilde{y}_{\sim i}).
\end{aligned}
$$

In the third step we have used for the transformation of the first expression the fact that $\tilde{Y}_{\sim i}$ constitutes a sufficient statistic for X_i given $Y_{\sim i}$ so that $X_i \to \tilde{Y}_{\sim i} \to Y_{\sim i}$. To transform the second term, note that $\tilde{Z}_{\sim i}$ is a function of $Z_{\sim i}$. Therefore, it suffices to consider $p_{Z_{\sim i}|X_i,Y_{\sim i},\tilde{Y}_{\sim i}}(z_{\sim i}\,|\,x_i,y_{\sim i},\tilde{y}_{\sim i})$. Note that $p_{X,Z_{\sim i}|Y_{\sim i}}(x,z_{\sim i}\,|\,y_{\sim i}) = p_{X|Y_{\sim i}}(x\,|\,y_{\sim i})\,p_{Z_{\sim i}|Y_{\sim i}}(z_{\sim i}\,|\,y_{\sim i})$ implies

$$p_{X_i,Z_{\sim i}|Y_{\sim i}}(x_i,z_{\sim i}\,|\,y_{\sim i}) = p_{X_i|Y_{\sim i}}(x_i\,|\,y_{\sim i})\,p_{Z_{\sim i}|Y_{\sim i}}(z_{\sim i}\,|\,y_{\sim i}).$$

We conclude that $X_i \to Y_{\sim i} \to Z_{\sim i}$. Since further $\tilde{Y}_{\sim i}$ is a function of $Y_{\sim i}$, it follows that, conditioned on $Y_{\sim i}$, $Z_{\sim i}$ is no longer a function of X_i. □

Lemma 4.80 (APP Processing Preserves Degradation). Let X be chosen with uniform probability from a binary linear code whose i-th position is proper. Let X be transmitted over a memoryless channel. Let Y be the output assuming that the i-th bit is transmitted through the BMS channel characterized by its L-density $\mathsf{a}_{\text{BMSC}_i}$, $i \in [n]$. Let Z denote the corresponding output if we assume that the i-th bit is transmitted over the BMS channel characterized by its L-density $\mathsf{b}_{\text{BMSC}_i}$, $i \in [n]$. Let a_i, respectively b_i, denote the density of Φ_i (as defined in Theorem 4.30) under these two cases, assuming that $X_i = 1$. If $\{\mathsf{a}_{\text{BMSC}_i}\} \prec \{\mathsf{b}_{\text{BMSC}_i}\}$ then $\{\mathsf{a}_i\} \prec \{\mathsf{b}_i\}$.

Discussion: In words the lemma states the intuitive fact that if we degrade all or just some of the observations then the output of the APP processor is degraded as well. A particularly simple case of degradation which appears frequently in practice is erasures of some observations.

Proof. We know from Theorem 4.29 that $X_i \to \Phi_i(Y_{\sim i}) \to Y_{\sim i}$, and by the same argument $X_i \to \Phi_i(Z_{\sim i}) \to Z_{\sim i}$ (in the statement of Theorem 4.29 we assume that all components are sent through the *same* channel but the proof applies verbatim to the more general case where the channels are possibly different). Applying Theorem 4.79 we conclude that $X_i \to \Phi_i(Y_{\sim i}) \to \Phi_i(Z_{\sim i})$.

Again by Theorem 4.29 we know that $p_{\Phi_i(Y_{\sim i}) | X_i}(\phi_i | x_i)$ and $p_{\Phi_i(Z_{\sim i}) | X_i}(\phi_i | x_i)$ represent BMS channels. By assumption the associated L-densities under the condition $X_i = 1$ are a_i and b_i, respectively. We conclude from $X_i \to \Phi_i(Y_{\sim i}) \to \Phi_i(Z_{\sim i})$ that $\mathsf{a}_i \prec \mathsf{b}_i$. □

§4.2. Message-Passing Decoder

All the decoders which we consider in the sequel are motivated by the decoder derived in Chapter 2. In particular, they are all of the *message-passing* type; i.e., the output sent along a particular edge only depends on the input along all *other* edges. It is convenient to introduce some degrees of freedom regarding the message alphabet as well as regarding the computation rules. These degrees of freedom allow a trade-off between performance and complexity. We start by defining the *class* of message-passing decoders that we consider.

The decoding proceeds by rounds of message exchanges. First, the incoming messages at the check nodes are processed and the outgoing messages are forwarded to the variable nodes. These messages are then processed at the variable nodes and messages are sent back along all edges to the check nodes. This constitutes *one round* of message passing. In general, decoding consists of several such rounds. As mentioned earlier, an important condition on the processing is that a message sent from a node along a particular edge must not depend on the message previously received along that edge. It is exactly this restriction that makes it possible to analyze the behavior of the decoder.

In order not to complicate the notation, we suppress in the sequel the dependency of the maps on the node degrees. Let $\mathcal{O}$ denote the alphabet of the received messages which, without loss of essential generality, we can assume to be equal to the channel output alphabet. Further, let $\mathcal{M}$ denote the message alphabet. Let $\Phi^{(\ell)} : \mathcal{M}^{\mathtt{r}-1} \to \mathcal{M}$, $\ell \geq 0$, denote the check-node message map as a function of $\ell \in \mathbb{N}$, and let $\Psi^{(\ell)} : \mathcal{O} \times \mathcal{M}^{\mathtt{l}-1} \to \mathcal{M}$, $\ell \geq 0$, denote the variable-node message map where $\mathtt{r}$ and $\mathtt{l}$ are the check- and variable-node degree, respectively. These functions represent the processing performed at the check nodes and variables nodes, respectively. Iteration $\ell = 0$ corresponds to the initialization and the "real" message-passing algorithm starts at $\ell = 1$. Because of the imposed restriction on the dependence of messages, the outgoing message only depends on $\mathtt{r} - 1$ incoming messages at a check node and $\mathtt{l} - 1$ incoming messages at a variable node. Also, we allow these maps to depend on the iteration number. We assume that each node of the same degree invokes the same message map for each edge and that all edges connected to such a node are treated equally. In other words, the maps are symmetric functions.

It is helpful to think of the messages (and the received values) in the following way. Each message represents an estimate of a particular codeword bit. The *sign* of the message indicates whether the transmitted bit is estimated to be -1 or $+1$ and the *absolute value* of the message is a measure of the *reliability* of this estimate. The sign of the particular value 0, which represents an *erasure*, is equally likely to be $+1$ or -1.

In the sequel we denote the *received* message by μ_0, the incoming messages to a check node of degree $\mathtt{r}$ by $\mu_1, \dots, \mu_{\mathtt{r}-1}$, and the incoming messages to a variable of degree $\mathtt{l}$ by $\mu_1, \dots, \mu_{\mathtt{l}-1}$. Our subsequent analysis and notation is greatly simplified by imposing the following symmetry conditions on the decoding algorithm.

Definition 4.81 (Message-Passing Symmetry Conditions).
Check-Node Symmetry:

$$\Phi^{(\ell)}(b_1\mu_1, \dots, b_{\mathtt{r}-1}\mu_{\mathtt{r}-1}) = \Phi^{(\ell)}(\mu_1, \dots, \mu_{\mathtt{r}-1}) \left(\prod_{i=1}^{\mathtt{r}-1} b_i \right)$$

for any ± 1 sequence $(b_1, \dots, b_{\mathtt{r}-1})$, i.e., signs factor out of the check-node message map.

Variable-Node Symmetry:

$$\ell = 0 \qquad \Psi^{(\ell)}(-\mu_0, \dots) = -\Psi^{(\ell)}(\mu_0, \dots),$$

$$\ell \geq 1 \qquad \Psi^{(\ell)}(-\mu_0, -\mu_1, \dots, -\mu_{\mathtt{l}-1}) = -\Psi^{(\ell)}(\mu_0, \mu_1, \dots, \mu_{\mathtt{l}-1}),$$

i.e., the initial message out of a variable node only depends on the value received at this node from the channel and sign inversion invariance of the variable-node message map holds. ▽

Example 4.82 (Gallager Algorithm A). Gallager's algorithm A is probably the simplest non-trivial message-passing algorithm applicable to general channels. Consider transmission over the BSC. (If the transmission takes place over a more general channel then apply a hard decision at the input.) The message alphabet is $\mathcal{M} = \{-1, 1\}$. The message maps are given by

$$\Phi^{(\ell)}(\mu_1, \dots, \mu_{\mathtt{r}-1}) = \prod_{i=1}^{\mathtt{r}-1} \mu_i, \qquad \Psi^{(\ell=0)}(\mu_0, \mu_1, \dots, \mu_{\mathtt{l}-1}) = \mu_0,$$

$$\Psi^{(\ell\geq 1)}(\mu_0, \mu_1, \dots, \mu_{l-1}) = \begin{cases} -\mu_0, & \text{if } \mu_1 = \mu_2 = \dots = \mu_{l-1} = -\mu_0, \\ \mu_0, & \text{otherwise.} \end{cases}$$

In words, check nodes send a message indicating the product of the incoming messages. The variable nodes send their received value unless the incoming messages are unanimous, in which case the sign indicated by these messages is sent. At iteration zero the messages emitted by the variable nodes depend only on the values received from the channel but not on the internal messages in the graph so that we do not need to specify the initial internal messages which enter the check nodes.

These message-processing rules are intuitive. At check nodes the product rule (applied to the ± 1-valued messages) reflects the modulo-2 structure of check constraints. At variable nodes we assume that the message received from the channel is more reliable than the internal messages. Only if all internal messages agree on a value do we trust them.

Let us give a short preview of the main characteristics which we encounter when using Gallager's algorithm A. Figure 4.83 shows the block (left) and bit (right) error probability of Gallager's algorithm A for the $(3, 6)$-regular ensemble when transmission takes place over the BSC. We will show in Section 4.7 that the algorithm has

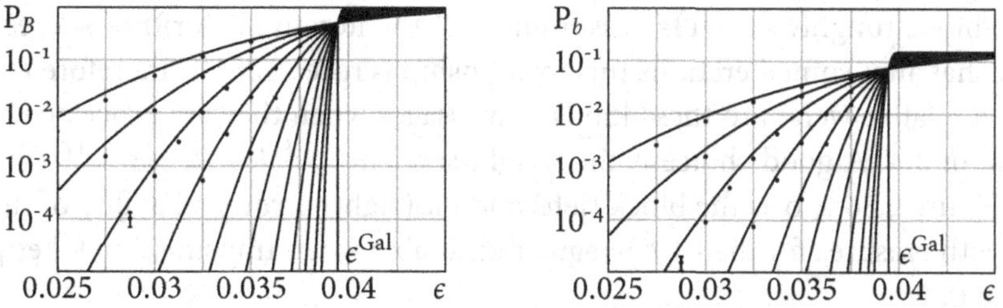

Figure 4.83: Performance of Gallager's algorithm A for the $(3, 6)$-regular ensemble when transmission takes place over the BSC. The blocklengths are $n = 2^i$, $i = 10, \dots, 20$. The left-hand graph shows the block error probability, whereas the right-hand graph concerns the bit error probability. The dots correspond to simulations. For most simulation points the 95% confidence intervals (see Problem 4.37) are smaller than the dot size. The lines correspond to the analytic approximation of the waterfall curves based on scaling laws (see Section 4.13).

a threshold and that the threshold for the $(3, 6)$-regular ensemble is $\epsilon^{\text{Gal}} \approx 0.0394$. We recognize in the figure the characteristic "waterfall" shape of the performance curve. (Since the figure shows the performance of expurgated ensembles, the "error floor" is not visible.) As the length increases the curves converge to a step function at the threshold. The solid lines correspond to scaling laws of essentially the same form as we discussed in Section 3.23 for the BEC. These scaling laws are discussed in Section 4.13. ◇

Example 4.84 (Decoder with Erasures). Assume that we extend the previous example by allowing "erasures" in the decoder, i.e., the alphabet is $\mathcal{M} = \{-1, 0, 1\}$. The message maps are specified by

$$\Phi^{(\ell)}(\mu_1, \dots, \mu_{\mathtt{r}-1}) = \prod_{i=1}^{\mathtt{r}-1} \mu_i,$$

$$\Psi^{(\ell=0)}(\mu_0, \mu_1, \dots, \mu_{\mathtt{l}-1}) = \mu_0,$$

$$\Psi^{(\ell\geq 1)}(\mu_0, \mu_1, \dots, \mu_{\mathtt{l}-1}) = \text{sgn}\,(w^{(\ell)}\mu_0 + \sum_{i=1}^{\mathtt{l}-1} \mu_i),$$

where $w^{(\ell)}$ is an appropriately chosen weight sequence. The motivation for this extension is simple. At the beginning of the decoding process, the received messages are more reliable than the computed messages sent from the check nodes to the variable nodes: consider the messages entering the variable nodes during the first iteration; if the initial error probability of a message is ϵ (where we think of ϵ as a small positive number) then at the output of a check node of degree $\mathtt{r}$ the error probability is roughly $(\mathtt{r} - 1)\epsilon$. Assuming that the iterative algorithm succeeds, it is clear that after some iterations this relationship is reversed. It is therefore natural to give a relative weight to these kinds of messages when they are processed at the variable nodes. A good choice is the weight sequence $w^{(1)} = 2$, and $w^{(\ell)} = 1$ for $\ell > 1$. Figure 4.86 shows the block (left) and bit (right) error probability of the decoder with erasures for the $(3, 6)$-regular ensemble when transmission takes place over the BSC.

The threshold for the $(3, 6)$-regular ensemble is $\epsilon^{\text{DE}} \approx 0.0708$, which is considerably larger than the threshold for Gallager's algorithm A. This shows that small increases in complexity can result in large gains in performance. Again, the waterfall part of the curves is well captured by a scaling law (see Section 4.13). ◇

Example 4.85 (Belief Propagation Decoder). The belief propagation (BP) decoder is the most powerful example, employing a "locally" optimal processing rule. Nodes act under the assumption that each message communicated to them represents a conditional probability on the bit, and that each message is conditionally independent of all others, i.e., the random variables on which the different messages are based are independent. This is the message-passing algorithm discussed in Chapter 2, i.e., the message-passing algorithm that results naturally from the factor graph approach.

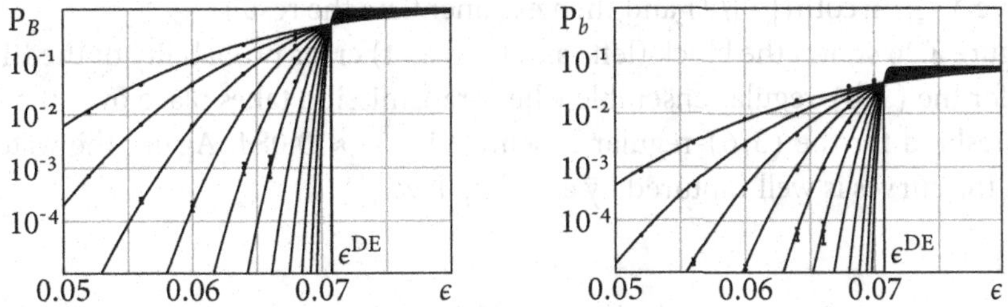

Figure 4.86: Performance of the decoder with erasures for the $(3,6)$-regular ensemble when transmission takes place over the BSC. The blocklengths are $n = 2^i$, $i = 10, \ldots, 20$. The left-hand graph shows the block error probability, whereas the right-hand graph concerns the bit error probability. The dots correspond to simulations. The lines correspond to the analytic approximation of the waterfall curves based on scaling laws (see Section 4.13).

As we have seen in Section 2.5.2, assuming that the messages (received from the channel) are in log-likelihood form, we have

$$\Phi^{(\ell)}(\mu_1, \ldots, \mu_{\mathtt{r}-1}) = \begin{cases} 0, & \ell = 0, \\ 2\tanh^{-1}\left(\prod_{i=1}^{\mathtt{r}-1} \tanh(\frac{\mu_i}{2})\right), & \ell \geq 1, \end{cases} \tag{4.87}$$

$$\Psi^{(\ell)}(\mu_0, \mu_1, \ldots, \mu_{\mathtt{l}-1}) = \mu_0 + \sum_{i=1}^{\mathtt{l}-1} \mu_i. \tag{4.88}$$

The processing rule on the check-node side can be written in several equivalent forms. We have

$$2\tanh^{-1}\Big(\prod_{i=1}^{\mathtt{r}-1} \tanh(\frac{\mu_i}{2})\Big) = 2\coth^{-1}\Big(\prod_{i=1}^{\mathtt{r}-1} \coth(\frac{\mu_i}{2})\Big) = g^{-1}\Big(\sum_{i=1}^{\mathtt{r}-1} g(\mu_i)\Big).$$

The equivalence of the first two forms is straightforward – instead of multiplying the terms $\tanh(\mu_i/2)$ we can multiply their inverses $1/\tanh(\mu_i/2) = \coth(\mu_i/2)$.

In the last expression we have used the map $g(l) = \big(\mathfrak{H}(l), \ln\coth(|l|/2)\big)$, which we introduced on page 180. With a slight abuse of notation we have assumed here that the received messages are already in log-likelihood ratio form so that the function takes as input l and not y. The "sum" $\sum_{i=1}^{\mathtt{r}-1} g(\mu_i)$ has to be interpreted in the following way. For $\mu \in \mathbb{R}$, $g(\mu)$ has the form (s, y), where the first component is the *sign* taking values in $\{\pm 1\}$ and the second is the *reliability*, which is a non-negative extended real number. We then have $\sum_{i=1}^{\mathtt{r}-1}(s_i, y_i) = (\prod_{i=1}^{\mathtt{r}-1} s_i, \sum_{i=1}^{\mathtt{r}-1} y_i)$. The equivalence of this last expression is easy to see: first, factor out the signs of the messages

μ_i. These signs multiply. Then, instead of taking the product $\prod_{i=1}^{r-1} \coth(|\mu_i|/2)$ we compute $\sum_{i=1}^{r-1} \ln \coth(|\mu_i|/2)$ and then exponentiate the result.

Figure 4.89 shows the block (left) and bit (right) error probability of the BP decoder for the $(3,6)$-regular ensemble when transmission takes place over the BSC. The threshold for the $(3,6)$-regular ensemble is $\epsilon^{\text{BP}} \approx 0.084$. Again, the waterfall part of the curves is well captured by a scaling law. ◇

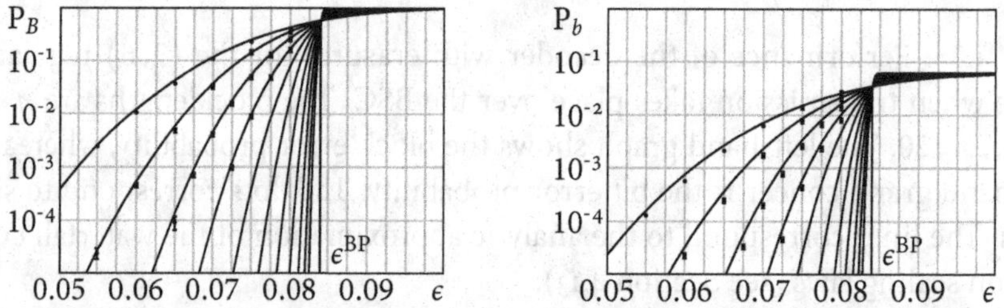

Figure 4.89: Performance of the BP decoder for the $(3,6)$-regular ensemble when transmission takes place over the BSC. The blocklengths are $n = 2^i$, $i = 10, \dots, 20$. The left-hand graph shows the block error probability, whereas the right-hand graph concerns the bit error probability. The dots correspond to simulations. The lines correspond to the analytic approximation of the waterfall curves based on scaling laws.

It is apparent that there is an infinite variety of decoders that fulfills the basic symmetry conditions described in Definition 4.81. This degree of freedom is important in practice. Although the BP decoder is the optimum choice, it requires infinite precision arithmetic. Therefore, in any digital implementation one must resort to quantized versions of BP. Typically one tries to mimic as much as possible the behavior of the BP decoder given the constraint that the messages take values only in a finite set. This provides a natural trade-off between achievable performance and complexity of the decoder.

§4.3. Two Basic Simplifications

The analysis proceeds in lock step with the one we have given for the BEC. Most fundamental properties have an analog in this general setting, although in many instances we have to be content with weaker statements. This is due to our current inability to prove the corresponding results and (presumably) not because these results are inherently not true.

As we have seen in the previous section, the class of message-passing algorithms is very broad. Some of the subsequent statements apply to all elements of this class.

Some other statements require an investigation of the specific chosen message-passing algorithm. We show how to proceed for Gallager's algorithm A as well as the BP decoder. Gallager's algorithm A is a simple example of a "quantized" decoder. The proofs for this algorithm are based on calculus. For the BP decoder on the other hand we can often make use of its "local" optimality and provide more conceptual proofs. In practice one very often deals with a "quantized" decoder with a relatively large number of quantization levels so that the resulting decoder closely mimics the BP decoder. In such a case one can either rely on the results for the BP decoder or proceed in a fashion similar to our analysis of Gallager's algorithm A.

§4.3.1. Restriction to the All-One Codeword

Lemma 4.90 (Conditional Independence of Error Probability). Let $\mathtt{G}$ be a binary Tanner graph representing a binary linear code C. Suppose that C is used to transmit over a BMS channel characterized by its L-density a_{BMSC} and suppose that the receiver performs message-passing decoding on $\mathtt{G}$ (we denote a generic such decoder by MP). Let $\text{P}^{\text{MP}}(\mathtt{G}, \mathsf{a}_{\text{BMSC}}, \ell, x)$ denote the conditional (bit or block) probability of error after the ℓ-th decoding iteration, assuming that x was sent, $x \in C$. If the decoder fulfills the symmetry conditions stated in Definition 4.81, then $\text{P}^{\text{MP}}(\mathtt{G}, \mathsf{a}_{\text{BMSC}}, \ell, x) = \frac{1}{|C|}\sum_{c \in C} \text{P}^{\text{MP}}(\mathtt{G}, \mathsf{a}_{\text{BMSC}}, \ell, c) = \text{P}^{\text{MP}}(\mathtt{G}, \mathsf{a}_{\text{BMSC}}, \ell)$. This means that $\text{P}^{\text{MP}}(\mathtt{G}, \mathsf{a}_{\text{BMSC}}, \ell, x)$ is independent of the transmitted codeword.

Proof. Recall from page 181 that a BMS channel can be modeled multiplicatively as

$$Y_t = x_t Z_t, \tag{4.91}$$

where $\{Z_t\}_t$ is a sequence of iid random variables with density a_{BMSC}. Let $x \in C$ and let Y denote the corresponding channel output, $Y = xZ$ (multiplication is component-wise and all three quantities are vectors of length n). Note that Z by itself is equal to the observation assuming that the all-one codeword was transmitted. We will now show that the messages sent during the decoding process for the cases where the received word is either xZ or Z are in one-to-one correspondence.

Let i be an arbitrary variable node and let j be one of its neighboring check nodes. Let $\mu_{ij}^{(\ell)}(y)$ denote the message sent from i to j in iteration ℓ assuming that the received value is y and let $\mu_{ji}^{(\ell)}(y)$ denote the corresponding message sent from j to i.

We have $\mu_{ij}^{(0)}(y) \overset{(4.91)}{=} \mu_{ij}^{(0)}(xz) = x_i \mu_{ij}^{(0)}(z)$, where the second step follows from the variable-node symmetry property stated in Definition 4.81. Assume that we have $\mu_{ij}^{(\ell)}(y) = x_i \mu_{ij}^{(\ell)}(z)$ for all (i, j) pairs and some $\ell \geq 0$. Let ∂j denote all variable nodes which are connected to check node j. Since x is a codeword, we have $\prod_{k\in\partial j} x_k = 1$.[3] From the check-node symmetry condition stated in Definition 4.81 we conclude that $\mu_{ji}^{(\ell+1)}(y) = x_i\mu_{ji}^{(\ell+1)}(z)$. Further, invoking once more the variable-node symmetry condition, it follows that $\mu_{ij}^{(\ell+1)}(y) = x_i\mu_{ij}^{(\ell+1)}(z)$ for all (i,j) pairs. Thus, by induction, all messages to and from variable node i, when y is received, are equal to the product of x_i and the corresponding message when z is received. Hence, both decoders proceed in lock step and commit exactly the same number of errors (if any), which proves the claim. □

§4.3.2. Concentration

The second major simplification comes from the fact that, rather than analyzing individual codes, it suffices to assess the ensemble average performance. This latter task is accomplished much more easily. This is true, since, as the next theorem asserts, the individual behavior of elements of an ensemble is with high probability close to the ensemble average.

Theorem 4.92 (Concentration around Ensemble Average). Let $\mathtt{G}$, chosen uniformly at random from LDPC(n,λ,ρ), be used for transmission over a BMS channel characterized by its L-density a_{BMSC}. Assume that the decoder performs ℓ rounds of message-passing decoding and let $\mathrm{P}_{\mathrm{b}}^{\mathrm{MP}}(\mathtt{G},\mathsf{a}_{\text{BMSC}},\ell)$ denote the resulting bit error probability. Then, for any given $\delta > 0$, there exists an $\alpha > 0$, $\alpha = \alpha(\lambda,\rho,\delta)$, such that

$$\mathbb{P}\{|\mathrm{P}_{\mathrm{b}}^{\mathrm{MP}}(\mathtt{G},\mathsf{a}_{\text{BMSC}},\ell) - \mathbb{E}_{\text{LDPC}(n,\lambda,\rho)}\left[\mathrm{P}_{\mathrm{b}}^{\mathrm{MP}}(\mathtt{G},\mathsf{a}_{\text{BMSC}},\ell)\right]| > \delta\} \le e^{-\alpha n}.$$

In words, the theorem asserts that all except an exponentially (in the blocklength) small fraction of codes behave within an arbitrarily small δ from the ensemble average. Therefore, assuming sufficiently large blocklengths, the ensemble average is a good indicator for the individual behavior and it seems a reasonable route to focus one's effort on the design and construction of ensembles whose average performance approaches the Shannon theoretic limit. The proof of the theorem is based on the so-called Hoeffding-Azuma inequality and can be found on page 487.

§4.4. Tree Channel and Convergence to Tree Channel

§4.4.1. Tree Channel

Definition 4.93 ($(\mathcal{T}_\ell, \mathsf{a}_{\text{BMSC}})$-Tree Channel). Given a BMS channel characterized by its L-density a_{BMSC} and a tree ensemble $\mathcal{T}_\ell = \mathcal{T}_\ell(\lambda,\rho)$, we define the associated

[3] In the case of parallel edges, x_k has to be counted according to the multiplicity of the edge.

$(\mathcal{T}_\ell, \mathsf{a}_{\text{BMSC}})$-tree channel. The channel takes binary input $X \in \{\pm 1\}$ with uniform probability. The output of the channel is constructed as follows. Given X, first pick $\mathtt{T}$ from $\mathcal{T}_\ell$ with probability according to the process defined by (λ, ρ). Next, pick a codeword from $C^X(\mathtt{T})$ uniformly at random. Transmit this codeword over the BMS channel defined by a_{BMSC}. Call the output Y. The receiver sees $(\mathtt{T}, Y)$ and estimates X.

Consider processing the tree by a MP decoder to form an estimate of X. Let $\mathrm{P}^{\text{MP}}_{\mathcal{T}_\ell}(\mathsf{a}_{\text{BMSC}})$ denote the resulting bit error probability.

As an important special case, consider BP decoding. Since we are operating on a tree we know that the BP algorithm computes the log-likelihood ratio of the root bit. More precisely, given $(\mathtt{T}, Y)$ the BP decoder computes

$$\ln \frac{p((\mathtt{T}, Y) \mid X = 1)}{p((\mathtt{T}, Y) \mid X = -1)}.$$

The distribution of this quantity conditioned on $X = 1$ is the L-density.

As an aid to analysis, it is convenient to consider also the following generalization. Assume that during transmission of the codeword all *internal* variable nodes are transmitted through a BMS channel characterized by a_{BMSC} but that all leaf nodes are transmitted through a BMS channel characterized by b_{BMSC}. We denote the corresponding tree channel by $(\mathcal{T}_\ell, \mathsf{a}_{\text{BMSC}}, \mathsf{b}_{\text{BMSC}})$ and the corresponding error probability of an MP decoder by $\mathrm{P}^{\text{MP}}_{\mathcal{T}_\ell}(\mathsf{a}_{\text{BMSC}}, \mathsf{b}_{\text{BMSC}})$. ▽

§4.4.2. Convergence to Tree Channel

Theorem 4.94 (Convergence to Tree Channel). For a given degree distribution pair (λ, ρ) consider the sequence of associated ensembles $\text{LDPC}(n, \lambda, \rho)$ of increasing blocklengths n under ℓ rounds of message-passing decoding. Then

$$\lim_{n\to\infty} \mathbb{E}_{\text{LDPC}(n,\lambda,\rho)}[\mathrm{P}^{\text{MP}}_{\mathrm{b}}(\mathtt{G}, \mathsf{a}_{\text{BMSC}}, \ell)] = \mathrm{P}^{\text{MP}}_{\mathcal{T}_\ell}(\mathsf{a}_{\text{BMSC}}), \quad \mathrm{P}^{\text{BP}}_{\mathcal{T}_\ell}(\mathsf{a}_{\text{BMSC}}) = \mathrm{P}^{\text{MAP}}_{\mathcal{T}_\ell}(\mathsf{a}_{\text{BMSC}}).$$

Proof. The proof of the convergence to the tree channel performance is virtually identical to the one for the BEC on page 94. Recall that the main idea of the proof is that almost surely the computation graph of a fixed depth is a tree. This gives us the identity between the two sides.

Further, we already know from Chapter 2 that on a tree BP decoding is equivalent to MAP decoding. □

§4.5. Density Evolution

Theorem 4.94 asserts that in the limit of large blocklengths, the average performance of an ensemble $\text{LDPC}(n, \lambda, \rho)$ converges to the performance of the corresponding tree channel. The performance of the tree channel is relatively easy to assess because of its recursive structure.

§4.5.1. Gallager Algorithm A

Theorem 4.95 (Performance of Tree Channel). Consider a degree distribution pair $(\lambda, \rho) \triangleq (L, R)$ and transmission over the BSC(ϵ). Define $x_0 = \epsilon$ and for $\ell \geq 1$ let

$$x_\ell = \epsilon(1 - p^+(x_{\ell-1})) + \bar{\epsilon} p^-(x_{\ell-1}), \tag{4.96}$$

where

$$p^+(x) = \lambda\Big(\frac{1+\rho(1-2x)}{2}\Big), \qquad p^-(x) = \lambda\Big(\frac{1-\rho(1-2x)}{2}\Big).$$

Then

$$\mathrm{P}^{\mathrm{Gal}}_{\vec{T}_\ell}(\epsilon) = x_\ell,$$
$$\mathrm{P}^{\mathrm{Gal}}_{\vec{T}_\ell}(\epsilon) = \epsilon\Big(1 - L\Big(\frac{1+\rho(1-2x_{\ell-1})}{2}\Big)\Big) + \bar{\epsilon}\, L\Big(\frac{1-\rho(1-2x_{\ell-1})}{2}\Big).$$

Proof. Consider $\mathrm{P}^{\mathrm{Gal}}_{\vec{T}_\ell}(\epsilon)$ for $\ell \in \mathbb{N}$. By definition of the algorithm, the initial variable-to-check message is equal to the received message. This message is in error with probability ϵ. It follows that $\mathrm{P}^{\mathrm{Gal}}_{\vec{T}_0}(\epsilon) = \epsilon$, as claimed. We proceed by induction over ℓ. Assume that $\mathrm{P}^{\mathrm{Gal}}_{\vec{T}_\ell}(\epsilon) = x_\ell$ for some $\ell \geq 0$. Let us derive the error probability of the check-to-variable message in the $(\ell+1)$-th iteration. Recall that a check-to-variable message emitted by a check node of degree $\mathtt{r}$ along a particular edge is the product of all the $\mathtt{r}-1$ incoming messages along all other edges. By assumption, each such message is in error with probability x_ℓ and all messages are statistically independent. The outgoing message is in error if an odd number of incoming messages is in error. This happens with probability (see Section D.3)

$$\sum_k \binom{\mathtt{r}-1}{2k-1} x_\ell^{2k-1}(1-x_\ell)^{\mathtt{r}-2k} = \frac{1-(1-2x_\ell)^{\mathtt{r}-1}}{2}.$$

Since an edge chosen uniformly at random is connected to a check node of degree $\mathtt{r}$ with probability $\rho_\mathtt{r}$, it follows that the expected probability of error of a randomly chosen check-to-variable message in the $(\ell+1)$-th iteration is equal to

$$\sum_\mathtt{r} \rho_\mathtt{r} \frac{1-(1-2x_\ell)^{\mathtt{r}-1}}{2} = \frac{1-\rho(1-2x_\ell)}{2}.$$

Now let us derive $\mathrm{P}^{\mathrm{Gal}}_{\vec{T}_{\ell+1}}(\epsilon)$, the error probability of the variable-to-check message in the $(\ell+1)$-th iteration. Consider an edge which is connected to a variable node of degree 1. The outgoing variable-to-check message along this edge in the $(\ell+1)$-th iteration is in error if either the received value is in error and at least one incoming message is in error or the received value is correct but all incoming messages are in error. The first event has probability

$$\epsilon\big(1-\big(1-\frac{1-\rho(1-2x_\ell)}{2}\big)^{1-1}\big)=\epsilon\big(1-\big(\frac{1+\rho(1-2x_\ell)}{2}\big)^{1-1}\big).$$

The second event has probability

$$\bar{\epsilon}\big(\frac{1-\rho(1-2x_\ell)}{2}\big)^{1-1}.$$

The claim now follows by averaging over the edge degree distribution $\lambda(\cdot)$.

The proof for $\mathrm{P}^{\mathrm{MP}}_{\mathring{\mathcal{T}}_\ell}(\epsilon)$ is very similar. It follows by performing the last step with respect to the *node* degree distribution $L(\cdot)$ instead of the edge degree distribution $\lambda(\cdot)$. In the preceding derivation we assume the following decision rule: the bit value is equal to the received value unless *all* internal incoming messages to the variable node agree. In the latter case we set the bit value equal to this common value. Of course, other decision rules are possible. □

§4.5.2. Belief Propagation

Theorem 4.97 (Performance of Tree Channel). Consider a degree distribution pair $(\lambda,\rho) \triangleq (L,R)$ and transmission over a BMS channel with associated L-density $\mathsf{a}_{\mathrm{BMSC}}$. Define $\mathsf{a}_0 = \mathsf{a}_{\mathrm{BMSC}}$ and for $\ell \geq 1$ let

$$\mathsf{a}_\ell = \mathsf{a}_{\mathrm{BMSC}} \circledast \lambda(\rho(\mathsf{a}_{\ell-1})), \tag{4.98}$$

where for an L-density a

$$\lambda(\mathsf{a}) = \sum_i \lambda_i \mathsf{a}^{\circledast(i-1)}, \qquad \rho(\mathsf{a}) = \sum_i \rho_i\, \mathsf{a}^{\boxast(i-1)}.$$

Then a_ℓ is the L density associated with the tree channel $(\mathcal{T}_\ell, \mathsf{a}_{\mathrm{BMSC}})$ and $\mathsf{a}_{\mathrm{BMSC}} \circledast L(\rho(\mathsf{a}_{\ell-1}))$ is the L-density associated with the tree channel $(\mathring{\mathcal{T}}_\ell, \mathsf{a}_{\mathrm{BMSC}})$. In particular,

$$\mathrm{P}^{\mathrm{BP}}_{\mathcal{T}_\ell} = \mathfrak{E}(\mathsf{a}_\ell), \qquad \mathrm{P}^{\mathrm{BP}}_{\mathring{\mathcal{T}}_\ell} = \mathfrak{E}\left(\mathsf{a}_{\mathrm{BMSC}} \circledast L(\rho(\mathsf{a}_{\ell-1}))\right).$$

Proof. Consider the distribution of the initial variable-to-check messages. By definition of the algorithm, these messages are equal to the received messages which have density $\mathsf{a}_{\mathrm{BMSC}}$ (under the still-running assumption that the all-one codeword is transmitted). It follows that $\mathsf{a}_0 = \mathsf{a}_{\mathrm{BMSC}}$, as claimed. We proceed by induction over

ℓ. Assume that the density of the variable-to-check messages in the ℓ-th iteration is a_ℓ as described in (4.98).

Consider a check node of degree $\mathtt{r}$. By definition of the algorithm, the outgoing message is the sum of the G-representations of $\mathtt{r}-1$ independent messages (see Example 4.85). It follows from the discussion on page 181 that the density of the outgoing message is

$$\mathsf{a}_\ell^{\boxast(\mathtt{r}-1)}.$$

Averaging over the right edge-degree distribution therefore gives

$$\mathsf{b}_{\ell+1} = \rho(\mathsf{a}_\ell) = \sum_i \rho_i \, \mathsf{a}_\ell^{\boxast(i-1)}.$$

Consider next a variable node of degree $\mathtt{l}$. By definition of the algorithm, the variable-to-check message is the sum (using L-representations) of the received message, which has density a_{BMSC}, and $\mathtt{l}-1$ check-to-variable messages, each of which has density $\mathsf{b}_{\ell+1}$. Since all messages are statistically independent, it follows that the density of the outgoing message is the convolution of the densities of the summands, i.e.,

$$\mathsf{a}_{\text{BMSC}} \circledast \mathsf{b}_{\ell+1}^{\circledast(\mathtt{l}-1)}.$$

Averaging again over the edge degree distribution we see that the outgoing density is equal to

$$\mathsf{a}_{\ell+1} = \mathsf{a}_{\text{BMSC}} \circledast \lambda(\mathsf{b}_{\ell+1}) = \mathsf{a}_{\text{BMSC}} \circledast \lambda(\rho(\mathsf{a}_\ell)). \qquad \square$$

In the preceding theorem we have assumed that $\mathsf{a}_0 = \mathsf{a}_{\text{BMSC}}$ but the recursions stay valid if we start with a general initial density a_0. In general, a_ℓ is the density associated with the tree channel $(\vec{\mathcal{T}}_\ell, \mathsf{a}_{\text{BMSC}}, \mathsf{a}_0)$.

Example 4.99 (Density Evolution for the BEC(ϵ)). The density evolution equation for the BEC(ϵ), which we derived in Chapter 3, is a particular case. Indeed, guided by Example 4.19 let us assume that for $\ell \geq 0$, $\mathsf{a}_\ell(y)$ has the form $\mathsf{a}_\ell(y) = x_\ell \Delta_0(y) + \bar{x}_\ell \Delta_{+\infty}(y)$. This is true for $\ell = 0$ with $x_0 = \epsilon$, since the initial density is the one corresponding to the BEC(ϵ). Consider the evolution of the densities at the check nodes. The G-density corresponding to $\mathsf{a}_\ell(y) = x_\ell \Delta_0(y) + \bar{x}_\ell \Delta_{+\infty}(y)$ is $a_\ell(\pm 1, y)$, where $a_\ell(1, y) = \bar{x}_\ell \Delta_0(y) + \frac{x_\ell}{2} \Delta_{+\infty}(y)$ and $a_\ell(-1, y) = \frac{x_\ell}{2} \Delta_{+\infty}(y)$. Assume that we convolve two densities of the preceding form. More precisely, we have the G-density $a(\pm 1, y)$, where $a(1, y) = \bar{\alpha} \Delta_0(y) + \frac{\alpha}{2} \Delta_{+\infty}(y)$ and $a(-1, y) = \frac{\alpha}{2} \Delta_{+\infty}(y)$ and the G-density $b(\pm 1, y)$ which has the same form but has parameter β in place of α. If we perform the convolution of these two G-densities, i.e., if we compute $a(\pm 1, y) \boxast b(\pm 1, y)$ as defined on page 181, then we see that the result is the G-density $c(\pm 1, y)$, where $c(1, y) = \bar{\gamma} \Delta_0(y) + \frac{\gamma}{2} \Delta_{+\infty}(y)$ and $c(-1, y) = \frac{\gamma}{2} \Delta_{+\infty}(y)$, with $\bar{\gamma} = \bar{\alpha}\bar{\beta}$. In words, convolution of such G-densities corresponds to a multiplication of the "dual" parameters. If we now perform this G-convolution according to the degrees present in the graph, take the corresponding weighted average, and, finally, convert back to L-densities, then we see that at the input of the variable nodes we have the density $(1-\rho(1-x_\ell))\Delta_0(y) + \rho(1-x_\ell)\Delta_{+\infty}(y)$.

Assume that we have two L-densities of the form

$$\alpha\Delta_0(y) + \bar{\alpha}\Delta_{+\infty}(y) \quad \text{and} \quad \beta\Delta_0(y) + \bar{\beta}\Delta_{+\infty}(y),$$

and that we convolve them at a variable node. The result is the L-density $\gamma\Delta_0(y) + \bar{\gamma}\Delta_{+\infty}(y)$, where $\gamma = \alpha\gamma$. In words, convolution of such L-densities corresponds to a multiplication of the parameters.

Combining the two steps, it follows that after one full iteration we go from the L-density $\mathsf{a}_\ell(y) = x_\ell\Delta_0(y) + \bar{x}_\ell\Delta_{+\infty}(y)$ to the L-density $\mathsf{a}_{\ell+1}(y) = x_{\ell+1}\Delta_0(y) + \bar{x}_{\ell+1}\Delta_{+\infty}(y) = \epsilon\lambda(1-\rho(1-x_\ell))\Delta_0(y) + (1-\epsilon\lambda(1-\rho(1-x_\ell)))\Delta_{+\infty}(y)$. In other words, we recover

$$x_{\ell+1} = \epsilon\lambda(1-\rho(1-x_\ell)). \qquad \diamond$$

Example 4.100 (Density Evolution for the BAWGNC(σ)). Consider the density evolution process for the BAWGNC(σ) and the rate one-half degree distribution pair

$$\begin{aligned}\lambda(x) &= 0.212332x + 0.197596x^2 + 0.0142733x^4 + 0.0744898x^5 + \\ &\quad 0.0379457x^6 + 0.0693008x^7 + 0.086264x^8 + 0.00788586x^{10} + \\ &\quad 0.0168657x^{11} + 0.283047x^{30}, \\ \rho(x) &= x^8.\end{aligned}$$

From Example 4.21 we know that $\mathsf{a}_{\text{BAWGNC}(\sigma)}(y) \sim \mathcal{N}(2/\sigma^2, 4/\sigma^2)$. Figure 4.101 shows the evolution of a_ℓ (the densities of messages emitted by the variable nodes in iteration ℓ) and $\mathsf{b}_{\ell+1}$ (the densities emitted by the check nodes at the subsequent iteration) for $\ell = 0, 5, 10, 50$, and 140 for $\sigma = 0.93$. Note that the densities a_ℓ "move to the right" as ℓ increases. Consequently, the error probability decreases with each iteration. $\diamond$

§4.6. Monotonicity

It is perhaps natural to expect that the performance of an iterative decoder will improve if the channel improves or if the number of iterations is increased, at least in the tree-like setting. For belief propagation this expectation is borne out. The optimality of the belief propagation in the tree-like setting supports and validates our

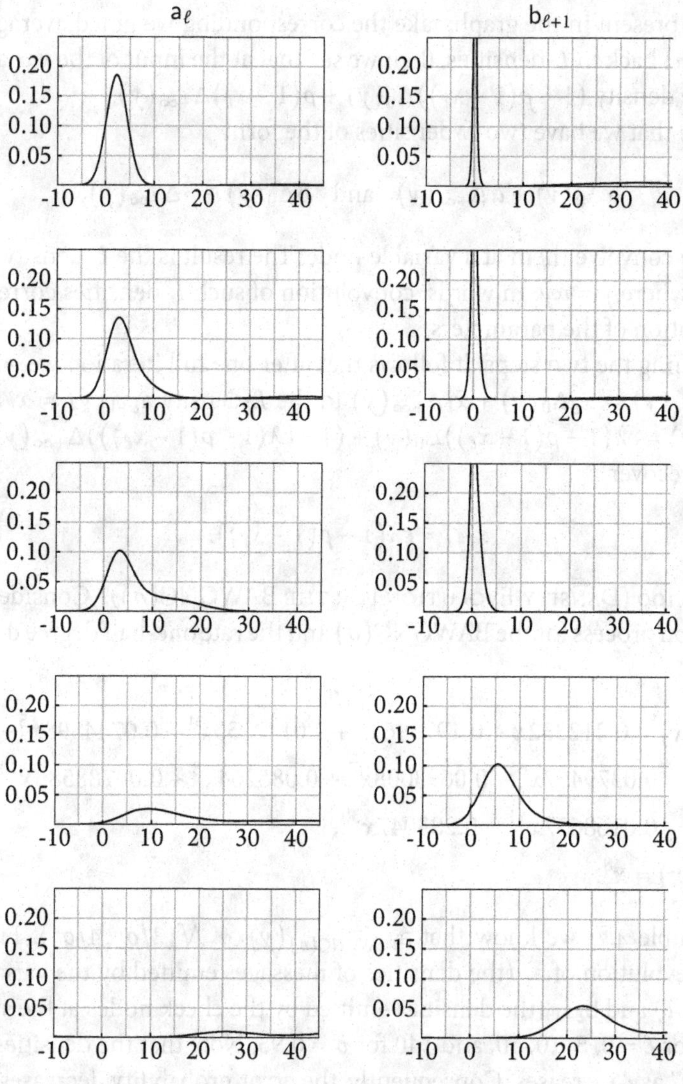

Figure 4.101: Evolution of a_ℓ (densities of messages emitted by variable nodes) and $b_{\ell+1}$ (densities of messages emitted from check nodes) for $\ell = 0, 5, 10, 50$, and 140 for the BAWGNC($\sigma = 0.93$) and the code given in Example 4.100. The densities "move to the right," indicating that the error probability decreases as a function of the number of iterations.

expectation. For message-passing algorithms in general, however, this expectation of monotonicity is not always borne out. In view of the fact that such a decoder may be highly suboptimal, it is actually unreasonable to assume that it should be monotonic, at least with regard to iteration. A good example is the min-sum decoder whose lack of monotonicity is discussed in Problem 4.42.

§4.6.1. Gallager Algorithm A

Lemma 4.102 (Monotonicity of $f(\cdot,\cdot)$). For a given degree distribution pair (λ,ρ) let $f(x,y) = x(1-p^+(y)) + \bar{x}p^-(y)$, where p^+ and p^- are defined in Theorem 4.95. Then $f(x,y)$ is increasing in both its arguments for $x, y \in [0, \frac{1}{2}]$.

Proof. We start by showing that $f(x,y)$ is an increasing function in x. Note that $f(x,y) = x(1-p^+(y)-p^-(y)) + p^-(y)$. Now $p^+(y)+p^-(y) = \lambda(1-q(y)) + \lambda(q(y))$, where $0 < q(y) = \frac{1-\rho(1-2y)}{2} \le \frac{1}{2}$, for $0 < y \le \frac{1}{2}$. Since $\lambda(1-z)+\lambda(z)$ is strictly decreasing in the range $0 \le z \le \frac{1}{2}$ (see Problem 4.35) and since $\lambda(1)+\lambda(0) = 1$ it follows that $1 - p^+(y) - p^-(y) > 0$ for $0 < y \le \frac{1}{2}$.

To show monotonicity in y we differentiate $f(x,y)$ with respect to y to get

$$\frac{\partial f(x,y)}{\partial y} = \left[x\lambda'\left(\frac{1+\rho(1-2y)}{2}\right) + (1-x)\lambda'\left(\frac{1-\rho(1-2y)}{2}\right)\right]\rho'(1-2y).$$

If $0 < x, y \le \frac{1}{2}$ then this expression is positive, which proves that f is an increasing function in its second argument as well. □

Lemma 4.103 (Monotonicity with Respect to Channel). Let (λ,ρ) be a degree distribution pair and let $\epsilon \in [0,\frac{1}{2}]$. If $0 \le \epsilon' \le \epsilon$ then $\mathrm{P}^{\mathrm{Gal}}_{\mathcal{T}_\ell}(\epsilon') \le \mathrm{P}^{\mathrm{Gal}}_{\mathcal{T}_\ell}(\epsilon)$. In particular, if $\mathrm{P}^{\mathrm{Gal}}_{\mathcal{T}_\ell}(\epsilon) \overset{\ell\to\infty}{\longrightarrow} 0$ then $\mathrm{P}^{\mathrm{Gal}}_{\mathcal{T}_\ell}(\epsilon') \overset{\ell\to\infty}{\longrightarrow} 0$.

Proof. We prove the claim for $\mathrm{P}^{\mathrm{Gal}}_{\vec{\mathcal{T}}_\ell}(\epsilon)$. The corresponding claim for $\mathrm{P}^{\mathrm{Gal}}_{\mathcal{T}_\ell}(\epsilon)$ can be shown in a nearly identical manner. Let $f(\cdot,\cdot)$ be defined as in Lemma 4.102. Recall from Theorem 4.95 that $\mathrm{P}^{\mathrm{Gal}}_{\vec{\mathcal{T}}_\ell}(\epsilon) = x_\ell(\epsilon)$, where $x_0(\epsilon) = \epsilon$ and $x_\ell(\epsilon) = f(\epsilon, x_{\ell-1}(\epsilon))$. Since $\epsilon' \le \epsilon$ it follows that $x_0(\epsilon') = \epsilon' \le \epsilon = x_0(\epsilon)$. Further, if for some $\ell \ge 0$, $x_\ell(\epsilon') \le x_\ell(\epsilon)$ then

$$x_{\ell+1}(\epsilon') = f(\epsilon', x_\ell(\epsilon')) \overset{\text{Lem. 4.102}}{\le} f(\epsilon, x_\ell(\epsilon)) = x_{\ell+1}(\epsilon).$$

By induction, $x_\ell(\epsilon') \le x_\ell(\epsilon)$ for $\ell \ge 0$. So if $x_\ell(\epsilon) \overset{\ell\to\infty}{\longrightarrow} 0$, then $x_\ell(\epsilon') \overset{\ell\to\infty}{\longrightarrow} 0$. □

Lemma 4.104 (Monotonicity with Respect to Iteration). For a given degree distribution pair (λ,ρ) define $f(\cdot,\cdot)$ as in Lemma 4.102. Let $\epsilon, x_0 \in [0,\frac{1}{2}]$. For

$\ell = 1, 2, \ldots$, define $x_\ell(x_0) = f(\epsilon, x_{\ell-1}(x_0))$. Then $x_\ell(x_0)$ is a monotone sequence converging to the *nearest* (in the direction of monotonicity) solution of the equation $x = f(\epsilon, x)$, $x \in [0, \frac{1}{2}]$.

Proof. If $x_0 = 0$ then $x_\ell = 0$ for $\ell \geq 1$ and the fixed point is $x = 0$. If for some $\ell \geq 1$, $x_\ell \geq x_{\ell-1}$ then $x_{\ell+1} = f(\epsilon, x_\ell) \overset{\text{Lem. 4.102}}{\geq} f(\epsilon, x_{\ell-1}) = x_\ell$, and the corresponding conclusion holds if $x_\ell \leq x_{\ell-1}$. This shows that the sequence $\{x_\ell\}$ is monotone.

For $\epsilon \in [0, \frac{1}{2}]$ we have $0 \leq f(\epsilon, x) \leq \frac{1}{2}$ for all $x \in [0, \frac{1}{2}]$. It follows that x_ℓ converges to an element of $[0, \frac{1}{2}]$, call it x_∞. By the continuity of f we have $x_\infty = f(\epsilon, x_\infty)$. It remains to show that x_∞ is the nearest (in the sense of monotonicity) fixed point. Consider a fixed point z such that $x_\ell(x_0) \leq z$ for some $\ell \geq 0$. Then $x_{\ell+1}(x_0) = f(\epsilon, x_\ell(x_0)) \overset{\text{Lem. 4.102}}{\leq} f(\epsilon, z) = z$, which shows that $x_\infty \leq z$. Similarly, if $x_\ell(x_0) \geq z$ for some $\ell \geq 0$ then $x_\infty \geq z$. This shows that x_ℓ cannot "jump" over any fixed point and in fact converges to the nearest one. □

§4.6.2. Belief Propagation

Lemma 4.105 (Monotonicity of $f(\cdot,\cdot)$). For a given degree distribution pair (λ, ρ) define the operator $f(\mathsf{a}, \mathsf{b}) = \mathsf{a} \circledast \lambda(\rho(\mathsf{b}))$, where a and b are two symmetric L-densities and where $\mathsf{a} \circledast \lambda(\rho(\mathsf{b}))$ is defined as in Theorem 4.97. If $\mathsf{a}' \prec \mathsf{a}$ and $\mathsf{b}' \prec \mathsf{b}$ then $f(\mathsf{a}', \mathsf{b}') \prec f(\mathsf{a}, \mathsf{b})$.

Proof. Looking back at Definition 4.93 we see that the density $f(\mathsf{a}, \mathsf{b})$ is the density associated with the tree channel $(\vec{\mathcal{T}}_1, \mathsf{a}, \mathsf{b})$. That is, it is the distribution of the log-likelihood ratio of X (the root bit) conditioned on $X = 1$.

By Lemma 4.80 we know that APP processing preserves the order implied by degradation, so that if we degrade the tree channel $(\vec{\mathcal{T}}_1, \mathsf{a}', \mathsf{b}')$ to the tree channel $(\vec{\mathcal{T}}_1, \mathsf{a}, \mathsf{b})$ then the density emitted at the root bit is degraded as well. □

Lemma 4.106 (Monotonicity with Respect to Channel). Let (λ, ρ) be a degree distribution pair and consider two BMS channels characterized by their L-densities $\mathsf{a}'_{\text{BMSC}}$ and a_{BMSC}, respectively. If $\mathsf{a}'_{\text{BMSC}} \prec \mathsf{a}_{\text{BMSC}}$ then $\mathrm{P}^{\text{BP}}_{\mathcal{T}_\ell}(\mathsf{a}'_{\text{BMSC}}) \leq \mathrm{P}^{\text{BP}}_{\mathcal{T}_\ell}(\mathsf{a}_{\text{BMSC}})$. In particular, if $\mathrm{P}^{\text{BP}}_{\mathcal{T}_\ell}(\mathsf{a}_{\text{BMSC}}) \overset{\ell\to\infty}{\longrightarrow} 0$, then $\mathrm{P}^{\text{BP}}_{\mathcal{T}_\ell}(\mathsf{a}'_{\text{BMSC}}) \overset{\ell\to\infty}{\longrightarrow} 0$.

Proof. The proof for $\mathrm{P}^{\text{BP}}_{\mathcal{T}_\ell}$ follows the same logic as the previous proof: degrade the tree channel $(\mathcal{T}_\ell, \mathsf{a}'_{\text{BMSC}})$ to the tree channel $(\mathcal{T}_\ell, \mathsf{a}_{\text{BMSC}})$ and use the fact that APP processing preserves the order imposed by degradation. □

Lemma 4.107 (Monotonicity with Respect to Iteration). For a given degree distribution pair (λ, ρ) define the operator $f(\cdot,\cdot)$ as in Lemma 4.105. Let a_{BMSC} and

a_0 denote two symmetric L-densities. For $\ell \geq 1$ define

$$a_\ell = f(a_{BMSC}, a_{\ell-1}).$$

If for some $\ell' \geq 0$ and $k \geq 1$ we have $a_{\ell'+k} \prec a_{\ell'}$ ($a_{\ell'} \succ a_{\ell'+k}$) then for $\ell'' \geq \ell'$, ℓ'' and k fixed, the sequence $a_{\ell''+k\ell}$ is monotone (with respect to degradation) in $\ell \geq 0$, and converges to a symmetric limit density. In particular, if $a_0 = a_{BMSC}$ then a_ℓ is monotonic (with respect to degradation) and converges to a symmetric limit $a_\infty \prec a_{BMSC}$.

Proof. We prove the statement for sequences which are monotonically upgraded. The case of sequences which are monotonically degraded can be handled in an identical fashion by reversing the direction of the arrows.

Assume that for some $\ell \geq 0$, $a_{\ell'+k(\ell+1)} \prec a_{\ell'+k\ell}$ so that the input to the tree channel $(\vec{\mathcal{T}}_k, a_{BMSC}, a_{\ell'+k(\ell+1)})$ is upgraded with respect to the input to the tree channel $(\vec{\mathcal{T}}_k, a_{BMSC}, a_{\ell'+k\ell})$. The conditions of the Lemma state that this is true for $\ell = 0$. This case serves as our anchor in the induction over ℓ.

From Lemma 4.80 we know that APP processing preserves the order imposed by degradation. Since the density associated to $(\vec{\mathcal{T}}_k, a_{BMSC}, a_{\ell'+k(\ell+1)})$ is $a_{\ell'+k(\ell+2)}$, whereas the density associated to $(\vec{\mathcal{T}}_k, a_{BMSC}, a_{\ell'+k\ell})$ is $a_{\ell'+k(\ell+1)}$, we conclude that $a_{\ell'+k(\ell+2)} \prec a_{\ell'+k(\ell+1)}$. From Lemma 4.75 we know that a sequence of symmetric densities ordered by degradation converges to a symmetric limit density.

By monotonicity of $f(\cdot,\cdot)$, if $a_{\ell'+k} \prec a_{\ell'}$ then $a_{(\ell'+1)+k} \prec a_{(\ell'+1)}$ and, more generally, $a_{(\ell'+j)+k} \prec a_{(\ell'+j)}$. By using the same argument as before with ℓ' replaced by $(\ell' + j)$ we conclude the general case.

Finally, if $a_0 = a_{BMSC}$ then we have $a_1 \prec a_0 = a_{BMSC}$. The statement follows from the previous case if we set $\ell' = 0$ and $k = 1$. □

Although $a_{\ell'+k\ell}$ converges in ℓ for *all* ℓ' we cannot conclude that a_ℓ converges. In general, the sequence could converge to a limit cycle of length k. Such a limit cycle has never been observed and we conjecture that limit cycles cannot occur and that a_ℓ in fact converges. For technical reasons we will later need the following result, which substitutes for the conjecture.

Corollary 4.108 (Monotonicity with Respect to Iteration). If in addition to the conditions of Lemma 4.107 we have $a_{\ell'+k+1} \prec a_{\ell'}$ ($a_{\ell'+k+1} \succ a_{\ell'}$) then a_ℓ converges to a limit density a_∞.

Proof. By Lemma 4.107 the sequence $a_{\ell'+(k+1)\ell}$ converges in ℓ. Since this sequence coincides with $a_{\ell'+j+k\ell}$ infinitely often the limits must be the same. □

§4.7. Threshold

Definition 4.109 (Threshold of a Degree Distribution Pair). Consider a family of BMS channels $\{\mathrm{a}_{\mathrm{BMSC}(\sigma)}\}_{\underline{\sigma}}^{\overline{\sigma}}$. Assume that for a fixed message-passing decoder MP

$$\lim_{\ell\to\infty} \mathrm{P}_{\mathcal{T}_\ell}^{\mathrm{MP}}(\mathrm{a}_{\mathrm{BMSC}(\underline{\sigma})}) = 0, \quad \text{but} \quad \lim_{\ell\to\infty} \mathrm{P}_{\mathcal{T}_\ell}^{\mathrm{MP}}(\mathrm{a}_{\mathrm{BMSC}(\overline{\sigma})}) > 0,$$

and that the decoder is *monotone* with respect to this channel family. Then there exists a supremum of the set of σ for which $\lim_{\ell\to\infty} \mathrm{P}_{\mathcal{T}_\ell}^{\mathrm{MP}}(\mathrm{a}_{\mathrm{BMSC}(\sigma)}) = 0$ so that for all smaller channel parameters the error probability is 0 and for all larger channel parameters the error probability is non-zero. This supremum is called the *threshold* and it is denoted by σ^{MP},

$$\sigma^{\mathrm{MP}}(\lambda,\rho) = \sup\{\sigma \in [\underline{\sigma},\overline{\sigma}] : \mathrm{P}_{\mathcal{T}_\ell}^{\mathrm{MP}}(\mathrm{a}_{\mathrm{BMSC}(\sigma)}) \xrightarrow{\ell\to\infty} 0\}. \qquad \triangledown$$

Example 4.110 ($\epsilon^{\mathrm{GAL}}(x^2,x^5)$). We want to determine $\epsilon^{\mathrm{Gal}}(x^2,x^5)$, i.e., the threshold of the degree distribution pair (x^2,x^5) for the family $\{\mathrm{BSC}(\epsilon)\}$ and Gallager's algorithm A. From Lemma 4.103 we know that this decoder is monotone with respect to the family $\{\mathrm{BSC}(\epsilon)\}$, so that the threshold is well defined. Figure 4.113 shows $\mathrm{P}_{\vec{\mathcal{T}}_\ell}^{\mathrm{Gal}}(\epsilon)$ as a function of ℓ for various values of ϵ. From this figure we see that $\epsilon^{\mathrm{Gal}}(3,6) \approx 0.03946365$. $\diamond$

Example 4.111 ($\sigma^{\mathrm{BP}}(3,6)$). In the same manner, we determine $\sigma_{\mathrm{BAWGNC}}^{\mathrm{BP}}(3,6)$ for the family $\{\mathrm{BAWGNC}(\sigma)\}$. From Lemma 4.106 we know that the BP decoder is monotone with respect to any channel family which is ordered by degradation, and from the remarks following Example 4.71 we know that the family $\{\mathrm{BAWGNC}(\sigma)\}$ is monotone. It follows that the threshold is well defined. Figure 4.114 shows $\mathrm{P}_{\vec{\mathcal{T}}_\ell}^{\mathrm{BP}}(\sigma)$ as a function of ℓ for various values of σ. From this figure we see that $\sigma^{\mathrm{BP}}(3,6) \approx 0.881$. $\diamond$

Table 4.115 lists thresholds for the family $\{\mathrm{BSC}(\epsilon)\}$ under Gallager's algorithm A and the BP algorithm as well as thresholds for the family $\{\mathrm{BAWGNC}(\sigma)\}$ under BP decoding. Also listed are the threshold values corresponding to Shannon capacity.

§4.8. Fixed Point Characterization of Threshold

As was the case for the BEC, the threshold can in many cases be characterized in terms of fixed points of the density evolution equation (but see Problem 4.39).

§4.8.1. Gallager Algorithm A

Theorem 4.112 (Fixed Point Characterization for Gallager A). For a given degree distribution pair (λ,ρ) define $f(\cdot,\cdot)$ as in Lemma 4.102 so that density evolution can be written as $x_\ell = f(\epsilon, x_{\ell-1})$, $\ell \geq 1$, $x_0 = \epsilon$.

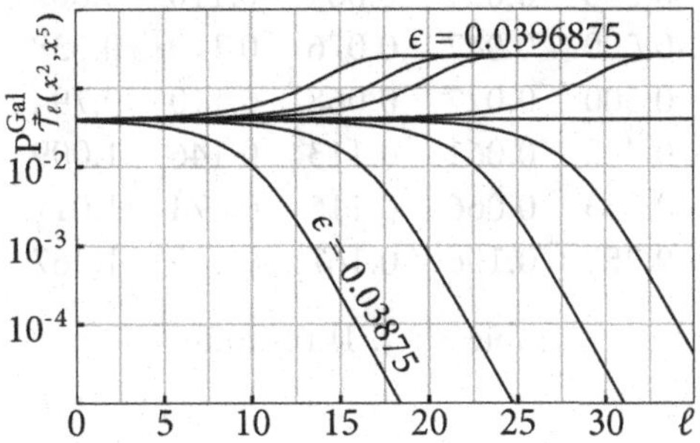

Figure 4.113: Evolution of $P^{Gal}_{\vec{\mathcal{T}}_\ell(x^2,x^5)}(\epsilon)$ as a function of ℓ for various values of ϵ. For $\epsilon = 0.03875, 0.039375, 0.0394531$, and 0.039462 the error probability converges to zero, whereas for $\epsilon = 0.039465, 0.0394922, 0.0395313$, and 0.0396875 the error probability converges to a non-zero value. For $\epsilon \approx 0.03946365$ the error probability stays constant. We conclude that $\epsilon^{Gal}(3,6) \approx 0.03946365$. Note that for $\epsilon > \epsilon^{Gal}(3,6)$, $P^{Gal}_{\vec{\mathcal{T}}_\ell(x^2,x^5)}(\epsilon)$ is an *increasing* function of ℓ, whereas below this threshold it is a *decreasing* function. In either case, $P^{Gal}_{\vec{\mathcal{T}}_\ell(x^2,x^5)}(\epsilon)$ is monotone as guaranteed by Lemma 4.104.

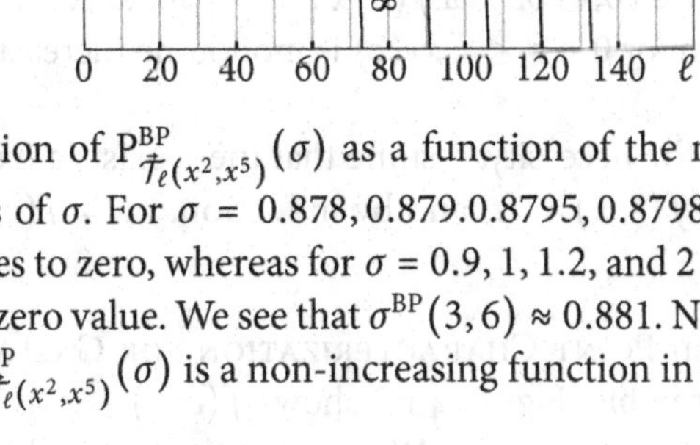

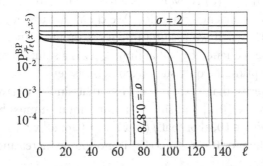

Figure 4.114: Evolution of $P^{BP}_{\vec{\mathcal{T}}_\ell(x^2,x^5)}(\sigma)$ as a function of the number of iterations ℓ for various values of σ. For $\sigma = 0.878, 0.879.0.8795, 0.8798$, and 0.88 the error probability converges to zero, whereas for $\sigma = 0.9, 1, 1.2$, and 2 the error probability converges to a non-zero value. We see that $\sigma^{BP}(3,6) \approx 0.881$. Note that, as predicted by Lemma 4.107, $P^{BP}_{\vec{\mathcal{T}}_\ell(x^2,x^5)}(\sigma)$ is a non-increasing function in ℓ.

l	r	rate	ϵ^{Gal}	ϵ^{BP}	ϵ^{Sha}	σ^{BP}	σ^{Sha}
3	6	0.500	0.039	0.084	0.110	0.881	0.979
4	8	0.500	0.047	0.076	0.110	0.838	0.979
5	10	0.500	0.027	0.068	0.110	0.794	0.979
3	5	0.400	0.061	0.113	0.146	1.009	1.149
4	6	0.333	0.066	0.116	0.174	1.011	1.297
3	4	0.250	0.106	0.167	0.215	1.267	1.550

Table 4.115: Thresholds.

[Convergence] For any $\epsilon \in [0, \frac{1}{2}]$, $x_\ell(\epsilon)$ converges to a solution of $x = f(\epsilon, x)$ with $x \in [0, \frac{1}{2}]$.

[Sufficiency] If $\epsilon^{\text{Gal}} > 0$ and if $x \neq f(\epsilon, x)$ for all $x \in (0, \epsilon]$, then $x_\ell(\epsilon)$ converges to zero as ℓ tends to infinity, hence, $\epsilon \leq \epsilon^{\text{Gal}}$.

[Necessity] If there exists an x, $x \in (0, \epsilon]$, such that $x = f(\epsilon, x)$, then $x_\ell(\epsilon) \geq x$ for all $\ell \geq 0$, hence $\epsilon^{\text{Gal}} \leq \epsilon$.

[Fixed Point Characterizations of the Threshold]

(i) $\epsilon^{\text{Gal}}(\lambda, \rho) = \sup\{\epsilon \in [0, \frac{1}{2}] : x = f(\epsilon, x) \text{ has no solution in } (0, \epsilon]\}$.

(ii) $\epsilon^{\text{Gal}}(\lambda, \rho) = \inf\{\epsilon \in [0, \frac{1}{2}] : x = f(\epsilon, x) \text{ has a solution in } (0, \epsilon]\}$.

Proof. We showed that $x_\ell(\epsilon)$ converges to a fixed point already in Lemma 4.104.

Let us show the sufficiency condition. We claim that under the stated assumptions $x_1 = f(\epsilon, \epsilon) < \epsilon$. If we assume this for a moment then it follows by the monotonicity of the sequence $\{x_\ell\}$ that $\lim_{\ell \to \infty} x_\ell$ is a solution of $x = f(\epsilon, x)$ with $x \in [0, \epsilon]$. By assumption the only such fixed point is zero.

It remains to verify that $f(\epsilon, \epsilon) < \epsilon$. Note that since we have no fixed points of $x = f(\epsilon, x)$ with $x \in (0, \epsilon]$ and since f is continuous it must be true that either $f(x, x) - x < 0$ for all $x \in (0, \epsilon]$ or that $f(x, x) - x > 0$ for all $x \in (0, \epsilon]$. If $f(x, x) - x > 0$ for all $x \in (0, \epsilon]$ then for $0 < \epsilon' \leq \epsilon$, $x_\ell(\epsilon)$ is monotone increasing. In other words, $\epsilon^{\text{Gal}} = 0$.

In order to show the necessity, assume that there exists a fixed point $x \in (0, \epsilon]$. Then $x_1 = f(\epsilon, \epsilon) \geq f(\epsilon, x) = x$, and by induction, $x_\ell = f(\epsilon, x_{\ell-1}) \geq f(\epsilon, x) = x$. □

Example 4.116 (Fixed Point Characterization for Gallager A). Consider the $(3, 3)$-regular ensemble. Figure 4.117 shows $f(\epsilon, x) - x$ as a function of x for $\epsilon = \epsilon^{\text{Gal}} \approx 0.22305$ (left-hand graph). We see that the threshold is determined by a single critical point. ◇

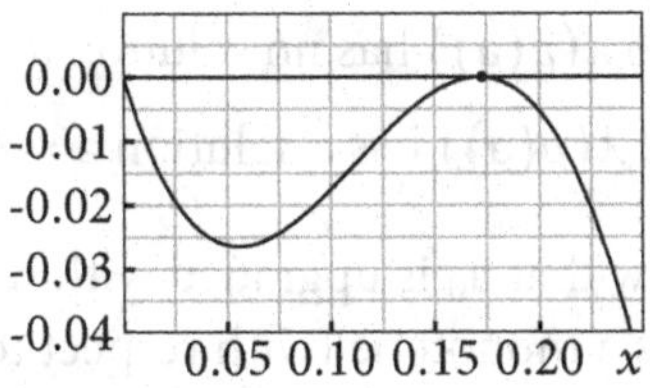

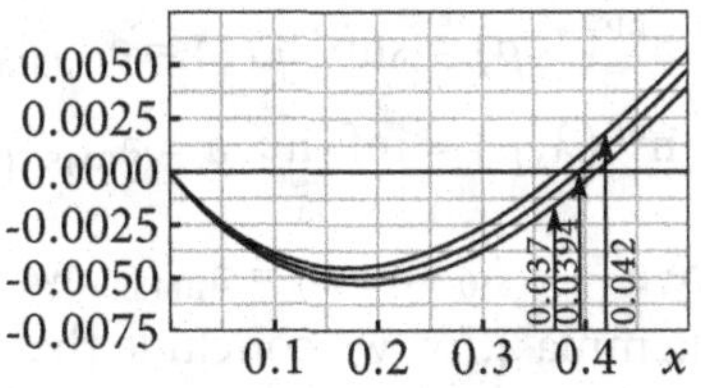

Figure 4.117: Left: $f(\epsilon, x) - x$ as a function of x for the $(3,3)$-regular ensemble and $\epsilon = \epsilon^{\text{Gal}} \approx 0.22305$. Right: $f(\epsilon, x) - x$ as a function of x for the $(3,6)$-regular ensemble and $\epsilon = 0.037$, $\epsilon = \epsilon^{\text{Gal}} \approx 0.394$, and $\epsilon = 0.042$.

Example 4.118 (Fixed Point Characterization for Gallager A). Consider the $(3,6)$-regular ensemble. In this case the threshold is determined by a fixed point at the beginning of the decoding process. Figure 4.117 shows $f(\epsilon, x) - x$ as a function of x for $\epsilon = 0.037$, $\epsilon = \epsilon^{\text{Gal}} \approx 0.03946$, and $\epsilon = 0.042$ (right-hand graph). Also shown is the corresponding starting point of the recursion $x = \epsilon$. If ϵ is smaller than the fixed point (crossing point with the horizontal axis) then the recursion converges to zero, otherwise it converges to a non-zero value. The threshold is characterized by the value of ϵ so that $f(\epsilon, \epsilon) = \epsilon$. ◇

§4.8.2. Belief Propagation Algorithm

Theorem 4.119 (Fixed Point Characterization for BP). Consider a given degree distribution pair (λ, ρ) and assume that transmission takes place over a BMS channel characterized by its L-density a_{BMSC}. Define $\mathsf{a}_{-1} = \Delta_0$ and for $\ell \geq 0$

$$\mathsf{a}_\ell = \mathsf{a}_{\text{BMSC}} \circledast \lambda(\rho(\mathsf{a}_{\ell-1})). \tag{4.120}$$

[Convergence] The sequence of densities a_ℓ converges to a symmetric density a_∞ which is a fixed point solution to (4.120).

[Sufficiency] If there does not exist a symmetric density $\mathsf{a} \neq \Delta_{+\infty}$ such that $\mathsf{a} = \mathsf{a}_{\text{BMSC}} \circledast \lambda(\rho(\mathsf{a}))$ then $\mathfrak{E}(\mathsf{a}_\ell)$ converges to zero as ℓ tends to infinity, or, equivalently, $\mathsf{a}_\infty = \Delta_{+\infty}$.

[Necessity] If there exists a symmetric density $\mathsf{a} \neq \Delta_{+\infty}$ such that $\mathsf{a} = \mathsf{a}_{\text{BMSC}} \circledast \lambda(\rho(\mathsf{a}))$ then $\mathfrak{E}(\mathsf{a}_\ell)$ does not converge to zero as ℓ tends to infinity, or, equivalently, $\mathsf{a}_\infty \neq \Delta_\infty$.

[Fixed Point Characterizations of the Threshold] Let $\{\mathsf{a}_{\text{BMSC}(\mathtt{h})}\}$ be a family of BMS channels ordered by degradation and parameterized by the real parameter $\mathtt{h}$. Then

(i) $h^{BP}(\lambda, \rho) = \sup\{h : a = a_{BMSC(h)} \circledast \lambda(\rho(a)) \text{ has no solution } a \neq \Delta_{+\infty}\}$

(ii) $h^{BP}(\lambda, \rho) = \inf\{h : a = a_{BMSC(h)} \circledast \lambda(\rho(a)) \text{ has a solution } a \neq \Delta_{+\infty}\}$.

Proof. Since $a_{-1} = \Delta_0$ and $a_0 = a_{BMSC}$, we know that a_0 is upgraded with respect to a_{-1}. From Lemma 4.107 we conclude that a_ℓ is a monotone (with respect to degradation) sequence that converges to a symmetric limit density a_∞.

The density a_∞ is a fixed point of (4.120) since the update equations are continuous under our notion of convergence. More precisely: $\{a_\ell\}$ converges in ℓ to a if the cumulative distributions converge at points of continuity of the cumulative distribution of a. Then, our assertion is simply that $a_{BMSC} \circledast \lambda(\rho(a_\ell))$ converges in ℓ to $a_{BMSC} \circledast \lambda(\rho(a))$ under the same notion of convergence.

If $a \neq \Delta_{+\infty}$ is a fixed point then the tree channel $(\mathcal{T}_\ell, a_{BMSC}, a)$ has associated density a. In particular $P^{BP}_{\mathcal{T}_\ell}(a_{BMSC}, a) = \mathfrak{E}(a)$. If we are below threshold then, for ℓ sufficiently large, we have $P^{BP}_{\mathcal{T}_\ell}(a_{BMSC}, \Delta_0) < \mathfrak{E}(a)$. But, since $a \prec \Delta_0$, this is a contradiction. Thus, below threshold there can be no fixed point other than $\Delta_{+\infty}$.

Above threshold we have $a_\infty \neq \Delta_{+\infty}$ as a fixed point. □

Example 4.121 (Fixed Point Density). Figure 4.122 shows the fixed point density for the $(3, 6)$-regular ensemble and the BAWGNC(σ) with $\sigma \approx 0.881$. ◇

§4.9. Stability

From the preceding section we see that the behavior of a message-passing decoder is often determined by its fixed points. One fixed point which is virtually always present is the one corresponding to perfect decoding (zero error rate). It is desirable that this fixed point is *stable*. That is, assuming the density has evolved to something "close" to perfect decoding, it should converge to the perfect decoding fixed point. One is also interested in the *rate of convergence* in this case. This stability property is amenable to analysis and this section is devoted to its study.

§4.9.1. Gallager Algorithm A

Let us look at the behavior of (4.96) in Theorem 4.95 for small values of x_ℓ. Some calculus shows that

$$x_\ell = (\epsilon\lambda'(1) - \epsilon\lambda'(0) + \lambda'(0))\rho'(1)x_{\ell-1} + O(x_{\ell-1}^2).$$

Clearly, for sufficiently small $x_{\ell-1}$ the convergence behavior is determined by the term linear in $x_{\ell-1}$. More precisely, the convergence depends on whether $(\epsilon\lambda'(1) - \epsilon\lambda'(0) + \lambda'(0))\rho'(1)$ is smaller or larger than 1. The precise statement is given in the following theorem.

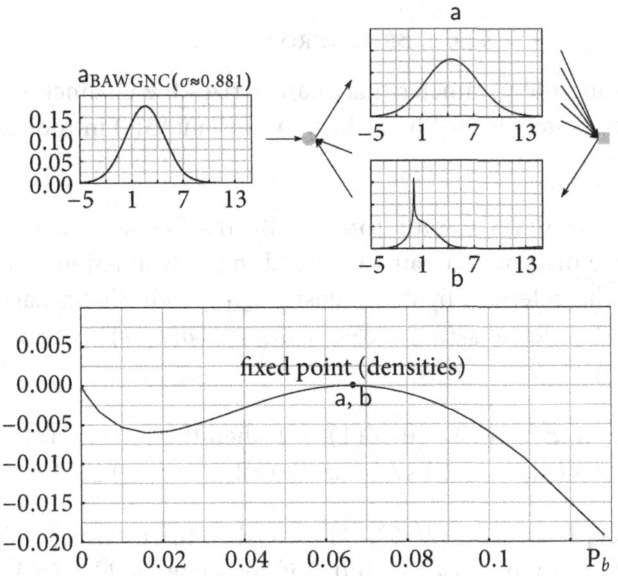

Figure 4.122: Progress per iteration (change of error probability) of density evolution for the $(3,6)$-ensemble and the BAWGNC(σ) channel with $\sigma \approx 0.881$ as a function of the bit error probability. In formulae: we plot $\mathfrak{E}(a_\ell) - \mathfrak{E}(a_{\ell-1})$ as a function of $P_b = \mathfrak{E}(a_{BAWGNC(\sigma)} \circledast L(\rho(a_{\ell-1})))$, where $L(x) = x^3$ and $\rho(x) = x^5$. For cosmetic reasons this discrete set of points was interpolated to form a smooth curve. The initial error probability is equal to $Q(1/0.881) \approx 0.12817$. At the fixed point the progress is zero. The associated fixed point densities are a (emitted at the variable nodes) and b (emitted at the check nodes).

THEOREM 4.123 (STABILITY CONDITION FOR GALLAGER ALGORITHM A). *Assume that we are given a degree distribution pair (λ, ρ) and a real number ϵ, $\epsilon \in [0, \frac{1}{2}]$. Define $f(\cdot,\cdot)$ as in Lemma 4.102 and let $x_\ell(\epsilon) = f(\epsilon, x_{\ell-1})$, $\ell \geq 1$, $x_0(\epsilon) = \epsilon$.*

[Necessity] If $\epsilon > \frac{1-\lambda'(0)\rho'(1)}{\lambda'(1)\rho'(1)-\lambda'(0)\rho'(1)}$, then there exists a strictly positive constant $\xi = \xi(\lambda, \rho, \epsilon)$ such that for all $\ell \in \mathbb{N}$, $x_\ell(\epsilon) > \xi$.

[Sufficiency] If $\epsilon < \frac{1-\lambda'(0)\rho'(1)}{\lambda'(1)\rho'(1)-\lambda'(0)\rho'(1)}$, then there exists a strictly positive constant $\xi = \xi(\lambda, \rho, \epsilon)$ such that if, for some $\ell \in \mathbb{N}$, $x_\ell(\epsilon) \leq \xi$, then $x_\ell(\epsilon)$ converges to zero as ℓ tends to infinity.

As an immediate consequence we get the upper bound on the threshold

$$\epsilon^{\text{Gal}} \leq \frac{1-\lambda'(0)\rho'(1)}{\lambda'(1)\rho'(1)-\lambda'(0)\rho'(1)}. \tag{4.124}$$

As discussed in Problem 4.56, this simple bound is often tight.

§4.9.2. Belief Propagation

Under BP decoding the parameter that characterizes the channel with respect to the stability of the system is the Bhattacharyya constant $\mathfrak{B}(\cdot)$ introduced in Definition 4.61.

Theorem 4.125 (Stability Condition for Belief Propagation). Assume we are given a degree distribution pair (λ, ρ) and that transmission takes place over a BMS channel characterized by its L-density a_{BMSC} with Bhattacharyya constant $\mathfrak{B}(\mathsf{a}_{\text{BMSC}})$. For $\ell \geq 1$ define $\mathsf{a}_\ell(\mathsf{a}_0) = \mathsf{a}_\ell = \mathsf{a}_{\text{BMSC}} \circledast \lambda(\rho(\mathsf{a}_{\ell-1}))$ with a_0 an *arbitrary* symmetric density.

> [Necessity] If $\mathfrak{B}(\mathsf{a}_{\text{BMSC}})\lambda'(0)\rho'(1) > 1$, then there exists a strictly positive constant $\xi = \xi(\lambda, \rho, \mathsf{a}_{\text{BMSC}})$ such that $\liminf_{\ell\to\infty} \mathfrak{E}(\mathsf{a}_\ell) \geq \xi$ for all $\mathsf{a}_0 \neq \Delta_{+\infty}$.
>
> [Sufficiency] If $\mathfrak{B}(\mathsf{a}_{\text{BMSC}})\lambda'(0)\rho'(1) < 1$, then there exists a strictly positive constant $\xi = \xi(\lambda, \rho, \mathsf{a}_{\text{BMSC}})$ such that if, for some $\ell \in \mathbb{N}$, $\mathfrak{E}(\mathsf{a}_\ell) \leq \xi$, then a_ℓ converges to $\Delta_{+\infty}$.

Example 4.126 (Stability Condition for BEC(ϵ)). We have

$$\mathfrak{B}(\mathsf{a}_{\text{BEC}(\epsilon)}) = \int [\epsilon\Delta_0 + \bar{\epsilon}\Delta_{+\infty}]\, \mathrm{e}^{-x/2} dx = \epsilon .$$

Therefore, the stability condition reads $\epsilon\lambda'(0)\rho'(1) < 1$, which agrees with the result of Section 3.13. Equivalently, we get $\epsilon^{\text{BP}}(\lambda, \rho) \leq \frac{1}{\lambda'(0)\rho'(1)}$. ◇

Example 4.127 (Stability Condition for BSC(ϵ)). We have

$$\mathfrak{B}(\mathsf{a}_{\text{BSC}(\epsilon)}) = \int [\epsilon\Delta_{-\ln\frac{\bar{\epsilon}}{\epsilon}} + \bar{\epsilon}\Delta_{\ln\frac{\bar{\epsilon}}{\epsilon}}]\, \mathrm{e}^{-x/2} dx = 2\sqrt{\epsilon\bar{\epsilon}} .$$

The stability condition for the BSC(ϵ) is therefore $2\sqrt{\epsilon\bar{\epsilon}}\lambda'(0)\rho'(1) < 1$. Formulated as an upper bound, $\epsilon^{\text{BP}}(\lambda, \rho) \leq \frac{1}{2}\left(1 - \sqrt{1 - \frac{1}{(\lambda'(0)\rho'(1))^2}}\right)$. ◇

Example 4.128 (Stability Condition for BAWGNC(σ)). We have

$$\mathfrak{B}(\mathsf{a}_{\text{BAWGNC}(\sigma)}) = \int \sqrt{\frac{\sigma^2}{8\pi}}\, \mathrm{e}^{-\frac{(x-\frac{2}{\sigma^2})^2\sigma^2}{8}}\, \mathrm{e}^{-x/2} dx = \mathrm{e}^{-\frac{1}{2\sigma^2}} .$$

Thus, the stability condition reduces to $\mathrm{e}^{-\frac{1}{2\sigma^2}}\lambda'(0)\rho'(1) < 1$. This gives rise to $\sigma^{\text{BP}}(\lambda, \rho) \leq \frac{1}{\sqrt{2\ln(\lambda'(0)\rho'(1))}}$. ◇

Proof of Necessity in Theorem 4.125. The proof is based on (the Erasure Decomposition) Lemma 4.78. Recall that for the BEC we observed that zero was a fixed point of the density evolution equation. By linearizing the recursion around this fixed point we were able to analyze its stability. For the general case we proceed along the same lines. Note that $\Delta_{+\infty}$ is the unique symmetric density a so that $\mathfrak{E}(\mathsf{a}) = 0$. Since $\mathsf{a}_{\text{BMSC}} \circledast \lambda(\rho(\Delta_{+\infty})) = \Delta_{+\infty}$, we see that $\Delta_{+\infty}$ is a fixed point of density evolution. To analyze local convergence we consider a linearization of the density evolution equation about this fixed point.

To that end, consider the (BEC) density $\mathsf{b}_0 = \mathsf{b}_0(\epsilon) = 2\epsilon\Delta_0 + (1-2\epsilon)\Delta_{+\infty}$. This density is symmetric and $\mathfrak{E}(2\epsilon\Delta_0 + (1-2\epsilon)\Delta_{+\infty}) = \epsilon$. After a complete iteration of density evolution, this density evolves to

$$\mathsf{b}_1 = 2\epsilon\lambda'(0)\rho'(1)\mathsf{a}_{\text{BMSC}} + (1 - 2\epsilon\lambda'(0)\rho'(1))\Delta_{+\infty} + O(\epsilon^2).$$

More generally, if we consider n iterations of density evolution we see that the density b_0 evolves to

$$\mathsf{b}_n = 2\epsilon(\lambda'(0)\rho'(1))^n \mathsf{a}_{\text{BMSC}}^{\circledast n} + (1 - 2\epsilon(\lambda'(0)\rho'(1))^n)\Delta_{+\infty} + O(\epsilon^2).$$

We are interested in the error probability associated with b_n, i.e., we are interested in $\mathfrak{E}(\mathsf{b}_n)$. Recall from Section 4.1.12, Lemma 4.66, that $\lim_{n\to\infty} \frac{1}{n} \log \mathfrak{E}(\mathsf{a}_{\text{BMSC}}^{\circledast n}) = \log(\mathfrak{B}(\mathsf{a}_{\text{BMSC}}))$. Therefore, if we assume that $\mathfrak{B}(\mathsf{a}_{\text{BMSC}})\lambda'(0)\rho'(1) > 1$, then there exists an integer N such that $(\lambda'(0)\rho'(1))^n\, \mathfrak{E}(\mathsf{a}_{\text{BMSC}}^{\circledast n}) > 1$ for $n \geq N$. We can then write

$$\mathfrak{E}(\mathsf{b}_n) = 2\epsilon(\lambda'(0)\rho'(1))^n\, \mathfrak{E}(\mathsf{a}_{\text{BMSC}}^{\circledast n}) + O(\epsilon^2) > 2\epsilon + O(\epsilon^2) > \epsilon \quad \text{if } \epsilon \leq \xi,$$

for $n \geq N$ where ξ is a strictly positive constant depending only on (λ, ρ) and a_{BMSC}. By (the Erasure Decomposition) Lemma 4.78 we see that if $\epsilon \leq \xi$ then b_N and b_{N+1} are degraded with respect to b_0. It follows from Corollary 4.108 that b_n converges to a fixed point $\mathsf{b}_\infty = \mathsf{b}_\infty(\epsilon)$ that is degraded with respect to $\mathsf{b}_0(\epsilon)$ and that satisfies $\mathfrak{E}(\mathsf{b}_\infty(\epsilon)) > \epsilon$.

Now let a_0 satisfy $\mathfrak{E}(\mathsf{a}_0) = \epsilon \in (0, \xi]$. Again by Lemma 4.78, a_0 is degraded with respect to $\mathsf{b}_0(\epsilon)$. It follows by induction on monotonicity, Lemma 4.105, that a_n is degraded with respect to $\mathsf{b}_n(\epsilon)$ and hence $\liminf_{n\to\infty} \mathfrak{E}(\mathsf{a}_n) \geq \mathfrak{E}(\mathsf{b}_\infty(\epsilon)) > \epsilon$.

Perhaps a little surprising, we can now conclude that $\mathsf{b}_\infty(\epsilon) = \mathsf{b}_\infty(\xi)$ for all $\epsilon \in (0, \xi]$. Indeed, we must have $\mathfrak{E}(\mathsf{b}_\infty(\epsilon)) > \xi$ or by the last paragraph it cannot be a fixed point (set $\mathsf{a}_0 = \mathsf{b}_\infty(\epsilon)$ and observe that $\mathsf{a}_n = \mathsf{b}_\infty(\epsilon)$ since $\mathsf{b}_\infty(\epsilon)$ is a fixed point). But this means that $\mathsf{b}_\infty(\epsilon)$ is degraded with respect to $\mathsf{b}_0(\xi)$. Since it is a fixed point it follows from monotonicity, Lemma 4.105, that $\mathsf{b}_\infty(\epsilon)$ must be degraded with respect to $\mathsf{b}_n(\xi)$ for all n and hence also with respect to $\mathsf{b}_\infty(\xi)$. Since $\mathsf{b}_\infty(\xi)$ is also degraded with respect to $\mathsf{b}_\infty(\epsilon)$ they must in fact be equal.

Now, if $\mathfrak{E}(\mathsf{a}_0) = \epsilon \in (0, \xi]$ then a_0 is degraded with respect to $\mathsf{b}_0(\epsilon)$ and so $\liminf_{n\to\infty} \mathfrak{E}(\mathsf{a}_n) \geq \mathfrak{E}(\mathsf{b}_\infty(\epsilon)) > \xi$. If $\mathfrak{E}(\mathsf{a}_0) > \xi$ then by Lemma 4.78 a_0 is degraded with respect to $\mathsf{b}_0(\xi)$ and the same conclusion holds. □

Proof of Sufficiency in Theorem 4.125. We give two proofs. The first proof is based on the idea of *extremes of information* combining (see Section 4.10.2). Let a_0 be a symmetric L-density at the output of the variable nodes. Consider density evolution, where a_ℓ denotes the densities at the output of the variable nodes and b_ℓ denotes the densities at the output of the check nodes. Define $x_\ell = \mathfrak{B}(\mathsf{a}_\ell)$ and $y_\ell = \mathfrak{B}(\mathsf{b}_\ell)$. By the multiplicativity of the Bhattacharyya constant at the variable nodes (Lemma 4.63) and the extremality property (iv) discussed in Problem 4.62 we have for $\ell \geq 1$

$$y_\ell \leq 1 - \rho(1 - x_{\ell-1}), \qquad x_\ell = \mathfrak{B}(\mathsf{a}_{\text{BMSC}})\lambda(y_\ell),$$

so that $x_\ell \leq \mathfrak{B}(\mathsf{a}_{\text{BMSC}})\lambda(1 - \rho(1 - x_{\ell-1}))$. If we expand this inequality around zero, this implies that $x_\ell \leq \mathfrak{B}(\mathsf{a}_{\text{BMSC}})\lambda'(0)\rho'(1)x_{\ell-1} + O(x_{\ell-1}^2)$. Since

$$\mathfrak{B}(\mathsf{a}_{\text{BMSC}})\lambda'(0)\rho'(1) < 1,$$

we can find $\eta > 0$ so that $\mathfrak{B}(\mathsf{a}_{\text{BMSC}})\lambda'(0)\rho'(1) + \eta < 1$. For a sufficiently small constant κ, $x_{\ell-1} \leq \kappa$ implies $x_\ell \leq (\mathfrak{B}(\mathsf{a}_{\text{BMSC}})\lambda'(0)\rho'(1) + \eta)x_{\ell-1} < x_{\ell-1}$. It follows that if, for some $\ell \in \mathbb{N}$, $x_\ell \leq \kappa$, then $x_\ell \to 0$.

Let $\xi = \kappa^2/4$. Then it follows by the right-hand side of (4.65) that if $\mathfrak{E}(\mathsf{a}_\ell) \leq \xi$ then $x_\ell = \mathfrak{B}(\mathsf{a}_\ell) \leq \kappa$. Hence, $x_\ell = \mathfrak{B}(\mathsf{a}_\ell) \to 0$. This in turn implies that $\mathsf{a}_\ell \to \Delta_{+\infty}$.

The second proof uses the idea of minimal codewords in a tree. You can find it on page 545 in Appendix E. □

§4.10. EXIT Charts

In the case of the BEC we saw that the EXIT chart was a useful and intuitive tool to visualize the density evolution process. Let us therefore discuss how to define EXIT functions and EXIT charts for general BMS channels.

Definition 4.129 (EXIT Function for BMS Channels). Let X be a vector of length n chosen with probability $p_X(x)$ from a binary code C. Assume that transmission takes place over the family $\{\text{BMSC}(\mathtt{h})\}$. Then

$$h_i(\mathtt{h}) = H(X_i \,|\, Y_{\sim i}(\mathtt{h})),$$

$$h(\mathtt{h}) = \frac{1}{n}\sum_{i=1}^{n} H(X_i \,|\, Y_{\sim i}(\mathtt{h})) = \frac{1}{n}\sum_{i=1}^{n} h_i(\mathtt{h}).$$

Lemma 4.130 (EXIT Function for Linear Codes via EXIT Functional). Let X be a vector of length n chosen uniformly at random from a proper binary linear code C and assume that transmission takes place over the family $\{\text{BMSC}(\mathtt{h})\}$. Define the *extrinsic* (i.e., based only on $Y_{\sim i}$ and not on the whole vector Y) MAP estimator of X_i,

$$\phi_i(y_{\sim i}) = \ln \frac{p_{X_i \mid Y_{\sim i}}(+1 \mid y_{\sim i})}{p_{X_i \mid Y_{\sim i}}(-1 \mid y_{\sim i})}, \tag{4.131}$$

and $\Phi_i = \phi_i(Y_{\sim i})$. Let a_i denote the density of Φ_i, assuming that the all-one codeword was transmitted, and let $\mathsf{a} = \frac{1}{n}\sum_{i=1}^{n} \mathsf{a}_i$, where for simplicity of notation we have suppressed the dependency of a_i and a on the channel parameter $\mathtt{h}$. Then

$$h_i(\mathtt{h}) = \mathrm{H}(\mathsf{a}_i), \qquad h(\mathtt{h}) = \mathrm{H}(\mathsf{a}).$$

Proof. From Theorem 4.29 we know that Φ_i constitutes a sufficient statistic for X_i given $Y_{\sim i}$. According to our discussion on page 29, this implies that $H(X_i \mid Y_{\sim i}) = H(X_i \mid \Phi_i)$. Further, since the code C is proper and $p_X(x)$ is uniform, we know from Theorem 4.29 that the channel $p_{\Phi_i \mid X_i}(\phi_i \mid x_i)$ is symmetric, that there is a uniform distribution on X_i, and that ϕ_i is in fact a log-likelihood ratio. We can therefore compute the entropy $H(X_i \mid \Phi_i)$ by applying the entropy operator to the distribution of Φ_i conditioned on $X_i = 1$. Since, again by Theorem 4.29, this distribution is the same as the distribution of Φ_i conditioned that the all-one codeword was transmitted and this distribution is denoted by a_i, the claim that $H(X_i \mid Y_{\sim i}) = \mathrm{H}(\mathsf{a}_i)$ follows. □

Example 4.132 ($[n, 1, n]$ Repetition Code). Assume that transmission takes place over a family of BMS channels characterized by their L-densities $\{\mathsf{a}_{\text{BMSC}(\mathtt{h})}\}$. By symmetry, $h(\mathtt{h}) = h_i(\mathtt{h})$, $i \in [n]$, and

$$h(\mathtt{h}) = \mathrm{H}(\mathsf{a}_{\text{BMSC}(\mathtt{h})}^{\circledast(n-1)}).$$

If we specialize to the family $\{\text{BSC}(\mathtt{h})\}$ we get

$$h(\mathtt{h}) = \sum_{i=0}^{n-1} \binom{n-1}{i} \epsilon^i \bar{\epsilon}^{n-1-i} h_2\Big(\frac{\epsilon^{|n-1-2i|}}{\epsilon^{|n-1-2i|} + \bar{\epsilon}^{|n-1-2i|}}\Big),$$

where $\epsilon = h_2^{-1}(\mathtt{h})$. For the Gaussian case we can express $h(\mathtt{h})$ in parametric form as

$$(\mathrm{H}(\mathsf{a}_{\text{BAWGNC}(\sigma)}), \mathrm{H}(\mathsf{a}_{\text{BAWGNC}(\sigma/\sqrt{n-1})})).$$

The two terms can be evaluated as discussed in Example 4.38. ◇

Example 4.133 ($[n, n-1, 2]$ Parity-Check Code – $\{\text{BSC}(\mathtt{h})\}$). Again by symmetry, $h(\mathtt{h}) = h_i(\mathtt{h})$, $i \in [n]$, and

$$h(\mathtt{h}) = \text{H}(\mathsf{a}_{\text{BSC}(\epsilon)}^{\boxast(n-1)}) = \text{H}(\mathsf{a}_{\text{BSC}(\frac{1-(1-2\epsilon)^{n-1}}{2})}) = h_2\left(\frac{1-(1-2\epsilon)^{n-1}}{2}\right),$$

where $\epsilon = h_2^{-1}(\mathtt{h})$. ◇

Example 4.134 ($[n, n-1, 2]$ Parity-Check Code – $\{\text{BAWGNC}(\mathtt{h})\}$). Formally, we have

$$h(\mathtt{h}) = \text{H}(\mathsf{a}_{\text{BAWGNC}(\mathtt{h})}^{\boxast(n-1)}).$$

Unfortunately, there is no elementary way to express the result of this convolution. For practical matters the following accurate approximation is available. Let $\psi(m)$ denote the function which gives the entropy of a symmetric Gaussian of mean m. We can compute $\psi(m)$ as discussed in Example 4.38 using the simple additional relationship $\sigma = \sqrt{2/m}$. The EXIT function according to the proposed approximation is

$$h(\mathtt{h}) = 1 - \psi\big((n-1)\psi^{-1}(1-\mathtt{h})\big).$$

The idea of this approximation is the following: we have seen in Theorem 3.77 that for the BEC we have the duality relationship: $h(\epsilon) = 1 - h^{\perp}(\bar{\epsilon})$. In words, to compute the EXIT function for a particular parameter ϵ and code C we can instead compute the EXIT function for the "dual" parameter $\bar{\epsilon}$ and the dual code $C^{\perp}$. This statement is easily generalized. In Problem 4.43 we discuss the Hartmann-Rudolph decoding rule. It states that an equivalent way of performing MAP decoding on a linear code is to use the dual code and to "dualize" the input to the code by taking the Fourier transform. This means that instead of operating with probabilities $(p(x=1 \mid y), p(x=-1 \mid y))$ we operate with the tuple $(p(x=1 \mid y) - p(x=-1 \mid y), p(x=1 \mid y) + p(x=-1 \mid y)) = (p(x=1 \mid y) - p(x=-1 \mid y), 1)$. Alternatively, this means that instead of working with L-densities we work with D-densities. For the BAWGNC($\mathtt{h}$) unfortunately the dual channel is not again a BAWGNC. Nevertheless, it has been observed that we get an accurate approximation if we perform the calculation on the dual code with input chosen from the BAWGNC($\mathtt{h}^{\perp} = 1 - \mathtt{h}$). To get the aforementioned approximation proceed as follows. First compute the dual parameter $1 - \mathtt{h}$. Then determine the corresponding mean by computing $\psi^{-1}(1-\mathtt{h})$. Since the means at a repetition code (the dual of the parity-check code) add, multiply the result by $n-1$. Finally, bring the resulting mean back to an entropy by applying $\psi(\cdot)$ and compute the dual parameter. ◇

Figure 4.135 shows the EXIT curves for the repetition as well as for the parity-check code for the families $\{\text{BEC}(\mathtt{h})\}$, $\{\text{BSC}(\mathtt{h})\}$, and also $\{\text{BAWGNC}(\mathtt{h})\}$. The EXIT curves for the various channel families are very similar. This observation is at the heart of the EXIT chart method for general channels. But note also that the EXIT function corresponding to the repetition code for the BEC is strictly smaller than the corresponding EXIT function for the BSC. This implies that the area theorem cannot be fulfilled in the general case. We will see in Section 4.12 how to remedy this situation.

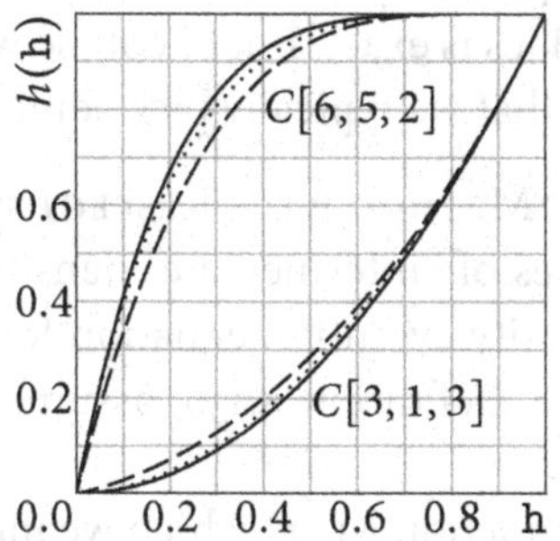

Figure 4.135: EXIT function of the $[3,1,3]$ repetition code and the $[6,5,2]$ parity-check code for the BEC (solid curve), the BSC (dashed curve), and also the BAWGNC (dotted curve).

§4.10.1. Thresholds and Optimization via EXIT Charts

Consider the density evolution process for BP stated in Theorem 4.97. We start with $\mathsf{a}_0 = \mathsf{a}_{\text{BMSC}}$, and for $\ell \geq 1$, $\mathsf{a}_\ell = \mathsf{a}_{\text{BMSC}} \circledast \lambda(\rho(\mathsf{a}_{\ell-1}))$. In general the "intermediate" densities a_ℓ do not have simple descriptions. This makes density evolution difficult to handle analytically.

If at each iteration ℓ we replace the intermediate density a_ℓ in the density evolution process with an "equivalent" density chosen from some suitable family of densities then we get the EXIT chart method. The most "faithful" equivalence rule is to choose the element of the channel family which has *equal entropy*.

Definition 4.136 (EXIT Chart Method). Consider a degree distribution pair (λ, ρ) and assume that transmission takes place over a BMS channel with L-density a_{BMSC}. Let $\{\mathsf{b}_\mathtt{h}\}$ and $\{\mathsf{a}_\mathtt{h}\}$ denote two families of symmetric L-densities. Let $\mathtt{h}_{c,\ell}$ ($\mathtt{h}_{v,\ell}$) denote the entropy of the density emitted at check nodes (variables nodes) at the ℓ-th iteration according to the EXIT chart method with respect to these channel families. More precisely, let these densities be $\mathsf{b}_{\mathtt{h}_{c,\ell}}$ and $\mathsf{a}_{\mathtt{h}_{v,\ell}}$, respectively. Then

$h_{v,0} = H(a_{BMSC})$ and for $\ell \geq 1$

$$h_{c,\ell} = H(\rho(a_{h_{v,\ell-1}})), \qquad h_{v,\ell} = H(a_{BMSC} \circledast \lambda(b_{h_{c,\ell}})),$$

where $\lambda(b_h) = \sum_i \lambda_i b_h^{\circledast(i-1)}$ and $\rho(a_h) = \sum_i \rho_i a_h^{\boxast(i-1)}$. We say that $h_{v,\ell}$ is the entropy emitted by the variable nodes in the ℓ-th iteration according to the EXIT chart method.

Discussion: If the chosen families $\{b_h\}$ and $\{a_h\}$ contain the actual densities encountered when computing density evolution then the EXIT chart method with respect to these channel families is exact. But of course we do not know these intermediate densities a priori so that we typically pick some "universal" family.

EXAMPLE 4.137 (EXIT CHART METHOD WITH RESPECT TO $\{a_{BAWGNC(h)}\}$). The preferred choice for both families of "intermediate" densities is $\{a_{BAWGNC(h)}\}$. Let us explicitly write down the density evolution equation for this case according to the EXIT chart method stated in Definition 4.136, assuming that transmission takes place over the BAWGNC($\hat{h}$).

In the sequel we use the function $\psi(m)$ which we introduced in Example 4.134. Let h denote the entropy entering a variable or check node. Define the two functions

$$v_{\hat{h}}(h) = \sum_i \lambda_i \psi\big((i-1)\psi^{-1}(h) + \psi^{-1}(\hat{h})\big),$$

$$c(h) = 1 - \sum_i \rho_i \psi\big((i-1)\psi^{-1}(1-h)\big).$$

As their names suggest, v/c describes the output entropy at a variable/check node as a function of the input entropy. We have encountered the function $c(h)$ (for the regular case) already in Example 4.134 as the (approximate) EXIT function of a parity-check code. In the current setting we consider irregular ensembles, which explains the extra averaging over the check-node degree distribution. That we have used an approximation of the output entropy rather than an exact expression does little harm. The approximation is very accurate and the EXIT chart method is an approximate method to start. The small additional error incurred by using the dual approximation is therefore easily outweighed by the advantage of being able to write down a simple analytic expression.

The expression for the variable-node side is easy to explain as well. We assume that the incoming internal messages have a symmetric Gaussian distribution with entropy h. Therefore $\psi^{-1}(h)$ gives the corresponding mean. The means of the inputs at a variable node add, whereby $\psi^{-1}(\hat{h})$ accounts for the mean of the received distribution.

Let $h_{v,\ell}$ denote the entropy at the output of the variable nodes at the end of the ℓ-th iteration. Then $h_{v,0} = \hat{h}$, and for $\ell \geq 1$, $h_{v,\ell} = v_{\hat{h}}(c(h_{v,\ell-1}))$. ◇

EXAMPLE 4.138 (EXIT CHART ANALYSIS FOR $(3,6)$-REGULAR ENSEMBLE). Assume that transmission takes place over the BAWGN($\hat{\mathtt{h}}$). Figure 4.140 shows the density evolution process according to the EXIT chart method for the two parameters $\hat{\mathtt{h}} \approx 0.3765$ ($\sigma \approx 0.816$) and $\hat{\mathtt{h}} \approx 0.427$ ($\sigma \approx 0.878$) and the $(3,6)$-regular ensemble. To construct this graph, plot $c(\mathtt{h})$, which describes the evolution at the check nodes and $v_{\hat{\mathtt{h}}}^{-1}(\mathtt{h})$ (for the chosen channel parameter $\hat{\mathtt{h}}$), which describes the process at the variable nodes. Rather than computing $v_{\hat{\mathtt{h}}}^{-1}(\mathtt{h})$, plot $v_{\hat{\mathtt{h}}}(\mathtt{h})$ but exchange the horizontal with the vertical axis. The density evolution progress is now easily read off from this picture by constructing a "staircase" function in the same manner as we have done for the case of transmission over the BEC in Figure 3.69. To recall: the initial entropy entering the check nodes is $\hat{\mathtt{h}}$. Let us consider the case where $\hat{\mathtt{h}} \approx 0.3765$. According to the EXIT chart method the entropy at the output of the check nodes is then $c(0.3765) \approx 0.8835$. We can construct this value graphically if we look for the intersection of the vertical line located at 0.3765 with the graph $c(\mathtt{h})$. This entropy now enters the variable nodes and according to the EXIT chart method the entropy at the output of the variable nodes is equal to $v_{\hat{\mathtt{h}}}(c(0.3765)) = v_{\hat{\mathtt{h}}}(0.8835) \approx 0.3045$. Since we have plotted the function $v_{\hat{\mathtt{h}}}^{-1}(\mathtt{h})$ we can graphically construct this value by looking for the intersection of the horizontal line at height 0.8835 with the function $v_{\hat{\mathtt{h}}}^{-1}(\mathtt{h})$. If we continue in this fashion, the corner points of the resulting staircase function describe the progress of density evolution according to the EXIT chart method.

We see from this figure that for $\hat{\mathtt{h}} \approx 0.3765$ the staircase function eventually reaches the point $(0,0)$, corresponding to successful decoding. On the other hand, for the value $\hat{\mathtt{h}} \approx 0.427$ the functions $c(\mathtt{h})$ and $v_{\hat{\mathtt{h}}}^{-1}(\mathtt{h})$ touch at some point (black dot in figure) and the staircase function converges to a non-zero fixed point. We conclude that the critical parameter according to the EXIT chart method is $\hat{\mathtt{h}} \approx 0.427$. This parameter differs only slightly from the true threshold value computed according to density evolution. In Table 6.24 we find the threshold listed as $\sigma^{\text{BP}} \approx 0.88$ (whereas $\hat{\mathtt{h}} \approx 0.427$ corresponds to $\sigma \approx 0.878$). ◇

EXAMPLE 4.139 (OPTIMIZATION FOR THE GAUSSIAN CHANNEL VIA EXIT CHARTS). From Example 4.137 we know that if the entropy at the output of the variable nodes is $\mathtt{h}$ then, after one further iteration, it becomes

$$\mathtt{h} \mapsto v_{\hat{\mathtt{h}}}(c(\mathtt{h})) = \sum_i \lambda_i \psi\big((i-1)\psi^{-1}(c(\mathtt{h})) + \psi^{-1}(\hat{\mathtt{h}})\big).$$

The condition for progress at each iteration is $v_{\hat{\mathtt{h}}}(c(\mathtt{h})) \leq \mathtt{h}$. This formulation is *linear* in the variable edge degree fractions λ_i. If we fix ρ, we can therefore optimize λ by linear programming techniques in the same manner as we have done in the case of the BEC (see Section 3.18). To give an example, if we assume that the channel

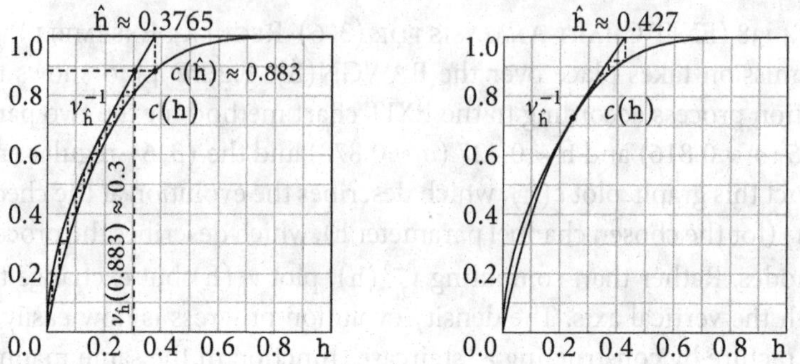

Figure 4.140: EXIT function of the $(3, 6)$-regular ensemble on the BAWGN channel. In the left-hand graph the parameter is $\hat{h} \approx 0.3765$ ($\sigma \approx 0.816$), whereas in the right-hand graph we chose $\hat{h} \approx 0.427$ ($\sigma = 0.878$).

parameter $\hat{h}$ is fixed and we take as objective function the rate of the code, then the corresponding (infinite) linear program reads

$$\max_{\lambda}\Big\{\sum_{i\geq 2}\lambda_i/i \mid \lambda_i \geq 0; \sum_{i\geq 2}\lambda_i = 1; \sum_{i\geq 2}\lambda_i\psi\big((i-1)\psi^{-1}(c(h)) + \psi^{-1}(\hat{h})\big) \leq h; h \in [0,1]\Big\}.$$

Further, the roles of λ and ρ can be interchanged. If h is the entropy entering a variable node then after one iteration this entropy has evolved to

$$h \mapsto c(\nu_{\hat{h}}(h)) = 1 - \sum_i \rho_i\psi\big((i-1)\psi^{-1}(1-\nu_{\hat{h}}(h))\big).$$

This formulation is now linear in ρ and so we can fix λ and optimize over ρ. Again, if our objective function is the rate of the code, we get the (infinite) linear program

$$\min_{\rho}\Big\{\sum_{i\geq 1}\rho_i/i \mid \rho_i \geq 0; \sum_{i\geq 2}\rho_i = 1; \sum_i \rho_i\psi\big((i-1)\psi^{-1}(1-\nu_{\hat{h}}(h))\big) \geq 1 - h; h \in [0,1]\Big\}.$$

Starting with a fixed pair (λ, ρ), an alternating application of the two preceding linear programs quickly leads to very good degree distributions.

The approximation underlying this optimization technique can be improved: in Example 4.138 we assumed that all intermediate densities are from the family $\{a_{\text{BAWGNC}(h)}\}$. But, as discussed in Definition 4.136, exactly the same procedure can be used with respect to *any* (complete) family of BMS densities. Further, we can use *distinct* families to represent the densities entering the variable and the check nodes, respectively. For alternative choices of the family of "interpolating" distributions we typically do not have succinct analytic characterizations of $\nu_{\hat{h}}(h)$ and $c(h)$. But we can compute these functions numerically once the intermediate channel families are specified.

Imagine now that for a given degree distribution pair (λ, ρ) we were able to "guess" the true intermediate densities which appear in the density evolution process and that out of this discrete set of densities we construct two complete channel families (e.g., by interpolating between successive elements of this sequence). If we were to apply the EXIT chart method with respect to *these* families of intermediate densities then the result would be *exact*. The idea of an improved approximation is to recursively try to find better and better guesses for these intermediate densities.

More precisely, start with some degree distribution pair (λ, ρ). Run density evolution, using the degree distribution pair (λ, ρ), and collect the sequence of actual intermediate densities. Construct out of this sequence of densities two complete channel families (one associated to the input at variable nodes and one associated to the input at check nodes). Next, employ the EXIT chart method and optimization technique outlined earlier *with respect to this constructed family of densities*. If the resulting optimized degree distribution pair differs only slightly from the original one, then there is good reason to believe that the true (according to the density evolution progress with respect to the new degree distribution pair) intermediate densities are "close" to the assumed such densities and that, therefore, the EXIT chart approximation is close to the true density evolution process. The procedure can be repeated a sufficient number of times until all quantities have (hopefully) converged.

This procedure was applied to find a rate one-half code for the BAWGNC channel with maximum left degree equal to 100. The result of this optimization is

$$\begin{aligned}\lambda(x) =& 0.169010x + 0.161244x^2 + 0.005938x^4 + 0.016799x^5 + 0.186455x^6 + \\ & 0.006864x^{13} + 0.025890x^{16} + 0.096393x^{18} + 0.010531x^{26} + \\ & 0.004678x^{27} + 0.079616x^{28} + 0.011885x^{38} + 0.224691x^{99} \\ \rho(x) =& x^{10}.\end{aligned}$$

The average right degree is 11. The threshold is $\mathtt{h}^{\text{BP}} \approx 0.4982$ ($\sigma^{\text{BP}} \approx 0.976$), which corresponds to a gap to capacity of 0.02370 dB. ◇

§4.10.2. Universal Lower Bound on Threshold

So far EXIT curves appeared as approximations to the true density evolution process. Surprisingly, they can be used to give strict bounds. The basis for such bounds is the following lemma.

Theorem 4.141 (Extremes of Information Combining). Let a and b represent two BMS channels and fix $H(b) = h$, $0 \leq h \leq 1$. Then

$$\begin{aligned} &\text{repetition code} && \overbrace{H(a \circledast b_{BEC(h)})}^{H(a)h} \leq H(a \circledast b) \leq H(a \circledast b_{BSC(h)}), \\ &\text{parity-check code} && H(a \boxast b_{BSC(h)}) \leq H(a \boxast b) \leq \underbrace{H(a \boxast b_{BEC(h)})}_{1-(1-H(a))(1-h)}. \end{aligned}$$

Discussion: Recall that $a \circledast b$ is the output density at a variable node of degree 3 assuming that the two input densities are a and b, respectively. Consider the following experiment. We fix one input density a and a parameter h, $h \in [0, 1]$. For all input densities b with entropy h we compute the entropy of the output density $a \circledast b$. The lemma asserts that the input densities which minimize/maximize the entropy of the output (and have themselves entropy h) are $b_{BEC(h)}$ and $b_{BSC(h)}$, respectively. These densities are sometimes called the *most* and *least informative* densities. The second assertion concerns the equivalent statement at a check node, since if a and b are the input densities at the input of a check node of degree three then $a \boxast b$ is the corresponding output density. Again $b_{BEC(h)}$ and $b_{BSC(h)}$ are the extremal densities but now the roles are *reversed*. Note that a can itself be the convolution (at either check or variable node) of any number of BMS channels so that the statement extends to the convolution of any number of densities. In particular the statement implies that any time we substitute at a variable node an input density with a density from the family $\{a_{BEC(h)}\}$ ($\{a_{BSC(h)}\}$) of equal entropy then the entropy of the output is decreased (increased). The reverse statements hold at a check node. Since the entropy operator is linear, it follows that if we take a symmetric density and "move it closer" to a BEC density (by taking the convex combination) then the entropy of the resulting output density at a variable node decreases. Problem 4.62 discusses several other settings in which the densities $\{b_{BSC(h)}\}$ and $\{b_{BEC(h)}\}$ can be shown to be extremal.

Proof of Theorem 4.141. If you believe one of the two claims there is an easy way to convince you of the other one: from Lemma 4.41 we know that $H(a \boxast b)$, the entropy at the output of a check node, whose inputs have symmetric densities a and b, respectively, is $H(a) + H(b) - H(a \circledast b)$. Therefore, for fixed entropies $H(a)$ and $H(b)$, the entropy of $a \boxast b$ is minimized/maximized if the entropy of $a \circledast b$ is maximized/minimized. In other words, the extremality for L-densities follows directly from the extremality for G-densities and vice versa.

We prove the extremality for check nodes. Consider a check node of degree 3 with two designated inputs and one designated output. Assume that the inputs experience the symmetric G-densities a and b, respectively. We are interested in the resulting output density. This density is given by $a \circledast b$. The corresponding entropy is $\text{H}(a \circledast b)$.

To start, assume that both input densities are from the BSC family. For the proof it is more convenient to parameterize the BSC channel in terms of the cross-over probability ϵ rather than the entropy h. Let $a = a_{\text{BSC}(\epsilon_a)}$ and $b = b_{\text{BSC}(\epsilon_b)}$. Their associated entropies are $h_2(\epsilon_a)$ and $h_2(\epsilon_b)$, respectively. The resulting density at the output corresponds also to a BSC and it has parameter $\epsilon_a\bar{\epsilon}_b + \epsilon_b\bar{\epsilon}_a$. This is true since the output is wrong if *exactly one* of the two inputs is wrong. The associated entropy is therefore $h_2(\epsilon_a\bar{\epsilon}_b + \epsilon_b\bar{\epsilon}_a)$. This is important: we can think of any BMS channel as the weighted sum (convex combination) of BSCs. More precisely, we have two density functions $w_a(\text{h})$ and $w_b(\text{h})$ so that

$$a(y) = \int_0^1 w_a(\text{h}) a_{\text{BSC}(h_2^{-1}(\text{h}))}(y)\text{dh}, \quad b(y) = \int_0^1 w_b(\text{h}) a_{\text{BSC}(h_2^{-1}(\text{h}))}(y)\text{dh}.$$

We can think of $w(\text{h})$ as yet another density (besides L-, D-, and G-densities) which characterizes the channel. Since the operator $\text{H}(\cdot)$ is linear we can compute $\text{H}(a \circledast b)$ by first conditioning on the "entropy" of each channel and then taking the expectation. We therefore get the representation

$$\begin{aligned}\text{H}(a \circledast b) &= \int_0^1 \int_0^1 w_a(\text{h}_a) w_b(\text{h}_b) h_2\big(\epsilon_a\bar{\epsilon}_b + \bar{\epsilon}_a\epsilon_b\big)\text{dh}_a\text{dh}_b \\ &= \int_0^1 w_a(\text{h}_a)\Big(\int_0^1 w_b(\text{h}_b) h_2\big(\epsilon_b(1-2\epsilon_a)+\epsilon_a\big)\text{dh}_b\Big)\text{dh}_a, \end{aligned} \tag{4.142}$$

where $\epsilon_{a/b} = h_2^{-1}(\text{h}_{a/b})$. We claim that for fixed $v \in [0, \frac{1}{2}]$ the function $h_2\big(h_2^{-1}(u)(1-2v)+v\big)$ is non-decreasing and convex-∪ in u, $u \in [0,1]$ (see the left-hand graph of Figure 4.143). To see that the function is non-decreasing note that $h_2^{-1}(u)(1-2v)+v$ is a non-decreasing function of u, $u \in [0,1]$, for any $v \in [0, \frac{1}{2}]$ and that it takes values in $[0, \frac{1}{2}]$. Further, $h_2(\cdot)$ is an increasing function of its argument in this range. Let us postpone the proof of the convexity for a moment and consider directly its implication. Recall that by assumption $\int_0^1 w_b(\text{h}_b)\text{h}_b\text{dh}_b = \text{h}$. Combined with the convexity this implies that $\int w_b(\text{h}_b) h_2\big(\epsilon_b(1-2\epsilon_a)+\epsilon_a\big)\text{dh}_b \geq h_2\big(h_2^{-1}(\text{h})(1-2\epsilon_a)+\epsilon_a\big)$. If we insert this into (4.142) we get $\text{H}(a \circledast b) \geq \text{H}(a \circledast b_{\text{BSC}(h_2^{-1}(\text{h}))})$. On the other hand, since $h_2\big(\epsilon_b(1-2\epsilon_a)+\epsilon_b\big)$ is convex-∪ in h_b and non-decreasing it follows that it is upper bounded by the straight line joining its two boundary values. For $\text{h}_b = 0$ the function takes on the value h_a, whereas for $\text{h}_b = 1$ it takes on the value 1. Therefore we get the upper bound

$$h_2\big(\epsilon_b(1-2\epsilon_a)+\epsilon_a\big) \leq \text{h}_a(1-\text{h}_b) + \text{h}_b = 1 - (1-\text{h}_a)(1-\text{h}_b).$$

If we insert this bound into (4.142) we get $\text{H}(a \circledast b) \leq 1 - (1-\text{H}(a))(1-\text{h}) = \text{H}(a \circledast b_{\text{BEC}(\text{h})})$.

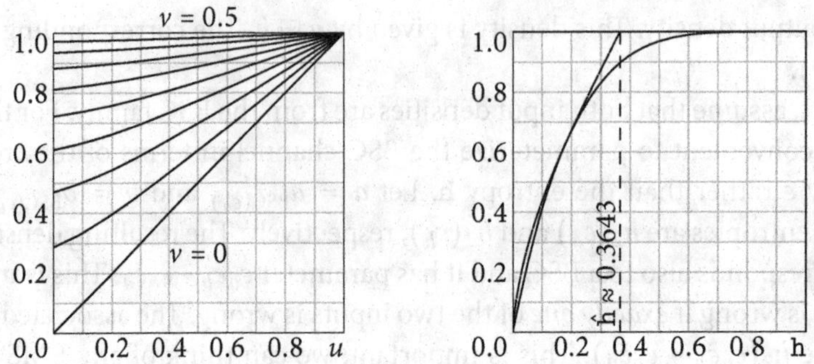

Figure 4.143: Left: For $v \in [0, \frac{1}{2}]$ the function $h_2(h_2^{-1}(u)(1-2v)+v)$ is non-decreasing and convex-∪ in u, $u \in [0,1]$. Right: Universal bound applied to the (3, 6)-regular ensemble.

Define $f(u,v) = h_2(h_2^{-1}(u)(1-2v)+v)$. It remains to be shown that, for a fixed $v \in [0, \frac{1}{2}]$, $f(u,v)$ is convex-∪ in u for $u \in [0,1]$. This requires that $\frac{\partial^2 f(u,v)}{\partial u^2} \geq 0$ for $u \in [0,1]$ and $v \in [0, \frac{1}{2}]$. A tedious computation shows that

$$\frac{\partial^2 f(u,v)}{\partial u^2} = \frac{(1-2v)\big[z(1-z)\log(\frac{1-z}{z}) - (1-2v)y(1-y)\log(\frac{1-y}{y})\big]}{\log(\frac{1-y}{y})^3},$$

where $y = h_2^{-1}(u)$, $y \in [0, \frac{1}{2}]$, and $z = v(1-y) + y(1-v)$. For the range of interest, this expression is non-negative if and only if the term between the square brackets is non-negative. If we take the derivative of that term with respect to y we get $(1-2v)\log(\frac{1-z}{z}) - \log(\frac{1-y}{y})$. Since $y \leq z$ and $1-2v \leq 1$, it follows that $\frac{\partial^2 f(u,v)}{\partial u^2}$ is decreasing as a function of y. Since for $y = \frac{1}{2}$ a direct check shows that $\frac{\partial^2 f(u,v)}{\partial u^2} = 0$, it follows that $\frac{\partial^2 f(u,v)}{\partial u^2} \geq 0$ in the range of interest. □

Extremal densities are useful in deriving universal bounds on thresholds. To be concrete, consider the (3, 6)-regular ensemble. The same idea applies to general ensembles as well. Assume that transmission takes place over a BMS channel with entropy $\mathfrak{h}$. What is the largest such entropy under which we can guarantee that BP decoding is successful (in the asymptotic limit), regardless of the distribution of the BMS channel? We proceed as follows. We employ the EXIT chart methodology but to get a bound we assume that the intermediate densities are the *least informative* ones. This means that we assume that the input densities to check nodes are elements of $\{a_{BEC(h)}\}$ and that the input densities at variable nodes (including the received one) are from the family $\{a_{BSC(h)}\}$. Consider Figure 4.143. The lower curve corresponds to the EXIT curve for the [6, 5, 2] single parity-check code with respect the channel family $\{\mathtt{a}_{\mathrm{BEC}(\mathtt{h})}\}$. In formulae it is given by $1-(1-\mathtt{h})^5$. This represents the extremal EXIT curve of the check-node side. The upper curve corresponds to the EXIT curve of a repetition code of length 4 assuming that one of the four inputs has density $\mathtt{a}_{\mathrm{BSC}(\hat{\mathtt{h}})}$ (the channel) and that the remaining inputs are from the family $\{\mathtt{a}_{\mathrm{BSC}(\mathtt{h})}\}$ (representing the internal messages). This is a slight generalization of the case discussed in Example 4.132. Explicitly the curve is given by

$$\mathrm{H}(\mathtt{a}_{\mathrm{BSC}(\mathtt{h})}^{\circledast 2} \circledast \mathtt{a}_{\mathrm{BSC}(\hat{\mathtt{h}})}) = (\bar{\epsilon}^2\bar{\hat{\epsilon}} + \epsilon^2\hat{\epsilon})h_2\Big(\frac{\bar{\epsilon}^2\bar{\hat{\epsilon}}}{\bar{\epsilon}^2\bar{\hat{\epsilon}} + \epsilon^2\hat{\epsilon}}\Big) + 2\epsilon\bar{\epsilon}h_2(\hat{\epsilon}) +$$
$$(\bar{\epsilon}^2\hat{\epsilon} + \epsilon^2\bar{\hat{\epsilon}})h_2\Big(\frac{\bar{\epsilon}^2\hat{\epsilon}}{\bar{\epsilon}^2\hat{\epsilon} + \epsilon^2\bar{\hat{\epsilon}}}\Big),$$

where $\mathtt{h} = h_2(\epsilon)$ and $\hat{\mathtt{h}} = h_2(\hat{\epsilon})$. This represents the extremal EXIT curve at the variable-node side.

The largest value of $\hat{\mathtt{h}}$ was chosen so that the two curves do not overlap and the critical parameter is $\hat{\mathtt{h}} \approx 0.3643$ (which corresponds to an error probability of 0.06957 for the BSC channel). We conclude that if we use the $(3,6)$-regular ensemble and BP decoding then we can transmit reliably over any BMS channel with entropy 0.3643 bits per channel use or less (this corresponds to a capacity of 0.6357 bits per channel use or more). Many variants of this idea are possible and useful. Instead of taking the least informative intermediate densities we can take the most informative ones. This gives a lower bound on the required entropy to transmit reliably using a particular ensembles and BP decoding (see Problem 4.57). Alternatively, we can use the method to derive a bound on the critical parameters for a particular family of input densities by only replacing the intermediate densities but by explicitly dealing with the specific input density. This is discussed in Problem 4.58.

§4.11. Gallager's Lower Bound on Density

Assuming that transmission takes place over the BEC, Theorem 3.93 states a lower bound on the gap to capacity as a function of the density of the parity-check matrix. This lower bound can be generalized to BMS channels.

Theorem 4.144 (Gallager's Inequality). Consider transmission over a BMS channel with L-density $\mathtt{a}_{\mathrm{BMSC}}$. Define $\mathrm{C}_{\mathrm{BMSC}} = \mathrm{C}(\mathtt{a}_{\mathrm{BMSC}})$. Assume that a proper binary linear code C of length n and dimension k is used and define $r = k/n$. Let H denote any full-rank parity-check matrix representing C and let $R(x)$ denote the normalized check-node degree distribution so that $n(1-r)R_i$ equals the number of check nodes of H of degree i. If X denotes the transmitted codeword, where we assume a uniform prior on X, and if Y denotes the received word, then

$$\frac{H(X \mid Y)}{n} \geq r - \mathrm{C}_{\mathrm{BMSC}} + \frac{1-r}{2\ln(2)} \sum_{k=1}^{\infty} \frac{R(\mathfrak{D}_{2k}(\mathtt{a}_{\mathrm{BMSC}}))}{k(2k-1)}, \quad (4.145)$$

where $\mathfrak{D}_k(\mathsf{a}_{\text{BMSC}})$ is the D-k-moment introduced in Definition 4.56 and where we measure the entropy in bits.

Remark: Since every term in the sum on the right-hand side of (4.145) is non-negative we are free to replace the infinite sum with a finite one. Any such choice gives a lower bound on the conditional entropy.

Proof. Start by expanding the mutual information $I(X;Y)$ as $H(X) - H(X \mid Y)$ as well as $H(Y) - H(Y \mid X)$. Equating the two expressions and rearranging terms yields

$$H(X \mid Y) = H(X) - H(Y) + H(Y \mid X). \tag{4.146}$$

Since the prior on X is uniform we know that $H(X) = nr$. Further, since the channel is memoryless,

$$H(Y \mid X) = \sum_{i=1}^{n} H(Y_i \mid X_i) = \sum_{i=1}^{n} H(Y_i) - \sum_{i=1}^{n} I(X_i; Y_i) \geq \sum_{i=1}^{n} H(Y_i) - n\mathrm{C}_{\text{BMSC}}.$$

If we plug these two (in)equalities into (4.146) then we get $H(X \mid Y) \geq nr - H(Y) + \sum_{i=1}^{n} H(Y_i) - n\mathrm{C}_{\text{BMSC}}$.

Recall from page 181 that we can describe the channel output Y equivalently as $(|Y|, S)$: $|Y|$ denotes the vector consisting of the *reliability* values (magnitudes), and S denotes the vector of signs. For our current purpose it is useful to let the sign take values in $\{0,1\}$ instead of the usual $\{\pm 1\}$. Now, $H(Y_i) = H(|Y_i|, S_i) = H(|Y_i|) + H(S_i \mid |Y_i|) = H(|Y_i|) + 1$. To see the last step, note that the code is assumed to be proper and that the distribution of X is the uniform one so that X_i has uniform distribution as well. Since the channel is symmetric, this implies that Y_i has a symmetric distribution around zero so that the sign has a uniform distribution conditioned on $|Y_i|$.

It remains to bound $H(Y)$. We have

$$H(Y) = H(|Y|, S) = H(|Y|) + H(S \mid |Y|) \leq \sum_{i=1}^{n} H(|Y_i|) + H(S \mid |Y|).$$

Pick an *information set* of the code, call it $\mathcal{I}$. More precisely, $\mathcal{I}$ is a subset of $[n]$ of cardinality k (the dimension of the code) so that each codeword takes on distinct values on $\mathcal{I}$ and that, conversely, every assignment to the positions of $\mathcal{I}$ corresponds to a valid codeword. For a given sign vector s, let $s_{\mathcal{I}}$ denote the subvector of s restricted to the components indexed by $\mathcal{I}$. Note that $s_{\mathcal{I}} \in \mathbb{F}_2^k$. Further, introduce the *syndrome* $\hat{s} = Hs^T$. Since H is by assumption a full-rank matrix and the code has length n and dimension k we have $\hat{s} \in \mathbb{F}_2^{n-k}$. We claim that there is a one-to-one correspondence between sign vectors s and $(\hat{s}, s_{\mathcal{I}})$. Clearly, both s and $(\hat{s}, s_{\mathcal{I}})$ can take on 2^n distinct values. Further, every element of s gives rise to a distinct tuple $(\hat{s}, s_{\mathcal{I}})$: two s which belong to distinct cosets of C map into different tuples $(\hat{s}, s_{\mathcal{I}})$ since their syndromes differ, and for each element s of the same coset there is exactly one which corresponds to a given $s_{\mathcal{I}}$. Therefore,

$$H(S\,|\,|Y|) = H(\hat{S}, S_{\mathcal{I}}\,|\,|Y|) \le H(S_{\mathcal{I}}\,|\,|Y|) + H(\hat{S}\,|\,|Y|) \le nr + \sum_{j=1}^{n(1-r)} H(\hat{S}_j\,|\,|Y|).$$

If we summarize our results so far we get

$$\frac{H(X\,|\,Y)}{n} \ge 1 - C_{\text{BMSC}} - \frac{1}{n}\sum_{j=1}^{n(1-r)} H(\hat{S}_j\,|\,|Y|). \tag{4.147}$$

Now note that $\hat{S}_j$ is a Bernoulli random variable. Consider a check node of degree $\mathtt{r}$. Then

$$\mathbb{P}\{\hat{S}_j = 1\,|\,|Y| = |y|\} = \frac{1}{2}\Big(1 - \prod_{j=1}^{\mathtt{r}}\Big(1 - \frac{2\mathrm{e}^{-|y_j|}}{1+\mathrm{e}^{-|y_j|}}\Big)\Big). \tag{4.148}$$

To see this, recall from page 181 that the probability that bit j was transmitted in error, call it p_j, given its reliability $|y_j|$ equals $\mathrm{e}^{-|y_j|}/(1+\mathrm{e}^{-|y_j|})$. Some thought then shows that the probability that the sum of $\mathtt{r}$ independent bits is in error (that the bits are independent follows from the fact that the channel is memoryless), call it p, is given by $(1-2p) = \prod_{j=1}^{\mathtt{r}}(1-2p_j)$, which leads to (4.148).

If we average the entropy of this Bernoulli random variable over $|Y|$ we get

$$\begin{aligned}
H(\hat{S}_j\,|\,|Y|) &= \int h_2\Big(\frac{1}{2}\Big(1 - \prod_{j=1}^{\mathtt{r}}\Big(1 - \frac{2\mathrm{e}^{-|y_j|}}{1+\mathrm{e}^{-|y_j|}}\Big)\Big)\Big)\prod_{j=1}^{\mathtt{r}} |\mathsf{a}_{\text{BMSC}}|(y_j)\,\mathrm{d}y_j \\
&= \int h_2\Big(\frac{1}{2}\Big(1 - \prod_{j=1}^{\mathtt{r}} \tanh(|y_j|/2)\Big)\Big)\prod_{j=1}^{\mathtt{r}} |\mathsf{a}_{\text{BMSC}}|(y_j)\,\mathrm{d}y_j \\
&= 1 - \frac{1}{2\ln(2)}\sum_{k=1}^{\infty}\frac{1}{k(2k-1)}\Big(\int_0^{+\infty}|\mathsf{a}_{\text{BMSC}}|(y)\tanh^{2k}(y/2)\,\mathrm{d}y\Big)^{\mathtt{r}} \\
&= 1 - \frac{1}{2\ln(2)}\sum_{k=1}^{\infty}\frac{1}{k(2k-1)}\Big(\int_{-\infty}^{+\infty}\mathsf{a}_{\text{BMSC}}(y)\tanh^{2k}(y/2)\,\mathrm{d}y\Big)^{\mathtt{r}} \\
&= 1 - \frac{1}{2\ln(2)}\sum_{k=1}^{\infty}\frac{(\mathfrak{D}_{2k}(\mathsf{a}_{\text{BMSC}}))^{\mathtt{r}}}{k(2k-1)}.
\end{aligned}$$

In the second-to-last step we have used for $0 \le x \le 1$ the series expansion

$$h_2(x) = 1 - \frac{1}{2\ln(2)}\sum_{k=1}^{\infty}\frac{(1-2x)^{2k}}{k(2k-1)}, \tag{4.149}$$

which allows us to break the $\mathtt{r}$-dimensional integral into a product of $\mathtt{r}$ independent integrals. This infinite series consists of non-negative terms so that we are justified to exchange the summation with the integral. The claim now follows by averaging over the degree distribution and by substituting this expression into (4.147). □

Theorem 4.150 (Lower Bound on Parity-Check Density). Assume we are given a BMS channel with capacity C_{BMSC} and $|L|$-density $|\mathsf{a}_{\text{BMSC}}|$. Let $\{C_N\}$ be a sequence of proper binary linear codes achieving a fraction $1-\delta$ of C_{BMSC} with vanishing bit error probability. Let $\{\Delta_N\}$ be the corresponding sequence of densities for an *arbitrary representation* of the binary linear block codes by full-rank parity-check matrices. Then

$$\liminf_{N\to\infty} \Delta_N \geq \frac{K_1 + K_2 \ln \frac{1}{\delta}}{1-\delta},$$

where

$$K_1 = K_2 \ln \frac{\xi(1-C_{\text{BMSC}})}{C_{\text{BMSC}}}, \quad K_2 = \frac{1-C_{\text{BMSC}}}{C_{\text{BMSC}} \ln\left(\frac{1}{\mathfrak{D}_2(\mathsf{a}_{\text{BMSC}})}\right)}, \quad \xi = \begin{cases} 1, & \text{BEC}, \\ \frac{1}{2\ln(2)}, & \text{otherwise}. \end{cases}$$

Proof. Using Jensen's inequality we have

$$\frac{1}{2\ln(2)} \sum_{k=1}^{\infty} \frac{R(\mathfrak{D}_{2k}(\mathsf{a}_{\text{BMSC}}))}{k(2k-1)} \geq \frac{1}{2\ln(2)} \sum_{k=1}^{\infty} \frac{\mathfrak{D}_{2k}(\mathsf{a}_{\text{BMSC}})^{\mathtt{r}_{\text{avg}}}}{k(2k-1)} \geq \frac{1}{2\ln(2)} \mathfrak{D}_2(\mathsf{a}_{\text{BMSC}})^{\mathtt{r}_{\text{avg}}},$$

where the second step is true since all terms are non-negative. We therefore can replace (4.145) with the slightly weaker but easier to handle inequality

$$\frac{H(X \mid Y)}{n} \geq r - C_{\text{BMSC}} + \frac{1-r}{2\ln(2)} \mathfrak{D}_2(\mathsf{a}_{\text{BMSC}})^{\mathtt{r}_{\text{avg}}}, \tag{4.151}$$

where all rates and entropies are measured in bits. By assumption the sequence $\{C_N\}$ achieves a fraction $1-\delta$ of capacity with vanishing bit error probability. Therefore, by Fano's inequality the sequence of normalized conditional entropies must converge to zero. This means that

$$\liminf_{N\to\infty} r_N - C_{\text{BMSC}} + \frac{1-r_N}{2\ln(2)} \mathfrak{D}_2(\mathsf{a}_{\text{BMSC}})^{\mathtt{r}_{\text{avg}_N}} \leq 0.$$

We know that the rate is asymptotically at least equal to $(1-\delta)C_{\text{BMSC}}$. If we solve the resulting equation for $\mathtt{r}_{\text{avg}_N}$ we get

$$\liminf_{N\to\infty} \mathtt{r}_{\text{avg}_N} \geq \frac{\log\left(\frac{1}{2\ln(2)}\left(1 + \frac{1-C_{\text{BMSC}}}{\delta C_{\text{BMSC}}}\right)\right)}{\log\left(\frac{1}{\mathfrak{D}_2(\mathsf{a}_{\text{BMSC}})}\right)}.$$

Finally, if we drop the 1 inside the log and recall that the density of the matrix is related to the average right degree by multiplying with $r/(1-r)$, the result for generic BMS channels follows.

Using a similar technique as before we can also show that (see Problem 4.54)

$$\frac{1}{2\ln(2)}\sum_{k=1}^{\infty}\frac{R(\mathfrak{D}_{2k}(\mathsf{a}_{\text{BMSC}}))}{k(2k-1)} \geq \mathrm{C}_{\text{BMSC}}^{\mathtt{r}_{\text{avg}}}.$$

We can therefore replace the $\frac{1}{2\ln(2)}\mathfrak{D}_2(\mathsf{a}_{\text{BMSC}})^{\mathtt{r}_{\text{avg}}}$ in the bound above with $\mathrm{C}_{\text{BMSC}}^{\mathtt{r}_{\text{avg}}}$. In particular, for the BEC we have $\mathfrak{D}_2(\mathsf{a}_{\text{BMSC}}) = \mathrm{C}_{\text{BEC}}$. This yields the tighter bound in this case. Since $\mathrm{C}_{\text{BEC}} = 1-\epsilon$, this bound is identical to the bound stated in Theorem 4.144. □

§4.12. GEXIT Function and MAP Performance

In the previous section we have seen several useful applications of EXIT functions in the context of general BMS channels. But EXIT functions are not without their shortcomings – most notably they do not fulfill the area theorem in the general setting. If we look back at Definition 4.129, we see that our generalization of EXIT functions is based on Definition 3.70. But as we have seen in Lemma 3.74, EXIT functions have several alternative characterizations. These characterization are equivalent only in the setting of transmission over the BEC. We next introduce GEXIT functions which are the natural extension of characterization (iii) in Lemma 3.74.

Definition 4.152 (GEXIT Function). Let X be a binary vector of length n chosen with probability $p_X(x)$ from a code C of length n. Assume that transmission takes place over the smooth family $\{\text{BMSC}(\mathtt{h})\}$. Then

$$g(\mathtt{h}) = \frac{\mathrm{d}H(X \mid Y(\mathtt{h}))}{n\mathrm{d}\mathtt{h}}. \tag{4.153}$$

More generally, assume that the channel from X to Y is memoryless and that the i-th bit is transmitted over the smooth family $\{\text{BMSC}(\mathtt{h}_i)\}$. Then

$$g_i(\mathtt{h}_1,\dots,\mathtt{h}_n) = \frac{\partial H(X \mid Y(\mathtt{h}_1,\dots,\mathtt{h}_n))}{\partial \mathtt{h}_i}.$$

If all these channels are parameterized in a smooth way by a common parameter ϵ, i.e., $\mathtt{h}_i = \mathtt{h}_i(\epsilon)$, then

$$g(\epsilon) = \frac{1}{n}\sum_{i=1}^{n} g_i(\mathtt{h}_1,\dots,\mathtt{h}_n)\frac{\partial \mathtt{h}_i(\epsilon)}{\partial \epsilon}. \tag{4.154}$$

Discussion: We see from this definition that the GEXIT function $g(\mathtt{h})$ measures the change of the conditional entropy as a function of a change in the channel entropy. If we only tweak the i-th channel parameter then we get the i-th GEXIT function $g_i(\mathtt{h}_1, \dots, \mathtt{h}_n)$. Consider the case where all channel families $\{\text{BMSC}(\mathtt{h}_i)\}$ are identical and where $\mathtt{h}_1(\epsilon) = \cdots = \mathtt{h}_n(\epsilon) = \epsilon$, so that $\frac{\partial \mathtt{h}_i(\epsilon)}{\partial \epsilon} = 1$, $i = 1, \dots, n$. Then, $g(\epsilon = \mathtt{h}) \overset{(4.154)}{=} \frac{1}{n}\sum_{i=1}^{n} \frac{\partial H(X \mid Y(\mathtt{h}_1,\dots,\mathtt{h}_n))}{\partial \mathtt{h}_i} = \frac{\mathrm{d} H(X \mid Y(\mathtt{h}))}{n \mathrm{d}\mathtt{h}} \overset{(4.153)}{=} g(\mathtt{h})$. We conclude that characterization (4.153) is consistent with, and indeed a special case of, characterization (4.154). With some abuse of notation we write in the sequel $g_i(\epsilon)$ to mean $g_i(\mathtt{h}_1, \dots, \mathtt{h}_n)$ in the case where all channels are parameterized by the common parameter ϵ.

An immediate consequence of Definition 4.152 is the General Area Theorem (GAT), which states that if we integrate the GEXIT function along a smooth path then the integral equals the difference of the conditional entropy at the two endpoints.

Corollary 4.155 (GEXIT Functions and the General Area Theorem). Let X be a vector of length n chosen with probability $p_X(x)$ from a binary code C. Assume that the channel from X to Y is memoryless and that the i-th bit is transmitted over the smooth family $\{\text{BMSC}(\mathtt{h}_i)\}$, where $\mathtt{h}_i = \mathtt{h}_i(\epsilon)$, $i \in [n]$, $\epsilon \in I$, i.e., all channels are parameterized by the common parameter ϵ in a smooth way. Then for $\underline{\epsilon}, \overline{\epsilon} \in I$,

$$\int_{\underline{\epsilon}}^{\overline{\epsilon}} g(\epsilon) \mathrm{d}\epsilon = \frac{1}{n}\big(H(X \mid Y(\mathtt{h}_1(\overline{\epsilon}), \dots, \mathtt{h}_n(\overline{\epsilon}))) - H(X \mid Y(\mathtt{h}_1(\underline{\epsilon}), \dots, \mathtt{h}_n(\underline{\epsilon})))\big).$$

The form of the GEXIT function as specified in Definition 4.152 is typically not convenient for computations. The following alternative characterization is much closer in spirit to the EXIT function and it is easier to handle.

Lemma 4.156 (Local Characterization of GEXIT Function). Under the conditions of Definition 4.152,

$$g(\epsilon) = \frac{1}{n}\sum_{i=1}^{n} \underbrace{\frac{\partial H(X_i \mid Y(\epsilon))}{\partial \mathtt{h}_i}}_{g_i(\epsilon)} \frac{\partial \mathtt{h}_i(\epsilon)}{\partial \epsilon}. \tag{4.157}$$

Proof. For $i \in [n]$, the entropy rule gives $H(X \mid Y) = H(X_i \mid Y) + H(X_{\sim i} \mid X_i, Y)$. Since the channel is memoryless this is equal to $H(X_i \mid Y) + H(X_{\sim i} \mid X_i, Y_{\sim i})$. The term $H(X_{\sim i} \mid X_i, Y_{\sim i})$ does not depend on the channel parameter $\mathtt{h}_i$. Therefore,

$$g_i(\epsilon) = \frac{\partial H(X \mid Y(\epsilon))}{\partial \mathtt{h}_i} = \frac{\partial H(X_i \mid Y(\epsilon))}{\partial \mathtt{h}_i}. \qquad \square$$

Lemma 4.158 (GEXIT Function for Proper Linear Codes). Let X be chosen uniformly at random from a binary linear code whose i-th position is proper and assume that the channel from X to Y is memoryless, where the i-th bit is transmitted over a smooth BMS channel family characterized by its L-density $\{\mathsf{a}_{\text{BMSC}(\mathtt{h}_i)}\}$. Define the *extrinsic* MAP estimator of X_i,

$$\phi_i(y_{\sim i}) = \ln \frac{p_{X_i \mid Y_{\sim i}}(+1 \mid y_{\sim i})}{p_{X_i \mid Y_{\sim i}}(-1 \mid y_{\sim i})}, \tag{4.159}$$

and $\Phi_i = \phi_i(Y_{\sim i})$. Let a_i denote the density of Φ_i, assuming that the all-one codeword was transmitted. Then

$$\begin{aligned} g_i(\mathtt{h}_1, \dots, \mathtt{h}_n) &= \frac{\partial \mathrm{H}(\mathsf{a}_{\text{BMSC}(\mathtt{h}_i)} \circledast \mathsf{a}_i)}{\partial \mathtt{h}_i} \\ &= G(\mathsf{a}_{\text{BMSC}(\mathtt{h}_i)}, \mathsf{a}_i) = \int \mathsf{a}_i(y) l^{\mathsf{a}_{\text{BMSC}(\mathtt{h}_i)}}(y) \mathrm{d}y, \end{aligned}$$

where $l^{\mathsf{a}_{\text{BMSC}(\mathtt{h})}}(y)$ is the GEXIT kernel introduced in Definition 4.43 and given by

$$l^{\mathsf{a}_{\text{BMSC}(\mathtt{h})}}(y) = \int \frac{\mathrm{d}\mathsf{a}_{\text{BMSC}(\mathtt{h})}(z)}{\mathrm{d}\mathtt{h}} \log_2(1 + \mathrm{e}^{-z-y}) \mathrm{d}z. \tag{4.160}$$

Proof. We know from Theorem 4.29 that $L_i + \Phi_i$ constitutes a sufficient statistic for X_i given Y. It follows that $H(X_i \mid Y) = H(X_i \mid L_i + \Phi_i)$. Further, again by Theorem 4.29, the prior on X_i is the uniform one, the channel $p_{\Phi_i \mid X_i}(\phi_i \mid x_i)$ is symmetric, and the distribution of Φ_i for $X_i = 1$ is the same as the distribution of Φ_i under the all-one codeword assumption. Let this (common) density be denoted by a_i so that the density of $L_i + \Phi_i$ under the all-one codeword assumption is given by $\mathsf{a}_{\text{BMSC}(\mathtt{h}_i)} \circledast \mathsf{a}_i$. We conclude that $H(X_i \mid L_i + \Phi_i) = \mathrm{H}(\mathsf{a}_{\text{BMSC}(\mathtt{h}_i)} \circledast \mathsf{a}_i)$, from which the result follows if we differentiate with respect to the i-th channel parameter $\mathtt{h}_i$. □

Example 4.161 ($[n, 1, n]$ Repetition Code). Let $\{\mathsf{a}_{\text{BMSC}(\mathtt{h})}\}$ characterize a smooth family of BMS channels and assume that all bits are transmitted over the same channel. If we use Definition 4.152 we see that the GEXIT function for the $[n, 1, n]$ repetition code is given by $g(\mathtt{h}) = \frac{1}{n}\frac{\mathrm{d}}{\mathrm{d}\mathtt{h}}\mathrm{H}(\mathsf{a}_{\text{BMSC}(\mathtt{h})}^{\circledast n})$. As a first example, we get $g_{\text{BEC}}(\mathtt{h}) = \mathtt{h}^{n-1} = h_{\text{BEC}}(\mathtt{h})$. As a second example, g_{BSC} is given in parametric form by

$$\Big(h_2(\epsilon), \frac{\sum_{j=\pm 1} j \sum_{i=1}^{n} \binom{n}{i} \epsilon^i \bar{\epsilon}^{n-i} \log\big(1 + (\epsilon/\bar{\epsilon})^{n-2i-j}\big)}{n \log(\bar{\epsilon}/\epsilon)}\Big),$$

with $\bar{\epsilon} = 1 - \epsilon$. From this representation it is not immediately obvious that for the $[n, 1, n]$ repetition code the area under $g(\mathtt{h})$ equals $1/n$, but the GAT asserts that this is indeed true. ◇

Example 4.162 ($[n, n-1, 2]$ Parity-Check Code). Some calculations show that g_{BSC} is given in parametric form by

$$\Big(h_2(\epsilon), 1-(1-2\epsilon)^{n-1}\frac{\log\big(\frac{1+(1-2\epsilon)^n}{1-(1-2\epsilon)^n}\big)}{\log\big(\frac{1-\epsilon}{\epsilon}\big)}\Big). \qquad \diamond$$

No simple analytic expressions are known for the case of transmission over the BAWGNC. Figure 4.163 compares the EXIT to the GEXIT function for some repetition and some single parity-check codes.

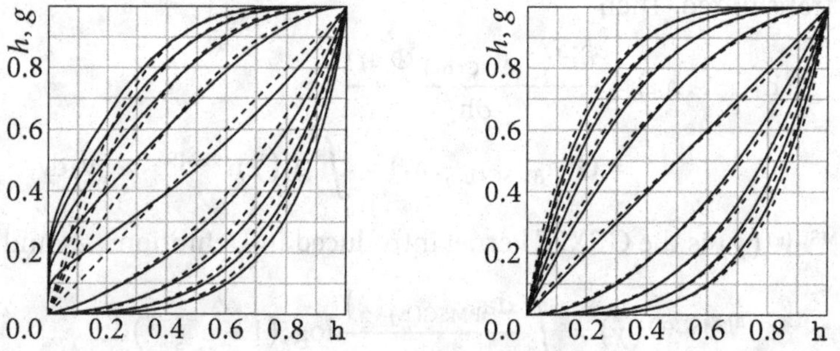

Figure 4.163: EXIT (solid) and GEXIT (dashed) function of the $[n, 1, n]$ repetition code and the $[n, n-1, 2]$ parity-check code assuming that transmission takes place over the BSC($\mathtt{h}$) (left) or the BAWGNC($\mathtt{h}$) (right), $n \in \{2, 3, 4, 5, 6\}$.

Our aim is to derive an easy-to-compute (hopefully tight) upper bound on the MAP threshold by mimicking the steps that we have applied in the case of transmission over the BEC. As a first ingredient we introduce the easy-to-compute BP GEXIT function.

Definition 4.164 (BP GEXIT Function for Linear Codes). Let X be chosen uniformly at random from a proper binary linear code and assume that the channel from X to Y is memoryless, where the i-th bit is transmitted over the smooth family $\{\text{BMSC}(\mathtt{h}_i)\}$. Let $\phi_i^{\text{BP},\ell}(y_{\sim i})$ denote the *extrinsic* BP estimator of X_i, where the number of iterations ℓ and the schedule have been fixed. Then

$$g_i^{\text{BP},\ell}(\epsilon) = \frac{\partial H(X_i \mid Y_i, \phi_i^{\text{BP},\ell}(Y_{\sim i}))}{\partial \mathtt{h}_i}\frac{\partial \mathtt{h}_i(\epsilon)}{\partial \epsilon}.$$

Discussion: If we compare this to Lemma 4.156 and in light of Lemma 4.158, we see that the only difference from the GEXIT function is that we have replaced the MAP estimate $\phi_i(Y_{\sim i})$ with the corresponding BP estimate $\phi_i^{\text{BP},\ell}(Y_{\sim i})$. With respect to the actual computation of $\phi_i^{\text{BP},\ell}$ the same remark as in the BEC case applies: to compute $\phi_i^{\text{BP},\ell}$ we input to the decoder the vector $Y_{\sim i}$, i.e., we erase the position Y_i. This ensures that $\phi_i^{\text{BP},\ell}$ is only a function of $y_{\sim i}$.

As a second ingredient we need to show that the BP GEXIT function is an upper bound on the GEXIT function. For the case of transmission over the BEC this was easily accomplished by using the data processing theorem. Surprisingly a very similar technique still works in this more general setting. We first show, more generally, that the GEXIT function preserves the order imposed by degradation. We state the result for BMS channels, since this is the context of interest, but the result remains valid as long as the channel is memoryless.

Lemma 4.165 (GEXIT Functions Preserve Degradation). Let X be chosen uniformly at random from a code C of length n. Let the channel from X to Y be memoryless, where Y_i is the result of passing X_i through the smooth family $\{\text{BMSC}(\mathtt{h}_i)\}$, $\mathtt{h}_i \in I_i$ (some interval contained in $[0,1]$), which is ordered by degradation. If $X_i \to Y_{\sim i} \to \Phi_i$ forms a Markov chain then

$$\frac{\partial H(X_i \mid Y)}{\partial \mathtt{h}_i} \leq \frac{\partial H(X_i \mid Y_i, \Phi_i)}{\partial \mathtt{h}_i}. \tag{4.166}$$

We relegated the proof to Appendix E since it is primarily an exercise in applying information-theoretic inequalities.

Recall that $\Phi_i^{\text{BP},\ell} = \phi_i^{\text{BP},\ell}(Y_{\sim i})$, i.e., it is a *function* of $Y_{\sim i}$, so that trivially $X_i \to Y_{\sim i} \to \Phi_i$. This gives us the following immediate consequence.

Corollary 4.167 (GEXIT versus BP GEXIT). Let X be chosen uniformly at random from a proper binary linear code C. Let the channel from X to Y be memoryless, where Y_i is the result of passing X_i through a smooth family $\{\text{BMSC}(\mathtt{h}_i)\}$, $\mathtt{h}_i \in [0,1]$, ordered by degradation. Assume that all individual channels are parameterized in a smooth (differentiable) way by a common parameter ϵ, i.e., $\mathtt{h}_i = \mathtt{h}_i(\epsilon)$, $i \in [n]$. Let $g_i(\epsilon)$ and $g_i^{\text{BP},\ell}(\epsilon)$ be as defined in Definitions 4.152 and 4.164. Then

$$g_i(\epsilon) \leq g_i^{\text{BP},\ell}(\epsilon).$$

So far all our statements concerned finite-length codes. Let us now consider the limit of large blocklengths.

Definition 4.168 (Asymptotic BP GEXIT Function). Consider a degree distribution pair (λ, ρ) and the corresponding sequence of ensembles LDPC (n, λ, ρ). Further, consider a smooth family $\{\text{BMSC}(\mathtt{h})\}$. Assume that all bits of X are sent through the channel BMSC$(\mathtt{h})$. For $\mathtt{G} \in \text{LDPC}(n, \lambda, \rho)$ and $i \in [n]$, let $g_i(\mathtt{G}, \epsilon)$ and $g_i^{\text{BP},\ell}(\mathtt{G}, \epsilon)$ denote the i-th MAP and BP GEXIT function associated to code $\mathtt{G}$. With some abuse of notation, define the asymptotic (and average) quantities

$$g(\mathtt{h}) = \limsup_{n\to\infty} \mathbb{E}_{\mathtt{G}}\Big[\frac{1}{n}\sum_{i\in[n]} g_i(\mathtt{G},\mathtt{h})\Big],$$

$$g^{\mathrm{BP},\ell}(\mathtt{h}) = \lim_{n\to\infty} \mathbb{E}_{\mathtt{G}}\Big[\frac{1}{n}\sum_{i\in[n]} g_i^{\mathrm{BP},\ell}(\mathtt{G},\mathtt{h})\Big],$$

$$g^{\mathrm{BP}}(\mathtt{h}) = \lim_{\ell\to\infty} g^{\mathrm{BP},\ell}(\mathtt{h}).$$

Discussion: For notational simplicity we suppress the dependence of the preceding quantities on the degree distribution pair and the channel family. We used the lim sup to define the asymptotic GEXIT function since it is difficult to assert the existence of the ordinary limit. On the contrary, the limit of the BP GEXIT functions is not only easy to assert but also easy to compute.

LEMMA 4.169 (ASYMPTOTIC BP GEXIT FUNCTION VIA GEXIT FUNCTIONAL). Consider a degree distribution pair (λ,ρ) and the corresponding sequence of ensembles LDPC (n,λ,ρ). Assume that for a fixed ℓ the expected fraction of computation graphs of height ℓ that are not proper is $o_n(1)$. Assume further that transmission takes place over the smooth family of BMS channels characterized by their L-densities $\{\mathsf{a}_{\mathrm{BMSC}(\mathtt{h})}\}$. Let $\mathsf{a}_0 = \mathsf{a}_{\mathrm{BMSC}(\mathtt{h})}$ and for $\ell \geq 1$, $\mathsf{a}_\ell = \mathsf{a}_{\mathrm{BMSC}(\mathtt{h})} \circledast \lambda(\rho(\mathsf{a}_{\ell-1}))$. Finally, let a_∞ denote the fixed point density to which density evolution converges. Then

$$g^{\mathrm{BP},\ell}(\mathtt{h}) = G(\mathsf{a}_{\mathrm{BMSC}(\mathtt{h})}, L(\rho(\mathsf{a}_{\ell-1}))),$$
$$g^{\mathrm{BP}}(\mathtt{h}) = G(\mathsf{a}_{\mathrm{BMSC}(\mathtt{h})}, L(\rho(\mathsf{a}_{\infty}))).$$

Proof. By assumption, if we fix ℓ and let n tend to infinity then most computation graphs are *proper*. This means that they are trees and that the set of codewords that are compatible with the local constraint is equal to the set of projections of the global codewords onto the computation tree. For such computation trees the BP GEXIT function is equal to the regular GEXIT function (on the tree) and is equal to $G(\mathsf{a}_{\mathrm{BMSC}(\mathtt{h})}, L(\rho(\mathsf{a}_{\ell-1})))$. The claim now follows since GEXIT functions are bounded so that the vanishing fraction of non-proper computation graphs only has a vanishing influence on the result. □

Discussion: From Lemma 3.47 we know that for regular ensembles the condition on the fraction of computation graphs which are non-proper is fulfilled.

In Figure 4.170 we plot the BP GEXIT function $g^{\mathrm{BP}}(\mathtt{h})$ for a few regular LDPC ensembles.

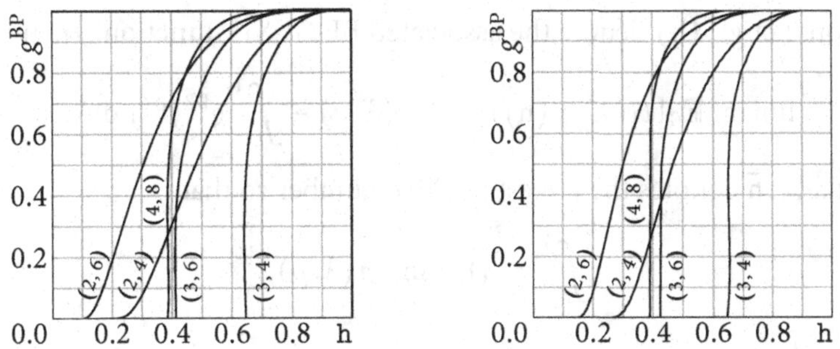

Figure 4.170: BP GEXIT curve for several regular LDPC ensembles for the BSC (left) and the BAWGNC (right).

DEFINITION 4.171 (MAP THRESHOLD). Consider a degree distribution pair (λ, ρ) and the smooth family $\{\mathrm{BMSC}(\mathtt{h})\}$, which is ordered by degradation. The MAP *threshold* $\mathtt{h}^{\mathrm{MAP}}$ is *defined* as

$$\mathtt{h}^{\mathrm{MAP}} = \inf\{\mathtt{h} \in [0,1] : \liminf_{n\to\infty} \mathbb{E}_{\mathrm{G}}[H(X \mid Y(\mathtt{h}))]/n > 0\}.$$

Discussion: Let us consider the operational meaning of the preceding definition. Let $\mathtt{h} < \mathtt{h}^{\mathrm{MAP}}$. Then by definition of the threshold, there exists a sequence of blocklengths $n_1, n_2, n_3, \dots$, so that the normalized (divided by the blocklength n) *average* conditional entropy converges to zero. Although we have not stated the corresponding (concentration) theorem, it is possible to show that this implies that *most* of the codes in the corresponding ensembles have a normalized conditional entropy less than any fixed constant. For sufficiently large blocklengths, a conditional entropy that grows sublinearly implies that the receiver can limit the set of hypothesis to a subexponential list, which with high probability contains the correct codeword. Therefore, in this sense reliable communication is possible.

On the other hand, assume that $\mathtt{h} > \mathtt{h}^{\mathrm{MAP}}$. In this case the normalized conditional entropy stays bounded away from zero by a strictly positive constant for all sufficiently large blocklengths. Again, by the omitted concentration theorem, this is not only true for the average over the ensemble but for most elements from the ensemble. It follows that with most elements from the ensemble reliable communication is not possible.

THEOREM 4.172 (UPPER BOUND ON MAP THRESHOLD). Consider a degree distribution pair (λ, ρ) whose asymptotic rate converges to the design rate $r(\lambda, \rho)$ (see Lemma 3.22) and which fulfills the conditions of Lemma 4.169. Assume further that transmission takes place over a smooth family $\{\mathrm{BMSC}(\mathtt{h})\}$, which is ordered by degradation. Let $g^{\text{BP}}(\mathtt{h})$ denote the associated BP GEXIT function. Then

$$(4.173)\qquad \liminf_{n\to\infty} \mathbb{E}_{\mathtt{G}}[H(X \mid Y(\mathtt{h}))]/n \geq r(\lambda,\rho) - \int_{\mathtt{h}}^{1} g^{\text{BP}}(\mathtt{h}')\, d\mathtt{h}'.$$

In particular, if $\bar{\mathtt{h}}$ denotes the largest positive number so that

$$\int_{\bar{\mathtt{h}}}^{1} g^{\text{BP}}(\mathtt{h})\, d\mathtt{h} = r(\lambda,\rho),$$

then $\mathtt{h}^{\text{MAP}} \leq \bar{\mathtt{h}}$.

Proof. Let $\mathtt{G}$ be chosen uniformly at random from the ensemble $\text{LDPC}(n,\lambda,\rho)$. Since by assumption the rate of the ensemble converges to the design rate we have

$$\begin{aligned}
& r(\lambda,\rho) - \liminf_{n\to\infty} \mathbb{E}_{\mathtt{G}}[H(X \mid Y(\mathtt{h}))]/n \\
& \qquad = \limsup_{n\to\infty} \frac{1}{n}\, \mathbb{E}_{\mathtt{G}}[H(X \mid Y(1)) - H(X \mid Y(\mathtt{h}))] \\
\text{Theorem 4.155} & \qquad = \limsup_{n\to\infty} \mathbb{E}_{\mathtt{G}}\Big[\int_{\mathtt{h}}^{1} g(\mathtt{G},\mathtt{h}')\, d\mathtt{h}'\Big] \\
\text{Fubini} & \qquad = \limsup_{n\to\infty} \int_{\mathtt{h}}^{1} \mathbb{E}_{\mathtt{G}}\big[g(\mathtt{G},\mathtt{h}')\, d\mathtt{h}'\big] \\
\text{Fatou-Lebesgue} & \qquad \leq \int_{\mathtt{h}}^{1} \limsup_{n\to\infty} \mathbb{E}_{\mathtt{G}}\big[g(\mathtt{G},\mathtt{h}')\, d\mathtt{h}'\big] \\
\text{Corollary 4.167} & \qquad \leq \int_{\mathtt{h}}^{1} \limsup_{n\to\infty} \mathbb{E}_{\mathtt{G}}\big[g^{\text{BP},\ell}(\mathtt{G},\mathtt{h}')\, d\mathtt{h}'\big] \\
\text{Definition 4.168} & \qquad = \int_{\mathtt{h}}^{1} g^{\text{BP},\ell}(\mathtt{h}')\, d\mathtt{h}'.
\end{aligned}$$

Since this is true for any $\ell \in \mathbb{Z}$, we get

$$\liminf_{n\to\infty} \mathbb{E}_{\mathtt{G}}[H(X \mid Y(\mathtt{h}))]/n \geq r(\lambda,\rho) - \int_{\mathtt{h}}^{1} g^{\text{BP}}(\mathtt{h}')\, d\mathtt{h}'.$$

Now note that the right-hand side of (4.173) is increasing in $\mathtt{h}$. Therefore,

$$\limsup_{n\to\infty} \mathbb{E}_{\mathtt{G}}[H(X \mid Y(\mathtt{h}))]/n$$

is bounded away from 0 for any $\mathtt{h} > \bar{\mathtt{h}}$ and the thesis follows from the definition of $\mathtt{h}^{\text{MAP}}$ given in Definition 4.171. □

Figure 4.174 shows the simple geometric interpretation for the construction of the upper bound for the $(3,6)$-regular ensemble: integrate the BP GEXIT curve from right to left until the area under the curve equals the design rate. This point is an upper bound on the MAP threshold.

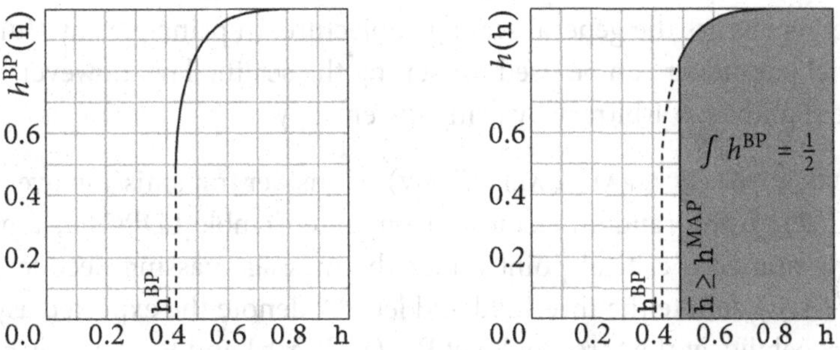

Figure 4.174: Left: BP GEXIT function $g^{\text{BP}}(\mathtt{h})$ for the $(3,6)$-regular ensemble; Right: Corresponding upper bound on GEXIT function $g(\mathtt{h})$ constructed according to Theorem 4.172.

EXAMPLE 4.175. The following table presents the upper bounds on the MAP threshold for transmission over the BSC$(\mathtt{h})$ as derived from Theorem 4.172 for a few regular ensembles.

$\mathtt{l}$	$\mathtt{r}$	$\mathtt{h}^{\text{BP}}$	$\overline{\mathtt{h}}$	$\tilde{\mathtt{h}}$	$\mathtt{h}^{\text{Sha}}$
3	4	0.6507(5)	0.7417(1)	0.743231	3/4
3	5	0.5113(5)	0.5800(3)	0.583578	3/5
3	6	0.4160(5)	0.4721(5)	0.476728	1/2
4	6	0.5203(5)	0.6636(2)	0.663679	2/3

Also shown is the result of an information-theoretic upper bound which is based on Gallager's inequality discussed in Theorem 4.150 (see Problem 4.51). This upper bound is denoted by $\tilde{\mathtt{h}}$. For the specific case of $(\mathtt{l},\mathtt{r})$-regular codes and transmission over the BSC the bound is given by $\tilde{\mathtt{h}} = h_2(\tilde{\epsilon})$, where $\tilde{\epsilon}$ is the unique positive root of the equation $\mathtt{r}h_2(\epsilon) = \mathtt{l}h_2((1-(1-2\epsilon)^{\mathtt{r}})/2)$. We see that for the considered cases $\tilde{\mathtt{h}}$ and $\overline{\mathtt{h}}$ are close. ◇

For the case of transmission over the BEC this simple upper bound was only the starting point. We could prove via a counting argument that in many instances this bound is tight. Finally, we could give an alternative graphical construction of the MAP threshold via the so-called Maxwell construction and then show that the Maxwell construction had an operational interpretation. The general case is technically more challenging and, although we conjecture that the same picture remains valid, currently no proofs for the general case are known.

§4.13. FINITE-LENGTH SCALING

Empirically, the contribution to the error probability due to "large" error events follows a scaling law very similar to what we discussed for the BEC, but no proofs are known. Let us discuss the general scaling conjecture. In principle any (function of the) channel parameter can be used for stating the scaling law; however, we make this choice slightly less arbitrary by using the entropy.

CONJECTURE 4.176 (GENERAL SCALING LAW). Consider transmission over the family $\{\text{BMSC}(\mathtt{h})\}$ using random elements from an ensemble LDPC (n, λ, ρ), which has a single non-zero critical point under the message-passing decoder MP. Let $\mathtt{h}^{\text{MP}} = \mathtt{h}^{\text{MP}}(\lambda, \rho)$ denote the threshold and let ν^{MP} denote the expected asymptotic bit error probability at the threshold. Let $\text{P}_{\text{b},\gamma}(n, \lambda, \rho, \mathtt{h})$ and $\text{P}_{\text{B},\gamma}(n, \lambda, \rho, \mathtt{h})$ denote the expected bit/block error probability *due to errors of size at least* $\gamma\nu^{MP}$, where $\gamma \in (0,1)$. Fix z to be $z = \sqrt{n}(\mathtt{h}^{\text{MP}} - \beta n^{-\frac{2}{3}} - \mathtt{h})$. Then as n tends to infinity,

$$\text{P}_{\text{B},\gamma}(n, \lambda, \rho, \mathtt{h}) = Q\left(z/\alpha\right)\left(1 + O\left(n^{-1/3}\right)\right),$$
$$\text{P}_{\text{b},\gamma}(n, \lambda, \rho, \mathtt{h}) = \nu^{\text{MP}} Q\left(z/\alpha\right)\left(1 + O\left(n^{-1/3}\right)\right),$$

where $\alpha = \alpha(\lambda, \rho, \text{BMSC}, \text{MP})$ and $\beta = \beta(\lambda, \rho, \text{BMSC}, \text{MP})$ are constants which depend on the ensemble, the channel BMS, as well as the decoder MP.

We have already seen this scaling law in Figures 4.83, 4.86, and 4.89 applied to three different message-passing decoders (Gallager A, decoder with erasures, and BP decoder). In all three cases we saw a good match of the predicted performance to the actual performance as measured by simulations. Let us give one further example. Figure 4.178 shows $\mathbb{E}_{\text{LDPC}(n,x^2,x^5)}[\text{P}_{\text{B}}(\mathtt{G}, \mathtt{h})]$, the average block error probability for the $(3,6)$-regular ensemble, assuming that transmission takes place over the BAWGNC$(\mathtt{h})$ and that we are using a (quantized) BP decoder. The elements of the ensemble were expurgated to ensure that indeed only large error events were counted. The two parameters α and β were fitted to the numerical data. We see again a good agreement of the empirical curves with the scaling laws. From an engineering point of view scaling laws can be used to quickly gauge what impact an increase of the blocklength would have on an existing system without having to simulate it again. More importantly, if it is possible to quickly compute the scaling parameters, scaling laws can be used to perform an efficient finite-length optimization.

§4.14. ERROR FLOOR UNDER MAP DECODING

The error floor under *MAP* decoding is relatively straightforward to determine using the results of Section 3.23. Let us phrase the result for the specific case of transmission over the BAWGNC$(\mathtt{h})$. Transmission over other BMS channels can be handled in a similar manner.

LEMMA 4.177 (ASYMPTOTIC ERROR FLOOR FOR LDPC (n, λ, ρ)). Consider the ensemble LDPC (n, λ, ρ) of rate r. Let s denote the expurgation parameter introduced

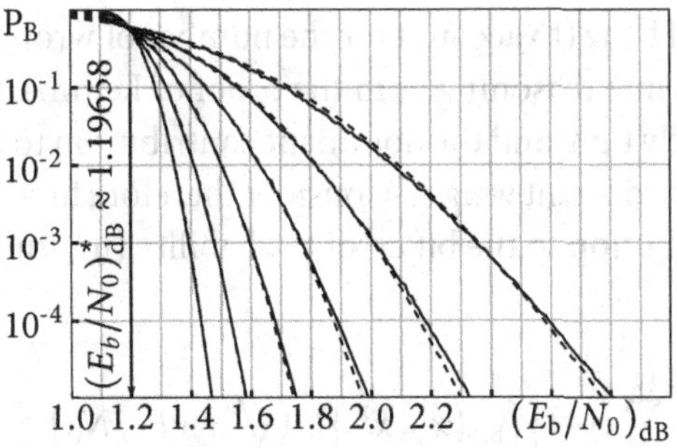

Figure 4.178: Scaling of $\mathbb{E}_{\text{LDPC}(n,x^2,x^5)}[P_B(G, h)]$ for transmission over the BAWGNC(h) and a quantized version of belief propagation decoding implemented in hardware. The threshold for this combination is $(E_b/N_0)^*_{dB} \approx 1.19658$. The block-lengths n are $n = 1000, 2000, 4000, 8000, 16,000$, and $32,000$, respectively. The solid curves represent the simulated ensemble averages. The dashed curves are computed according to the scaling law of Conjecture 4.176 with scaling parameters $\alpha = 0.8694$ and $\beta = 5.884$. These parameters were fitted to the empirical data.

in Definition 3.143. Define $\mu = \lambda'(0)\rho'(1)$ and

$$\hat{P}_s(x) = \sum_{w \geq s} \frac{(\mu x)^w}{2w}, \qquad \hat{P}_{b,s}(x) = \frac{1}{2} \sum_{w \geq s} (\mu x)^w.$$

Assume that transmission takes place over the channel family BAWGNC(E_b/N_0). Then for $E_b/N_0 > (E_b/N_0)^{BP}$

$$\lim_{n \to \infty} \mathbb{E}_{\text{ELDPC}(n,s,\lambda,\rho)}[P_B^{\text{MAP}}(G, E_b/N_0)] = 1 - e^{-\frac{1}{\pi}\int_0^{\frac{\pi}{2}} \hat{P}_s\left(e^{-\frac{rE_b/N_0}{\sin^2(\theta)}}\right) d\theta}, \tag{4.179}$$

$$\lim_{n \to \infty} n \, \mathbb{E}_{\text{ELDPC}(n,s,\lambda,\rho)}[P_b^{\text{MAP}}(G, E_b/N_0)] = \frac{1}{\pi} \int_0^{\frac{\pi}{2}} \hat{P}_{b,s}\left(e^{-\frac{rE_b/N_0}{\sin^2(\theta)}}\right) d\theta. \tag{4.180}$$

Discussion: As in Lemma 3.166 we state the error floor only for channel parameters below the BP threshold but we conjecture it to be true for all channel parameters up to the MAP threshold.

Proof. The proof follows the same path as the proof of Lemma 3.166, which gives the corresponding expression for transmission over the BEC(ϵ). We are therefore brief and concentrate on the new elements of the proof.

We start with the bit error probability. From Lemma 3.164) we know that $\hat{P}_{b,s}(x)$ is the generating function counting the expected number of minimal codewords of size at least s in the limit as $n \to \infty$, where the term corresponding to codewords of weight w is multiplied by w (to account for the number of wrong bits caused by such a codeword). By the same reasoning as in the proof of Lemma 3.166 a simple union bound is asymptotically tight and the dominant contribution to the error probability is due to codewords of constant weight. Consider therefore first codewords of weight up to W. Their contribution to the bit error probability (multiplied by n) converges to

$$\sum_{w=s}^{W} \text{coef}\{\hat{P}_{b,s}(x), x^w\} Q\big(\sqrt{2rwE_b/N_0}\big). \tag{4.181}$$

The claim follows if in (4.181) we can take the limit of W tending to infinity. That this limiting expression is equal to (4.180) follows by an application of *Craig's* formula discussed in Problem 4.63.

Let us now consider the block error probability. First restrict the attention to the contribution of the block error probability due to codewords that are entirely composed of minimal codewords of individual weight at most W, where W is a fixed integer. The number of such codewords and their composition is given by a Poisson distribution. We make a mistake if we make a mistake in at least one of the components. The corresponding expression is

$$1 - \sum_{a_s,\dots,a_W} \prod_{w=s}^{W} \frac{\mu_w^{a_w} e^{-\mu_w}}{a_w!} \big(1 - Q\big(\sqrt{2rwE_b/N_0}\big)\big)^{a_w}$$
$$= 1 - e^{-\sum_{w=s}^{W} \mu_w Q\left(\sqrt{2rwE_b/N_0}\right)} = 1 - e^{-\sum_{w=s}^{W} \frac{\mu^w}{2w} Q\left(\sqrt{2rwE_b/N_0}\right)}.$$

The claim now follows by letting W tend to infinity and by applying Craig's formula. □

Discussion: Similar expressions can also be derived for transmission over other channels by replacing the $Q(\cdot)$ function with the appropriate function that expresses the pairwise error probability as a function of the Hamming distance.

In general, the error floor under message-passing decoding is *higher* than the one under MAP decoding. Similar to the case of transmission over the BEC, there are so-called *pseudo codewords* which are not necessarily codewords and which cause the decoder to make erroneous decisions because of the suboptimality of the decoder. For some decoders (e.g., the linear programming decoder, which we do not discuss in these notes, and to some extent also the min-sum message-passing decoder of Section 2.5.4) the set of pseudo codewords is well understood. Much less is known, however, about the nature of pseudo codewords under general message-passing decoding.

Notes

Gallager introduced iterative decoding in his seminal thesis [26]. With two notable exceptions, iterative decoding was all but forgotten for a period of roughly 30 years. The first exception concerns the papers by Pinsker and Zyablov [101, 102] which contain an analysis of iterative decoding over the erasure channel and an analysis of the so-called *flipping algorithm* (see Chapter 8), respectively. The second important exception is Tanner [82], who generalized Gallager's construction by introducing more complex subcodes (not only single parity-check nodes) and by systematically exploiting the *graphical representation* (Tanner graphs) of codes in terms of bipartite graphs.

Iterative coding experienced a spectacular renaissance with the introduction of *turbo codes* by Berrou, Glavieux, and Thitimajshima [11]. The turbo code construction and the associated decoding algorithm were not inspired by the work of Gallager (which at that point was basically forgotten) but by the idea of a mechanical machine (turbo engine) in which partial output of the decoder is used to "boost" the decoding performance further. In the paper, which was presented at the ICC'93 in Geneva, the authors showed a practical (in terms of complexity) coding scheme which was capable of approaching capacity within less than 1 dB at a bit error probability of 10^{-6}, a sensational jump forward in the practice of coding. Initial disbelief quickly changed to astonishment. The significance of this contribution cannot be overstated.

Much credit should also be given to Lodge, Young, Hoeher, and Hagenauer [45] who introduced a very similar iterative decoding technique. As fate would have it, the two papers appeared at the same conference. Lodge and his co-authors considered a less powerful class of codes and concentrated on shorter blocklengths. Therefore the final coding gains were slightly less impressive.

The story does not stop here. The concept of iterative decoding and the principle of sparse graph codes must have been in the air around the early 1990s. Several other groups, working in entirely different communities and unaware of each other's work, converged to very similar concepts. Let us trace their stories here.

In 1989 Sourlas published a paper in which he observed that codes can be phrased in the language of spin systems [78]. If the graphical model describing this spin system is taken to be sparse, then one can use the methods of statistical physics to analyze the resulting codes. Sourlas considered what now would be called low-density generator-matrix (LDGM) codes. He observed that such codes show a large error floor. He therefore concluded that such sparse codes are not very useful and did not pursue them further. Although Sourlas was not performing iterative decoding, a standard method of analysis for such dilute (sparse) spin systems is to consider the so-called *Bethe* free energy. This is an approximation of the *free energy* introduced by Bethe in the 1930s. It is much easier to compute than the free energy itself. To compute the Bethe free energy one has to find the stationary points of a system of equations. Here is the connection to iterative decoding: Yedidia, Weiss, and Freeman showed that these stationary points are fixed points of density evolution for the BP algorithm [99, 100].

At around the same time turbo codes were introduced, and completely independent from the previous work, MacKay started his exploration of iterative decoding techniques inspired by papers by Meier and Staffelbach [62] as well as Mihaljević and Golić [65] and initiated by questions from Anderson. It is interesting to note that all of these papers deal with questions of cryptography and not communications. His initial explorations led him to construct LDGM codes and to use an iterative decoder inspired by a variational approach [50]. MacKay and Neal then introduced MN codes (see Chapter 7) and rediscovered Gallager's class of LPDC codes. Also, they implemented a BP decoder and found that it had superior performance [51, 52, 53, 54, 55]. The resulting codes showed very good performance, even outperforming turbo codes in some cases. A collaboration with McEliece and Cheng then led to [59], which describes the turbo decoding algorithm as an instance of BP.

Starting in the fall of 1993, Sipser and Spielman set out to find constructions for probabilistically checkable proofs (a complexity class even more powerful than NP). One construction they considered was based on expanders. Although this construction turned out to be not useful in the original context, Spielman realized that it gave rise to error-correcting codes. While looking for prior art, Spielman found Gallager's book "Low-Density Parity-Check Codes" in the MIT library. During the summer of 1994, Spielman met Luby at Bell Labs and first discussions concerning error correction for the erasure channel started. This gives the connection to a further independent thread that started with the work of Alon and Luby who were interested in the transmission of video over the Internet. Their original construction was based on Reed-Solomon codes [1]. Since for video transmission the blocklength can be very large, the complexity of Reed-Solomon decoding can become a bottleneck. The question was therefore if it was possible to construct codes that allow transmission over the Internet (the erasure channel) at rates close to capacity but that have a much lower complexity. At Berkeley, Luby, Mitzenmacher, Shokrollahi, Spielman, and Stemann started to collaborate on the problem in the period of 1995 – 1996. As was mentioned in the notes of Chapter 3 starting on page 156, while studying iterative decoding for transmission over the BEC, this group introduced [46, 47, 48, 49] a variety of new tools and concepts. Many of the results for general channels are directly inspired by these papers.

In the period from 1994 – 1996 two further developments took place that had a strong subsequent impact.

Wiberg, working with Loeliger and Kötter at Linköping University, showed in his thesis that the turbo decoding algorithm and the standard iterative algorithm used by Gallager can both be seen as an instance of the "sum-product" algorithm applied to a graphical model [94, 95]. This was an important conceptual step forward. He also systematically developed the use of graphical models, echoing Tanner's work from his 1981 paper [82], which he only discovered late into his thesis work.

The second development concerns weight distributions. It is difficult to compute the weight distribution of an individual turbo code. But Benedetto and Montorsi realized that it is easy to compute the *average* weight distribution of an ensemble, where the ensemble is defined by taking a uniform probability distribution over the set of all permutations. This led to the notion of a "uniform interleaver" [8, 9, 10]. Most results to date concern ensemble averages.

The preceding description traces the general developments until early 1997. Most of the groups mentioned worked essentially in isolation and these and other threads continued to stay independent for some time. Let us now come back to the specific material contained in this chapter.

The various representations of BMS channels and their associated distributions (Sections 4.1.1-4.1.4) were introduced by Richardson, and Urbanke [72]. In the same paper the authors introduced the class of message-passing decoders which we have considered (Section 4.2); they discussed the simplifications that arise when restricting one's attention to this class of message-passing decoders (Section 4.3); they extended the concentration theorem to general channels (Section 4.3.2); and they introduced the notion of density evolution for general channels (Section 4.5), the notion of degradation and monotonicity of channel families (Sections 4.1.14 and 4.6), and the notion of a threshold (Section 4.7).

The symmetry of distributions for BMS channels (Definition 4.11), (the Channel Equivalence) Lemma 4.28, the stability condition (Section 4.9), a partial fixed point characterization of the threshold (Section 4.8), and some optimization techniques are due to Richardson, Shokrollahi, and Urbanke [71]. The full proof of the stability condition and a complete description of the fixed point characterization can be found in the paper by Richardson and Urbanke [73]. This paper also contains the bounds on the Bhattacharyya constants stated in Lemmas 4.64 and 4.66. The short proof of the sufficiency of the stability condition that we present on page 234 is due to Khandekar and McEliece [42]. One can think of this proof as an instance of the "extremes of information combining" approach applied to the Bhattacharyya constant.

From a historical perspective it is interesting to note that Elias proposed in the mid-1950s the following coding scheme. In modern language, he proposed a product code with simple single-error-correcting components in each row. The proposed decoder was a *one-pass* decoding procedure (i.e., the decoder would process each component in a predetermined order as in the iterative decoding procedure but only in a single pass). He showed that for a proper choice of block sizes such a scheme can provide arbitrarily high reliability at a positive rate.

The density evolution equations (Theorem 4.95) as well as some thresholds for the Gallager algorithm A are already contained in Gallager's thesis [26]. The material relating to monotonicity (Lemma 4.102), stability (Theorem 4.123), the fixed point characterization (Section 4.8), and the optimization of this algorithm are due to Bazzi, Richardson, and Urbanke [7].

The idea that the performance of a sparse graph code can be upper bounded in terms of its degrees (Section 4.11) is due to Gallager. In his thesis Gallager states such a bound for the BSC [25]. This idea was first extended by Burshtein, Krivelevich, Litsyn, and Miller to general channels [13] by quantizing a general BMS channel to a BSC with the same error probability. Theorem 4.150 in Section 4.11 is a strengthened version and it is due to Sason and Wiechmann [96]. We decided to denote this more general result still as Gallager's inequality. The remaining material in Section 4.11 and Problem 4.53 relating to the trade-off between complexity and gap to capacity is originally due to Sason and Urbanke [75] and the strengthened form we present is due to Sason and Wiechman [76, 96, 97]. In the case of transmission over the BEC we saw that the implied bounds are tight by constructing an iterative decodable ensemble (the check-concentrated ensemble) which achieved the upper bound. For general BMS channels no capacity-achieving ensembles are known, but it is tempting to conjecture that such ensembles exist and that the best achievable trade-off follows again the information-theoretic upper bound.

EXIT charts (Section 4.10) were pioneered by ten Brink who initially introduced them in the context of choosing a suitable mapper and a suitable constellation in an iterative demapping and decoding scheme [83, 84]. Soon thereafter he applied the same technique to the analysis of turbo codes [85, 86]. A Gaussian approximation as a one-dimensional approximation to density evolution for LDPC codes was suggested by Chung, Richardson, and Urbanke [19, 20]. At the same conference El Gamal and Hammons introduced an equivalent concept for turbo codes [27]. These Gaussian approximations differed from the EXIT chart method in the choice of the one-dimensional parameter which is used to characterize the density. Experiments have shown that the entropy criterion used by EXIT charts is indeed the most faithful parameter and it has long been the preferred choice.

The optimization technique presented in Section 4.10.1 is very similar to the one introduced by Chung in his thesis [17] (see also [20] and the paper by Chung, Forney, Richardson, and Urbanke [18]). The main difference is that Chung matched the mean of the Gaussian instead of its entropy and that the presented method allows for general interpolating channel families to improve the approximation. This form of the optimization was introduced in the thesis [2] of Amraoui for the optimization of point-to-point channels, and it was also employed by Amraoui, Dusad, and Urbanke in [3] to optimize degree distributions for multiple-access channels. A very readable description of optimization techniques applied to RA codes was given by Roumy, Guemghar, Caire, and Verdú [74].

Chung conjectured in his thesis various forms of extremality of the BEC and the BSC in the setting of iterative decoding [17]. The notion of "extremals for information combining" was coined by Huettinger and Hueber [35, 36]. The proof of Theorem 4.141 follows the one given by Land, Hoeher, Huettinger, and Huber [44], as well as Sutskover, Shamai, and Ziv [79, 80]. The main step in the proof, namely that $f(u,v)$ is convex-$\cup$ in u, is identical to the the main step in the proof of the so-called "Mrs. Gerber's Lemma" due to Wyner and Ziv [98]. An alternative proof using Tchebycheff system theory was proposed by Jiang, Ashikhmin, Kötter, and Singer [37].

There are other ways of bounding the performance of iterative decoding systems. A previous such bound was derived by Burshtein and Miller [14], except that this bound was derived by propagating the mean of the distributions and not the corresponding entropy. And the inequalities of Khandekar and McEliece [42] for the Bhattacharyya constant can also be seen as early instances of this approach.

The material on GEXIT functions which we present in Section 4.12 is due to Méasson, Montanari, Richardson, and Urbanke [60]. In many ways GEXIT charts are the natural extension of EXIT charts to general channels. In particular they fulfill the so-called Area Theorem (Theorem 4.155). This conservation law allows one to give upper bounds on the MAP threshold of iterative decoding systems. In the BEC case we were able to show that the implied bounds are in fact tight. The same result is conjectured to be true in the general case. An entirely different (but in terms of the result equivalent) approach was introduced by Montanari [68]. (See also the extension by Macris, Korada, and Kudekar [57, 58].) This approach is based on Guerra's interpolation technique [28] (see also the paper by Franz, Leone, and Toninelli [24]). Yet another approach of proving bounds on the GEXIT function using correlation inequalities was pointed out by Macris [56]. The observation that the GEXIT functional preserves the partial ordering implied by degradation (Lemma 4.77) is due to Méasson, Montanari, and Urbanke [61]. The same bounds on the MAP threshold were first computed using the (non-rigorous) replica method from statistical physics [66]. In [68], they were shown to be upper bounds for $\mathtt{r}$ even, using an interpolation technique. The present proof applies also to the case of odd $\mathtt{r}$. It can be proved that the three characterizations of the threshold are indeed equivalent; i.e., they give *exactly* the same value. If one computes the GEXIT function for a $[2,1,2]$ repetition code in Gaussian noise one recovers a beautiful connection between the derivative of the conditional entropy and the minimum mean-square error detector which was found before the introduction of GEXIT functions by Guo, Shamai, and

Verdú [29, 30]. Based on this observation, an MSE-based EXIT chart analysis was proposed by Bhattad and Narayanan [12].

The finite-length scaling conjecture presented in Section 4.13 is due to Amraoui, Montanari, Richardson, and Urbanke [4]. It is the natural extension of the scaling law which we presented for the BEC in Section 3.23. Scaling laws for LDPC ensembles decoded with the Gallager algorithm A were considered by Ezri, Montanari, and Urbanke [21].

In our exposition we consider ensembles, not individual codes. This makes an analysis possible. We have seen that there is a strong concentration of individual instances with respect to the performance in the waterfall region. But the error floor does not concentrate and significant differences can arise among the individual members of an ensemble. It is therefore important to be able to construct or identify "good" elements of an ensemble. Let us give here a few references to the literature concerning this topic. The progressive edge growth method was proposed by Hu, Eleftheriou, and Arnold [33] to generate graphs which do not contain small cycles. The ACE algorithm, which has a similar goal, is due to Tian, Jones, Villasenor, and Wesel [87]. Halford and Chugg [31] present methods to efficiently count short cycles in a bipartite graph. A method to compute the minimum distance was presented by Hu, Fossorier, and Eleftheriou [34]. Methods to reduce the complexity of the decoding were discussed by Chen, Dholakia, Eleftheriou, Fossorier, and Hu [15]. The minimum distance or the size of the smallest cyle are unfortunately only proxies for the error floor performance under iterative decoding. In the case of transmission over the BEC we were able to give a tight characterization of the contribution to the erasure probability that stems from small weaknesses in the graph, under both iterative and MAP decoding. In particular, we saw that stopping sets played under iterative decoding the same role that codewords do under MAP decoding. For general channels a complete such characterization is currently still open. A very promising avenue is the theory of pseudo codewords. This was initially discussed by Wiberg in his thesis [94]. An exact characterization can be given in the setting of the linear-programming decoder. This was done by Feldman in his thesis [22]. Further results can be found in the work of Vontobel and Kötter [43, 92].

There is also a considerable body of literature on the connection of coding and statistical physics. As mentioned before, this connection is based on the observation by Sourlas that codes can be phrased in the language of spin systems [78]. For references regarding the weight distribution problem see the Notes in Chapter 3. References concerning *turbo codes* are given in the Notes in Chapter 6. Let us mention the papers by Kanter and Saad [41], Murayama, Kabashima, Saad, and Vicente [69], Kabashima, Murayama, and Saad [38], Montanari [66, 67, 68], Franz, Leone, Montanari, and Ricci-Tersenghi [23], Saad and Kabashima [88], Vicente and Saad [91], and Kabashima, Sazuka, Nakamura, and Saad [40]. The books by Nishimori [70],

Mézard, Parisi, and Virasoro [64], Talagrand [81], as well as Mézard and Montanari [63] and the review paper by Kabashima and Saad [39] are good starting points.

If you compare our exposition for BMS channels to the one for the BEC you will notice that the broad outline is very much the same but that on several occasions we had to be content with weaker statements (or no statement at all). Let us review some of the biggest open questions.

In Section 3.19 we saw that for the BEC no matter what order we increase the number of iterations and the blocklength, as long as both tend to infinity, we always get the same threshold. The same statement is conjectured to be true for BMS channels. The theory becomes easy if we first let n tend to infinity and then increase ℓ afterwards. This is how we proceeded. In practice, we proceed closer the the converse: for a given blocklength we typically iterate until no further progress is achieved and we are interested in the behavior of the performance as the blocklength increases. Empirically, we observe the same threshold that we found analytically also for this limit. It would be nice to be able to formulate a proof of this important fact.

Although we have shown several bounds on thresholds, these bounds are too loose to prove that there are capacity-achieving ensembles. By optimizing degree distributions numerically and computing their threshold we can seemingly get arbitrarily close to capacity, but neither does this constitute a proof that capacity-achieving degree distributions exist, nor does it give much insight into their structure.

The connection between the belief propagation (BP) decoder and the MAP decoder make for another interesting topic. For the BEC the set of fixed points of density evolution forms a smooth one-dimensional manifold, which we called the extended BP EBP curve, and this curve encodes both the performance of the BP decoder as well as the one of the MAP decoder via the Maxwell construction. Experimentally the same holds for the general case but no formal justification is known.

For practical purposes the finite-length performance of codes is of crucial importance. To this end, we would like to justify the scaling approach which we introduced in Section 4.13 and find ways of computing the scaling parameters in an efficient manner. Equally important, we need to find ways of determining the error floor of either ensembles or fixed codes.

Problems

4.1 (Symmetric Distributions – Montanari [68]). Let a be a symmetric L-density and let X have density a. Show that

$$\mathbb{E}[\tanh^{2i-1}(X/2)] = \mathbb{E}[\tanh^{2i}(X/2)] = \mathbb{E}[\frac{\tanh^{2i}(X/2)}{1+\tanh(X/2)}],\ i \geq 1,$$

$$\mathbb{E}[\ln(1+\tanh(X/2))] = \sum_{i=1}^{\infty}\Big(\frac{1}{2i-1}-\frac{1}{2i}\Big)\mathbb{E}[\tanh^{2i}(X/2)].$$

Hint: Recall that since a is symmetric, $\mathbb{E}[f(X)] = \mathbb{E}[f(X)+f(-X)e^{-X}]/2$ for any function $f(\cdot)$ for which the expectations exist. Choose $f_1(x) = \tanh^{2i-1}(x/2)$ and $f_2(x) = \tanh^{2i}(x/2)$, respectively, and compare the results.

4.2 (Symmetry in D- and G-Representation). Show that the symmetry condition in the D-representation reads as stated in (4.13) and that the symmetry in the G-representation is indeed the one stated in (4.18).

4.3 (Symmetry for Densities). The aim of this problem is to verify the equivalence of conditions stated in Definition 4.11.

Show that if a is an L-density with $a(x) = e^{x}a(-x)$ for $x \in \mathbb{R}$ then

$$\int f(x)\mathrm{d}A(x) = \int e^{-x} f(-x)\mathrm{d}A(x)$$

for all bounded continuous functions $f(x)$ so that $f(-x)e^{-x}$ is bounded.

Conversely, show that if for an L-density a we have

$$\int f(x)a(x)\mathrm{d}x = \int e^{-x} f(-x)a(x)\mathrm{d}x$$

for all bounded continuous functions $f(x)$ so that $f(-x)e^{-x}$ is bounded then

$$\int (a(x) - a(-x)e^{x})^2 \mathrm{d}x = 0.$$

If a is continuous then it is meaningful to talk about the value of $a(x)$ at the point x and the preceding assertion implies $a(x) = e^{x}a(-x)$ for $x \in \mathbb{R}$.

4.4 ($L(Y)$ for Cauchy and Laplace Channel). The binary Cauchy and the binary Laplace channels are defined by $Y_t = X_t + Z_t$, where $\{Z_t\}_t$ is a sequence of iid random variables with density $p_{\text{BCC}(\lambda)}(z) = \lambda/(\pi(\lambda^2+z^2))$ and $p_{\text{BLC}(\lambda)}(z) = e^{-|z|/\lambda}/(2\lambda)$, respectively. Determine the L-densities $a_{\text{BCC}(\lambda)}(y)$ and $a_{\text{BLC}(\lambda)}(y)$ and verify that they are symmetric.

4.5 (Channel Equivalence for BAWGN(σ)). Let a denote the L-density associated to the channel BAWGNC(σ), conditioned that $X = 1$. Show by explicit computation that the symmetric channel with transition probability $p(y \mid x = 1) = a(y)$ has L-density equal to a.

4.6 (Stronger Version of Symmetry under MAP Decoding). Show that Theorem 4.29 stays correct under the weaker assumption that transmission does not necessarily take place over a BMS channel but that only

$$p_{Y_{\sim i}\,|\,X}(y_{\sim i}\,|\,w) = p_{Y_{\sim i}\,|\,X}(w_{\sim i}y_{\sim i}\,|\,\underline{1}) \tag{4.182}$$

for all $w \in C$.

4.7 (Alternative Proof of Symmetry under BP Decoding). Consider the repetition code $C[n,1,n]$ and transmission over a BMS channel with L-density a. Express the L-density under MAP bit decoding in terms of a and show directly that it is symmetric. Repeat the same steps for the parity-check code $C[n, n-1, 2]$.

Next show that the mixture (convex combination) of symmetric densities is symmetric. Now argue that the preceding results give an (alternative) proof that all densities encountered in the iterative decoding of a tree channel are symmetric.

4.8 (Symmetry and Channel Capacity). Show that the optimal input distribution for a BMS channel is the uniform one.

4.9 (Error Probability Functional). Apply the error probability functional $\mathfrak{E}(\cdot)$ to the densities $\mathsf{a}_{\text{BBC}(\lambda)}$ and $\mathsf{a}_{\text{BLC}(\lambda)}$.

4.10 (Error Probability Functional in D-Domain). Let a be a symmetric L-density and $\mathfrak{a}$ be the corresponding D-density. Show that in the D-domain the error probability functional $\mathfrak{E}(\cdot)$ has the following alternative characterizations:

$$\mathfrak{E}(\mathsf{a}) = \int_0^1 \mathfrak{a}(z)\frac{1-z}{1+z}\text{d}z = \frac{1}{2} - \frac{1}{2}\int_0^1 |\mathfrak{a}|(z)z\text{d}z = \frac{1}{2} - \frac{1}{2}\int_{-1}^1 \mathfrak{a}(z)|z|\text{d}z.$$

4.11 (Series Expansion for Capacity of BAWGNC). Derive the series expansion for the capacity of the BAWGNC given in Example 4.38.

Hint: Represent the range $(-\infty,+\infty)$ as $(-\infty,0)\cup[0,+\infty)$ and consider both parts separately. For example, for $x \geq 0$, use

$$\ln(1+\text{e}^{-x}) = -\sum_{i\geq 1}\frac{(-1)^i\text{e}^{-ix}}{i}.$$

4.12 (Asymptotic Expansion of Capacity of BAWGNC). Let $m = \frac{2}{\sigma^2}$. Show that

$$\begin{aligned}
\text{C}_{\text{BAWGNC}}(m) &= \ln(2) - \frac{1}{2\sqrt{\pi m}}\int \text{e}^{-\frac{(z-m)^2}{4m}}\ln\left(1+\text{e}^{-z}\right)\text{d}z \;\text{ nats}\\
&= \ln(2) - \sqrt{\frac{\pi}{m}}\text{e}^{-\frac{m}{4}}\sum_{k=0}^{\infty}\frac{(-1)^k c_k}{m^k} \;\text{ nats}
\end{aligned}$$

where

$$c_k = \frac{1}{2^{2k+1}\pi(k!)} \int z^{2k} e^{\frac{z}{2}} \ln(1 + e^{-z})\, dz.$$

We have $c_0 = 1$, $c_1 = \frac{8+\pi^2}{4}$, $c_2 = \frac{384+48\pi^2+5\pi^4}{32}$, and so on. We conclude that

$$C_{\text{BAWGNC}}(m) = \ln(2) - \sqrt{\pi} e^{-\frac{m}{4}} \left(m^{-\frac{1}{2}} - \frac{8+\pi^2}{4} m^{-\frac{3}{2}} + O(m^{-\frac{5}{2}}) \right) \text{ nats,}$$

which is a good approximation for large values of m. On the other hand, for small m expand the function $\ln(1 + e^{-z})$ around zero and show that you get the approximation

$$C_{\text{BAWGNC}}(m) = \frac{m}{4} - \frac{m^2}{16} + \frac{m^3}{48} + O(m^4) \text{ nats.}$$

4.13 (Capacity of BCC(λ)). Determine the capacity of the BCC(λ), call it $C_{\text{BCC}(\lambda)}$ (see Problem 4.4).

4.14 (Capacity of BLC(λ)). Show that the capacity of the BLC(λ) (see Problem 4.4) is given by

$$C_{\text{BLC}(\lambda)} = \frac{-\pi + 4\arctan(e^{-1/\lambda})}{2e^{1/\lambda}\ln(2)} - \log_2\big((1 + e^{-2/\lambda})/2\big) \text{ bits per channel use.}$$

4.15 (Threshold Values for Rate One-Half Codes). Determine the critical value, call it ϵ^{Sha}, so that $C_{\text{BEC}}(\epsilon^{\text{Sha}}) = 1/2$. Repeat the equivalent calculation for the BSC(ϵ), the BAWGNC(σ), and the AWGNC(σ).

4.16 (GEXIT Kernel for BEC). Consider the family $\{\text{BEC}(\mathtt{h})\}$ with the associated family of L-densities $\{\mathsf{a}_{\text{BEC}(\mathtt{h})}\}$. Show that the GEXIT kernel in the $|D|$-domain is $|d|^{\mathsf{a}_{\text{BEC}(\mathtt{h})}}(z) = h_2((1+z)/2)$.

4.17 (GEXIT Kernel for BSC). Consider the family $\{\text{BSC}(\mathtt{h})\}$ with the associated family of L-densities $\{\mathsf{a}_{\text{BSC}(\mathtt{h})}\}$. Show that the GEXIT kernel in the $|D|$-domain is given by

$$|d|^{\mathsf{a}_{\text{BSC}(\mathtt{h})}}(z) = 1 + \frac{z}{\log(\bar{\epsilon}/\epsilon)} \log\left(\frac{1 + 2\epsilon z - z}{1 - 2\epsilon z + z}\right),$$

where $\epsilon = h_2^{-1}(\mathtt{h})$. Give the limits for $\mathtt{h} \to 1$ and $\mathtt{h} \to 0$.

4.18 (GEXIT KERNEL FOR BAWGNC). Consider the family $\{\text{BAWGNC}(\mathtt{h})\}$ with the associated family of L-densities $\mathtt{a}_{\text{BAWGNC}(\mathtt{h})}$. Define σ explicitly by $\mathtt{h} = \mathtt{h}(\sigma)$. Use Example (4.47) and (4.49) to write the kernel in the $|D|$-domain as

$$|d|^{\mathtt{c}_{\text{BAWGNC}(\mathtt{h})}}(z) = \sum_{i\in\{-1,+1\}} \frac{\int \frac{(1-z^2)e^{-\frac{(w-\sigma^2)^2}{4\sigma^2}}}{(1+iz)+(1-iz)e^w} dw}{\int \frac{2e^{-\frac{(w-\sigma^2)^2}{4\sigma^2}}}{1+e^w} dw}.$$

Give the limits for $\mathtt{h} \to 1$ and $\mathtt{h} \to 0$.

4.19 (ALTERNATIVE REPRESENTATION OF GEXIT KERNEL FOR BAWGNC). Consider the family $\text{BAWGNC}(\mathtt{h} = \mathtt{h}(\sigma))$ modeled as $Y = X + Z$, where X takes values in $\{-1, +1\}$ and Z is Gaussian with zero mean and variance σ^2. We will derive several different *equivalent* expressions for the GEXIT kernel in the L-domain.

(a) What is the L-density $\mathtt{a}(w)$ associated with $\text{BAWGNC}(\sigma)$? Using the parameter $m = \frac{2}{\sigma^2}$, express $\frac{\partial \mathtt{a}(w)}{\partial m}$ in terms of $\frac{\partial^i \mathtt{a}(w)}{\partial w^i}$, $i = 1, 2$. Start with (4.44) and use integration by parts twice to show that the GEXIT kernel is given by

$$l^{\mathtt{a}_{\text{BAWGNC}(\mathtt{h})}}(z) = \left(e^{-z} \int \frac{e^{-\frac{(w\sigma^2-2)^2}{8\sigma^2}}}{(\cosh(\frac{w-z}{2}))^2} dw\right) \Big/ \left(\int \frac{e^{-\frac{(w\sigma^2-2)^2}{8\sigma^2}}}{(\cosh(\frac{w}{2}))^2} dw\right).$$

In the remainder of this exercise, Φ denotes a further observation of X conditionally independent of Y: It is the result of passing X through a symmetric channel and is assumed to be in log-likelihood ratio form (if we use coding, Φ represents the extrinsic estimate of X).

(b) Show that $\tanh(\frac{w+z}{2}) = \mathbb{E}[X \,|\, \frac{2Y}{\sigma^2} = w, \Phi = z]$.

(c) Show that, if $f(y)$ is even, then $\mathbb{E}[f(Y)] = \mathbb{E}[f(Y) \,|\, X = +1]$.

(d) Use (b), (c), and Example 4.48 to show that the following expression represents an *equivalent* kernel:

$$l'^{\mathtt{c}_{\text{BAWGNC}(\mathtt{h})}}(z) = \frac{1 - \mathbb{E}[\mathbb{E}[X \,|\, Y, \Phi = z]^2]}{1 - \mathbb{E}[\mathbb{E}[X \,|\, Y]^2]}.$$

(e) Problem 4.1 shows that $\mathbb{E}[\tanh(Y/(\sigma^2)) \,|\, X = +1] = \mathbb{E}[\tanh^2(Y/(\sigma^2)) \,|\, X = +1]$. Can you find this result directly using the definition of the conditional expectation (which gives $\mathbb{E}[X\,\mathbb{E}[X \,|\, Y]] = \mathbb{E}[\mathbb{E}[X \,|\, Y]^2]$)?

(f) Starting with (d), use (e) and Example 4.48 to prove that

$$l''^{\text{CBAWGNC}(\mathtt{h})}(z) = \frac{1 - \mathbb{E}[\mathbb{E}[X \mid Y, \Phi = z] \mid X = 1]}{1 - \mathbb{E}[\mathbb{E}[X \mid Y] X = 1]}$$

is an equivalent kernel.

4.20 (Bhattacharyya Constant Directly from $p_{Y|X}(y|x)$). Starting with Definition 4.61, show that the Bhattacharyya constant associated to a BMS channel characterized by its transition probability $p_{Y|X}(y|x)$ can also be computed as

$$\int \sqrt{p_{Y|X}(y| + 1) p_{Y|X}(y| - 1)} \mathrm{d}y.$$

4.21 (Bhattacharyya Constant versus Error Probability). Show that the left inequality in (4.65) is tight for the BEC and that the right inequality is tight for the BSC.

4.22 (Uniform Bound on $\mathfrak{E}(\mathsf{a}^{\circledast n})$). Let a denote a symmetric L-density with strictly positive Bhattacharyya constant $\mathfrak{B}(\mathsf{a})$. Show that for any $0 < \mathfrak{B}' < \mathfrak{B}(\mathsf{a})$ we have the uniform (in n bound)

$$\alpha (\mathfrak{B}')^n \leq \mathfrak{E}(\mathsf{a}^{\circledast n}) \leq \mathfrak{B}(\mathsf{a})^n,$$

where α is a strictly positive constant. Show that a valid choice for α is

$$\alpha = \frac{(\mathfrak{B}(\mathsf{a})\mathrm{e})^{\frac{3}{2}} \sqrt{2 \ln \frac{\mathfrak{B}(\mathsf{a})}{\mathfrak{B}'}}}{9\pi}.$$

Hint: Start with Lemma 4.66.

4.23 (Alternative Derivation of Bhattacharyya Constant). Let a denote a symmetric L-density and consider the sum $\sum_{i=1}^n Y_i$, where the Y_i are iid samples distributed according to a. Recall from the discussion of Bernstein's inequality in Section C.2 that for any $s > 0$

$$\frac{1}{n} \log \mathfrak{E}(\mathsf{a}^{\circledast n}) \leq \frac{1}{n} \log \mathbb{P}\{\sum_{i=1}^n Y_i \leq 0\} \leq \frac{1}{n} \log \mathbb{E}[\mathrm{e}^{-s \sum_{i=1}^n Y_i}] = \log \int \mathsf{a}(y) \mathrm{e}^{-sy} \mathrm{d}y.$$

Justify the following steps:

$$\begin{aligned} \inf_{s>0} \int \mathsf{a}(x)\, \mathrm{e}^{-sx} \mathrm{d}x &= \inf_{s>0} \frac{1}{2} \int \mathsf{a}(x)\, \mathrm{e}^{-x/2} [\mathrm{e}^{-x(-s+1/2)} + \mathrm{e}^{x(-s+1/2)}] \mathrm{d}x \\ &= \inf_{s>0} \int \mathsf{a}(x)\, \mathrm{e}^{-x/2} \cosh((-s+1/2)x) \mathrm{d}x \end{aligned}$$

$$= \int \mathsf{a}(x) e^{-x/2} dx = \mathfrak{B}(\mathsf{a}).$$

This shows that $\mathfrak{E}(\mathsf{a}^{\circledast n}) \leq \mathfrak{B}(\mathsf{a})^n$. We have already seen in Lemma 4.66 and in Problem 4.22 that this bound is essentially tight.

4.24 (Alternative Derivation of $\mathfrak{B}(\mathsf{a}_{\text{BEC}(\epsilon)})$). Derive the Bhattacharyya constant $\mathfrak{B}(\mathsf{a}_{\text{BEC}(\epsilon)})$ by explicitly computing $\mathsf{a}^{\circledast n}_{\text{BEC}(\epsilon)}$ and then applying the error probability functional $\mathfrak{E}(\cdot)$.

4.25 (Alternative Derivation of $\mathfrak{B}(\mathsf{a}_{\text{BSC}(\epsilon)})$). Derive the Bhattacharyya constant $\mathfrak{B}(\mathsf{a}_{\text{BSC}(\epsilon)})$ by explicitly computing $\mathsf{a}^{\circledast n}_{\text{BSC}(\epsilon)}$ and then applying the error probability functional $\mathfrak{E}(\cdot)$.

4.26 (Alternative Derivation of $\mathfrak{B}(\mathsf{a}_{\text{BAWGNC}(\sigma)})$). Derive the Bhattacharyya constant $\mathfrak{B}(\mathsf{a}_{\text{BAWGNC}(\sigma)})$ by explicitly computing $\mathsf{a}^{\circledast n}_{\text{BAWGNC}(\sigma)}$ and then applying the error probability functional $\mathfrak{E}(\cdot)$.

Hint: It might be helpful in this respect to recall that if X is a Gaussian random variable with mean μ and variance σ^2 then $\mathbb{P}\{X \geq \alpha\} = Q((\alpha - \mu)/\sigma)$, where $Q(\cdot)$ is the so-called Q-function. One has $Q(x) = 1 - Q(-x)$ and for x tending to infinity $Q(x) \sim e^{-x^2/2}/(\sqrt{2\pi}x)$.

4.27 (Degradation of $\{\text{BCC}(\lambda)\}$). Prove that the channel family $\{\text{BCC}(\lambda)\}$ (see Problem 4.4) is ordered by degradation.

Hint: Show that the concatenation of a $\text{BCC}(\lambda_1)$ with a channel whose additive noise has Cauchy distribution with parameter λ_2 is equal to the channel $\text{BCC}(\lambda_1 + \lambda_2)$. It is helpful to know that the Fourier transform of the function $1/(a^2 + t^2)$ is $\pi e^{-a|f|}/a$ for all a with positive real part.

4.28 (Degradation of $\{\text{BLC}(\lambda)\}$). Prove that the channel family $\{\text{BLC}(\lambda)\}$ (see Problem 4.4) is ordered by degradation.

Hint: Let $\lambda' > \lambda$ and look at the concatenation of the $\text{BLC}(\lambda)$ with an additive memoryless channel whose noise distribution has the density

$$p(z) = (\lambda/\lambda')^2 \Delta(z)(1 - (\lambda/\lambda')^2)/(2\lambda')e^{-|z|/\lambda'}.$$

4.29 (Degradation via Symmetric Channel). In Lemma 4.73 it is claimed that if $p \twoheadrightarrow q$ and both p and q are memoryless symmetric then there exists a memoryless *symmetric* degrading channel r. Prove this claim by showing that if r is a degrading channel then the *symmetrized* version,

$$\frac{1}{2}\big(r_{Y \mid Y'}(y|y') + r_{Y \mid Y'}(-y| - y')\big),$$

is also a degrading channel.

4.30 (Degradation of Two Mixture Channels). Consider the two BMS channels characterized by their L-densities

$$\mathsf{a} = \alpha \mathsf{a}_{\text{BSC}(\epsilon_1)} + \bar{\alpha} \mathsf{a}_{\text{BSC}(\epsilon_3)}, \qquad \mathsf{b} = \beta \mathsf{a}_{\text{BSC}(\epsilon_2)} + \bar{\beta} \mathsf{a}_{\text{BSC}(\epsilon_4)},$$

with $0 \leq \epsilon_1 < \epsilon_2 < \epsilon_3 < \epsilon_4 \leq \frac{1}{2}$.

(i) Draw $|\mathfrak{A}|$ and $|\mathfrak{B}|$ assuming that $\bar{\beta} \geq \bar{\alpha}$.

(ii) Show that in this case $\mathsf{a} \prec \mathsf{b}$.

(iii) Explicitly construct the degrading channel for this case.

(iv) We have seen in (ii) that $\bar{\beta} \geq \bar{\alpha}$ is a sufficient condition for $\mathsf{a} \prec \mathsf{b}$. Is it also necessary?

4.31 (Kernels That Reverse the Partial Order). Consider the L-domain kernels y^i, $i \in \mathbb{N}$. Show that they *reverse* the partial order implied by degradation. Consider next the L-domain kernels $\tanh(y/2)^i$, $i \in \mathbb{N}$. Show that the $|D|$-domain kernel corresponding to even i, $i = 2j$, equals the one for odd i, $i = 2j - 1$, and is given by w^{2j} (see Problem 4.1). Show that these kernels reverse the partial order implied by degradation.

4.32 (GEXIT Kernels Preserve Ordering). Show that the $|D|$-domain GEXIT kernel for a family of BMS channels given by their $|D|$-densities $\{|\mathsf{a}|_{\text{BMSC}(\mathtt{h})}\}$ can be written as

$$\int_0^1 \frac{\mathrm{d}|\mathsf{a}_{\text{BMSC}(\mathtt{h})}|(z)}{\mathrm{d}\mathtt{h}} \Big(\sum_{i,j=\pm 1} \frac{(1+iz)(1+jw)}{4} \log_2\Big(1 + \frac{1-iz}{1+iz}\frac{1-jw}{1+jw}\Big)\Big) \mathrm{d}z.$$

Use this representation and Theorem 4.74 to prove Lemma 4.77.

4.33 (Alternative Proof of Erasure Decomposition Lemma). Let p be a BMS channel characterized by its L-density a. Lemma 4.78 shows that p is degraded with respect to $\text{BEC}(2\,\mathfrak{E}(\mathsf{a}))$. In this problem we are interest in an explicit construction of the degrading channel.

Construct a ternary-input memoryless output-symmetric channel, say q, such that p can be represented as the concatenation of $\text{BEC}(2\,\mathfrak{E}(\mathsf{a}))$ and q.

Hint: Let q denote a channel whose input alphabet is $\{-1, ?, 1\}$. Further, let q have real output y and set $q_{Y|X}(y\,|\,?) = \frac{1}{2\epsilon} e^{-|y|} \mathsf{a}(|y|)$, $q_{Y|X}(y\,|\,1) = \frac{1}{1-2\epsilon}(1 - e^{-y}) \mathbb{1}_{\{y \geq 0\}} \mathsf{a}(y)$, and $q_{Y|X}(y\,|-1) = \frac{1}{1-2\epsilon}(1 - e^{y}) \mathbb{1}_{\{y \leq 0\}} \mathsf{a}(-y)$.

4.34 (Equivalence of Convergence). Let a characterize a BMS channel and consider iterative decoding using a message-passing decoder. Show that $\mathrm{P}^{\mathrm{MP}}_{\dot{\mathcal{T}}_\ell}(\mathtt{a}) \xrightarrow{\ell\to\infty} 0$ if and only if $\mathrm{P}^{\mathrm{MP}}_{\mathcal{T}_\ell}(\mathtt{a}) \xrightarrow{\ell\to\infty} 0$.

4.35 ($\lambda(z)+\lambda(1-z)$ is Decreasing). Show that $\lambda(x)+\lambda(1-x)$ is strictly decreasing on $x \in [0, \frac{1}{2}]$ for any degree distribution $\lambda(x)$.

4.36 (Gallager Algorithm A). Determine the threshold of the $(4,5)$-regular degree distribution pair for the Gallager decoding algorithm A.

4.37 (Confidence Interval). In Figure 4.83 you see 95% confidence intervals indicated. Here is how they are computed. Consider a sequence of iid random variables $\{X_i\}$. Assume that we know that they have a Gaussian distribution but that we know neither their mean μ nor their variance σ^2. We form the estimate of the mean $\bar{X} = \frac{1}{n}\sum_{i=1}^n X_i$, which is itself a random variable. We want to determine an interval so that

$$\mathbb{P}\{\mu \in [\bar{X} - \underline{\delta}(X_1,\dots,X_n), \bar{X} + \overline{\delta}(X_1,\dots,X_n)]\} \geq 0.95.$$

The interval itself is a function of the data $X_1,\dots,X_n$ and the randomness resides in the realization of this data and not in μ which is considered fixed. Define $S^2 = \frac{1}{n-1}\sum_{i=1}^n (X_i - \bar{X})^2$. Define $T_{n-1} = \frac{\bar{X}-\mu}{S/\sqrt{n}}$. Show that regardless of the values of μ and σ^2 the distribution of T_{n-1} is equal to the distribution of

$$\frac{\frac{1}{n}\sum_{i=1}^n Y_i}{\sqrt{\frac{1}{n-1}\sum_{j=1}^n (Y_j - \frac{1}{n}\sum_{i=1}^n Y_i)^2}},$$

where the Y_i are iid Gaussian with zero mean and unit variance. This is the key reason why T_{n-1} can be used to derive confidence intervals.

This distribution is called the Student's t-distribution with $n-1$ degrees of freedom. The associated density is

$$f_{n-1}(t) = \frac{\Gamma(\frac{n}{2})}{\sqrt{(n-1)\pi}\,\Gamma(\frac{n-1}{2})}(1 + t^2/(n-1))^{-n/2}.$$

The exact density of T_{n-1} is cumbersome to deal with. But for n sufficiently large ($n \geq 61$) we have

$$\mathbb{P}\{-2 \leq T_{n-1} \leq 2\} \geq 0.95. \tag{4.183}$$

Argue that this implies that we can choose $\underline{\delta} = \overline{\delta} = 2S/\sqrt{n}$.

For our application the X_i are typical not Gaussian. For the block error probability the X_i are zero-one valued. Nevertheless, for n sufficiently large the stated formula gives a good approximation to the exact confidence interval.

4.38 (Decoder with Erasures). Consider the decoder with erasures introduced in Example 4.84. Derive the density evolution equations for this decoder. For the simple degree distribution pair (x^2, x^5) write down the fixed point equation. Show that there is no fixed point of this system of equations for $\epsilon \leq 0.0708$, but that for ϵ exceeding this value there is at least one such fixed point. Check, by running the density evolution recursions with the fixed weight sequence $w_\ell = 1$ that indeed $\epsilon^{\text{DE}} \approx 0.0708$.

4.39 (Decoder With Erasures – Threshold versus Fixed Points). This problem shows that the relationship between threshold and fixed points of density evolution can be complex. Consider the $(4, 6)$ regular ensemble, transmission over the $\text{BSC}(\epsilon = 0.087)$, and decoding using the decoder with erasures introduced in Example 4.84.

(i) Assume we initialize density evolution with the probabilities $\mu(-1)$, $\mu(0)$, and $\mu(1)$, where the individual terms are non-negative and sum up to 1. Fix the weight sequence to $w_\ell = 1$. Check empirically for which pairs $(\mu(0), \mu(1))$ the error probability converges to zero. You should get the gray area indicated in Figure 4.184. This gray area does *not* include the "natural" starting point for this channel which is $(\mu(0) = 0, \mu(1) = 1 - \epsilon = 0.913)$.

(ii) Run density evolution for the weight sequence $w_\ell = 3, 1, 1, 1, \dots, \ell \geq 1$. You should get the sequence of circles indicated in Figure 4.184. The sequence converges to a fixed point but this is not the desired fixed point. This shows that for $\epsilon = 0.087$ a non-trivial fixed point exists.

(iii) Run density evolution for the weight sequence $w_\ell = 3, 2, 2, 2, 2, 2, 1, 1, \dots$, $\ell \geq 1$. You should get the sequence of points shown as squares in Figure 4.184. For this weight sequence the densities eventually reach the gray area, and so density evolution subsequently converges with the weight set to 1.

We conclude that the existence of a non-trivial fixed point does not necessarily imply that density evolution converges to this point. Some fixed points are only reachable with a particular choice of the weight sequence and some fixed points cannot be reached at all.

4.40 (Density Evolution for Min-Sum Decoder – Chung [17], Anastasopoulos [5], Wei and Akansu [93], and Chen and Fossorier [16]). Consider a degree distribution pair (λ, ρ) and a BMS channel given in terms of its L-density a_{BMSC}. Show that the density evolution equations under min-sum decoding are

$$\mathsf{a}_0(y) = \mathsf{a}_{\text{BMSC}}(y),$$

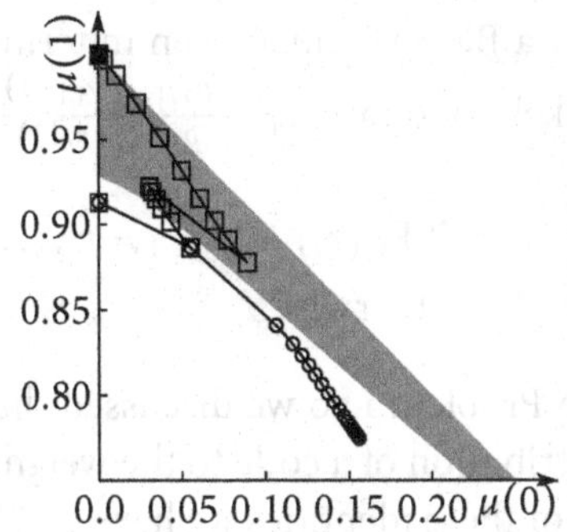

Figure 4.184: Region of convergence for the all-one weight sequence (indicated in gray).

$$\begin{aligned}
\mathsf{a}_\ell(y) &= \mathsf{a}_{\text{BMSC}}(y) \circledast \lambda(\mathsf{b}_\ell(y)), \quad \ell \geq 1,\\
\mathsf{b}_\ell(y) &= \sum_i \rho_i \frac{i-1}{2}\Big[(\mathsf{a}_{\ell-1}(y) + \mathsf{a}_{\ell-1}(-y))\Big(\int_{|y|}^{+\infty}(\mathsf{a}_{\ell-1}(x) + \mathsf{a}_{\ell-1}(-x))\mathrm{d}x\Big)^{i-2} +\\
&\quad (\mathsf{a}_{\ell-1}(y) - \mathsf{a}_{\ell-1}(-y))\Big(\int_{|y|}^{+\infty}(\mathsf{a}_{\ell-1}(x) - \mathsf{a}_{\ell-1}(-x))\mathrm{d}x\Big)^{i-2}\Big], \quad \ell \geq 1.
\end{aligned}$$

4.41 (Threshold for Min-Sum Decoder). Show that density evolution for min-sum decoding has a threshold for the channel family $\{\text{BSC}(\epsilon)\}$, call it ϵ^{MinSum}. More precisely, show that if $\epsilon < \epsilon^{\text{MinSum}}$ then $\lim_{\ell\to\infty} \mathrm{P}^{\text{MinSum}}_{\mathcal{T}_\ell}(\epsilon) = 0$, whereas for $\epsilon > \epsilon^{\text{MinSum}}$, $\lim_{\ell\to\infty} \mathrm{P}^{\text{MinSum}}_{\mathcal{T}_\ell}(\epsilon) > 0$.

Hint: Good luck with this one. We do not know of a proof ourselves.

4.42 (Min-Sum Decoder Is not Monotone – Rathi). Consider transmission over the $\text{BSC}(\epsilon = \frac{2}{10})$ and the degree-distribution pair $(\lambda(x) = \frac{1}{10}x + \frac{9}{10}x^2, \rho(x) = x^5)$. Explicitly perform density evolution for the first few iterations under min-sum decoding and determine the corresponding error probability. Show that the error probability takes on the values 0.2, 0.3139, 0.3374, 0.2818, 0.2646, etc. This shows that the min-sum decoder is in general *not* monotone with respect to iterations.

4.43 (MAP Decoding via Dual Code – Hartmann and Rudolph [32] (see also Battail, Decouvelaere, and Godlewski [6])). Consider a binary linear code $C[n, k, d]$ with components in $\{\pm 1\}$ and a uniform prior on the set of codewords. Define the multivariate polynomial

$$\mathrm{P}_C(z_1, \dots, z_n) = \sum_{x\in C}\prod_{i=1}^n z_i^{(1-x_i)/2}.$$

Consider transmission over a BMS channel given in terms of its transition probability $p(y \mid x)$. Define the likelihood ratio $r_i = \frac{p_{Y_i \mid X_i}(y_1 \mid -1)}{p_{Y_i \mid X_i}(y_i \mid 1)}$. Show that

$$\hat{x}_i^{\text{MAP}}(r) = \begin{cases} 1, & \text{if } \mathrm{P}_C(r_1, \dots, r_{i-1}, -r_i, \dots, r_n) > 0, \\ -1, & \text{otherwise.} \end{cases}$$

Let $C^{\perp}$ be the dual code. In Problem 1.20 we discussed the MacWilliams identities which relate the weight distribution of a code to the weight distribution of the dual code. By a slight extension of these identities we have

$$\mathrm{P}_C(z_1, \dots, z_n) = \frac{1}{|C^{\perp}|} \mathrm{P}_{C^{\perp}}\left(\frac{1-z_1}{1+z_1}, \cdots, \frac{1-z_n}{1+z_n}\right) \prod_{i=1}^{n} (1+z_i).$$

Define $\hat{r}_i = \frac{p_{Y_i \mid X_i}(y_i \mid 1) - p_{Y_i \mid X_i}(y_i \mid -1)}{p_{Y_i \mid X_i}(y_i \mid 1) + p_{Y_i \mid X_i}(y_i \mid -1)} = \frac{1-r_i}{1+r_i}$. Now show that, equivalently, one can decode via the dual code by using the decoding rule

$$\hat{x}_i^{\text{MAP}}(\hat{r}) = \begin{cases} 1, & \text{if } \hat{r}_i \mathrm{P}_{C^{\perp}}(\hat{r}_1, \dots, \hat{r}_{i-1}, 1/\hat{r}_i, \dots, \hat{r}_n) < 0, \\ -1, & \text{otherwise.} \end{cases}$$

4.44 (Isotropic Codes – Boutros and Vialle [89, 90]). Call a linear binary code C *isotropic* if the density of Φ_i, the extrinsic information defined in (4.30), is independent of the position i, assuming that the all-one codeword is transmitted through a fixed BMS channel. Show that (i) repetition codes, (ii) single-parity check codes, (iii) Hamming codes, and (iv) binary BCH codes are isotropic.

Hint: Express the extrinsic output as a function obtained from the weight enumerators of the subcode associated with i. Show that, if an isotropic code has as many codewords of weight $2w$ with 1 at a given position as it has codewords of weight $2w-1$ with 0 at the same position, then the extended code is also isotropic (v). This property can be used to show that the extension of a binary primitive BCH code is isotropic.

4.45 (Stability Condition for Gallager A). Explicitly determine the stability condition for the $(3, 4)$ as well as the $(4, 6)$ ensemble under decoding with Gallager A. How does the critical noise value given by the stability condition compare to the threshold?

4.46 (Geometric Convergence Close to Fixed Point). Consider a degree distribution pair (λ, ρ), transmission over the BSC(ϵ), and decoding via the Gallager algorithm A. Assume that $\epsilon < \epsilon^{\text{Gal}}(\lambda, \rho)$. Show that $\mathrm{P}^{\text{Gal}}_{\vec{\mathcal{T}}_{\ell+1}}(\epsilon) / \mathrm{P}^{\text{Gal}}_{\vec{\mathcal{T}}_{\ell}}(\epsilon)$ converges to $(\epsilon\lambda'(1) + (1-\epsilon)\lambda'(0))\rho'(1)$ as ℓ tends to infinity.

What would you expect the equivalent statement to be for the BP decoder?

4.47 (Stability Condition for BLC(λ)). Determine the stability condition for $\{\text{BLC}(\lambda)\}$ (see Problem 4.4).

Hint: Compute $\mathfrak{B}$ according to the characterization given in Problem 4.20.

4.48 (Stability Condition for BCC(λ)). Determine the stability condition for $\{\text{BCC}(\lambda)\}$ (see Problem 4.4).

4.49 (Gaussian Approximation – Stability). Consider the BAWGNC(σ^2). Show that the stability condition implied by the Gaussian approximation discussed in Example 4.139 coincides with the stability condition under BP decoding, i.e., it reads $\lambda'(0)\rho'(1)e^{-1/(2\sigma^2)} < 1$.

Hint: Use the two expansions of the capacity function (and therefore also of the function $\psi(m)$) discussed in Problem 4.12.

4.50 (Gaussian Approximation – Optimization). Use the Gaussian approximation to find a degree distribution pair $(\lambda(x), \rho(x))$ of rate one-half and maximum degree 10 with as large a threshold for the BAWGNC(σ) as possible.

4.51 (Gallager's Upper Bound on MAP Threshold). Consider $(\mathtt{l}, \mathtt{r})$-regular codes. Starting with Gallager's inequality given in Theorem 4.144, show that for transmission over the BEC(ϵ) the MAP threshold for the $(\mathtt{l}, \mathtt{r})$-regular ensemble is upper bounded by the unique positive solution of the equation $\mathtt{l} - \mathtt{r}\epsilon = \mathtt{l}(1-\epsilon)^{\mathtt{r}}$. How does this compare to the result that you get if you start with (3.95)?

In the same manner show that for transmission over the BSC(ϵ), Gallager's inequality implies the upper bound on the MAP threshold which is the unique positive solution of the equation $\mathtt{r}h_2(\epsilon) = \mathtt{l}h_2((1-(1-2\epsilon)^{\mathtt{r}})/2)$.

Hint: Explicitly evaluate $\mathfrak{D}_{2k}$ in both cases and then evaluate (4.145) using (4.149).

4.52 (Bad News about Cycle-Free Codes). Let C be a *binary* linear code of rate r that admits a binary Tanner graph which is a tree. Assume we use this code for transmission over a BMS channel characterized by its L-density a. Let $P_b^{\text{MAP}}(C)$ denote the bit error probability of this code under MAP decoding. Prove that

$$P_b^{\text{MAP}} \geq (2r-1)\,\mathfrak{E}(\mathsf{a} \circledast \mathsf{a}). \tag{4.185}$$

4.53 (Bad News about Codes with Few Cycles). The bound stated in Problem 4.52 is non-trivial only for rates above one-half. Let us discuss an information-theoretic bound which is non-trivial for all rates and that applies more generally to the case of codes with cycles.

Recall from Definition 3.91 the notion of the density of a parity-check matrix H. Consider a parity-check matrix H whose Tanner graph is a tree. As mentioned in the preceding proof, if the code has length n and rate r then there are $(2-r)n-1 <$

$(2-r)n$ edges in the graph. From this follows that the density of the graph is (at most) $\frac{2-r}{r}$. Denote this particular density corresponding to a tree by Δ^*. Let C be a binary linear code of rate r which admits a binary Tanner graph of density Δ. Assume that the codewords of C are used equally likely and that transmission takes place over a BMS channel with capacity C_{BMSC} and L-density a.

Prove that, under any decoding algorithm, the bit error probability P_b satisfies the lower bound

$$h_2(\text{P}_\text{b}) \geq r - C_{\text{BMSC}} + \frac{1-r}{2\ln(2)} \mathfrak{D}_2^{\frac{\Delta}{\Delta^*}\frac{2-r}{1-r}},$$

where $\mathfrak{D}_2 = \mathfrak{D}_2(\text{a}_{\text{BMSC}})$.

4.54 (Inequality on Moments of L-Density). Show the inequality

$$\frac{1}{2\ln(2)} \sum_{k=1}^{\infty} \frac{R(\mathfrak{D}_{2k})}{k(2k-1)} \geq C_{\text{BMSC}}^{\mathtt{r}_{\text{avg}}},$$

which is used in the proof of Theorem 4.150. Follow your own path or justify the following steps

$$\begin{aligned}
&\frac{1}{2\ln(2)} \sum_{k=1}^{\infty} \frac{\mathfrak{D}_{2k}^{\mathtt{r}_{\text{avg}}}}{k(2k-1)} \\
&\overset{\text{(i)}}{=} \frac{1}{2\ln(2)} \sum_{k=1}^{\infty} \frac{1}{k(2k-1)} \left(\int_0^{+\infty} \text{a}_{\text{BMSC}}(y)(1+\text{e}^{-y}) \tanh^{2k}(y/2)\text{d}y \right)^{\mathtt{r}_{\text{avg}}} \\
&\overset{\text{(ii)}}{\geq} \left(\int_0^{+\infty} \text{a}_{\text{BMSC}}(y)(1+\text{e}^{-y}) \sum_{k=1}^{\infty} \frac{1}{2\ln(2)} \frac{1}{k(2k-1)} \tanh^{2k}(y/2)\text{d}y \right)^{\mathtt{r}_{\text{avg}}} \\
&\overset{\text{(iii)}}{=} \left(\int_0^{+\infty} \text{a}_{\text{BMSC}}(y)(1+\text{e}^{-y}) \left[1 - h_2\left(\frac{1}{2}(1-\tanh(y/2))\right)\right]\text{d}y \right)^{\mathtt{r}_{\text{avg}}} \\
&\overset{\text{(iv)}}{=} \left(\int_0^{+\infty} \text{a}_{\text{BMSC}}(y)(1+\text{e}^{-y})[1 - h_2(\frac{1}{1+\text{e}^{y}})]\text{d}y \right)^{\mathtt{r}_{\text{avg}}} \\
&\overset{\text{(v)}}{=} C_{\text{BMSC}}^{\mathtt{r}_{\text{avg}}}.
\end{aligned}$$

4.55 (Upper Bound on the Decoding Error Probability of BP). Consider transmission over a BMS channel. Assume that we perform a hard decision at the receiver. This means that we map the observation Y taking on real values to the set $\{\pm 1\}$ according to the sign of Y. The channel seen by the decoder therefore corresponds to a BSC(ϵ).

In the case of a BAWGNC with variance σ^2, what is the corresponding cross-over probability ϵ after hard decision? Show that the channel after hard decision is a degraded version of the original channel.

4.56 (Upper Bound on Threshold for Gallager Algorithm A). For a given degree distribution pair (λ, ρ), let τ denote the smallest positive real root of the polynomial $p(x) = xp^{+}(x) + (x-1)p^{-}(x)$. Show that

$$\epsilon^{\text{Gal}}(\lambda, \rho) \leq \min\Big\{\frac{1-\lambda_2\rho'(1)}{\lambda'(1)\rho'(1)-\lambda_2\rho'(1)}, \tau\Big\}.$$

4.57 (Universal Lower Bound on Capacity for Reliable Transmission). In Section 4.10.2 we have used the least informative intermediate densities in the EXIT chart methodology to derive an upper bound on the required capacity for reliable transmission under BP decoding. Now use instead the most informative intermediate densities to derive in this way a lower bound. For the $(3,6)$-regular ensemble what is the required capacity? Is this bound useful for this concrete example?

4.58 (Lower Bound on Capacity for Reliable Transmission). In Section 4.10.2 we have used the least informative densities in the EXIT chart methodology not only for the intermediate densities but also for the input density. This gives rise to an *universal* bound on the threshold which only depends on the capacity of the input distribution (and the degree distribution of course). We get tighter bounds if we take into account the specific input distribution. Consider transmission over the channel BAWGNC(σ). Replace the intermediate densities by their least informative counterparts but explicitly compute the effect of the input density. What is the lower bound on the critical noise value σ^{BP} according to this method when using the $(3,6)$-regular ensemble?

4.59 (Alternative Representation of Common Functionals). Show that the functionals $\mathrm{H}(\cdot)$, $\mathfrak{E}(\cdot)$, and $\mathfrak{B}(\cdot)$ can be expressed in terms of the $|D|$-density $|\mathsf{a}|$ and in terms of the densities $w_a(x)$ discussed in the proof of Theorem 4.141 as

$$\begin{aligned}
\mathfrak{B}(|\mathsf{a}|) &= \int_0^1 |\mathsf{a}|(w)\sqrt{1-w^2}\,\mathrm{d}w &&= \int_0^1 w_a(x)2\sqrt{h_2^{-1}(x)(1-h_2^{-1}(x))}\,\mathrm{d}x,\\
\mathfrak{E}(|\mathsf{a}|) &= \int_0^1 |\mathsf{a}|(w)\frac{1}{2}(1-w)\,\mathrm{d}w &&= \int_0^1 w_a(x)h_2^{-1}(x)\,\mathrm{d}x,\\
\mathrm{H}(|\mathsf{a}|) &= \int_0^1 |\mathsf{a}|(w)h_2\Big(\frac{1-w}{2}\Big)\mathrm{d}w &&= \int_0^1 w_a(x)x\,\mathrm{d}x.
\end{aligned}$$

Use the preceding expressions to rederive (4.65).

4.60 (Extremal Densities of Fixed $\mathrm{H}(\cdot)$). Let a denote a symmetric L-density. Show that for a fixed $\mathrm{H}(\mathsf{a})$ the symmetric density which minimizes/maximizes $\mathfrak{E}(\mathsf{a})$ (maximize/minimize) $\mathfrak{B}(\mathsf{a})$) is from the family $\{\text{BSC}(\epsilon)\}/\{\text{BEC}(\epsilon)\}$.

Hint: Use the (right) representation of the functionals $\mathrm{H}(\cdot)$, $\mathfrak{E}(\cdot)$, and $\mathfrak{B}(\cdot)$ given in Problem 4.59.

4.61 (Extremal Densities of Fixed $\mathfrak{E}(\cdot)$). Let a denote a symmetric L-density. Show that for a fixed $\mathfrak{E}(\mathsf{a})$ the symmetric density that minimizes/maximizes $\mathrm{H}(\mathsf{a})$ (maximize/minimize) $\mathfrak{B}(\mathsf{a})$) is from the family $\{\mathrm{BSC}(\epsilon)\}/\{\mathrm{BEC}(\epsilon)\}$.

Hint: Use the (left) representation of the functionals $\mathrm{H}(\cdot)$, $\mathfrak{E}(\cdot)$, and $\mathfrak{B}(\cdot)$ given in Problem 4.59.

Conclude from this and Problem 4.60 that for a fixed $\mathfrak{B}(\mathsf{a})$ the symmetric density that minimizes/maximizes $\mathrm{H}(\mathsf{a})$ (maximize/minimize $\mathfrak{E}(\mathsf{a})$) is from the family $\{\mathrm{BSC}(\epsilon)\}/\{\mathrm{BEC}(\epsilon)\}$.

4.62 (More on Extremal Densities). In the following we write $\mathsf{a} \circledast \mathsf{b}$ to denote the convolution of two symmetric L-densities (variable node) and $a \boxast b$ to denote the convolution of two symmetric G-densities (check node).

(i) Let a and b denote two symmetric L-densities and assume each of the following three cases: $\mathfrak{E}(\mathsf{a}) = \epsilon_a$ and $\mathfrak{E}(\mathsf{b}) = \epsilon_b$ *or* $\mathfrak{B}(\mathsf{a}) = \beta_a$ and $\mathfrak{B}(\mathsf{b}) = \beta_b$ *or* $\mathrm{H}(\mathsf{a}) = \mathfrak{h}_a$ and $\mathrm{H}(\mathsf{b}) = \mathfrak{h}_b$. Determine for each case bounds on $\mathfrak{B}(\mathsf{a} \circledast \mathsf{b})$ and the corresponding extremal densities.

(ii) Let a and b denote two symmetric G-densities and assume each of the following three cases: $\mathfrak{E}(a) = \epsilon_a$ and $\mathfrak{E}(b) = \epsilon_b$ *or* $\mathrm{H}(a) = \mathfrak{h}_a$ and $\mathrm{H}(b) = \mathfrak{h}_b$ *or* $\mathfrak{B}(a) = \beta_a$ and $\mathfrak{B}(b) = \beta_b$. Determine for each case bounds on $\mathfrak{E}(a \boxast b)$ and the corresponding extremal densities.

(iii) Let a and b denote two symmetric L-densities with $\mathfrak{E}(\mathsf{a}) = \epsilon_a$ and $\mathfrak{E}(\mathsf{b}) = \epsilon_b$. Show that

$$\epsilon_a \epsilon_b \leq \mathfrak{E}(\mathsf{a} \circledast \mathsf{b}) \leq \min\{\epsilon_a, \epsilon_b\}$$

and determine the corresponding extremal densities.

Hint: Start by showing that if a and b are from the family $\{\mathrm{BSC}(\epsilon)\}$ then

$$\mathfrak{E}(\mathsf{a} \circledast \mathsf{b}) = \min\{\epsilon_a, \epsilon_b\},$$

and that this function, for a fixed ϵ_b, is convex-$\cap$ and non-decreasing in ϵ_a.

(iv) Let a and b denote two symmetric G-densities with $\mathfrak{B}(a) = \beta_a$ and $\mathfrak{B}(b) = \beta_b$. Show that

$$\sqrt{\beta_a^2 + \beta_b^2 - \beta_a^2 \beta_b^2} \leq \mathfrak{B}(a \boxast b) \leq \beta_a + \beta_b - \beta_a \beta_b,$$

and determine the corresponding extremal densities. More generally, show that if ρ is the edge-perspective degree distribution at the check nodes and if a is a symmetric G-density then $\mathfrak{B}(\rho(a)) \leq 1 - \rho(1 - \mathfrak{B}(a))$.

Hint: Start by showing that if a and b are from the family $\{\text{BSC}(\epsilon)\}$ then

$$\mathfrak{B}(a \circledast b) = \sqrt{\beta_a^2 + \beta_b^2 - \beta_a^2\beta_b^2},$$

and that this function, for a fixed β_b, is convex-∪ and increasing in β_a.

(v) Let a and b denote two symmetric L-densities with $\mathfrak{D}(\mathsf{a}) = d_a$ and $\mathfrak{D}(\mathsf{b}) = d_b$. Show that

$$d_a + d_b - 2d_a d_b \leq \frac{d_a + d_b - 2d_a d_b}{1 - d_a d_b} \leq \mathfrak{D}(\mathsf{a} \circledast \mathsf{b}) \leq d_a + d_b - d_a d_b$$

and determine the corresponding extremal densities.

4.63 (Craig's Formula – Simon and Divsalar [77]). Show that for $x \geq 0$

$$Q(x) = \frac{1}{\pi}\int_0^{\frac{\pi}{2}} \mathrm{e}^{-\frac{x^2}{2\sin^2(\theta)}}\,\mathrm{d}\theta, \qquad Q^2(x) = \frac{1}{\pi}\int_0^{\frac{\pi}{4}} \mathrm{e}^{-\frac{x^2}{2\sin^2(\theta)}}\,\mathrm{d}\theta.$$

Hint: Let X and Y be two independent Gaussian random variables distributed according to $\mathcal{N}(0,1)$ and compute $\mathbb{P}\{X \geq 0, Y \geq y\} = Q(0)Q(y)$. Explicitly evaluate the left-hand side using polar coordinates. The second identity can be proved in a similar way.

References

[1] N. Alon and M. Luby, *A linear time erasure-resilient code with nearly optimal recovery*, IEEE Trans. Inform. Theory, 42 (1996), pp. 1732–1736. [262]

[2] A. Amraoui, *Asymptotic and finite-length optimization of LDPC codes*, PhD thesis, EPFL, Lausanne, Switzerland, 2006. Number 3558. [264]

[3] A. Amraoui, S. Dusad, and R. Urbanke, *Achieving general points in the 2-user Gaussian MAC without time-sharing or rate-splitting by means of iterative coding*, in Proc. of the IEEE Int. Symposium on Inform. Theory, Lausanne, Switzerland, June 2002, p. 334. [265, 313]

[4] A. Amraoui, A. Montanari, T. Richardson, and R. Urbanke, *Finite-length scaling for iteratively decoded LDPC ensembles*, in Proc. of the Allerton Conf. on Commun., Control, and Computing, Monticello, IL, USA, Oct. 2003. [159, 266, 500]

[5] A. Anastasopoulos, *A comparison between the sum-product and the min-sum iterative detection algorithms based on density evolution*, in Proc. of GLOBECOM, San Antonio, TX, USA, Nov. 2001, pp. 1021–1025. [276]

[6] G. Battail, M. Decouvelaere, and P. Godlewski, *Replication decoding*, IEEE Trans. Inform. Theory, 25 (1979), pp. 332–345. [67, 277]

[7] L. Bazzi, T. Richardson, and R. Urbanke, *Exact thresholds and optimal codes for the binary-symmetric channel and Gallager's decoding algorithm A*, IEEE Trans. Inform. Theory, 50 (2004), pp. 2010–2021. [162, 264]

[8] S. Benedetto and G. Montorsi, *Performance evaluation of turbo-codes*, Electron. Lett., 31 (1995), pp. 163–165. [263]

[9] ——, *Design of parallel concatenated convolutional codes*, IEEE Trans. Commun., 44 (1996), pp. 591–600. [263, 367]

[10] ——, *Unveiling turbo codes: Some results on parallel concatenated coding schemes*, IEEE Trans. Inform. Theory, 42 (1996), pp. 409–428. [263, 367]

[11] C. Berrou, A. Glavieux, and P. Thitimajshima, *Near Shannon limit error-correcting coding and decoding*, in Proc. of ICC, Geneva, Switzerland, May 1993, pp. 1064–1070. [261, 366]

[12] K. Bhattad and K. R. Narayanan, *An MSE-based transfer chart for analyzing iterative decoding schemes using a Gaussian approximation*, IEEE Trans. Inform. Theory, 53 (2007), pp. 22–38. [266]

[13] D. Burshtein, M. Krivelevich, S. L. Litsyn, and G. Miller, *Upper bounds on the rate of LDPC codes*, IEEE Trans. Inform. Theory, 48 (2002), pp. 2437–2449. [264]

[14] D. Burshtein and G. Miller, *Bounds on the performance of belief propagation decoding*, IEEE Trans. Inform. Theory, 48 (2002), pp. 112–122. [265]

[15] J. Chen, A. Dholakia, E. Eleftheriou, M. P. C. Fossorier, and X.-Y. Hu, *Reduced-complexity decoding of LDPC codes*, IEEE Trans. Commun., 53 (2005), pp. 1288–1299. [266]

[16] J. Chen and M. P. C. Fossorier, *Near optimum universal belief propagation based decoding of low-density parity check codes*, IEEE Trans. Commun., 50 (2002), pp. 406–414. [276]

[17] S.-Y. Chung, *On the Construction of Some Capacity-Approaching Coding Schemes*, PhD thesis, MIT, 2000. [264, 265, 276]

[18] S.-Y. Chung, G. D. Forney, Jr., T. Richardson, and R. Urbanke, *On the design of low-density parity-check codes within 0.0045 dB of the Shannon limit*, IEEE Commun. Lett., 5 (2001), pp. 58–60. [264, 477]

[19] S.-Y. Chung, T. Richardson, and R. Urbanke, *Gaussian approximation for sum-product decoding of low-density parity-check codes*, in Proc. of the IEEE Int. Symposium on Inform. Theory, Sorrento, Italy, June 2000, p. 318. [264]

[20] ——, *Analysis of sum-product decoding of low-density parity-check codes using a Gaussian approximation*, IEEE Trans. Inform. Theory, 47 (2001), pp. 657–670. [264]

[21] J. Ezri, A. Montanari, and R. Urbanke, *Finite-length scaling for Gallager A*, in 44th Allerton Conf. on Communication, Control, and Computing, Monticello, IL, USA, Oct. 2006. [266]

[22] J. Feldman, *Decoding Error-Correcting Codes via Linear Programming*, PhD thesis, MIT, 2003. [266]

[23] S. Franz, M. Leone, A. Montanari, and F. Ricci-Tersenghi, *The dynamic phase transition for decoding algorithms*, Phys. Rev. E, 66 (2002). E-print: cond-mat/0205051. [266]

[24] S. Franz, M. Leone, and F. L. Toninelli, *Replica bounds for diluted non-Poissonian spin systems*, J. Phys. A, 36 (2003), pp. 10967–10985. [265]

[25] R. G. Gallager, *Low-density parity-check codes*, IRE Trans. Inform. Theory, 8 (1962), pp. 21–28. [33, 66, 159, 264, 419, 434]

[26] ——, *Low-Density Parity-Check Codes*, MIT Press, Cambridge, MA, USA, 1963. [156, 158, 164, 261, 264, 420]

[27] H. E. Gamal and A. R. Hammons, Jr., *Analyzing the turbo decoder using the Gaussian approximation*, in Proc. of the IEEE Int. Symposium on Inform. Theory, Sorrento, Italy, June 2000, p. 319. [264]

[28] F. Guerra and F. L. Toninelli, *Quadratic replica coupling in the Sherrington-Kirkpatrick mean field spin glass model*, J. Math.Phys., 43 (2002), pp. 3704–3716. [265]

[29] D. Guo, S. Shamai, and S. Verdú, *Mutual information and MMSE in Gaussian channels*, in Proc. of the IEEE Int. Symposium on Inform. Theory, Chicago, IL, USA, June 27–July 2, 2004, p. 349. [266]

[30] D. Guo, S. Verdú, and S. Shamai, *Mutual information and conditional mean estimation in Poisson channels*, in Proc. of the IEEE Inform. Theory Workshop, Austin, TX, USA, Oct. 24–Oct. 29, 2004, pp. 265–270. [266]

[31] T. R. Halford and K. M. Chugg, *An algorithm for counting short cycles in bipartite graphs*, IEEE Trans. Inform. Theory, 52 (2006), pp. 287–292. [266]

[32] C. Hartmann and L. Rudolph, *An optimum symbol-by-symbol decoding rule for linear codes*, IEEE Trans. Inform. Theory, 22 (1976), pp. 514–517. [277]

[33] X.-Y. Hu, E. Eleftheriou, and D. M. Arnold, *Regular and irregular progressive edge-growth tanner graphs*, IEEE Trans. Inform. Theory, 51 (2005), pp. 386–398. [266]

[34] X.-Y. Hu, M. P. C. Fossorier, and E. Eleftheriou, *Approximate algorithms for computing the minimum distance of low-density parity-check codes*, in Proc. of the IEEE Int. Symposium on Inform. Theory, Chicago, IL, USA, June 2004, p. 475. [266]

[35] S. Huettinger and J. B. Huber, *Design of "multiple-turbo codes" with transfer characteristics of component codes*, in Proc. of Conf. on Inform. Sciences and Systems (CISS), Princeton, NJ, USA, Mar. 2002. [265]

[36] ——, *Information processing and combining in channel coding*, in Proc. of the Int. Conf. on Turbo Codes and Related Topics, Brest, France, Sept. 2003, pp. 95–102. [265]

[37] Y. Jiang, A. Ashikhmin, R. Kötter, and A. C. Singer, *Extremal problems of information combining*, in Proc. of the IEEE Int. Symposium on Inform. Theory, Adelaide, Australia, Sept. 4–9, 2005, pp. 1730–1743. [265]

[38] Y. Kabashima, T. Murayama, and D. Saad, *Typical performance of Gallager-type error-correcting codes*, Phys. Rev. Lett., 84 (2000), pp. 1355–1358. [266]

[39] Y. Kabashima and D. Saad, *Statistical mechanics of low-density parity check codes*, J. Phys. A, 37 (2004), pp. R1–R43. Invited paper. [267]

[40] Y. Kabashima, N. Sazuka, K. Nakamura, and D. Saad, *Evaluating zero error noise thresholds by the replica method for Gallager code ensembles*, in Proc. of the IEEE Int. Symposium on Inform. Theory, Lausanne, Switzerland, June 2002, IEEE, p. 255. [160, 266]

[41] I. Kanter and D. Saad, *Error-correcting codes that nearly saturate Shannon's bound*, Phys. Rev. Lett., 83 (1999), pp. 2660–2663. [266, 418, 419]

[42] A. K. Khandekar, *Graph-Based Codes and Iterative Decoding*, PhD thesis, Caltech, Pasadena, CA, USA, 2002. [263, 265]

[43] R. Kötter and P. O. Vontobel, *Graph covers and iterative decoding of finite-length codes*, in Proc. of the Int. Conf. on Turbo Codes and Related Topics, Brest, France, Sept. 1–5, 2003, pp. 75–82. [266, 419]

[44] I. Land, P. Hoeher, S. Huettinger, and J. B. Huber, *Bounds on information combining*, in Proc. of the Int. Conf. on Turbo Codes and Related Topics, Brest, France, Sept. 2003, pp. 39–42. [265]

[45] J. Lodge, R. Young, P. Hoeher, and J. Hagenauer, *Separable MA, USAP-filters for the decoding of product and concatenated codes*, in Proc. of ICC, Geneva, Switzerland, May 1993, pp. 1740–1745. [261]

[46] M. Luby, M. Mitzenmacher, A. Shokrollahi, and D. A. Spielman, *Analysis of low density codes and improved designs using irregular graphs*, in Proc. of the 30th Annual ACM Symposium on Theory of Computing, 1998, pp. 249–258. [157, 262, 500]

[47] ——, *Efficient erasure correcting codes*, IEEE Trans. Inform. Theory, 47 (2001), pp. 569–584. [157, 262, 500]

[48] ——, *Improved low-density parity-check codes using irregular graphs*, IEEE Trans. Inform. Theory, 47 (2001), pp. 585–598. [157, 262, 500]

[49] M. Luby, M. Mitzenmacher, A. Shokrollahi, D. A. Spielman, and V. Stemann, *Practical loss-resilient codes*, in Proc. of the 29th annual ACM Symposium on Theory of Computing, 1997, pp. 150–159. [157, 262, 434, 456]

[50] D. J. C. MacKay, *A free energy minimization framework for inference problems in modulo 2 arithmetic*, in Fast Software Encryption (Proceedings of 1994 K. U. Leuven Workshop on Cryptographic Algorithms), B. Preneel, ed., no. 1008 in Lect. Notes Comput. Sci., Springer, 1995, pp. 179–195. [262]

[51] ——, *Good error-correcting codes based on very sparse matrices*, in Proc. of the IEEE Int. Symposium on Inform. Theory, Ulm, Germany, 1997, p. 113. [262]

[52] ——, *Good error correcting codes based on very sparse matrices*, IEEE Trans. Info. Theory, 45 (1999), pp. 399–431. [262]

[53] D. J. C. MacKay and R. M. Neal, *Good codes based on very sparse matrices*, in Cryptography and Coding. 5th IMA, USA Conf., LNCS 1025, C. Boyd, ed., Springer, Berlin, 1995, pp. 100–111. [262]

[54] ——, *Near Shannon limit performance of low density parity check codes*, Electron. Lett., 32 (1996), pp. 1645–1646. Reprinted in *Electron. Lett.*, **33** (1997), pp. 457–458. [262]

[55] ——, *Near Shannon limit performance of low density parity check codes*, Electron. Lett., 32 (1996), pp. 1645–1646. [262]

[56] N. Macris, *Correlation inequalities: A useful tool in the theory of LDPC codes*, in Proc. of the IEEE Int. Symposium on Inform. Theory, Adelaide, Australia, Sept. 2005, pp. 2369–2373. [265]

[57] N. Macris, S. Korada, and S. Kudekar, *Exact solution for the conditional entropy of Poissonian LDPC codes over the binary erasure channel*, in Proc. of the IEEE Int. Symposium on Inform. Theory, Nice, France, Sept. 2007. [265]

[58] N. Macris and S. Kudekar, *Sharp bounds for MAP decoding of general irregular LDPC codes*, in Proc. of the IEEE Int. Symposium on Inform. Theory, Seattle, WA, USA, Sept. 2006. [265]

[59] R. J. McEliece, D. J. C. MacKay, and J.-F. Cheng, *Turbo decoding as an instance of Pearl's 'belief propagation' algorithm*, IEEE J. Sel. Area. Commun., 16 (1998), pp. 140–152. [66, 262]

[60] C. Méasson, A. Montanari, T. Richardson, and R. Urbanke, *Life above threshold: From list decoding to area theorem and MSE*, in Proc. of the IEEE Inform. Theory Workshop, San Antonio, TX, USA, Oct. 2004. E-print: cs.IT/0410028. [158, 265]

[61] C. Méasson, A. Montanari, and R. Urbanke, *Maximum a posteriori deocoding and turbo codes for general memoryless channels*, in Proc. of the IEEE Int. Symposium on Inform. Theory, Adelaide, Australia, Sept. 2005, pp. 1241–1245. [265]

[62] W. Meier and O. Staffelbach, *Fast correlation attacks on certain stream ciphers*, J. Cryptology, 1 (1989), pp. 159–176. [262]

[63] M. Mézard and A. Montanari, *Information, Physics, and Computation*, Clarendon Press, Oxford, 2007. [267]

[64] M. Mézard, G. Parisi, and M. A. Virasoro, *Spin-Glass Theory and Beyond*, World Scientific Publ., 1987. [267]

[65] M. J. Mihaljević and J. D. Golić, *A comparison of cryptanalytic principles based on iterative error-correction*, Lect. Notes Comput. Sci., 547 (1991), pp. 527–531. [262]

[66] A. Montanari, *The glassy phase of Gallager codes*, Eur. Phys. J. B, 23 (2001), pp. 121–136. E-print: cond-mat/0104079. [159, 265, 266]

[67] ———, *Why "practical" decoding algorithms are not as good as "ideal" ones?*, in Proc. DIMA, USACS Workshop on Codes and Complexity, Rutgers University, Piscataway, NJ, USA, Dec. 4–7, 2001, pp. 63–66. [158, 266]

[68] ———, *Tight bounds for LDPC and LDGM codes under MAP decoding*, IEEE Trans. Inform. Theory, 51 (2005), pp. 3221–3246. [265–267]

[69] T. Murayama, Y. Kabashima, D. Saad, and R. Vicente, *Statistical physics of regular low-density parity-check error-correcting codes*, Phys. Rev. E, 62 (2000), pp. 1577–1591. [266]

[70] H. Nishimori, *Statistical Physics of Spin Glasses and Information Processing: An Introduction*, Oxford Science Publ., 2001. [266]

[71] T. Richardson, A. Shokrollahi, and R. Urbanke, *Design of capacity-approaching irregular low-density parity-check codes*, IEEE Trans. Inform. Theory, 47 (2001), pp. 619–637. [263]

[72] T. Richardson and R. Urbanke, *The capacity of low-density parity check codes under message-passing decoding*, IEEE Trans. Inform. Theory, 47 (2001), pp. 599–618. [263]

[73] ———, *Fixed points and stability of density evolution*, Commun. Inform. Syst., 4 (2004), pp. 103–116. [263, 418]

[74] A. Roumy, S. Guemghar, G. Caire, and S. Verdú, *Design methods for irregular repeat-accumulate codes*, IEEE Trans. Inform. Theory, 50 (2004), pp. 1711–1727. [265]

[75] I. Sason and R. Urbanke, *Parity-check density versus performance of binary linear block codes over memoryless symmetric channels*, IEEE Trans. Inform. Theory, 49 (2003), pp. 1611–1635. [158, 264]

[76] I. Sason and G. Wiechman, *On achievable rates and complexity of LDPC codes over parallel channels: Bounds and applications*, IEEE Trans. Inform. Theory, 53 (2007), pp. 580–598. [264]

[77] M. K. Simon and D. Divsalar, *Some new twists to problems involving the Gaussian probability integral*, IEEE Trans. Commun., 46 (1998), pp. 200–210. [283]

[78] N. Sourlas, *Spin-glass models as error-correcting codes*, Nature, 339 (1989), pp. 693–695. [160, 261, 266, 367]

[79] I. Sutskover, S. Shamai, and J. Ziv, *Extremes of information combining*, in Proc. of the Allerton Conf. on Commun., Control, and Computing, Monticello, IL, USA, Oct. 2003. [265]

[80] ——, *Extremes of information combining*, IEEE Trans. Inform. Theory, 51 (2005), pp. 1313–1325. [265]

[81] M. Talagrand, *Spin Glasses: A Challenge for Mathematicians: Cavity and Mean Field Models*, Springer, New York, NY, USA, 2003. [267]

[82] R. M. Tanner, *A recursive approach to low complexity codes*, IEEE Trans. Inform. Theory, 27 (1981), pp. 533–547. [65, 161, 261, 263]

[83] S. ten Brink, *Convergence of iterative decoding*, Electron. Lett., 35 (1999), pp. 806–808. [158, 264]

[84] ——, *Iterative decoding for multicode CDMA, USA*, in Proc. IEEE VTC, vol. 3, May 1999, pp. 1876–1880. [264]

[85] ——, *Iterative decoding trajectories of parallel concatenated codes*, in Proc. 3rd IEEE/ITG Conf. Source Channel Coding, Münich, Germany, Jan. 2000, pp. 75–80. [264]

[86] ——, *Convergence behavior of iteratively decoded parallel concatenated codes*, IEEE Trans. Inform. Theory, 49 (2001), pp. 1727–1737. [264]

[87] T. Tian, C. R. Jones, J. D. Villasenor, and R. D. Wesel, *Selective avoidance of cycles in irregular LDPC code constructions*, IEEE Trans. Commun., 52 (2004), pp. 1242–1247. [266]

[88] J. van Mourik, D. Saad, and Y. Kabashima, *Critical noise levels for LDPC decoding*, Phys. Rev. E, 66 (2002). [160, 266]

[89] S. Vialle, *Construction et analyse de nouvelles structures de codage de canal adaptées au traitement itératif*, PhD thesis, ENST, Paris, France, Dec. 2000. [278]

[90] S. Vialle and J. Boutros, *Performance limits of concatenated codes with iterative coding*, in Proc. of the IEEE Int. Symposium on Inform. Theory, Sorrento, Italy, 2000, p. 150. [278]

[91] R. Vicente, D. Saad, and Y. Kabashima, *Low Density Parity Check Codes–A Statistical Physics Perspective*, vol. 125 of Advances in Imaging and Electron Phys., Elsevier Science, 2002. In press. [266]

[92] P. O. Vontobel and R. Kötter, *On the relationship between linear programmming decoding and min-sum algorithm decoding*, in Proc. of the Int. Symposium on Inform. Theory and its Applications, Parma, Italy, Oct. 2004, pp. 991–996. [266, 420]

[93] X. Wei and A. N. Akansu, *Density evolution for low-density parity check codes under Max-Log-MA, USAP decoding*, Electron. Lett., 37 (2001), pp. 1225–1226. [276]

[94] N. Wiberg, *Codes and Decoding on General Graphs*, PhD thesis, Linköping University, S-581 83, Linköping, Sweden, 1996. [65, 66, 263, 266, 366]

[95] N. Wiberg, H.-A. Loeliger, and R. Kötter, *Codes and iterative decoding on general graphs*, Eur. Trans. Telecomm. (ETT), 6 (1995), pp. 513–526. [65, 263]

[96] G. Wiechman and I. Sason, *On the parity-check density and achievable rates of LDPC codes for memoryless binary-input output-symmetric channels*, in Proc. of the Allerton Conf. on Commun., Control, and Computing, Monticello, IL, USA, Sept. 2005, pp. 1747–1758. [264]

[97] ———, *Parity-check density versus performance of binary linear block codes: New bounds and applications*, IEEE Trans. Inform. Theory, 53 (2007), pp. 550–579. [264]

[98] A. D. Wyner and J. Ziv, *A theorem on the entropy of certain binary sequences and applications: Part I*, IEEE Trans. Inform. Theory, 19 (1973), pp. 769–772. [265]

[99] J. S. Yedidia, W. T. Freeman, and Y. Weiss, *Constructing free-energy approximations and generalized belief propagation algorithms*, IEEE Trans. Inform. Theory, 51 (2005), pp. 2282–2312. [262]

[100] J. S. Yedidia, Y. Weiss, and W. T. Freeman, *Understanding belief propagation and its generalizations*, Exploring Artificial Intelligence in the New Millennium, (2003), pp. 239–236. Chap. 8. [262]

[101] V. Zyablov and M. Pinsker, *Decoding complexity of low-density codes for transmission in a channel with erasures*, Problemy Peredachi Informatsii, 10 (1974), pp. 15–28. [157, 261]

[102] ———, *Estimation of the error-correction complexity of Gallager low-density codes*, Problemy Peredachi Informatsii, 11 (1975), pp. 23–36. [261, 434, 529]

Chapter 5

GENERAL CHANNELS

We now look at a select list of applications beyond the simple model of *binary memoryless symmetric* channels. We formalize each problem and point out how the system can be analyzed. Rather than discussing applications in their full generality, we limit ourselves to interesting special cases. In the same spirit, we do not present highly tuned solutions but explain how each system can be optimized. This keeps the exposition simple. The generalizations are quite routine. Since the Forney-style factor graph (FSFG) of a large system is the composition of the individual FSFGs of its components, it suffices for the most part to study those components in isolation. Real transmission scenarios typically involve combinations of the various components discussed in the following.

Each example introduces one new ingredient. A simple model of a *fading* channel is discussed in Section 5.1. We next discuss the prototypical *asymmetric* channel (the so-called Z channel) in Section 5.2. We then turn in Section 5.3 to an *information-theoretic* application of factor graphs - computing information rates of channels with *memory*. We also discuss how to code over channels with *memory*. In Section 5.4 we see how to construct systems with *high spectral efficiency* from simple binary ones. Very similar in spirit is the discussion on *multiple-access* channels in Section 5.5.

§5.1. Fading Channel

Consider the following simple model of a *fading* channel. We have

$$Y_t = A_t X_t + Z_t,$$

where $X_t \in \{\pm 1\}$, $Z_t \sim \mathcal{N}(0, \sigma^2)$, and A_t is Rayleigh, i.e., $p_A(a) = 2a\mathrm{e}^{-a^2}$. We further assume that both $\{A_t\}$ and $\{Z_t\}$ are iid and that they are jointly independent. We refer to this model as the *binary Rayleigh fading* (BRAYF) channel.

There are two interesting scenarios: we can assume that the fading coefficients are *known* or that they are *unknown* at the receiver. In both cases they are unknown at the transmitter. In the first case we talk about the *known side information* (KSI) case and the channel output at time t is (Y_t, A_t), whereas in the second case we say we have *unknown side information* (USI), and the channel output at time t is Y_t. Note that the channel in the case of USI is *degraded* with respect to the KSI case since part of the output is erased. In both cases the uniform input distribution is capacity-achieving.

Consider first the KSI case. The log-likelihood ratio function is

$$l(y,a) = \ln \frac{p_{Y,A|X}(y,a\,|\,1)}{p_{Y,A|X}(y,a\,|-1)} = \frac{2ay}{\sigma^2}.$$

It follows that conditioned on $A = a$ and $X = 1$, the L-density is $\mathcal{N}(\frac{2a^2}{\sigma^2}, \frac{4a^2}{\sigma^2})$. The L-density conditioned only on $X = 1$ is therefore

$$\mathsf{a}^{\text{KSI}}_{\text{BRAYFC}(\sigma)}(y) = \int_0^\infty \underbrace{2a\mathrm{e}^{-a^2}}_{p_A(a)} \underbrace{\frac{\sigma}{\sqrt{8\pi a^2}}\mathrm{e}^{-\frac{(y-2a^2/\sigma^2)^2}{8a^2/\sigma^2}}}_{\sim\mathcal{N}\left(\frac{2a^2}{\sigma^2},\frac{4a^2}{\sigma^2}\right)} da = \frac{\sigma^2}{2\sqrt{1+2\sigma^2}}\mathrm{e}^{-|y|\frac{\sqrt{1+2\sigma^2}}{2}}\mathrm{e}^{y/2}.$$

The symmetry of this density is apparent from this last expression (see the factor $\mathrm{e}^{y/2}$) but it also follows from the fact that $\mathsf{a}^{\text{KSI}}_{\text{BRAYFC}(\sigma)}(y)$ is the weighted sum of symmetric densities. From this expression we can compute the Bhattacharyya constant according to (4.62) and the stability condition according to Theorem 4.125. We have

(5.1) $$\mathfrak{B}(\mathsf{a}^{\text{KSI}}_{\text{BRAYFC}(\sigma)}) = \int \mathsf{a}^{\text{KSI}}_{\text{BRAYFC}(\sigma)}(y)\mathrm{e}^{-y/2}dy = \frac{2\sigma^2}{1+2\sigma^2} \overset{\text{stability}}{<} 1/(\lambda'(0)\rho'(1)).$$

Consider next the USI case. If $X = \pm 1$ then $Y = \pm A + Z$, so that the density of Y is the convolution of a (possibly negated) Rayleigh with a Gaussian with variance σ^2. Since the density of Z is symmetric, it follows from this representation that $p_{Y|X}(y\,|\,1) = p_{Y|X}(-y\,|-1)$, i.e., the channel is symmetric. After a tedious calculation we get

$$p_{Y|X}(y\,|\,1) = \int_{\mathbb{R}} \underbrace{2a\mathrm{e}^{-a^2}}_{p_A(a)} \underbrace{\frac{1}{\sqrt{2\pi\sigma^2}}\mathrm{e}^{-\frac{(y-a)^2}{2\sigma^2}}}_{\sim\mathcal{N}(a,\sigma^2)} da$$

$$= \frac{\sqrt{2}\mathrm{e}^{-\frac{y^2}{1+2\sigma^2}}}{\sqrt{\pi(1+2\sigma^2)}}\left(\sigma\mathrm{e}^{-\frac{y^2}{2\sigma^2(1+2\sigma^2)}} + \frac{\sqrt{2\pi}y}{\sqrt{1+2\sigma^2}}Q(-y/(\sigma^2(1+2\sigma^2)))\right),$$

where $Q(x) = \frac{1}{\sqrt{2\pi}}\int_x^\infty \mathrm{e}^{-\frac{z^2}{2}}dz$, the standard Q-function. The density $\mathsf{a}^{\text{USI}}_{\text{BRAYFC}(\sigma)}$ can be expressed in parametric form in terms of $p_{Y|X}(y\,|\,1)$ as

$$\left(\ln\frac{p_{Y|X}(y\,|\,1)}{p_{Y|X}(-y\,|\,1)}, \frac{p_{Y|X}(y\,|\,1)}{\left|\frac{p'_{Y|X}(y|1)}{p_{Y|X}(y|1)} + \frac{p'_{Y|X}(-y|1)}{p_{Y|X}(-y|1)}\right|}\right).$$

Figure 5.2 shows $\mathsf{a}^{\text{KSI}}_{\text{BRAYFC}(\sigma)}$ and $\mathsf{a}^{\text{USI}}_{\text{BRAYFC}(\sigma)}$ for $\sigma = 1$. The stability condition can be

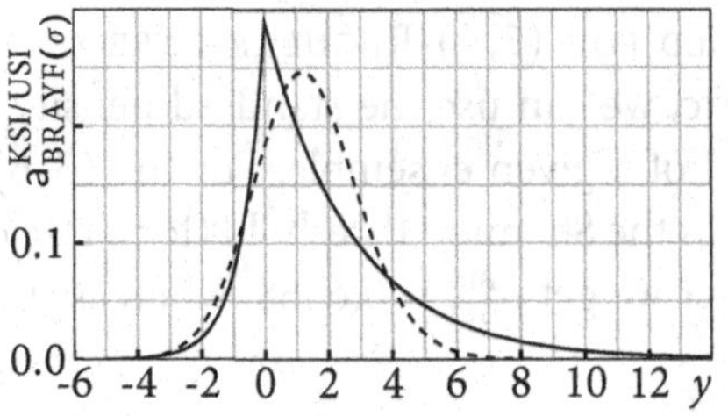

Figure 5.2: L-densities $a^{KSI}_{BRAYFC(\sigma)}$ (solid curve) and $a^{USI}_{BRAYFC(\sigma)}$ (dashed curve) for $\sigma = 1$.

evaluated (at least numerically) according to Theorem 4.125 or using the formulation of Problem 4.20. Figure 5.3 compares $1/\mathfrak{B}$, the upper bound on $\lambda'(0)\rho'(1)$, for the two cases. Not surprisingly, the KSI case can tolerate a higher value of $\lambda'(0)\rho'(1)$ than the USI case for a fixed σ. In fact, since the USI case is degraded with respect to the KSI case we know from our discussion on page 207 that this must be the case. A plot of the corresponding capacities $C^{KSI}_{BRAYFC}(\sigma)$ and $C^{USI}_{BRAYFC}(\sigma)$, both computed

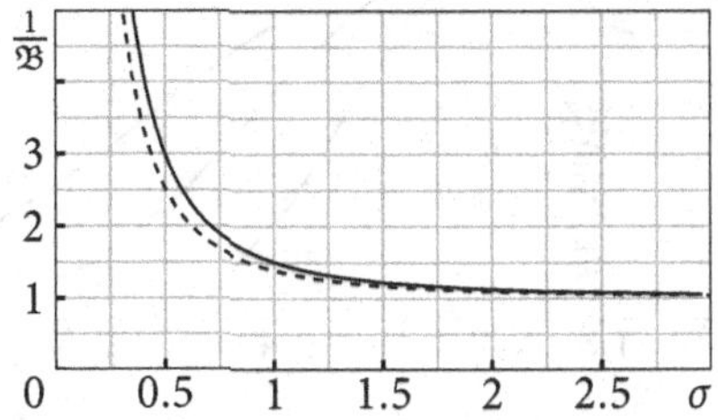

Figure 5.3: Upper bound on $\lambda'(0)\rho'(1)$, i.e., $1/\mathfrak{B}$, for the KSI case (solid curve), computed according to (5.1), and for the USI case (dashed curve).

via the general expression given in Lemma 4.35, is shown in Figure 5.4.

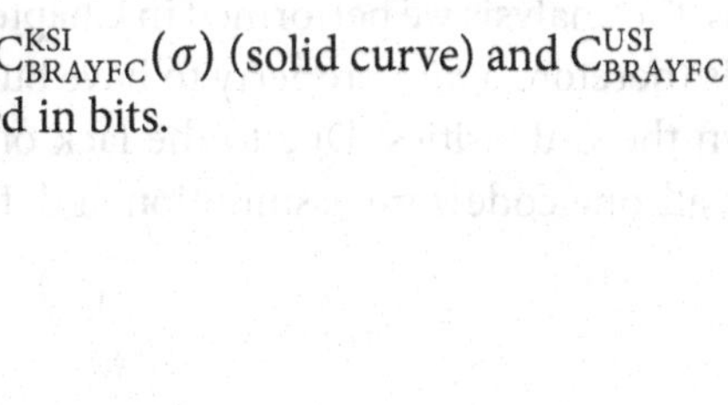

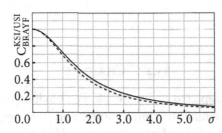

Figure 5.4: Capacities $C^{KSI}_{BRAYFC}(\sigma)$ (solid curve) and $C^{USI}_{BRAYFC}(\sigma)$ (dashed curve) as a function of σ measured in bits.

Example 5.5 (Threshold for (3,6)-Regular Ensemble). Since in both cases the channel is symmetric, we can use the standard density evolution approach to determine the threshold of a given ensemble. For the $(3,6)$-regular ensemble we get $\sigma_{\text{KSI}}^{\text{BP}} \approx 0.7016$, whereas the Shannon threshold (for a rate one-half code) is $\sigma_{\text{KSI}}^{\text{Sha}} \approx 0.81$. In the same manner we get $\sigma_{\text{USI}}^{\text{BP}} \approx 0.6368$, whereas the Shannon threshold is $\sigma_{\text{USI}}^{\text{Sha}} \approx 0.74366$. ◇

Example 5.6 (Optimization of Degree Distributions). As usual, better thresholds can be achieved by allowing irregular degree distributions. To give just one example, the following rate one-half ensemble has $\sigma_{\text{KSI}}^{\text{BP}} \approx 0.8028$, which is quite close to the Shannon threshold $\sigma_{\text{KSI}}^{\text{Sha}} \approx 0.81$. The corresponding degree distribution is $\lambda(x) = 0.194x + 0.207x^2 + 0.092x^6 + 0.112x^7 + 0.014x^8 + 0.114x^{14} + 0.004x^{48} + 0.263x^{49}$ and $\rho(x) = 0.347x^8 + 0.645x^9 + 0.008x^{10}$. Figure 5.7 shows $\mathbb{E}_{\text{LDPC}(n,\lambda,\rho)}[\text{P}_\text{b}(\text{G}, E_b/N_0)]$. ◇

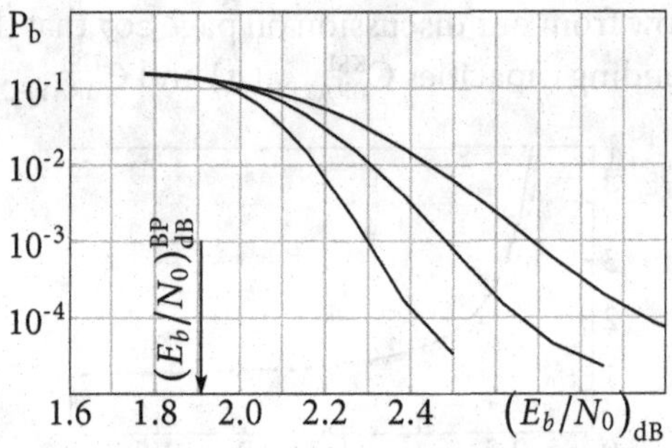

Figure 5.7: $\mathbb{E}_{\text{LDPC}(n,\lambda,\rho)}[\text{P}_\text{b}(\text{G}, E_b/N_0)]$ for the optimized ensemble stated in Example 5.6 and transmission over the BRAYF(E_b/N_0) with KSI and belief-propagation decoding. As stated in Example 5.6, the threshold for this combination is $\sigma_{\text{KSI}}^{\text{BP}} \approx 0.8028$ which corresponds to $(E_b/N_0)_{\text{dB}} \approx 1.90785$. The blocklengths/expurgation parameters n/s are $n = 8192/10$, $16384/10$, and $32768/10$, respectively.

§5.2. Z Channel

Consider the channel depicted in Figure 5.8. For obvious reasons it is called the Z channel (ZC). This channel has binary input and it is memoryless but it is *not* symmetric. Nevertheless, the analysis we performed in Chapter 4 can still be applied to this case. Symmetry is therefore a *nice* property to have but it is *not essential.* Let us start by writing down the L-densities. Due to the lack of symmetry we are no longer able to make the all-one codeword assumption and, therefore, we need the

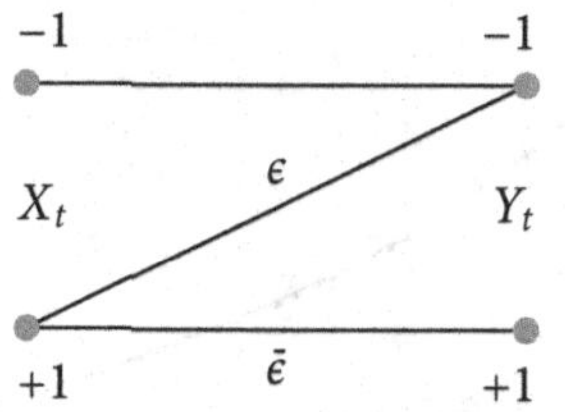

Figure 5.8: Z channel with parameter ϵ.

L-density for both $X = 1$ and $X = -1$. We have

$$\begin{cases} \mathsf{a}^{+}_{\text{ZC}(\epsilon)}(y) = \epsilon \Delta_{\ln(\epsilon)}(y) + \bar{\epsilon}\Delta_{\infty}(y), & X = 1, \\ \mathsf{a}^{-}_{\text{ZC}(\epsilon)}(y) = \Delta_{\ln(\epsilon)}(y), & X = -1. \end{cases}$$

Consider the capacity of this channel. Assuming that we use the input distribution $p_X(1) = \alpha$, the output distribution satisfies

$$(p_{Y|X}(1\,|\,1), p_{Y|X}(-1\,|\,1)) = (\bar{\epsilon}, \epsilon),$$
$$(p_{Y|X}(1\,|-1), p_{Y|X}(-1\,|-1)) = (0, 1),$$
$$(p_Y(1), p_Y(-1)) = (\alpha\bar{\epsilon}, 1 - \alpha\bar{\epsilon}),$$

so that the information rate $I_\alpha(X;Y) = I(X;Y)\,|_{p_X(1)=\alpha}$ for a fixed α equals

(5.9) $$I_\alpha(X;Y) = H(Y) - H(Y\,|\,X) = h(\alpha\bar{\epsilon}) - \alpha h(\epsilon).$$

Some calculus reveals (see Problem 5.1) that the optimal choice for α is

(5.10) $$\alpha(\epsilon) = \frac{\epsilon^{\epsilon/\bar{\epsilon}}}{1 + \bar{\epsilon}\epsilon^{\epsilon/\bar{\epsilon}}},$$

so that

$$\text{C}_{\text{ZC}(\epsilon)} = h(\alpha(\epsilon)\bar{\epsilon}) - \alpha(\epsilon)h(\epsilon) = \log(1 + \bar{\epsilon}\epsilon^{\frac{\epsilon}{\bar{\epsilon}}}),$$

where the last step requires several lines of calculus. Figure 5.11 compares $\text{C}_{\text{ZC}(\epsilon)}$ with $I_{\alpha=\frac{1}{2}}(X;Y)$. This is the highest rate that can be achieved with a uniform input distribution. Only little is lost by insisting on the uniform input distribution. As discussed in more detail in Problem 5.2, the rate which is achievable by using a uniform input distribution is at least a fraction $\frac{1}{2}\mathsf{e}\ln(2) \approx 0.942$ of capacity over the entire range of ϵ (with equality when ϵ approaches 1): from this perspective the Z channel is the extremal case – the information rate of any binary-input memoryless channel when the input distribution is the uniform one is at least a fraction $\frac{1}{2}\mathsf{e}\ln(2)$ of its

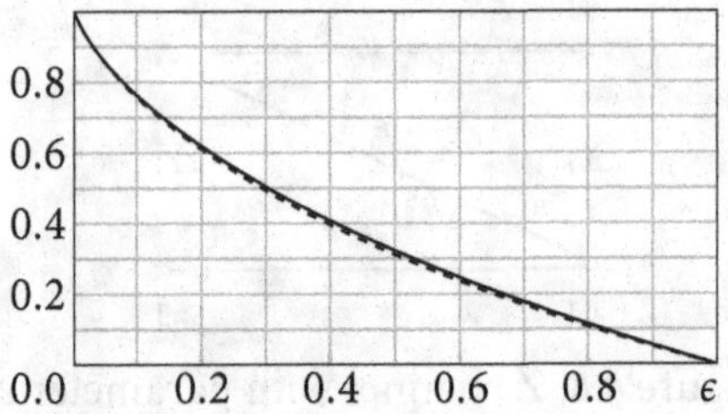

Figure 5.11: Comparison of $C_{ZC(\epsilon)}$ (solid curve) with $I_{\alpha=\frac{1}{2}}(X;Y)$ (dashed curve), both measured in bits.

capacity. From the preceding discussion we conclude that, when dealing with asymmetric channels, not much is lost if we use a binary linear coding scheme (inducing a uniform input distribution).

Consider the density evolution analysis. It seems at first that we have to analyze the behavior of the decoder with respect to each codeword. Fortunately this is not necessary. First note that, because we consider an ensemble average, only the "type" of the codeword matters. More precisely, let us say that a codeword has type τ if the fraction of zeros and ones is τ and $\bar{\tau}$, respectively. For $x \in C$, let $\tau(x)$ be its type. Let us assume that we use a low-density parity-check (LDPC) ensemble whose dominant type is one-half. This means that typical codewords contain roughly as many zeros as ones. Although it is possible to construct degree distributions which violate this constraint, most degree distributions that we encounter do fulfill it (see proof of Lemma 3.22). Under this assumption there exists some strictly positive constant γ such that

$$\mathbb{P}\{\tau(X) \notin [1/2 - \delta/\sqrt{n}, 1/2 + \delta/\sqrt{n}]\} \leq e^{-\delta^2 \gamma}. \tag{5.12}$$

We can therefore analyze the performance of such a system in the following way: determine the error probability assuming that the type of the transmitted codeword is close to the typical one. Since sublinear changes in the type do not figure in the density analysis, this task can be accomplished by a straightforward density evolution analysis. Now add to this the probability that the type of a random codeword deviates significantly from the typical one. The second term can be made arbitrarily small (see right-hand side of (5.12)) by choosing δ sufficiently large.

Consider therefore the density evolution with respect to the typical type. This means that half the nodes have initial density $\mathsf{a}^{+}_{ZC(\epsilon)}(y)$ and the remaining nodes have initial density $\mathsf{a}^{-}_{ZC(\epsilon)}(y)$. Proceed with a density evolution analysis which has two types of messages (namely those that are connected to a variable node with transmitted value +1 and those that are connected to a variable node with transmitted value −1). Fortunately we can do even better. We can "factor out" the sign of the received message. More precisely, assume that for all nodes with associated value -1, call them "minus nodes," we flip the sign of the received message. By using the symmetry of the processing rules as discussed in Section 4.1.9, one can check that the signs of all those messages which enter or exit minus nodes are flipped as well (with respect to the identical decoder which is fed with the original input) but that their magnitude is identical. Further, for this modified decoder the message densities flowing into the variable nodes are the same regardless of the sign of the variable node. In short, density evolution for an asymmetric channel with respect to the typical type is equivalent to density evolution with respect to the "symmetrized" channel

$$(\mathsf{a}^{+}_{\text{ZC}(\epsilon)}(y) + \mathsf{a}^{-}_{\text{ZC}(\epsilon)}(-y))/2.$$

That this density is indeed symmetric is quickly checked by direct computations. More generally, as discussed in Problem 5.3, this is the case for any binary-input memoryless channel.

From this observation the derivation of the stability condition as well as the methods of optimization follow in a straightforward fashion. It is the goal of Problem 5.4 to show that (under the uniform input distribution) the Bhattacharyya constant associated with this channel is $\mathfrak{B}(\mathsf{a}_{\text{ZC}(\epsilon)}) = \sqrt{\epsilon}$, so that the stability condition for this channel reads

$$\lambda'(0)\rho'(1) < 1/\sqrt{\epsilon}.$$

As a final remark: if it is crucial to approach capacity very closely, so that a uniform input distribution is not sufficient, one can combine the linear code with a non-linear mapper in order to induce a non-uniform input distribution.

§5.3. Channels with Memory

Consider an instance in which the factor graph methodology can help in answering an information-theoretic question. We want to compute the information rate (maximal rate at which information can be transmitted reliably for a given input distribution) of a channel with memory. Assuming that the memory has a Markov structure, this problem can be solved in a computationally efficient manner using the factor graph framework. This is of interest in itself, but it also forms the starting point in our investigation of low-complexity coding schemes for channels with memory. More precisely, assume we are interested in computing the information rate

$$\lim_{n\to\infty} I(X_1^n; Y_1^n)/n$$

between the input process $\{X_t\}_{t\geq 1}$ and the output process $\{Y_t\}_{t\geq 1}$. Let us write X_1^n as a shorthand for the set of random variables $X_1, \dots, X_n$. We assume that the input process takes values in a finite alphabet and that there exists a state sequence $\{\Sigma_t\}_{t\geq 0}$, taking values in a finite alphabet, such that the joint probability distribution factors in the form

$$p_{X_1^n, Y_1^n, \Sigma_0^n}(x_1^n, y_1^n, \sigma_0^n) = p_{\Sigma_0}(\sigma_0) \prod_{i=1}^{n} p_{X_i, Y_i, \Sigma_i \mid \Sigma_{i-1}}(x_i, y_i, \sigma_i \mid \sigma_{i-1}). \tag{5.13}$$

We also assume that the state sequence $\{\Sigma_t\}_{t\geq 0}$ is ergodic. This means that if $p_\Sigma(\sigma)$ is the stationary distribution on the state and if f is bounded then for any initial distribution on the state

$$\lim_{T\to\infty} \frac{1}{T} \sum_{t=1}^{T} f(\Sigma_t) \overset{\text{almost surely}}{=} \sum_{\sigma} f(\sigma) p_\Sigma(\sigma).$$

In words, the time average is almost surely equal to the ensemble average. For our purpose it suffices to know the following fact: (since we assumed that the state space is finite) the state sequence is ergodic if and only if from any state σ_i we can go to any other state σ_j in a finite number of steps with strictly positive probability. The FSFG that corresponds to (5.13) is shown in Figure 5.14.

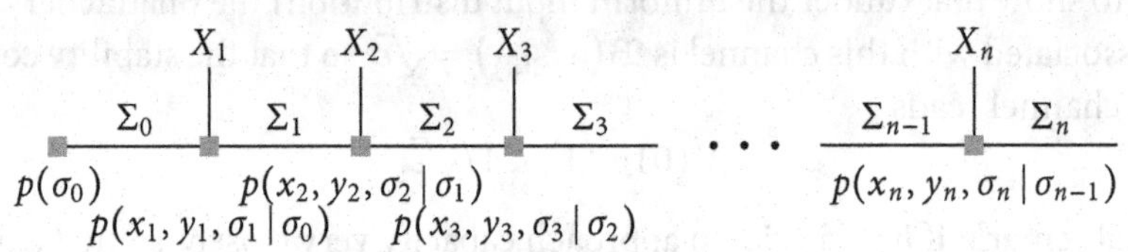

Figure 5.14: FSFG corresponding to (5.13).

Example 5.15 (Intersymbol Interference Channel (IIC)). The IIC is defined by

$$Y_t = \sum_{k=0}^{d} h_k X_{t-k} + Z_t, \tag{5.16}$$

where $\{Z_t\}$ is an iid sequence of zero-mean Gaussian random variables with variance σ^2 and $\{h_t\}_{t=0}^{d}$ represents the channel response. We assume that the channel response is *causal* and of finite length and that it is known at the receiver. Although more general cases can be dealt with in the same framework, assume that the input sequence $\{X_t\}_{t\geq 1}$ is iid, taking values in $\{\pm 1\}$ with uniform probability. Define the *state* Σ_t, $t \geq 0$, as $\Sigma_t = (X_{t-1}, \dots, X_{t-d})$, where X_t is defined as zero for $t \leq 0$. We see that $p_{X_1^n, Y_1^n, \Sigma_0^n}(x_1^n, y_1^n, \sigma_0^n)$ factors in the form (5.13). Also, the state takes values in a finite alphabet and the state sequence is ergodic: we can go from any state to any other state in at most d steps and since the X_t are iid these steps have a positive probability. ◇

EXAMPLE 5.17 (GILBERT-ELLIOTT CHANNEL). In Section 5.1 we discussed a model of a simple fading channel. In this discussion we assumed that the fading coefficients are independent. In order to arrive at a more realistic fading model we have to consider correlations between subsequent fading coefficients. A possible avenue is to assume that the channel can be in one of several states, each state being associated with a channel of a particular quality. In the simplest case there are exactly two states as shown in Figure 5.18. This is the original Gilbert-Elliott channel (GEC) model. Assume that $\{X_t\}_{t\geq 1}$ is iid, taking values in $\{\pm 1\}$ with uniform probability. The channel is either in a *good* state, denoted by G, or in a *bad* state, B. In either state

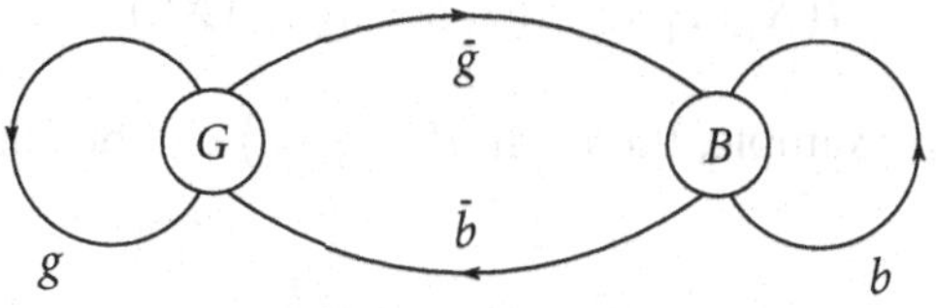

Figure 5.18: Gilbert-Elliott channel with two states.

the channel is a BSC. Let the cross-over probability in the good state be ϵ_G and in the bad state be ϵ_B, with $0 \leq \epsilon_G < \epsilon_B \leq 1/2$. Let P be the 2×2 matrix

$$P = \begin{pmatrix} g & \bar{b} \\ \bar{g} & b \end{pmatrix}$$

which encodes the transition probabilities between states (columns correspond to the present state and rows indicate the next state). Define the *steady-state* probability vector $p = (p_G, p_B)$, i.e., the vector which fulfills $Pp^T = p^T$. This means that in steady-state the system spends a fraction p_G of the time in state G and a fraction p_B of the time in state B. If we consider, e.g., the state G then in steady-state it must be true that $p_G\bar{g} = p_B\bar{b}$. From this we get $p = (\bar{b}/(\bar{g}+\bar{b}), \bar{g}/(\bar{g}+\bar{b}))$.

More generally, let us assume that we have s states, $s \in \mathbb{N}$, and that the channel in state i, $i \in [s]$, is the BSC(ϵ_i). Let P be the $s \times s$ matrix encoding the transition probabilities between these states. Let p denote the steady-state probability distribution vector. Assuming that $(I - P^T + E)$ is invertible, we claim that $p = e(I - P^T + E)^{-1}$, where e is the all-one vector of length s, I is the $s \times s$ identity matrix, and E is the $s \times s$ all-one matrix. To see this, note that $p = e(I - P^T + E)^{-1}$ is equivalent to $p(I - P^T + E) = e$. Expanding and canceling common terms we get $p - pP^T = 0$, the steady-state equation.

Note that also in this case the state sequence is ergodic as long as the Markov chain is recurrent, i.e., as long as there is a path of strictly positive probability from any state to any other state. An equivalent condition is that the steady-state probability distribution must exist. In the original Gilbert-Elliott model this is true as long as $0 < g, b < 1$. ◇

Consider the computation of the information rate. For sake of definiteness let us assume that the output Y takes values in $\mathbb{R}$ and has a density. In this case the mutual information is the difference of two *differential* entropies. (In the case of probability mass functions, replace the two differential entropies with two regular entropy expressions.) In order to keep the notational burden to a minimum we maintain this notation even for the GEC where the output is discrete. We then have

$$I(X_1^n; Y_1^n) = h(Y_1^n) - h(Y_1^n \mid X_1^n).$$

As shown in the next example, the term $h(Y_1^n \mid X_1^n)$ can be computed analytically for the IIC.

Example 5.19 (Computation of $h(Y_1^n \mid X_1^n)$: IIC). Note that for each realization of $X_1^n \in \{\pm 1\}^n$ the mean of Y_1^n is fixed and that the randomness in Y_1^n is exclusively due to Z_1^n. We conclude that

$$h(Y_1^n \mid X_1^n) = h(Z_1^n) = \sum_{i=1}^{n} h(Z_i) = \frac{n}{2} \log\left(2\pi e \sigma^2\right),$$

where in the middle step we used the fact that the random variables Z_i are independent. The last step follows since a Gaussian with variance σ^2 has differential entropy $\frac{1}{2}\log\left(2\pi e\sigma^2\right)$, i.e., if $p_Z(z) = 1/\sqrt{2\pi\sigma^2}\exp(-z^2/(2\sigma^2))$, then

$$-\int p_Z(z) \log p_Z(z) \mathrm{d}z = \frac{1}{2}\log\left(2\pi e\sigma^2\right). \qquad ◇$$

Let us see how we can compute $\lim_{n\to\infty} h(Y_1^n)/n$. Because of the ergodicity assumption on the state sequence, $-\frac{1}{n}\log p_{Y_1^n}(y_1^n)$ converges with probability 1 to $\lim_{n\to\infty} h(Y_1^n)/n$. It follows that if we can compute $-\frac{1}{n}\log p_{Y_1^n}(y_1^n)$ for a very large sequence, then with high probability the value will be close to the desired entropy rate. Instead of computing $p_{Y_1^n}(y_1^n)$, let us compute $p_{\Sigma_n, Y_1^n}(\sigma_n, y_1^n)$. From this we trivially get our desired quantity by summing

$$p_{Y_1^n}(y_1^n) = \sum_{\sigma_n} p_{\Sigma_n, Y_1^n}(\sigma_n, y_1^n).$$

We have

$$p_{\Sigma_n, Y_1^n}(\sigma_n, y_1^n) = \sum_{x_n, \sigma_{n-1}} p_{X_n, \Sigma_{n-1}, \Sigma_n, Y_1^n}(x_n, \sigma_{n-1}, \sigma_n, y_1^n)$$

$$= \sum_{x_n,\sigma_{n-1}} \underbrace{p_{X_n,\Sigma_n,Y_n\,|\,\Sigma_{n-1}}(x_n,\sigma_n,y_n\,|\,\sigma_{n-1})}_{\text{kernel}} \underbrace{p_{\Sigma_{n-1},Y_1^{n-1}}(\sigma_{n-1},y_1^{n-1})}_{\text{message}}.$$

Consider the FSFG of Figure 5.14. We claim that the preceding recursion produces the messages of the belief propagation (BP) decoder which are sent from "left to right." Indeed, if we denote the message sent along the edge corresponding to Σ_n by $\alpha_{\Sigma_n}(\sigma_n)$ then by the standard message-passing rules

$$\alpha_{\Sigma_n}(\sigma_n) = \sum_{x_n,\sigma_{n-1}} p_{X_n,\Sigma_n,Y_n\,|\,\Sigma_{n-1}}(x_n,\sigma_n,y_n\,|\,\sigma_{n-1})\alpha_{\Sigma_{n-1}}(\sigma_{n-1}).$$

In other words, $\alpha_{\Sigma_n}(\sigma_n) = p_{\Sigma_n,Y_1^n}(\sigma_n,y_1^n)$, so that

$$\lim_{n\to\infty} h(Y_1^n)/n = -\lim_{n\to\infty} \log\Big(\sum_{\sigma_n} \alpha_{\Sigma_n}(\sigma_n)\Big)/n. \tag{5.20}$$

We can therefore get a good estimate of $\lim_{n\to\infty} h(Y_1^n)/n$ by computing the messages $\alpha_{\Sigma_n}(\sigma_n)$ for a very long trellis, summing them up, taking the log, and normalizing the result by the length n.

From a practical perspective it is typically more convenient to pass *normalized* messages $\tilde{\alpha}_{\Sigma_n}(\sigma_n)$ so that $\sum_\sigma \tilde{\alpha}_{\Sigma_n}(\sigma_n) = 1$. The first message $\alpha_{\Sigma_0}(\sigma_0) = p_{\Sigma_0}(\sigma_0)$ is a probability distribution and, hence, $\tilde{\alpha}_{\Sigma_0}(\sigma_0) = \alpha_{\Sigma_0}(\sigma_0)$. Compute $\alpha_{\Sigma_1}(\sigma_1)$ and let $\lambda_1 = \sum_{\sigma_1} \alpha_{\Sigma_1}(\sigma_1)$. Define $\tilde{\alpha}_{\Sigma_1}(\sigma_1) = \alpha_{\Sigma_1}(\sigma_1)/\lambda_1$. Note that all subsequent messages are also scaled by this factor. Therefore, if λ_n denotes the normalization constant by which we have to divide at step i so as to normalize the message then $\tilde{\alpha}_{\Sigma_n}(\sigma_n) = \alpha_{\Sigma_n}(\sigma_n)/(\prod_{i=1}^n \lambda_i)$. It follows that

$$\begin{aligned}\lim_{n\to\infty} h(Y_1^n)/n &= -\lim_{n\to\infty} \log\Big(\sum_{\sigma_n} \alpha_{\Sigma_n}(\sigma_n)\Big)/n \\ &= -\lim_{n\to\infty} \log\Big(\Big(\prod_{i=1}^n \lambda_i\Big)\sum_{\sigma_n} \tilde{\alpha}_{\Sigma_n}(\sigma_n)\Big)/n = -\lim_{n\to\infty}\Big(\sum_{i=1}^n \log(\lambda_i)\Big)/n.\end{aligned}$$

The preceding observation and the derivation contained in Example 5.19 allow us to determine $I(X_1^n; Y_1^n)$ for the IIC case. For the GEC, where we do not have an analytic expression of $h(Y_1^n \,|\, X_1^n)$, we write $h(Y_1^n \,|\, X_1^n)/n = h(Y_1^n, X_1^n)/n - h(X_1^n)/n$. The second part is trivial since the inputs are binary, uniform, and iid by assumption, hence $h(X_1^n)/n = 1$. For the term $h(Y_1^n, X_1^n)/n$ we use the same technique as for the computation of $h(Y_1^n)/n$. Due to the ergodicity assumption $-\frac{1}{n}\log p_{Y_1^n,X_1^n}(y_1^n, x_1^n)$ converges with probability 1 to $\lim_{n\to\infty} h(Y_1^n, X_1^n)/n$. Write $p_{Y_1^n,X_1^n}(y_1^n, x_1^n)$ in the form $\sum_{\sigma_n} p_{\Sigma_n,Y_1^n,X_1^n}(\sigma_n, y_1^n, x_1^n)$ and use the factorization

$$p_{\Sigma_n,Y_1^n,X_1^n}(\sigma_n, y_1^n, x_1^n) = \sum_{\sigma_{n-1}} p_{\Sigma_{n-1},\Sigma_n,Y_1^n,X_1^n}(\sigma_{n-1}, \sigma_n, y_1^n, x_1^n)$$

$$= \sum_{\sigma_{n-1}} \underbrace{p_{X_n,\Sigma_n,Y_n \mid \Sigma_{n-1}}(x_n, \sigma_n, y_n \mid \sigma_{n-1})}_{\text{kernel}} \cdot \underbrace{p_{\Sigma_{n-1},Y_1^{n-1},X_1^{n-1}}(\sigma_{n-1}, y_1^{n-1}, x_1^{n-1})}_{\text{message}}.$$

In words, we generate a random instance X_1^n and Y_1^n and run the forward pass on the FSFG shown in Figure 5.14 assuming that *both* Y_1^n *and* X_1^n are "frozen." Taking the logarithm, multiplying by -1, and normalizing by $1/n$ gives us an estimate of $\lim_{n\to\infty} h(Y_1^n, X_1^n)/n$.

Now that we can compute the information rates, let us consider coding over the IIC or the GEC. The decoding function of the optimal bit-wise decoder is

$$\begin{aligned}
\hat{x}_i &= \text{argmax}_{x_i \in \{\pm 1\}} p_{X_i \mid Y_1^n}(x_i \mid y_1^n) \\
&= \text{argmax}_{x_i \in \{\pm 1\}} \sum_{\sim x_i} p_{X_1^n, Y_1^n, \Sigma_0^n}(x_1^n, y_1^n, \sigma_0^n) \\
&= \text{argmax}_{x_i \in \{\pm 1\}} \sum_{\sim x_i} p_{\Sigma_0}(\sigma_0) \prod_{j=1}^{n} p_{X_j, Y_j, \Sigma_j \mid \Sigma_{j-1}}(x_j, y_j, \sigma_j \mid \sigma_{j-1}) \mathbb{1}_{\{x \in C\}}.
\end{aligned}$$

In words, the FSFG in Figure 5.14 describes also the factorization for the iterative decoder if we add to it the factor node describing the definition of the code. As always, this factor graph together with the initial messages stemming from the channel completely specify the message-passing rules, except for the message-passing schedule. Let us agree that we alternate one round of decoding with one round of channel estimation. No claim as to the optimality of this scheduling rule is made.

It is not hard to see that, as in the case of binary memoryless symmetric channels, for a fixed number of iterations, the decoding neighborhood is again asymptotically tree-like (this assumes windowed decoding for the channel estimation part). In the case of the GEC the channel is also symmetric and we can proceed assuming that the all-one codeword was transmitted. In the case of the IIC the overall channel is in general not symmetric but we have already seen in Section 5.2 when discussing the Z channel how we can deal with this situation: symmetrize the channel and analyze the system relative to the typical codeword type. Therefore, in both cases we can employ the technique of density evolution to determine thresholds and to optimize the ensembles.

Example 5.21 (GEC: State Estimation). For the case of transmission over the GEC the iterative decoder implicitly estimates the state of the channel. Let us demonstrate this by means of the following example. We pick a GEC with three states. Let

$$P = \begin{pmatrix} 0.99 & 0.005 & 0.02 \\ 0.005 & 0.99 & 0.02 \\ 0.005 & 0.005 & 0.96 \end{pmatrix},$$

which has a steady-state probability vector of $p \approx (0.4444, 0.4444, 0.1112)$. Let the channel parameters of the BSC in these three states be $(\epsilon_1, \epsilon_2, \epsilon_3) = (0.01, 0.11, 0.5)$. This corresponds to an *average* error probability of $\epsilon_{\text{avg}} = \sum_{i=1}^{3} p_i \epsilon_i \approx 0.108889$. Using the methods described earlier, the maximum rate at which information can be transmitted reliably over this channel assuming iid inputs is $\text{C} \approx 0.583$ bits per channel use. This is markedly higher than $1 - h(\epsilon_{\text{avg}}) \approx 0.503444$, which is the capacity of the BSC(ϵ_{avg}), the channel that we experience if we ignore the Markov structure.

Assume we use the ensemble

$$\text{LDPC}\left(n, \lambda(x) = 0.245x + 0.4585x^2 + 0.1183x^5 + 0.1782x^7, \rho(x) = x^6\right),$$

which has a design rate of $r \approx 0.5498$. Figure 5.22 shows the evolution of the densities for this case. The pictures on the right correspond to the channel state estimates. Note that 5 clear peaks emerge after 10 iterations: at $\pm\ln(0.99/0.01) \approx \pm 4.595$, $\pm\ln(0.9/0.1) \approx \pm 2.197$, $\pm\ln(0.5/0.5) = 0$. They correspond to the received likelihoods in the three possible channel states. The emergence of the peaks indicates that at this stage the system has identified the channel states with high reliability. ◇

§5.4. Coding for High Spectral Efficiency

Let us reconsider the additive white Gaussian noise channel, $Y_t = X_t + Z_t$, where $\{Z_t\}$ is an iid sequence of zero-mean Gaussian random variables with variance σ^2. If we limit the input X_t to be an element of $\{\pm 1\}$, then the rate is upper bounded by 1, regardless of the noise variance σ^2. Particularly for small values of σ^2, higher rates are achievable by allowing X_t to lie in an extended constellation.

Most commonly such constellations are chosen to be one- or two-dimensional. Some standard two-dimensional constellations are discussed in Problem 5.5. As a concrete example let us consider the pulse-amplitude modulation (PAM) constellation on 4 points shown in Figure 5.24. The scaling is chosen so that the average signal energy is equal to 1 (assuming a uniform probability distribution on the points).

Assume that we decided on the signal constellation, call it $\mathcal{S}$. For convenience we assume that $\mathcal{S}$ contains 2^k points. It is then natural to label these points by k bits, call them $x = \left(x^{[1]}, \dots, x^{[k]}\right)$. We formally specify this labeling by introducing a one-to-one map, $\psi : \mathbb{F}_2^k \to \mathcal{S}$, which maps each k-tuple of bits $x = \left(x^{[1]}, \dots, x^{[k]}\right)$ into a distinct point of the constellation $\mathcal{S}$. We assume a uniform prior on the bits

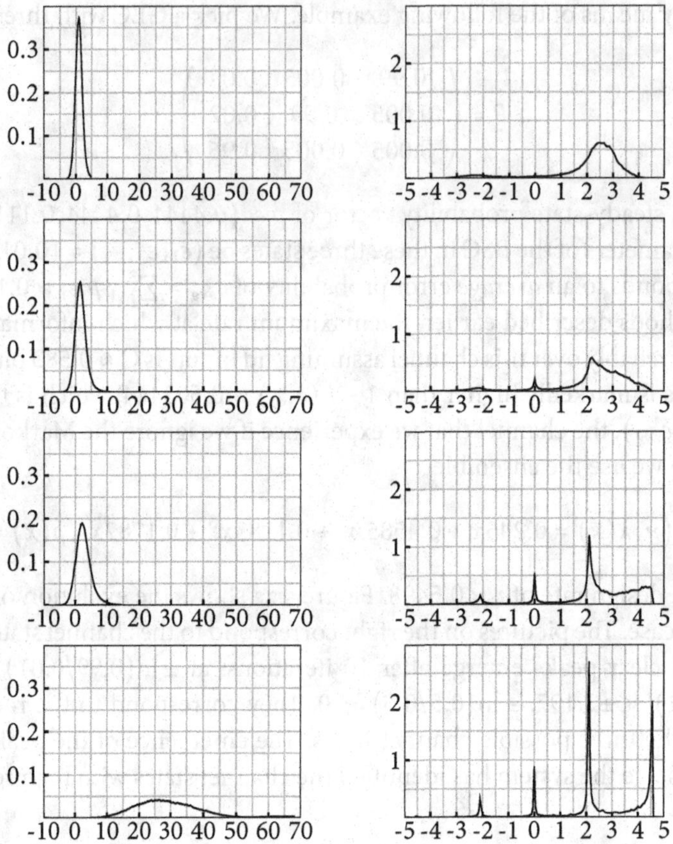

Figure 5.22: L-densities of density evolution at iteration 1, 2, 4, and 10. The left pictures show the densities of the messages which are passed from the code toward the part of the FSFG which estimates the channel state. The right-hand side shows the density of the messages which are the estimates of the channel state and which are passed to the part of the FSFG corresponding to the code.

which induces a uniform prior on the elements of $\mathcal{S}$. There are $2^k!$ possible such maps and we will see shortly how our choice affects the system performance. Note that

$$I(X;Y) = I(X^{[1]}, \ldots, X^{[k]}; Y) = \sum_{i=1}^{k} I(X^{[i]}; Y \mid X^{[1]}, \ldots, X^{[i-1]}), \tag{5.23}$$

where the second step is the well-known *chain rule* of mutual information (it follows from (1.44), the chain rule of entropies). It is crucial to notice that the mutual information $I(X;Y)$ is *independent* of the map ψ but that the *split* of the mutual information into subterms *does* depend on ψ. Note that the i-th term on the right-hand side expresses the mutual information between the i-th bit and the received symbol Y, given the previous $i-1$ bits. For the 4-PAM example two possible maps are shown in Figure 5.24. Equation (5.23) can be given an operational meaning which

Figure 5.24: Two specific maps ψ for the 4-PAM constellation.

one can use to design a transmission system. Consider the case $k=2$ (which would apply to 4-PAM). Equation (5.23) then reads

$$I(X^{[1]},X^{[2]};Y) = I(X^{[1]};Y) + I(X^{[2]};Y\,|\,X^{[1]}). \tag{5.25}$$

The first term is the mutual information between bit $X^{[1]}$ and the received symbol Y considering channel input $X^{[2]}$ as noise, i.e., as part of the channel. More precisely, the channel seen by bit $X^{[1]}$ (assuming that $X^{[2]}$ is treated as noise and has uniform prior) has transition probability

$$p_{Y\,|\,X^{[1]}}(y\,|\,x^{[1]}) = \frac{1}{2}\left(p_{Y\,|\,X^{[1]},X^{[2]}}(y\,|\,x^{[1]},0) + p_{Y\,|\,X^{[1]},X^{[2]}}(y\,|\,x^{[1]},1)\right).$$

This transition probability depends on the map ψ. The transition probabilities for the two choices of ψ depicted in Figure 5.24 are shown in Figure 5.26.

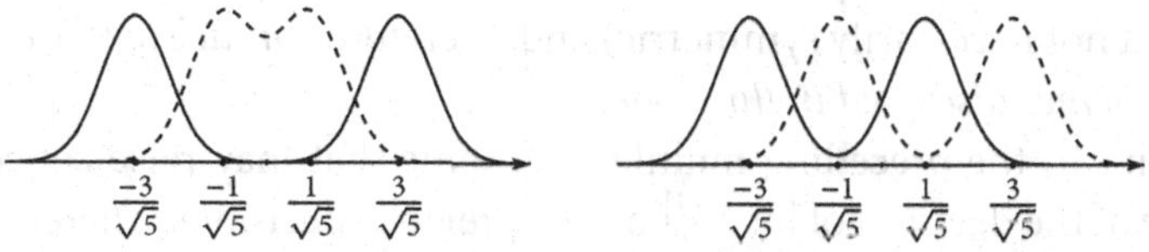

Figure 5.26: Transition probabilities $p_{Y\,|\,X^{[1]}}(y\,|\,x^{[1]})$ for $\sigma \approx 0.342607$ as a function of $x^{[1]} = 0/1$ (solid/dashed). The two cases correspond to the two maps ψ shown in Figure 5.24.

The second term has a similar interpretation, except that now at the receiver we have *side information* $X^{[1]}$, i.e., the term is the mutual information between bit $X^{[2]}$ and the channel output Y given that $X^{[1]}$ is available at the receiver. If we consider, e.g., the map shown on the left-hand side of Figure 5.24, and if we assume that $X^{[1]} =$

0 then the channel "seen" by the second decoder is a binary channel with inputs located at $\{\pm 3/\sqrt{5}\}$ and additive Gaussian noise with variance σ^2. Figure 5.27 shows the corresponding receiver diagram. Because the decoding process is done in (two) levels, the scheme is called a *multilevel* decoding scheme.

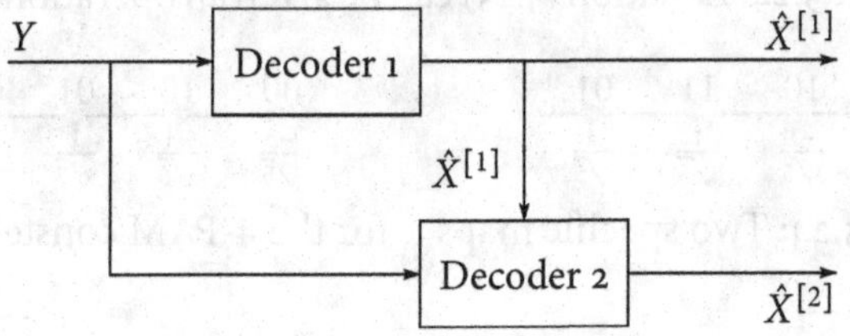

Figure 5.27: Multilevel decoding scheme. The two decoding parts correspond to the two parts of (5.25).

EXAMPLE 5.28 (MULTILEVEL SCHEME FOR 4-PAM). Consider the multilevel scheme for 4-PAM and the two maps shown in Figure 5.24. We know that both maps lead to the same overall capacity. For $\sigma \approx 0.342607$ this sum capacity is 3/2 bits per channel use. But this sum rate is *split* in different ways between the two *subchannels*. We have

$$I(X^{[1]}; Y) + I(X^{[2]}; Y \mid X^{[1]}) = 0.666549 + 0.833455 = 3/2, \qquad \text{left map},$$
$$I(X^{[1]}; Y) + I(X^{[2]}; Y \mid X^{[1]}) = 0.518198 + 0.981806 = 3/2, \qquad \text{right map}.$$

◇

The preceding interpretation gives rise to the following *multilevel* scheme. For a given constellation of size 2^k, choose a map ψ. This gives rise to k channels, where the i-th channel has capacity $I(X^{[i]}; Y \mid X^{[1]}, \ldots, X^{[i-1]})$. Note that each channel is *binary* (albeit not necessarily symmetric) and, therefore, *on each level we can employ the coding schemes discussed in this book.*

One point in the preceding multilevel scheme that may raise concerns is the dependence of the decision of bit $X^{[i]}$ on the previous decisions. Therefore, once an error is made at level i it is likely that this error will adversely affect all following levels. This is called *error propagation*. In order to limit error propagation, one has to ensure that levels are decoded highly reliably. This usually means large latency.

This issue can be circumvented at a small cost in transmission rate by using a *bit-interleaved coded modulation* (BICM) scheme. The basic idea of BICM is straightforward. We have

$$I(X; Y) = I(X^{[1]}, \ldots, X^{[k]}; Y) = \sum_{i=1}^{k} I(X^{[i]}; Y \mid X^{[1]}, \ldots, X^{[i-1]}) \geq \sum_{i=1}^{k} I(X^{[i]}; Y).$$

The interpretation of the inequality is immediate. Rather than first decoding bit $X^{[1]}$ and then using this information as side information for decoding bit $X^{[2]}$ and so on, decode all bits in *parallel*. This is shown in Figure 5.29. This obviously avoids the

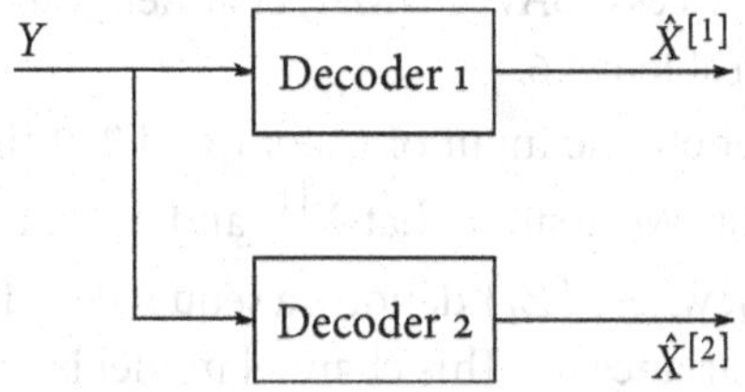

Figure 5.29: BICM decoding scheme. The two decoding parts correspond to $I(X^{[1]}; Y)$ and $I(X^{[2]}; Y)$, respectively.

latency and error propagation problems. On the other hand, each term $I(X^{[i]}; Y)$ is now, in general, *strictly* smaller than the corresponding term

$$I(X^{[i]}; Y \mid X^{[1]}, \ldots, X^{[i-1]});$$

i.e., the overall transmission rate achievable by BICM is, in general, *strictly* less than the optimal multilevel scheme. How much is lost crucially depends on the mapping ψ. The optimal BICM mapping ψ is the one which maximizes $\sum_{i=1}^{k} I(X^{[i]}; Y)$. As a rule of thumb, the so-called *Gray* mapping is typically a good choice. A Gray mapping of a constellation is a binary labeling so that the labels of any two *nearest* neighbors (in the Euclidean sense) differ in exactly one position. For example, if we look back at Figure 5.24 we see that the left mapping is a Gray mapping but that the right one is not. It turns out that, for those constellations most frequently used, surprisingly little is lost by employing BICM as opposed to the more complicated multilevel scheme.

Example 5.30 (BICM for 4-PAM). Consider BICM for 4-PAM and the two mappings ψ shown in Figure 5.24. If we choose again $\sigma \approx 0.342607$ then

$$I(X^{[1]}; Y) = 0.666549, \qquad I(X^{[2]}; Y) = 0.832902, \qquad \text{left map},$$
$$I(X^{[1]}; Y) = 0.518198, \qquad I(X^{[2]}; Y) = 0.832902, \qquad \text{right map}.$$

It follows that for the map on the left the achievable sum rate is 1.49945, which is only negligibly less than the full sum rate of 1.5. On the other hand, for the map on the right the sum rate is 1.3511, which is noticeably smaller. ◇

§5.5. Multiple-Access Channel

The factor graph approach can deal with multiple-user scenarios with ease equal to that for the single-user case. Let us consider the binary-input additive white-Gaussian noise multiple-access (BAWGNMA) channel. A "noiseless" multiple-access channel is the topic of Problem 5.6.

Let $X_t^{[1]}$ and $X_t^{[2]}$ denote the input of users 1 and 2 at time t, and let Y_t denote the corresponding output. We assume that $X_t^{[1]}$ and $X_t^{[2]}$ are elements of $\{\pm 1\}$. We have $Y_t = X_t^{[1]} + X_t^{[2]} + Z_t$, where $\{Z_t\}$ denotes a sequence of iid zero-mean Gaussian random variables with variance σ^2. This channel model is shown in Figure 5.31. We

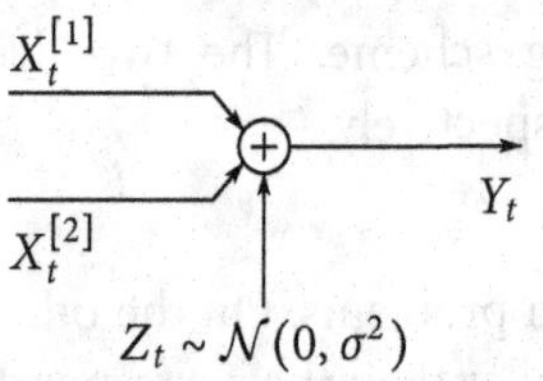

Figure 5.31: BAWGNMA channel with two users.

assume that the two users can not coordinate their transmissions. Mathematically we model this by assuming that the input distribution has product form:

$$p_{X^{[1]},X^{[2]}}(x^{[1]}, x^{[2]}) = p_{X^{[1]}}(x^{[1]})p_{X^{[2]}}(x^{[2]}).$$

Let $r^{[1]}$ and $r^{[2]}$ denote the transmission rate of users 1 and 2, respectively. It is intuitive, and in fact correct, that any achievable rate pair $(r^{[1]}, r^{[2]})$ is upper bounded by

$$\begin{aligned} r^{[1]} &\le I(X^{[1]}; Y \mid X^{[2]}), \\ r^{[2]} &\le I(X^{[2]}; Y \mid X^{[1]}), \\ r^{[1]} + r^{[2]} &\le I(X^{[1]}, X^{[2]}; Y). \end{aligned}$$

The first two inequalities state an upper bound on the individual rates assuming that the signal from the other user is known at the receiver. The third inequality states a bound on the sum rate of both users. It is more surprising, and a cornerstone of information theory, that all rate tuples within this pentagon are indeed achievable. Figure 5.32 shows this capacity region for the particular choice $\sigma \approx 0.778$.

Consider rate tuples $(r^{[1]}, r^{[2]})$ which have maximal sum. This means that $r^{[1]} + r^{[2]} = I(X^{[1]}, X^{[2]}; Y)$. The set of such rate tuples is called the *dominant face* and is typically denoted by $\mathcal{D}$ (see Figure 5.32). Without loss of generality we can restrict our attention to $\mathcal{D}$. This is true, since all other achievable rate tuples are *dominated*

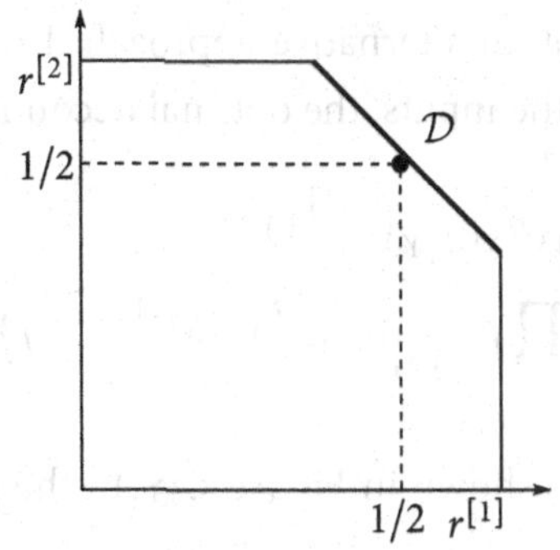

Figure 5.32: Capacity region for $\sigma \approx 0.778$. The dominant face $\mathcal{D}$ (thick diagonal line) is the set of rate tuples of the capacity region of maximal sum rate.

(component-wise) by at least one of those in $\mathcal{D}$. Consider a "corner" point of $\mathcal{D}$. One such corner point corresponds to the rate tuple

$$(r^{[1]}, r^{[2]}) = (I(X^{[1]}; Y), I(X^{[2]}; Y \mid X^{[1]})),$$

and the second one has the same form with the roles of the two users reversed. This is essentially the situation which we encountered in Section 5.4 when discussing the multilevel coding scheme. More precisely, we see that this rate tuple can be achieved by two binary single-user codes. The first is operating at rate $I(X^{[1]}; Y)$ over a channel with transition probability

$$p_{Y \mid X^{[1]}}(y \mid x^{[1]}) = \frac{1}{2\sqrt{2\pi\sigma^2}}\left(e^{-\frac{(y-x^{[1]}-1)^2}{2\sigma^2}} + e^{-\frac{(y-x^{[1]}+1)^2}{2\sigma^2}}\right).$$

The second user is transmitting at rate $I(X^{[2]}; Y \mid X^{[1]})$ over the channel with transition probability

$$p_{Y \mid X^{[1]}, X^{[2]}}(y \mid x^{[1]}, x^{[2]}) = \frac{1}{\sqrt{2\pi\sigma^2}} e^{-\frac{\left(y-x^{[1]}-x^{[2]}\right)^2}{2\sigma^2}},$$

where $x^{[1]}$ is assumed to be known.

How about general points in $\mathcal{D}$? One technique is to use time-sharing between the two corner points with an appropriate time-sharing constant. This achieves any point in $\mathcal{D}$ with essentially the same complexity as the two corner points. A slight disadvantage is that this scheme requires at least two codes as well as coordination between the two users in terms of synchronization.

Let us therefore look at an alternative approach. Factor graphs to the rescue. Assuming equal priors on the inputs, the optimal decoding rule for bit $x_i^{[1]}$ reads

$$\hat{x}_i^{[1]}(y) = \operatorname{argmax}_{x_i^{[1]} \in \{\pm 1\}} p_{X_i^{[1]} \mid Y}(x_i^{[1]} \mid y)$$
$$= \operatorname{argmax}_{x_i^{[1]} \in \{\pm 1\}} \sum_{\sim x_i^{[1]}} \Big(\prod_j p_{Y_j \mid X_j^{[1]}, X_j^{[2]}}(y_j \mid x_j^{[1]}, x_j^{[2]}) \Big) \mathbb{1}_{\{x^{[1]} \in C^{[1]}\}} \mathbb{1}_{\{x^{[2]} \in C^{[2]}\}}.$$

The corresponding FSFG is shown in Figure 5.33. In the preceding derivation and

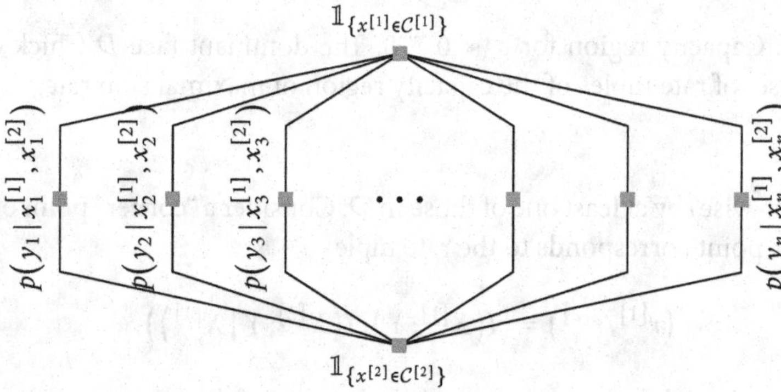

Figure 5.33: FSFG corresponding to decoding on the BAWGNMA channel.

in Figure 5.33 we have assumed that the transmissions of the two users are *block* aligned, i.e., the two users employ the same blocklength and the boundaries of the codewords are aligned. In fact this is not necessary. Nothing essential changes if we only have *bit* alignment, i.e., the codewords might be of different lengths or simply are not aligned but bits are aligned. For the following discussion we will assume that the decoder interleaves one decoding round for user 1 with one decoding round of user 2. No claim with regards to the optimality of such a scheme is made.

Under the aforementioned conditions, it is quickly verified that the computation graph for the decoding of a specific bit is again asymptotically tree-like if we consider a fixed number of iterations and if we let the blocklength grow to infinity. We can therefore employ the density evolution method. Further, by using the symmetries in the problem, we can assume that one user transmits the all-one codeword and that the second user employs a codeword of type one-half. As for the Z channel in Section 5.2, we can "symmetrize" the remaining problem and reduce the density evolution analysis to a single density that we need to track.

Example 5.34 (Optimized Degree Distribution for Bit-Aligned Scenario). Assume we want to achieve a sum rate of one bit per channel use. One can check that the corresponding threshold as given by the Shannon capacity, call it σ^{Sha}, is equal to

$\sigma^{\text{Sha}} \approx 0.7945$. We want to achieve the equal rate point, i.e, we are looking for a pair of ensembles, each of rate one-half, with an iterative decoding threshold σ^{BP} as close as possible to σ^{Sha}. To simplify matters, assume that both users employ identical ensembles LDPC (n, λ, ρ). Finally, we assume bit alignment and assume that the matching of the variable nodes of the two users at the receiver is done randomly. Consider the degree distribution pair

$$\lambda(x) = 0.315434x + 0.242272x^2 + 0.0988336x^{16} + 0.05x^{17} + 0.293462x^{99},$$
$$\rho(x) = x^7.$$

The density evolution analysis shows that $\sigma^{\text{BP}}(\lambda, \rho) \approx 0.778$ which is only 0.18 dB away from capacity. Figure 5.32 shows the achievable rate pair $(1/2, 1/2)$ and the capacity region for $\sigma \approx 0.778$. ◇

Consider the number of minimal codewords in the product $C^{[1]} \times C^{[2]}$ and let $\hat{P}(x^{[1]}, x^{[2]})$ be its asymptotic generating function. Since the two codes are independent and since we are looking at minimal codewords, it follows that $\hat{P}(x^{[1]}, x^{[2]})$ is the sum of the individual generating functions. More precisely, if the degree distribution pair associated to $C^{[j]}$ is $(\lambda^{[j]}, \rho^{[j]})$ let

$$\mu^{[j]} = \Big(\frac{d\lambda^{[j]}(x)}{dx}\Big|_{x=0}\Big)\Big(\frac{d\rho^{[j]}(x)}{dx}\Big|_{x=1}\Big).$$

Then

$$\hat{P}(x^{[1]}, x^{[2]}) = -\frac{1}{2}\sum_{j=1}^{2} \ln\Big(1 - \mu^{[j]} x^{[j]}\Big).$$

The stability condition follows from this in the usual manner.

Lemma 5.35 (Stability Condition for Multiple-Access Channel). Consider transmission over the two-user BAWGNMAC(σ). Assume user j, $j = 1, 2$, uses the ensemble LDPC $(n, \lambda^{[j]}, \rho^{[j]})$ and that a BP decoder is used. Let $\mathfrak{B}(\mathsf{a}_{\text{BAWGNC}(\sigma)})$ denote the Bhattacharyya constant associated to the BAWGNC(σ) as computed in Example 4.26. Then the desired fixed point corresponding to correct decoding is stable if and only if for each j, $j = 1, 2$,

$$\mu^{[j]}\, \mathfrak{B}(\mathsf{a}_{\text{BAWGNC}(\sigma)}) < 1. \tag{5.36}$$

In words, the stability condition for the multiple-access case reduces to the two single-user stability conditions.

Notes

The material of Section 5.1 concerning fading channels is taken from Hou, Siegel, and Milstein [41]. The exact density for the USI case was worked out by Flegbo and Méasson [27].

The Z channel discussed in Section 5.2 is probably the simplest non-trivial asymmetric channel. It was shown by Majani and Rumsey [51] that for the class of binary-input discrete memoryless channels at most $1 - \frac{1}{2}e\ln(2) \approx 5.8$ percent of capacity is lost if we use the uniform input distribution instead of the optimal input distribution. This result was later strengthened by Shulman and Feder [67], who showed that the Z channel is extremal in this respect (see also [48]). Iterative decoding for the Z channel was first discussed by McEliece [53], who pointed out the fortunate consequence of the insensitivity of the capacity with respect to the input distribution for the setting of iterative decoding. The approach we presented, reducing the analysis to the symmetric case by regarding density evolution with respect to the dominant codeword type, is not the only possible approach. An alternative path, which considers the *average* of the density evolution analysis with respect to each codeword, was put forward by Wang, Kulkarni, and Poor [73]. Both approaches lead to the same result. The stability condition for the asymmetric case was first considered by Bennatan and Burshtein [11].

The first papers regarding iterative decoding over channels with memory (see Section 5.3) are from Douillard, Picart, Jézéquel, Didier, Berrou, and Glavieux [23], Hagenauer, Bauch, and Khorram [8, 38], and Garcia-Frias and Villasenor [28, 29, 30, 31, 32]. Kschischang and Eckford considered iterative decoding for the GEC [47, 25]. The code presented in Example 5.21 is taken from their work and the densities shown in Figure 5.22 were computed by Neuberg and Méasson [55]. A partial ordering (in the sense of thresholds) of GECs under iterative decoding is discussed by Eckford, Kschischang, and Pasupathy [26]. Iterative decoding for the IIC was investigated by Kavčić, Ma, and Mitzenmacher [46] as well as Doan and Narayanan [22]. The approach of computing the information rate of a channel with memory by means of a factor graph was proposed independently and more or less simultaneously by Arnold and Loeliger [4, 5], Sharma and Singh [66], Pfister, Soriaga, and Siegel [58, 57], as well as Kavčić [45] (see also the work by Holliday, Goldsmith, and Glynn [40].) Extensions of this method are discussed by Arnold, Loeliger, and Vontobel [6], Zhang, Duman, and Kurtas [76], and Dauwels and Loeliger [21].

In a landmark paper, Ungerboeck showed how one could construct bandwidth-efficient coding and modulation schemes, called *trellis-coded* modulation, built from binary (convolutional) codes and suitable signal constellations and maps [68, 69]. The first investigation into high spectral efficiency coding using iterative techniques (see Section 5.4) was performed by Le Goff, Glavieux, and Berrou [35] exactly one year after the publication of the turbo coding concept. This was soon followed by the work of Benedetto, Divsalar, and Montorsi [9], Robertson and Woerz [61, 62, 63], and Blackert and Wilson [14], as well as Benedetto, Divsalar, Montorsi, and Pollara [10]. An overview can be found in Robertson [60].

At the same time as Ungerboeck's set-partitioning scheme, the paper by Imai and Hirakawa appeared [43]. They proposed the so-called *multilevel* scheme. The multilevel scheme in principle can achieve capacity (see [70]) but it needs as component codes powerful binary codes. This is the reason why, until the advent of turbo codes, multilevel coding played a less prominent role. It was suggested by Wachsmann and Huber to build bandwidth-efficient coding schemes using the multilevel scheme [70, 71]. With the advent of powerful binary codes this was possible.

As we have discussed, BICM is closely related to multilevel schemes. It was introduced by Zehavi [75] as a pragmatic and flexible approach. A good summary can be found in the book by Jamali and Le-Ngoc [44]. A systematic study of BICM schemes can be found in the paper of Caire, Taricco, and Biglieri [18].

Consider now the multiple-access channel discussed in Section 5.5. The capacity region of the multiple-access channel was determined by Ahlswede [1], Liao [49], Bergmans and Cover [12], as well as Wyner [74]. An alternative to the time-sharing technique as well as the joint-iterative method discussed in the section is the rate-splitting method developed by Rimoldi and Urbanke [59] (for the Gaussian case) and Grant, Rimoldi, Urbanke, and Whiting [37] (for discrete memoryless channels). The first papers on the use of joint iterative techniques applied to the Gaussian multiple-access channel are the ones by Ibrahim and Kaleh [42], as well as by Chayat and Shamai [19]. Palanki, Khandekar, and McEliece [56] considered the binary adder multiple-access channel discussed in Problem 5.6. An optimization of degree distributions for multiple-access channels was performed by Amraoui, Dusad, and Urbanke [2]. The same problem was also discussed by Roumy, Declercq, and Fabre [64]. Coding for the MIMO broadcast channel was considered by Amraoui, Kramer, and Shamai [3]. LDPC codes for the fading Gaussian broadcast channels were studied by Berlin and Tuninetti [13].

The list of problems which are amenable to the factor graph approach is much larger than what we could present in the limited space. We mention just a few further applications. The source-coding problem for memoryless sources using iterative schemes was addressed by Garcia-Frias and Zhao [34]. It was then shown by Caire, Shamai, and Verdú [16, 17] how to compress stationary ergodic sources with memory. Further related references are the papers by Hagenauer, Barros, and Schaefer [39] as well as Dütsch and Hagenauer [24]. Iterative solutions to the Slepian-Wolf coding problem of correlated sources were proposed by Bajcsy and Mitran [7], Garcia-Frias and Zhao [33], Liveris, Xiong, and Georghiades [50], Murayama [54], as well as Schonberg, Ramchandran, and Pradhan [65]. The K-SAT problem was investigated by Braunstein, Mézard, and Zecchina [15] through the introduction of the so-called *survey* propagation algorithm. It was then shown by Maneva, Mossel, and Wainwright [52] and also Braunstein and Zecchina [15] that the survey propagation algorithm can be reformulated as a BP algorithm. Finally, the survey propagation algorithm was applied by Ciliberti, Mézard, and Zecchina [20] as well as Maneva and Wainwright [72] to the source-coding problem.

Problems

5.1 (Z-Channel: Optimal Input Distribution – Golay [36]). Consider the Z channel discussed in Section 5.2. Show that the choice of α that maximizes the information rate given in (5.9) is the one stated in (5.10).

5.2 (Z-Channel: Extremal Property – Majani and Rumsey [51]). Show that the uniform input distribution achieves at least a fraction $\frac{1}{2}\text{e}\ln(2) \approx 0.942$ of the channel capacity of the Z channel for all channel parameters ϵ.

Hint: Show that $I_\alpha/(\alpha \ln \frac{1}{\alpha})$ is a monotone increasing function in α for $\alpha \in [0,1]$. Now use the fact that for $\alpha \in [0,1]$, $\alpha \ln \frac{1}{\alpha} \leq \frac{1}{\text{e}}$.

Note: Without proof we mention that the Z channel is extremal in this sense. This means that for any binary-input memoryless channel the fraction of capacity that can be achieved using the uniform input distribution is at least $\frac{1}{2}\text{e}\ln(2)$.

5.3 (Symmetry of "Symmetrized Density"). Consider a binary-input memoryless channel (not necessarily symmetric) and let $\mathsf{a}^+_{\text{BM}}(y)$ and $\mathsf{a}^-_{\text{BM}}(y)$ denote the corresponding L-densities assuming that $X = \pm 1$, respectively. Define the "symmetrized" density

$$\mathsf{a}(y) = (\mathsf{a}^+_{\text{BM}}(y) + \mathsf{a}^-_{\text{BM}}(-y))/2.$$

Show that this density is symmetric.

5.4 (Z-Channel: The Bhattacharyya Constant and Stability Condition). Using the symmetrized density, compute the Bhattacharyya constant for the ZC(ϵ) according to (4.62). Show that the resulting stability condition reads $\lambda'(0)\rho'(1) < 1/\sqrt{\epsilon}$.

5.5 (Two-Dimensional Modulation Schemes). In the two-dimensional case, X_t is assumed to be an element of $\mathbb{R}^2$ with average energy equal to $2E$ and Z_t is a two-dimensional zero-mean Gaussian with independent components, each of variance σ^2. This normalization ensures that the energy and the noise variance *per dimension* are the same as in the one-dimensional case. Let X_t take values from the *signal set* $\mathcal{S} = \{s_1, \ldots, s_{2^k}\}$, $s_i \in \mathbb{R}^2$, where for convenience we have assumed that the number of signal points is a power of 2. Figure 5.37 shows several standard constellations:

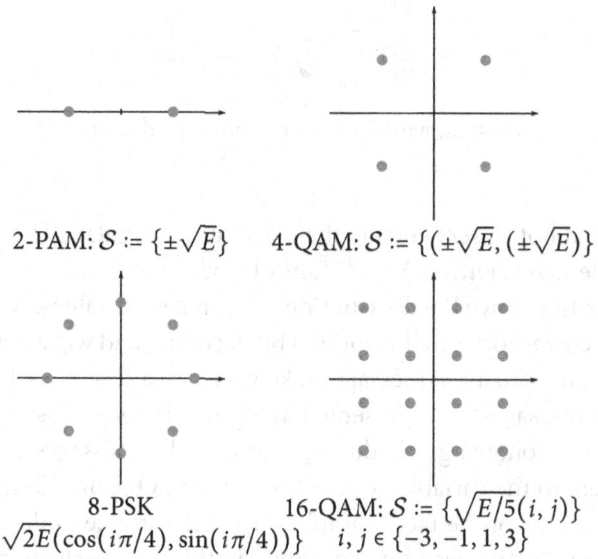

Figure 5.37: Four standard signal constellations: 2-PAM (top left), 4-QAM (top right), 8-PSK (bottom left), and 16-QAM (bottom right). In all cases it is assumed that the prior on $\mathcal{S}$ is uniform. The signal constellations are scaled so that the average energy per dimension is E.

2-PAM, which we discussed beforehand, 4-QAM, and 16-QAM, where the signal points are arranged on a square grid, and 8-PSK, in which the signals are equally spaced along a circle. In each case a uniform prior on the signals is assumed.

Show that the capacity for a given constellation $\mathcal{S}$ can be written as

$$\begin{aligned} C_{\mathcal{S}} &= I(X;Y) = h(Y) - h(Y\,|\,X) = h(Y) - h(Z) \\ &= -\int p_Y(y)\log p_Y(y)\mathrm{d}y - \log(2\pi e\sigma^2), \end{aligned}$$

where

$$p_Y(y) = \frac{1}{2\pi\sigma^2|\mathcal{S}|}\sum_{s\in\mathcal{S}} \mathrm{e}^{-\frac{\|y-s\|^2}{2\sigma^2}}.$$

Plot the capacities of the four constellations shown in Figure 5.37 and compare them with the Shannon capacity (using Gaussian inputs).

5.6 (Analysis of Noiseless Multiple-Access Binary-Input Adder Channel – Palanki, Khandekar, and McEliece [56]). Consider the following noiseless two-user multiple-access channel shown in Figure 5.38. We have $Y_t = X_t^{[1]} + X_t^{[2]}$, where $X_t^{[j]} \in \{0,1\}$, $j = 1, 2$, and $Y_t \in \{0, 1, 2\}$.

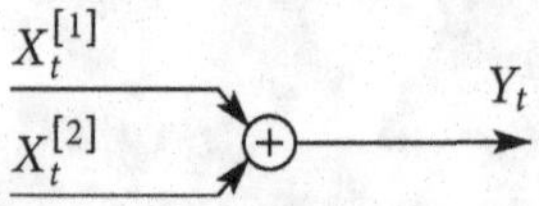

Figure 5.38: Multiple-access binary adder channel.

Assume that both users encode their data using regular LDPC codes from the same ensemble LDPC $\left(n, \lambda(x) = x^{\mathtt{l}-1}, \rho(x) = x^{\mathtt{r}-1}\right)$.

Start by writing down the distribution of the received values. Next, draw the factor graph that corresponds to the optimal bit decoding and write down the message-passing rules. Show that the messages take values in a discrete set. What is this set assuming that messages are represented as log-likelihood ratios?

Consider the following scheduling. Initially, the messages received from the channel are sent to the variable nodes of both users. One decoding iteration is performed in each part of the factor graph. Then the variable nodes of the two users exchange messages. The decoding continues in this way, performing in parallel the operations of both users. We call this the *symmetric* schedule. Show that for a fixed number of iterations the computation graph is indeed a tree with probability converging to one as the blocklength tends to infinity. Assuming symmetric scheduling, argue that the average error probability $\mathrm{P}^{\mathrm{BP}}_{\mathcal{T}_\ell}$ can be expressed in the form of a one dimensional recursion. Write down this recursion.

What is the degree distribution pair $(\mathtt{l}, \mathtt{r})$ corresponding to the highest rate that you can find such that, for infinite length, the average error probability after decoding on this factor graph tends to zero.

Assume that, for any erasure probability, we can find degree distribution pairs that achieve capacity on a BEC. Devise a method that enables the construction of codes that are suited for our binary-input adder channel from one of these capacity-achieving codes.

What is the capacity region of this multiple-access channel?

References

[1] R. Ahlswede, *Multi-way communication channels*, in Proc. of the IEEE Int. Symposium on Inform. Theory, Tsahkadsor, Armenian SSSR., 1971, pp. 23–52. [313]

[2] A. Amraoui, S. Dusad, and R. Urbanke, *Achieving general points in the 2-user Gaussian MAC without time-sharing or rate-splitting by means of iterative coding*, in Proc. of the IEEE Int. Symposium on Inform. Theory, Lausanne, Switzerland, June 2002, p. 334. [265, 313]

[3] A. Amraoui, G. Kramer, and S. Shamai, *Coding for the MIMO broadcast channel*, in Proc. of the IEEE Int. Symposium on Inform. Theory, Tokyo, Japan, June 29–July 4, 2003, p. 296. [313]

[4] D. M. Arnold, *Computing Information Rates of Finite-State Models with Application to Magnetic Recording*, PhD thesis, ETHZ, Zürich, Switzerland, 2003. [312]

[5] D. M. Arnold and H.-A. Loeliger, *On the information rate of binary-input channels with memory*, in Proc. 2001 IEEE Int. Conf. on Commun., vol. 9, Helsinki, Finnland, June 11–14, 2001, pp. 2692–2695. [312]

[6] D. M. Arnold, H.-A. Loeliger, and P. O. Vontobel, *Computation of information rates from finite-state source/channel models*, in Proc. of the Allerton Conf. on Commun., Control, and Computing, Monticello, IL, USA, Oct. 2002, pp. 457–466. [312]

[7] J. Bajcsy and P. Mitran, *Coding for the Slepian-Wolf problem with turbo codes*, in Proc. of GLOBECOM, San Antonio, TX, USA, Nov. 25–29, 2001, IEEE, pp. 1400–1404. [313]

[8] G. Bauch, H. Khorram, and J. Hagenauer, *Iterative equalization and decoding in mobile communications systems*, in Proc. of GLOBECOM, 1997. [312]

[9] S. Benedetto, D. Divsalar, and G. Montorsi, *Bandwidth efficient parallel concatenated coding schemes*, Electron. Lett., 31 (1995), pp. 2067–2069. [313]

[10] S. Benedetto, D. Divsalar, G. Montorsi, and F. Pollara, *Parallel concatenated trellis coded modulation*, in Proc. of ICC, June 1996, pp. 974–978. [313]

[11] A. Bennatan and D. Burshtein, *Iterative decoding of LDPC codes over arbitrary discrete-memoryless channels*, in Proc. of the Allerton Conf. on Commun., Control, and Computing, Monticello, IL, USA, Oct. 2003, pp. 1416–1425. [312]

[12] P. P. Bergmans and T. M. Cover, *Cooperative broadcasting*, IEEE Trans. Inform. Theory, IT–20 (1974), pp. 317–324. [313]

[13] P. Berlin and D. Tuninetti, *LDPC codes for fading Gaussian broadcast channels*, IEEE Trans. Inform. Theory, 51 (2005), pp. 2173–2182. [313]

[14] W. J. Blackert and S. G. Wilson, *Turbo trellis coded modulation*, in Proc. of CISS'96, Mar. 1996. [313]

[15] A. Braunstein, M. Mézard, and R. Zecchina, *Survey propagation: algorithm for satisfiability*. E-print: cs.CC//0212002. [314]

[16] G. Caire, S. Shamai, and S. Verdú, *Lossless data compression with error correcting code*, in Proc. of the IEEE Int. Symposium on Inform. Theory, Yokohama, Japan, June 29–July 4, 2003, p. 22. [313]

[17] ——, *A new data compression algorithm for sources with memory based on error correcting codes*, in Proc. of the IEEE Inform. Theory Workshop, Paris, France, Mar. 2003, pp. 291–295. [313]

[18] G. Caire, G. Taricco, and E. Biglieri, *Bit-interleaved coded modulation*, IEEE Trans. Inform. Theory, 44 (1998), pp. 927–946. [313]

[19] N. Chayat and S. Shamai, *Convergence properties of iterative soft onion peeling*, in Proc. of IEEE ITW'99, Kruger National Park, South Africa, June 1999, pp. 9–11. [313]

[20] S. Ciliberti, M. Mézard, and R. Zecchina, *Lossy data compression with random gates*, Phys. Rev. Lett., 95 (2005). [314]

[21] J. Dauwels and H.-A. Loeliger, *Computation of information rates by particle methods*, in Proc. of the IEEE Int. Symposium on Inform. Theory, Chicago, IL, USA, June 27–July 2, 2004, p. 178. [312]

[22] D. N. Doan and K. R. Narayanan, *Design of good low-rate coding schemes for ISI channels based on spectral shaping*, IEEE Trans. Wirel. Commun., 4 (2005), pp. 2309–2317. [312]

[23] C. Douillard, A. Picart, M. Jézéquel, P. Didier, C. Berrou, and A. Glavieux, *Iterative correction of intersymbol interference: Turbo-equalization*, Eur. Trans. Telecomm. (ETT), 6 (1995), pp. 507–511. [312]

[24] N. Dütsch and J. Hagenauer, *Combined incremental and decremental redundancy in joint source-channel coding*, in Proc. Int. Symposium on Inform. Theory and its Applications (ISITA 2004), Parma, Italy, Oct. 2004, pp. 775–779. [313]

[25] A. W. Eckford, *Low-Density Parity-Check Codes for Gilbert-Elliot and Markov-Modulated Channels*, PhD thesis, University of Toronto, 2004. [312]

[26] A. W. Eckford, F. R. Kschischang, and S. Pasupathy, *On partial ordering of Markov modulated channels under LDPC decoding*, in Proc. of the IEEE Int. Symposium on Inform. Theory, Tokyo, Japan, June 29–July 4, 2003, p. 295. [312]

[27] F.-E. Flegbo, *Iterative coding over the fading channel.* EPFL, Semester Project (Supervisor: Cyril Méason), 2002/2003. [312]

[28] J. Garcia-Frias and J. D. Villasenor, *Combining hidden Markov source models and parallel concatenated codes*, IEEE Commun. Lett., 1 (1997), pp. 111–113. [312]

[29] ——, *Exploiting binary Markov channels with unknown parameters in Turbo decoding*, in Proc. of GLOBECOM, Sydney, Australia, Nov. 1998, pp. 3244–3249. [312]

[30] ——, *Joint source channel coding and estimation of hidden Markov structures*, in Proc. of the IEEE Int. Symposium on Inform. Theory, Cambridge, MA, USA, Aug. 1998, p. 201. [312]

[31] ——, *Turbo decoders for Markov channels*, IEEE Commun. Lett., 2 (1998), pp. 257–259. [312]

[32] ——, *Joint turbo decoding and estimation of hidden Markov sources*, IEEE J. Sel. Area. Commun., 19 (2001), pp. 1671–1679. [312]

[33] J. Garcia-Frias and Y. Zhao, *Compression of correlated binary sources using turbo codes*, IEEE Commun. Lett., 5 (2001), pp. 417–419. [313]

[34] ———, *Compression of correlated binary sources using punctured turbo codes*, IEEE Commun. Lett., 6 (2002), pp. 394–396. [313]

[35] S. L. Goff, A. Glavieux, and C. Berrou, *Turbo-codes and high spectral efficiency modulation*, in Proc. of ICC, New Orleans, LA, May 1994, pp. 645–649. [312, 366]

[36] S. W. Golomb, *The limiting behavior of the Z-channel*, IEEE Trans. Inform. Theory, 26 (1980), p. 372. [314]

[37] A. Grant, B. Rimoldi, R. Urbanke, and P. Whiting, *Rate-splitting multiple access for discrete memoryless channel*, IEEE Trans. Inform. Theory, 47 (2001), pp. 873–890. [313]

[38] J. Hagenauer, *The turbo principle–tutorial introduction and state of the art*, in Proc. of the Int. Conf. on Turbo Codes and Related Topics, Brest, France, Sept. 1997, pp. 1–11. [312, 366]

[39] J. Hagenauer, J. Barros, and A. Schaefer, *Lossless turbo source coding with decremental redundancy*, in Proc. 5th International ITG Conference on Source and Channel Coding (SCC'04), Erlangen, Germany, Jan. 2004, pp. 333–340. [313]

[40] T. Holliday, A. Goldsmith, and P. Glynn, *Capacity of finite state Markov channels with general inputs*, in Proc. of the IEEE Int. Symposium on Inform. Theory, Yokohama, Japan, June 29–July 4, 2003, IEEE, p. 289. [312]

[41] J. Hou, P. H. Siegel, and L. B. Milstein, *Performance analysis and code optimization of low density parity-check codes on rayleigh fading channels*, IEEE J. Sel. Area. Commun., 19 (2001), pp. 924–934. [312]

[42] N. Ibrahim and G. K. Kaleh, *Iterative decoding and soft interference cancellation for the Gaussian multiple access channel*, in Int. Symp. on Signals, Systems, and Electron., Pisa, Italy, 1998, pp. 156–161. [313]

[43] H. Imai and S. Hirakawa, *A new multilevel coding method using error-correcting codes*, IEEE Trans. Inform. Theory, 23 (1977), pp. 371–377. [313]

[44] S. H. Jamali and T. Le-Ngoc, *Coded Modulation Techniques for Fading Channels*, Kluwer Academic Publ., New York, NY, USA, 1994. [313]

[45] A. Kavčić, *On the capacity of Markov sources over noisy channels*, in Proc. of GLOBECOM, San Antonio, TX, USA, Nov. 25–29, 2001, IEEE, pp. 2997–3001. [312]

[46] A. Kavčić, X. Ma, and M. Mitzenmacher, *Binary intersymbol interference channels: Gallager codes, density evoluton, and code performance bounds*, IEEE Trans. Inform. Theory, 49 (2003), pp. 1636–1652. [312]

[47] F. R. Kschischang and A. W. Eckford, *Low-density parity-check codes for the Gilbert-Elliot channel*, in Proc. of the Allerton Conf. on Commun., Control, and Computing, Monticello, IL, USA, Oct. 2003. [312]

[48] X.-B. Liang, *On a conjecture of Majani and Rumsey*, in Proc. of the IEEE Int. Symposium on Inform. Theory, June 2004, p. 62. [312]

[49] H. Liao, *Multiple access channels*, PhD thesis, University of Hawaii, Honolulu, HI, USA, 1972. [313]

[50] A. D. Liveris, Z. Xiong, and C. N. Georghiades, *Compression of binary sources with side information at the decoder using LDPC codes*, IEEE Commun. Lett., 6 (2002), pp. 440–442. [313]

[51] E. E. Majani and H. Rumsey, Jr., *Two results on binary-input discrete memoryless channels*, in Proc. of the IEEE Int. Symposium on Inform. Theory, June 1991, p. 104. [312, 314]

[52] E. Maneva, E. Mossel, and M. J. Wainwright, *A new look at survey propagation and its generalizations*, in SODA, Vancouver, Canada, 2005. [314]

[53] R. J. McEliece, *Are turbo-like codes effective on nonstandard channels?*, IEEE Inform. Theory Soc. Newslett., 51 (2001), pp. 1–8. [312]

[54] T. Murayama, *Statistical mechanics of the data compression theorem*, J. Phys. A: Math. Gen., 35 (2002). [313]

[55] C. Neuberg, *Gilbert-Elliott channel and iterative decoding.* EPFL, Semester Project (Supervisor: Cyril Méason), 2004. [312]

[56] R. Palanki, A. K. Khandekar, and R. J. McEliece, *Graph-based codes for synchronous multiple-access channels*, in Proc. of the Allerton Conf. on Commun., Control, and Computing, Monticello, IL, USA, Oct. 2001. [313, 315]

[57] H. D. Pfister, *On the Capacity of Finite State Channels and the Analysis of Convolutional Accumulate-m Codes*, PhD thesis, UCSD, San Diego, CA, USA, 2003. [312, 367, 373]

[58] H. D. Pfister, J. B. Soriaga, and P. H. Siegel, *On the achievable information rates of finite-state ISI channels*, in Proc. of GLOBECOM, vol. 5, San Antonio, TX, USA, Nov. 25–29, 2001, IEEE, pp. 2992–2996. [312]

[59] B. Rimoldi and R. Urbanke, *A rate-splitting approach to the Gaussian multiple-access channel*, IEEE Trans. Inform. Theory, 42 (1996), pp. 364–375. [313]

[60] P. Robertson, *An overview of bandwidth efficient turbo coding schemes*, in Proc. of the Int. Conf. on Turbo Codes and Related Topics, Brest, France, Sept. 1997, pp. 103–110. [313]

[61] P. Robertson and T. Woerz, *Coded modulation scheme employing turbo codes*, Electron. Lett., 31 (1995), pp. 1546–1547. [313]

[62] ——, *A novel bandwidth efficient coding scheme employing turbo codes*, in Proc. of ICC, June 1996, pp. 962–967. [313]

[63] ——, *Extensions of turbo trellis coded modulation to high bandwidth efficiencies*, in Proc. of ICC, vol. 3, June 1997, pp. 1251–1255. [313]

[64] A. Roumy, D. Declercq, and E. Fabre, *Low Complexity Code Design for the 2-user Gaussian Multiple Access Channel*, in Proc. of the IEEE Int. Symposium on Inform. Theory, June 2004, p. 483. [313]

[65] D. Schonberg, K. Ramchandran, and S. S. Pradhan, *Distributed code constructions for the entire Slepian-Wolf rate region for arbitrarily correlated sources*, in DCC, Utah, UT, USA, 2004, pp. 292–301. [313]

[66] V. Sharma and S. K. Singh, *Entropy and channel capacity in the regenerative setup with applicatons to Markov channels*, in Proc. of the IEEE Int. Symposium on Inform. Theory, Washington, DC, USA, June 24–29, 2001, IEEE, p. 283. [312]

[67] N. Shulman and M. Feder, *The uniform distribution as a universal prior*, IEEE Trans. Inform. Theory, 50 (2004), pp. 1356–1362. [312]

[68] G. Ungerboeck, *Trellis-coded modulation with redundant signal sets part I: Introduction*, IEEE Commun. Mag., 25 (1987), pp. 5–11. [312]

[69] ——, *Trellis-coded modulation with redundant signal sets part II: State of the art*, IEEE Commun. Mag., 25 (1987), pp. 12–21. [312]

[70] U. Wachsmann, F. H. Fischer, and J. B. Huber, *Multilevel codes: Theoretical concepts and practical design rules*, IEEE Trans. Inform. Theory, 45 (1999), pp. 1361–1391. [313]

[71] U. Wachsmann and J. B. Huber, *Power and bandwidth efficient digital communication using turbo-codes in multilevel codes*, Eur. Trans. Telecomm. (ETT), 6 (1995), pp. 557–567. [313]

[72] M. J. Wainwright and E. Maneva, *Lossy source coding via message-passing and decimation over generalized codewords of LDGM codes*, in Proc. of the IEEE Int. Symposium on Inform. Theory, Adelaide, Australia, Sept. 2005, pp. 1493–1497. [314]

[73] C.-C. Wang, S. R. Kulkarni, and H. V. Poor, *Density evolution for asymmetric memoryless channels*, IEEE Trans. Inform. Theory, 51 (2005), pp. 4216–4236. [312]

[74] A. D. Wyner, *Recent results in the Shannon theory*, IEEE Trans. Inform. Theory, IT–20 (1974), pp. 2–10. [313]

[75] E. Zehavi, *8-PSK trellis codes for a Rayleigh channel*, IEEE Trans. Commun., 40 (1992), pp. 873–884. [313]

[76] Z. Zhang, T. M. Duman, and E. M. Kurtas, *Information rates of binary-input intersymbol interference channels with signal-dependent media noise*, IEEE Trans. Magnet., 39 (2003), pp. 599–607. [312]

Chapter 6
TURBO CODES

Turbo codes played a decisive role in the development of iterative decoding. Therefore, they deserve special attention. There are many variations on the theme. We start with a discussion of convolutional codes, the building blocks of turbo codes. We then introduce the two main flavors of turbo codes: parallel and serially concatenated convolutional codes with systematic recursive rate one-half convolutional encoders as component codes. We formulate most theorems only for the parallel case and leave the extensions to the serial case as problems. In Section 6.10 we briefly mention some of the many generalizations.

§6.1. Convolutional Codes

The encoding function of block codes maps *blocks* of data into (longer) blocks of data. *Convolutional* codes, on the other hand, are codes in which the encoder maps (in principle continuous and infinite) *streams* of data into (more) streams of data. The mapping (encoding) is realized by sending the input streams over linear filters. The name convolutional code/encoder stems from the fact that this filtering operation can be expressed as a convolution.

Depending on the nature of the filter, we distinguish between different types of convolutional *encoders*. The most common case is when all operations are over $\mathbb{F}_2$. We then talk about *binary* convolutional encoders. The filter can be feed-forward or recursive. We then talk about *non-recursive* or *recursive* encoders. The filter can have a single or multiple input stream(s). Finally, if the input streams appear unaltered among the output streams we talk about *systematic* encoders.

The most important convolutional encoders in the context of iterative decoding are *binary systematic recursive convolutional* encoders since they are the defining components for standard *turbo codes*. We discuss their structure and the associated decoding algorithm in detail.

A *binary systematic recursive convolutional encoder of memory m and rate one-half* is defined in terms of a binary rational function $G(D) = p(D)/q(D)$ of memory m, $m = \max\{\deg(p), \deg(q)\}$. In order for this rational function to represent a well-defined filter we require that $q_0 = 1$ (see the discussion starting on page 505 in Appendix D). Note that we allow filters also with $\deg(p) > \deg(q)$ and/or $\gcd(p, q) \neq 1$. It is convenient for future discussions to introduce the following shorthand to denote binary rational functions: with some abuse of notation, we characterize the polynomial $p(D) = \sum_i p_i D^i$ with binary coefficients p_i by p which is equal to $\sum_i p_i 2^i$ in *octal* notation. For example, the octal number 21 is equal in binary notation to 010001. In both cases the least significant digit is on the right and each octal digit corresponds to a binary triple: e.g., 2 corresponds to 010 and 1 corresponds to 001. The polynomial with binary coefficients corresponding to 21 is therefore $1 + D^4$. In this sense we write $G = p/q$, where p and q are the octal numbers which characterize $p(D)$ and $q(D)$, respectively. To give some more examples, $G = 21/37$ corresponds to $G(D) = (1 + D^4)/(1 + D + D^2 + D^3 + D^4)$ and $G = 1/13$ corresponds to $G(D) = 1/(1 + D + D^3)$. Figure 6.1 depicts the particular example $G = 7/5$.

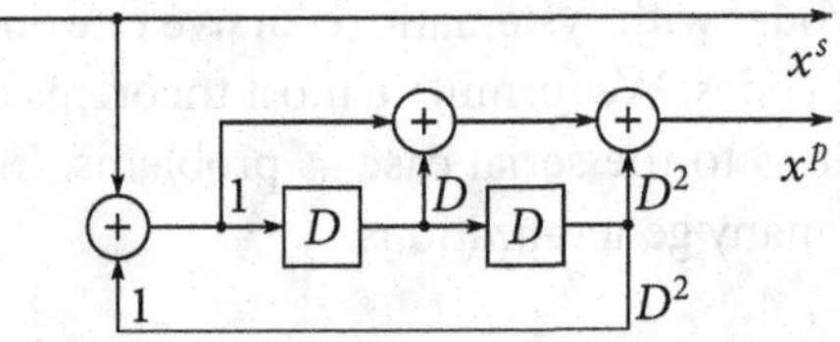

Figure 6.1: Binary systematic recursive convolutional encoder of memory $m = 2$ and rate one-half defined by $G = 7/5$. The two square boxes are delay elements. The 7 corresponds to $1 + D + D^2$. These are the coefficients of the "forward" branch (the top branch of the filter) with 1 corresponding to the leftmost coefficient. In a similar manner, 5 corresponds to $1 + D^2$, which represents the coefficients of the "feedback" branch. Again, the leftmost coefficient corresponds to 1.

Although the "natural" setting is to consider encoders that map semi-infinite streams into semi-infinite streams, for our application it is more convenient to consider "terminated" schemes that have a fixed blocklength. For a given rational function G and a given length n, $n \in \mathbb{N}$, we define a code $C(G, n)$ as follows: let the input x^s (*systematic* bits) be

$$x^s = \big(x_1^s, \ldots, x_n^s, \underbrace{0, \ldots, 0}_{m \text{ times}}\big),$$

where the first n components are arbitrary elements of $\mathbb{F}_2$ and the last m components are zero. Associated with each input x^s is an output $x^p = \big(x_1^p, \ldots, x_{n+m}^p\big)$ (*parity* bits). It is the result of passing x^s through a linear filter. For the first n steps the filter is $G(D)$, whereas for the last m steps the filter is $\bar{G}(D) = p(D)$, i.e., we *remove the feedback*. This is not necessarily the best termination scheme for a particular situation, but it is universally applicable and simplifies our discussion. The Notes at the end of this chapter contain a summary of the many alternative termination schemes proposed in the literature. As discussed in Problem 6.18, it is typically quite straightforward to adapt the calculations of the weight distribution given in the following to alternative termination schemes.

A convenient way of describing this map is in terms of a so-called *state-space* model: let x_i^s denote the i-th component of the input, where $i = 1, \ldots, n+m$. Further, let σ_i denote the *state* of the system at time i, $i = 0, \ldots, n+m$. We define the state σ_{i-1} to be equal to the content of the shift register just *before* the i-th bit x_i^s has been input to the system and σ_i to be the resulting content of the shift register. We therefore have the sequence $(\sigma_{i-1}, x_i^s) \mapsto (\sigma_i, x_i^p)$. The state is a binary m-tuple. By convention, the initial state is the *all-zero* state, i.e., we have $\sigma_0 = (0, \ldots, 0)$. For $1 \le i \le n+m$, the evolution of the system is described by

$$(6.2) \qquad \sigma_i = \sigma_{i-1}A + x_i^s C, \qquad\qquad x_i^p = \sigma_{i-1}B^T + x_i^s p_0,$$

where A is an $m \times m$ binary matrix and B and C are $1 \times m$ binary matrices. More specifically, in terms of $p(D)$ and $q(D)$ these matrices are

$$A = \begin{pmatrix} q_1 & 1 & 0 & \cdots & 0 \\ q_2 & 0 & 1 & \cdots & 0 \\ \vdots & \vdots & \vdots & \ddots & \vdots \\ q_{m-1} & 0 & 0 & \cdots & 1 \\ q_m & 0 & 0 & \cdots & 0 \end{pmatrix}, \quad B^T = \begin{pmatrix} p_1 + p_0 q_1 \\ p_2 + p_0 q_2 \\ \vdots \\ p_{m-1} + p_0 q_{m-1} \\ p_m + p_0 q_m \end{pmatrix}, \quad C^T = \begin{pmatrix} 1 \\ 0 \\ \vdots \\ 0 \\ 0 \end{pmatrix},$$

where for the last m steps we have $q(D) = 1$. By removing the feedback in the last m steps we guarantee that the final state is again the all-zero state, i.e., $\sigma_{n+m} = (0, \ldots, 0)$.

Let the encoding map be denoted by $x^p = \gamma(x^s)$. For simplicity we suppress the dependency of γ on G and n in our notation. In terms of γ, the code $C(G, n)$ is defined as

$$C(G,n) = \Big\{ \big(x^s, \gamma(x^s)\big) : x^s = \big(x_1^s, \ldots, x_n^s, \underbrace{0, \ldots, 0}_{m \text{ times}}\big), x_i^s \in \mathbb{F}_2 \Big\}.$$

Because of the m appended zeros the rate is slightly less than one-half but the difference is of order $1/n$ and we ignore this issue in the following. It is important to note that, as in the case of block codes, there are many different encoders which generate the same code. For iterative decoding systems, the choice of encoder is typically at least as important as the choice of code.

Example 6.3 (Encoding and State Sequence for $G = 7/5$). Consider the encoder depicted in Figure 6.1 where $G = 7/5$ and $n = 5$. We have $m = \max\{\deg(1 + D + D^2), \deg(1 + D^2)\} = 2$. Table 6.4 lists the state sequence σ as well as the output x^p associated with the input $x^s = (x_1^s, x_2^s, x_3^s, x_4^s, x_5^s, x_6^s, x_7^s) = (1, 0, 1, 1, 0, 0, 0)$. ◇

Assume that a codeword of C is transmitted over a binary memoryless symmetric (BMS) channel with transition probability $p(y \mid x)$. In order to avoid having to

i	x_i^s	σ_i	x_i^p	i	x_i^s	σ_i	x_i^p
0		(00)		4	1	(10)	1
1	1	(10)	1	5	0	(01)	1
2	0	(01)	1	6	0	(00)	1
3	1	(00)	1	7	0	(00)	0

Table 6.4: State sequence σ as well as the output x^p associated with the input $x^s = (x_1^s, x_2^s, x_3^s, x_4^s, x_5^s, x_6^s, x_7^s) = (1,0,1,1,0,0,0)$ for the code $C(G = 7/5, n = 5)$.

switch notation frequently let us refer in this chapter to the logical values of the input, i.e., we assume that x takes values in $\{0,1\}$ (rather than the usual $\{\pm 1\}$, which corresponds to the physical values the input takes on). Let (Y^s, Y^p), taking values (y^s, y^p), be the observation at the output of the channel assuming (X^s, X^p) was transmitted. We are interested in the optimal decoder. We start with the bit-wise MAP decoder. We denote its decoding function by $\hat{x}_i^s = \hat{x}_i^s(y^s, y^p)$:

$$\begin{aligned}
\hat{x}_i^s &= \text{argmax}_{x_i^s \in \{0,1\}} p(x_i^s \mid y^s, y^p) \\
&= \text{argmax}_{x_i^s \in \{0,1\}} \sum_{\sim x_i^s} p(x^s, x^p, \sigma \mid y^s, y^p) \\
&= \text{argmax}_{x_i^s \in \{0,1\}} \sum_{\sim x_i^s} p(y^s, y^p \mid x^s, x^p, \sigma) p(x^s, x^p, \sigma) \\
&= \text{argmax}_{x_i^s \in \{0,1\}} \sum_{\sim x_i^s} p(\sigma_0) \prod_{j=1}^{n+m} \underbrace{p(y_j^s \mid x_j^s) p(y_j^p \mid x_j^p)}_{\text{channel}} \\
&\qquad \underbrace{p(x_j^s)}_{\text{prior}} \underbrace{p(x_j^p, \sigma_j \mid x_j^s, \sigma_{j-1})}_{\text{allowed transitions}}.
\end{aligned}$$

Due to our convention on the initial state, $p(\sigma_0)$ is a zero-one function: it is 1 for the all-zero state, and zero otherwise. In the preceding derivation we encountered an important new ingredient in finding efficient representations. The crucial innovation is to introduce the *state* sequence σ, even though this state sequence is "hidden" and cannot be observed. Figure 6.5 shows the corresponding Forney-style factor graph (FSFG). Note that the factor node p_{Σ_0} at the bottom of the figure enforces the constraint that all paths start in the zero state. No such node is needed at the top of the figure: the constraint that all paths end in the zero straight is implicitly contained in the factor nodes $p_{X_i^p, \Sigma_i \mid X_i^s, \Sigma_{i-1}}$ for the last m steps. This FSFG is a tree and therefore the message-passing algorithm is *exact*. In the conventional coding literature the associated message-passing algorithm applied to the decoding of convolutional

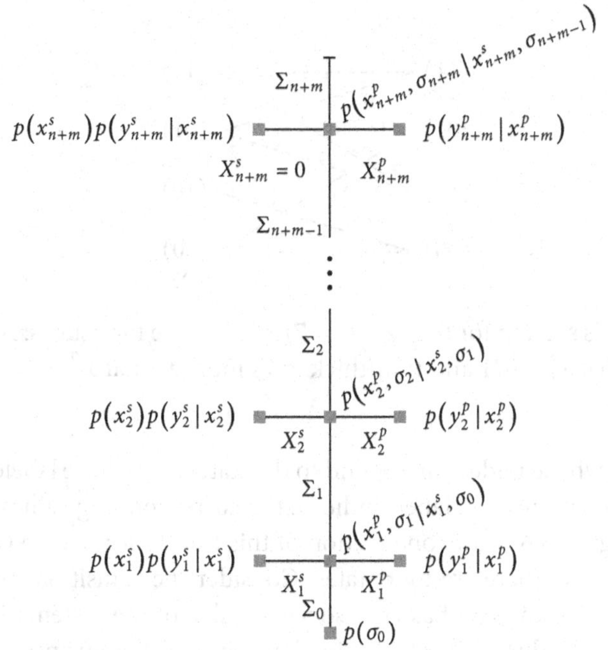

Figure 6.5: FSFG for the MAP decoding of $C(G, n)$.

codes is known as the *BCJR algorithm*. The graph is essentially a line. Therefore, for the message-passing algorithm there are essentially two flows of information. In our representation this is the flow which starts at the bottom and goes toward the top and the reverse flow, starting at the top and eventually reaching the bottom. These two flows of messages correspond to what in the standard literature are called the *α-recursion* and *β-recursion*, respectively. Finally, consider the decision step. It is usually called the *γ-step*. In principle we could at this point simply refer to the message-passing rules summarized in Figure 2.12. But since the BCJR algorithm is the core ingredient that makes *turbo codes* perform well, it is worth discussing this special case in some more detail.

Consider one *section* of the factor graph. This section consists of two consecutive states, call them σ_{i-1} and σ_i, the two associated variables x^s_i and x^p_i, and three factor nodes: two of these factor nodes correspond to $p(x^s_i)p(y^s_i \mid x^s_i)$ and $p(y^p_i \mid x^p_i)$, respectively. They describe the effect of the channel as well as the prior on x^s_i. The third factor node is associated with $p(x^p_i, \sigma_i \mid x^s_i, \sigma_{i-1})$. This is a $\{0, 1\}$-valued function: given that the encoder is in state σ_{i-1} and that the input to the encoder is x^s_i, the function describes which is the next state σ_i and what is the corresponding output bit x^p_i. A useful graphical way to describe this function is in form of a *trellis*, a

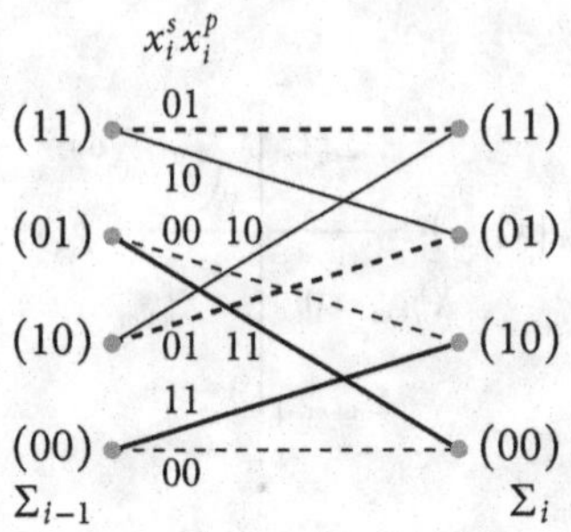

Figure 6.6: Trellis section for the case $G = 7/5$. There are four states. A dashed/solid line indicates that $x_i^s = 0/1$ and thin/thick lines indicate that $x_i^p = 0/1$.

directed graph, whose nodes correspond to the states and whose labeled edges indicate allowed transitions. The labels indicate the corresponding values of x_i^s and x_i^p, respectively. Figure 6.6 shows one section of this trellis for the case $G = 7/5$. Since for this case $m = 2$, there are four states. Consider the transition from state (10) to state (11). This transition has an associated value of the systematic bit of 1 and the parity bit output during this transition has the value 0. In principle this graph is directed. The natural evolution is from left to right corresponding to the encoding operation. As we will soon see, the decoding also requires us to pass this graph from right to left. It is common however not to draw any arrows.

We start with the α-recursion corresponding to the flow of messages from bottom to top in Figure 6.5. Consider the message sent along the edge representing Σ_i. Let us call this message $\alpha_{\Sigma_i}(\sigma_i)$. By the message-passing rules (see (2.21)) this message is equal to

$$\sum_{x_i^s, x_i^p, \sigma_{i-1}} \underbrace{p(x_i^p, \sigma_i \mid x_i^s, \sigma_{i-1})}_{\text{kernel}} \underbrace{p(x_i^s) p(y_i^s \mid x_i^s) p(y_i^p \mid x_i^p) \alpha_{\Sigma_{i-1}}(\sigma_{i-1})}_{\text{product of incoming messages}}.$$

We claim that $\alpha_{\Sigma_i}(\sigma_i) = p(\sigma_i, y_1^s, \dots, y_i^s, y_1^p, \dots, y_i^p)$. This is true for $\alpha_{\Sigma_0}(\sigma_0)$, since the emitted message at a leaf node is equal to the function itself, i.e., $\alpha_{\Sigma_0}(\sigma_0) = p_{\Sigma_0}(\sigma_0)$. To prove the claim for $i > 0$, use induction and write

$$\begin{aligned}
&p(\sigma_i, y_1^s, \dots, y_i^s, y_1^p, \dots, y_i^p) \\
&\quad = \sum_{x_i^s, x_i^p, \sigma_{i-1}} p(\sigma_{i-1}, y_1^s, \dots, y_{i-1}^s, y_1^p, \dots, y_{i-1}^p, x_i^s, x_i^p, y_i^s, y_i^p, \sigma_i) \\
&\quad = \sum_{x_i^s, x_i^p, \sigma_{i-1}} p(\sigma_{i-1}, y_1^s, \dots, y_{i-1}^s, y_1^p, \dots, y_{i-1}^p) p(x_i^s) \\
&\qquad\qquad p(x_i^p, \sigma_i \mid x_i^s, \sigma_{i-1}) p(y_i^s \mid x_i^s) p(y_i^p \mid x_i^p)
\end{aligned}$$

$$= \sum_{x_i^s, x_i^p, \sigma_{i-1}} p(x_i^p, \sigma_i \mid x_i^s, \sigma_{i-1}) p(x_i^s) p(y_i^s \mid x_i^s) p(y_i^p \mid x_i^p) \alpha_{\Sigma_{i-1}}(\sigma_{i-1}),$$

where in the last line we have used the induction hypothesis. In other words, the message along Σ_i is proportional to the probability that $\Sigma_i = \sigma_i$, conditioned on the the observations $y_1, \dots, y_i$.

So far we have considered the computation from the point of view of the FSFG. Alternatively, we can accomplish this recursive computation on the *trellis*. We initialize the boundary values to $\alpha_{\Sigma_0}(\sigma_0) = \mathbb{1}_{\{\sigma_0=0\}}$, which encodes the constraint that all paths must start in the all-zero state. Further, we associate with each edge in the i-th trellis section the weight $p(x_i^s)p(y_i^s \mid x_i^s)p(y_i^p \mid x_i^p)$: note that for a given edge, x_i^s, x_i^p, y_i^s, and y_i^s are known quantities so that the preceding label is well defined. We can compute the values $\alpha_{\Sigma_i}(\sigma_i)$ recursively, starting with $i = 1$, based on the values $\alpha_{\Sigma_{i-1}}(\sigma_{i-1})$, which we assume have been stored in the nodes corresponding to $\Sigma_{i-1} = \sigma_{i-1}$ (recall that the values $\alpha_{\Sigma_0}(\sigma_0)$ are already determined via the boundary condition). To compute $\alpha_{\Sigma_i}(\sigma_i)$, run over all incoming edges of the state $\Sigma_i = \sigma_i$, and compute the sum of the product of the corresponding $\alpha_{\Sigma_{i-1}}(\sigma_{i-1})$ times the edge weights. Store this value in the node corresponding to $\Sigma_i = \sigma_i$. This is the traditional way of phrasing the BCJR algorithm.

Let us switch back to the FSFG point of view and interpret the flow of messages starting at Σ_{n+m} and propagating down to Σ_0. Let us call the corresponding message $\beta_{\Sigma_i}(\sigma_i)$. If we apply the message-passing rules to $\beta_{\Sigma_{m+n-1}}(\sigma_{m+n-1})$ we get

$$\beta_{\Sigma_{m+n-1}}(\sigma_{m+n-1}) = \sum_{x_{m+n-1}^s, x_{m+n-1}^p, \sigma_{m+n-1}} \underbrace{p(x_{m+n-1}^p, \sigma_{m+n-1} \mid x_{m+n-1}^s, \sigma_{m+n-2})}_{\text{kernel}}$$
$$\underbrace{p(x_{m+n-1}^s) p(y_{m+n-1}^s \mid x_{m+n-1}^s) p(y_{m+n-1}^p \mid x_{m+n-1}^p)}_{\text{product of incoming messages}}$$
$$= p(y_{n+m}^s, y_{n+m}^p \mid \sigma_{m+n-1}).$$

We claim that, more generally, for $i = m + n - 1, \dots, 0$,

$$\beta_{\Sigma_i}(\sigma_i) = p(y_{i+1}^s, \dots, y_{n+m}^s, y_{i+1}^p, \dots, y_{n+m}^p \mid \sigma_i).$$

As for the α-recursion we can prove this by induction: we start with the case $i = m + n - 1$ and work our way down. Using the general message-passing rules and the induction hypothesis,

$$\beta_{\Sigma_i}(\sigma_{i-1}) = \sum_{x_i^s, x_i^p, \sigma_i} \underbrace{p(x_i^p, \sigma_i \mid x_i^s, \sigma_{i-1})}_{\text{kernel}} \underbrace{p(x_i^s) p(y_i^s \mid x_i^s) p(y_i^p \mid x_i^p) \beta_{\Sigma_i}(\sigma_i)}_{\text{product of incoming messages}}$$

$$= \sum_{x_i^s, x_i^p, \sigma_i} p(x_i^p, \sigma_i \mid x_i^s, \sigma_{i-1}) p(x_i^s) p(y_i^s \mid x_i^s) p(y_i^p \mid x_i^p) \cdot$$

$$\cdot\, p(y_{i+1}^s, \dots, y_{n+m}^s, y_{i+1}^p, \dots, y_{n+m}^p \mid \sigma_i)$$

$$= p(y_i^s, \dots, y_{n+m}^s, y_i^p, \dots, y_{n+m}^p \mid \sigma_{i-1}).$$

Last but not least, let us consider the final (decision) step. Typically one is interested in the decision regarding X_i^s. This decision metric is a function of x_i^s and, by the message-passing rules, it is equal to

$$p(x_i^s) p(y_i^s \mid x_i^s) \sum_{\sim x_i^s} p(x_i^p, \sigma_i \mid x_i^s, \sigma_{i-1}) p(y_i^p \mid x_i^p) \beta_{\Sigma_i}(\sigma_i) \alpha_{\Sigma_{i-1}}(\sigma_{i-1}).$$

If we insert the expressions for $\alpha_{\Sigma_{i-1}}(\sigma_{i-1})$ and $\beta_{\Sigma_i}(\sigma_i)$, we see that this is equal to

$$p(x_i^s, y_1^s, \dots, y_{n+m}^s, y_1^p, \dots, y_{n+m}^p),$$

as one would expect: up to a global normalization, this decision metric is the a posteriori function of x_i^s given the observation (y^s, y^p).

Example 6.7 (BCJR for $C(G = 7/5, n = 5)$). Consider the simple convolutional code $C(G = 7/5, n = 5)$. Assume that transmission takes place over the BSC$(\epsilon = \frac{1}{4})$ and that the received word is $(y^s, y^p) = (1001000, 1111100)$. By definition of the code we have $p_{\Sigma_0}(0) = p_{\Sigma_7}(0) = 1$. The top row in Figure 6.9 shows the trellis with branch labels corresponding to $p(y_i^s \mid x_i^s) p(y_i^p \mid x_i^p)$. (We did not include in these branch labels the prior $p(x_i^s)$ since it is uniform and so does not influence the computation.) The middle row shows the α-recursion, and the bottom row shows the β-recursion. Finally, the decoded sequence is indicated at the very bottom and it is $(\hat{x}^s, \hat{x}^p) = (1011000, 1111100)$.

This is important: this sequence does *not* constitute a valid codeword. This might seem strange at first. But it is a consequence of the fact that bit-wise MAP decoding optimizes the decision for each bit individually, instead of optimizing the decision of the whole sequence jointly (which is done by the block-wise MAP decoder).

As the index i increases, the metric α_{Σ_i} decreases (and, conversely, β_{Σ_i} decreases as i decreases). This can lead to numerical problems when performing the computation on long graphs. As we discussed already in Section 2.5.2, the *normalization* of the messages plays no role. In practice it is therefore common to normalize both α_{Σ_i} and β_{Σ_i} with respect to the maximum component. ◇

Example 6.8 (Performance of Convolutional Codes). Figure 6.10 shows the performance of the rate one-half code $C(G = 21/37, n = 2^{16})$ over the BAWGNC(σ) when decoded via the BCJR algorithm. We show the performance only for a single

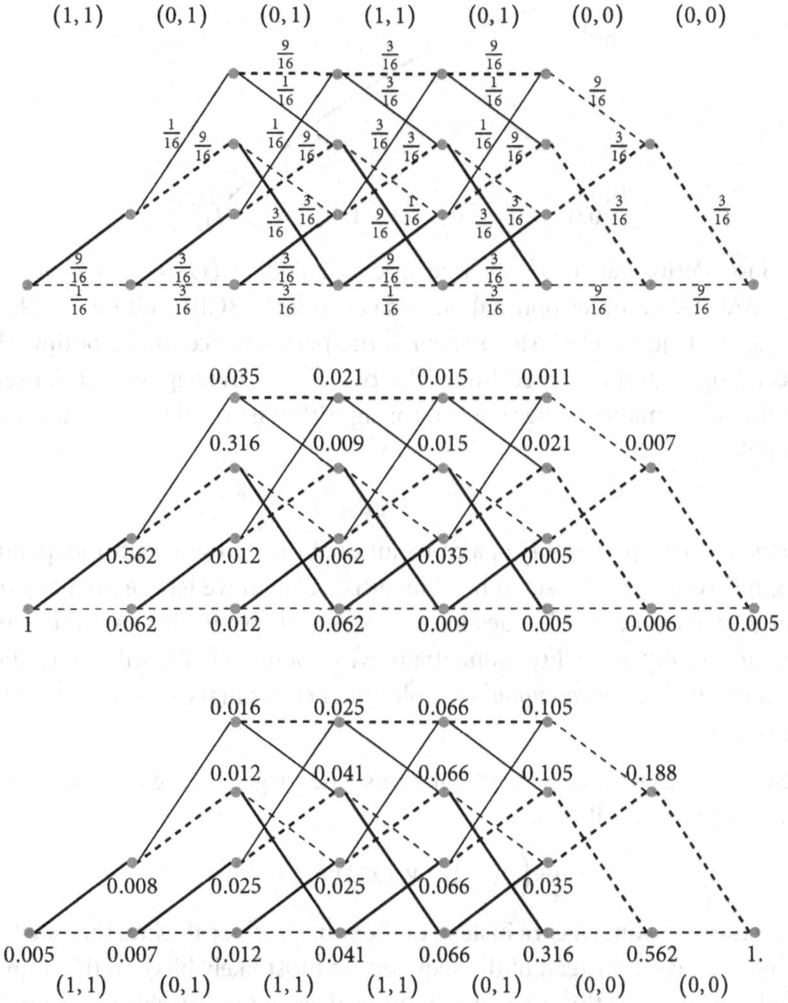

Figure 6.9: BCJR algorithm applied to the code $C(G = 7/5, n = 5)$ assuming transmission takes place over the BSC$(\epsilon = 1/4)$. The received word is equal to $(y^s, y^p) = (1001000, 1111100)$. The top figure shows the trellis with branch labels corresponding to the received sequence. We have *not* included the prior $p(x_i^s)$, since it is uniform. The middle and bottom figures show the α- and the β- recursion, respectively. On the very bottom, the estimated sequence is shown.

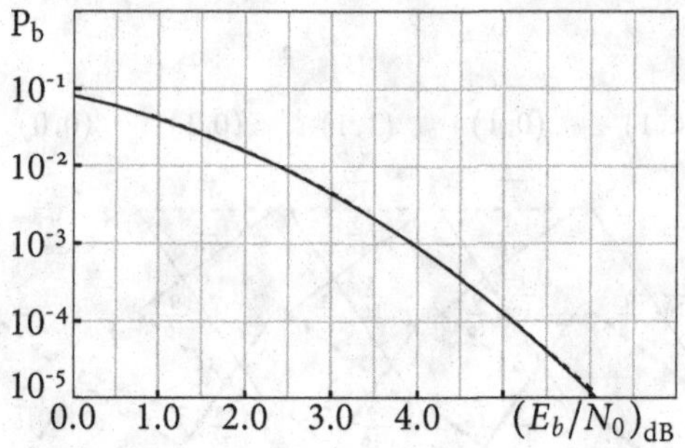

Figure 6.10: Performance of the rate one-half code $C(G = 21/37, n = 2^{16})$ over the BAWGNC under optimal bit-wise decoding (BCJR, solid line). Note that $(E_b/N_0)_{\text{dB}} = 10 \log_{10} \frac{1}{2r\sigma^2}$. Also shown is the performance under optimal block-wise decoding (Viterbi, dashed line). The two curves overlap almost entirely. Although the performance under the Viterbi algorithm is strictly worse the difference is negligible.

n since the bit error probability of a convolutional code is essentially independent of the length (except for very small blocklengths). Only if we let the memory m tend to infinity do convolutional codes have a threshold, but in this case the complexity tends to infinity as well (exponentially with memory). We will see in the next section that suitably *concatenated* convolutional codes make excellent iterative decoding systems. ◇

In Section 2.5.5 we have seen that if we use the (negative) log of the usual branch labels, i.e., for our case if we use

$$-\log\big(p(x_i^s)p(y_i^s \mid x_i^s)p(y_i^p \mid x_i^p)\big),$$

and apply the min-sum algebra instead of the sum-product algebra, then we find the *most likely codeword* (instead of the sequence of most likely bits). In the framework of convolutional codes this algorithm is equivalent to the *Viterbi* algorithm. For iterative decoding it plays a less important role. There is a slight difference between the Viterbi algorithm (as it is usually defined) and the message-passing algorithm which we get by applying the min-sum algebra. The difference lies in the *schedule.* For the Viterbi algorithm we typically perform one *forward* recursion, applying the min-sum rules once from the beginning of the trellis until the end. In a second step we *back-trace* from the end of the trellis until the beginning to "read off" the best path. A straightforward application of the message-passing algorithm, on the other hand, involves three steps: a forward pass, a backward pass, and a decision step. It is the topic of Problem 6.2 to see why the Viterbi algorithm can get away with only two passes.

EXAMPLE 6.11 (VITERBI FOR $C(G = 7/5, n = 5)$). Consider again the code $C(G = 7/5, n = 5)$ as in Example 6.7 and transmission via the BSC($\epsilon = 1/4$). As before, the received word is equal to $(y^s, y^p) = (1001000, 1111100)$. Figure 6.12 shows the application of the message-passing algorithm using the min-sum algorithm to this problem. The decoded sequence of bits is indicated at the very bottom and it is equal to $(\hat{x}^s, \hat{x}^p) = (1011000, 1111110)$. For our example, optimum bit and optimum block decoding *do not* lead to the same result. ◇

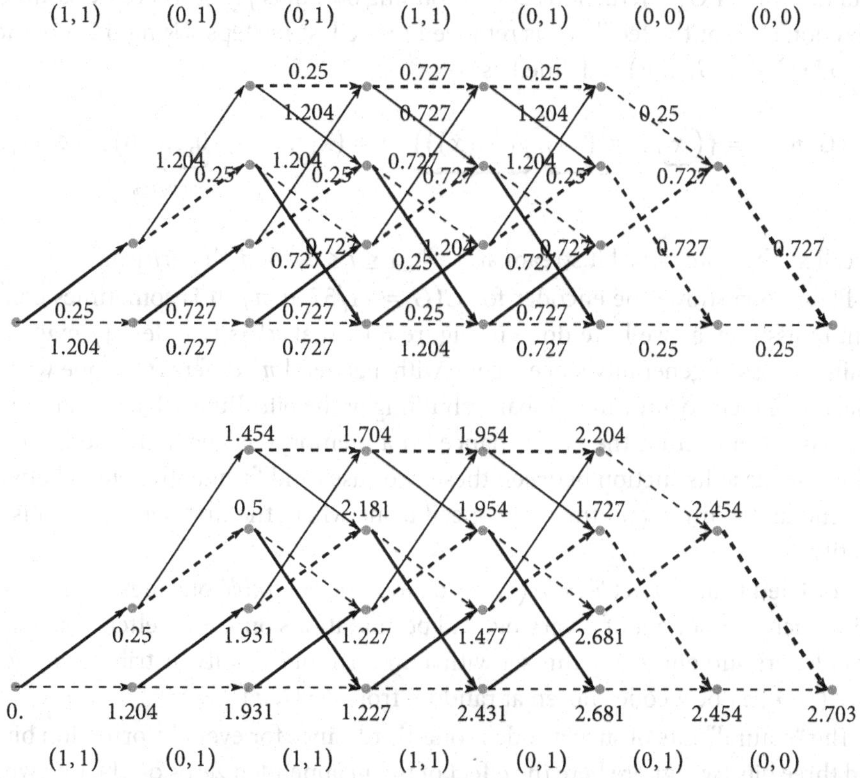

Figure 6.12: Viterbi algorithm applied to the code $C(G = 7/5, n = 5)$ assuming transmission takes place over the BSC($\epsilon = 1/4$). The received word is $(y^s, y^p) = (1001000, 1111100)$. The top figure shows the trellis with branch labels corresponding to $-\log_{10}\left(p(y_i^s \mid x_i^s)p(y_i^p \mid x_i^p)\right)$. Since we have a uniform prior we can take out the constant $p(x_i^s)$. These branch labels are easily derived from Figure 6.9 by applying the function $-\log_{10}$. The bottom figure show the workings of the Viterbi algorithm. On the very bottom the estimated sequence is shown.

§6.2. Structure and Encoding

A standard *parallel* concatenated turbo code with two component codes is defined as follows. Fix a binary rational function $G(D) = p(D)/q(D)$ of memory m with $q_0 = 1$ and a length n. Further, let $\pi = (\pi^1, \pi^2)$, where $\pi^i : \{1, \ldots, n+m\} \rightarrow \{1, \ldots, n+m\}$, $1 \leq i \leq 2$, is a permutation on $n+m$ letters which fixes the last m letters. Recall from Section 6.1 that we denote the encoding map associated with a rational function $G(D)$ by γ. More precisely, if $x = (x_1, \ldots, x_n, 0, \ldots, 0)$ denotes the input to the filter $G(D)$, then the corresponding output is $\gamma(x)$, where we assume as in Section 6.1 that the feedback is removed in the last m steps. Using this notation, the code $C = C(G, n, \pi)$ is defined as

$$C(G,n,\pi) = \{(\underbrace{x}_{x^s}, \underbrace{\gamma(\pi^1(x))}_{x^{p_1}}, \underbrace{\gamma(\pi^2(x))}_{x^{p_2}}) : x = (x_1, \ldots, x_n, \underbrace{0, \ldots, 0}_{m \text{ times}}), x_i \in \mathbb{F}_2\}.$$

We call x^s the *systematic* bit stream and x^{p_i}, $1 \leq i \leq 2$, the i-th *parity* bit stream.

Figure 6.13 shows the encoder for $C(G = 21/37, n, \pi)$. It is sometimes convenient to assume, as we have done in Figure 6.13, that π^1 is the identity map. This entails no loss of generality since a code with a general π^1 differs from one with π^1 equal to the identity only in a global reshuffling of the bits. But such a global reshuffling has no impact on the performance on a memoryless channel. Therefore, we will not make a distinction between these two cases and frequently switch between the general definition and the more specific one where the first permutation is the identity.

For fixed G and n, let $\mathcal{P} = \mathcal{P}(G, n)$ denote the *ensemble* of codes generated by letting each component of π vary over all permutations on $n+m$ letters that fix the last m letters and endowing this set with a uniform probability distribution. We let $C = C(G, n, \pi)$ be a code chosen at random from $\mathcal{P}(G, n)$.

The "natural" rate of such a code is one-third, since for every information bit we send three bits (we ignore here the effect of the m appended zeros on the rate which vanishes with $\Theta(1/n)$). Often one is interested in *punctured* turbo codes to adjust the rate. If we puncture (delete), e.g., every second bit of both x^{p_1} and x^{p_2} then we get a rate one-half code. More generally, by picking a suitable puncturing pattern the rate r can be varied in the range $1/3 \leq r \leq 1$. From the point of view of analysis, *random* puncturing is particularly appealing: to achieve rate r, puncture each bit of each parity stream with probability $(3r-1)/(2r)$. This gives rise to an ensemble of punctured codes whose *average* rate is r. We denoted this ensemble by $\mathcal{P}(G, n, \pi, r)$. By standard concentration results (see Appendix C) we know that most elements in $\mathcal{P}(G, n, \pi, r)$ have rate $r \pm O(1/\sqrt{n})$. We denote a punctured code of *design rate* r by $C(G, n, \pi, r)$, where the exact nature of the puncturing will be clear from context.

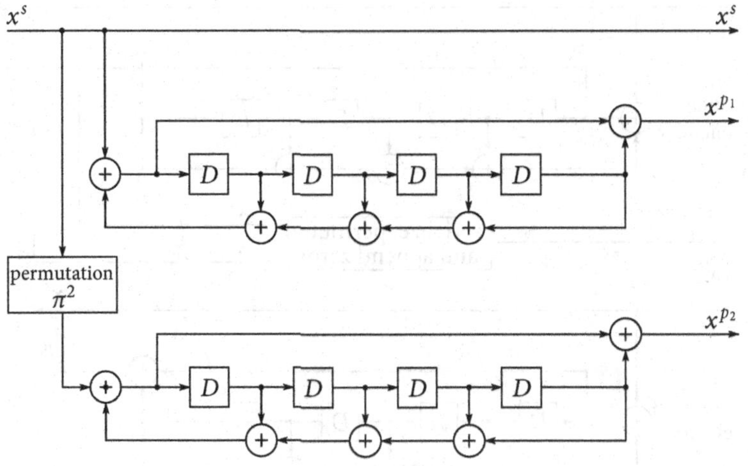

Figure 6.13: Encoder for $C(G = 21/37, n, \pi = (\pi^1, \pi^2))$, where π^1 is the identity permutation.

We can also puncture the m extra systematic bits appended to the n information bits since they are known to be zero.

Standard *serially* concatenated codes are defined in an analogous way. Fix a binary rational function G^o of memory m^o, a binary rational function G^i of memory m^i, and a length n (the superscripts o and i stand for outer and inner, respectively). Further, let π, $\pi : \{1, \dots, 2(n+m^o)+m^i\} \to \{1, \dots, 2(n+m^o)+m^i\}$, be a permutation on $2(n+m^o)+m^i$ letters which fixes the last m^i letters. The associated code $C = C(G^o, G^i, n, \pi)$ is defined as

$$C(G^o, G^i, n, \pi) = \{(\pi(x \cdot \gamma^o(x) \cdot \underbrace{0, \dots, 0}_{m^i \text{ times}}), \gamma^i(\pi(x \cdot \gamma^o(x) \cdot \underbrace{0, \dots, 0}_{m^i \text{ times}}))) :$$
$$x = (x_1, \dots, x_n, \underbrace{0, \dots, 0}_{m^o \text{ times}}), x_i \in \mathbb{F}_2\},$$

where "$\cdot$" denotes concatenation of sequences. We call x^{s_o} the systematic bits of the *outer* code and $x^{p_o} = \gamma^o(x^{s_o})$ the corresponding parity bits. Further, we denote by $x^{s_i} = \pi(x^{s_o} \cdot x^{p_o} \cdot 0, \dots, 0)$ the systematic bits of the *inner code* and $x^{p_i} = \gamma^i(x^{s_i})$ the corresponding parity bits. We append m^i zeros to the output of the first encoder so that the input to the second encoder has the proper form. Figure 6.14 shows the encoder for $C(G^o = 21/37, G^i = 21/37, n, \pi)$. For fixed G^o, G^i, and n let $\mathcal{S} = \mathcal{S}(G^o, G^i, n)$ be the ensemble of codes generated by varying π over *all permutations* on $2(n+m^o)+m^i$ letters which fix the last m^i positions and endowing this set

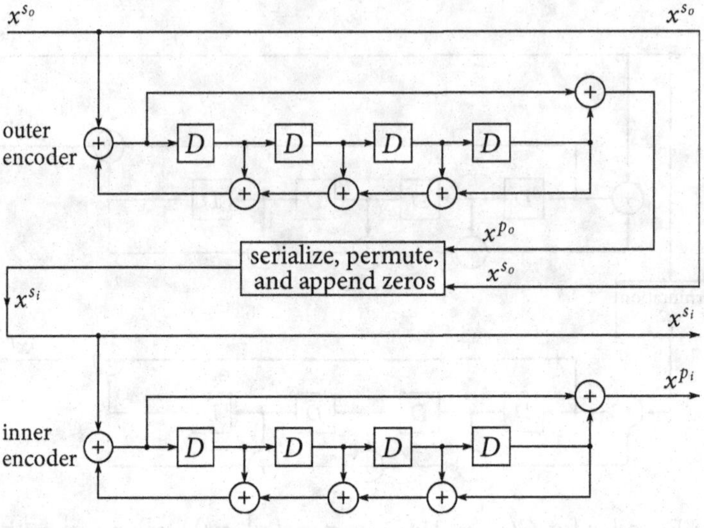

Figure 6.14: Encoder for $C(G^o = 21/37, G^i = 21/37, n, \pi)$.

with a uniform probability distribution. We let $C = C(G^o, G^i, n, \pi)$ be a code chosen at random from $\mathcal{S}(G^o, G^i, n)$. The "natural" rate of such a code is one-quarter (ignoring again the vanishing effect of the additional appended zeros). We can increase the rate to r, where $1/4 \le r \le 1$, by an appropriate puncturing of the output. There is a considerable degree of freedom in the choice of bits that are punctured and different choices lead, in general, to different performances. We denote a punctured code by $C = C(G^o, G^i, n, \pi, r)$ and the corresponding ensemble by $\mathcal{S}(G^o, G^i, n, r)$, where the exact nature of the puncturing is again understood from the context.

From the preceding description it is hopefully clear that the *encoding* operation is *linear* in the blocklength and that it can be implemented efficiently. We can therefore proceed directly to the decoding problem.

§6.3. Decoding

Assume that transmission takes place over a BMS channel. We denote the channel input by X and the channel output by Y. Consider the ensemble $\mathcal{P}(G, n)$ and optimal bit-wise decoding. For the moment, we assume that none of the bits are punctured. Recall from Section 6.1 that the mapping of a convolutional encoder at time j is completely determined by the *current state*, call it σ_{j-1}, and the current *input* x_j. Let $x^{s_1} = \pi^1(x^s)$ and $x^{s_2} = \pi^2(x^s)$, i.e., x^{s_1} and x^{s_2} represent the permuted versions of the systematic bits. Let σ^1 and σ^2 denote the state sequence of encoders 1 and 2, respectively. Then the bit MAP decoder has the form

$$\hat{x}_i^{\text{MAP}}(y^s, y^{p_1}, y^{p_2})$$

$$
\begin{aligned}
&= \operatorname{argmax}_{x_i^s \in \{0,1\}} p(x_i^s \mid y^s, y^{p_1}, y^{p_2}) \\
&= \operatorname{argmax}_{x_i^s \in \{0,1\}} \sum_{\sim x_i^s} p(x^s, x^{p_1}, x^{p_2}, \sigma^1, \sigma^2, y^s, y^{p_1}, y^{p_2}) \\
&= \operatorname{argmax}_{x_i^s \in \{0,1\}} \sum_{\sim x_i^s} \Big(\prod_{j=1}^{n+m} \underbrace{p(x_j^s)}_{\text{prior}} \underbrace{p(y_j^s \mid x_j^s) p(y_j^{p_1} \mid x_j^{p_1}) p(y_j^{p_2} \mid x_j^{p_2})}_{\text{channel}} \Big) \\
&\qquad p(\sigma_0^1) p(\sigma_0^2) \Big(\prod_{j=1}^{n+m} \underbrace{p(x_j^{p_1}, \sigma_j^1 \mid x_j^{s_1}, \sigma_{j-1}^1)}_{\text{code 1}} \underbrace{p(x_j^{p_2}, \sigma_j^2 \mid x_j^{s_2}, \sigma_{j-1}^2)}_{\text{code 2}} \Big).
\end{aligned}
$$

The corresponding FSFG is shown in Figure 6.15. The message-passing rules stated

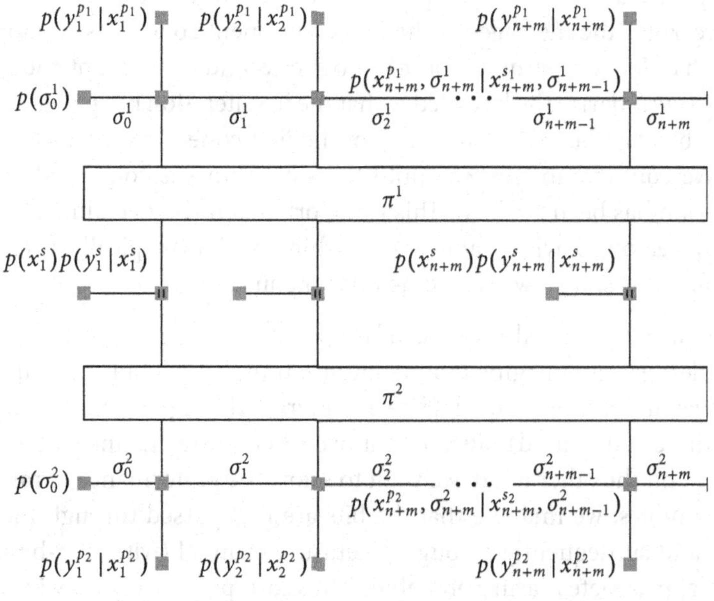

Figure 6.15: FSFG for the optimum bit-wise decoding of an element of $\mathcal{P}(G, n)$.

in Figure 2.12 completely specify the iterative decoder, except for the *scheduling* of the messages. Since the FSFG of the present example is *not* a tree, different schedules can result in different performances. Based on our discussion on factor graphs, the *natural* schedule is to initialize all messages in the graph and then in each "iteration" to recompute the outgoing messages from the incoming ones. If we use log-likelihood ratios as messages, the natural initialization is to set all messages to zero (this acts like the neutral value).

If you look closer at Figure 6.15, you see that it is composed of FSFGs of two convolutional codes (compare to Figure 6.5). This is not surprising. After all, we defined elements of $\mathcal{P}(G, n)$ exactly in this way as a concatenation of two convolutional codes. Let us call each of the two convolutional codes a *component*. If we think of turbo codes as being composed of two components, we arrive at a second natural schedule. We call it the *turbo schedule*. In the turbo schedule, we also initialize all messages by setting them to zero (assuming again that the messages are in log-likelihood ratio form). In each "iteration" we "freeze" (do not change) the messages in one component and decode the other component by running one complete iteration of the BCJR algorithm as discussed in Section 6.1. To be specific: in iteration 1 we "freeze" the messages in the second component and run the BCJR algorithm on the first component. (The messages which flow via the permuter π^2 into the first component are initially zero.) In iteration 2 we exchange the roles of component one and component two and run one complete BCJR algorithm on the second component while "freezing" the messages in the first component code. This is important: the messages that flow via permutation π^1 into the second component code are no longer zero, but summarize the knowledge that the first iteration has gathered about the systematic bits via the code constraints of the first code. This information acts like a prior. We continue in this way until the algorithm has converged or some stopping criterion has been reached. This is the original turbo decoding algorithm as proposed by Berrou, Glavieux, and Thitimajshima and it is typically a very good choice. In what follows, we always assume this schedule.

So far we have considered the case of no puncturing, but the general case is not much harder: we can imagine that *all* bits are transmitted, but that part of the bits are sent over an erasure channel. If we are interested in a particular puncturing pattern then these (punctured) bits are sent over an erasure channel with erasure probability 1. If, on the other hand, we want to know the performance of randomly punctured ensembles, we imagine that the bits are first passed through the given BMS channel and subsequently through an erasure channel (with non-binary inputs) with the appropriate erasure probability. In short, puncturing only influences the local functions $p(y_j^{s/p_1/p_2} \mid x_j^{s/p_1/p_2})$, but the overall form of the FSFG remains unchanged.

Let $\mathrm{P_b}(C, E_b/N_0)$ denote the bit error probability of a code C assuming that transmission takes place over a BAWGNC with parameter $(E_b/N_0)_{\mathrm{dB}}$. Figure 6.16 shows $\mathbb{E}_{\mathcal{P}(G=21/37,n,r=1/2)}[\mathrm{P_b}(C, E_b/N_0)]$ for $n = 2^{11}, \dots, 2^{16}$. Recall that the *block-length* is equal to $(1 + 2r)(n + m)$ which equals $2(n + 4)$ for the particular case. As indicated, the ensemble $\mathcal{P}(G = 21/37, n, r = 1/2)$ has a rate of one-half. This is achieved by an alternating puncturing pattern, puncturing every second parity bit. Fifty iterations using the turbo schedule (i.e., 25 iterations for each component code) as discussed earlier were performed. As usual, these plots show two distinct behaviors: the *waterfall* region in which the error probability drops off steeply as a function of the channel parameter, and the *error-floor* region, with a much gentler slope. From these plots one can see that for increasing lengths the codes show again a threshold behavior. In the next section we discuss how this threshold can be determined and we describe analytic approximations (the dashed curves in Figure 6.16) of the error-floor in Lemma 6.52. This is important: if we do not pick random permu-

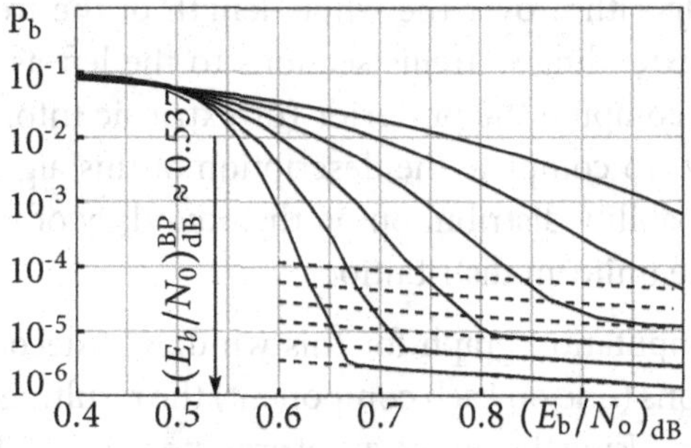

Figure 6.16: $\mathbb{E}_{\mathcal{P}(G=21/37,n,r=1/2)}[P_b(C, E_b/N_0)]$ for an alternating puncturing pattern (identical on both branches), $n = 2^{11}, \ldots, 2^{16}$, 50 iterations, and transmission over the BAWGNC(E_b/N_0). The arrow indicates the position of the threshold $(E_b/N_0)^{BP}_{dB} \approx 0.537$ ($\sigma^{BP} \approx 0.94$) which we compute in Section 6.5. The dashed curves are analytic approximations of the error floor discussed in Lemma 6.52.

tations but design the permutations carefully, then we can get a significant reduction in the error floor. The Notes at the end of this chapter give points to the literature which describe the so-called "interleaver design" problem.

Iterative decoding and the FSFG for serial concatenated ensembles are discussed in Problem 6.6. Generally speaking, serially concatenated ensembles show a better error floor behavior than parallel concatenated ensembles, but at the cost of slightly worse thresholds.

§6.4. Basic Simplifications

The asymptotic analysis proceeds according to the standard recipe: decoder symmetry, all-one codeword assumption, concentration, asymptotically tree-like computation graph for a fixed number of iterations, and – finally – density evolution.

We start with the symmetry of the decoder. Each (part of the) decoder performs MAP decoding on a linear code (BCJR on trellis). Therefore, by Theorem 4.29 the output density is symmetric assuming that all input densities are symmetric. It follows that if transmission takes place over a BMS channel (so that the received densities are symmetric) then we can make the all-one codeword assumption.

Consider the ensemble $\mathcal{P}(G, n, r)$. Focus on one bit and consider one round of the decoding algorithm (according to the turbo schedule): a little thought shows that the computation graph for one iteration includes *all* variables, *regardless* of the size of the code. Therefore, even in the asymptotic case, the computation graph of this naive decoder does not become tree-like and we cannot employ our standard density-evolution analysis.

One way to proceed is to consider a *windowed* decoding algorithm: instead of running the BCJR algorithm over the whole length of the trellis, we construct a symmetric window extending w trellis sections to the left as well as to the right of any given bit and compute the posterior (or extrinsic information) based only on this local window. To complete the description of this algorithm we also need to decide on the probability distribution on the boundary of these windows – it is convenient to pick the uniform distribution.

Consider the computation graph for this windowed decoding algorithm. For $w = 1$ and two iterations (one on each component) the resulting computation graph is shown in Figure 6.17. Using the same type of argument as for LDPC ensembles (see Section 3.8.2), we see that, for a fixed number of iterations, this computation graph is a tree with probability converging to one as the blocklength tends to infinity. Next,

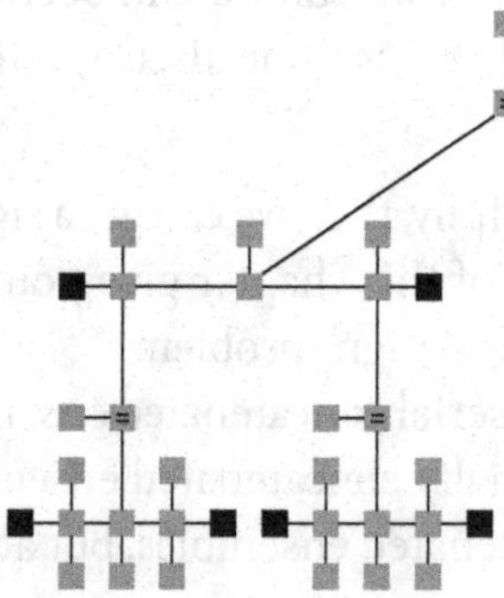

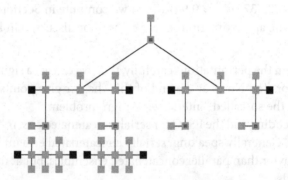

Figure 6.17: Computation graph corresponding to windowed ($w = 1$) iterative decoding of a parallel concatenated code for two iterations. The black factor nodes indicate the end of the decoding windows and represent the prior which we impose on the boundary states.

for a fixed window size and a fixed number of iterations, it is rather routine to check the concentration of the performance as a function of the permutation used and the observations received. We can therefore again use an ensemble average approach. The computation graph for the serial case is the topic of Problem 6.8.

§6.5. Density Evolution

Let us consider density evolution for general BMS channels. From our previous discussion, we know that for a fixed window size w we can use density evolution to compute

$$\lim_{\ell\to\infty}\lim_{n\to\infty}\mathbb{E}_{\mathcal{P}(G,n)}[\mathrm{P_b}(C,\mathsf{a},\ell)],$$

where a characterizes the BMS channel and where ℓ denotes the number of iterations. If we now let the window size tend to infinity we get

$$\lim_{w\to\infty}\lim_{\ell\to\infty}\lim_{n\to\infty}\mathbb{E}_{\mathcal{P}(G,n)}[\mathrm{P_b}(C,\mathsf{a},\ell)].$$

From a practical point of view it is convenient to exchange the two outer limits and to compute

$$\lim_{\ell\to\infty}\lim_{w\to\infty}\lim_{n\to\infty}\mathbb{E}_{\mathcal{P}(G,n)}[\mathrm{P_b}(C,\mathsf{a},\ell)].$$

This is the limit we consider in the following. The proof that the exchange of limits is admissible is technical and so we skip it. In fact, it is conjectured that all three limits can be taken in any order without changing the result. This is quite important from a practical perspective since in a typical scenario we use a fixed length, we use an unbounded window, and we run as many iterations as we can afford from the perspective of complexity.

Define the following maps acting on pairs of densities as depicted in Figure 6.18. There is a bi-infinite trellis generated by $G(D)$. For the sake of definiteness, we assume that all messages are in log-likelihood ratio form. The systematic bits experience a channel with L-density a, whereas the parity bits are passed through a channel with L-density b. We are interested in the *outgoing message densities* as indicated in Figure 6.18. Let us denote them by c and d, respectively, and the map between

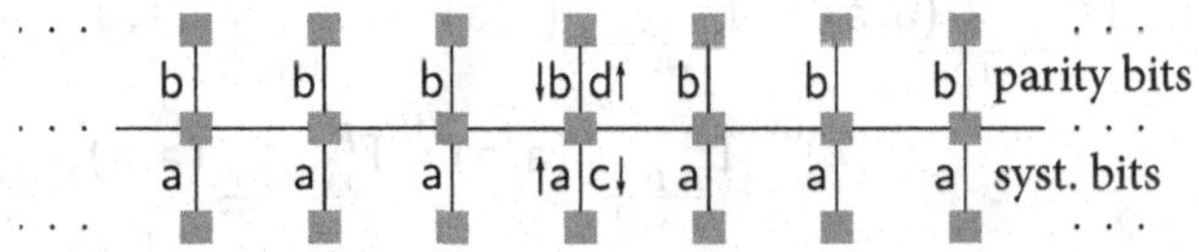

Figure 6.18: Definition of the maps $\mathsf{c} = \Gamma_G^s(\mathsf{a},\mathsf{b})$ and $\mathsf{d} = \Gamma_G^p(\mathsf{a},\mathsf{b})$. We are given a bi-infinite trellis defined by a rational function $G(D)$. Associated with all systematic variables are iid samples from a density a, whereas the parity bits experience the channel b. The resulting densities of the outgoing messages are denoted by c and d, respectively.

incoming and outgoing messages by

$$\mathsf{c} = \Gamma^s_G(\mathsf{a}, \mathsf{b}), \qquad \mathsf{d} = \Gamma^p_G(\mathsf{a}, \mathsf{b}).$$

We write $\Gamma_G(\cdot,\cdot)$ if we mean either of the two maps and we write $\Gamma(\cdot,\cdot)$ if G is understood from context. Before discussing how the map $\Gamma(\cdot,\cdot)$ can be computed, let us state several equivalence relationships. These relations significantly decrease the number of cases that one has to investigate when trying to optimize the threshold of iterative systems with convolutional components.

Consider a binary polynomial $p(D)$ of degree $\deg(p)$ such that $p_0 = 1$. Associate with such a polynomial the *time-reversed and shifted* polynomial

$$\hat{p}(D) = D^{\deg(p)} p(1/D).$$

We extend this definition to a binary rational function $G(D) = \frac{p(D)}{q(D)}$ so that $p_0 = q_0 = 1$ by defining $\hat{G}(D) = \frac{\hat{p}(D)}{\hat{q}(D)}$. The proof of the following theorem is the topic of Problem 6.4.

Theorem 6.19 (Equivalence of Encoders). Consider a convolutional encoder defined by a binary rational function $G(D) = \frac{p(D)}{q(D)}$ with $q_0 = 1$. Then, for any pair of symmetric densities a and b and $j \geq 1$,

$$\text{(i) } \Gamma_{G(D^j)}(\mathsf{a},\mathsf{b}) = \Gamma_{G(D)}(\mathsf{a},\mathsf{b}), \qquad \text{(ii) } \Gamma_{G(D)}(\mathsf{a},\mathsf{b}) = \Gamma_{D^jG(D)}(\mathsf{a},\mathsf{b}),$$
$$\text{(iii) } \Gamma_G(\mathsf{a},\mathsf{b}) = \Gamma_{\hat{G}}(\mathsf{a},\mathsf{b}), \text{if } p_0 = 1, \quad \text{(iv) } \Gamma^s_G(\mathsf{a},\mathsf{a}) = \Gamma^p_{G^{-1}}(\mathsf{a},\mathsf{a}), \text{if } p_0 = 1.$$

Example 6.20 (Application of Equivalence Relationships). We claim that

$$\Gamma^s_{\frac{D+D^2}{1+D^2+D^3}}(\mathsf{a},\mathsf{a}) = \Gamma^p_{\frac{1+D^2+D^6}{1+D^2}}(\mathsf{a},\mathsf{a}).$$

Indeed,

$$\Gamma^s_{\frac{D+D^2}{1+D^2+D^3}}(\mathsf{a},\mathsf{a}) \overset{\text{(ii)}}{=} \Gamma^s_{\frac{1+D}{1+D^2+D^3}}(\mathsf{a},\mathsf{a}) \overset{\text{(iii)}}{=} \Gamma^s_{\frac{1+D}{1+D+D^3}}(\mathsf{a},\mathsf{a})$$
$$\overset{\text{(iv)}}{=} \Gamma^p_{\frac{1+D+D^3}{1+D}}(\mathsf{a},\mathsf{a}) \overset{\text{(i)}}{=} \Gamma^p_{\frac{1+D^2+D^6}{1+D^2}}(\mathsf{a},\mathsf{a}). \qquad \diamond$$

As a slight generalization, if we consider punctured ensembles then we will assume that the functional $\Gamma(\cdot,\cdot)$ is computed according to this puncturing and the nature of the concatenation. To be concrete: assume that every second parity symbol is punctured and that we are interested in a parallel concatenation. In this case we assume that the input to every second parity node is Δ_0 (instead of b). If we consider randomly punctured ensembles where each parity symbol is punctured with probability p, we replace each input density b of each parity symbol with probability p with the density Δ_0. The equivalent statements are true regarding puncturing of systematic bits. The assumed puncturing will be understood from the context and we do not explicitly include it in our notation. Under some puncturing schemes the above equivalence relations remain valid but in some cases they will no longer hold: under random puncturing (of either the systematic or the parity bits or both) statements (i)–(iii) are still correct; on the other hand, statement statement (i) is incorrect for j even if we puncture every second parity bit.

Let us now return to the description of density evolution. Since density evolution concerns the performance of ensembles in the limit as n tends to infinity we drop the n from the description of ensembles in what follows and simply write $\mathcal{P}(G, r)$. We can "read off" from Figure 6.15 the following relations concerning the density evolution process (see Problem 6.7 for the serial case.)

THEOREM 6.21 (DENSITY EVOLUTION FOR PARALLEL CONCATENATED CODES). Consider density evolution for the ensembles $\mathcal{P}(G, r)$ when transmission takes place over a BMS channel with L-density a_{BMSC} and the turbo schedule is used. Let c_ℓ denote the density emitted from the trellis toward the systematic bits in the ℓ-th iteration. Then $\mathsf{c}_0 = \Delta_0$, and for $\ell \geq 1$

$$\mathsf{c}_\ell = \Gamma_G^s(\mathsf{a}_{\text{BMSC}} \circledast \mathsf{c}_{\ell-1}, \mathsf{a}_{\text{BMSC}}).$$

Unfortunately, the map $\Gamma(\cdot, \cdot)$ is in general not easily computed since intermediate distributions (the distributions on the states which are computed by the α- and β- recursions) "live" in $2^m - 1$ dimensions, where m is the memory of the rational function $G(D)$. In practice, the outgoing densities c and d are therefore most often determined by sampling: for given input densities a and b we run the BCJR algorithm on a very long trellis which we initialize with independent samples from the given distributions. By collecting a large number of output samples we estimate the densities c and d.

EXAMPLE 6.22 (DENSITY EVOLUTION FOR $\mathcal{P}(G = 21/37, r = 1/2)$). Assuming transmission over the BAWGN channel, the densities $\mathsf{c}_\ell(y)$, for $\ell = 1, \ldots, 25$, are shown in Figure 6.23. ◇

Exactly as in the case of density evolution of LDPC ensembles, depending on the channel parameter, the densities either converge to a fixed point density with a positive associated error probability, or the error probability converges to zero. Since density evolution is a process on a tree and corresponds to the performance of a MAP decoder, if we assume that the channel family is degraded then we can conclude that there exists a threshold: it is that parameter of the channel which separates the values for which density evolution converges to the desired fixed point

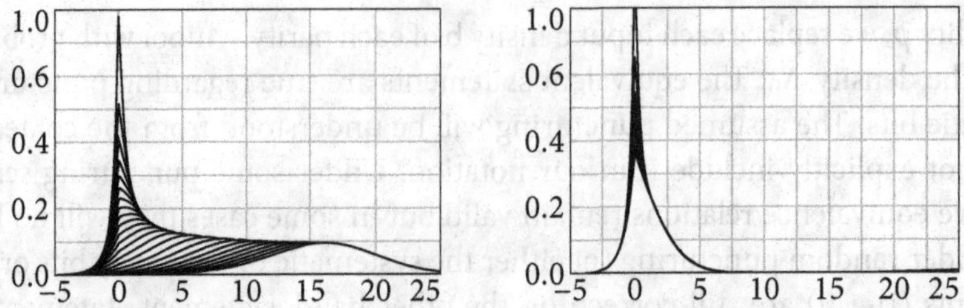

Figure 6.23: Evolution of c_ℓ for $\ell = 1, \ldots, 25$ for the ensemble $\mathcal{P}(G = 21/37, r = 1/2)$, an alternating puncturing pattern of the parity bits, and transmission over the BAWGNC(σ). In the left picture $\sigma = 0.93$ ($E_b/N_0 \approx 0.63$ dB). For this parameter the densities keep moving "to the right" toward Δ_∞. In the right picture $\sigma = 0.95$ ($E_b/N_0 \approx 0.446$ dB). For this parameter the densities converge to a fixed point density.

(Δ_∞) from those where the error probability stays bounded away from zero. Table 6.24 lists thresholds for some ensembles $\mathcal{P}(G, r = 1/2)$.

m	$G = p/q$	σ^{BP}	$(E_b/N_0)_{\text{dB}}$
2	5/7	0.883	1.08
3	13/15, 15/13	0.93	0.63
4	21/37, 27/37, 35/37	0.94	0.537
5	37/55, 47/41, 63/41, 71/41	0.94	0.537
6	41/167, 103/177, 51/153	0.94	0.537

Table 6.24: Thresholds of some ensembles $\mathcal{P}(G, r = 1/2)$ for transmission over the BAWGNC(σ) and alternating puncturing. The Shannon threshold for rate one-half codes is $\sigma^{\text{Sha}} \approx 0.979$ ($E_b/N_0 \approx 0.18$ dB).

§6.6. Stability Condition

The stability condition plays a fundamental role in the analysis of low-density parity-check (LDPC) ensembles. It is therefore natural to see whether an equivalent condition can be derived for turbo codes. In the present section we need to refer to some notions and results concerning the weight distribution of turbo codes. These are collected in Section 6.9.

Conjecture 6.25 (Stability Condition for Parallel Concatenated Ensemble). Consider the ensemble $\mathcal{P}(G, r)$ and let $D(x, y) = \sum_i d_i(y) x^i$ denote the detour generating function associated with G (taking the appropriate puncturing into account). Assume that transmission takes place over a BMS channel with L-density a. Then the fixed point corresponding to correct decoding is stable if and only if

$$2\mathfrak{B}(\mathsf{a})d_2(\mathfrak{B}(\mathsf{a})) < 1. \tag{6.26}$$

We state the stability condition as a conjecture since we do not present a proof of the preceding result but take a considerable shortcut by making use of an analogy with respect to standard LDPC ensembles. Recall from (3.165) that for standard LDPC ensembles with degree distribution pair (λ, ρ) the generating function counting the number of minimal codewords in the asymptotic limit is equal to $\hat{P}(x) = -\frac{1}{2}\log(1 - \lambda'(0)\rho'(1)x)$. More precisely, this means that, if $\hat{P}(x) = \sum_w \hat{p}_w x^w$, then $\hat{p}_w$ denotes the expected number of minimal codewords of weight w in the limit of blocklengths tending to infinity. Apply the union bound to this asymptotic weight distribution: if we have a codeword of weight w, the probability that a MAP decoder decides on this codeword instead on the transmitted all-zero sequence (considering only these two alternatives) is given by $\mathfrak{E}(\mathsf{a}^{\circledast w})$. From Problem 4.22 we know that for any $0 < \mathfrak{B}' < \mathfrak{B}(\mathsf{a})$ there exists a strictly positive α so that for all $w \in \mathbb{N}$

$$\alpha(\mathfrak{B}')^w \leq \mathfrak{E}(\mathsf{a}^{\circledast w}) \leq \mathfrak{B}(\mathsf{a})^w.$$

This gives us the bounds

$$\alpha\hat{P}(\mathfrak{B}') = \alpha\sum_w \hat{p}_w(\mathfrak{B}')^w \leq \sum_w \hat{p}_w\,\mathfrak{E}(\mathsf{a}^{\circledast w}) \leq \sum_w \hat{p}_w\,\mathfrak{B}(\mathsf{a})^w = \hat{P}(\mathfrak{B}(\mathsf{a})).$$

Roughly speaking, taking the union bound amounts to replacing the indeterminate x in the generating function by the Bhattacharyya constant $\mathfrak{B}(\mathsf{a})$. For the standard irregular LDPC ensemble where $\hat{P}(x) = -\frac{1}{2}\log(1 - \lambda'(0)\rho'(1)x)$ we conclude that the union bound converges if and only if

$$\mathfrak{B}(\mathsf{a})\lambda'(0)\rho'(1) < 1.$$

The latter condition is exactly the stability condition.

Let us apply the same reasoning to parallel concatenated turbo codes. From Lemma 6.49 we know that $\hat{P}(x) = -\frac{1}{2}\log(1 - 4x^2 d_2^2(x))$. Convergence of the union bound therefore requires that $4\mathfrak{B}(\mathsf{a})^2 d_2^2(\mathfrak{B}(\mathsf{a})) < 1$. This is equivalent to $2\mathfrak{B}(\mathsf{a})d_2(\mathfrak{B}(\mathsf{a})) < 1$. As always, the stability condition gives an upper bound on the threshold. In some cases this bound is tight; for others it is not.

Example 6.27 (Stability of $\mathcal{P}(G = 7/5, r = 1/3)$). For the ensemble $\mathcal{P}(G = 7/5, r = 1/3)$ we know from Example 6.41 that $d_2(y) = \frac{y^3}{1-y}$. Let x^{Sta} be the unique positive solution to the equation $2xd_2(x) = 2x\frac{x^3}{1-x} = 1$. We have $x^{\text{Sta}} \approx 0.647799$.

If we assume that transmission takes place over the BEC(ϵ) we have $\mathfrak{B}(\epsilon) = \epsilon$, so that $\epsilon^{\text{Sta}} \approx 0.647799$. If transmission takes place over the BAWGNC(σ), so that $\mathfrak{B}(\sigma) = e^{-\frac{1}{2\sigma^2}}$, then $\sigma^{\text{Sta}} \approx 1.07313$ ($E_b/N_0 \approx 1.148$ dB). By running density evolution we see that in both cases the stability condition determines the belief propagation (BP) threshold. From the discussion in Figure 6.34 we know that the MAP threshold is upper bounded by the same number (and it is of course always lower bounded by the BP threshold). It follows that all three thresholds (stability, BP, and MAP) are the same. ◇

Example 6.28 (Stability of $\mathcal{P}(G = 21/37, r = 1/2)$ and Alternating Puncturing). For the ensemble $\mathcal{P}(G = 21/37, r = 1/2)$ with alternating puncturing we know from Example 6.42 that $d_2(y) = \frac{y^2(3+y^2)}{2-2y^2}$. Let x^{Sta} be the unique positive solution to the equation $2xd_2(x) = 1$. We have $x^{\text{Sta}} \approx 0.582645$. If we assume transmission over the BEC(ϵ) we have $\mathfrak{B}(\epsilon) = \epsilon$ so that $\epsilon^{\text{Sta}} \approx 0.582645$. If we consider transmission over the BAWGNC(σ) then we have $\sigma^{\text{Sta}} \approx 0.962093$ ($E_b/N_0 \approx 0.336$ dB). For both channels the stability condition differs from the BP threshold. ◇

§6.7. EXIT Charts

Consider again the density evolution process introduced in Section 6.5. According to Theorem 6.21 we start with $\mathsf{c}_0 = \Delta_0$, and for $\ell \geq 1$, $\mathsf{c}_\ell = \Gamma_G^s(\mathsf{a}_{\text{BMSC}} \circledast \mathsf{c}_{\ell-1}, \mathsf{a}_{\text{BMSC}})$. Unfortunately, the "intermediate" densities c_ℓ do not have simple analytic descriptions in general.

If at each iteration ℓ we replace the intermediate density c_ℓ in the density evolution process with an "equivalent" density chosen from some "suitable" family of densities then we get an approximate density evolution process. If we choose an intermediate density which has *equal entropy* then we get the EXIT chart method. Although other choices are possible and sometimes also useful, the preferred choice for the family of intermediate densities is $\{\mathsf{a}_{\text{BAWGNC}(\mathtt{h})}\}$.

Definition 6.29 (EXIT Chart Method with Respect to $\{\mathsf{a}_{\text{BAWGNC}(\mathtt{h})}\}$). The density evolution process according to the EXIT chart method with respect to the channel family $\{\mathsf{a}_{\text{BAWGNC}(\mathtt{h})}\}$ can be specified as follows. Let $\mathtt{h}_0 = 1$. For $\ell \geq 1$, define

$$\mathtt{h}_\ell = H(\Gamma_G^s(\mathsf{a}_{\text{BMSC}} \circledast \mathsf{a}_{\text{BAWGNC}(\mathtt{h}_{\ell-1})}, \mathsf{a}_{\text{BMSC}})).$$

We then say that $\mathtt{h}_\ell$ is the entropy emitted in the ℓ-th iteration according to the EXIT chart method. The condition for convergence reads

$$H(\Gamma^s(\mathsf{a}_{\text{BMSC}} \circledast \mathsf{a}_{\text{BAWGNC}(\mathtt{h})}, \mathsf{a}_{\text{BMSC}})) < \mathtt{h}, \mathtt{h} \in (0,1). \tag{6.30}$$

Discussion: In words, at each stage of the decoding process the entropy $\mathtt{h}$ has to decrease. If we plot the entropy after one decoding round as a function of the entropy of the input, call it $\mathtt{h}$, then the corresponding graph must lie below the line $\mathtt{h}$.

Example 6.31 (EXIT Chart Method for Ensemble $\mathcal{P}(G = 21/37, r = 1/2)$). The left-hand picture in Figure 6.32 shows the decoding process according to the EXIT chart method for $\sigma = 0.93$ ($E_b/N_0 \approx 0.63$ dB). This is the same parameter used in Figure 6.23, where we considered true density evolution. The interpretation of these curves is essentially the same as discussed in Sections 3.14 and 4.10 for the case of LDPC ensembles. The only difference is that for LDPC ensembles the two curves correspond to the "action" of the variable nodes and check nodes, respectively, whereas in the present context the two curves correspond to the "action" of the two component codes. Since we have assumed identical component codes, the two curves are identical, which explains why we get a symmetric picture. In the right-hand picture the parameter has been increased to $\sigma \approx 0.941$ ($E_b/N_0 \approx 0.528$ dB). We see from this figure that $\sigma \approx 0.941$ is the critical parameter according to the EXIT chart method. This parameter differs only slightly form the true parameter σ^{BP} computed according to density evolution. It is listed in Table 6.24 as $\sigma^{\text{BP}} \approx 0.94$ ($E_b/N_0 \approx 0.537$ dB). ◇

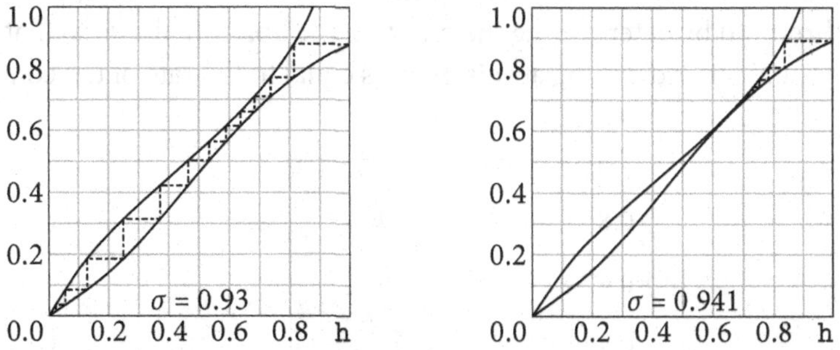

Figure 6.32: EXIT chart method for the ensemble $\mathcal{P}(G = 21/37, r = 1/2)$ with alternating puncturing on the BAWGN channel. In the left-hand picture the parameter is $\sigma = 0.93$, whereas in the right-hand picture we chose $\sigma = 0.941$.

§6.8. GEXIT Function and MAP Performance

In the preceding description we have used EXIT curves as a convenient (albeit approximate) tool to visualize the density evolution process. But we can also plot the EXIT curve of the *overall code* in response to a channel family $\{\mathsf{a}_{\text{BMSC}(\mathtt{h})}\}$. As we have seen for LDPC ensembles, in this case it is more useful to plot the GEXIT curve. Typically, it is difficult to determine the actual such GEXIT curve since no efficient procedure is known to compute the MAP performance. It is much easier to find the corresponding BP GEXIT curve, i.e., we use the densities at the output of the iterative decoder, instead of the true MAP densities.

LEMMA 6.33 (BP GEXIT CURVE FOR PARALLEL CONCATENATED ENSEMBLE). Consider the ensemble $\mathcal{P}(G, r)$. Assume that transmission takes place over a smooth family of BMS channels ordered by degradation characterized by their L-densities $\{\mathsf{a}_{\text{BMSC}(\mathtt{h})}\}$, where $\mathtt{h}$ denotes the entropy of the channel. For each $\mathtt{h}$, let $\mathsf{c}_{\mathtt{h}}$ denote the fixed point density of the density evolution process, i.e.,

$$\mathsf{c}_{\mathtt{h}} = \Gamma^{s}(\mathsf{a}_{\text{BMSC}(\mathtt{h})} \circledast \mathsf{c}_{\mathtt{h}}, \mathsf{a}_{\text{BMSC}(\mathtt{h})}).$$

Then the BP GEXIT curve is given in parametric form as

$$\Big(\mathtt{h}, \int \big(r\mathsf{c}_{\mathtt{h}} \circledast \mathsf{c}_{\mathtt{h}} + \bar{r}\Gamma^{p}(\mathsf{a}_{\text{BMSC}(\mathtt{h})} \circledast \mathsf{c}_{\mathtt{h}}, \mathsf{a}_{\text{BMSC}(\mathtt{h})})\big) l^{\text{BMSC}}(y)\mathrm{d}y\Big),$$

where $l^{\text{BMSC}(\mathtt{h})}(y)$ is the GEXIT kernel associated with the BMS channel.

Figure 6.34 shows the BP GEXIT curve for the ensemble $\mathcal{P}(G = 7/5, r = 1/3)$. Transmission takes place over the BAWGNC($\mathtt{h}$).

As for LDPC ensembles, we get from the BP GEXIT curve an upper bound on the MAP threshold by determining the point on the horizontal axis so that the integral starting at $\mathtt{h} = 1$ and ending at this point is equal to the rate of the code. For the

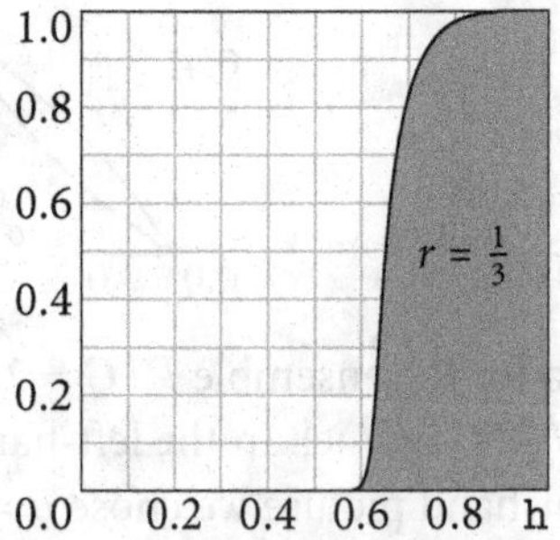

Figure 6.34: BP GEXIT curve for the ensemble $\mathcal{P}(G = 7/5, r = 1/3)$ assuming that transmission takes place over the BAWGNC($\mathtt{h}$). The BP and the MAP thresholds coincide and both thresholds are given by the stability condition. We have $\mathtt{h}^{\text{MAP/BP}} \approx 0.559$ ($\sigma^{\text{MAP/BP}} \approx 1.073$).

ensemble $\mathcal{P}(G = 7/5, r = 1/3)$ the BP threshold coincides with the MAP threshold. Both are given by the stability condition that is discussed in the previous section.

§6.9. Weight Distribution and Error Floor

Weight distributions played a pivotal role in the development of the theory of turbo codes. We start by determining the weight distribution of a convolutional code.

§6.9.1. Convolutional Codes

Let $C(G, n)$ denote the convolutional code of input length n defined by the binary rational function G. Let $a_{i,o,n}$ count the number of codewords of *input* weight i and *output* weight o. More precisely, i counts the weight of the information bits and o counts the weight of the parity bits, so that $i + o$ is the (*total/regular*) weight of a codeword. Define the associated generating function

$$A(x, y, z) = \sum_{i,o,n} a_{i,o,n} x^i y^o z^n.$$

We call $A(x, y, z)$ the *input-output* weight-generating function of the (sequence of) code(s) $C(G, n)$. It is our aim to compute $A(x, y, z)$.

Example 6.35 (The Encoder $G = 7/5$). We use the binary rational function $G = 7/5$ as our running example. A trellis section for this code is shown in Figure 6.6 on page 328. ◇

We employ the *transfer matrix* method. Encode the effect of the state transitions at each step in matrix form as

$$M(x,y) = \begin{array}{c} \\ (00) \\ (10) \\ (01) \\ (11) \end{array} \begin{array}{c} \begin{array}{cccc} (00) & (10) & (01) & (11) \end{array} \\ \begin{pmatrix} 1 & xy & 0 & 0 \\ 0 & 0 & y & x \\ xy & 1 & 0 & 0 \\ 0 & 0 & x & y \end{pmatrix} \end{array}.$$

More precisely, the rows of the matrix are associated with the current state, whereas the columns are associated with the state after the transition. For our running example we get $M_{(00),(10)} = xy$, since the transition from state (00) to state (10) requires an input of 1 and results in a parity bit of 1 as well. This matrix encodes the transitions corresponding to paths of length 1. Consider now paths of length 2. Such paths are the concatenation of two paths of length 1 whose intermediate states coincide. Let $M^{(2)}$ denote the matrix which encodes such paths of length 2. Let us associate integers to each state in the natural way, i.e., state (01) corresponds to 2, whereas (11) is identified with 3. Since the intermediate state, call it k, must coincide we have

$$M^{(2)}_{i,j} = \sum_k M_{i,k} M_{k,j}.$$

This means that $M^{(2)}(x,y) = M^2(x,y)$, a simple matrix multiplication. For our example we get

$$M^{(2)}(x,y) = \begin{matrix} \\ (00) \\ (10) \\ (01) \\ (11) \end{matrix} \begin{matrix} (00) & (10) & (01) & (11) \\ \end{matrix} \begin{pmatrix} 1 & xy & xy^2 & x^2y \\ xy^2 & y & x^2 & xy \\ xy & x^2y^2 & y & x \\ x^2y & x & xy & y^2 \end{pmatrix}.$$

The same reasoning applies to paths of length n: we can think of them as paths of length $n-1$ concatenated with paths of length 1 that share the same intermediate state. Therefore, $M^{(n)}(x,y) = M^n(x,y)$ by induction.

Recall that in the last m steps we eliminate the feedback. For these m steps the transition matrix, call it $\bar{M}$, corresponds to the encoder $p/1$ (instead of p/q). Further, all the input bits are zero so that $\bar{M} = \bar{M}(y)$. For our example we have

$$\bar{M}(y) = \begin{matrix} & (00) & (10) & (01) & (11) \\ (00) & 1 & 0 & 0 & 0 \\ (10) & 0 & 0 & y & 0 \\ (01) & y & 0 & 0 & 0 \\ (11) & 0 & 0 & 1 & 0 \end{matrix}.$$

By assumption the codewords start and end in the zero state, where for the first n steps we use $G = p/q$, whereas for the last m steps we set $G = p/1$. Therefore, the codewords of $C(G,n)$ are encoded by $\left[M^n(x,y)\bar{M}^m(y)\right]_{0,0}$. It follows that (see Problem 6.14)

$$A(x,y,z) = \sum_n \left[M^n(x,y)\bar{M}^m(y)\right]_{0,0} z^n = \left[(I - zM(x,y))^{-1}\bar{M}^m(y)\right]_{0,0}.$$

Example 6.36 (Weight Distribution of $C(G=7/5,n)$). To accomplish the preceding calculations for a specific example it is convenient to use a symbolic computer algebra system. For $C(G=7/5,n)$ we find that $A(x,y,z)$ equals

$$\frac{1 + z\left(-(x^2z^2) - y(1+z) + xy^3(1+z) - xy^4z(1+z) + y^2z(x^2+z+x^3z)\right)}{1 - (1+y)z + \left(y(1+y) - x^2(1+y^3)\right)z^3 + \left(x^2 - (1+x^4)y^2 + x^2y^4\right)z^4}.$$

The first few terms (up to z^2 corresponding to $n + m = 2 + 2 = 4$) of the expansion are (see Problem 6.9)

$$A(x,y,z) = 1 + (1 + xy^3)z + (1 + x^2y^2 + 2xy^3)z^2 + O(z)^3.$$

This means that there is 1 codeword (namely the empty one) contained in the length zero code, there are 2 codewords (the all-zero codeword and a codeword with 1 non-zero input bit and 3 non-zero output bits) contained in the length 1 code, there are 4 codewords in the length two 2, and so on. ◇

Let us define the *regular* (as compared to the input-output) weight distribution of the code $C(G, n)$. We have

$$A(x, z) = A(x, y = x, z) = \sum_{w,n} a_{w,n} x^w z^n$$

with coefficients $a_{w,n} = \sum_{i,o:i+o=w} a_{i,o,n}$. The coefficient $a_{w,n}$ counts the number of codewords of length n and weight w in the sequence of codes $C(G, n)$.

Example 6.37 (Regular Weight Distribution for $C(G = 7/5, n)$). Specializing the expression in Example 6.36 we get

$$A(x, z) = \frac{xz\left(((x-2)z-1)x^3+z+1\right)-1}{x\,(x^4-1)\,z^3+(x+1)z-1}. \qquad ◇$$

Example 6.38 (Asymptotic Weight Distribution for $C(G = 7/5, n)$). Figure 6.39 shows $\frac{1}{n}\log_2(a_{w,n})$ as a function of the normalized weight w/n for $n = 64, 128$, and 256 (recall that the blocklength of the code is $2(n+m)$). Also shown is $\lim_{n\to\infty}\frac{1}{n}\log_2(a_{w,n})$. It is the topic of Problem 6.17 to see how this limit can be computed efficiently. ◇

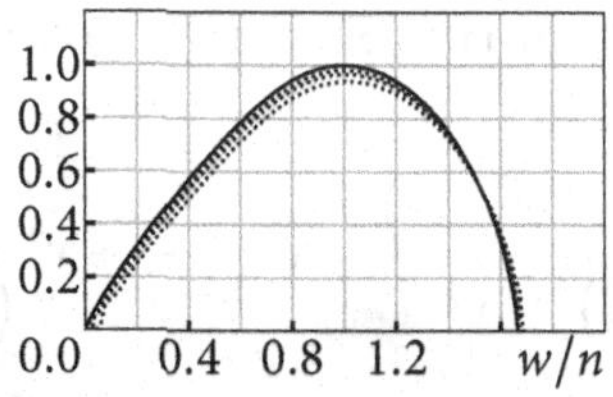

Figure 6.39: Exponent $\frac{1}{n}\log_2(a_{w,n})$ of the regular weight distribution of the code $C(G = 7/5, n)$ as a function of the normalized weight w/n for $n = 64, 128$, and 256 (dashed curves). Also shown is the asymptotic limit (solid line).

As discussed in Problems 6.10 and 6.11, it is not much harder to deal with punctured ensembles.

Before we discuss how to compute the weight distribution of concatenated ensembles, let us see how we can compute the generating function counting *detours*. A detour is a codeword that starts in state zero at time zero, diverges in the first transition from the zero state, and stops the first time it returns to the zero state. We have already seen that detours (of input weight 2) play the key role in determining the stability condition. We will soon discuss that the error floor is determined by such detours as well.

LEMMA 6.40 (DETOUR GENERATING FUNCTION). Consider the binary rational function G and let $M(x, y)$ be the corresponding transfer matrix where x encodes the input weight and y encodes the output weight. Let $M^{\bullet}(x, y)$ be equal to $M(x, y)$ except for entry $(0, 0)$ which we set equal to zero. Let $D(x, y)$ be the generating function counting detours. Then we have

$$D(x, y) = 1 - \frac{1}{\left[(I - M^{\bullet}(x, y))^{-1}\right]_{0,0}}.$$

Proof. By setting the entry $(0, 0)$ equal to zero we do not allow transitions from the zero state to the zero state. Therefore, $\left[(I - M^{\bullet}(x, y))^{-1}\right]_{0,0}$ counts all paths (irrespective of their length) that start and end in the zero state and have no zero transition. Such paths are the concatenation of an arbitrary number of detours. If the generating function for detours is $D(x, y)$ then the generating function for paths which are the concatenation of two detours is $D^2(x, y)$ and the generating function of paths which are the concatenation of i detours is $D^i(x, y)$. We have

$$\left[(I - M^{\bullet}(x, y))^{-1}\right]_{0,0} = \sum_i D^i(x, y) = \frac{1}{1 - D(x, y)},$$

from which the claim follows. □

EXAMPLE 6.41 (DETOUR GENERATING FUNCTION FOR $G = 7/5$). For $G = 7/5$ we find

$$D(x, y) = \frac{x^2 y^2 (x^2 + y - y^2)}{(1 - y)^2 - x^2} = x^2 \frac{y^3}{1 - y} + x^4 \frac{y^2}{(1 - y)^3} + O(x^6). \qquad \diamond$$

We are primarily interested in codewords of small weight since these are the ones responsible for the error floor that we have observed in Figure 6.16. Therefore, we write

$$D(x, y) = \sum_{i \geq 2} d_i(y) x^i.$$

Note that $i \geq 2$ since a detour requires an input of weight at least 2: one non-zero input to drive the encoder out of the zero state and a second one to force it back to the zero state. We will soon see that $d_2(y)$ plays the key role in the asymptotic analysis of the weight distribution for small weights. In many cases $d_2(y)$ has the form

$$d_2(y) = \frac{y^{\alpha}}{1 - y^{\beta}}.$$

In these cases α and β have the following interpretation: α is the weight of the lowest weight detour due to inputs of weight 2; β is the weight of the unique cycle in the state diagram corresponding to zero input (where the transition from the zero state to the zero state is forbidden). A simple "atypical" example is $G(D) = \frac{1+D+D^2}{1+D}$, for which a direct calculation leads to

$$d_2(y) = \frac{y^2}{1-y} + y^3.$$

How to compute the detour generating function for punctured codes is the topic of Problem 6.12.

EXAMPLE 6.42 (DETOUR GENERATING FUNCTION FOR $G = 21/37$ WITH ALTERNATING PUNCTURING). Specializing the general expressions in Problem 6.12 to the specific case $G = 21/37$ we get

$$d_2(y) = \frac{y^2\left(3+y^2\right)}{2-2y^2}. \qquad \diamond$$

§6.9.2. CONCATENATED ENSEMBLES

We have seen how to compute the weight distribution of convolutional codes. It is now only a small step to derive the weight distribution of parallel and serially concatenated ensembles. More precisely, we compute the expected weight distribution where the expectation is over the choice of interleavers.

LEMMA 6.43 (EXPECTED WEIGHT DISTRIBUTION OF $\mathcal{P}(G,n,r)$). Consider the ensemble $\mathcal{P}(G,n,r)$. Let $A_{\frac{3r-1}{2r}}(x,y,z) = \sum_{i,o,n} a_{i,o,n} x^i x^o z^n$ denote the generating function of the input-output weight distribution of $C(G,n)$ under puncturing of rate $\frac{3r-1}{2r}$ so that the overall rate of the ensemble $\mathcal{P}$ is r (see discussion on page 334). Let $P(x,y,z) = \sum_{i,o,n} p_{i,o,n} x^i y^o z^n$ denote the generating function of the expected input-output weight distribution of $\mathcal{P}(G,n,r)$, where the expectation is taken over all elements of the ensemble (i.e., all interleavers). Then

$$p_{i,o,n} = \frac{\sum_j a_{i,j,n} a_{i,o-j,n}}{\binom{n}{i}}. \tag{6.44}$$

The generating function of the regular weight distribution is $P(x,z) = P(x,y = x,z)$.

Proof. Consider a codeword of the underlying convolutional code with systematic weight i and parity weight o. The permuted information sequence also has weight i.

Since we impose a uniform probability distribution over all interleavers, the permuted input sequence is mapped into any of the possible $\sum_o a_{i,o,n}$ sequences of input weight i with equal probability. This explains the "convolution" with respect to the parity weight o. The factor $\binom{n}{i}$ ensures the proper normalization: we have $\sum_o a_{i,o,n} = \binom{n}{i}$, since there are $\binom{n}{i}$ inputs of weight i, each leading to one distinct codeword. □

Example 6.45 (Expected Weight Distribution for $\mathcal{P}(G = 7/5, n, r = 1/3)$). Consider the ensemble $\mathcal{P}(G = 7/5, n, r = 1/3)$. In Example 6.36 we computed $A(x, y, z)$ for $C(G = 7/5, n)$. Let us compute the first few terms of the associated regular weight distribution for $n = 64$, 128, and 256 via (6.44):

$$
\begin{aligned}
&P(x,z) \approx \\
&\dots \\
&\left(1 + 0.0005\,x^6 + 0.126\,x^7 + 2.2255\,x^8 + 4.2853\,x^9 + 6.34\,x^{10} + \dots\right)z^{64} + \\
&\dots \\
&\left(1 + 0.0001\,x^6 + 0.0627\,x^7 + 2.111\,x^8 + 4.1337\,x^9 + 6.1395\,x^{10} + \dots\right)z^{128} + \\
&\dots \\
&\left(1 + 0.0313\,x^7 + 2.0551\,x^8 + 4.0647\,x^9 + 6.0622\,x^{10} + 8.0525\,x^{11} + \dots\right)z^{256} + \\
&O(z^{257}).
\end{aligned}
$$

Figure 6.46 shows the exponent $\frac{1}{n}\log_2(p_{w,n})$ as a function of the normalized weight w/n for the ensemble $\mathcal{P}(G = 7/5, n, r = 1/3)$ and $n = 64$, 128, and 256. ◇

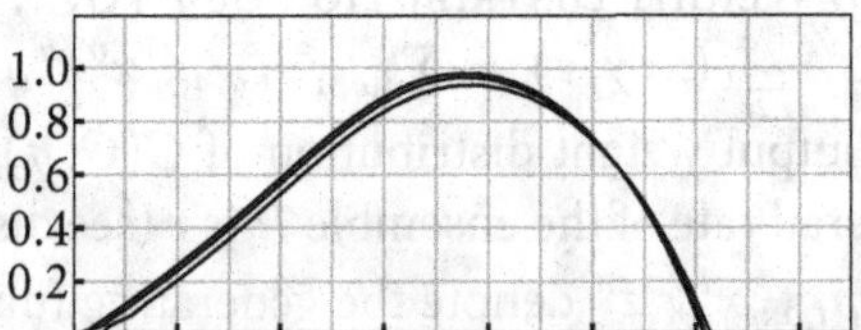

Figure 6.46: Exponent $\frac{1}{n}\log_2(p_{w,n})$ as a function of the normalized weight w/n for the ensemble $\mathcal{P}(G = 7/5, n, r = 1/3)$ and $n = 64$, 128, and 256. The normalization of the weight is with respect to n, not the blocklength.

Consider the first few terms of $P(x, z)$ in Example 6.45. These terms give the *expected* number of codewords of low weight for $n = 64$, 128, and 256. These low-weight codewords are the cause of the error floor (see, e.g., the performance curve in Figure 6.16).

We will now show that the expected number of codewords of *fixed weight* converges to a limit as the blocklength increases (see Example 6.45). We will also show how this limit can be computed. From this result we will be able to derive the expected number of *minimal* codewords of fixed weight in the asymptotic limit and we will show that these minimal codewords are distributed according to a Poisson distribution. This will allow us to compute the probability that a random element from the ensemble has a given minimum distance.

LEMMA 6.47 (ASYMPTOTIC EXPECTED NUMBER OF CODEWORDS OF FIXED WEIGHT). Consider the ensemble $\mathcal{P}(G,n,r)$. Let $d_2(y)$ count detours of $G(D)$ that have input weight 2. Let $P(x,y,z) = \sum_{i,o,n} p_{i,o,n} x^i y^o z^n$ denote the generating function of the input-output weight distribution of the ensemble $\mathcal{P}(G,n,r)$ and let $\bar{P}(x,y) = \sum_{i,o} \bar{p}_{i,o} x^i y^o$ be the generating function of the *asymptotic* input-output weight distribution. More precisely, $\bar{p}_{i,o} = \lim_{n\to\infty} p_{i,o,n}$. Then

$$\bar{P}(x,y) = \frac{1}{\sqrt{1-4x^2 d_2^2(y)}}. \tag{6.48}$$

Proof. Let $D(x,y) = \sum_{i,o} d_{i,o} x^i y^o$ denote the generating function counting detours of $G(D)$. We want to count *minimal* codewords of $C(G,n)$ of "type" (i,o), i.e., minimal codewords that have input weight i and output weight o. We claim that this number, divided by n, converges for increasing n to $d_{i,o}$. This is true since, ignoring boundary effects, each detour of type (i,o) can be shifted to start at n positions to create a minimal codeword of $C(G,n)$. Conversely, every minimal codeword is a shift of a detour.

Consider now a generic codeword of $C(G,n)$. Such a codeword is the union of minimal codewords. In the large n limit, the number of codewords of fixed type (i,o) which consist of l minimal codewords converges to the coefficient of $x^i y^o z^l$ of $\sum_l \binom{n}{l} D^l(x,y) z^l$. This is true since the l detours give rise to $\binom{n}{l}$ degrees of freedom of choosing their starting value. Define $e_{i,o,l}$ as the coefficient in front of $x^i y^o z^l$ of $\sum_l \binom{n}{l} D^l(x,y) z^l$. Note that $e_{i,o,l}$ is proportional to n^l.

Look at the weight distribution of the parallel concatenated code. We know from Lemma 6.43 that the asymptotic weight distribution is equal to

$$\bar{p}_{i,o} = \lim_{n\to\infty} \sum_l \frac{\sum_j e_{i,o-j,l} e_{i,j,l}}{\binom{n}{i}}.$$

A minimal codeword has input weight at least 2. Therefore, a codeword consisting of l minimal codewords has input weight at least $2l$. If its weight is strictly larger than $2l$ then its contribution to the preceding sum will vanish like n^{2l-i} (recall that $e_{i,o,l}$ is proportional to n^l). The only surviving terms are therefore from those minimal codewords that have input weight exactly 2 and their contribution will converge to a constant. We can restrict ourselves therefore to $\sum_l \binom{n}{l}(x^2 d_2(y))^l z^l$ instead of $\sum_l \binom{n}{l} D^l(x,y) z^l$. We get

$$\bar{P}(x,y) = \sum_i \binom{2i}{i} d_2^{2i}(y) x^{2i} = \frac{1}{\sqrt{1-4x^2 d_2^2(y)}},$$

where the term $\binom{2i}{i}$ appears because $\binom{2i}{i} = \lim_{n\to\infty} \frac{\binom{n}{i}^2}{\binom{n}{2i}}$. □

Lemma 6.49 (Expected Number of Minimal Codewords of Fixed Weight in the Asymptotic Limit). Consider the ensemble $\mathcal{P}(G,n,r)$. Let $\bar{P}(x,y)$ denote the generating function of the asymptotic input-output weight distribution of $\mathcal{P}(G,n,r)$ and let $\hat{P}(x,y)$ denote the corresponding generating function counting *minimal* codewords. Then

$$\hat{P}(x,y) = \log(\bar{P}(x,y)) = -\frac{1}{2}\log\big(1-4x^2 d_2^2(y)\big). \tag{6.50}$$

Let $\hat{P}(x) = \hat{P}(x, y = x) = \sum_w \hat{p}_w x^w$. Let C denote a random element of $\mathcal{P}(G,n,r)$. For any $d > 0$, $d \in \mathbb{N}$,

$$\lim_{n\to\infty} \mathbb{P}\{d_{\min}(C) > d\} = \mathrm{e}^{-\sum_{w=1}^{d} \hat{p}_w}.$$

Further, if $(W_1, W_2, \ldots, W_d)$ denotes the random vector counting the number of minimal codewords of weight up to d for a random sample C, then the distribution of this vector converges to a vector of independent Poisson distributed random variables with mean equal to $(\hat{p}_1, \hat{p}_2, \ldots, \hat{p}_d)$.

Proof. We proceed as in the case of DLPC ensembles. Recall that in the realm of LDPC codes, codewords/stopping sets of small weight are (asymptotically) due to cycles in the bipartite graph which involve exclusively nodes of degree 2. For turbo codes, the role of variable nodes of degree 2 is played by detours of input weight 2. For LDPC ensembles we were able to reduce the question of the number and distribution of cycles to a well-known random graph problem by converting the bipartite graph into a standard graph (see Lemma C.37). For the ensemble of turbo codes no such reduction is known. Therefore we have to perform the necessary computations explicitly. Our main tools are Theorem C.33 and Fact C.35.

Let X_w denote the number of minimal codewords of weight w. As we have seen previously, in the limit of large blocklengths, all such codewords are due to combinations of detours of input weight 2, which are mapped via the interleaver again to combinations of detours of input weight 2. The minimality condition implies that such a codeword cannot be decomposed into two smaller codewords. Consider now $(X_w)_r$, the factorial moment of X_w defined in (C.34). We can represent X_w as $X_w = \sum_i Z_{w,i}$, where $Z_{w,i}$ is an indicator random variable which corresponds to a particular choice of placement of up to w input bits. If this choice of input bits leads to a minimal codeword and if this codeword has weight w then $Z_{w,i}$ takes the value 1, otherwise it takes the value 0. The randomness resides in the choice of the interleaver. From Fact C.35 we know that we can write $(X_w)_r$ as $(X_w)_r = \sum_{(i_1,\dots,i_r)} Z_{w,i_1} \dots Z_{w,i_r}$. Consider the contribution of the sum on the right-hand side stemming from all those choices of $(i_1, \dots, i_r)$ which correspond to non-overlapping bits. Since each bit pattern has some finite (small) weight and n tends to infinity, almost all placements have the property that they do not overlap. Further, for non-overlapping patterns the involved indicator random variables become asymptotically independent. We conclude that the portion of the sum $\sum_{(i_1,\dots,i_r)} Z_{w,i_1} \dots Z_{w,i_r}$ corresponding to non-overlapping patterns tends to $\prod_{i=1}^m \mu_i^{r_i}$.

It remains to look at those summands which involve overlapping patterns. We claim that their contributions vanish like $O(1/n)$. For this step to hold it is important that we look at *minimal* codewords as it would not hold for codewords themselves. We skip the tedious verification of this claim.

Now since we know that the distribution of the minimal codewords is Poisson (in the asymptotic limit), what is the relationship between $\hat{P}(x,y)$ (counting minimal codewords) and $\bar{P}(x,y)$ (counting regular codewords)? We claim that we have $\hat{P}(x,y) = \log \bar{P}(x,y)$ as stated in the lemma. This is equivalent to

$$\bar{P}(x,y) = e^{\hat{P}(x,y)} = \sum_k \frac{\hat{P}(x,y)^k}{k!}.$$

The interpretation is similar to the one on page 155 where we derived the block error probability from the distribution of minimal codewords. Consider one term of $\hat{P}(x,y)$, call it $\hat{p}_{i,j}x^i y^j$. It corresponds to the occurrence of a minimal codeword of "type" $x^i y^j$. From our preceding discussion we know that this type has a Poisson distribution. Therefore,

$$e^{\hat{p}_{i,j}x^i y^j} = \sum_k \frac{\hat{p}_{i,j}^k x^{ik} y^{ik}}{k!}$$

represents the generating function counting regular codewords which are composed entirely of minimal codewords of this particular type. Why can we simply add the weights of the individual codewords? This is true since, if we have k minimal codewords that form a regular codeword and if we consider the blocklength n as very large, then with high probability these minimal codewords do not overlap – their weights therefore add. We get the final formula $\bar{P}(x,y) = e^{\hat{P}(x,y)}$ by recalling that the joint distribution of different types is of product form and of Poisson type as well. □

Example 6.51 ($\bar{P}(x,y)$ and $\hat{P}(x,y)$ for $\mathcal{P}(G = 7/5, r = 1/3)$). In Example 6.41 we determined $d_2(y)$ for this case. Inserting this into (6.48) and (6.50) we get

$$\bar{P}(x,y) = \frac{1}{\sqrt{1-(2xy^3/(1-y))^2}}, \quad \hat{P}(x,y) = -\frac{1}{2}\log(1-(2xy^3/(1-y))^2).$$

Let us expand out the first few terms. We get

$$\bar{P}(x) = 1 + 2x^8 + 4x^9 + 6x^{10} + 8x^{11} + O(x^{12}) = \hat{P}(x).$$

The first few terms of $\bar{P}(x,y)$ and $\hat{P}(x)$ are the same but they differ as soon as we look at weights starting from twice the minimum distance.

Compare this to the finite-length weight distribution which we computed in Example 6.45: e.g., we have $p_{w=8,n=64} = 2.2255$, $p_{w=8,n=128} = 2.111$, and $p_{w=8,n=256} = 2.0551$. From earlier, we know that the asymptotic value is $\bar{p}_{w=8} = 2$. We conclude that the finite-length values converge to the asymptotic limit reasonably quickly (at the speed of $1/n$). This implies that the asymptotic quantities are of practical value. From Lemma 6.49, we know that for an element C chosen uniformly at random from the ensemble we have

$$\lim_{n\to\infty} \mathbb{P}\{d_{\min}(C) \geq 8\} = 1, \quad \lim_{n\to\infty} \mathbb{P}\{d_{\min}(C) \geq 9\} = e^{-2},$$
$$\lim_{n\to\infty} \mathbb{P}\{d_{\min}(C) \geq 10\} = e^{-6}, \quad \lim_{n\to\infty} \mathbb{P}\{d_{\min}(C) \geq 11\} = e^{-12}. \qquad \diamond$$

§6.9.3. Error Floor under MAP Decoding

Now since we know the distribution of low-weight codewords, it is only a small step to the error floor under MAP decoding. We could start from the finite-length weight distribution. However, as we have seen, the difference from the asymptotic case is typically small and the asymptotic expressions are easier to handle. We skip the proof of the following lemma since it is similar to the proof of Lemmas 3.166 and 4.177.

Lemma 6.52 (Asymptotic Error Floor for $\mathcal{P}(G,n,r)$). Consider the ensemble $\mathcal{P}(G,n,r)$ and assume that transmission takes place over the BAWGNC(E_b/N_0). Let $\hat{P}(x,y)$ be the generating function counting the asymptotic input-output weight distribution of minimal codewords, where we assume that the appropriate puncturing has been performed. Define

$$\hat{P}(x) = \hat{P}(x, y = x) = -\frac{1}{2}\log(1 - 4x^2 d_2^2(x)),$$

$$\hat{P}_{\rm b}(x) = \frac{x\partial\hat{P}(x,y)}{\partial x}\Big|_{y=x} = \frac{4d_2^2(x)x^2}{1-4d_2^2(x)x^2}.$$

Let $\sigma^{\rm BP}$ denote threshold under BP decoding and define $(E_b/N_0)^{\rm BP} = \frac{1}{2r\sigma^{\rm BP}}$. Then for $E_b/N_0 > (E_b/N_0)^{\rm BP}$

(6.53) $$\lim_{n\to\infty} \mathbb{E}_{\mathcal{P}(G,n,r)}[\mathrm{P}_{\rm B}^{\rm MAP}(C, E_b/N_0)] = 1 - \mathrm{e}^{-\frac{1}{\pi}\int_0^{\frac{\pi}{2}} \hat{P}\left(\mathrm{e}^{-\frac{rE_b/N_0}{\sin^2(\theta)}}\right)\mathrm{d}\theta},$$

(6.54) $$\lim_{n\to\infty} n\,\mathbb{E}_{\mathcal{P}(G,n,r)}[\mathrm{P}_{\rm b}^{\rm MAP}(C, E_b/N_0)] = \frac{1}{\pi}\int_0^{\frac{\pi}{2}} \hat{P}_{\rm b}\left(\mathrm{e}^{-\frac{rE_b/N_0}{\sin^2(\theta)}}\right)\mathrm{d}\theta.$$

Discussion: If we consider expurgated ensembles then instead of $\hat{P}(x)$ and $\hat{P}_{\rm b}(x)$ use the corresponding generating functions where the summation ranges from the expurgation parameter to infinity.

Example 6.55 (Error Floor of $\mathcal{P}(G = 7/5, n, r = 1/3)$). From Example 6.51 we have

$$\hat{P}(x) = \hat{P}(x,y)\,|_{y=x} = -\frac{1}{2}\log\left(1-\left(2x^4/(1-x)\right)^2\right),$$

$$\hat{P}_{\rm b}(x) = \frac{x\partial\hat{P}(x,y)}{\partial x}\Big|_{y=x} = \frac{4x^8}{1-2x+x^2-4x^8}.$$ ◇

Example 6.56 (Error Floor of Berrou-Glavieux-Thitimajshima Code). Using the result of Example 6.42 we get

$$\hat{P}(x) = -\frac{1}{2}\log\left(1-\left(2x^3(3+x^2)/(2-2x^2)\right)^2\right),$$

$$\hat{P}_{\rm b}(x) = \frac{x^6(3+x^2)^2}{1-2x^2+x^4-9x^6-6x^8-x^{10}}.$$

The error floor which is predicted if we insert $\hat{P}_{\rm b}(x)$ into (6.54) and numerically perform the integration is shown in Figure 6.16 together with the curve corresponding to simulations. As you can see they match well. ◇

§6.9.4. Minimum Distance

Our previous investigation concerned the behavior of *typical* elements of the ensemble. What is the minimum distance of the *best* element of the ensemble? From the result concerning the Poisson distribution we can conclude that there exist turbo codes with arbitrary large minimum distance (assuming the blocklength is chosen sufficiently large). But what about the relative minimum distance? Without proof we state that a more careful variation of the previous probabilistic argument shows that we can let d grow as fast as $\Theta(\log^{1/4}(n))$. This shows the existence of turbo codes whose minimum distance grows like $\Theta(\log^{1/4}(n))$. We will now show, conversely, that no element of the ensemble has minimum distance larger than $\Theta(\log(n))$.

Assume that we feed the input $1 + x^d$ to a convolutional encoder with $G = p/q$. This means, we input a 1 at time zero, followed by another 1, exactly d time instances later. All other inputs are 0. We claim that for a proper choice of d the output has small weight, even if n tends to infinity. This is true since for any polynomial $q(x)$ with $q_0 = 1$ there exists an integer d such that $q(x)$ divides $1+x^d$ (see Problem 6.15). Assuming a sufficiently large n (so that the termination no longer plays a role), the output is $(1 + x^d)\frac{p(x)}{q(x)}$. But if $q(x)$ divides $1 + x^d$ then this is a polynomial and has, hence, fixed weight regardless of n. We call the smallest such integer d the *period* of the convolutional encoder.

Note that $1 + x^{2d} = 1 + x^d + x^d(1 + x^d)$ and, more generally, $1 + x^{di}$ is the sum of i shifted copies of $1 + x^d$. By linearity of the code, we conclude that, for a fixed $i \in \mathbb{N}$, the output corresponding to the input $1 + x^{di}$ has small weight, even if n tends to infinity: indeed, there exist two integers, α and β, so that the weight of the codeword corresponding to the input $1 + x^{di}$ is upper bounded by $\alpha + \beta i$. Often, but not always (see the discussion on page 353), α is the weight of the lowest weight detour due to inputs of weight 2 and β is the weight of the unique cycle in the state diagram corresponding to input zero.

Theorem 6.57 (Logarithmic Upper Bound on Minimum Distance). Consider the ensemble $\mathcal{P}(G, n, r)$. Let d denote the period of the convolutional encoder and let α and β be as discussed earlier. Then for every $\Delta \in \mathbb{N}$, $1 \le \Delta \le \sqrt{\frac{n+1}{d}}$, and every $C \in \mathcal{P}(G, n, r)$,

$$d_{\min}(C) < 2\frac{\log(n(\Delta-1)^2/\Delta)}{\log(\Delta-1)}(\alpha + \Delta\beta).$$

Discussion: If we choose Δ to be a constant then the preceding expression shows that the minimum distance grows at most logarithmically with the blocklength.

Proof. First note that puncturing only decreases the minimum distance. Although better bounds can be derived by taking the effect of puncturing explicitly into account, a valid upper bound is derived by considering the unpunctured ensemble $\mathcal{P}(G, n)$.

Consider the set $[n] = \{1, \dots, n\}$, corresponding to the n bits of the input, and pick an integer Δ, $1 \le \Delta \le \sqrt{\frac{n+1}{d}}$. We claim that we can partition $[n]$ into non-overlapping subsets $\mathcal{S}_j$, $j = 1, \dots, \lfloor n/\Delta \rfloor$, so that $\bigcup_j \mathcal{S}_j = [n]$ and that each subset $\mathcal{S}_j$ has the form $\mathcal{S}_j = \{0, d, \dots, (\Delta-1)d\} + s_j$ or $\mathcal{S}_j = \{0, d, \dots, \Delta d\} + s_j$. In both cases the *shift* s_j is an integer. That such a partition can always be found under the given conditions can be seen as follows. First partition $[n]$ into the d subsets $\mathcal{N}_i = [n] \bigcap (\mathbb{Z}_d + i)$, $i = 0, \dots, d-1$. Now partition each $\mathcal{N}_i$ into $\lfloor |\mathcal{N}_i|/\Delta \rfloor$ subsets: the first $\lfloor |\mathcal{N}_i|/\Delta \rfloor(\Delta+1) - |\mathcal{N}_i|$ have size Δ, the remaining have size $\Delta + 1$. The condition $\Delta \leq \sqrt{\frac{n+1}{d}}$ guarantees that $\lfloor |\mathcal{N}_i|/\Delta \rfloor(\Delta+1) - |\mathcal{N}_i| \geq 0$.

To be concrete, consider the case $n = 19$, $d = 2$, and $\Delta = 3$. We have $3 = \Delta \leq \sqrt{(n+1)/d} \approx 3.16$. We get

$$\mathcal{N}_0 = \{2,4,6,8,10,12,14,16,18\}, \qquad \mathcal{N}_1 = \{1,3,5,7,9,11,13,15,17,19\}.$$

We partition $\mathcal{N}_0$ into the subsets $\{2,4,6\}$, $\{8,10,12\}$, and $\{14,16,18\}$, and $\mathcal{N}_1$ into the subsets $\{1,3,5\}$, $\{7,9,11\}$, and $\{13,15,17,19\}$.

Consider a particular input x^{s_1}. Assume that this input enters the first encoder directly and that it is permuted via a permutation π before being fed to the second encoder. Call the permuted sequence x^{s_2}, $x^{s_2} = \pi(x^{s_1})$. Group the bits of x^{s_1} as well as x^{s_2} according to the partition $\{\mathcal{S}_j\}_j$. Construct a bipartite graph in the following way. The nodes of the bipartite graph correspond to the groups of the partition for each encoder. For the preceding example the bipartite graph would have in total 12 nodes, 6 on each side. For example, node 1 (on each side of the bipartite graph) represents the input bits 2, 4, and 6; node 2 represents the input bits 8, 10, 12, and so on. The edges are drawn according to the permutation π: if a bit of x^{s_1} belongs to, let's say, group $\mathcal{S}_k$ and is mapped via the permutation to a position which belongs to group $\mathcal{S}_l$ in x^{s_2} then we draw an edge in the bipartite graph from node k (on the left) to node l (on the right). Multiple edges between pairs of nodes are possible. Consider our example and the permutation

$$\pi = \{12,1,4,18,17,4,3,19,5,13,2,6,16,7,15,14,9,8,10\}.$$

This means that the permutation sends 1 to 12 and so on. The corresponding bipartite graph is shown in Figure 6.58. Note that 1 belongs to $\mathcal{S}_4$ and that 12 belongs to $\mathcal{S}_2$. There is therefore an edge from $\mathcal{S}_4$ (on the left) to $\mathcal{S}_2$ (on the right).

This bipartite graph has $\lfloor n/\Delta \rfloor$ nodes on each side and each node has degree at least Δ (more precisely degree Δ or degree $(\Delta+1)$). Such a bipartite graph must have a *cycle* of length strictly less than

$$4\left(\frac{\log(n/\Delta)}{2\log(\Delta-1)}+1\right) = 2\frac{\log(n(\Delta-1)^2/\Delta)}{\log(\Delta-1)}.$$

This means the graph must contain a path of length strictly less than this number which starts and ends at the same node. The argument is identical to the one used in Problem 3.25 to determine the girth of the Tanner graph of an LDPC code.

Consider a cycle of length $2c$, which involves c left nodes and c right nodes. One such node corresponds in the encoder to a set of at most $\Delta + 1$ bits which

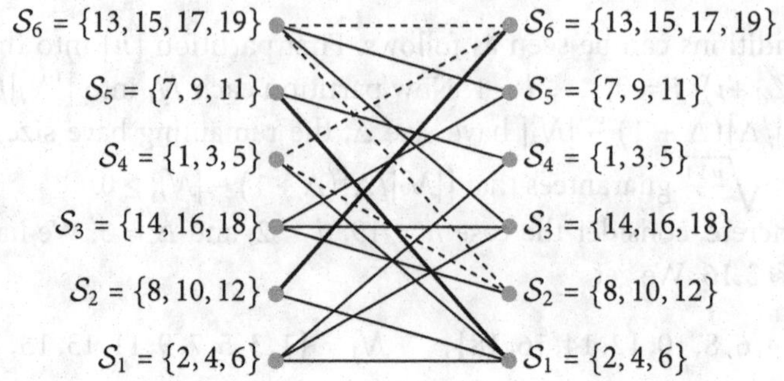

Figure 6.58: Bipartite graph corresponding to the parameters $n = 19$, $d = 2$, $\Delta = 3$, and $\pi = \{12,1,4,18,17,4,3,19,5,13,2,6,16,7,15,14,9,8,10\}$. Double edges are indicated by thick lines. The cycle of length 4, formed by (starting on the left) $\mathcal{S}_6 \to \mathcal{S}_6 \to \mathcal{S}_4 \to \mathcal{S}_2$, is shown as dashed lines.

are always exactly d positions apart. Assume at first that the bipartite graph has no double edges. For each node in the cycle there are then exactly two bits that correspond to the two outgoing edges. For both x^{s_1} as well as x^{s_2} set these bits to 1 and all other ones to 0. Note that this definition guarantees that $x^{s_2} = \pi(x^{s_1})$. We have constructed an input of weight $2c$ (for both the first encoder and the encoder with the permitted input).

By construction, x^{s_1} as well as x^{s_2} give rise to c detours, each of input weight 2 and each corresponding to an input of the form $1+x^{di}$ for some i. The corresponding codeword consists therefore of $2c$ such detours. Since the length of each such detour is limited to at most Δd, by our previous discussion the weight of each such detour is at most $\alpha + \Delta\beta$. From this the indicated upper bound on the total length follows.

For our example there is a cycle of length 4 formed by (starting on the left) $\mathcal{S}_6 \to \mathcal{S}_6 \to \mathcal{S}_4 \to \mathcal{S}_2$. On the left we set the bits 1 and 5 (corresponding to the two edges in the cycle entering $\mathcal{S}_4 = \{1,3,5\}$) as well as 15 and 19 (corresponding to the two edges in the cycle entering $\mathcal{S}_6 = \{13,15,17,19\}$) to 1. On the right we set the bits 10, 12, 15, and 17 to 1. By construction, the second word is the permutation (under π) of the first one. Both inputs give rise to two detours, each of input weight 2. The separation of the four detours is (here $d = 2$) $2d$, $2d$, d, and d, respectively. We therefore have an upper bound on the total weight of $4\alpha + 6\beta$. The argument used in the theorem is weaker and uses the upper bound $4(\alpha + \Delta\beta) = 4\alpha + 12\beta$.

If the bipartite graph has double edges then we have some degree of freedom in placing the non-zero bits, but the argument still applies. □

§6.10. Variations on the Theme

As mentioned before, there are many variations on the basic theme of turbo codes. The aim of these variations is to introduce some extra degree of irregularity in their design and to choose this irregularity in such a way as to increase the performance or to decrease the complexity. Let us mention some of the more prominent variations.

First, in our definition of admissible encoders $G = p/q$ we imposed no restriction on the degree of p with respect of the degree q. In particular, we can choose $\deg(p) > \deg(q)$. This means that the feedback does not extend over the whole range of the shift register. In a classical setting, where we are interested in the minimum distance corresponding to G, it is not hard to show that such a choice yields no advantage and that, to the contrary, it is detrimental in terms of complexity. However, in the iterative setting, the resulting codes (called *big numerator* codes) are sometimes superior to classical choices.

Asymmetric turbo codes represent another interesting variation. These are turbo codes which employ different component encoders. That such a choice might be advantageous is easily understood from the EXIT chart analysis. Consider a parallel concatenated ensemble and consider the EXIT chart as shown in Figure 6.32. If the two component encoders G are identical then the two component EXIT curves are symmetric around the 45 degree line. The threshold is determined by the condition that each such curve just touches the 45 degree line. Clearly, a potentially better "matching of the curves" can be achieved if we allow different component codes. In this case each individual curve is allowed to cross the 45 degree line – as long as the two curves do not intersect.

Example 6.59 (Asymmetric Big Numerator Parallel Concatenated Ensemble). Figure 6.60 shows the EXIT curves for transmission over the BEC($\mathtt{h} \approx 0.6481$) for an asymmetric big-numerator parallel concatenated ensemble. The two component encoders are $G_1 = 11/3$ and $G_2 = 11/13$. Note that G_1 is a big-numerator encoder. The threshold of $\mathtt{h}^{\text{BP}} \approx 0.6481$ is larger than the threshold of any symmetric memory-three parallel concatenated ensemble. ◇

As a final generalization consider the ensemble of *irregular* turbo codes. Let us first represent a standard parallel concatenated code with two component codes in the following alternative form. The encoder structure is shown in Figure 6.61. As before, let x^s denote the information sequence of length $n + m$, where the last m bits are assumed to be zero. This sequence is again sent directly over the channel. To generate the parity bits, first repeat each bit of x^s once to generate a sequence of the form $(x_1^s, x_1^s, x_2^s, x_2^s, \dots, x_{n+m}^s, x_{n+m}^s)$. Next, permute this sequence. Assume at first that the permutation is constrained so that the permuted sequence has the form $(x^s, \pi(x^s))$, where π is a permutation on $n + m$ letters fixing the last m positions.

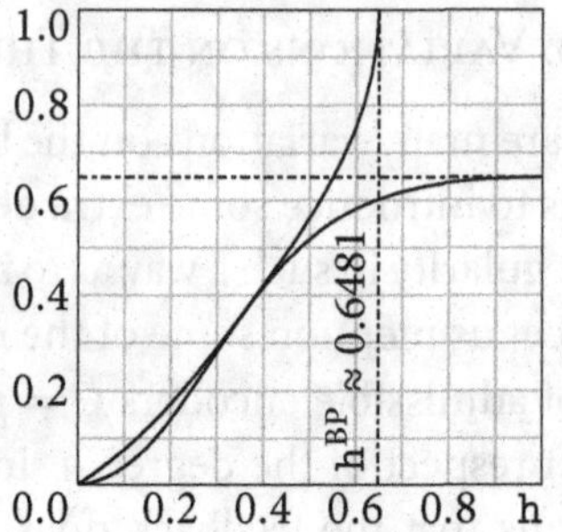

Figure 6.60: EXIT chart for transmission over the BEC(h ≈ 0.6481) for an asymmetric (big-numerator) parallel concatenated ensemble.

Input this permuted sequence of length $2(n+m)$ into the convolutional encoder. Because we have m zeros at the end of x^s as well as at the end of $\pi(x^s)$ the output of the convolutional encoder is simply the concatenation of the output of the standard parallel concatenated code with two component codes, i.e., $x^p = (x^{p_1}, x^{p_2})$. We therefore see that this new encoder structure is simply an alternative way of

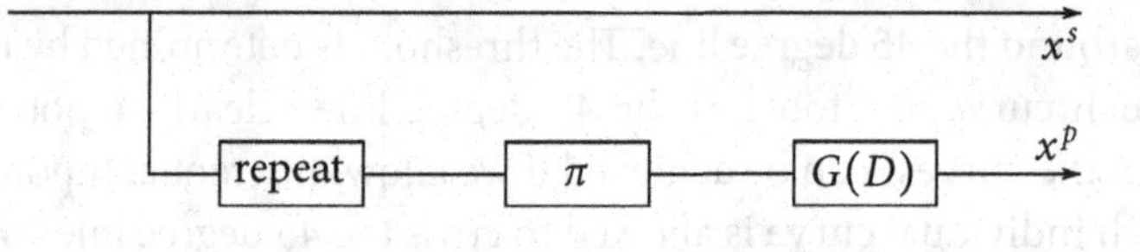

Figure 6.61: Alternative view of an encoder for a standard parallel concatenated code.

accomplishing the encoding. Assume next that we lift the restriction on the permutation on $2(n+m)$ letters (except that the last m positions should be fixed). The corresponding FSFG is shown in Figure 6.15. A moment's thought shows that the density evolution analysis for this new (larger) ensemble of codes is identical to the one of the corresponding parallel concatenated code, i.e., the thresholds are identical. In this FSFG every systematic bit has degree exactly 2, i.e., it appears exactly twice at the input to the encoder. This corresponds to the fact that the code is composed of two concatenated component codes. It is now straightforward to generalize this picture by allowing a varying degree of repetition. This is shown in Figure 6.63. As always, let $L(x)$ denote the (normalized) degree distribution from a node perspective. If we assume that we puncture a fraction p of all parity bits we see that the design rate fulfills the relation

$$\frac{1}{r} = 1 + \bar{p}L'(1).$$

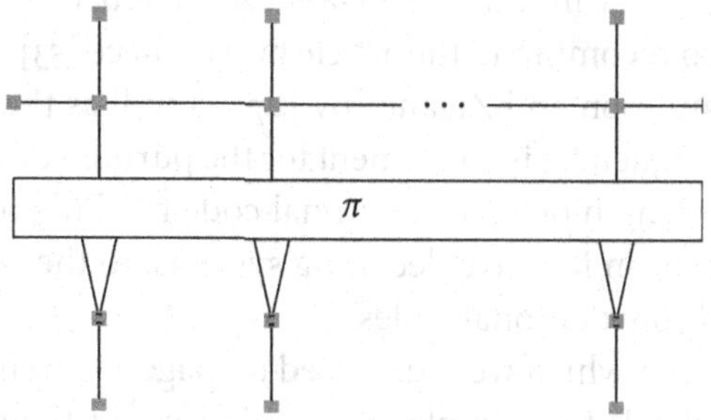

Figure 6.62: Alternative view of the FSFG of a standard parallel concatenated code.

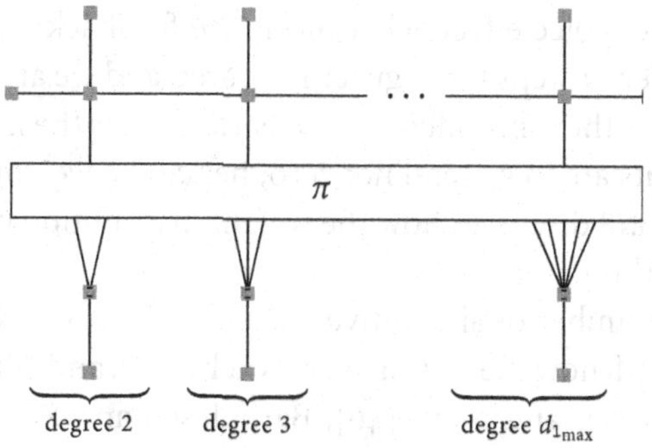

Figure 6.63: FSFG of an irregular parallel concatenated turbo code.

Example 6.64 (Irregular Turbo Ensemble). Consider the turbo code ensemble with a degree distribution from an edge perspective equal to $\lambda(x) = 55/100x + 45/100x^9$. This correspond to a node degree distribution of $L(x) = 55/64x^2 + 9/64x^{10}$. Further assume that the puncturing is random and that we puncture parity bits at a rate $p = 68/100$. From the previous formula we can then compute the rate of the code to be equal to $\frac{1}{r} = 1 + (1 - \frac{68}{100})\frac{25}{8}$, which yields $r = 1/2$. To complete the specification of the code we choose $G = 11/13$.

A density evolution analysis for the case that transmission takes place over the BEC(ϵ) shows that $\epsilon^{\text{BP}} \approx 0.4825$, which is close to the Shannon threshold of one-half. Further, we can compute the upper bound on the threshold $\bar{\epsilon}^{\text{MAP}} \approx 0.495$. ◇

Notes

Our approach to convolutional codes presented in Section 6.1 is non-standard. Usually, a convolutional code is defined as the set of all output streams resulting from the set of all input streams, where these streams are taken to be either semi-infinite, starting at time zero, or bi-infinite. The classical reference is the work by Forney [30, 31, 32, 33]. We also recommend the article by McEliece [53]. In-depth references are the book by Johanneson and Zigangirov [47] as well as the book by Piret [64]. Considering infinite sequences is convenient for the purpose of analysis, but it is also meaningful since the length of a convolutional code has little impact on its performance. For applications in iterative decoding systems, to the contrary, one usually considers *terminated* convolutional codes.

The *termination rule* which we introduced on page 324 (removing the feedback for the last m steps) leads to a simple analysis but is not the best in terms of the resulting error floor. A popular solution proposed by Divsalar and Pollara [28] is the following: for the last m steps the input is chosen in such a way that the first encoder terminates. This can be accomplished by choosing the input equal to the feedback signal. This choice effectively cancels the feedback and pushes zeros into the shift register. After m steps the register is cleared and we are back to the all-zero state. This method has the same effect on the parity bits as the method we presented but the systematic bits are in general not zero; hence the weight of the codeword is increased. Problem 6.18 discusses how the weight distribution can be calculated for this termination method.

There is a large number of alternatives. They differ in which of the component encoders is terminated, how the termination is achieved, and whether or not tail bits are transmitted: Joerssen and Meyr [46]; Barbulescu and Pietrobon [5]; Blackert, Hall, and Wilson [14]; Reed and Pietrobon [65]; Hattori, Murayama, and McEliece [42]; Khandani [49]; van Dijk, Egner, and Motwani [81]; Le Dantec and Piret [26]; Tanner [79]; McEliece, Le Dantec, and Piret [55]; Huang, Vucetic, Feng, and Tan [84]; Le Bars, Le Dantec, and Piret [6]; and Anderson and Hladik [3]. A summary and comparison of many of the proposed termination schemes was written by Hokfelt, Edfors, and Maseng [43].

It is worth pointing out that in classical coding the *code* itself plays the central role, whereas for iterative systems more prominence is due to the *encoder*.

Turbo codes were introduced together with the corresponding turbo decoding algorithm by Berrou, Glavieux, and Thitimajshima [12]. This paper set off a revolution in coding and, more generally, communications and led to the rediscovery of Gallager's thesis. The importance of the introduction of turbo codes on the development of coding cannot be overstated. Very soon after the introduction of their original scheme for binary transmission, Le Goff, Glavieux, and Berrou showed [40] that the same principle can be applied for non-binary modulation and it quickly became clear that the *turbo principle*, as Hagenauer termed it in [41], had wide applicability.

Wiberg introduced the notion of a support tree, which is our computation tree [82]. An early paper that recognized the role of the *computation tree* is by Gelblum, Calderbank, and Boutros [39]. The method of analysis (concentration around ensemble average and asymptotic analysis via density evolution) presented in Section 6.4 is due to Richardson and Urbanke [67]. An alternative route for the analysis is the geometric interpretation of the turbo decoding algorithm introduced by Richardson [66] and followed up by Agrawal and Vardy [2] and Kocarev, Lehmann, Maggio, Scanavino, Tasev, and Vardy [50]. Turbo codes can also be analyzed from a *statistical mechanics* point of view. This was accomplished by Montanari and Sourlas [58]. The basis for this analysis is the observation by Sourlas that codes can be phrased in the language of spin-glass systems [75].

The *weight distribution* of turbo codes has been studied by a large set of authors. Probably one of the most important steps was the realization by Benedetto and Montorsi that, although the weight distribution of individual codes is hard to determine, the *average* weight distribution of the ensemble is relatively easy to compute [10, 11]. This was the beginning of the ensemble average analysis. With few exceptions, most analytic results for iterative decodes are due to this ensemble point of view. Much of what we discuss in Section 6.9 has been the topic of investigation in [11]. Similar concepts were discussed around the same time by Perez, Seghers, and Costello [60]. In particular, [10] as well as [60] both contain the error floor expressions for the parallel case. The average value analysis was refined by Baligh and Khandani [4] as well as Richardson and Urbanke [68] to include the *distribution* of low-weight codewords. The limiting *Poisson distribution* of the number of minimal codewords stated in Lemma 6.49 is similar in spirit to the distribution of cycles in the turbo graph which was studied by Ge, Eppstein, and Smyth [38]. In the derivation of the average weight distribution of convolutional codes in Section 6.9.1 we follow the description of McEliece [54] and employ the *transfer matrix* method. It was shown by Sason, Telatar, and Urbanke [73, 74] how to efficiently compute the *growth rate* of the weight distribution. In this respect it is interesting to note that Bender, Richmond, and Williamson showed [8] that one can also derive central and local limit theorems for the growth of the components of the power of a matrix. This allows one in principle to apply Hayman-like techniques to the problem of the weight distribution of turbo codes in a similar manner as we discuss in Appendix D for the weight distribution of LDPC ensembles. Much material on the weight distribution problem (including Problem 6.17) can be found in the Ph.D. thesis by Pfister [62].

The average weight distribution was used by a large number of authors to derive *upper bounds* on the performance of maximum-likelihood decoders. Since the corresponding list of references is considerable we cite Sason and Shamai [72], where the reader can find an extensive literature review.

The *minimum distance* was investigated by Kahale and Urbanke in a *probabilistic* setting [48]. In particular, it was shown that the minimum distance of typical parallel concatenated turbo codes with two component codes grows at most logarithmically. The first *worst case* upper bound was the $O(\sqrt{n})$ bound proposed by Breiling and Huber [18]. It has since been improved to $O(n^{1/3})$ and, finally, to $O(\log n)$ [7, 19]. This is the result we present in Section 6.9.4. A similar logarithmic bound was also shown by Perotti and Benedetto [61]. The material in Section 6.9.4 follows closely the results of Breiling and Huber. Conversely, Truhachev, Lentmaier, Wintzell, and Zigangirov give [80] an explicit construction of a permutation that leads to a minimum distance that grows like $\Theta(\log n)$ (see also the work of Boutros and Zémor [16]). This is quite pleasing since the combination of the upper bound on the lower bound shows that the optimum growth of the minimum distance as a function of the blocklength is $\Theta(\log n)$. A very general and fundamental bound which relates the complexity of the encoding process with the resulting minimum distance is due to Bazzi and Mitter [7]. Algorithms to compute the low-weight terms of the weight distribution for a specific turbo code were discussed by Perez, Seghers, and Costello [60]; Breiling and Huber [17]; Berrou and Vaton [13]; Garello, Pierleoni, and Benedetto [37]; Garello and Casado [36]; Crozier, Guinand, and Hunt [23, 24]; and Rosnes and Ytrehus [69].

The *stability* of specific turbo code ensembles was first studied by Montanari and Sourlas [58]. The general stability condition as stated in Conjecture 6.25 is due to Richardson and Urbanke [68]. Lentmaier, Truhachev, Zigangirov, and Costello state [51] a sufficient condition for a turbo code that the bit error probability converges to zero by tracking the evolution of the Bhattacharyya parameter.

An important practical question which we have not discussed is the choice of the permutation(s) π. This is typically referred to as the *interleaver design* problem. It is probably fair to say that there are more papers written on this issue than on any other topic in iterative decoding. We will not even attempt to summarize the state of affairs, but simply give a number of pointers to the literature. The aim of most interleaver designs is to construct permutations with a large minimum distance. Under MAP decoding such an interleaver guarantees a low error floor. A popular class of interleavers are the so-called S-random interleavers which require that for the permutation π the sum $|i - j| + |\pi(i) - \pi(j)|$ is lower bounded for all distinct pairs (i, j) (there are several variants of this definition in the literature). The intuition for this restriction is simple. Consider a convolutional encoder and inputs of weight 2. If the two non-zero positions are in a general position then the encoder produces an output weight which is proportional to the length of the encoded sequence and we do not need to worry. However, if we place the inputs a multiple of the period of the encoder apart, then the corresponding output has a weight which is proportional to the length $|i - j|$. The same argument is true for the permuted tuple $(\pi(i), \pi(j))$. The lower bound on $|i - j| + |\pi(i) - \pi(j)|$ therefore translates directly into a lower bound on the weight due to inputs of weight 2. As we have seen in our discussion on the weight distribution, for a "random" permutation the number of low-weight codewords is asymptotically determined by such input pairs of weight 2. It is therefore intuitive that the preceding restriction should lead to an increased (as compared to random interleavers) minimum distance. The condition on the permutation does not depend on the component codes (in particular the period of the component codes). S-random permuters are therefore *universal.* On the other hand, they are not optimized with respect to the particular components that are used. S-random permuters were introduced by Divsalar and Pollara [28], with follow-up work by Fragouli and Wesel [34], Crozier [21], Dinoi and Benedetto [27], Breiling, Peeters, and Huber [20], and Sadjadpour, Sloane, Nebe, and Salehi [71]. Dithered relative prime interleavers were introduced by Crozier and Guinand [22]. Quadratic interleavers were proposed by Sun and Takeshita [76].

Alternatively, one can directly optimize the minimum distance for a given component code. As samples of this approach we refer to the papers by Yuan, Vucetic, and Feng [83]; Daneshgaran and Mondin [25]; Le Ruyet, Sun, and Thien [70]; Abbasfar and Yao [1]; and Ould-Cheikh-Mouhamedou, Crozier, and Kabal [59].

Explicit constructions were given by Takeshita and Costello [78] and Le Bars, Le Dantec, and Piret [6].

A word of warning: Under *iterative* decoding the error floor may not be dominated by the contributions due to codewords (i.e., pseudo codewords might dominate).

As discussed in the text, there are uncountable many flavors of the basic theme. *Serially concatenated* codes were introduced by Benedetto, Divsalar, Montorsi, and Pollara [9]. *Asymmetric* turbo codes were suggested by Takeshita, Collins, Massey, and Costello [77]. *Big numerator* codes were introduced by Massey, Takeshita, and Costello [52] and independently discovered by Huettinger, ten Brink, and Huber [45]. It was suggested by Hokfelt and Maseng [44] to optimize the energy which is assigned to the various streams (systematic and parity) of turbo codes. The same idea was investigated also by Duman and Salehi [29]. *Irregular* turbo codes were first suggested by Frey and MacKay [35] and optimized versions were presented by Boutros, Caire, Viterbo, Sawaya, and Vialle [15]. Examples 6.59 and 6.64 are due to Méasson and appeared in [56, 57].

Implementation issues concerning turbo codes are discussed by Pietrobon [63].

Problems

6.1 (More General Convolutional Encoders). Consider the convolutional encoder in Figure 6.65. This convolutional encoder is a rate one-half *non-recursive* encoder that has $m = 2$. For any memory m, let σ_i denote the state vector at time i and let $x_i^{p,q} = (x_i^p, x_i^q)$ be the length 2 output vector at time i. Write down the general form of the state equations (6.2) and specify the corresponding matrices in terms of the filter coefficients. Draw one labeled trellis section.

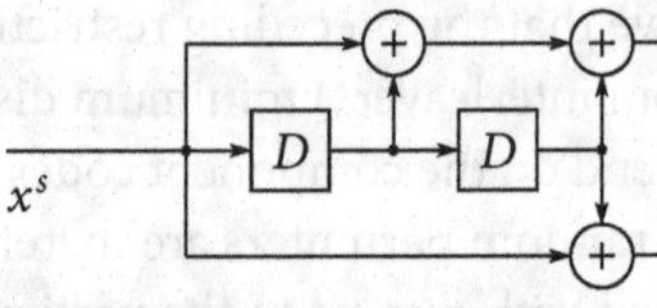

Figure 6.65: Binary feed-forward convolutional encoder of memory $m = 2$ and rate one-half defined by $(p(D) = 1 + D + D^2, q(D) = 1 + D^2)$.

6.2 (Viterbi versus BCJR). Explain why the Viterbi algorithm only needs two passes (running over the FSFG once from let's say left to right and then backtracking), whereas in the BCJR we needed three passes (one forward, one backward, and one decision step)?

6.3 (State Probability in Trellis). Consider the trellis shown in Figure 6.9. Using the results of the α- and β- recursion shown – what is $p_{\Sigma_3 \mid Y^s, Y^p}((0,0) \mid y^s, y^p)$?

6.4 (Proof of Equivalence Relationships). Prove Theorem 6.19.

6.5 (Symmetric Turbo Schedule for Parallel Concatenation). Consider iterative decoding of an element from the ensemble $\mathcal{P}(G, n, r)$ using the following "symmetric" turbo schedule: one decoding round consists of running in parallel one complete BCJR algorithm *on each* component code and then exchanging the messages via the interleaver.

Draw the computation tree for one iteration assuming windowed decoding with $w = 1$.

Prove that the threshold for this symmetric turbo schedule is identical to the threshold for the standard turbo schedule discussed in the main text.

Hint: Show that the computation tree for the symmetric turbo schedule is included in the computation tree of the standard turbo schedule for a sufficiently large number of iterations, and vice versa. Use the fact that on the computation tree iterative decoding is optimal.

6.6 (Iterative Decoding and FSFG for $\mathcal{S}(G^o, G^i, n, r)$). Consider the ensemble $\mathcal{S}(G^o, G^i, n, r)$ and optimal bit-wise decoding. Write the decision map

$$\hat{x}_j^{\text{MAP}}(y^{s_i}, y^{p_i})$$

in factorized form. Show that the corresponding FSFG is the one depicted in Figure 6.66. Label the various elements of the FSFG with the corresponding factors.

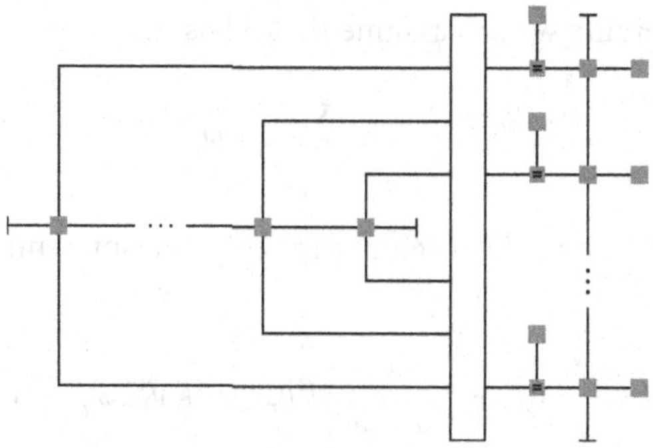

Figure 6.66: FSFG for the optimal bit-wise decoding of $\mathcal{S}(G^o, G^i, n, r)$.

6.7 (Density Evolution for Serially Concatenated Ensembles). Consider the density evolution equations for the ensemble $\mathcal{S}(G^o, G^i, n, r)$ when transmission takes place over a BMS channel with L-density a_{BMSC}. Assume that no puncturing is performed on the output of the outer encoder. Consider the following *schedule*: in each *even* iteration we perform one complete decoding pass on the outer code (left part in Figure 6.66); in each *odd* iteration we perform one complete decoding pass on the inner code (right part in Figure 6.66). In between iterations, the messages are exchanged via the interleaver. Let b_ℓ denote the density emitted from the trellis in the ℓ-th iteration of density evolution and let $\mathsf{c}_\ell = \mathsf{a}_{\text{BMSC}} \circledast \mathsf{b}_\ell$. Argue that $\mathsf{b}_0 = \Delta_0$, and that for $\ell \geq 1$

$$\mathsf{b}_\ell = \begin{cases} \Gamma^s_{G^i}(\mathsf{c}_{\ell-1}, \mathsf{a}_{\text{BMSC}}), & \ell \text{ odd}, \\ \frac{1}{2}\left(\Gamma^s_{G^o}(\mathsf{c}_{\ell-1}, \mathsf{c}_{\ell-1}) + \Gamma^p_{G^o}(\mathsf{c}_{\ell-1}, \mathsf{c}_{\ell-1})\right), & \ell \text{ even}. \end{cases}$$

6.8 (Computation Graph of $\mathcal{S}(G^o, G^i, n, r)$). Draw the computation graph for $\mathcal{S}(G^o, G^i, n, r)$ corresponding to windowed ($w = 1$) iterative decoding for two iterations and the schedule indicated in Problem 6.7.

6.9 (Recursion for Computation of Weight Distribution). The generating function $A(x, y, z)$ introduced on page 349 completely specifies the weight distribution $a_{i,o,n}$ and in principle we can derive from it any desired term by long division. This method quickly runs into computational limitations though. A more efficient way of determining $a_{i,o,n}$ is to note that the rational function gives rise to a recursion. Assume that we have

$$A(x, y, z) = \frac{\alpha(x, y, z)}{\beta(x, y, z)},$$

so that

$$A(x, y, z)\beta(x, y, z) = \alpha(x, y, z). \tag{6.67}$$

Without loss of generality we can assume that β has the form

$$\beta(x, y, z) = 1 + \sum_{i,o,n\geq 1} \beta_{i,o,n} x^i y^o z^n.$$

If we expand the left-hand side of (6.67) explicitly as a sum and compare terms we see that we have

$$a_{i,o,n} = \alpha_{i,o,n} - \sum_{i',o',n'\geq 1} \beta_{i',o',n'} a_{i-i',o-o',n-n'},$$

where the recursion is valid for any triple (i, o, n) if we assume that all coefficients are zero if any of the indices is negative. Assuming that we are interested in the weight distribution for a particular value of n, the complexity of the recursion is $O(n^2)$ in space and $O(n^3)$ in time.

Use this method to compute the terms of $A(x, y, z)$ up to and including z^2 for the encoder $C(G = 7/5, n)$ (see Example 6.36).

6.10 (Weight Distribution of $C(G = 7/5, n)$ and Alternating Puncturing). Consider the weight distribution of $C(G = 7/5, n)$ where we puncture all even parity bit positions. Assuming that we only look at codes of even length, write down a general expression for $A(x, y, z)$. Show that the first few terms of $A(x, y, z)$ are

$$A(x, y, z) = 1 + \left(1 + 2x y^2 + x^2 y^2\right) z^2 +$$
$$\left(1 + x y \left(1 + 3y + x\left(\left(1 + y\right)\left(1 + 2y\right) + x\left(2 + y^2\left(1 + x + y\right)\right)\right)\right)\right) z^4 + O(z^6).$$

6.11 (Weight Distribution of $C(G, n)$ and Puncturing). Let $A(x, y, z)$ be the generating function of the input-output weight distribution of the code $C(G, n)$. Let $A_\alpha(x, y, z)$ denote the corresponding generating function for the punctured ensemble in which each parity bit is punctured independently with probability α. Prove that $A_\alpha(x, y, z) = A(x, y\bar{\alpha} + \alpha, z)$.

6.12 (Detour Generating Function for Punctured Codes). Consider the detour generating function for punctured ensembles. Show that for random puncturing we have $D_\alpha(x, y) = D(x, y\bar{\alpha} + \alpha)$, whereas for alternating puncturing we have

$$2(1 - D(x, y)) = \frac{1}{\left[(I - M^\bullet(x, y) M^\bullet(x, y = 1))^{-1}(I + M^\bullet(x, y))\right]_{0,0}} + \frac{1}{\left[(I - M^\bullet(x, y = 1) M^\bullet(x, y))^{-1}(I + M^\bullet(x, y = 1))\right]_{0,0}}.$$

6.13 (Expected Weight Distribution of $\mathcal{S}(G^i, G^o, n, r)$). Consider the ensemble $\mathcal{S}(G^i, G^o, n, r)$. Let $A^i(x, y, z)$ and $A^o(x, y, z)$ denote the generating functions corresponding to the input-output weight distribution of the inner and outer code, respectively, where we assume that the appropriate puncturing has been performed. Let $S(x, y, z) = \sum_{i,o,n} s_{i,o,n} x^i x^o z^n$ denote the generating function corresponding to the expected input-output weight distribution of $\mathcal{S}(G^i, G^o, n, r)$, where the expectation is taken over all interleavers. Show that

$$s_{i,o,n} = \frac{\sum_j a^o_{i,j,n} a^i_{j,o,n}}{\binom{n}{j}}.$$

6.14 (Matrix Series). Consider the formal power sum $\sum_{n\geq 0} M(x)^n y^n$, where $M(x)$ is a symbolic square matrix in the variable x over some field $\mathbb{F}$ (see Appendix D). Show that over the ring of formal power sums we have

$$\sum_{n\geq 0} M(x)^n y^n = (I - yM(x))^{-1},$$

where I is the identity matrix.

Note: In Appendix D we assume that the coefficients of a power sum are elements of a field. In fact, all that is needed is that they are elements of a ring.

6.15 ($q(x)$ divides $1+x^d$). Let $q(x)$ be a polynomial over $\mathbb{F}_2$ with $q_0 = 1$. Show that there exists an integer d such that $q(x)$ divides $1 + x^d$. Show that $d \leq 2^{\deg(q)} - 1$.

Hint: Consider the set $x^i \mod q(x)$ for $i = 0, \cdots, 2^{\deg(q)} - 1$.

6.16 (Parallel Turbo Code Ensemble with k Components). Define a parallel concatenated turbo code ensemble with k components, $k \in \mathbb{N}$, analogous to the case $k = 2$ discussed on page 334. Draw the encoder diagram equivalent (Figure 6.13 for $k = 2$) and the FSFG for the optimum bit-wise decoding (Figure 6.15 for $k = 2$). What is the "natural" rate of this ensemble assuming no puncturing? Consider a decoding schedule in which we perform in parallel one complete decoding pass on each branch and then exchange the messages via the interleavers. Write down the density evolution equations for this schedule. Finally, derive the stability condition.

6.17 (Asymptotic Growth Rate of Weight Distribution - Pfister [62]). In Figure 6.39 we plotted $\lim_{n\to\infty} \frac{1}{n}\log_2(a_{w,n})$. Let us discuss how this limit can be computed in an efficient manner.

Note that when the blocklength tends to infinity the effect of the specific termination scheme vanishes. Therefore, we only need to compute

$$\lim_{n\to\infty} \frac{1}{n} \log \operatorname{coef}\{[M^n(x)]_{0,0}, x^{n\omega}\}.$$

Let 2^m be the number of states. Assume that the matrix $M(x)$ is diagonalizable so that we can write the identity

$$M^n(x) = \sum_{i=1}^{2^m} \lambda_i^n(x) L_i(x),$$

where $\lambda_i(x)$ is the i-th eigenvalue and $L_i(x)$ is the matrix formed by the outer product of the corresponding i-th left and right eigenvector. Suppose that we have the ordering $\lambda_1(x) > |\lambda_2(x)| \geq \cdots \geq |\lambda_{2^m}(x)|$ for all $x \in \mathbb{R}$, i.e., $\lambda_1(x)$ is the strictly largest eigenvalue (it has to be real).

Argue that

$$\lim_{n\to\infty} \frac{1}{n} \log \operatorname{coef}\{[M^n(x)]_{0,0}, x^{n\omega}\} = \lim_{n\to\infty} \frac{1}{n} \log \operatorname{coef}\{\lambda_1^n(x), x^{n\omega}\}.$$

Therefore, if $\lambda_1(x)$ is Hayman admissible (see Definition D.5 on page 510) then we can use the Hayman method to compute the asymptotic growth rate.

Apply this idea to compute $\lim_{n\to\infty} \frac{1}{n} \log_2(a_{w,n})$ for the code $C(G = 1/3, n)$.

For larger examples it is in general difficult to compute $\lambda_1(x)$ symbolically as a function of x but it is easy to compute it numerically for a given value of x. In order to determine the growth rate via the Hayman method we need not only $\lambda_1(x)$ but also its derivative $\lambda_1'(x)$. Show that this derivative can be computed in the following way: let $p(x,z)$ be the characteristic polynomial, $p(x,z) = \det(M(x) - zI)$. For example, for the code $C(G = 1/3, n)$ we get $p(x,z) = x - x^3 - z - xz + z^2$. Let $\mathrm{d}p(x,z) = a(x,z)\mathrm{d}x + b(x,z)\mathrm{d}z$ denote the total derivative of $p(x,z)$, where $a(x,z) = \partial p(x,z)/\partial x$ and $b(x,z) = \partial p(x,z)/\partial z$. Set $\mathrm{d}p(x,z) = 0$, and solve for $\mathrm{d}z/\mathrm{d}x$ so that $\mathrm{d}z/\mathrm{d}x = -a(x,z)/b(x,z)$. For our specific example we get $\mathrm{d}z/\mathrm{d}x = \frac{1-3x^2-z}{1+x-2z}$. Show that $\lambda_1'(x) = -a(x,z)/b(x,z)|_{z=\lambda_1(x)}$. This method of computing $\lambda_1'(x)$ works often even if we cannot compute $\lambda_1(x)$ symbolically. This is true since the characteristic polynomial can be computed symbolically for relatively large matrices.

6.18 (Weight Distribution for Alternative Termination Scheme). Consider a convolutional code with rational function G and the alternative termination discussed in the Notes. Instead of terminating the feedback for the last m steps and setting the input to zero, we keep the feedback and set the input in the last m steps to be equal to the feedback signal.

Using the matrix transfer method as introduced on page 349, express the weight distribution for this case in terms of a generating function. Apply your general formula to the example $G = 7/5$.

References

[1] A. Abbasfar and K. Yao, *Interleaver design for turbo codes by distance spectrum shaping*, in Proc. of WCNC, Atlanta, GA, USA, Mar. 2004, pp. 1616–1619. [369]

[2] D. Agrawal and A. Vardy, *On the phase trajectories of the turbo decoding algorithm*. IMA, USA 1999 Summer Program: Codes, Systems and Graphical Models. [367]

[3] J. B. Anderson and S. M. Hladik, *Tailbiting MA, USAP decoders*, IEEE J. Sel. Area. Commun., 16 (1998), pp. 297–302. [366]

[4] H. Baligh and A. K. Khandani, *Asymptotic effect of interleaver structure on the performance of turbo codes*, in Proc. of Conf. on Inform. Sciences and Systems (CISS), Princeton, NJ, USA, Mar. 2002. [367]

[5] S. A. Barbulescu and S. S. Pietrobon, *Terminating the trellis of turbo-codes in the same state*, Electron. Lett., 31 (1995), pp. 22–23. [366]

[6] P. L. Bars, C. L. Dantec, and P. M. Piret, *Bolt interleavers for turbo codes*, IEEE Trans. Inform. Theory, 49 (2003), pp. 391–400. [366, 369]

[7] L. Bazzi and S. Mitter, *Endcoding complexity versus minimum distance*, IEEE Trans. Inform. Theory, 50 (2005), pp. 2010–2021. [368]

[8] E. A. Bender, L. B. Richmond, and S. G. Williamson, *Central and local limit theorems applied to asymptotic enumeration III: Matrix recursions*, J. Combin. Theory, A 35 (1983), pp. 263–278. [367, 530]

[9] S. Benedetto, D. Divsalar, G. Montorsi, and F. Pollara, *Serial concatenation of interleaved codes: Performance analysis, design, and iterative decoding*, IEEE Trans. Inform. Theory, 44 (1998), pp. 909–926. [369]

[10] S. Benedetto and G. Montorsi, *Design of parallel concatenated convolutional codes*, IEEE Trans. Commun., 44 (1996), pp. 591–600. [263, 367]

[11] ———, *Unveiling turbo codes: Some results on parallel concatenated coding schemes*, IEEE Trans. Inform. Theory, 42 (1996), pp. 409–428. [263, 367]

[12] C. Berrou, A. Glavieux, and P. Thitimajshima, *Near Shannon limit error-correcting coding and decoding*, in Proc. of ICC, Geneva, Switzerland, May 1993, pp. 1064–1070. [261, 366]

[13] C. Berrou and S. Vaton, *Computing the minimum distance of linear codes by the error impulse method*, in Proc. of the IEEE Int. Symposium on Inform. Theory, Lausanne, Switzerland, June 2002, IEEE, p. 5. [368]

[14] W. J. Blackert, E. K. Hall, and S. G. Wilson, *Turbo code termination and interleaver conditions*, Electron. Lett., 31 (1995), pp. 2082–2084. [366]

[15] J. Boutros, G. Caire, E. Viterbo, H. Sawaya, and S. Vialle, *Turbo code at 0.03 dB from capacity limit*, in Proc. of the IEEE Int. Symposium on Inform. Theory, Lausanne, Switzerland, June 30–July 5, 2002, p. 56. [369]

[16] J. Boutros and G. Zémor, *On quasi-cyclic interleavers for parallel turbo codes*, IEEE Trans. Inform. Theory, 52 (2006), pp. 1732–1739. [368]

[17] M. Breiling and J. B. Huber, *A method for determining the distance profile of turbo codes*, in Proc. 3rd ITG Conf. Sourve Channel Coding, Münich, Germany, Jan. 2000, pp. 219–224. [368]

[18] ———, *Combinatorial analysis of the minimum distance of turbo codes*, IEEE Trans. Inform. Theory, 47 (2001), pp. 2737–2750. [368]

[19] ———, *A logarithmic upper bound on the minimum distance of turbo codes*, IEEE Trans. Inform. Theory, 50 (2004), pp. 1692–1710. [368]

[20] M. Breiling, S. Peeters, and J. B. Huber, *Interleaver design using backtracking and spreading methods*, in Proc. of the IEEE Int. Symposium on Inform. Theory, Sorrento, Italy, June 25–30, 2000, p. 451. [369]

[21] S. N. Crozier, *New high-spread high-distance interleavers for turbo codes*, in Proc. Biennal Symp. Commun., Kingston, ON, Canada, May 2000, pp. 3–7. [369]

[22] S. N. Crozier and P. Guinand, *High-performance low-memory interleaver banks for turbo codes*, in Proc. of the Vehicular Technol. Conf., Atlantic City, NJ, USA, Oct. 2001. [369]

[23] S. N. Crozier, P. Guinand, and A. Hunt, *Estimating the minimum distance of turbo codes unsing double and triple impulse methods*, IEEE Commun. Lett., 9 (2005), pp. 631–633. [368]

[24] ———, *Estimating the minimum distance of large-block turbo codes using iterative multiple-impulse methods*, in Proc. of the Int. Conf. on Turbo Codes and Related Topics, Munich, Germany, Apr. 3–7, 2006. [368]

[25] F. Daneshgaran and M. Mondin, *Design of interleaver for turbo codes: Iterative interleaver growth algorithms of polynomial complexity*, IEEE Trans. Inform. Theory, 45 (2003), pp. 1845–1859. [369]

[26] C. L. Dantec and P. M. Piret, *Algebraic and combinatorial methods producing good interleavers*, in Proc. of the Int. Conf. on Turbo Codes and Related Topics, Brest, France, Sept. 2000, pp. 271–274. [366]

[27] L. Dinoi and S. Benedetto, *Variable-size interleaver design for parallel turbo decoder architectures*, IEEE Trans. Commun., 53 (2005), pp. 1833–1840. [369]

[28] D. Divsalar and F. Pollara, *Turbo codes for PCS applications*, in Proc. of ICC, Seattle, WA, USA, June 1995, pp. 54–59. [366, 369]

[29] T. M. Duman and M. Salehi, *On optimal power allocation for turbo codes*, in Proc. of the IEEE Int. Symposium on Inform. Theory, Ulm, Germany, June 1997, IEEE, p. 104. [369]

[30] G. D. FORNEY, JR., *Convolutional codes I: Algebraic structure*, IEEE Trans. Inform. Theory, 16 (1970), pp. 720–738. [33, 366]

[31] ——, *Correction to 'Convolutional codes I: Algebraic structure'*, IEEE Trans. Inform. Theory, 17 (1971), p. 360. [33, 366]

[32] ——, *Convolutional codes II: Maximum-likelihood decoding*, Inform. Contr., 25 (1974), pp. 222–266. [34, 366]

[33] ——, *Convolutional codes III: Sequential decoding*, Inform. Contr., 25 (1974), pp. 267–297. [366]

[34] C. FRAGOULI AND R. D. WESEL, *Semi-random interleaver design criteria*, in Proc. of GLOBECOM, Rio de Janeiro, Brazil, Dec. 1999, pp. 2352–2356. [369]

[35] B. J. FREY AND D. J. C. MACKAY, *Irregular turbo codes*, in Proc. of the IEEE Int. Symposium on Inform. Theory, Sorrento, Italy, June 2000, p. 121. [369]

[36] R. GARELLO AND A. CASADO, *The all-zero iterative decoding algorithm for turbo code minimum distance computation*, in Proc. of ICC, June 2004, pp. 361–364. [368]

[37] R. GARELLO, P. PIERLEONI, AND S. BENEDETTO, *Computing the free distance of turbo codes and serially concatenated codes with interleavers: Algorithms and applications*, IEEE J. Sel. Area. Commun., 19 (2001), pp. 800–812. [368]

[38] X. GE, D. EPPSTEIN, AND P. SMYTH, *The distribution of loop lengths in graphical models for turbo codes*, IEEE Trans. Inform. Theory, 47 (2001), pp. 2549–2553. [367]

[39] E. A. GELBLUM, A. R. CALDERBANK, AND J. BOUTROS, *Understanding serially concatenated codes from a support tree approach*, in Proc. of the Int. Conf. on Turbo Codes and Related Topics, Brest, France, Sept. 1997, pp. 271–274. [366]

[40] S. L. GOFF, A. GLAVIEUX, AND C. BERROU, *Turbo-codes and high spectral efficiency modulation*, in Proc. of ICC, New Orleans, LA, May 1994, pp. 645–649. [312, 366]

[41] J. HAGENAUER, *The turbo principle–tutorial introduction and state of the art*, in Proc. of the Int. Conf. on Turbo Codes and Related Topics, Brest, France, Sept. 1997, pp. 1–11. [312, 366]

[42] M. HATTORI, J. MURAYAMA, AND R. J. MCELIECE, *Pseudorandom and self-terminating interleavers for turbo codes*, in Proc. of the IEEE Inform. Theory Workshop, San Diego, CA, USA, Feb. 1998. [366]

[43] J. HOKFELT, O. EDFORS, AND T. MASENG, *On the theory and performance of trellis termination methods for turbo codes*, IEEE J. Sel. Area. Commun., 19 (2001), pp. 838–847. [366]

[44] J. HOKFELT AND T. MASENG, *Optimizing the energy of different bitstreams of turbo codes*, in Turbo Coding Seminar Proceedings, Lund, Sweden, Aug. 1996, pp. 59–63. [369]

[45] S. Huettinger, S. ten Brink, and J. B. Huber, *Turbo-code representation of RA codes and DRS codes for reduced decoding complexity*, in Proc. Conf. on Inform. Sciences and Systems, Baltimore, MD, USA, Mar. 2001, pp. 201–210. [369]

[46] O. Joerssen and H. Meyr, *Terminating the trellis of turbo codes*, Electron. Lett., 30 (1994), pp. 1285–1286. [366]

[47] R. Johannesson and K. S. Zigangirov, *Fundamentals of Convolutional Coding*, IEEE Press, Piscataway, NJ, USA, 1999. [33, 34, 366]

[48] N. Kahale and R. Urbanke, *On the minimum distance of parallel and serially concatenated codes*, in Proc. of the IEEE Int. Symposium on Inform. Theory, Boston, MA, USA, Aug. 1998, p. 31. [367]

[49] A. K. Khandani, *Design of turbo code interleaver using Hungarian method*, Electron. Lett., 34 (1998), pp. 63–65. [366]

[50] L. Kocarev, F. Lehmann, G. M. Maggio, B. Scanavino, Z. Tasev, and A. Vardy, *Nonlinear dynamics of iterative decoding systems: Analysis and applications*, IEEE Trans. Inform. Theory, 52 (2006), pp. 1366–1384. [367]

[51] M. Lentmaier, D. V. Truhachev, K. S. Zigangirov, and D. J. Costello, Jr., *Turbo codes and Shannon's condition for reliable communication*, in Proc. of the IEEE Int. Symposium on Inform. Theory, Chicago, IL, USA, June 27–July 2, 2004, p. 442. [368]

[52] P. C. Massey, O. Y. Takeshita, and D. J. Costello, Jr., *Contradicting a myth: Good turbo codes with large memory order*, in Proc. of the IEEE Int. Symposium on Inform. Theory, Sorrento, Italy, June 2000, p. 122. [369]

[53] R. J. McEliece, *The algebraic theory of convolutional codes*, in Handbook of Coding Theory, V. S. Pless and W. C. Huffman, eds., vol. 1, Elsevier Science, Amsterdam, Holland, 1998, pp. 1065–1138. [366]

[54] ——, *How to compute weight enumerators for convolutional codes*, in Communications and Coding (P. G. Farrell 60th birthday celebration), M. Darnell and B. Honoray, eds., Wiley, New York, NY, USA, 1998, pp. 121–141. [367]

[55] R. J. McEliece, C. L. Dantec, and P. M. Piret, *Permutations preserving divisibility*, IEEE Trans. Inform. Theory, 47 (2001), pp. 1206–1207. [366]

[56] C. Méasson and R. Urbanke, *Asymptotic analysis of turbo codes over the binary erasure channel*, in Proc. of the 12th Joint Conference on Communications and Coding, Saas Fee, Switzerland, Mar. 2002. [369]

[57] ——, *Further analytic properties of EXIT-like curves and applications*, in Proc. of the IEEE Int. Symposium on Inform. Theory, Yokohama, Japan, June 29–July 4, 2003, p. 266. [369]

[58] A. Montanari and N. Sourlas, *The statistical mechanics and turbo codes*, in Proc. of the Int. Conf. on Turbo Codes and Related Topics, Brest, France, Sept. 2000, pp. 63–66. [367, 368]

[59] Y. Ould-Cheikh-Mouhamedou, S. N. Crozier, and P. Kabal, *Distance measurement method for double binary turbo codes and a new interleaver design for DVB-RCS*, in Proc. of GLOBECOM, Dallas, TX, USA, Nov. 2004, pp. 172–178. [369]

[60] L. C. Perez, J. Seghers, and D. J. Costello, Jr., *A distance spectrum interpretation of turbo codes*, IEEE Trans. Inform. Theory, 42 (1996), pp. 1698–1709. [367, 368]

[61] A. Perotti and S. Benedetto, *New upper bound on the minimum distance of turbo codes*, IEEE Trans. Inform. Theory, 50 (2004), pp. 2985–2997. [368]

[62] H. D. Pfister, *On the Capacity of Finite State Channels and the Analysis of Convolutional Accumulate-m Codes*, PhD thesis, UCSD, San Diego, CA, USA, 2003. [312, 367, 373]

[63] S. S. Pietrobon, *Implementation and performance of a turbo/MAP decoder*, Int. J. Sat. Commun., 16 (1998), pp. 23–46. [369]

[64] P. M. Piret, *Convolutional Codes: An Algebraic Approach*, MIT Press, 1988. [34, 366]

[65] M. C. Reed and S. S. Pietrobon, *Turbo-code termination schemes and a novel alternative for short frames*, in IEEE Int. Sympo. on Personal, Indoor and Mobile Radio Commun., Taipai, Taiwan, Oct. 1996, pp. 354–358. [366]

[66] T. Richardson, *The geometry of turbo-decoding dynamics*, IEEE Trans. Inform. Theory, 46 (2000), pp. 9–23. [367]

[67] T. Richardson and R. Urbanke, *Thresholds for turbo codes*, in Proc. of the IEEE Int. Symposium on Inform. Theory, Sorrento, Italy, June 2000, p. 317. [367]

[68] ——, *On the distribution of low-weight codewords for turbo codes*, in Proc. of the Allerton Conf. on Commun., Control, and Computing, Monticello, IL, USA, Sept. 2004. [367, 368]

[69] E. Rosnes and Y. Ytrehus, *Improved algorithms for the determination of turbo-code weight distributions*, IEEE Trans. Commun., 51 (2005), pp. 20–26. [368]

[70] D. L. Ruyet, H. Sun, and H. V. Thien, *An interleaver design algorithm based on a cost matrix for turbo codes*, in Proc. of the IEEE Int. Symposium on Inform. Theory, Sorrento, Italy, June 25–30, 2000, p. 452. [369]

[71] H. Sadjadpour, N. J. Sloane, G. Nebe, and M. Salehi, *Interleaver design for turbo codes*, in Proc. of the IEEE Int. Symposium on Inform. Theory, Sorrento, Italy, June 25–30, 2000, p. 453. [369]

[72] I. Sason and S. Shamai, *Improved upper bounds on the ML decoding error probability of parallel and serial concatenated turbo codes via their ensemble distance spectrum*, IEEE Trans. Inform. Theory, 46 (2000), pp. 24–47. [367]

[73] I. Sason, E. Telatar, and R. Urbanke, *The asymptotic input-output weight distribution of convolutional encoders*, in Proc. of the Allerton Conf. on Commun., Control, and Computing, Monticello, IL, USA, Oct. 2000. [367]

[74] ——, *On the asymptotic input-output weight distributions and thresholds of convolutional and turbo-like encoders*, IEEE Trans. Inform. Theory, 48 (2002), pp. 3052–3061. [367, 530]

[75] N. Sourlas, *Spin-glass models as error-correcting codes*, Nature, 339 (1989), pp. 693–695. [160, 261, 266, 367]

[76] J. Sun and O. Y. Takeshita, *Interleavers for turbo codes using permutation polynomials over integer rings*, IEEE Trans. Inform. Theory, 51 (2005), pp. 101–119. [369]

[77] O. Y. Takeshita, O. M. Collins, P. C. Massey, and D. J. Costello, Jr., *A note on asymmetric turbo-codes*, IEEE Commun. Lett., 3 (1999), pp. 69–71. [369]

[78] O. Y. Takeshita and D. J. Costello, Jr., *New deterministic interleaver designs for turbo codes*, IEEE Trans. Inform. Theory, 46 (2000), pp. 1988–2006. [369]

[79] R. M. Tanner, *Toward an algebraic theory for turbo codes*, in Proc. of the Int. Conf. on Turbo Codes and Related Topics, Brest, France, Sept. 2000, pp. 17–25. [366]

[80] D. V. Truhachev, M. Lentmaier, O. Wintzell, and K. S. Zigangirov, *On the minimum distance of turbo codes*, in Proc. of the IEEE Int. Symposium on Inform. Theory, Lausanne, Switzerland, June 2002, IEEE, p. 84. [368]

[81] M. van Dijk, S. Egner, and R. Motwani, *Simultaneous zero-tailling of parallel convolutional codes*, in Proc. of the IEEE Int. Symposium on Inform. Theory, Sorrento, Italy, June 2000, p. 368. [366]

[82] N. Wiberg, *Codes and Decoding on General Graphs*, PhD thesis, Linköping University, S-581 83, Linköping, Sweden, 1996. [65, 66, 263, 266, 366]

[83] J. Yuan, B. Vucetic, and W. Feng, *Combined turbo codes and interleaver design*, in Proc. of the IEEE Int. Symposium on Inform. Theory, Cambridge, MA, USA, Aug. 1998, p. 176. [369]

[84] J. Yuan, B. Vucetic, W. Feng, and M. Tan, *Design of cyclic shift interleavers for turbo codes*, Ann. Télécommun., 56 (2001), pp. 384–393. [366]

Chapter 7

GENERAL ENSEMBLES

In practical LDPC design one is invariably interested in achieving the best possible performance. Although we have used the framework of irregular low-density parity-check (LDPC) ensembles throughout this book, the notion of "irregularity" we have employed is not the most general possible and the best performing structures are not properly captured in that framework.

We start by presenting in Section 7.1 a generalization of irregular LDPC ensembles called *multi-edge-type LDPC ensembles*. In principle, any of the ensembles suggested to date can be represented in this form and we demonstrate this discussing a few examples.

In Section 7.2 we review the theory surrounding these ensembles. It is largely the same as for the standard irregular case, which is why we have avoided using this more complex notion so far. The multi-edge-type generalization enjoys several advantages. With multi-edge-type LDPC codes one can achieve better performance at lower complexity. The generalization is especially useful under extreme conditions where standard irregular LDPC codes do not fare so well. Examples of these conditions include very low code rates, high rate codes that target very low bit error rates, and codes used in conjunction with turbo equalization schemes.

We discuss in Section 7.3 an alternative and complementary way of describing structured ensembles. These ensembles are the result of *lifting* a very small graph to a large graph by first replicating the structure of the small graph a large number of times and then by choosing the connections between these copies in a random way. The advantage of this construction is that both the encoder as well as the decoder can exploit the underlying structure.

Section 7.4 introduces *non-binary* LDPC codes. To keep the exposition simple we focus on non-binary codes for transmission over binary channels, but these codes are of course also of interest for transmission over non-binary channels.

Finally, we discuss in Section 7.5 the dual of LDPC codes, namely *low-density generator-matrix* (LDGM) codes. On their own these codes have limited performance since by their very definition they necessarily contain some very low-weight codewords. But if we *precode* them to clean up the error floor the performance can be remarkably good. More importantly, an important generalization of precoded LDGM codes leads to the concept of *rateless* codes. Rateless codes are ideally suited for the purpose of broadcasting or in situations where the channel quality varies widely and a (very-low-rate) feedback channel is available.

§7.1. Multi-Edge-Type LDPC Code Ensembles

§7.1.1. Introduction

The basic idea of multi-edge-type LDPC ensembles is conveyed by the modifier *multi-edge-type*. In standard irregular LDPC ensembles the graph connectivity is constrained only by the node degrees. In the multi-edge-type setting we define several edge (equivalence) classes and every node is characterized by its number of sockets in each class. *Only sockets of the same class can be connected by an edge.* This gives considerable control over the graph. In fact, as an extreme case it is possible, in this framework, to specify a particular graph. That such a further specification can potentially improve the performance is easily seen. Recall from the analysis of standard LDPC ensembles that optimized (with respect to the threshold) degree distributions contain a significant fraction of degree 2 edges. Given the importance of degree 2 edges, it is natural to conjecture that degree 1 edges could bring further benefits. In the standard irregular framework, because of the random nature of the edge placements, such degree 1 edges inadvertently lead to error probabilities bounded away from zero (see Problem 3.6). If, on the other hand, we carefully place these degree 1 edges (e.g., by assuming that each check node contains at most one connection into a degree 1 edge) then we can reap the benefits and at the same time circumvent the related problems. The multi-edge-type framework provides a convenient language for expressing and optimizing such constraints.

§7.1.2. Parameterization

A multi-edge-type ensemble is comprised of a finite number m_{e} of *edge types.* The *degree type* of a *check* node is a vector of (non-negative) integers of length m_{e}; the i-th entry of this vector records the number of sockets of the i-th edge type. We shall also refer to this vector as an *edge degree.* The *degree type* of a *variable* node has two parts. These are of length m_{e} and $m_{\mathrm{r}} + 1$, respectively. The first part specifies the *edge degree*: it plays the same role as for check nodes – edges of a given type can only connect to sockets of the same type. The second part relates to the *received* distribution: different node types may have different received distributions, i.e., the associated bits may go through different channels. This is convenient, e.g., if we look at punctured bits in which case the corresponding channel is the binary erasure channel with erasure probability 1. To each i, $i = 0, \dots, m_{\mathrm{r}}$, we associate a binary memoryless symmetric (BMS) channel with L-density $\mathsf{a}_{\mathrm{BMSC}_i}$. The channel for $i = 0$ plays a special role, we assume that $\mathsf{a}_{\mathrm{BMSC}_0} = \Delta_0$. In words, this channel corresponds to *punctured* bits.

To represent the structure of the graph we introduce a node-perspective multinomial representation; we interpret degrees as exponents. Let $\underline{d} = (d_1, \dots, d_{m_{\mathrm{e}}})$ be a multi-edge-type degree and let $\underline{x} = (x_1, \dots, x_{m_{\mathrm{e}}})$ denote a vector of variables. We use $\underline{x}^{\underline{d}}$ to denote $\prod_{i=1}^{m_e} x_i^{d_i}$. Similarly, let $\underline{b} = (b_0, \dots, b_{m_r})$ be a received degree and let $\underline{r} = (r_0, \dots, r_{m_r})$ denote variables corresponding to received distributions. By $\underline{r}^{\underline{b}}$ we mean $\prod_{i=0}^{m_r} r_i^{b_i}$. We assume in the sequel that received degrees $\underline{b}$ have *exactly one* entry set to 1 and the rest set to 0. The seemingly more general case in which a node has several observations is easily cast in this framework since every *combination* of BMS channels is again a BMS channel.

A graph ensemble is specified through two multinomials, one associated to variable nodes and the other associated to check nodes. With some abuse of notation, we denote these multinomials by

$$L(\underline{r}, \underline{x}) = \sum L_{\underline{b},\underline{d}} \underline{r}^{\underline{b}} \underline{x}^{\underline{d}}, \qquad R(\underline{x}) = \sum R_{\underline{d}} \underline{x}^{\underline{d}},$$

where coefficients $L_{\underline{b},\underline{d}}$ and $R_{\underline{d}}$, are non-negative reals. We will now interpret the coefficients. Let n be the block-length, i.e., the *length of the codeword*. This is in general *not equal* to the number of variable nodes in the *graph* since *punctured* bits are not included in the codeword but are included in the graph. For each variable-node degree type $(\underline{b}, \underline{d})$ the quantity $L_{\underline{b},\underline{d}} n$ is the number of variable nodes of type $(\underline{b}, \underline{d})$ in the graph. Similarly, for each check-node degree type $\underline{d}$ the quantity $R_{\underline{d}} n$ equals the number of check nodes of type $\underline{d}$ in the graph. Further, define

$$L_{r_i}(\underline{r}, \underline{x}) = \frac{\mathrm{d}L(\underline{r}, \underline{x})}{\mathrm{d}r_i}, \qquad L_{x_i}(\underline{r}, \underline{x}) = \frac{\mathrm{d}L(\underline{r}, \underline{x})}{\mathrm{d}x_i}, \qquad R_{x_i}(\underline{x}) = \frac{\mathrm{d}R(\underline{x})}{\mathrm{d}x_i}.$$

The coefficients of L and R are constrained to ensure that the number of sockets of *each type* is the same on both (variable and check) sides of the graph. This gives rise to m_e linear conditions on the coefficients of L and R as follows. First, we have the *socket count* equality constraints

$$L_{x_i}(\underline{1}, \underline{1}) = R_{x_i}(\underline{1}), \qquad i = 1, \dots, m_e. \tag{7.1}$$

(We use $\underline{1}$ to denote a vector of all 1's, the length being clear from the context.) If we assume that the i-th channel is used for a fraction (with respect to n) ξ_i of the transmitted bits, then we have the *received* constraints

$$L_{r_i}(\underline{1}, \underline{1}) = \xi_i, \qquad i = 1, \dots, m_r, \tag{7.2}$$

where the ξ_i are given positive constants satisfying $\sum_{i=1}^{m_r} \xi_i = 1$. In the special case of a single channel this constraint reduces to $L_{r_1}(\underline{1}, \underline{1}) = 1$. Note that L_{r_0} is not directly constrained; we are free to introduce punctured nodes The *design rate* is given by

$$r = L(\underline{1}, \underline{1}) - R(\underline{1}). \tag{7.3}$$

To see that this is the design rate, note that if we multiply r by n then we get the number of variable nodes in the graph minus the number of check nodes. This is, nominally, the number of independent variable nodes. Dividing this by n, the length of the codeword, is, by definition, the design rate. Fixing the design rate amounts to imposing a linear constraint on the parameters.

As we have seen, all key functionals and constraints on the parameters are linear. Thus, even after fixing the design rate, the space of admissible parameters is a convex polytope. Extreme points may be easily found using linear programming. This aspect is useful when optimizing the parameters.

Consider enumerating the sockets on both sides of the graph in some arbitrary manner. Let the total number of sockets on either side of the graph be $K = K_1 + K_2 + \cdots + K_{m_e}$, where K_i is the number of edges of type i. A particular graph is determined by a permutation π on K letters such that socket i on the variable-node side is connected to socket $\pi(i)$ on the check-node side. Further, we constrain socket $\pi(i)$ to be of the same type as socket i. Thus, $\pi = (\pi_1, \ldots, \pi_{m_e})$ can be decomposed into m_e distinct permutations where π_i is a permutation on K_i elements.

The *ensemble* of graphs under consideration is defined by viewing π_i as a random variable distributed uniformly over the set of all permutations on K_i elements, where the permutations for different edge types are independent.

§7.1.3. Some Examples

Example 7.4 (Standard Irregular Codes). Consider a standard irregular LDPC ensemble characterized by its edge degree distribution pair (λ, ρ). Assume that all bits are transmitted over the same BMS channel and that no bits are punctured. In this case we have only a single edge type, so that $\underline{x} = x_1$. Also, there is only a single BMS channel, so that $\underline{r} = r_1$. We therefore get

$$L(\underline{r}, \underline{x}) = L(r_1, x_1) = r_1 \frac{\int_0^{x_1} \lambda(z)\mathrm{d}z}{\int \lambda}, \qquad R(\underline{x}) = R(x_1) = \frac{\int_0^{x_1} \rho(z)\mathrm{d}z}{\int \lambda}.$$

Note the normalization of $R(x_1)$. This is so that $R(1) = (1-r)$, where r is the design rate, and $R(1)n$ gives the number of check nodes in the graph.

We have seen examples of Tanner graphs of LDPC codes throughout this book. Nevertheless, to facilitate the comparison with more general ensembles we depict a further such Tanner graph in Figure 7.6. ◇

Example 7.5 (Repeat-Accumulate Codes). Repeat-accumulate (RA) codes were originally introduced as a "stripped-down" version of turbo codes (see Notes at the end of this chapter). It has since been recognized that these are powerful codes themselves. Since their introduction a staggering number of variants on the basic theme have been proposed, a select few of which we discuss.

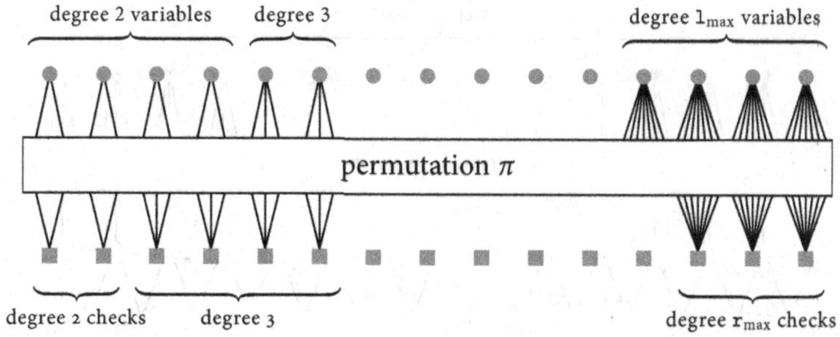

Figure 7.6: Tanner graph of a standard irregular LDPC code.

The codes are defined as follows. Let $x_1^s, \ldots, x_m^s$ be a sequence of m information bits. These bits form the *systematic* part of the codeword. To generate the *parity* part, *repeat* each bit $\mathtt{l}$ times to obtain a sequence of $\mathtt{l}m$ bits. Order this sequence according to a permutation π. Encode the thus ordered sequence using a $1/(1+D)$ convolutional encoder (accumulate). An additional bit is appended at the end to force the encoder into the '0' state. If no bits are punctured then the resulting codeword has length $m + \mathtt{l}m + 1$. If all the systematic bits are punctured then a codeword is of length $\mathtt{l}m+1$. These are the two most common versions, but of course more general puncturing schemes are possible. The encoding scheme is shown in Figure 7.7.

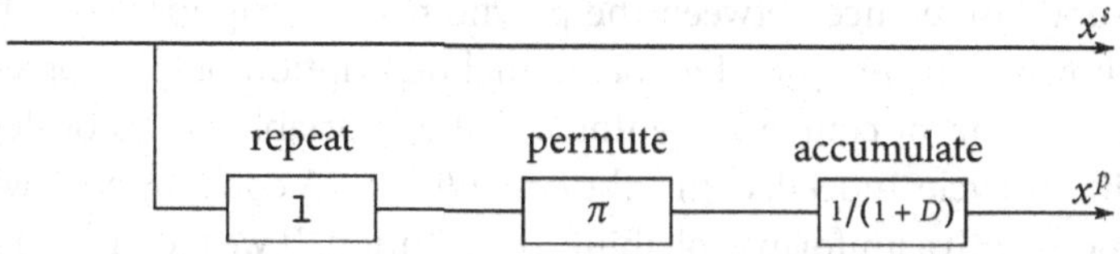

Figure 7.7: Encoder for an RA code. Each systematic bit is repeated $\mathtt{l}$ times; the resulting vector is permuted and fed into a filter with response $1/(1+D)$ (accumulate).

The code may also be represented as an LDPC code as shown on the right-hand side of Figure 7.8 in which each check node (except for the two "end" nodes) has degree 3. The check nodes are connected by a chain of nodes of degree 2, and the systematic bits are nodes of degree $\mathtt{l}$. From this point of view an RA code is a check 3-regular LDPC code with variable degrees 2 and $\mathtt{l}$ and where all variable nodes of degree 2 are arranged "in a chain." Because of the particular structure we can think of the variable nodes of degree 2 as the *parity* bits and the degree $\mathtt{l}$ ones as the *systematic* bits. In this construction it makes sense to consider a slight modification in which the chain is extended to a ring so that all check nodes have degree 3. The additional two edges are shown as dashed lines in Figure 7.8.

There are two edge types, those connected to systematic bits and those con-

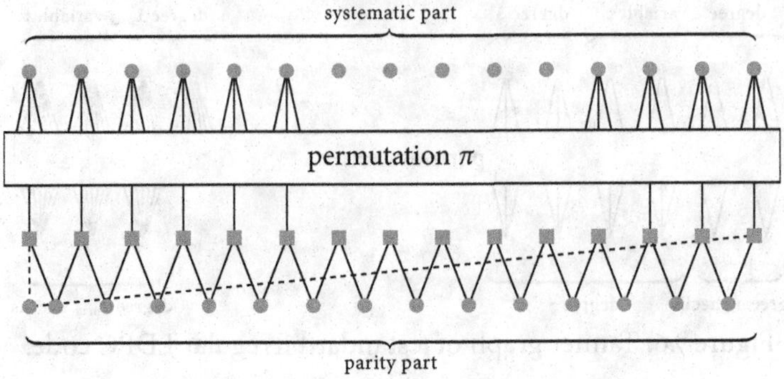

Figure 7.8: Tanner graph of an RA code with $\mathtt{l} = 3$.

nected to check bits. We introduce two variables, x_1 and x_2, accordingly. The polynomials for this ensemble are

$$L(\underline{r}, \underline{x}) = r_1 x_1^2 + \frac{1}{\mathtt{l}} r_0 x_2^{\mathtt{l}}, \qquad R(\underline{x}) = x_1^2 x_2,$$

if we consider the punctured case. For the unpunctured case this changes to

$$L(\underline{r}, \underline{x}) = \frac{1}{1+\mathtt{l}} r_1 x_1^2 + \frac{\mathtt{l}}{1+\mathtt{l}} r_1 x_2^{\mathtt{l}}, \qquad R(\underline{x}) = \frac{\mathtt{l}}{1+\mathtt{l}} x_1^2 x_2.$$

There is a subtle difference between the polynomial description we just gave and the graph shown in Figure 7.8. The polynomial description only enforces the *local* connectivity but can of course not enforce that all variable nodes of degree 2 are connected in *one* cycle. Indeed, define the *ensemble* of RA codes by endowing the set of permutations with a uniform probability distribution. If we pick a code at random from this ensemble then for increasing lengths the permutation between the variable nodes of degree 2 and the check nodes will break into typically $\Theta(\ln(n))$ cycles so that in average each cycle has length $\Theta(n/\ln(n))$. The difference to the case of a single cycle can be significant if we are interested in the behavior of the error floor but it has no influence on the result of density evolution. Similar remarks apply to many of the other ensembles which we discuss in the sequel. ◇

Example 7.9 (Irregular Repeat-Accumulate Codes). In the previous example we have seen that RA codes are check 3-regular LDPC codes, with variable nodes of degree 2 and $\mathtt{l}$, where the variable nodes of degree 2 are arranged in a cycle. Irregular repeat-accumulate (IRA) codes are a natural generalization of this concept in which we allow general degree distributions but still insist that the variable nodes of degree 2 form a ring. The corresponding Tanner graph is shown in Figure 7.10. Again two edge types are needed, namely one for edges connected to variable nodes in the ring and one for the remaining edges. Consider the case of no puncturing. If we let

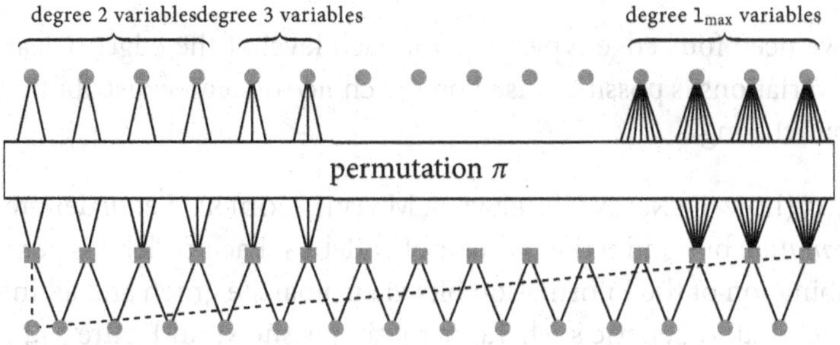

Figure 7.10: Tanner graph corresponding to an IRA code.

(λ, ρ) denote the degree distribution pair which describes the "LDPC part" of the IRA ensemble, i.e., the part of the graph excluding the ring, then we get

$$L(\underline{r}, \underline{x}) = \frac{\int \rho}{\int \lambda + \int \rho} r_1 x_1^2 + \sum_i \frac{\lambda_i}{i(\int \lambda + \int \rho)} r_1 x_2^i,$$
$$R(\underline{x}) = \sum_i \frac{\rho_i}{i(\int \lambda + \int \rho)} x_1^2 x_2^i.$$

More generally, some of the variable nodes (either in the ring or outside the ring) may be punctured which changes the degree distributions accordingly. ◇

Example 7.11 (Accumulate-Repeat-Accumulate Codes). Another natural extension of RA ensembles is the ensemble of accumulate-repeat-accumulate (ARA) codes. The Tanner graph of this ensemble is shown in Figure 7.12. Here two "rings"

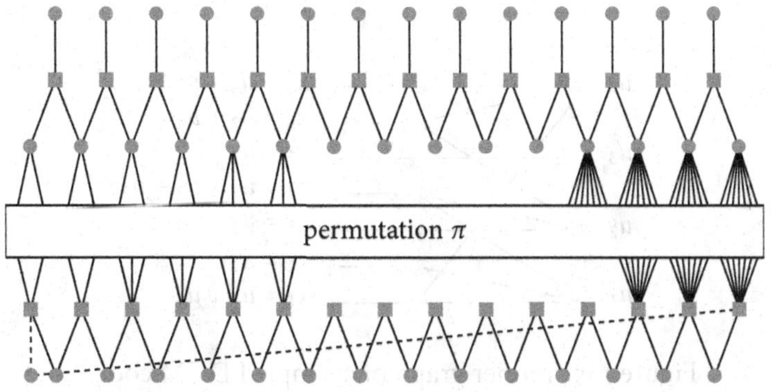

Figure 7.12: Tanner graph of an ARA code.

are added which explains the double "accumulate" in the name. To describe this ensemble we need four edge types, one for each level of the edges. Clearly, a large number of variations is possible based on the choice of degree distributions and the choice of puncturing. ◇

EXAMPLE 7.13 (LOW-DENSITY GENERATOR MATRIX CODES). Let m denote the number of *information* bits and n the number of codebits. Each codebit is generated as a linear combination of the information bits via a bipartite graph and all information bits are punctured. A generic such Tanner graph is shown in Figure 7.14 and a specific case is drawn in Figure 7.15. Two distinct edge types are needed. For the specific case we have $n = 7$ and the degree distribution is given by

$$L(\underline{r}, \underline{x}) = \frac{4}{7} r_0 x_1^3 + r_1 x_2, \qquad R(\underline{x}) = \frac{3}{7} x_1 x_2 + \frac{3}{7} x_1^2 x_2 + \frac{1}{7} x_1^3 x_2.$$

The *design rate* is $r = \frac{m}{n}$. Contrary to the case of LDPC codes, the design rate for

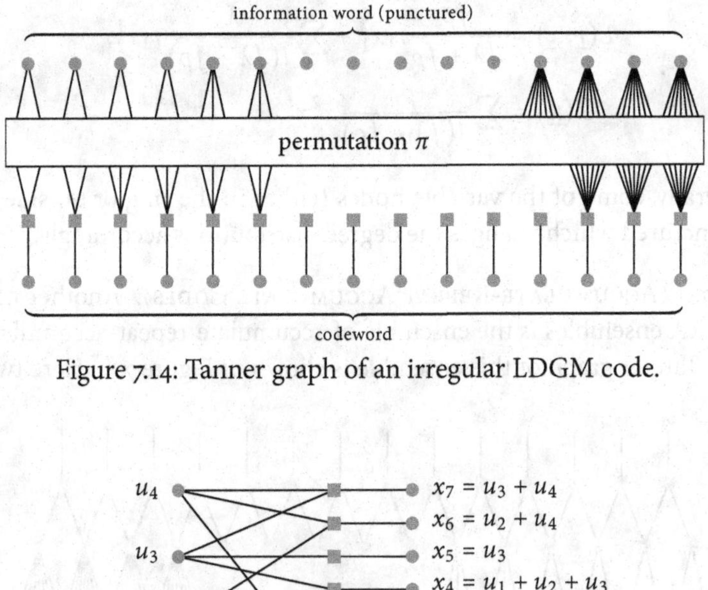

Figure 7.14: Tanner graph of an irregular LDGM code.

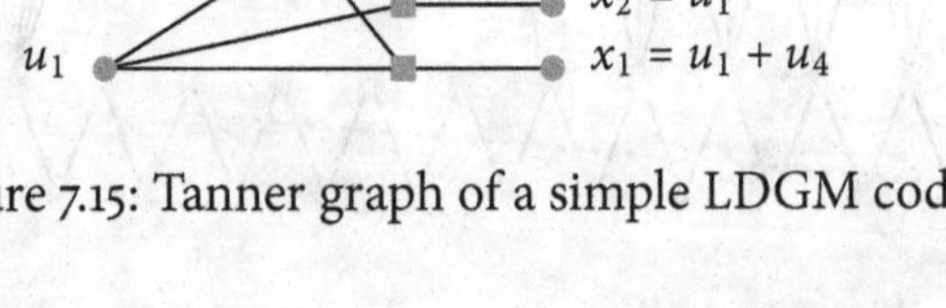

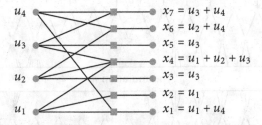

Figure 7.15: Tanner graph of a simple LDGM code.

LDGM codes is always an *upper bound*. We will have much more to say about LDGM codes in Section 7.5. ◇

EXAMPLE 7.16 (MN OR COMPOUND CODES). MN codes are the fusion of an LDGM with an LDPC code. This is best seen in the Tanner graph which is shown in Figure 7.17. The LDPC code is of rate 1 and the corresponding matrix must be invertible. To separate the two subgraphs more clearly, imagine instead of the one layer of check

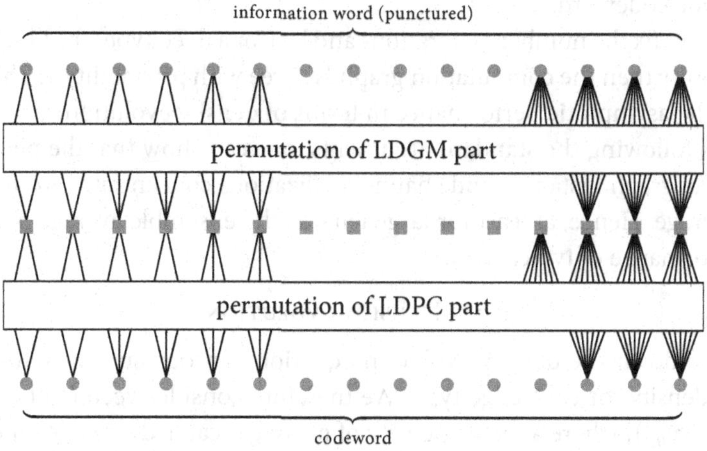

Figure 7.17: Tanner graph of an MN code.

nodes which has inputs from both top and bottom – two layers of check nodes with an intermediate layer of variable nodes. Each of these intermediate variable nodes is connected to exactly one of the top and exactly one of the bottom check nodes. Let s be the vector that corresponds to the intermediate variable nodes.

With this picture in mind, the information word, call it x^s, is first translated into the intermediate word s by means of an LDGM code, i.e., we have $s = x^s G$, where G is the sparse matrix that corresponds to the LDGM code. This intermediate word s is then used as a syndrome for the subsequent LDPC code. More precisely, we have the relationship $H(x^p)^T = s^T$, where H is the sparse matrix that corresponds to the LDPC code. For each given s we want the second relationship to yield a unique x^p. This implies that the matrix H must be invertible and so the LDPC code is of rate 1. The global relationship between input and output is $x^p = x^s G(H^{-1})^T$. Superficially, looking at the last relationship, it looks as if the resulting code is again an LDGM code. But in general $(H^{-1})^T$ is not sparse even if H has this property. ◇

§7.2. MULTI-EDGE-TYPE LDPC CODES: ANALYSIS

§7.2.1. ALL-ONE CODEWORD ASSUMPTION, TREE-LIKE COMPUTATION TREE, AND CONCENTRATION

Drawing from our experience with LDPC ensembles we can quickly dispense some of the preliminary steps of the asymptotic analysis.

First, assuming that all channels are BMS channels and that the message-passing decoder obeys the symmetry condition specified in Definition 4.81, Lemma 4.90 (which still applies) shows us that the conditional probability of error is independent of the transmitted codeword. Hence, we can analyze the performance with respect to the all-one codeword.

Next, if we fix the number of iterations and let (for a fixed type) the blocklength n tend to infinity then the computation graph is a tree with probability 1. This enables us to find the asymptotic performance in terms of density evolution.

Finally, following the standard arguments, one can show that the performance of individual graph instances and channel realizations concentrates around the ensemble average. Hence, at least for large lengths the ensemble average is indicative of the performance of typical codes.

§7.2.2. Density Evolution

Let us write down the density evolution equations. In our current setting we require one density for each edge type. We therefore consider vectors of L-densities $\underline{\mathsf{a}} = (\mathsf{a}_1, \dots, \mathsf{a}_{m_e})$, where a_i is the density of messages carried on edge type i. By $\underline{\Delta}_0$ we mean a vector of densities where each density is Δ_0. Similarly, by $\underline{\Delta}_\infty$ we mean a vector of densities where each density is $\Delta_{+\infty}$. We introduce the following multinomials:

$$\text{(7.18)}\quad \begin{aligned} \underline{\lambda}(\underline{r}, \underline{x}) &= \left(\frac{L_{x_1}(\underline{r}, \underline{x})}{L_{x_1}(\underline{1}, \underline{1})}, \frac{L_{x_2}(\underline{r}, \underline{x})}{L_{x_2}(\underline{1}, \underline{1})}, \dots, \frac{L_{x_{m_e}}(\underline{r}, \underline{x})}{L_{x_{m_e}}(\underline{1}, \underline{1})} \right), \\ \underline{\rho}(\underline{x}) &= \left(\frac{R_{x_1}(\underline{x})}{R_{x_1}(\underline{1})}, \frac{R_{x_2}(\underline{x})}{R_{x_2}(\underline{1})}, \dots, \frac{R_{x_{m_e}}(\underline{x})}{R_{x_{m_e}}(\underline{1})} \right). \end{aligned}$$

Example 7.19 (Standard Irregular LDPC Codes). Consider a standard irregular LDPC code with degree distribution pair (λ, ρ). We then get

$$\underline{\lambda}(\underline{r}, \underline{x}) = (r_1 \lambda(x_1)), \qquad \underline{\rho}(\underline{x}) = (\rho(x_1)).$$

This agrees with our standard notation, except that we have introduced a variable representing received distributions into the arguments of λ. Previously, the received distribution was not an argument of λ. ◇

Example 7.20 (RA Codes). For unpunctured RA ensembles discussed in Example 7.5 we have

$$\underline{\lambda}(\underline{r}, \underline{x}) = (r_1 x_1, r_1 x_2^{1-1}), \qquad \underline{\rho}(\underline{x}) = (x_1 x_2, x_1^2). \qquad ◇$$

In order to describe density evolution in a compact way it is useful to introduce a notation where we replace the vectors of variables $\underline{r}$ and $\underline{x}$ with vectors of densities

$\underline{a}_{\text{BMSC}}$ and $\underline{a}$, respectively. The meaning of this notation is analogous to the one which we introduced for standard LDPC ensembles: multiplication of variables is replaced by convolution of densities, where for $\underline{\lambda}$ we use the regular convolution $\circledast$, but for $\underline{\rho}$ we use the convolution $\boxast$. To be more explicit, note that

$$\underline{\lambda}(\underline{r}, \underline{x}) = (\underline{\lambda}_1(\underline{r}, \underline{x}), \dots, \underline{\lambda}_{m_e}(\underline{r}, \underline{x})), \qquad \underline{\rho}(\underline{x}) = (\underline{\rho}_1(\underline{x}), \dots, \underline{\rho}_{m_e}(\underline{x})),$$

where the components $\underline{\lambda}_i(\underline{r}, \underline{x})$ and $\underline{\rho}_i(\underline{x})$ are polynomials. Consider the previous example, $\underline{\lambda}_1(\underline{r}, \underline{x}) = r_1 x_1$, $\underline{\lambda}_2(\underline{r}, \underline{x}) = r_1 x_2^{1-1}$, $\underline{\rho}_1(\underline{x}) = x_1 x_2$, and $\underline{\rho}_2(\underline{x}) = x_1^2$; in this case $\underline{\lambda}_1(\underline{a}_{\text{BMSC}}, \underline{b}) = \underline{a}_{\text{BMSC}} \circledast \underline{b}_1$, $\underline{\lambda}_2(\underline{a}_{\text{BMSC}}, \underline{b}) = \underline{a}_{\text{BMSC}} \circledast \underline{b}_2^{\circledast(1-1)}$, $\underline{\rho}_1(\underline{a}) = \underline{a}_1 \boxast \underline{a}_2$, and $\underline{\rho}_1(\underline{a}) = \underline{a}_1^{\boxast 2}$.

Lemma 7.21 (Density Evolution). Consider a multi-edge-type ensemble. Let $\underline{a}_{\text{BMSC}}$ denote the vector of L-densities corresponding to channels over which information is received. Each of the involved channels is assumed to be a BMS channel. Let $\underline{a}_\ell$ denote the vector of densities passed from variable nodes to check nodes in iteration ℓ assuming that $\underline{a}_0 = \underline{\lambda}(\underline{a}_{\text{BMSC}}, \underline{\Delta}_0)$ (corresponding to each variable node sending out its received value), and let $\underline{b}_\ell$ denote the vector of densities passed from check nodes to variable nodes in iteration ℓ. Then for $\ell \geq 1$, $\underline{b}_\ell = \underline{\rho}(\underline{a}_{\ell-1})$ and $\underline{a}_\ell = \underline{\lambda}(\underline{a}_{\text{BMSC}}, \underline{b}_\ell)$, so that

$$\underline{a}_\ell = \underline{\lambda}(\underline{a}_{\text{BMSC}}, \underline{\rho}(\underline{a}_{\ell-1})). \tag{7.22}$$

Further, the density of the *decision* at time ℓ is given by $L(\underline{a}_{\text{BMSC}}, \underline{\rho}(\underline{a}_{\ell-1}))$.

Example 7.23 (RA Codes). For unpunctured RA ensembles discussed in Example 7.5 we have $\underline{a}_0 = (\Delta_0, a_{\text{BMSC}})$ and for $\ell \geq 1$

$$\underline{b}_\ell = \underline{\rho}(\underline{a}_{\ell-1}) = (a_{1,\ell-1} \boxast a_{2,\ell-1}, a_{1,\ell-1}^{\boxast 2}),$$
$$\underline{a}_\ell = \underline{\lambda}(\underline{a}_{\text{BMSC}}, \underline{b}_\ell) = (a_{\text{BMSC}} \circledast b_{1,\ell}, a_{\text{BMSC}} \circledast b_{2,\ell}^{\circledast(1-1)}),$$

where $a_{i,\ell}$ denotes the i-th component of $\underline{a}_\ell$. ◇

§7.2.3. Fixed Points, Convergence, and Monotonicity

In the multi-edge-type setting certain degeneracies can arise that are not possible in the standard irregular LDPC setting. There are two obvious degeneracies which we rule out a priori: punctured degree 1 variable nodes and received distributions equal to $\Delta_{+\infty}$. A punctured degree 1 variable node effectively eliminates its neighboring constraint from the decoding process, and a variable node with received distribution $\Delta_{+\infty}$ can be removed from the graph without effect.

We say that $\underline{b}$ is *degraded* with respect to $\underline{a}$, denoted $\underline{a} \prec \underline{b}$, if $a_i \prec b_i$, i.e., if b_i is degraded with respect to a_i for $i = 1, \dots, m_e$. We assume a parameterized family of input distributions $\underline{a}_{\text{BMSC}(\sigma)}$ ordered by degradation where increasing σ signifies a degradation of the input distribution.

We need to generalize density evolution to allow different values for $\underline{a}_0$. In general $\underline{a}_\ell(\underline{a}_0)$ denotes the vector density in the ℓ-th iteration assuming $\underline{a}_0$ in the 0-th iteration.

THEOREM 7.24 (MONOTONICITY). If, for some $\underline{b}$ and some $k \geq 0$, we have $\underline{a}_k(\underline{b}) \prec (\succ)\underline{b}$, then $\underline{a}_{k\ell}(\underline{b}) \prec (\succ)\underline{a}_{k(\ell-1)}(\underline{b})$ for all $\ell > 1$ and $\underline{a}_{k\ell}(\underline{b}) \xrightarrow{\ell\to\infty} \underline{f}$ for some vector density $\underline{f}$. If, in addition $\underline{a}_{k+1}(\underline{b}) \prec (\succ)\underline{b}$, then $\underline{a}_\ell(\underline{b}) \xrightarrow{\ell\to\infty} \underline{f}$.

Proof. Monotonicity, that $\underline{a}_{k\ell}(\underline{b}) \prec (\succ)\underline{a}_{k(\ell-1)}(\underline{b})$ under the stated assumptions, follows from the tree channel argument. Convergence of $\underline{a}_{k\ell}(\underline{b})$ in ℓ follows from sequential compactness of the space of symmetric distributions and completeness of functionals monotonic under degradation.

The extension in the final statement is analogous to Corollary 4.108 and can be proved the same way. □

As a special case, the preceding theorem implies that $\underline{a}_\ell(\underline{a}_{\text{BMSC}(\sigma)})$ always converges to a well-defined limit density. Usually we are interested in knowing when the output bit error probability goes to zero, i.e., we are interested in knowing when

$$\lim_{\ell\to\infty} L(\underline{a}_{\text{BMSC}(\sigma)}, \underline{\rho}(\underline{a}_\ell)) = \underline{\Delta}_\infty.$$

Let $\{\underline{a}_{\text{BMSC}(\sigma)}\}$ denote a family of distributions ordered by degradation: $\underline{a}_{\text{BMSC}(\sigma)} \prec \underline{a}_{\text{BMSC}(\sigma')}$ if $\sigma \leq \sigma'$. We define the *threshold* as

$$\sigma^{\text{BP}} = \sup\{\sigma : \lim_{\ell\to\infty} L(\underline{a}_{\text{BMSC}(\sigma)}, \underline{\rho}(\underline{a}_\ell)) = \underline{\Delta}_\infty\}.$$

In the setting of standard irregular LDPC ensembles, if $\sigma < \sigma^{\text{BP}}$ then $a_\ell \xrightarrow{\ell\to\infty} \Delta_{+\infty}$. In the multi-edge-type setting it is not always obvious what the fixed point limit of $\underline{a}_\ell$ might be. It is clear, however, that such a fixed point must be *perfectly decodable.*

A density vector $\underline{a}$ is a perfectly decodable fixed point of density evolution if it is a fixed point and $L(\underline{a}_{\text{BMSC}}, \underline{\rho}(\underline{a})) = \underline{\Delta}_\infty$. A multi-edge-type structure may have more than one perfectly decodable fixed point. Indeed, consider the simple example

$$L(r_0, r_1, x_1, x_2) = r_0 x_1 x_2, \qquad \text{and} \qquad R(x_1, x_2) = x_1 x_2,$$

where r_0 corresponds to Δ_0. Then *any* vector density $\underline{a} = (a_1, a_2)$ satisfying $a_1 = \Delta_{+\infty}$ or $a_2 = \Delta_{+\infty}$ is a perfectly decodable fixed point for density evolution.

LEMMA 7.25 (PERFECTLY DECODABLE FIXED POINT). Let $\underline{a}$ be a perfectly decodable fixed point and define $\underline{b} = \underline{\rho}(\underline{a})$. Then for each i we have either $a_i = \Delta_{+\infty}$ or $b_i = \Delta_{+\infty}$ (or both).

Proof. Assume that $\underline{a}$ is perfectly decodable and that for some i, $a_i \neq \Delta_{+\infty}$. Consider a variable-node type which contains the edge type i so that for this variable-node type the outgoing density along this edge, call it $\tilde{a}_i$, is not $\Delta_{+\infty}$. Note that a_i is the weighted convex sum of possibly several such densities and so if $\tilde{a}_i \neq \Delta_{+\infty}$ then $a_i \neq \Delta_{+\infty}$. For this variable node type the density of the decision can be written as $\tilde{a}_i \circledast b_i$. By assumption this must be equal to $\Delta_{+\infty}$, which is only true if $b_i = \Delta_{+\infty}$. □

An important corollary of the preceding lemma is that, for any perfectly decodable fixed point $\underline{a}$, every check-node type can have degree at most one in $\{i : a_i \neq \Delta_{+\infty}\}$.

One of the complicating features of the multi-edge-type framework is the possibility of having degree 1 variable nodes. As a consequence, the message distributions associated with certain edge types are strictly bounded away from $\Delta_{+\infty}$: any edge type that is connected to a degree 1 variable node never carries the distribution $\Delta_{+\infty}$ in the variable-to-check direction. Any other edge type that connects to a check node which is connected to a degree 1 variable node also never carries the distribution $\Delta_{+\infty}$ in the check-to-variable direction. This "bounded away from $\Delta_{+\infty}$" property propagates through the graph in a manner similar to erasure decoding with the roles of check nodes and variable nodes reversed.

Consider $\underline{a}_\ell(\underline{a}_{\text{BMSC}}, \underline{\Delta}_\infty)$; this is the density emitted in the ℓ-th iteration by the tree channel, in which the leaf nodes have density $\underline{\Delta}_\infty$, but the intermediate nodes have density $\underline{a}_{\text{BMSC}}$. Theorem 7.24 implies that this sequence of vector densities converges and that it is monotonic with respect to degradation. This is true since any vector density is degraded with respect to $\underline{\Delta}_\infty$. Let $\mathcal{E}^{\text{VC}}[\ell]$ denote $\{i : \underline{a}_{i,\ell}(\underline{a}_{\text{BMSC}}, \underline{\Delta}_\infty) \neq \Delta_{+\infty}\}$. Similarly, let $\mathcal{E}^{\text{CV}}[\ell]$ denote $\{i : \underline{\rho}_i(\underline{a}_\ell(\underline{a}_{\text{BMSC}}, \underline{\Delta}_\infty)) \neq \Delta_{+\infty}\}$. It is easy to see inductively that $\mathcal{E}^{\text{VC}}[\ell]$ and $\mathcal{E}^{\text{CV}}[\ell]$ are monotonically non-decreasing and that at some finite ℓ both reach their maximal value $\mathcal{E}^{\text{VC}}$ and $\mathcal{E}^{\text{CV}}$.

To be more explicit, we construct the sets recursively. Note that for this initial condition, densities are degrading each iteration so we need only find the first point at which a density ceases to be $\Delta_{+\infty}$. Initially $\mathcal{E}^{\text{VC}}[0] = \emptyset$ and $\mathcal{E}^{\text{CV}}[1] = \emptyset$. Clearly, the set $\mathcal{E}^{\text{VC}}[1]$ consists of all edge types that have positive probability of being connected to a degree one variable node. Since no received distribution is $\Delta_{+\infty}$ it follows that none of the distributions on these edge types are $\Delta_{+\infty}$. Tracing the propagation and bearing in mind that no received distribution is $\Delta_{+\infty}$ it follows that $\mathcal{E}^{\text{VC}}[\ell]$ and $\mathcal{E}^{\text{CV}}[\ell]$ can be defined by iterating the following two steps for ℓ starting with $\ell = 2$.

(i) Let $\mathcal{E}^{\text{CV}}[\ell]$ be the union of $\mathcal{E}^{\text{CV}}[\ell-1]$ and those edge types $i \notin \mathcal{E}^{\text{CV}}[\ell-1]$ that have a positive probability of being connected to a check node that has another edge in $\mathcal{E}^{\text{VC}}[\ell-1]$.

(ii) Let $\mathcal{E}^{\text{VC}}[\ell]$ be the union of $\mathcal{E}^{\text{VC}}[\ell-1]$ and those edge types $i \notin \mathcal{E}^{\text{VC}}[\ell-1]$ that have positive probability of connecting to a variable node having all other edges in $\mathcal{E}^{\text{CV}}[\ell]$.

For all sufficiently large ℓ we have $\mathcal{E}^{\text{VC}}[\ell] = \mathcal{E}^{\text{VC}}[\ell-1]$ and $\mathcal{E}^{\text{CV}}[\ell] = \mathcal{E}^{\text{CV}}[\ell-1]$ and we let $\mathcal{E}^{\text{VC}}$ and $\mathcal{E}^{\text{CV}}$ denote these limiting sets.

The edge types $\mathcal{E}^{\text{CV}}$ and $\mathcal{E}^{\text{VC}}$ arise from degree one variable nodes. Lemma 7.25 implies that in order to have a perfectly decodable fixed point that we must have $\mathcal{E}^{\text{CV}} \cup \mathcal{E}^{\text{VC}} = \emptyset$. This observation together with our example of a graph with multiple perfectly decodable fixed points nearly exhausts all possibilities. The following result summarizes this. This result can be proved with a few pages of fairly tight but unilluminating reasoning, the proof has therefore been omitted.

Theorem 7.26 (Degeneracy of Perfectly Decodable Fixed Points). Given a multi-edge-type structure, there exists a perfectly decodable fixed point if and only if $\mathcal{E}^{\text{VC}} \cap \mathcal{E}^{\text{CV}} = \emptyset$. Given $\mathcal{E}^{\text{VC}} \cap \mathcal{E}^{\text{CV}} = \emptyset$, the perfectly decodable fixed point is unique or there exist non-empty disjoint edge types $\mathcal{E}_1$ and $\mathcal{E}_2$, both disjoint from $\mathcal{E}^{\text{VC}} \cup \mathcal{E}^{\text{CV}}$, such that

A. No node type has degree more than 1 in $\mathcal{E}_1$.

B. All variable nodes with degree 1 in $\mathcal{E}_1$ are state variable nodes with all other edges in $\mathcal{E}_2$.

C. All check nodes with positive degree in $\mathcal{E}_2$ have degree 1 in $\mathcal{E}_1$.

Furthermore, if edge types satisfying A, B, and C, exist, then either there are multiple perfectly decodable fixed points or no perfectly decodable fixed point is reachable under standard density evolution.

We will call a multi-edge-type structure *non-degenerate* if $\mathcal{E}^{\text{VC}} \cap \mathcal{E}^{\text{CV}} = \emptyset$ and there do not exist non-empty disjoint edge types $\mathcal{E}_1$ and $\mathcal{E}_2$, both disjoint from $\mathcal{E}^{\text{VC}} \cup \mathcal{E}^{\text{CV}}$ satisfying A,B, and C, of Theorem 7.26. Otherwise, the structure is called *degenerate*. In practice, degenerate structures are of little or no interest.

§7.2.4. Stability

The stability analysis of density evolution examines the (asymptotic) behavior of the decoder when it is close to successful decoding. The basic question of interest is when can one be certain that the error probability is converging to zero, assuming it has gotten sufficiently close. In the standard irregular setting one can look at the error probability of the edge message distribution and ask whether it is tending to zero. In the multi-edge-type setting this notion is inadequate because, as we saw in the previous section, messages on certain edge types may not converge to zero error probability even though the output error probability converges to zero. Nevertheless, a more-or-less complete stability theory can be developed.

We generally limit our presentation to ensembles without degree 1 nodes, indicating some extension to that case at the end of this section. Define the matrices $\mathtt{L} = \mathtt{L}(\underline{r})$ and $\mathtt{P}$ by

$$\mathtt{L}_{i,j} = \frac{\mathrm{d}\underline{\lambda}_i(\underline{r},\underline{x})}{\mathrm{d}x_j}\Big|_{\underline{x}=\underline{0}}, \qquad \mathtt{P}_{i,j} = \frac{\mathrm{d}\underline{\rho}_i(\underline{x})}{\mathrm{d}x_j}\Big|_{\underline{x}=\underline{1}}.$$

THEOREM 7.27 (STABILITY FOR MULTI-EDGE). Assume a multi-edge-type structure with no degree 1 variable nodes. If $\mathtt{L}(\mathfrak{B}(\underline{\mathtt{a}}_{\text{BMSC}}))\mathtt{P}$ is a stable matrix (spectral radius less than one) then there exists $\xi > 0$ such that $\underline{\mathtt{a}}_\ell(\underline{\mathtt{a}}_{\text{BMSC}}, \underline{\mathtt{a}}) \xrightarrow{\ell\to\infty} \underline{\Delta}_\infty$ for any $\underline{\mathtt{a}}$ satisfying $\max_i\{\mathfrak{E}(\underline{\mathtt{a}}_i)\} < \xi$.

Discussion: If the structure is degenerate, possessing edge type subsets $\mathcal{E}_1$ and $\mathcal{E}_2$ as in Theorem 7.26, then the matrix product $\mathtt{L}(\mathfrak{B}(\underline{\mathtt{a}}_{\text{BMSC}})\mathtt{P}$ is, at best, marginally stable, i.e., it has 0 as an eigenvalue.

The converse to Theorem 7.27 is a little trickier to state. It is not difficult to concoct structures where setting the density on one edge type to $\Delta_{+\infty}$ causes all densities to eventually converge to $\Delta_{+\infty}$. Thus, assuming that $\mathtt{L}(\mathfrak{B}(\underline{\mathtt{a}}_{\text{BMSC}})\mathtt{P}$ is unstable, we cannot say, as we did in the standard irregular case, that if $\underline{\mathtt{a}} \neq \underline{\Delta}_\infty$ then $\underline{\mathtt{a}}_\ell(\underline{\mathtt{a}}_{\text{BMSC}}, \underline{\mathtt{a}})$ will not approach $\underline{\Delta}_\infty$. We do, however, have the following.

THEOREM 7.28. Assume a non-degenerate multi-edge-type structure with no degree 1 variable nodes. Assume that $\mathtt{L}(\mathfrak{B}(\underline{\mathtt{a}}_{\text{BMSC}})\mathtt{P}$ is an unstable matrix (spectral radius greater than one). Let $\underline{\mathtt{a}}$ satisfy $\min_i\{\mathfrak{E}(\underline{\mathtt{a}}_i)\} > 0$. Then

$$\liminf_{\ell\to\infty} \mathfrak{E}(L(\underline{\rho}(\underline{\mathtt{a}}_\ell(\underline{\mathtt{a}}_{\text{BMSC}}, \underline{\mathtt{a}})))) > 0.$$

Proof. Assume that $\mathtt{L}(\mathfrak{B}(\underline{\mathtt{a}}_{\text{BMSC}}))\mathtt{P}$ is unstable. By the Perron-Frobenius theorem there exists a non-negative (real) eigenvector $\underline{x}$ of $\mathtt{L}(\mathfrak{B}(\underline{\mathtt{a}}_{\text{BMSC}}))\mathtt{P}$ with real positive eigenvalue $\xi > 1$ whose magnitude equals the spectral radius. Without loss of generality we assume that $\underline{x}_i \leq 1$ for $i = 1, \dots, m_{\mathsf{e}}$.

Let $\epsilon = \min_i\{\mathfrak{E}(\underline{\mathtt{b}}_i)\} > 0$ and consider the density vector $\underline{\mathtt{c}}(\eta)$, where $\underline{\mathtt{c}}_i(\eta) = \eta\underline{x}_i\Delta_0 + (1-\eta\underline{x}_i)\Delta_{+\infty}$. For all $\eta \leq 2\epsilon$ we apply Lemma 4.78 to note that $\underline{\mathtt{a}}$ is degraded with respect to $\underline{\mathtt{c}}(\eta)$. Now, for any fixed ℓ, it is not hard to see that $\underline{\mathtt{c}}(\eta)$ has evolved to $\eta\big(\mathtt{L}(\mathfrak{B}(\underline{\mathtt{a}}_{\text{BMSC}})\mathtt{P}\big)^{\ell}\underline{x} + O(\eta^2)$ where the multiplication of matrix elements represents the variable domain convolution. If we apply the Bhattacharyya functional to this density we get $\eta\xi^\ell \underline{x} + O(\eta^2)$. We know from Lemma 4.66 that

$$\mathfrak{E}\left(\mathrm{L}(\mathfrak{B}(\underline{\mathsf{a}}_{\mathrm{BMSC}})\mathrm{P})^\ell \underline{x} \geq c\frac{\mathsf{e}}{3\pi}\sqrt{\frac{\mathfrak{B}_{\min}}{\ell + m_{\mathrm{r}}}}\xi^\ell \underline{x},\right.$$

where c is a positive constant independent of ℓ and the inequality is component-wise. Since $\xi > 1$, we can choose $\ell = L$ sufficiently large so that $\mathfrak{E}(\underline{\mathsf{a}}_L(\underline{\mathsf{a}}_{\mathrm{BMSC}}, \underline{\mathsf{c}}(\eta))) \geq \eta\underline{x} + O(\eta^2)$ component-wise. Assuming now that η is sufficiently small we have

$$\mathfrak{E}(\underline{\mathsf{a}}_k(\underline{\mathsf{a}}_{\mathrm{BMSC}}, \underline{\mathsf{c}}(\eta))) \geq \frac{1}{2}\eta\underline{x}.$$

for $k = L$ and $k = L+1$. It now follows from Lemma 4.78 that $\underline{\mathsf{a}}_k(\underline{\mathsf{a}}_{\mathrm{BMSC}}, \underline{\mathsf{c}}(\eta)) \prec \underline{\mathsf{c}}(\eta)$ for $k = L$ and $k = L+1$. By Theorem 7.24 we have $\underline{\mathsf{a}}_\ell(\underline{\mathsf{a}}_{\mathrm{BMSC}}, \underline{\mathsf{c}}) \xrightarrow{\ell\to\infty} \underline{\mathsf{f}}$ for some fixed point $\underline{\mathsf{f}}$ satisfying $\underline{\mathsf{f}} \prec \underline{\mathsf{c}}(\eta)$. Clearly, $\underline{\mathsf{f}}$ is not perfectly decodable. □

§7.2.5. Stability with Degree-one Variable Nodes

A complete and general stability theory including degree one variable nodes is somewhat complicated. The cases of most interest, however, are rather straightforward. In this section we briefly discuss the simplest case, which can easily be extended once the main idea is understood.

We consider codes that are essentially a non-degenerate multi-edge type LDPC code having no degree-one variable nodes together with its serial concatenation with a multi-edge type low-density generator matrix (LDGM) code. The Tanner graph of such a code includes the Tanner graph of the LDPC code having no degree-one variable nodes, which we will refer to as the core LDPC graph. In addition there is one check node, each with a single neighboring associated degree-one variable node, for each bit in the LDGM code. Thus, the code consists of the original LDPC code and additional bits formed directly as parities of the bits in the core LDPC code. If E_1 denotes the set of edge types in the core LDPC graph then, in addition, the full Tanner graph has edge types $E_{1,2}$ connecting variable nodes in the core LDPC graph to the check nodes associated to the LDGM code and edge types E_2 that connect the degree-one variable nodes to the check nodes associated to the LDGM code (exactly one edge for each such check node type).

It is easy to see that such a graph has a unique perfectly decodable fixed point with $\Delta_{+\infty}$ carried on each edge type in $E_{1,2}$ in the variable-to-check direction and $\Delta_{+\infty}$ carried on each edge type in E_2 in the check-to-variable direction. Consider the densities carried on edge types in $E_{1,2}$ in the check-to-variable direction at the perfectly decodable fixed point. From the perspective of the core LDPC graph these densities can be viewed as an enhancement, i.e. a physical upgrading, of the received distribution(s) of the core graph. The main result here is that if the perfectly decodable fixed point of the core LDPC graph is stable with this enhanced received distribution, then the fixed point associated to the full graph is stable.

A complete proof is simple but tedious. The main idea is to observe that convergence to the perfectly decodable fixed point in the core is preserved under a sufficiently slight degradation of the densities in the check-to-variable direction on edge types in $E_{1,2}$ and then to note that the degradation, with respect to the fixed point, of these messages can be controlled by the degradation of the densities in the variable-to-check direction on edge types in $E_{1,2}$. Since these densities are all converging to $\Delta_{+\infty}$, convergence to the perfectly decodable fixed point follows.

§7.3. Structured Codes

In a practical system there is motivation to impose further structure on the graph beyond the degree structure. This can reduce implementation cost, enable parallelism of the decoder, and/or simplify encoding.

§7.3.1. Liftings of LDPC Codes and Graphs

A *lifting* is conveniently visualized from the Tanner graph perspective. Consider a "small" graph corresponding to a short LDPC code as illustrated in Figure 7.29. This small graph goes by various names: *protograph, base graph*, or *projected graph*. The first two names convey that this graph forms the foundation for building a larger graph and the name "projected graph" conveys that this smaller graph is a projection (a map with the property that this map applied twice is equal to the map itself) of the latter constructed graph. Make *m copies* of the base graph as shown in the left-hand

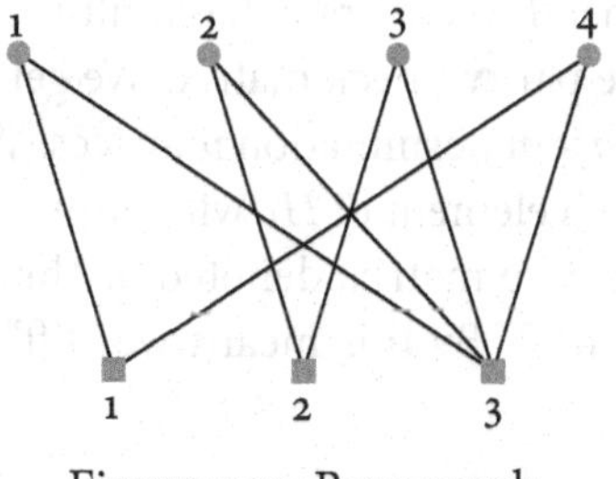

Figure 7.29: Base graph.

side of Figure 7.30. Identify like nodes (variable and check) as well as like edges into *clusters* of size m each. This means that for each node in the original graph there is now a cluster of size m and the same is true for each edge in the original graph.

To construct the lifted graph take each edge cluster and apply a permutation to this cluster so that different copies of the base graph become connected to each other.

The result is shown on the right-hand side of Figure 7.30 where all permutations are cyclic. In standard graph theory this construction is known as a *cover*, more specific

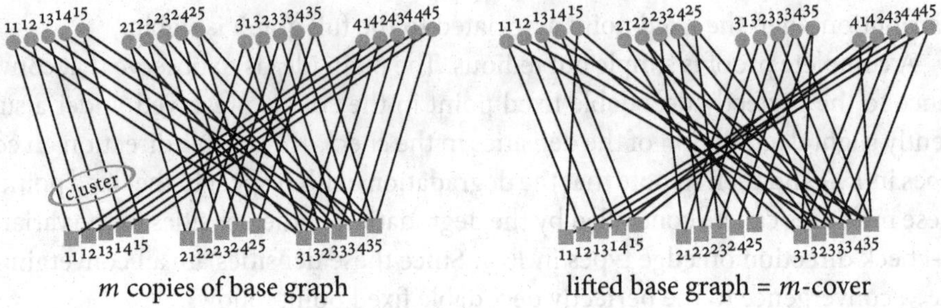

Figure 7.30: Left: m copies of base graph with $m = 5$. Right: Lifted graph resulting from applying permutations to the edge clusters.

an m-cover. This is important: density evolution is the same if we are working on the base graph or on an m-cover. *Locally* they are identical, and that is all that a message-passing decoder can tell.

For another perspective, consider the lifting of the parity-check matrix. Thus, let H_b be the parity-check matrix of the base graph. For our example we have

$$H_b = \begin{pmatrix} 1 & 0 & 0 & 1 \\ 0 & 1 & 1 & 0 \\ 1 & 1 & 1 & 1 \end{pmatrix}. \tag{7.31}$$

It often occurs in practice that the base graph has multiple edges. This is not captured in the description using the parity-check matrix. We get back to this point shortly. Let $\mathcal{P}$ denote the group of $m \times m$ permutation matrices. We form an m-times-larger LDPC code by replacing each element of H_b with an $m \times m$ matrix. The 0 elements of H_b are replaced with the zero matrix, denoted 0. The 1 elements of H_b are each replaced with a matrix from $\mathcal{P}$. By this means we "lift" an LDPC code to one m times larger:

$$H = \begin{pmatrix} P^2 & 0 & 0 & P^4 \\ 0 & P^4 & P^1 & 0 \\ P^3 & P^5 & P^2 & P^5 \end{pmatrix}$$

For the specific case depicted in Figure 7.30, P^i corresponds to a cyclic permutation of the set $\{1, \dots, m\}$ with shift $i - 1$: the columns correspond to the variables and the rows to the checks. For example, P^1 denotes the identity matrix and

$$P^2 = \begin{pmatrix} 0 & 0 & 0 & 0 & 1 \\ 1 & 0 & 0 & 0 & 0 \\ 0 & 1 & 0 & 0 & 0 \\ 0 & 0 & 1 & 0 & 0 \\ 0 & 0 & 0 & 1 & 0 \end{pmatrix},$$

which means that variable 11 is connected to check 12 and variable 12 is connected to check 13 and so on until finally variable 15 is connected to check 11. If we write a codeword of the lifted code in the form

$$x = (x_{11}, x_{12}, \dots, x_{1m}, x_{21}, \dots, x_{2m}, \dots, x_{mm}),$$

then the code is defined by the standard condition $Hx^T = 0^T$. For our example the condition implied by the first row of the parity-check matrix is $x_{15} + x_{44} = 0$.

The permutations are in general chosen according to their suitability with respect to the implementation of the encoder and decoder. For example, if we are interested in a hardware implementation then we might pick permutations that are supported by an Ω-*network*. Such a network accomplishes a permutation on 2^k el-

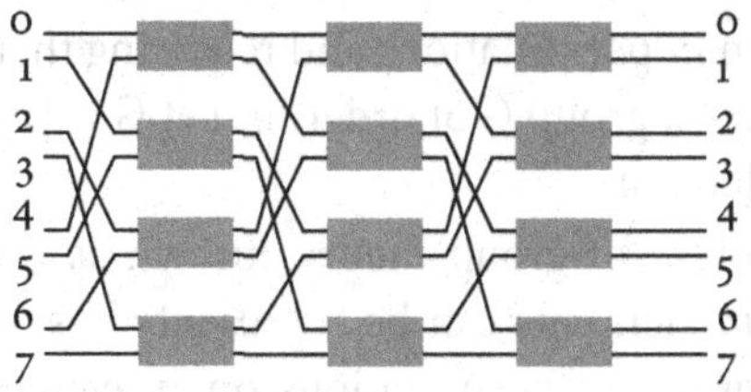

Figure 7.32: Ω-network for 8 elements. It has $\log_2(8) = 3$ stages, each consisting of a perfect shuffle.

ements, $k \in \mathbb{N}$. To see what such a network looks like consider the Ω-network for $k = 3$, i.e., for 8 elements, as shown in Figure 7.32. It has $\log_2(8) = 3 = k$ stages, each consisting of a *perfect shuffle*. A perfect shuffle of the numbers 0 to $2^k - 1$ is most easily described by representing these numbers in binary form with the leftmost bit being the most significant one: a perfect shuffle corresponds to mapping the number i (represented in binary form) into the number which results by performing a cyclic left-hand shift of the bits, i.e., a shift where the most significant bit becomes the least significant one. For example, the number 3 has binary representation 011 and it is mapped to the number 6 since this has binary representation 110. There is also a visual representation of this permutation from which the name "perfect shuffle" derives. To be concrete, consider the case $k = 3$. Think of the numbers 0 up to 7 as 8 ordered cards stacked on top of each other, with 0 being the bottom card. Split the deck into two equal sized stacks – the cards 0 to 3 make up one stack and the cards 4 to 7 are contained in the other stack. Lay down the two stacks side by side and merge them. This means we take the bottom card of the first stack (this is the card 0) and put on top of it the bottom card of the second stack (this is the card 4), then we put on top of this the second card of the first stack and so on.

Each box in Figure 7.32 represents a binary *switch*. The two inputs are either forwarded straight to the output or they are exchanged. For each setting of the $2^{k-1}k$ switches we get a different permutation. Without proof we state that for a proper position of the switches an Ω-network can accomplish permutations of the form $i \mapsto pi + c \mod 2^k$, where $p, c \in \mathbb{N}$ and p is odd.

§7.3.2. Matched Liftings

Matched liftings are characterized by requiring that the set of permutations used in the lifting form a group of order m, the size of the lifting. In this case a simplifying mathematical structure emerges which can be exploited in both encoding and decoding. We begin by discussing the underlying structure.

Group Rings

Rather than starting with m permutations and requiring that this set forms a group (under composition), take a group G of order m. Let $\text{G} = \{\text{g}_1, \dots, \text{g}_m\}$. By convention, g_1 is the identity element.

A fundamental theorem of group theory, known as *Cayley's* theorem, states that *any group* G with m elements can be represented as a subgroup of the symmetric group Σ_m, the group of permutations on m elements. Let $\{\pi_1, \dots, \pi_m\}$, $\pi_i : \{1, \dots, m\} \to \{1, \dots, m\}$, denote the set of m permutations which represent G. How do we find this subgroup given G? Here is how it works: set $\pi_i(u) = v$ if $\text{g}_i \text{g}_u = \text{g}_v$. Let us check that $\{\pi_1, \dots, \pi_m\}$ faithfully represents G under composition of maps. More precisely, we need to check that if $\text{g}_i \text{g}_j = \text{g}_k$ then $\pi_i \circ \pi_j = \pi_k$. Note that $\pi_j(u) = v$ if $\text{g}_j \text{g}_u = \text{g}_v$. Further, $p_i(v) = w$ if $\text{g}_i \text{g}_v = \text{g}_w$. Therefore, $(\pi_i \circ \pi_j)(u) = w$ if $\text{g}_i \text{g}_j \text{g}_u = \text{g}_w$. Since, $\pi_k(u) = w$ if $\text{g}_k \text{g}_u = \text{g}_w$ we conclude that we must have $\text{g}_i \text{g}_j = \text{g}_k$, as claimed.

A permutation on m elements can also be represented by an $m \times m$ permutation matrix. Let $P^1, P^2, \dots, P^m$ denote the m permutation matrices, where

$$P^i_{\pi_i(l),l} = 1, \quad l = 1, \dots, m,$$

and all other entries are 0. The permutation π_i on a vector x of length m is then effected by the multiplication $P^i x^T$.

Let us summarize. The group G is represented either by its elements g_i, by permutations π_i, or by permutation matrices P^i, and we have

$$g_i g_j = g_k \quad \Leftrightarrow \quad \pi_i \circ \pi_j = \pi_k \quad \Leftrightarrow \quad P^i P^j = P^k.$$

Consider again the Tanner graph representation. We described the lifting construction earlier, starting with m copies of the base graph. Label those copies with the elements of the group $g_1, \ldots, g_m$. Prior to permuting within the edge clusters, a variable node in copy g_k connects only to a check node in copy g_k. Suppose that to each edge cluster we now associate an element g_i of the group G, with the prescription that in this edge cluster the variable node in copy g_k connects to the check node in copy $g_i g_k$ via its edge in the given edge cluster. From the parity-check matrix perspective, the permutation matrix representing this lifted edge cluster is P^i.

Using such a group one can define a ring, the so-called *group ring* $\mathbb{F}_2[G]$, whose elements can be represented as binary vectors of length m.[1] Here, a binary vector $u = (u_1, \ldots, u_m)$ is identified with a formal sum $\sum_{i=1}^m u_i g_i$. Let us specify the rules of addition and multiplication in this ring. Addition is straightforward. Given two binary vectors u and v the sum $u + v$ is defined as component-wise addition of u and v over $\mathbb{F}_2$. The multiplication $u \cdot v$ is defined by

$$\begin{aligned}\Big(\sum_{i=1}^m u_i g_i\Big)\Big(\sum_{j=1}^m v_j g_j\Big) &= \sum_{i=1}^m \sum_{j=1}^m u_i v_j g_i g_j = \sum_{k=1}^m \Big(\sum_{(i,j):g_i g_j = g_k} u_i v_j\Big) g_k \\ &= \sum_{k=1}^m \Big(\sum_{(i,j):\pi_i(j)=k} u_i v_j\Big) g_k = \sum_{k=1}^m \Big(\sum_{i=1}^m u_i v_{\pi_i^{-1}(k)}\Big) g_k,\end{aligned}$$

where the arithmetic in $\big(\sum_{(i,j):g_i g_j = g_k} u_i v_j\big)$ is over $\mathbb{F}_2$.

By the correspondence of g_i with P^i (an $m \times m$ permutation matrix) we can interpret the sum $\sum_{i=1}^m u_i g_i$ also as a matrix sum $M(u) = \sum_{i=1}^m u_i P^i$. Here, $M(u)$ is a matrix over $\mathbb{F}_2$. This sum is invertible, i.e., $\sum_{i=1}^m u_i P^i$ uniquely determines u (in fact a single column or row determines u). By assumption g_1 is the identity element so that $M(e_1) = I$, where $e_1 = (1, 0, \ldots, 0)$ and I is the $m \times m$ identity matrix.

Given two binary vectors u and v, representing elements of $\mathbb{F}_2[G]$, their product $w = u \cdot v$ then corresponds to the regular matrix product $M(w) = M(u)M(v)$. Note, moreover, that $w^T = M(u)v^T$. Indeed,

$$(M(u)v^T)_k = \sum_{i=1}^m u_i (P^i v^T)_k = \sum_{i=1}^m u_i \sum_{j=1}^m P^i_{kj} v_j = \sum_{i=1}^m u_i v_{\pi_i^{-1}(k)}.$$

[1] Actually, since $\mathbb{F}_2$ is a field, this is a group algebra.

Matched Lifted LDPC Codes

We are now ready to describe matched lifted LDPC codes. To continue our example, assume at first that the base graph does not have multiple connections. In this case start with the binary parity-check matrix of the base code; call it H_b. To be concrete, consider H_b as given in (7.31). Now replace each element of H_b with a binary vector of length m. More precisely, replace 0 with the the all-zero m-tuple, and each element 1 with a binary m-tuple which contains exactly one 1. For our running example and a specific choice of binary vectors we get the matrix

$$\begin{pmatrix} (0,1,0,0,0) & (0,0,0,0,0) & (0,0,0,0,0) & (0,0,0,1,0) \\ (0,0,0,0,0) & (0,0,0,1,0) & (1,0,0,0,0) & (0,0,0,0,0) \\ (0,0,1,0,0) & (0,0,0,0,1) & (0,1,0,0,0) & (0,0,0,0,1) \end{pmatrix}.$$

Now apply to each component (i.e., vector) of this matrix the map $M(\cdot)$ to get the parity-check matrix of the lifted code. Assume that the underlying permutation matrices P^i correspond to cyclic permutations, where P^1 is the identity matrix and P^i corresponds to a cyclic permutation with shift $i-1$. These matrices form a group as required. In this case the lifted code corresponds to the code described on the right-hand side of Figure 7.30.

Let us now lift the restriction that the base graph cannot have multiple edges. In the general case, start with a matrix H with elements over $\mathbb{F}_2[\mathrm{G}]$and apply to it the map $M(\cdot)$ to get the parity-check matrix of the lifted code. The matrix $M(H)$ is a block matrix with each block of the form $M(H_{ij})$ where H_{ij} is a binary vector of length m. The weight of H_{ij}, is equal to the number of edges between variable node j and check node i in the base graph. As an example, consider the matrix

$$\begin{pmatrix} (0,1,1,0,0) & (0,0,0,0,0) & (0,0,0,0,0) & (0,0,0,1,0) \\ (0,0,0,0,0) & (0,0,0,1,0) & (1,0,0,0,0) & (0,0,0,0,0) \\ (0,0,1,0,0) & (0,0,0,0,1) & (0,1,0,0,0) & (0,0,0,0,1) \end{pmatrix},$$

which differs from our previous example only in the top-leftmost entry which now has weight 2. This corresponds to the base graph shown on the left-hand side of Figure 7.33. It has a double edge between variable 1 and check 1. The lifted graph is shown on the right-hand side of Figure 7.33.

The resulting lifted binary LDPC code is the set of solutions x to the equation $M(H)x^T = 0^T$. Writing $x = (x^1, \dots, x^N)^T$, where each x^i is a binary vector of length m, we can also write this equation as $Hx^T = 0^T$, where now each x^i is interpreted as an element of $\mathbb{F}_2[\mathrm{G}]$. Thus, the binary LDPC code associated to $M(H)$ can be identified with the LDPC code over $\mathbb{F}_2[\mathrm{G}]$ associated to H.

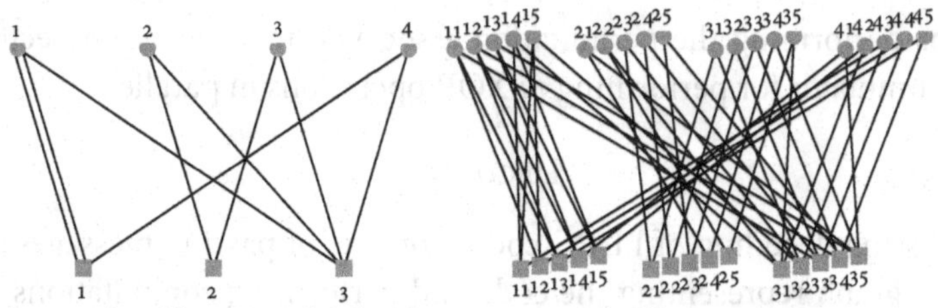

base graph with one multiple edge lifted base graph

Figure 7.33: Left: Base graph with a multiple edge between variable node 1 to check node 1. Right: Lifted graph.

Encoding

One of the most useful aspects of the matched lifting structure is the general encoding method it supports.

We use the general encoding method discussed in Appendix A but, in effect, we perform the approximate upper triangulation on the base parity-check matrix. The encoding is then performed by interpreting matrix operations as arithmetic over $\mathbb{F}_2[G]$ rather than as binary arithmetic. In practical settings, where the size of the lifting is sufficiently large, one can usually enforce $g = 1$ with no discernible loss in performance. In this case the matrix $\phi = ET^{-1}A + C$ is an element of $\mathbb{F}_2[G]$, and, since it is invertible, so is ϕ^{-1}. To keep complexity small, diagonal elements of T should have weight 1, i.e., correspond to single (not multiple) edges in the base graph. In this case one can assume without loss of generality that the diagonal elements are each the identity.

In the natural parallel implementations of the encoder, one implements the permutations associated to G directly. Thus, the complexity of performing a multiplication $u \cdot v$ depends on the weight of u. Most elements of H will typically have low weight, e.g., 1. The element ϕ^{-1}, on the other hand, is typically dense, having weight near $m/2$. Often, however, dense elements can be decomposed so that fewer parallel operations are required.

Let us describe a parallel encoder in a little more detail. Let m be the size of the lifting. Input bits are stored in a memory configured as an $L \times m$ array of bits for some L. Each column of m bits can be interpreted as an element of $\mathbb{F}_2[G]$. When memory is accessed, it is accessed a column at a time; thus, in effect, elements of $\mathbb{F}_2[G]$ are read from memory. Multiplication of such a vector by an element of weight 1 in $\mathbb{F}_2[G]$ amounts to performing a permutation from G on the vector. Multiplication of such a vector by an element of weight w of $\mathbb{F}_2[G]$ amounts to performing w permutations on the vector and XORing the results bit-wise. Thus, multiplication by elements of the ring and addition of elements in the ring is easily implemented given a device for performing the permutation associated to G on m-bit vectors and a parallel accumulator for performing m XOR operations in parallel.

Decoding

Message-passing decoding of LDPC codes consists of passing messages along the edges of the graph representing the code and performing computations based on those messages at the nodes of the graph.

Given a "lifted" LDPC code one can vectorize (parallelize) the decoding process as follows. Decode the base LDPC code in an edge serial fashion but, instead of single edges, process m parallel edges and/or nodes at once. Each parallel path requires identical execution. The paths mix only in the permutation within edge clusters.

Invariance

The group structure of matched liftings guarantees certain invariance. Consider the parity-check structure as defined over $\mathbb{F}_2[G]$. If we have

$$Hy = s$$

and $u \in \mathbb{F}_2[G]$ then

$$Hyu = su\,.$$

When $w(u) = 1$ then multiplying a binary vector of length m on the right by u simply permutes the vector. Note that if G is Abelian then the permutation is one of the lifting permutations and if G is not Abelian then the permutation need not be (it corresponds to right multiplication in the group rather than left multiplication). There are m such permutations.

The presence of the group structure in the graph facilitates optimization of the graph. Neighborhood structure is repeated m times so, in examining the graph, one obtains a speed-up of a factor of m.

Product Liftings

Suppose the lifting group is a product group $G_1 \times G_2$. Then the graph lifting may be interpreted as a lifting of G_1, G_2, or $G_1 \times G_2$. This can be quite useful in practice when different terminals in a communications system operate at different speeds or with different throughputs.

For example with $m = 64$ we can use the product of $2 \times 4 \times 8$ and there are then factor groups of size 2, 4, 8, 16, 32, and 64, any of which may be used for parallelism in an implementation. At the same time, the graph description always enjoys the compression of 64 times parallelism.

§7.4. Non-Binary Codes

So far we have been concerned exclusively with binary codes. Theoretically this is all that is needed. Assuming we have mastered the binary case, more general scenarios can typically be dealt with by using binary codes as building blocks in conjunction with suitable mappers as we have discussed in Section 5.4.

In some cases it is simply more natural to apply non-binary codes directly. Also, as we will see now, employing non-binary codes can lead to improved performance even when transmitting over binary channels. However, this improved performance comes at the cost of increased complexity. This section assumes that you are familiar with the basic material on finite fields as it is described in most classical coding books. There are many possible ways of viewing non-binary codes. We could describe them within the framework of matched liftings discussed in the previous section. To keep things simple we start from scratch. Also, as a main difference to the previous section we focus now on the question of an efficient implementation of *non-binary* message-passing schemes, whereas in the previous section we used the structure only to simplify the *scheduling* and processing of the messages while still considering the code a binary code.

Consider transmission using an element of the ensemble LDPC (n, λ, ρ) over $\mathbb{F}_q$. This ensemble is defined in an analogous way as in the binary case except that the non-zero entries of the parity-check matrix H are chosen uniformly at random from $\mathbb{F}_q^*$, the non-zero elements of $\mathbb{F}_q$. Figure. 7.34 shows the Forney-style factor graph (FSFG) of a simple code over $\mathbb{F}_4$ together with its corresponding parity-check matrix. If we are interested in transmission over a BMS channel it is natural to as-

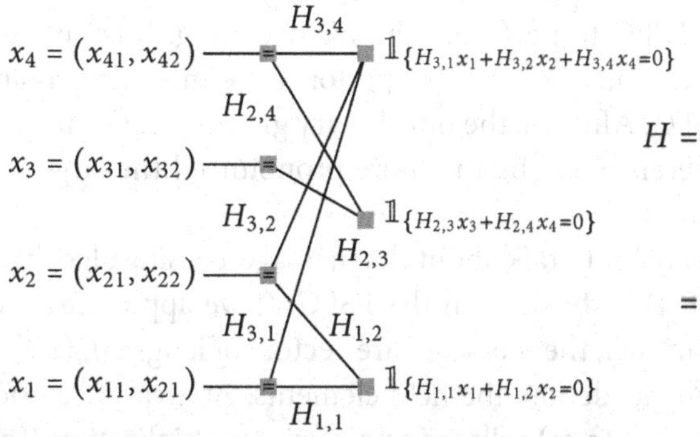

$$H = \begin{pmatrix} H_{1,1} & H_{1,2} & 0 & 0 \\ 0 & 0 & H_{2,3} & H_{2,4} \\ H_{3,1} & H_{3,2} & 0 & H_{3,4} \end{pmatrix} = \begin{pmatrix} 1 & 1+z & 0 & 0 \\ 0 & 0 & z & z \\ 1+z & z & 0 & 1 \end{pmatrix}$$

Figure 7.34: FSFG of a simple code over $\mathbb{F}_4$ and its associated parity-check matrix H. The primitive polynomial generating $\mathbb{F}_4$ is $p(z) = 1 + z + z^2$.

sume that $\mathbb{F}_q = \mathbb{F}_{2^m}$. Let $x = (x_1, \dots, x_n)$ denote the codeword over $\mathbb{F}_{2^m}$. Since we can think of each symbol of $\mathbb{F}_{2^m}$ as a binary m-tuple, we can equivalently think of the codeword as a binary codeword of length nm, $x = (x_{11}, \dots, x_{1m}, \dots, x_{n1}, \dots, x_{nm})$.

In fact, we can also translate the constraints over $\mathbb{F}_{2^m}$ into binary constraints. As

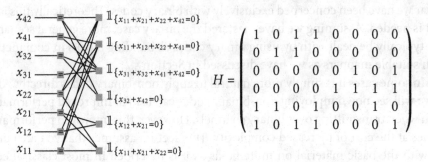

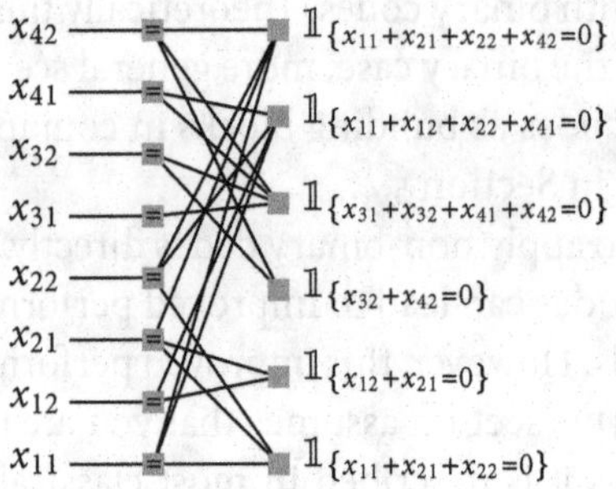

$$H = \begin{pmatrix} 1 & 0 & 1 & 1 & 0 & 0 & 0 & 0 \\ 0 & 1 & 1 & 0 & 0 & 0 & 0 & 0 \\ 0 & 0 & 0 & 0 & 0 & 1 & 0 & 1 \\ 0 & 0 & 0 & 0 & 1 & 1 & 1 & 1 \\ 1 & 1 & 0 & 1 & 0 & 0 & 1 & 0 \\ 1 & 0 & 1 & 1 & 0 & 0 & 0 & 1 \end{pmatrix}$$

Figure 7.35: FSFG of a simple code over $\mathbb{F}_4$ and its associated parity-check matrix H. The primitive polynomial generating $\mathbb{F}_4$ is $p(z) = 1 + z + z^2$.

an example, consider the constraint, $x_1 + (1+z)x_2 = 0$ over $\mathbb{F}_4$, assuming that the primitive polynomial generating $\mathbb{F}_4$ is $p(z) = 1 + z + z^2$. Write $x_1 = x_{11} + x_{12}z$ and $x_2 = x_{21} + x_{22}z$ so that $x_1 + (1+z)x_2 = 0$ becomes $(x_{11}+x_{12}z)+(1+z)(x_{21}+x_{22}z) = 0$. Expand out this expression and use the relationship $z^2 = 1 + z$ which is implied by the primitive polynomial $p(z)$. We get $(x_{11} + x_{21} + x_{22}) + (x_{12} + x_{21})z = 0$. We conclude that the original constraint $x_1 + (1+z)x_2 = 0$ over $\mathbb{F}_4$ is equivalent to the two binary constraints $x_{11} + x_{21} + x_{22} = 0$ and $x_{12} + x_{21} = 0$. The binary FSFG and the binary parity-check matrix are shown in Figure 7.35.

Assume that the binary components are transmitted over a BMS channel and that the received word is $y = (y_{11}, \dots, y_{1m}, \dots, y_{n1}, \dots, y_{nm})$. Given the received word y we can use this binary FSFG to perform iterative decoding. It is not surprising though that we do significantly better if we perform the message-passing algorithm on the non-binary FSFG. After all, the non-binary graph contains in general many fewer loops. The difference will be the more pronounced the larger we choose the field size.

In the sequel it is more convenient to talk about the message-passing algorithm on the standard factor graph (FG) instead of on the FSFG. If we apply message-passing directly on the non-binary FG, the messages are vectors of length q, $(\mu(x = \alpha_0), \dots, \mu(x = \alpha_{q-1}))$, where the α_i denote the field elements. At a variable node the message passing rule expressed in (2.21) calls for a pointwise multiplication. This causes no further headaches and the decoding complexity scales linearly with the alphabet size q.

The next component in the decoder is to take into account the effect of the edge labels. This leads to a permutation of the components of the vector. More precisely, if the edge label is the field element α, then the message vector $(\mu(x = \alpha_0), \dots, \mu(x = \alpha_{q-1}))$ gets permuted to the message vector $(\mu(x = \alpha^{-1}\alpha_0), \dots, \mu(x = \alpha^{-1}\alpha_{q-1}))$.

This is true since the message which relates to a specific field element β corresponds after the multiplication with the label α to the message which corresponds to the field element $\alpha\beta$. This transformation applies in both directions.

Finally, let us consider a check node. To start, consider a check node of degree 3. Let x, y, and z be the connected variables and consider the outgoing message along the edge to z as a function of the incoming messages along the edges of x and y. The outgoing message toward variable z is then

$$\begin{aligned}\mu_z(z=\gamma) &= \sum_{\alpha,\beta\in\mathbb{F}_q} \mathbb{1}_{\{\alpha+\beta+\gamma=0\}}\mu_x(x=\alpha)\mu_y(y=\beta)\\ &= \sum_{\alpha\in\mathbb{F}_q} \mu_x(x=\alpha)\mu_y(y=-\gamma-\alpha). \end{aligned} \tag{7.36}$$

In a brute force manner, the preceding evaluation (for all values $\gamma\in\mathbb{F}_q$) can be accomplished with complexity $O(q^2)$. But one can do much better. Note that $\mu_z(z=-\gamma)$ is given in terms of a *convolution* of two message vectors, where the index calculations are done with respect to the additive group of $\mathbb{F}_q$. It should therefore not be surprising that we can use Fourier transforms to accomplish this convolution in an efficient manner. Since $q=p^m$ for some prime p and natural number m, the Fourier transform is particularly simple. Let ξ be a p-th root of unity, e.g., $\xi=e^{2\pi j/p}$. Write an element of $\mathbb{F}_q$ as an m-tuple with components in $\mathbb{F}_p$, $\alpha=(\alpha_1,\dots,\alpha_m)$. Let $x=(x_\alpha)_{\alpha\in\mathbb{F}_p}$, denote a vector, whose components are indexed by the elements of $\mathbb{F}_p$, taking values in $\mathbb{C}$, and let $X=(X_\alpha)_{\alpha\in\mathbb{F}_p}$ denote its Fourier transform. The Fourier transform pair is

$$X_\alpha = \sum_\beta x_\beta \xi^{-\alpha\beta^T}, \qquad x_\alpha = \frac{1}{p^m}\sum_\beta X_\beta\xi^{\alpha\beta^T}.$$

In words, arrange the input vector in an m-dimensional box of side length p according to $\alpha=(\alpha_1,\dots,\alpha_m)$ (the i-th component α_i gives the position in the i-th dimension). The Fourier transform is then the conventional m-dimensional Fourier transform, taking the standard cyclic Fourier transform of length p along each dimension. We see that we can accomplish the convolution in (7.36) by taking the Fourier transform of both $\mu_x(x=\alpha)$ and $\mu_y(y=\alpha)$, multiplying the result, and taking the inverse Fourier transform. The result of this operation equals $\mu_z(z=-\gamma)$. This means that we need to rearrange the vector to get $\mu_z(z=-\gamma)$ (if the field has characteristic 2 we are done since in this case $\gamma=-\gamma$.)

Example 7.37 (Fourier Transform over $\mathbb{F}_4$). The case $\mathbb{F}=\mathbb{F}_{2^m}$ is particularly simple since then $\xi=-1$. Consider, e.g., the Fourier transform over $\mathbb{F}_4$. Arrange the input vector $(x_{(0,0)},x_{(1,0)},x_{(0,1)},x_{(1,1)})$ into a square box of side length 2:

$$\begin{pmatrix} x_{(0,0)} & x_{(0,1)} \\ x_{(1,0)} & x_{(1,1)} \end{pmatrix}.$$

Running a two-point Fourier transform on each row results in

$$\begin{pmatrix} x_{(0,0)} + x_{(0,1)} & x_{(0,0)} - x_{(0,1)} \\ x_{(1,0)} + x_{(1,1)} & x_{(1,0)} - x_{(1,1)} \end{pmatrix}.$$

We get the desired Fourier transform by next running a two-point Fourier transform on each column:

$$\begin{pmatrix} x_{(0,0)} + x_{(0,1)} + x_{(1,0)} + x_{(1,1)} & x_{(0,0)} - x_{(0,1)} + x_{(1,0)} - x_{(1,1)} \\ x_{(0,0)} + x_{(0,1)} - x_{(1,0)} - x_{(1,1)} & x_{(0,0)} - x_{(0,1)} - x_{(1,0)} + x_{(1,1)} \end{pmatrix}.$$

Figure 7.38 shows the performance of the $(2,3)$-regular ensemble over $\mathbb{F}_{2^m}$, for

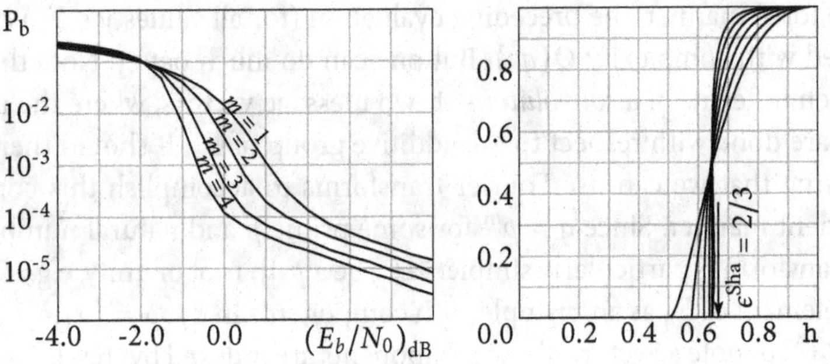

Figure 7.38: Left: Performance of the $(2,3)$-regular ensemble over $\mathbb{F}_{2^m}$, $m = 1,2,3,4$ of binary length 4320 over the BAWGNC(σ). Right: EXIT curves for the $(2,3)$-regular ensembles over $\mathbb{F}_{2^m}$ for $m = 1,2,3,4,5,6$, and transmission over the BEC(ϵ).

$m = 1,2,3,4$, over the BAWGNC(σ). In order to arrive at a fair comparison the *binary* length of each code was fixed to 4320. We see that the performance improves dramatically as the alphabet size is increased. From this figure one might get the impression that increasing the alphabet size always results in increased performance. This is unfortunately not the case. For the $(3,6)$-regular ensemble the performance gets strictly worse if we go to a non-binary alphabet.

Let us now consider the analysis. It is the task of Problem 7.2 to show that if the transmission takes place over a BMS channel then one can assume the all-zero codeword for the analysis. For the case of transmission over the binary erasure channel (BEC) the density evolution equations can again be written down in analytic form. Consider the result of the density evolution analysis for, e.g., the cycle-code ensemble LDPC (n, x, x^2) of rate one-third. In the binary case the threshold is given by the stability condition and is equal to one-half. Table 7.39 shows the evolution of this threshold as a function of the alphabet size. Also shown are the thresholds that

m	ϵ^{BP}	ϵ^{MAP}	ϵ^{Sta}
1	0.5000	0.5000	0.5000
2	0.5775	0.5775	0.5811
3	0.6183	0.6209	0.6510
4	0.6369	0.6426	0.7075
5	0.6446	0.6540	0.7518

Table 7.39: Thresholds for the $(2,3)$-regular ensemble over $\mathbb{F}_{2^m}$ for $m = 1,2,3,4,5$ for transmission over the BEC(ϵ).

one would get from the stability condition which is discussed in the following. As we can see, the thresholds improve significantly and, for higher alphabet sizes, they are no longer given by the stability condition. The EXIT curves of the overall code for various choices of m are shown on the right-hand side in Figure 7.38. The maximum a posteriori (MAP) threshold was computed according to the principle of Theorem 3.120. As we can see from these data, the MAP threshold seems to converge to the Shannon threshold at the speed $O(2^{-m})$. The BP threshold, on the other hand, first quickly increases toward the Shannon threshold but it actually peaks for some value of m and later starts decreasing again. For example, the BP threshold for $m = 15$ is only 0.616.

For the general case, an analysis in terms of density evolution is in principle possible but practically difficult. Even for codes over $\mathbb{F}_4$ densities already "live" in $\mathbb{R}^3$. The BP threshold can still be computed numerically by Monte Carlo methods in the same way as is done for computing the density evolution for convolutional codes.

Analytically, we can determine at least the stability condition. You are asked in Problem 7.3 to verify that the generating function counting the number of minimal codewords in the limit of infinite blocklengths is equal to

$$\tilde{P}(x) = -\frac{1}{2}\log\left(1 - \lambda'(0)\rho'(1)\frac{(1+x)^m - 1}{2^m - 1}\right). \tag{7.40}$$

Arguing as in the case of turbo codes (see Section 6.6) we arrive at the following conjecture.

Conjecture 7.41 (Stability Condition for Non-Binary Ensembles). Consider the ensemble LDPC (n, λ, ρ). Assume that transmission takes place over a BMS channel with L-density a and associated Bhattacharyya constant $\mathfrak{B}(\mathsf{a})$. Then the fixed point corresponding to correct decoding is stable if and only if

$$\lambda'(0)\rho'(1)\frac{(1+\mathfrak{B}(\mathsf{a}))^m - 1}{2^m - 1} < 1.$$

§7.5. Low-Density Generator Codes and Rateless Codes

§7.5.1. Low-Density Generator Codes

Most of this book is dedicated to low-density *parity-check* codes. Why have we not considered low-density *generator-matrix* (LDGM) codes so far, i.e., codes that are defined by a random sparse *generator* matrix? An immediate answer is that LDGM codes necessarily exhibit large error floors. To be more precise, assume that all rows of G have degree d and that the codeword X is generated by $X = UG$, where U denotes the information word. Assume further that X is sent over a BMS channel with L-density a_{BMSC}. We want to find U given Y, the observation at the channel output. Without loss of generality assume that we want to estimate U_1 and that $X_1, \dots, X_d$ are the d codebits connected to information bit U_1. We get a lower bound on the probability of error of estimating U_1 if we assume that $U_{\sim 1}$ is known perfectly and that in addition we have the observations $Y_1, \dots, Y_d$. If the code is proper (so that none of the components of the codeword are known a priori with certainty) we get the lower bound $\mathfrak{E}(\mathsf{a}_{\text{BMSC}}^{\circledast d})$.

So why would we nevertheless discuss LDGM codes? As we will see in the next section, the error-floor problem is easily solved if we apply *precoding* techniques, i.e., if we pick the information word to be itself an element of a (very-high-rate) code which can be used to "clean-up" the error floor. More importantly, as we will discuss shortly, by choosing a specific degree distribution we can use LDGM codes in a *rateless* fashion in a way that allows a receiver to accumulate simultaneously information from a variety of uncoordinated transmitters. This degree of freedom allows many new transmission scenarios which cannot be addressed with standard codes.

§7.5.2. Analysis of LDGM Ensembles

In Example 7.13 we introduced LDGM ensembles as a special case of multi-edge-type LDPC ensembles. We gave expressions for the generic degree distribution and for density evolution. Rather than proceeding along these lines, it is more convenient to revert to a more traditional notation and to introduce a standard degree distribution pair (λ, ρ), where $\lambda(x)$ is the degree distribution (from an edge perspective) corresponding to the information bits and $\rho(x)$ is the degree distribution (from an edge perspective) corresponding to the generator nodes *counting only the edges toward the information variable nodes.* For the example shown in Figure 7.15 we have

$$\lambda(x) = x^2, \qquad \rho(x) = \frac{1}{4} + \frac{1}{2}x + \frac{1}{4}x^2.$$

The design rate of the ensemble is $r = \frac{\int\lambda}{\int\rho}$. For our specific example we get $r = \frac{\frac{1}{3}}{\frac{1}{4}+\frac{1}{4}+\frac{1}{12}} = \frac{4}{7}$. Recall that all information bits are punctured and that the codebits are sent through a BMS channel with L-density a_{BMSC}. Let a_ℓ denote the variable-to-generator density emitted in the ℓ-th iteration and let b_ℓ denote the generator-to-variable density emitted in the ℓ-th iteration. Then we have $b_0 = \Delta_0$, and for $\ell \geq 1$

$$\mathsf{a}_\ell = \lambda(\mathsf{b}_{\ell-1}), \qquad \mathsf{b}_\ell = \mathsf{a}_{\text{BMSC}} \boxast \rho(\mathsf{a}_\ell),$$

where $\lambda(\mathsf{b}) = \sum_i \lambda_i \mathsf{b}^{\circledast(i-1)}$ and $\rho(\mathsf{a}) = \sum_i \rho_i \mathsf{a}^{\boxast(i-1)}$. In summary, $\mathsf{b}_\ell = \mathsf{a}_{\text{BMSC}} \boxast \rho(\lambda(\mathsf{b}_{\ell-1}))$. Remark: If you look at the equations carefully you will see that the process never gets started; i.e., all densities are Δ_0. We will soon discuss this in more detail and see how to eliminate this problem.

Example 7.42 (Density Evolution for BEC). Let x_ℓ denote the erasure probability of the variable-to-generator messages and let y_ℓ denote the erasure probability of the generator-to-variable messages. Then we have $y_0 = 1$, and for $\ell \geq 1$

$$x_\ell = \lambda(y_{\ell-1}), \qquad y_\ell = 1 - (1-\epsilon)\rho(1 - x_\ell),$$

so that $y_\ell = 1 - (1-\epsilon)\rho(1 - \lambda(y_{\ell-1}))$.

Assume that we make the following change of variables: we exchange the roles of λ and ρ, and we let $\epsilon \mapsto 1-\epsilon$, $x \mapsto 1-y$, and $y \mapsto 1-x$. Then the preceding equations are transformed to $x_0 = 0$, and for $\ell \geq 1$

$$y_\ell = 1 - \rho(1 - x_{\ell-1}), \qquad x_\ell = \epsilon\lambda(y_\ell),$$

so that $x_\ell = \epsilon\lambda(1 - \rho(1 - x_{\ell-1}))$. These are the standard density evolution equations except that here $x_0 = 0$ instead of $x_0 = \epsilon$, reflecting the fact that for LDGM ensembles the recursion proceeds "backwards." The close connection is not really surprising: we know that a generator matrix is the dual of a parity-check matrix and we have seen in Theorem 3.77 that the dual of the channel BEC(ϵ) (in the sense of going from the L-distribution to the G-distribution) is the channel BEC($1-\epsilon$).

Given this observation we might hope that the design of good degree distributions for LDGM ensembles amounts to not much more than translating results we have obtained for LDPC ensembles. This is unfortunately not quite true, although, as we will see shortly, at least when it comes to capacity-achieving ensembles we do not have to start from scratch. To see what goes wrong note that the standard fixed point condition for LDPC ensembles

$$x \geq \epsilon\lambda(1 - \rho(1-x)), \quad x \in (0,1),$$

translates via the previous transformation into the condition

$$y \leq 1-(1-\epsilon)\rho(1-\lambda(y)), \quad y \in (0,1).$$

Unfortunately, the inequality in the condition goes in the wrong direction. ◇

EXIT CHARTS

Let us visualize density evolution in terms of EXIT charts. To simplify matters we focus on the case of transmission over the BEC. To apply EXIT charts to general BMS channels just follow the steps outlined in Section 3.14.

Consider a degree distribution pair (λ, ρ). As we have just discussed, the condition for successful decoding (in the asymptotic limit) reads $y \geq 1-(1-\epsilon)\rho(1-\lambda(y))$, $y \in (0,1)$. Recall that $\rho(x)$ has an inverse (since it is a power series with non-negative coefficients) so that we can rewrite this condition as $1-\rho^{-1}((1-y)/(1-\epsilon)) \geq \lambda(y)$, $y \in (0,1)$. Here, the left-hand side represents the "action" of the generator nodes, whereas the right-hand side represents the "action" of the variable nodes.

EXAMPLE 7.43 (EXIT CHART). Pick $\lambda(x) = \frac{1}{2}x^4 + \frac{1}{2}x^5$ and $\rho(x) = \frac{2}{5}x + \frac{1}{5}x^2 + \frac{2}{5}x^8$. The design rate of this ensemble is $\frac{\int \lambda}{\int \rho} = \frac{33}{56} \approx 0.59$. Define $v(y) = \lambda(y)$ and $g_\epsilon^{-1}(y) = 1-\rho^{-1}((1-y)/(1-\epsilon))$. Figure 7.44 shows v and g_ϵ^{-1} where $\epsilon = 0.35$. A few remarks

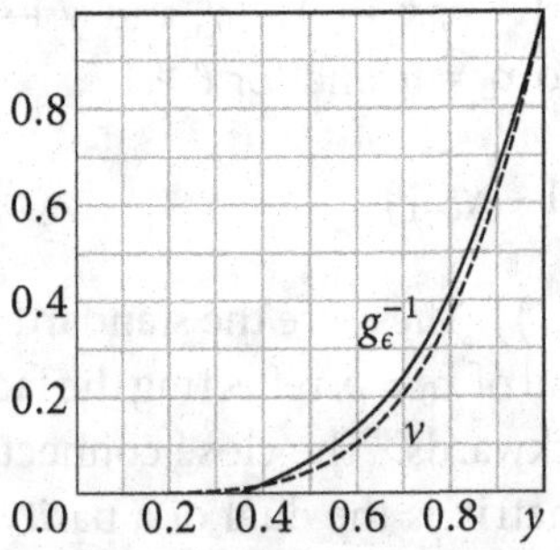

Figure 7.44: EXIT chart for the LDGM ensemble with $\lambda(x) = \frac{1}{2}x^4 + \frac{1}{2}x^5$ and $\rho(x) = \frac{2}{5}x + \frac{1}{5}x^2 + \frac{2}{5}x^8$ and transmission over the BEC($\epsilon = 0.35$).

are in order. First consider the start of the decoding process. This corresponds to the right- and topmost point on the plot. As we can see, without some help the process never starts since $y = 1$ is a fixed point of density evolution. This phenomenon is true for all LDGM ensembles that do not have degree 1 generator nodes. Fortunately it is easily overcome by leaving a strictly positive but arbitrarily small fraction of the information bits unpunctured.

Next note that close to $y = 1$ the "tunnel" between the two curves initially widens. For $y = 1$ the derivative of the curve $v(y)$ equals $\lambda'(1)$, whereas the derivative of the curve $g_\epsilon^{-1}(y)$ is seen to be equal to $1/((1-\epsilon)\rho'(0))$. Since $v(1) = 1 = g_\epsilon^{-1}(1)$ the condition for an initial widening of the tunnel reads $(g_\epsilon^{-1})'(1) < v'(1)$ which translates to $(1-\epsilon)\lambda'(1)\rho'(0) > 1$. We will see in Section 7.5.2 that this is the *stability condition* for the special case of transmission over the BEC.

Finally, observe that the two curves cross again at some point. This is not due to a bad choice of the degree distribution: a little thought shows that a crossing must occur for some $0 < y < 1$. To see this, first note that $v(y) = \lambda(y)$ is strictly positive for all $y > 0$. On the other hand, a direct check shows that $g_\epsilon(\epsilon) = 0$, so that there must be a crossing for some $y \geq \epsilon$. By a proper choice of the degree distributions, one can make the probability of error of the final decision very small. ◇

Capacity-Achieving Degree Distributions for the BEC

When we consider capacity-achieving LDGM ensembles we can reuse our knowledge gathered during the construction of capacity-achieving LDPC ensembles: the starting point is a pair of (un-normalized) degree distributions which fulfill density evolution with equality (so that the dual equations are also fulfilled with equality).

Lemma 7.45 (Capacity-Achieving LDGM Ensembles). Consider transmission over the BEC(ϵ). Let $N \in \mathbb{N}$ and $\mu > 0$ and set

$$\alpha = \frac{\mu + H(N) + 1}{1-\epsilon}. \tag{7.46}$$

Then, in the limit of infinite blocklengths, the ensemble LDGM $\left(n, \lambda_\alpha(x), \rho^{(N)}(x)\right)$, where

$$\lambda_\alpha(x) = \mathrm{e}^{-\alpha(1-x)}, \qquad \rho^{(N)}(x) = \frac{\mu + \sum_{i=2}^{N} \frac{x^{i-1}}{(i-1)} + \frac{(N+1)x^N}{N}}{\mu + H(N) + 1},$$

enables transmission at rate $r = (1-\epsilon)\frac{1-\mathrm{e}^{-\alpha}}{1+\mu}$ with a bit error probability not exceeding $\frac{1}{N-1}$.

Proof. We start with the heavy-tail Poisson distribution from Example 3.90,

$$\lambda_\alpha(x) = \mathrm{e}^{-\alpha(1-x)}, \qquad \rho(x) = -\ln(1-x) = \sum_{i=1}^{\infty} \frac{x^i}{i},$$

where we have exchanged the roles of λ and ρ as compared to the original example.

The variable degree distribution is already normalized (but of course it contains degrees of all orders), but the generator degree distribution has infinite mean. Let us therefore consider the modified degree distribution

$$\rho^{(N)}(x) = \frac{\mu + \sum_{i=2}^{N} \frac{x^{i-1}}{(i-1)} + \frac{(N+1)x^N}{N}}{\mu + H(N) + 1},$$

where $H(N) = \sum_{i=1}^{N} \frac{1}{i}$, the N-th harmonic number. This is a well-defined degree distribution since all coefficients are non-negative and $\rho^{(N)}(1) = 1$. There are two modifications with respect to the original degree distribution. First, the degree of $\rho^{(N)}(x)$ is bounded. Second, $\rho^{(N)}(x)$ contains a small fraction of degree 1 nodes. This second modification is related to the stability problem at the beginning of the decoding process. The design rate is quickly determined to be

$$r = \frac{\int \lambda}{\int \rho} = \frac{(\mu + H(N) + 1)(1 - e^{-\alpha})}{\alpha(\mu + 1)} \overset{(7.46)}{=} (1 - \epsilon)\frac{1 - e^{-\alpha}}{\mu + 1}.$$

Therefore, if we let μ tend to zero and let N tend to infinity (so that $1 - e^{-\alpha}$ tends to 1), then we can achieve rates arbitrarily close to capacity. Note that if $0 \le x < 1 - \frac{1}{N}$, then

$$\sum_{i=N+1}^{\infty} \frac{x^i}{i} < \frac{1}{N+1} \sum_{i=N+1}^{\infty} x^i = \frac{1}{N+1} \frac{x^{N+1}}{1-x} < x^N,$$

where the last step follows explicitly by calculations, taking into account the given range of x. It follows that for $0 \le x < 1 - \frac{1}{N}$

$$\rho^{(N)}(x) = \frac{\mu - \ln(1-x) + x^N - \sum_{i=N+1}^{\infty} \frac{x^i}{i}}{\mu + H(N) + 1} > \frac{\mu - \ln(1-x)}{\mu + H(N) + 1}.$$

Write the density evolution equations in terms of x (the erasure probability from the variable to the generator nodes) instead of y. We have for $\frac{1}{N} \le x \le 1$

$$\lambda_\alpha(1 - (1-\epsilon)\rho^{(N)}(1-x)) = e^{-\alpha(1-\epsilon)\rho^{(N)}(1-x)} \le e^{-\alpha(1-\epsilon)\frac{\mu - \ln(x)}{\mu + H(N) + 1}}$$
$$\overset{(7.46)}{=} e^{-\mu} x < x.$$

We see that the density evolution equations are fulfilled in this range.

It remains to check the implied bound on the probability of bit erasure. The normalized (to one) degree distribution from the node perspective of the variable nodes is

$$\frac{\lambda_\alpha(x) - e^{-\alpha}}{1 - e^{-\alpha}} \overset{(7.46)}{\le} \frac{N}{N-1} \lambda_\alpha(x).$$

The claim now follows by observing that the density evolution recursion is valid at least until $\lambda_\alpha(y) \le \frac{1}{N}$. □

Stability Condition

Let us consider the stability condition for LDGM ensembles. Whereas the stability condition for LDPC ensembles concerns the end of the decoding process, the stability condition for LDGM ensembles relates to the beginning of the decoding process. To render the system stable at the beginning we have essentially two options. We can either "move" the curves, e.g., if we add some degree 1 generator nodes the curve g_ϵ^{-1} no longer goes through the point $(1,1)$ and the system is stable at the beginning. The other option is to adjust the "slopes" of the two EXIT curves. A real system likely involves a combination of the two methods – add some degree 1 nodes to ensure that the system gets started and control the "slopes" to ensure that the two EXIT curves do not cross right away. The following theorem is concerned only with the slopes; i.e., we assume that there are no degree 1 generator nodes.

Theorem 7.47 (Stability Condition for LDGM Ensembles). Assume we are given a degree distribution pair (λ, ρ), where ρ contains no degree 1 nodes, and assume that transmission takes place over a BMS channel characterized by its L-density a_{BMSC} with D-mean $\mathfrak{D}(\mathsf{a}_{\text{BMSC}})$ (see Definition 4.56).

Let a_0 be a symmetric L-density, and for $\ell \geq 1$, define $\mathsf{a}_\ell = \mathsf{a}_{\text{BMSC}} \circledast \rho(\lambda(\mathsf{a}_{\ell-1}))$. Assume that $\mathfrak{D}(\mathsf{a}_{\text{BMSC}})\lambda'(1)\rho'(0) < 1$. Then there exists a strictly positive constant $\xi = \xi(\lambda, \rho, \mathsf{a}_{\text{BMSC}})$ such that if $\mathfrak{E}(\mathsf{a}_0) \geq \frac{1}{2} - \xi$ then $\mathfrak{E}(\mathsf{a}_\ell)$ converges to $\frac{1}{2}$ for ℓ tending to infinity.

If $\mathfrak{D}(\mathsf{a}_{\text{BMSC}})\lambda'(1)\rho'(0) > 1$, then there exists a strictly positive constant $\xi = \xi(\lambda, \rho, \mathsf{a}_{\text{BMSC}})$ such that if $\mathfrak{E}(\mathsf{a}_0) < \frac{1}{2}$ then $\liminf_{\ell\to\infty} \mathfrak{E}(\mathsf{a}_\ell) \leq \frac{1}{2} - \xi$.

Proof. Recall that the Bhattacharyya constant played a key role in the derivation of the stability condition for LDPC ensembles. The equivalent role for LDGM ensembles is played by the D-mean $\mathfrak{D}(\mathsf{a}_{\text{BMSC}})$. Recall from Lemma 4.60 that the D-mean is multiplicative at check nodes. Further, in Problem 4.62 we saw that for small values of the D-mean we have tight upper and lower bounds on the D-mean at the output of a variable node. Let a_0 be a symmetric L-density at the output of the variable nodes and consider the evolution of the density under density evolution, where a_ℓ denotes the densities at the output of the variable nodes and b_ℓ denotes the densities at the output of the generator nodes. Let $x_\ell = \mathfrak{D}(\mathsf{a}_\ell)$ and $y_\ell = \mathfrak{D}(\mathsf{b}_\ell)$.

Then by the multiplicativity of the D-densities at check nodes and the bounds at the variable nodes we have

$$y_\ell = \rho(x_{\ell-1})\,\mathfrak{D}(\mathsf{a}_{\text{BMSC}}), \ell \geq 1,$$

$$\frac{1}{2}(1-\lambda(1-2y_\ell)) \leq x_\ell \leq 1-\lambda(1-y_\ell). \tag{7.48}$$

This implies that

$$x_\ell \leq 1-\lambda(1-\rho(x_{\ell-1})\,\mathfrak{D}(\mathsf{a}_{\text{BMSC}})) = \lambda'(1)\rho'(0)\,\mathfrak{D}(\mathsf{a}_{\text{BMSC}})x_{\ell-1} + O(x_{\ell-1}^2).$$

Therefore, if $\lambda'(1)\rho'(0)\,\mathfrak{D}(\mathsf{a}_{\text{BMSC}}) < 1$ then there exits a strictly positive constant ξ' so that if $x_0 = \mathfrak{D}(\mathsf{a}_0) < \xi'$ then $\lim_{\ell\to\infty} x_\ell = \lim_{\ell\to\infty} \mathfrak{D}(\mathsf{a}_\ell) = 0$. For any strictly positive ξ' there exists a strictly positive ξ so that if $\mathfrak{D}(\mathsf{a}_0) < \xi'$ then $\mathfrak{E}(\mathsf{a}_0) > \frac{1}{2} - \xi$. Combining the two preceding statements shows that $\mathfrak{E}(\mathsf{a}_\ell)$ converges to $\frac{1}{2}$.

The converse follows by essentially the same argument if we use the lower bound in (7.48) instead of the upper bound. We skip the details. □

§7.5.3. Rateless Codes: LT-Codes and Raptor Codes

Assume that we make the special choice $\lambda(x) = e^{\mathtt{l}_{\text{avg}}(1-x)}$; i.e., we pick the variable degree distribution to be the Poisson one. Strictly speaking this degree distribution falls outside the validity of our analysis since $\lambda(x)$ is not bounded. But we will take a short-cut and simply assume that our analysis remains valid. In the Notes we list some references in which you can find a more careful analysis. The final results agree.

Consider the ensemble LDGM $\left(m, \lambda(x) = e^{\mathtt{l}_{\text{avg}}(1-x)}, \rho\right)$. By our standard assumption elements of the ensemble are created by first generating m variable and n generator nodes (both with the right degree distribution) with labeled sockets. As a second step the connections are chosen according to a random permutation on $\Lambda'(1)$ elements. To create a codeword, we choose the source word u and create from this the codeword w.

We get to the punch line. Rather than creating a codeword by first creating the whole code and then performing the encoding, we can create for a given source word u a code and the resulting codeword w one bit at a time. Further, we do not need to decide on the length of the codeword in advance but we simply continue the procedure as long as we please.

Let us look at this procedure in detail. We are given the length m, a source word u of length m, and the desired degree distribution pair (λ, ρ), where λ is a Poisson with average degree $\mathtt{l}_{\text{avg}}$. This average degree $\mathtt{l}_{\text{avg}}$ is only defined a posteriori once we decide how many codebits we generate from the given source bits. To create the k-th bit of the codeword proceed as follows. Pick a generator degree according to the distribution $\frac{\rho_i}{i \sum_j \rho_j/j}$. If the chosen degree is d pick a d-tuple of bits from u. This can be done either by picking one of the $\binom{n}{d}$ distinct d-tuples or by picking d elements of the m-tuple u with repetitions. For any reasonable length m the difference between these two models is negligible and we will therefore ignore it in the sequel. Now encode your choice of positions $(i_1, \dots, i_d)$ and send this information together with the resulting k-th bit of the codeword $w_k = u_{i_1} + \cdots + u_{i_d}$ to the receiver.

There are several obvious questions that arise. First, from the preceding description it seems doubtful that the scheme is of much use since for every bit we need to convey the choice of d-tuple. This requires on the order of $\mathtt{l}_{\text{avg}} \log(m)$ bits in average. This problem is easily fixed by, e.g., breaking up the data into many blocks and encoding several blocks *with the same choice of combinations.* In this case the overhead of $\mathtt{l}_{\text{avg}} \log(m)$ is amortized over many codewords and by choosing the number of blocks large enough we can make the overhead negligible. Alternatively, sender and receiver can use a shared source of randomness or use pseudorandom generators.

The second question concerns the relationship between this ensemble and the ensemble LDGM $\left(m, \lambda(x) = \mathtt{e}^{\mathtt{l}_{\text{avg}}(1-x)}, \rho\right)$. Since for every codebit we choose the degree *independently* according to the node-degree distribution we see that for increasing blocklengths the generator degree distribution converges to $\rho(x)$. Further, the variable-node degree distribution is the result of throwing $n\mathtt{r}_{\text{avg}}$ balls into m boxes, where the choices are independent and where $1/\mathtt{r}_{\text{avg}} = \int \rho$. On average there are therefore $\frac{n}{m}\mathtt{r}_{\text{avg}}$ balls in each box. In other words, $\mathtt{l}_{\text{avg}} = \frac{n}{m}\mathtt{r}_{\text{avg}}$. Since the choice of each box is made independently, the resulting degree distribution converges to a Poisson distribution with mean $\mathtt{l}_{\text{avg}}$.

If we choose $\mathtt{l}_{\text{avg}}$ to be a constant (lets say $\mathtt{l}_{\text{avg}} = 4$) and increase the blocklength m then a given source bit is not covered by any of the linear combinations with positive probability. For m increasing to infinity, this probability converges to the probability that a Poisson random variable with mean $\mathtt{l}_{\text{avg}}$ takes on the value 0; i.e., it converges to $\mathtt{e}^{-\mathtt{l}_{\text{avg}}}$. This implies a high error floor. One solution is to choose $\mathtt{l}_{\text{avg}}$ as a function of n. A necessary and sufficient growth is $\mathtt{l}_{\text{avg}}(n) = c\log(n)$, where c is a strictly positive constant. Codes generated in this way are called *LT* codes and they represent the historically first class of rateless codes. It is hopefully clear at this point why the codes are called *rateless*. The rate of the code is not specified a priori but is only fixed once the transmission for a particular block stops. Different data blocks encoded by the same procedure might end up being transmitted using a different code rate.

An even better solution is to keep $\mathtt{l}_{\text{avg}}$ constant but to *precode* the information vector u (e.g., with an LDPC code). If u denotes the original information vector and if $\tilde{u}$ is the result of the precoding then we proceed as follows. We first try to recover $\tilde{u}$ using the LDGM code. Since $\mathtt{l}_{\text{avg}}$ is a constant we know that there will be a fraction of the bits in $\tilde{u}$ that we cannot recover. By choosing $\mathtt{l}_{\text{avg}}$ sufficiently large we can make this unrecovered fraction as small as desired. We can now use the error-correcting capability of the precode to "clean up" the error floor. For the specific instance in which the precoding is done using an LDPC code the resulting rateless codes are called *Raptor* codes.

The real power of rateless codes appears once we realize that it is not necessary that the information is received from the *same* transmitter. Consider a broadcasting scenario where a large amount of data needs to be broadcast to a large set of users that "see" channels of highly varying conditions. We can imagine scenarios in which many transmitters (maybe satellites) broadcast the same information in a rateless fashion. The receiver can then accumulate information from any of the transmitters that are "within reach." The receiver can even switch on and off as he wishes. The only condition for success at the end is that the total amount of information that he has received suffices for decoding. Since the equations (codebits) generated at different transmitters are independent the received information is additive at the receiver.

Notes

Multi-edge-type LDPC ensembles appeared in the paper by Richardson and Urbanke [54]. The definition and parameterization of these ensembles (Section 7.1.2) as well as their asymptotic analysis (Section 7.2) is taken from this paper. The proof of the stability condition follows closely the proof for standard irregular LDPC ensembles which was given by Richardson and Urbanke [53].

RA codes were introduced by Divsalar, Jin, and McEliece [11] as a slimmed-down version of *turbo codes* to facilitate their analysis. (Chains of nodes of degree 2 appear also in the construction of Kanter and Saad [26].) Surprisingly, despite their simple structure these codes themselves are quite good and their basic structure in turn has been generalized ad infinitum. If you are interested in finding out more on the cycle distribution of random permutations a good reference is Wilf [70]. We have discussed *IRA* codes introduced by Jin, Khandekar, and McEliece [24]. In the same paper it was shown that, at least for the BEC, this class of ensembles is sufficiently powerful to achieve capacity. It was shown by Pfister, Sason, and Urbanke [49] that by a proper selection of the degree distribution the capacity of the BEC can be achieved with *bounded complexity*, i.e., a fixed number of computations per information bit. This is in contrast to standard LDPC ensembles where we proved in Theorem 3.85 that to achieve capacity the number of required operations must tend to infinity at least as the log of the gap to capacity. *ARA* codes were invented by Abbasfar, Divsalar, and Kung [1, 2]. It was shown by Pfister and Sason [48] that their asymptotic analysis can be reduced to the asymptotic analysis of IRA ensembles. The chain of generalizations of RA codes does not stop here. To mention just one such generalization which we have not discussed in this book, there are *accumulate-repeat-accumulate-accumulate* codes (see Divsalar, Dolinar, and Thorpe [10]).

Let us mention a few further classes of codes. Concatenated tree codes were proposed by Ping [50]. Boutros, Pothier, and Zémor showed that a random graph whose local codes are Hamming codes are asymptotically good and they simulated the performance of such codes under iterative decoding [6]. The same result was shown by Lentmaier and Zigangirov [33]. More general, it was shown by Barg and Zémor that random graphs with component codes whose minimum distance is at least 3 are asymptotically good [4]. A fairly general construction using *base codes* was put forth by Tillich [63] with follow-up work by Andriyanova, Tillich, and Carlach [3]. This construction contains as special cases many of the codes discussed in this book. Low-density convolutional codes were introduced by Engdahl and Zigangirov [15]. In a similar spirit, braided block codes where proposed by Truhachev, Lentmaier, and Zigangirov [64].

Low-density generator codes were already known to Gallager [19]. At first, these codes appeared more as a mathematical curiosity. But with the advent of *rateless* codes and the interest in the source-coding problem this class has become of much interest on its own.

MN codes were introduced by MacKay and Neal [41]. This is also the origin of the name MN. There are several variations on the theme and we only discussed one particular flavor. Properties of these codes were discussed by MacKay [39] as well as Kanter and Saad [26]. The same basic structure is not only of interest for the channel coding problem but seems also well suited for the source-coding problem (or a combination thereof). We refer the reader to the work of Martinian and Wainwright [42, 43], as well as Dimakis, Wainwright, and Ramchandran [9].

The idea of using a lifting for LDPC codes was first presented by Tanner [61]. A patent was granted to Richardson [52]. An important aspect of lifted codes is that one performs the encoding operation essentially on the base graph by assuming that encoding takes place over $\mathbb{F}_2[G]$ rather than as binary arithmetic. Very similar to liftings is the idea of *protographs* put forward by Thorpe, Andrews, and Dolinar [62]. Protographs with linear minimum distance were constructed by Divsalar, Jones, Dolinar, and Thorpe [12]. The idea of lifting is also strongly connected to a *graph cover*, a standard concept in graph theory. Such graph covers in turn play an important role in the analysis of *pseudo codewords* as was pointed out by Kötter and Vontobel [31]. Also the class of non-binary ensembles as discussed in Section 7.4 can be seen as a lifting of the underlying binary graph.

A good reference for Ω-networks, or more generally, non-blocking switching networks is the book by Hwang [23] (see also Knuth [30]).

All ensembles discussed so far are based on a small number of constraints (degree distributions and edge types) and a large amount of randomness. On the other end of the spectrum there are constructions of sparse graph codes which contain no or little randomness. The first such constructions are due to Tanner [61]. Codes based on finite-geometries were discussed by Kou, Lin, and Fossorier [32]. Combinatorial code designs were put forth by Vasic and Milenkovic [66], Djordjevic and Vasic [13, 14], as well as Vasic, Kurtas, and Kuznetsov [65]. Other constructions are by Smarandache and Vontobel [58], Greferath, O'Sullivan, and Smarandache [21], and Sridhara, Kelley, and Rosenthal [60].

The aforementioned ensembles contain many good codes. The main reason we have excluded these ensembles from our description is that with the methods described in this book it seems difficult to analyze the behavior of these codes under message-passing decoding. An analysis under MAP decoding is of course always possible using weight distributions, but this is of less interest in our context. As a first logical step one can investigate the pseudo-codeword distribution. This is the appropriate measure under linear-programming decoding but it also has some relevance to message-passing decoding as was shown by Vontobel and Kötter [68]. Such investigations have been undertaken by Kashyap and Vardy [28]; Vontobel and Kötter [67]; Kelley, Sridhara, Xu, and Rosenthal [29]; Vontobel, Smarandache, Kiyavash, Teutsch, and Vukobratovic [69]; Schwartz and Vardy [55]; and Etzion [18].

The question of how to accomplish an efficient decoding in quasi-cyclic constructions with high redundancy was addressed by Johnson and Weller [25] as well as Li, Chen, Zeng, Lin, and Fong [34].

In his thesis Gallager already mentioned the possibility of non-binary codes [20] which is the topic of Section 7.4. The fact that, by using non-binary alphabets, the performance of LDPC codes over BMS channels can be improved was first reported by Davey and Mackay [8]. They showed by specific examples that non-binary LDPC codes can perform significantly better than their binary counterparts for the BMS channels. Another reference on non-binary codes from the same time period is the one by Berkmann [5]. The first explicit mention of the use of Fourier transforms to perform the check-node computation efficiently can be found in the article by Mackay and Davey [40]. This approach has since been rediscovered a countless number of times. Hu showed that even the performance of cycle codes can be improved considerably with non-binary alphabets [22]. In [59], Sridhara and Fuja have designed codes over certain rings and groups for coded modulation based on the principle of non-binary LDPC codes. Much of the material presented in Section 7.4 is taken from Rathi and Urbanke [51]. In this paper it is shown that the analysis is simplified if the edge labels are taken from the set of invertible $m \times m$ binary matrices (rather than the subset of these matrices that represent finite field elements in $\mathbb{F}_{2^m}$).

Mitzenmacher as well as Luby and Mitzenmacher proposed to use codes over large alphabets as *verification codes* [38, 46]. Such codes are well suited if we consider *packets of bits* instead of bits. To see what a verification code is, consider transmission over a q-ary symmetric channel, where q denotes the alphabet size. Assume that for a particular check the sum of its neighbors is zero. Then, by the symmetry of the channel, the probability that the received value of any of the involved packets (variable nodes) is incorrect is at most $1/q$, a negligible value for any reasonable packet size (if the packet consists of b bytes then $q = 8^b$). We then say that these variable nodes have been verified. In a similar way we can correct and verify a node if all of its neighbors (with respect to a particular check) have already been verified. This gives rise to a decoding algorithm reminiscent of the erasure decoder.

The first mention of rateless codes is due to Byers, Luby, Mitzenmacher, and Rege [7]. The first explicit constructions of rateless rates are due to Luby and are named LT-codes (Luby transform codes) in his honor [35, 36, 37]. As only briefly mentioned in this text, these are essentially LDGM codes with an average degree which grows logarithmically in the blocklengths. This logarithmic growth is necessary to ensure that all input bits are "covered" at least once (are part of at least one linear combination). But this logarithmic growth causes two problems. First, it increases the complexity, and second, we can no longer apply our standard analysis which assumes that the degrees are fixed and that the length tends to infinity. (Although, surprisingly, a formal such application of the standard techniques *does* give the right result.) To remedy these problems, Shokrollahi introduced Raptor codes [56, 57]. Lemma 7.45 is taken from the journal paper. The idea of Raptor codes is to precode the information word. The LDGM part of the code is used to bring the error probability down to a small number. It is then the task of the code used in the precoding to eliminate the remaining errors. (See also the papers by Maymounkov [44] as well as Maymounkov and Mazieres [45].)

The stability condition presented in Section 7.5.2 is due to Etesami and Shokrollahi [17]. We have not presented a finite-length analysis of LT codes. But such an analysis has been accomplished by Karp, Luby, and Shokrollahi [27]. The use of Raptor codes on symmetric channels was investigated by Etesami, Molkaraie, and Shokrollahi [16] as well as Palanki and Yedidia [47].

Several standards have already adopted Raptor codes and this number is likely to increase further in the future.

Problems

7.1 (Socket Count Equality Constraint). Consider the RA ensemble discussed in Example 7.5. Check that the socket count equality constraints are fulfilled.

7.2 (All-One Codeword for Non-Binary Case). Consider transmission over a BMS channel using a non-binary LDPC ensemble and BP decoding. Prove that the error probability is independent of the transmitted codeword so that in the analysis we can make the all-one codeword assumption.

7.3 (Minimal Codewords for Non-Binary Codes). Show that the generating function counting the number of minimal codewords in a non-binary LDPC ensemble in the limit of infinite blocklengths is given by (7.40).

References

[1] A. Abbasfar, D. Divsalar, and Y. Kung, *Accumulate repeat accumulate codes*, in Proc. of the IEEE Int. Symposium on Inform. Theory, Chicago, IL, USA, June 27–July 2, 2004, IEEE, p. 505. [418]

[2] ——, *Accumulate repeat accumulate codes*, in Proc. of GLOBECOM, Nov. 2004, pp. 509–513. [418]

[3] I. Andriyanova, J.-P. Tillich, and J.-C. Carlach, *Asymptotically good codes with high iterative decoding performances*, in Proc. of the IEEE Int. Symposium on Inform. Theory, Adelaide, Australia, Sept. 4–9, 2005, pp. 850–854. [419]

[4] A. Barg and G. Zémor, *Distance properties of expander codes*, IEEE Trans. Inform. Theory, 52 (2006), pp. 78–90. [418, 434]

[5] J. Berkmann, *On turbo decoding of nonbinary codes*, IEEE Commun. Lett., 2 (1998), pp. 94–96. [420]

[6] J. Boutros, O. Pothier, and G. Zémor, *Generalized low-density (Tanner) codes*, in Proc. of ICC, Vancouver, Canada, June 1999, pp. 441–445. [418]

[7] J. Byers, M. Luby, M. Mitzenmacher, and A. Rege, *A digital fountain approach to reliable distribution of bulk data*, in Proc. of ACM SIGCOMM, Aug. 1998. [421]

[8] M. C. Davey and D. J. C. MacKay, *Low density parity check codes over GF(q)*, IEEE Commun. Lett., 2 (1998), pp. 165–167. [420]

[9] A. G. Dimakis, M. J. Wainwright, and K. Ramchandran, *Lower bounds on the rate-distortion function of LDGM codes*, in Information Theory Workshop 2007, Lake Tahoe, CA, USA, Sept. 2007, IEEE. [419]

[10] D. Divsalar, S. Dolinar, and J. Thorpe, *Accumulate-repeat-accumulate-accumulate codes*, in Proc. of the Vehicular Technol. Conf., Sept. 26–29, 2004, pp. 2292–2296. [418]

[11] D. Divsalar, H. Jin, and R. J. McEliece, *Coding theorems for "turbo-like" codes*, in Proc. of the Allerton Conf. on Commun., Control, and Computing, Monticello, IL, USA, Sept. 1998, pp. 201–210. [418]

[12] D. Divsalar, C. R. Jones, S. Dolinar, and J. Thorpe, *Protograph based LDPC codes with minimum distance linearly growing with block size*, in Proc. of GLOBECOM, Nov. 2005, pp. 1152–1156. [419]

[13] I. B. Djordjevic and B. Vasic, *Projective geometry LDPC codes for ultralong-haul WDM high-speed transmission*, IEEE Phot. Technol. Lett., 15 (2003), pp. 784–786. [419]

[14] ——, *Iteratively decodable codes from orthogonal arrays for optical communication systems*, IEEE Commun. Lett., 9 (2005), pp. 924–926. [419]

[15] K. Engdahl and K. S. Zigangirov, *On the theory of low-density convolutional codes I*, Probl. Peredach. Inform., 35 (1999), pp. 12–28. [419]

[16] O. Etesami, M. Molkaraie, and A. Shokrollahi, *Raptor codes on symmetric channels*, in Proc. of the IEEE Int. Symposium on Inform. Theory, June 2004, p. 38. [421]

[17] O. Etesami and A. Shokrollahi, *Raptor codes on binary memoryless symmetric channels*, IEEE Trans. Inform. Theory, 52 (2006), pp. 2033–2051. [421]

[18] T. Etzion, *On the stopping redundancy of Reed-Muller codes*, IEEE Trans. Inform. Theory, 52 (2006), pp. 4867–4879. [420]

[19] R. G. Gallager, *Low-density parity-check codes*, IRE Trans. Inform. Theory, 8 (1962), pp. 21–28. [33, 66, 159, 264, 419, 434]

[20] ——, *Low-Density Parity-Check Codes*, MIT Press, Cambridge, MA, USA, 1963. [156, 158, 164, 261, 264, 420]

[21] M. Greferath, M. E. O'Sullivan, and R. Smarandache, *Construction of good LDPC codes using dilation matrices*, in Proc. of the IEEE Int. Symposium on Inform. Theory, Chicago, IL, USA, 2004, p. 237. [419]

[22] X.-Y. Hu, *Low-Delay Low-Complexity Error-Correcting Codes on Sparse Graphs*, PhD thesis, EPFL, Lausanne, Switzerland, 2002. [420]

[23] F. K. Hwang, *The Mathematical Theory of Nonblocking Switching Networks*, vol. 15 of Series on Applied Math., World Scientific Publ., 2 ed., 2004. [419]

[24] H. Jin, A. K. Khandekar, and R. J. McEliece, *Irregular repeat-accumulate codes*, in Proc. of the Int. Conf. on Turbo Codes and Related Topics, Brest, France, Sept. 2000, pp. 1–8. [418]

[25] S. J. Johnson and S. R. Weller, *A family of irregular LDPC codes with low encoding complexity*, IEEE Commun. Lett., 7 (2003), pp. 79–81. [420]

[26] I. Kanter and D. Saad, *Error-correcting codes that nearly saturate Shannon's bound*, Phys. Rev. Lett., 83 (1999), pp. 2660–2663. [266, 418, 419]

[27] R. M. Karp, M. Luby, and A. Shokrollahi, *Finite-length analysis of LT-codes*, in Proc. of the IEEE Int. Symposium on Inform. Theory, June 2004, p. 39. [421]

[28] N. Kashyap and A. Vardy, *Stopping sets in codes from designs*, in Proc. of the IEEE Int. Symposium on Inform. Theory, Yokohama, Japan, June 2003, p. 122. [420]

[29] C. Kelley, D. Sridhara, J. Xu, and J. Rosenthal, *Pseudocodeword weights and stopping sets*, in Proc. of the IEEE Int. Symposium on Inform. Theory, Chicago, IL, USA, June 2004, p. 67. [420]

[30] D. E. Knuth, *The Art of Computer Programming: Sorting and Searching*, vol. 3, Addison-Wesley, Reading, MA, USA, 1973. [419]

[31] R. Kötter and P. O. Vontobel, *Graph covers and iterative decoding of finite-length codes*, in Proc. of the Int. Conf. on Turbo Codes and Related Topics, Brest, France, Sept. 1–5, 2003, pp. 75–82. [266, 419]

[32] Y. Kou, S. Lin, and M. P. C. Fossorier, *Low-density parity-check codes based on finite geometries: A rediscovery and new results*, IEEE Trans. Inform. Theory, 47 (2001), pp. 2711–2736. [166, 419]

[33] M. Lentmaier and K. S. Zigangirov, *On generalized low-density parity-check codes based on hamming component codes*, IEEE Commun. Lett., 3 (1999), pp. 248–260. [418]

[34] Z. Li, L. Chen, L. Zeng, S. Lin, and W. H. Fong, *Efficient encoding of quasi-cyclic low-density parity-check codes*, IEEE Trans. Inform. Theory, 54 (2006), pp. 71–81. [420]

[35] M. Luby, *Information additive code generator and decoder for communication systems.* US Patent Number 6,307,487, Oct. 2001. [421]

[36] ——, *Information additive code generator and decoder for communication systems.* US Patent Number 6,373,406, Apr. 2002. [421]

[37] ——, *LT-codes*, in Proc. of STOC, 2002, pp. 271–280. [421]

[38] M. Luby and M. Mitzenmacher, *Verification-based decoding for packet-based low-density parity-check codes*, IEEE Trans. Inform. Theory, 51 (2005), pp. 120–127. [420]

[39] D. J. C. MacKay, *Good error correcting codes based on very sparse matrices*, IEEE Trans. Inform. Theory, 45 (1999), pp. 399–431. [419]

[40] D. J. C. MacKay and M. C. Davey, *Evaluation of Gallager Codes for Short Block Length and High Rate Applications*, vol. 123 of IMA, USA Volumes in Math. and its Applications, Springer, New York, NY, USA, 2000. [420]

[41] D. J. C. MacKay and R. M. Neal, *Good codes based on very sparse matrices*, in Cryptography and Coding. 5th IMA, USA Conference, C. Boyd, ed., no. 1025 in Lect. Notes Comput. Sci, Springer, Berlin, Germany, 1995, pp. 100–111. [419]

[42] E. Martinian and M. J. Wainwright, *Analysis of LDGM and compound codes for lossy compression and binning*, in Proc. of the IEEE Inform. Theory Workshop, San Diego, CA, USA, Feb. 2006. [419]

[43] ——, *Low-density constructions can achieve the Wyner-Ziv and Gelfand-Pinsker bounds*, in Proc. of the IEEE Int. Symposium on Inform. Theory, Seattle, WA, USA, July 2006, pp. 484–488. [419]

[44] P. Maymounkov, *Online codes.* Technical Report TR2002-833, Nov. 2002. [421]

[45] P. Maymounkov and D. Mazieres, *Rateless codes and big downloads*, in Peer-to-Peer Systems II: Second International Workshop, IPTPS 2003, Berkeley, CA, February 21-22, no. 2735 in Lect. Notes Comput. Sci., Springer, Berlin, Germany, 2003, pp. 247–255. [421]

[46] M. Mitzenmacher, *Verification codes for deletions*, in Proc. of the IEEE Int. Symposium on Inform. Theory, Tokyo, Japan, June 29–July 4, 2003, p. 217. [420]

[47] R. Palanki and J. S. Yedidia, *Rateless codes on noisy channels*, in Proc. of the IEEE Int. Symposium on Inform. Theory, June 2004, p. 37. [421]

[48] H. D. Pfister and I. Sason, *Accumulate-repeat-accumulate codes: Capacity-achieving ensembles of systematic codes for the erasure channel with bounded complexity*, IEEE Trans. Inform. Theory, 53 (2007), pp. 2088–2115. [418]

[49] H. D. Pfister, I. Sason, and R. Urbanke, *Capacity-achieving ensembles for the binary erasure channel with bounded complexity*, IEEE Trans. Inform. Theory, 51 (2005), pp. 2352–2379. [418]

[50] L. Ping, *Concatenated tree codes: A low-complexity high-performance approach*, IEEE Trans. Inform. Theory, 47 (2001), pp. 791–799. [418]

[51] V. Rathi and R. Urbanke, *Density evolution, threshold and the stability condition for non-binary LDPC codes*, IEE Proc. Commun., 152 (2005), pp. 1069–1074. [420]

[52] T. Richardson, *Method and apparatus for performing low-density parity-check (LDPC) code operations using a multi-level permutation.* US Patent Number 6,450,245, Sept. 2004. [419]

[53] T. Richardson and R. Urbanke, *Fixed points and stability of density evolution*, Commun. Inform. Syst., 4 (2004), pp. 103–116. [263, 418]

[54] ——, *Multi-edge type LDPC codes.* submitted IEEE IT, 2004. [418]

[55] M. Schwartz and A. Vardy, *On the stopping distance and the stopping redundancy of codes*, IEEE Trans. Inform. Theory, 52 (2006), pp. 922–932. [420]

[56] A. Shokrollahi, *Raptor codes*, in Proc. of the IEEE Int. Symposium on Inform. Theory, Chicago, IL, USA, June 2004, p. 36. [421]

[57] ——, *Raptor codes*, IEEE Trans. Inform. Theory, 52 (2006), pp. 2551–2567. [421]

[58] R. Smarandache and P. O. Vontobel, *On regular quasicyclic LDPC codes from binomials*, in Proc. of the IEEE Int. Symposium on Inform. Theory, Chicago, IL, USA, 2004, p. 274. [419]

[59] D. Sridhara and T. E. Fuja, *Low density parity check codes over groups and rings*, in Proc. of the IEEE Inform. Theory Workshop, Banglore, India, Oct. 2002, pp. 163–166. [420]

[60] D. Sridhara, C. Kelley, and J. Rosenthal, *Tree-based construction of LDPC codes*, in Proc. of the IEEE Int. Symposium on Inform. Theory, Adelaide, Australia, 2005, pp. 845–849. [419]

[61] R. M. Tanner, *A transform theory for a class of group-invariant codes*, IEEE Trans. Inform. Theory, 34 (1988), pp. 725–775. [419]

[62] J. Thorpe, K. Andrews, and S. Dolinar, *Methodologies for designing LDPC codes using protographs and circulants*, in Proc. of the IEEE Int. Symposium on Inform. Theory, Chicago, IL, USA, June 2004. pp. 238. [419]

[63] J.-P. Tillich, *Habilitation á diriger des recherches.* INRIA Rocquencourt, France, 2006. [419]

[64] D. V. Truhachev, M. Lentmaier, and K. S. Zigangirov, *On braided block codes*, in Proc. of the IEEE Int. Symposium on Inform. Theory, Tokyo, Japan, June 29–July 4, 2003, p. 32. [419]

[65] B. Vasic, E. M. Kurtas, and A. V. Kuznetsov, *LDPC codes based on mutually orthogonal latin rectangles and their application in perpendicular magnetic recording*, IEEE Trans. Magnet., 38 (2002), pp. 2346–2348. [419]

[66] B. Vasic and O. Milenkovic, *Combinatorial constructions of low-density parity-check codes for iterative decoding*, IEEE Trans. Inform. Theory, 50 (2004), pp. 1156–1176. [419]

[67] P. O. Vontobel and R. Kötter, *Lower bounds on the minimum pseudo-weight of linear codes*, in Proc. of the IEEE Int. Symposium on Inform. Theory, Chicago, IL, USA, June 2004, p. 70. [420]

[68] ——, *On the relationship between linear programmming decoding and min-sum algorithm decoding*, in Proc. of the Int. Symposium on Inform. Theory and its Applications, Parma, Italy, Oct. 2004, pp. 991–996. [266, 420]

[69] P. O. Vontobel, R. Smarandache, N. Kiyavash, J. Teutsch, and D. Vukobratovic, *On the minimal pseudo-codewords of codes from finite geometries*, in Proc. of the IEEE Int. Symposium on Inform. Theory, Adelaide, Australia, 2005, pp. 980–984. [420]

[70] H. S. Wilf, *Generatingfunctionology*, Academic Press, 2 ed., 1994. [418, 529, 532]

Chapter 8

EXPANDER CODES AND FLIPPING ALGORITHM

Rather than looking at *message-passing* decoders and density evolution, one can use the *expansion* of a bipartite graph to guarantee an error-correcting capability of the associated code. One advantage of this approach is that it applies directly to finite-size graphs. On the negative side, the bounds derived by this method are typically pessimistic and do not reflect the true error-correcting potential observed in practice. The most prominent decoder used in conjunction with expansion arguments is the *flipping* algorithm.

§8.1. Building Codes from Expanders

It is said that one has to work in probability theory for at least 20 years before coming across the name *Chebyshev* spelled in exactly the same way twice. In a similar way, hardly any two authors agree on a definition of the *expansion* of a graph. Intuitively, in a graph with large expansion a subset of nodes has a large set of neighbors. Degree distributions play a minor role in our present investigation since the (analytically predictable) performance is only a function of the lowest and highest degree. We therefore consider $(\mathtt{l}, \mathtt{r})$-regular ensembles.

Definition 8.1 (Expansion of Bipartite Graph). Let $\mathtt{G}$ be an $(\mathtt{l}, \mathtt{r})$-regular bipartite graph with n *variable* nodes of degree $\mathtt{l}$ and $\frac{\mathtt{l}}{\mathtt{r}}n$ *check nodes* of degree $\mathtt{r}$. We say that $\mathtt{G}$ is an $(\mathtt{l}, \mathtt{r}, \alpha, \gamma)$ *expander* if for every subset $\mathcal{V}$ of at most αn variable nodes, the set of check nodes which are connected to $\mathcal{V}$ is at least $\gamma|\mathcal{V}|\mathtt{l}$.

The idea is clear. A set $\mathcal{V}$ of variable nodes has $|\mathcal{V}|\mathtt{l}$ outgoing edges and can hence be connected to *at most* $|\mathcal{V}|\mathtt{l}$ check nodes. Therefore, γ represents the minimum such *fraction* which is achieved by the given graph (where the minimum is over all non-empty sets $\mathcal{V}$ of cardinality at most αn). If the expansion is sufficiently large, the minimum distance of the associated code is a linear fraction of the blocklength.

Theorem 8.2 (Expansion and Minimum Distance). Let $\mathtt{G}$ be an $(\mathtt{l}, \mathtt{r}, \alpha, \gamma)$ expander with $\gamma > \frac{1}{2}$. Then the associated code has minimum distance greater than αn.

Proof. We proceed by contraposition. We assume that $\gamma > \frac{1}{2}$ and that there exists a non-zero codeword of size at most αn. Let $\mathcal{V}$ denote the set of variable nodes corresponding to the non-zero positions of this codeword. The constraints require that each check-node neighbor of $\mathcal{V}$ is connected to $\mathcal{V}$ an even number of times. In particular, each such neighbor must be connected at *least twice*. Thus, the number of neighbors is at most $\frac{1}{2}|\mathcal{V}|\mathtt{l}$. Since the graph is a $(\mathtt{l}, \mathtt{r}, \alpha, \gamma)$ expander it follows on the other hand that the number of neighbors is at least $\gamma|\mathcal{V}|\mathtt{l}$. This leads to $\frac{1}{2}|\mathcal{V}|\mathtt{l} \geq \gamma|\mathcal{V}|\mathtt{l}$, a contradiction since $\gamma > \frac{1}{2}$. □

§8.2. Flipping Algorithm

The *flipping algorithm* is the algorithm of choice if we consider codes built from expanders. It comes in many flavors. We limit our discussion to its generic version – the so-called *sequential* flipping algorithm.

Definition 8.3 (Sequential Flipping Algorithm). Associate with every variable node a current *estimate* of the corresponding bit. Initially this estimate is equal to the received value. In each iteration exactly one of these estimates is flipped until either a valid codeword is reached or until the decoding procedure stops. To decide which estimate to flip one proceeds as follows. Given the current estimates, call a check node *satisfied* if the modulo-2 sum of the estimates of its connected variable nodes is zero and *unsatisfied* otherwise. Choose a variable node that is connected to more unsatisfied constraints than satisfied constraints and flip its estimate. If no such variable node exists, stop.

Remark: By flipping such a variable node we *strictly* decrease the number of unsatisfied check nodes.

For graphs with sufficient expansion the flipping algorithm corrects a linear fraction of errors.

Theorem 8.4 (Expansion and Decoding Radius). Let G be a $(\mathtt{l}, \mathtt{r}, \alpha, \gamma)$ expander with $\gamma > \frac{3}{4}$. Then the sequential flipping algorithm correctly decodes all error patterns of weight up to $\frac{\alpha}{2}n$.

Proof. We call a variable node *good* if its current estimate is correct and *bad* otherwise. Let b_ℓ denote the number of bad variable nodes at the beginning of the ℓ-th iteration. Let s_ℓ and u_ℓ denote the number of satisfied and unsatisfied check nodes *which are connected to bad variable nodes* in the ℓ-th iteration. Note that *every* unsatisfied check node is connected to at least one bad variable node. Since by definition of the algorithm the number of unsatisfied check nodes is strictly decreasing it follows that the sequence u_ℓ is strictly decreasing until the algorithm terminates. By assumption the number of errors in the received word is at most $\frac{\alpha}{2}n$, i.e.,

$$b_1 \leq \frac{\alpha}{2}n.$$

We now show that if $0 < b_\ell < \alpha n$ then the algorithm flips a variable node in the ℓ-th decoding step and that $0 \leq b_{\ell+1} < \alpha n$. Since u_ℓ is strictly decreasing (and finite), it follows that the algorithm does not terminate until u_ℓ has reached zero. To see this claim assume that $0 < b_\ell < \alpha n$. From the expansion property we know that

$$s_\ell + u_\ell > \frac{3}{4}\mathtt{l} b_\ell.$$

Further, a check node which is connected to a bad variable node but which is satisfied must be connected to at least two bad variable nodes. Since there are exactly $\mathtt{l} b_\ell$ edges emanating from the set of bad variable nodes, it follows that

$$2s_\ell + u_\ell \leq \mathtt{l} b_\ell.$$

Combining the last two inequalities we conclude that

$$u_\ell > \frac{\mathtt{l}}{2} b_\ell. \tag{8.5}$$

In words, the average number of unsatisfied check nodes which are connected to a bad variable node is bigger than $\frac{\mathtt{l}}{2}$. We conclude that there must exist at least one (bad) variable node which is connected to more unsatisfied check nodes than satisfied check nodes. It follows that in the ℓ-th step the algorithm flips a variable node.

Let us now show that $0 \leq b_\ell < \alpha n$ throughout the whole decoding process. Using again (8.5) and the fact that u_ℓ is strictly decreasing we conclude that $\frac{\mathtt{l}}{2} b_\ell < u_\ell < u_1 \leq \mathtt{l} b_1 \leq \frac{\mathtt{l}\alpha}{2} n$. This shows that $b_\ell < \alpha n$.

It remains to check that the algorithm converges to the *correct* codeword. Without loss of generality we may assume that the zero codeword was sent. As we have seen, at no point in the algorithm does b_ℓ exceed αn. But from Theorem 8.2, we see that there are no codewords of weight less or equal to αn, so that the codeword found by the algorithm must indeed be the correct one. □

§8.3. Bound on Expansion of a Graph

In the previous section we have seen how the error-correction capability of a code can be lower bounded in terms of the expansion of its associated bipartite graph. Unfortunately, the problem of determining the expansion of a given bipartite graph is hard (NP-complete to be precise).

We therefore limit ourselves to stating a lower bound on the expansion of a given bipartite graph which can be determined efficiently. Unfortunately, as we will see, this bounding technique is limited to proving expansions of at most $\frac{1}{2}$.

Let $\mathtt{G}$ be an $(\mathtt{l}, \mathtt{r})$-regular bipartite graph with n variable nodes and $\frac{\mathtt{l}}{\mathtt{r}} n$ check nodes. Without essential loss of generality we assume that the graph contains no multiple edges. Let H be the *incidence* matrix of this bipartite graph, i.e., H_{ij} is equal to 1 if the i-th check node is connected to the j-th variable node and zero otherwise. In other words, H is the parity-check matrix of the associated binary code but considered as a $\{0,1\}$ matrix over the reals. The matrix $H^T H$ is an $n \times n$ real symmetric non-negative definite matrix. It is therefore diagonalizable and it has real non-negative eigenvalues and orthogonal eigenvectors. Let $\lambda_1 \geq \lambda_2 \geq \cdots \geq \lambda_n$ denote its ordered eigenvalues and let $e_1, \dots, e_n$ denote its associated orthogonal eigenvectors.

Theorem 8.6 (Tanner's Lower Bound on Expansion via Eigenvalues). Let $\mathtt{G}$ have second-largest eigenvalue λ_2. Then for any $\alpha \in \mathbb{R}^+$, $\mathtt{G}$ is an $(\mathtt{1}, \mathtt{r}, \alpha, \gamma(\alpha))$ expander with

$$\gamma(\alpha) \geq \frac{\mathtt{1}}{(\mathtt{1r} - \lambda_2)\alpha + \lambda_2}.$$

Proof. Let $\mathcal{V}$ denote a subset of the variable nodes and $\mathcal{C}$ denote that subset of check nodes that are connected to nodes in $\mathcal{V}$. Let x denote the characteristic vector associated to $\mathcal{V}$, i.e., x is a $\{0,1\}$-valued vector of length n which is 1 at position i if $i \in \mathcal{V}$, and 0 otherwise. Define $z = Hx^T$. By definition, z has exactly $|\mathcal{C}|$ non-zero entries and the sum of its entries is $|\mathcal{V}|\mathtt{1}$. Hence, by convexity of the function $f(x) = x^2$, we have

$$\|z\|_2^2 = \sum_{i=1}^{n} |z_i|^2 \geq \Big(\sum_{i=1}^{n} \frac{z_i}{|\mathcal{C}|}\Big)^2 |\mathcal{C}| = \frac{(|\mathcal{V}|\mathtt{1})^2}{|\mathcal{C}|}.$$

It follows that $\frac{|\mathcal{C}|}{|\mathcal{V}|\mathtt{1}} \geq \frac{|\mathcal{V}|\mathtt{1}}{\|z\|_2^2}$. Therefore, if we can find an upper bound on $\|z\|_2^2$ then we have a lower bound on the expansion. Since every row of H contains exactly $\mathtt{r}$ 1's and every column of H contains exactly $\mathtt{1}$ 1's we conclude that

$$(H^T H)(1,\dots,1)^T = H^T (H(1,\dots,1)^T) = H^T (1,\dots,1)^T \mathtt{r} = (1,\dots,1)^T \mathtt{1r}.$$

In other words, $(1,\dots,1)$ is an eigenvector of $H^T H$ with eigenvalue $\mathtt{1r}$. We claim that no eigenvalue can be larger: each row and column of $H^T H$ has L_1 norm $\mathtt{1r}$. It follows that for any vector x we have

$$\|(H^T H)x^T\|_1 \leq \sum_i \sum_j |(H^T H)_{ij} x_j| = \sum_j \Big\{\sum_i (H^T H)_{ij}\Big\} |x_j| = \sum_j \mathtt{1r}|x_j| = \mathtt{1r}\|x\|_1,$$

from which the claim follows. Therefore we have $e_1 = (1,\dots,1)/\sqrt{n}$ and $\lambda_1 = \mathtt{1r}$. Expand x, the characteristic vector associated to $\mathcal{V}$, in terms of the orthogonal basis spanned by the eigenvectors, $x = \sum_{i=1}^n \epsilon_i e_i$. Then we have $(H^T H)x^T = \sum_{i=1}^n \lambda_i \epsilon_i e_i^T$. Multiply the last equality from the left by $x = \sum_{i=1}^n \epsilon_i e_i$. By using the orthogonality of the eigenvectors, we have

$$\begin{aligned}
\|z\|_2^2 &= x(H^T H)x^T = \sum_{i=1}^{n} \lambda_i \epsilon_i^2 \\
&= \mathtt{lr}\frac{|\mathcal{V}|^2}{n} + \sum_{i=2}^{n} \lambda_i \epsilon_i^2 \le \mathtt{lr}\frac{|\mathcal{V}|^2}{n} + \lambda_2 \sum_{i=2}^{n} \epsilon_i^2 \\
&= (\mathtt{lr} - \lambda_2)\frac{|\mathcal{V}|^2}{n} + \lambda_2 \|x\|_2^2 = (\mathtt{lr} - \lambda_2)\frac{|\mathcal{V}|^2}{n} + \lambda_2 |\mathcal{V}|.
\end{aligned}$$

Minimizing over all non-empty sets $\mathcal{V}$ such that $\frac{|\mathcal{V}|}{n} \le \alpha$, the expansion satisfies

$$\begin{aligned}
\gamma(\alpha) &= \min_{\mathcal{V}} \frac{|\mathcal{C}|}{|\mathcal{V}|\mathtt{l}} \ge \min_{\mathcal{V}} \frac{|\mathcal{V}|\mathtt{l}}{(\mathtt{lr} - \lambda_2)\frac{|\mathcal{V}|^2}{n} + \lambda_2 |\mathcal{V}|} \\
&= \min_{\mathcal{V}} \frac{\mathtt{l}}{(\mathtt{lr} - \lambda_2)\frac{|\mathcal{V}|}{n} + \lambda_2} \ge \frac{\mathtt{l}}{(\mathtt{lr} - \lambda_2)\alpha + \lambda_2}.
\end{aligned}$$

□

§8.4. Expansion of a Random Graph

Let us state here without proof that the preceding eigenvalue argument can assert an expansion of at most $\frac{1}{2}$. Unfortunately this is not sufficient to prove good distance properties of a code (and far from sufficient to show that the code can correct a linear fraction of errors). But as we will see, in suitably chosen ensembles of graphs almost every element has large expansion.

Theorem 8.7 (Expansion of Ensembles). Let $\mathtt{G}$ be chosen uniformly at random from LDPC $(n, x^{\mathtt{l}-1}, x^{\mathtt{r}-1})$. Choose $\gamma \in [0, 1 - 1/\mathtt{l})$. Let $\alpha_{\max}$ be the positive solution of the equation

$$\frac{\mathtt{l}-1}{\mathtt{l}} h_2(\alpha) - \frac{1}{\mathtt{r}} h_2(\alpha\gamma\mathtt{r}) - \alpha\gamma\mathtt{r} h_2\left(\frac{1}{\gamma\mathtt{r}}\right) = 0.$$

Then for $0 < \alpha < \alpha_{\max}$ and $\beta = \mathtt{l}(1-\gamma) - 1$

$$\mathbb{P}\{\mathtt{G} \text{ is an } (\mathtt{l}, \mathtt{r}, \alpha, \gamma) \text{ expander}\} \ge 1 - O(n^{-\beta}).$$

Discussion: As discussed in Problem 8.2, it is straightforward to see that an $(\mathtt{l}, \mathtt{r})$-regular graph can not have expansion exceeding $1 - 1/\mathtt{l}$. The theorem asserts that with high probability a random graph has expansion close to this upper bound.

If $\mathtt{l} \ge 3$ then (since $\frac{1}{2} < 1 - 1/\mathtt{l} = 2/3$) with probability 1 (in the limit of large n) a random graph has expansion strictly bigger than $\frac{1}{2}$ for some strictly positive α (and therefore the associated code has linear minimum distance). If $\mathtt{l} \geq 5$ then (since $\frac{3}{4} < 1 - 1/\mathtt{l} = 4/5$) with probability one (in the limit of large n) a random graph has expansion strictly bigger than $\frac{3}{4}$ for some strictly positive α (and therefore the associated code has a linear error-correcting radius under the flipping decoder).

Example 8.8 (α versus γ). Figure 8.9 shows α as a function of γ for $\mathtt{l} = 2, 3, 4, 5$ and $\mathtt{r} = 6$. ◇

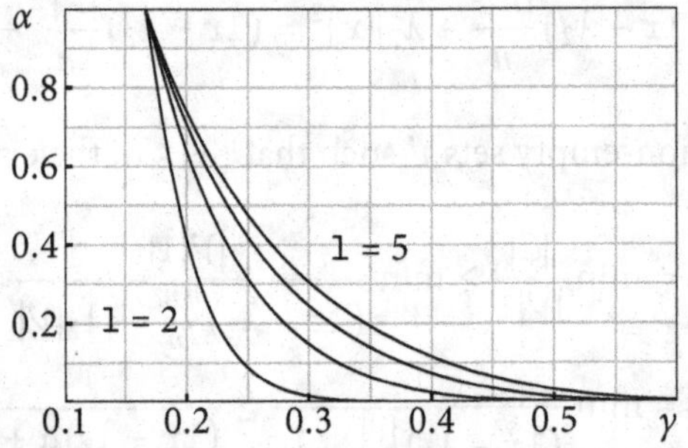

Figure 8.9: Value of α as a function of γ for $\mathtt{l} = 2, 3, 4, 5$ and $\mathtt{r} = 6$.

Proof. Consider a set of variable nodes $\mathcal{V}$ of size v. There are $\binom{n\mathtt{l}}{v\mathtt{l}}$ distinct ways of choosing the $v\mathtt{l}$ check sockets to which the edges emanating from $\mathcal{V}$ connect. We claim that at most $\binom{n\frac{\mathtt{l}}{\mathtt{r}}}{v\mathtt{l}\gamma}\binom{v\mathtt{l}\gamma\mathtt{r}}{v\mathtt{l}}$ of those are expanding by γ or less: a set $\mathcal{V}$ of size v that expands by at most γ is connected to at most $v\mathtt{l}\gamma$ distinct check nodes. Therefore, choose $v\mathtt{l}\gamma$ out of all $n\frac{\mathtt{l}}{\mathtt{r}}$ check nodes. This accounts for the first factor. Now choose the $v\mathtt{l}$ sockets out of the available $v\mathtt{l}\gamma\mathtt{r}$. This is represented by the second factor. Note that some constellations are counted repeatedly, so that the preceding expression is an upper bound. Since all constellations in the ensemble LDPC $\left(n, x^{\mathtt{l}-1}, x^{\mathtt{r}-1}\right)$ are equally likely, the probability that a random constellation on v variable nodes is expanding by at most γ is upper bounded by $\binom{n\frac{\mathtt{l}}{\mathtt{r}}}{v\mathtt{l}\gamma}\binom{v\mathtt{l}\gamma\mathtt{r}}{v\mathtt{l}}\binom{n\mathtt{l}}{v\mathtt{l}}^{-1}$. Since there are $\binom{n}{v}$ subsets of variable nodes of size v, an application of (the Markov inequality) (C.2) and the bounds (1.58) as well as (1.59) shows that the probability

$$\mathbb{P}\{\mathtt{G} \text{ contains } \mathcal{V} : |\mathcal{V}| = v \text{ and } |\mathcal{C}| \leq v\mathtt{l}\gamma\}$$

is upper bounded by

$$(8.10) \qquad \binom{n}{v}\binom{n\frac{\mathtt{l}}{\mathtt{r}}}{v\mathtt{l}\gamma}\binom{v\mathtt{l}\gamma\mathtt{r}}{v\mathtt{l}}\binom{n\mathtt{l}}{v\mathtt{l}}^{-1}$$

$$(8.11) \qquad \leq (n\mathtt{l}+1)2^{-n\mathtt{l}\left[\frac{\mathtt{l}-1}{\mathtt{l}}h_2(v/n)-\frac{1}{\mathtt{r}}h_2((v/n)\gamma\mathtt{r})-(v/n)\gamma\mathtt{r}h_2\left(\frac{1}{\gamma\mathtt{r}}\right)\right]}$$

(8.12) $$=(n\mathtt{l}+1)\exp\big(-n\mathtt{l}[\frac{\beta}{\mathtt{l}}\frac{v}{n}\ln\frac{n}{v}+O(v/n)]\big).$$

Fix $0 \leq \gamma < 1 - 1/\mathtt{l}$ so that $\beta = \mathtt{l}(1-\gamma) - 1 > 0$. We see from (8.12) that there exists a positive range for v/n such that the exponent in the bound is positive. The exact range is determined by (8.11): we get $v/n < \alpha_{\max}$.

Fix $\alpha < \alpha_{\max}$ and sum (8.10) for $v = 1, \dots, \alpha n$. We want to show that this sum tends to zero as n tends to infinity. This is best done in two steps. For small v we bound (8.10) using (1.56) and (1.57) by

(8.13) $$\Big(\frac{v}{cn}\Big)^{\beta v} e^{\frac{v^2 \mathtt{l}}{n}},$$

where c is a strictly positive constant, $c = c(\mathtt{l}, \mathtt{r}, \gamma)$. Set $\delta = \frac{\beta}{2(1+\beta)}$. The sum over (8.13) for $v = 1, \dots, n^\delta$ can be bounded by

$$\sum_{v=1}^{n^\delta}\Big(\frac{v}{cn}\Big)^{\beta v} e^{\frac{v^2\mathtt{l}}{n}} \leq e^{n^{2\delta-1}\mathtt{l}}\sum_{v=1}^{n^\delta}\Big(\frac{v}{cn}\Big)^{\beta v} \overset{n\geq\frac{1}{c}}{\leq} e^{n^{2\delta-1}\mathtt{l}}(cn)^{-\beta}\Big(1+n^{\delta\beta}\sum_{v=2}^{n^\delta}\Big(\frac{v}{cn}\Big)^{\beta(v-1)}\Big)$$
$$\overset{(i)}{\leq} (cn)^{-\beta}e^{n^{2\delta-1}-1}\big(1+n^{\delta\beta+\delta-\beta}(2/c)^\beta\big) \leq O(n^{-\beta}),$$

where in step (i) we assumed that $n \geq (\frac{1}{c})^{\frac{1}{1-\delta}}$ so that $\frac{v}{cn} \leq 1$. This implies that $\big(\frac{v}{cn}\big)^{\beta(v-1)}$ takes on its maximum value for $v = 2$. For the sum $v = n^\delta, \dots, \alpha n$ we use (8.12), which leads to the bound

$$O(\exp(-\beta(1-\delta)n^\delta \log n)) \leq O(n^{-\beta}).$$ □

Example 8.14 (Expansion of (5, 6)-Regular Ensemble). According to Theorem 8.7

$$\mathbb{P}\{\mathtt{G} \text{ is an } (\mathtt{l}=5, \mathtt{r}=6, \alpha=0.0302946, \gamma=\frac{1}{2}) \text{ expander}\} \geq 1 - O(n^{-\frac{1}{4}}).$$

Let us combine this with Theorem 8.2, taking into account that the relationship between α and γ is continuous, so that a small increase in γ only requires a small decrease in α. Therefore, a randomly chosen element $\mathtt{G}$ from the (5, 6)-regular ensemble has a minimum distance of (at least) 3 % of the blocklength with high probability. If we are asking for an expansion of $\gamma = \frac{3}{4}$ then the corresponding α is less than $3.375 \cdot 10^{-11}$, and by Theorem 8.4 the guaranteed error-correcting radius under the sequential flipping algorithm is half of this amount. Fortunately, in practice the observed performance is considerably better than that. The flipping algorithm is in general not as good as the BP algorithm but it typically can correct a reasonable fraction of the errors that the BP can correct. ◇

Notes

The flipping algorithm was introduced by Gallager [7]. The first analysis of this algorithm is due to Pinsker and Zyablov [16]. The analysis of the flipping algorithm in terms of the expansion of the underlying Tanner graph, as stated in Theorem 8.2 and Theorem 8.4, is due to Sipser and Spielman [14]. Improved constants were given by Burshtein [5].

The lower bound on the expansion based on the eigenvalue separation of the associated adjacency matrix is due to Tanner [15]. It was shown by Kahale in his thesis [10] that the eigenvalue method can certify an expansion of at most $\frac{1}{2}$. This is unfortunately not good enough to assert that there are expander codes of good minimum distance and it is far from sufficient to assert a good performance under the flipping algorithm.

The fact that suitable random bipartite graphs are expanders with high probability was discovered independently by several authors. Such a statement is contained in the thesis of Spielman, where the key idea is attributed to Alon. A slightly different proof was given by Luby, Mitzenmacher, Shokrollahi, Spielman, and Stemann [11]. Theorem 8.7 and its proof are taken from Burshtein and Miller [6].

Although we have only considered random graphs, it is possible to explicitly *construct* expanders to guarantee a certain error-correcting radius. The first such construction is due to Margulis [12]. For the binary case, the construction of expander codes with the currently largest minimum distance as a function of the rate is due to Barg and Zémor [3, 4]. The expanders with the largest error exponent for large rates as well as low rates (over the binary symmtric channel) are also due to these authors [2, 3]. For non-binary alphabets, nearly-MDS expander codes (which also admit a linear-time encoding) were presented by Guruswami and Indyk [8] as well as Roth and Skachek [13]. For a comparison of various constructions in terms of decoding characteristics versus gap to capacity see the paper by Ashikhmin and Skachek [1].

If you are looking for a survey on expander graphs see Hoory, Linial, and Wigderson [9].

From the material we presented it might appear that there are two completely separate (coding) universes: a *message-passing* one and an *expanding* one which we use when analyzing the flipping algorithm. It is possible to analyze message-passing decoders in terms of expansion of the underlying Tanner graph. This surprising result is due to Burshtein and Miller [6]. More precisely, if the underlying Tanner graph has expansion exceeding $\frac{3}{4}$ then suitably quantized versions of the BP decoder have a strictly positive error-correcting radius and, hence, an associated strictly positive error exponent.

Problems

8.1 (Tanner's Lower Bound on Expansion: Special Case). Consider the bound in Theorem 8.6. What happens if for an $(\mathtt{l}, \mathtt{r})$-regular graph $\mathtt{G}$ you have $\lambda_2 = \lambda_1$?

8.2 (Upper Bound on Expansion). Consider an $(\mathtt{l}, \mathtt{r})$-regular graph. Show that its expansion is at most $1 - 1/\mathtt{l} + \frac{1}{\mathtt{l}v_k}$ for sets of cardinality v_k, where $v_k = 1 + \sum_{i=1}^{k} (\mathtt{l} - 1)^{i-1}(\mathtt{r} - 1)^i$, $k \in \mathbb{N}$.

Hint: First consider the node-perspective computation graph of depth $\ell = 1$ of a variable node. Show that its expansion is at most $\frac{1+(\mathtt{l}-1)(\mathtt{r}-1)}{1+\mathtt{l}(\mathtt{r}-1)} = 1 - 1/\mathtt{l} + \frac{1}{\mathtt{l}(1+\mathtt{l}(\mathtt{r}-1))} = 1 - 1/\mathtt{l} + \frac{1}{\mathtt{l}v_1}$. Then generalize.

8.3 (Expansion of a Graph). Consider the Tanner graph of an $(\mathtt{l}, \mathtt{r})$-regular LDPC code of length n. Show that if $\mathtt{G}$ is an $(\mathtt{l}, \mathtt{r}, \alpha, \gamma)$ expander with $\gamma > \frac{1}{2}$ then the graph does not contain stopping sets of size αn or smaller.

References

[1] A. Ashikhmin and V. Skachek, *Decoding of expander codes at rates close to capacity*, IEEE Trans. Inform. Theory, 52 (2006), pp. 5475–5485. [434]

[2] A. Barg and G. Zémor, *Error exponents of expander codes under linear-complexity decoding*, SIAM J. Discrete Math., USATH., 17 (2004), pp. 426–445. [434]

[3] ——, *Multilevel expander codes*, in Algebraic Coding Theory and Information Theory, A. Ashikhmin and A. Barg, eds., vol. 68 of DIMA, USACS Series in Discrete Math. and Theoretical Computer Science, American Math. Soc., 2005, pp. 69–84. [434]

[4] ——, *Distance properties of expander codes*, IEEE Trans. Inform. Theory, 52 (2006), pp. 78–90. [418, 434]

[5] D. Burshtein, *On the error correction of regular LDPC codes using the flipping algorithm*, in Proc. of the IEEE Int. Symposium on Inform. Theory, Nice, France, July 2007, pp. 226–230. [434]

[6] D. Burshtein and G. Miller, *Expander graph arguments for message-passing algorithms*, IEEE Trans. Inform. Theory, 47 (2001), pp. 782–790. [434]

[7] R. G. Gallager, *Low-density parity-check codes*, IRE Trans. Inform. Theory, 8 (1962), pp. 21–28. [33, 66, 159, 264, 419, 434]

[8] V. Guruswami and P. Indyk, *Near-optimal linear-time codes for unique decoding and new list-decodable codes over smaller alphabets*, in STOC, Montreal, Quebec, Canada, May 2002, pp. 812–821. [434]

[9] S. Hoory, N. Linial, and A. Widgerson, *Expander graphs and their applications*, Bulletin of the Amercian Mathematical Society, 43 (2006), pp. 439–561. [434]

[10] N. Kahale, *Expander Graphs*, PhD thesis, MIT, 1993. [434]

[11] M. Luby, M. Mitzenmacher, A. Shokrollahi, D. A. Spielman, and V. Stemann, *Practical loss-resilient codes*, in Proc. of the 29th annual ACM Symposium on Theory of Computing, 1997, pp. 150–159. [157, 262, 434, 456]

[12] G. A. Margulis, *Explicit consctructions of graphs without short cycles and low density codes*, Combinatorica, 2 (1982), pp. 71–78. [434]

[13] R. Roth and V. Skachek, *Improved nearly-MDS expander codes*, IEEE Trans. Inform. Theory, 52 (2006), pp. 3650–3661. [434]

[14] M. Sipser and D. A. Spielman, *Expander codes*, IEEE Trans. Inform. Theory, 42 (1996), pp. 1710–1722. [434, 456]

[15] R. M. Tanner, *Explicit concentrators from generalized N-gons*, SIAM J. Alg. Disc. Meth., 5 (1984), pp. 287–293. [434]

[16] V. Zyablov and M. Pinsker, *Estimation of the error-correction complexity of Gallager low-density codes*, Problemy Peredachi Informatsii, 11 (1975), pp. 23–36. [261, 434, 529]

Appendix A

ENCODING LOW-DENSITY PARITY-CHECK CODES

Low-density parity-check (LDPC) codes are unusual in that their encoding method is not an intrinsic part of their definition. In this chapter we discuss how to efficiently encode generic LDPC codes. The method we present is *general*, applicable to any LDPC code. In practice, an LDPC code is often constrained further to yield a simple encoding. The general method presented here can always be used as a starting point in such constructions – controlling the complexity of the resulting encoder.

§A.1. Encoding Generic LDPC Codes

Assume we are given an LDPC code of length n and dimension k specified by its $(n-k) \times n$ parity-check matrix H; i.e., the code consists of the set of n-tuples x such that

$$Hx^T = 0^T.$$

Assuming that H has full row rank, we can accomplish the encoding if we can find a decomposition (permuting columns of H if necessary)

$$H = \begin{pmatrix} H_p & H_s \end{pmatrix}$$

such that H_p is square and *invertible*. Split the vector x into a *systematic* part s, $s \in \mathbb{F}^k$, and a *parity* part p, $p \in \mathbb{F}_2^{n-k}$, such that $x = (p, s)$. To encode we fill s with the k desired information bits and solve for p using $H_p p^T = H_s s^T$.

A straightforward construction of such an encoder is to use Gaussian elimination and column permutations to bring H into an equivalent[1] upper triangular form as shown in Figure A.1. In this way we not only find the decomposition $H = \begin{pmatrix} H_p & H_s \end{pmatrix}$, but we in effect premultiply H so that H_p is upper triangular. Using this equivalent form, we can solve for p, the $(n-k)$ parity-check bits, using *back substitution*. More precisely, for $l \in [n-k]$, starting with $l = n-k$, set

$$p_l = -\sum_{j=l+1}^{n-k} H_{l,j} p_j - \sum_{j=1}^{k} H_{l,j+n-k} s_j.$$

[1]We say equivalent because Gaussian elimination is right multiplication by an invertible matrix, hence the code is unchanged.

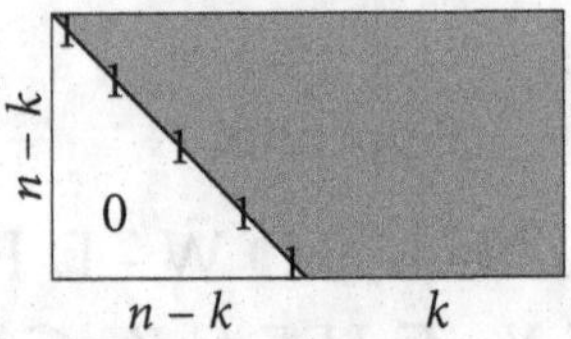

Figure A.1: H in upper triangular form.

What is the complexity of such an encoding scheme? Gaussian elimination with column swapping to bring the matrix H into the desired form requires $O(n^3)$ operations. Since this need only be done once, the result being usable in all encodings, we view this as *preprocessing*. The number of operations required for each encoding is $O(n^2)$ since, in general, after the preprocessing the matrix is no longer sparse. More precisely, we expect that we need about $n^2 \frac{r(1-r)}{2}$ operations to accomplish this encoding, where r is the rate of the code.

Given that the original parity-check matrix H is sparse, of $O(n)$ density, one might wonder if encoding can be accomplished in complexity $O(n)$. As we will see, for codes that allow transmission at rates close to capacity, linear time encoding is indeed possible. And for those codes for which the following encoding scheme leads to quadratic encoding complexity, the constant factor in front of the n^2 term is typically small so that the encoding complexity stays manageable up to large blocklengths.

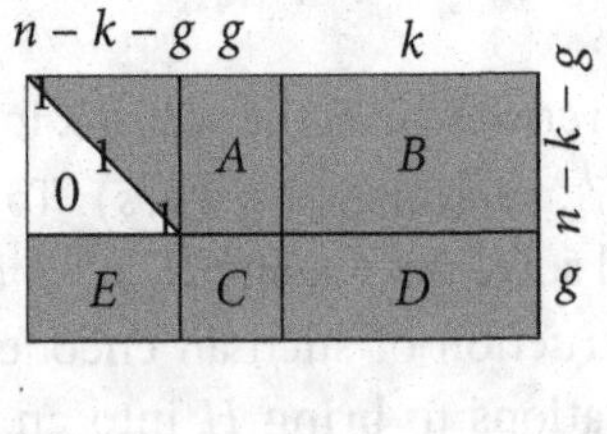

Figure A.2: H in approximate upper triangular form.

The general encoder is motivated by the preceding example. It is not restricted to the case $\mathbb{F}_2$. We therefore keep track of signs. The first step is to manipulate the parity-check matrix *using row and column permutations only* (but no algebraic operations) to put H into the form indicated in Figure A.2 and expressed in the equation

$$\begin{pmatrix} T & A & B \\ E & C & D \end{pmatrix}. \tag{A.3}$$

This is the decomposition $H = \begin{pmatrix} H_p & H_s \end{pmatrix}$ with $H_p = \begin{pmatrix} T & A \\ E & C \end{pmatrix}$, and the special requirement that T be square and upper triangular. We say that H is in *approximate upper triangular form*. This is important: since this transformation was accomplished solely by permutations, the matrix is *still sparse*. Hence, if we can make T to be of dimension $n-k$ then encoding with complexity $O(n)$ is possible. More generally, as we will show in the sequel, if we can make T of dimension $n-k-g$ then we can encode with complexity $O(n+g^2)$. Our goal in constructing such a decomposition is therefore to make the dimension of T as close to $n-k$ as possible. The height of $\begin{pmatrix} E & C & D \end{pmatrix}$ is called the *gap* and it is denoted by g. This parameter measures the distance of H from an upper triangular decomposition. The dimensions of the various submatrices are as follows:

$$
\begin{aligned}
&T : (n-k-g) \times (n-k-g), && A : (n-k-g) \times g, \\
&B : (n-k-g) \times k, && E : g \times (n-k-g), \\
&C : g \times g, && D : g \times k.
\end{aligned}
$$

Assuming that we have brought our parity-check matrix H into the form (A.3), encoding consists of solving the system of equations $H(p,s)^T = 0^T$ for p, given s. Multiply H from the left by

$$
\begin{pmatrix} I & 0 \\ -ET^{-1} & I \end{pmatrix}. \tag{A.4}
$$

This yields

$$
\begin{pmatrix} T & A & B \\ 0 & C - ET^{-1}A & D - ET^{-1}B \end{pmatrix}.
$$

In words, by premultiplying we have eliminated E, effectively performing the first step of Gaussian elimination. Rather than completing Gaussian elimination, we solve the implied system directly. Decompose $x = (p,s)$ further into $x = (p_1, p_2, s)$, where p_1 and p_2 combined denote the parity part; p_1 has length $(n-k-g)$, and p_2 has length g. The equation $Hx^T = 0^T$ splits into two equations, namely

$$
Tp_1^T + Ap_2^T + Bs^T = 0^T, \text{ and} \tag{A.5}
$$

$$
(C - ET^{-1}A)p_2^T + (D - ET^{-1}B)s^T = 0^T. \tag{A.6}
$$

Define $\phi = C - ET^{-1}A$, and assume for the moment that ϕ is invertible. This is equivalent to assuming that H_p is invertible. From (A.6)

$$
p_2^T = -\phi^{-1}(D - ET^{-1}B)s^T.
$$

Operation	Complexity
Bs^T	$O(n)$
$T^{-1}\left[Bs^T\right]$	$O(n)$
$-E\left[T^{-1}Bs^T\right]$	$O(n)$
Ds^T	$O(n)$
$\left[Ds^T\right]-\left[ET^{-1}Bs^T\right]$	$O(n)$
$-\phi^{-1}\left[Ds^T-ET^{-1}Bs^T\right]$	$O(g^2)$

Table A.7: Efficient computation of $p_2^T = -\phi^{-1}(D-ET^{-1}B)s^T$.

Operation	Complexity
Ap_2^T	$O(n)$
Bs^T	$O(n)$
$\left[Ap_2^T\right]+\left[Bs^T\right]$	$O(n)$
$-T^{-1}\left[Ap_2^T+Bs^T\right]$	$O(n)$

Table A.8: Efficient computation of $p_1^T = -T^{-1}(Ap_2^T + Bs^T)$.

Hence, if we precompute the $g \times k$ matrix $-\phi^{-1}(D-ET^{-1}B)$ then the determination of p_2 can be accomplished in complexity $O(g \times k)$ by performing a multiplication with this (generically dense) matrix. This complexity, however, can be further reduced as shown in Table A.7. Rather than precomputing $-\phi^{-1}(D - ET^{-1}B)$ and then multiplying with s^T, we precompute ϕ^{-1}, which is $g \times g$. Then we determine p_2 by first computing $(D-ET^{-1}B)s^T$ and then by multiplying with $-\phi^{-1}$. We claim that the first step can be done in $O(n)$ operations: we first determine Bs^T, which has complexity $O(n)$ since B is sparse. Next, we multiply the result by T^{-1}. Since $T^{-1}[Bs^T] = y^T$ is equivalent to the system $[Bs^T] = Ty^T$, this can also be accomplished in $O(n)$ by back-substitution (recall that T is upper triangular and sparse). The remaining steps are fairly straightforward. It follows that the overall complexity of determining p_2 is $O(n + g^2)$. In a similar manner, noting from (A.5) that $Tp_1^T = -(Ap_2^T + Bs^T)$, we can solve for p_1 in complexity $O(n)$ as shown step by step in Table A.8.

A summary of the proposed encoding procedure is given in Table A.9. It entails two steps: a *preprocessing* step and the actual *encoding* step. In the preprocessing step we first perform row and column permutations to bring the parity-check matrix into approximate upper triangular form with g as small as possible. We will see in the next section how this can be accomplished efficiently. Next we premultiply from the left by the matrix given in (A.4) to find ϕ. As discussed earlier, this premultiplication

PREPROCESSING:

In: Non-singular parity-check matrix H.

Out: An equivalent parity-check matrix of the form $\begin{pmatrix} T & A & B \\ E & C & D \end{pmatrix}$ such that $C - ET^{-1}A$ is non-singular.

1. [Triangulation] Perform row and column permutations to bring the parity-check matrix H into approximate upper triangular form

$$H = \begin{pmatrix} T & A & B \\ E & C & D \end{pmatrix},$$

with as small a gap g as possible.

2. [Check Rank] Use Gaussian elimination to perform the premultiplication

$$\begin{pmatrix} I & 0 \\ -ET^{-1} & I \end{pmatrix} \begin{pmatrix} T & A & B \\ E & C & D \end{pmatrix} = \begin{pmatrix} T & A & B \\ 0 & C - ET^{-1}A & D - ET^{-1}B \end{pmatrix}.$$

Check that $\phi = C - ET^{-1}A$ is non-singular – perform further column permutations if necessary to ensure this property. (Singularity of H is detected at this point.)

ENCODING:

In: Parity-check matrix of the form $\begin{pmatrix} T & A & B \\ E & C & D \end{pmatrix}$ such that $C - ET^{-1}A$ is non-singular and a vector $s \in \mathbb{F}^k$.

Out: The vector $x = (p_1, p_2, s)$, $p_1 \in \mathbb{F}^{n-k-g}$, $p_2 \in \mathbb{F}^g$, such that $Hx^T = 0^T$.

1. Determine p_2 as shown in Table A.7.

2. Determine p_1 as shown in Table A.8.

Table A.9: Summary of the proposed encoding procedure. It entails two steps: a preprocessing step and the actual encoding step.

is most efficiently done by running the Gaussian elimination algorithm to clear the matrix E. At this point we see the matrix ϕ. In order to check whether ϕ is singular we continue Gaussian elimination until the diagonal has been extended over all $n-k$ rows. Three cases can arise. First, Gaussian elimination succeeds without the need for column permutations. In this case we know that ϕ is non-singular. Second, Gaussian elimination succeeds but requires some column permutations. In this case we apply these column permutations to the matrix in approximate upper triangular form. This guarantees that in this new (and equivalent) matrix ϕ is non-singular. Finally, if no column permutation exists so that Gaussian elimination succeeds then the original matrix did not have full rank.

The actual encoding entails the steps listed in Tables A.7 and A.8. Note that the preprocessing step need only be carried out once and that the result can be used for all subsequent encodings.

Example A.10 (Parity-Check Matrix of (3, 6)-Regular Code of Length 12). Assume we are given the parity-check matrix

$$
\begin{array}{c}
\\ 2 \\ 1 \\ 4 \\ 3 \\ 5 \\ 6
\end{array}
\begin{array}{c}
\begin{array}{cccccccccccc}
1 & 4 & 10 & 3 & 5 & 6 & 7 & 8 & 9 & 2 & 11 & 12
\end{array} \\
\left(\begin{array}{cccc|cc|cccccc}
\mathbf{1} & 1 & 0 & 0 & 0 & 1 & 1 & 0 & 1 & 0 & 1 & 0 \\
0 & \mathbf{1} & 1 & 1 & 0 & 1 & 0 & 1 & 0 & 0 & 1 & 0 \\
0 & 0 & \mathbf{1} & 1 & 1 & 0 & 1 & 0 & 1 & 1 & 0 & 0 \\
0 & 0 & 0 & \mathbf{1} & 0 & 1 & 1 & 1 & 0 & 1 & 0 & 1 \\
\hline
1 & 1 & 0 & 0 & 1 & 0 & 0 & 1 & 1 & 0 & 0 & 1 \\
1 & 0 & 1 & 0 & 1 & 0 & 0 & 0 & 0 & 1 & 1 & 1
\end{array}\right)
\end{array}.
$$

(This matrix is the result of a reordering of the rows and columns of the parity-check matrix H discussed in the subsequent Example A.12. The original row and column labels have been maintained.) The gap is $g = 2$. Let us find ϕ. We perform Gaussian elimination to set E to zero. The result is

$$
\begin{array}{c}
\\ 2 \\ 1 \\ 4 \\ 3 \\ 5{+}2 \\ 6{+}2{+}1{+}3
\end{array}
\begin{array}{c}
\begin{array}{cccccccccccc}
1 & 4 & 10 & 3 & 5 & 6 & 7 & 8 & 9 & 2 & 11 & 12
\end{array} \\
\left(\begin{array}{cccc|cc|cccccc}
\mathbf{1} & 1 & 0 & 0 & 0 & 1 & 1 & 0 & 1 & 0 & 1 & 0 \\
0 & \mathbf{1} & 1 & 1 & 0 & 1 & 0 & 1 & 0 & 0 & 1 & 0 \\
0 & 0 & \mathbf{1} & 1 & 1 & 0 & 1 & 0 & 1 & 1 & 0 & 0 \\
0 & 0 & 0 & \mathbf{1} & 0 & 1 & 1 & 1 & 0 & 1 & 0 & 1 \\
\hline
0 & 0 & 0 & 0 & 1 & 1 & 1 & 1 & 0 & 0 & 1 & 1 \\
0 & 0 & 0 & 0 & 1 & 1 & 0 & 0 & 1 & 0 & 1 & 0
\end{array}\right)
\end{array}.
$$

We see that $\phi = \begin{pmatrix} 1 & 1 \\ 1 & 1 \end{pmatrix}$, which is singular. In order to make ϕ invertible we swap columns 6 and 7 so that ϕ becomes $\begin{pmatrix} 1 & 1 \\ 1 & 0 \end{pmatrix}$. Therefore, the appropriate approximate upper triangular matrix for the encoding is

$$\begin{array}{c} \\ 2 \\ 1 \\ 4 \\ 3 \\ 5 \\ 6 \end{array} \begin{array}{c} \begin{array}{cccccccccccc} 1 & 4 & 10 & 3 & 5 & 7 & 6 & 8 & 9 & 2 & 11 & 12 \end{array} \\ \left(\begin{array}{cccc|cc|cccccc} \mathbf{1} & 1 & 0 & 0 & 0 & 1 & 1 & 0 & 1 & 0 & 1 & 0 \\ 0 & \mathbf{1} & 1 & 1 & 0 & 0 & 1 & 1 & 0 & 0 & 1 & 0 \\ 0 & 0 & \mathbf{1} & 1 & 1 & 1 & 0 & 0 & 1 & 1 & 0 & 0 \\ 0 & 0 & 0 & \mathbf{1} & 0 & 1 & 1 & 1 & 0 & 1 & 0 & 1 \\ \hline 1 & 1 & 0 & 0 & 1 & 0 & 0 & 1 & 1 & 0 & 0 & 1 \\ 1 & 0 & 1 & 0 & 1 & 0 & 0 & 0 & 0 & 1 & 1 & 1 \end{array}\right) \end{array}.$$

Assume we choose $s = (1,0,0,0,0,0)$. To determine p_2 we follow the steps listed in Table A.7 (keep in mind for the sequel that we work in $\mathbb{F}_2$ so that signs do not matter). We get $Bs^T = (1,1,0,1)^T$, $T^{-1}[Bs^T] = (0,1,1,1)^T$, $E[T^{-1}Bs^T] = (1,1)^T$, $Ds^T = (0,0)^T$, $[Ds^T]+[ET^{-1}Bs^T] = (1,1)^T$, and $\phi^{-1}[Ds^T + ET^{-1}Bs^T] = (1,0)^T = p_2^T$. In a similar manner we execute the steps listed in Table A.8 to determine p_1. We get $Ap_2^T = (0,0,1,0)^T$, $[Ap_2^T] + [Bs^T] = (1,1,1,1)^T$, and $T^{-1}[Ap_2^T + Bs^T] = (1,0,0,1)^T = p_1^T$. Therefore, the codeword is

$$x = (p_1, p_2, s) = (1,0,0,1,1,0,1,0,0,0,0,0).$$

A quick check verifies that $Hx^T = 0^T$, as required. ◇

§A.1.1. LDPC Codes over Rings

LDPC codes may be defined over arbitrary fields or, more generally, over rings. In Chapter 7 we saw how certain structured LDPC codes – matched liftings – can be interpreted as LDPC codes over rings. The encoding procedure described here uses only linear algebra so it generalizes to LDPC codes over rings. The only point which needs to be emphasized is that diagonal elements of T need to be invertible (in the case of matched liftings the diagonal elements should have weight 1) and invertibility of ϕ is over the ring.

§A.2. Greedy Upper Triangulation

In the preceding section we have shown how to encode once the parity-check matrix H is in approximate upper triangular form. In this section we address the problem of finding column and row permutations to bring H into such a form. The goal is to make the gap g small. We address the problem in the context of irregular LDPC ensembles.

In general one could ask the problem: given H what is the smallest possible g? We do not know how to solve this problem efficiently in general, and it is likely to be NP-complete. In practice, however, it is virtually always possible to find a sufficiently small g. The algorithm we use in this chapter is simple and greedy and the analysis is stochastic, much like the analysis of the decoding process.

The algorithm operates on H, permuting rows and columns. We introduce t and g, two integer parameters that are initially set to zero and increase as the algorithm proceeds. The parameter t represents time (measured in the number of steps the algorithm has taken) and also the dimensions of the $t \times t$ matrix T as it is constructed. The parameter g represents the number of rows of the matrix assigned to the gap and increases to reach the actual gap at the end of the algorithm. Let H_t denote the matrix after step t with $H_0 = H$. At the end of the algorithm we have $t + g = n - k$. This assumes that H has full row rank (we always assume this in the sequel). The submatrix of H_t consisting of row range $[t + 1, n - k - g]$ and column range $[t+1, n]$ will be referred to as the *residual* parity-check matrix and the corresponding Tanner graph as the *residual* graph. If a column of the residual parity-check matrix contains exactly d non-zero elements then we say that the associated variable node has *residual* degree d. The *minimum residual degree* at time t is the minimum non-zero degree of all variable nodes if at least one node has non-zero residual degree, and it is zero otherwise.[2] The algorithm is listed in Table A.11.

At the completion of the preceding algorithm, the top leftmost $t \times t$ submatrix of H_t, call it T, is upper triangular with non-zero elements along the diagonal.

Example A.12 (Encoder Construction). Let $H_0 = H$ be as follows:

$$
\begin{array}{c}
\\ 1 \\ 2 \\ 3 \\ 4 \\ 5 \\ 6
\end{array}
\begin{array}{c}
\begin{array}{cccccccccccc} 1 & 2 & 3 & 4 & 5 & 6 & 7 & 8 & 9 & 10 & 11 & 12 \end{array} \\
\left(\begin{array}{cccccccccccc}
0 & 0 & 1 & 1 & 0 & 1 & 0 & 1 & 0 & 1 & 1 & 0 \\
1 & 0 & 0 & 1 & 0 & 1 & 1 & 0 & 1 & 0 & 1 & 0 \\
0 & 1 & 1 & 0 & 0 & 1 & 1 & 1 & 0 & 0 & 0 & 1 \\
0 & 1 & 1 & 0 & 1 & 0 & 1 & 0 & 1 & 1 & 0 & 0 \\
1 & 0 & 0 & 1 & 1 & 0 & 0 & 1 & 1 & 0 & 0 & 1 \\
1 & 1 & 0 & 0 & 1 & 0 & 0 & 0 & 0 & 1 & 1 & 1
\end{array}\right)
\end{array}.
$$

In the description of the general algorithm it was convenient to refer to the columns and rows of H_t in their natural order (the physical position if we write down the matrix). For the example, however, it is more convenient to label the rows and columns once and for all and to *always refer to the original* labels.

At the beginning of the algorithm $t = g = 0$. Therefore, the residual degree of all variable nodes is equal to their original degree. For our example all variable nodes have (residual) degree 3 (at time $t = 0$). We proceed to step Choose. We choose a column at random. Assume we pick column 1. This column has 1s in rows 2, 5,

[2] In fact, one can check that if the matrix H_0 has full rank then the minimum residual degree at time $t < n - k - g$ is never zero.

INITIALIZE: Set $H_0 = H$ and $t = g = 0$. Go to CONTINUE.

CONTINUE: If $t = n-k-g$ then stop and output H_t. Otherwise, if the minimum residual degree is 1 go to EXTEND, else go to CHOOSE.

EXTEND: Choose uniformly at random a column $\mathtt{c}$ of residual degree 1 in H_t. Let $\mathtt{r}$ be the row (in the range $[t+1, n-k-g]$) of H_t that contains the (residual) non-zero entry in column $\mathtt{c}$. Swap column $\mathtt{c}$ with column $t+1$ and row $\mathtt{r}$ with row $t+1$. (This places the non-zero element at position $(t+1, t+1)$, extending the diagonal by 1.) Call the resulting matrix H_{t+1}. Increase t by 1 and go to CONTINUE.

CHOOSE: Choose uniformly at random a column $\mathtt{c}$ in H_t with minimum positive residual degree, call the degree d. Let $\mathtt{r}_1, \mathtt{r}_2, \dots, \mathtt{r}_d$ denote the rows of H_t in the range $[t+1, n-k-g]$ which contain the d residual non-zero entries in column $\mathtt{c}$. Swap column $\mathtt{c}$ with column $t+1$. Swap row $\mathtt{r}_1$ with row $t+1$ and move rows $\mathtt{r}_2, \mathtt{r}_3, \dots, \mathtt{r}_d$ to the bottom of the matrix. Call the resulting matrix H_{t+1}. Increase t by 1 and increase g by $d-1$. Go to CONTINUE.

Figure A.11: Greedy algorithm to perform approximate upper triangulation.

and 6. Since the chosen column is already in the leftmost position we do not need to swap it. Next we swap row 2 with row 1. Rows 5 and 6 are already at the bottom of the matrix. The parameters are now $t = 1$ and $g = 2$ and the resulting matrix H_1 is

$$
\begin{array}{c}
 \\ 2 \\ 1 \\ 3 \\ 4 \\ 5 \\ 6
\end{array}
\begin{array}{c}
\begin{array}{cccccccccccc} 1 & 2 & 3 & 4 & 5 & 6 & 7 & 8 & 9 & 10 & 11 & 12 \end{array} \\
\left(\begin{array}{c|ccccccccccc}
\mathbf{1} & 0 & 0 & 1 & 0 & 1 & 1 & 0 & 1 & 0 & 1 & 0 \\ \hline
0 & 0 & 1 & 1 & 0 & 1 & 0 & 1 & 0 & 1 & 1 & 0 \\
0 & 1 & 1 & 0 & 0 & 1 & 1 & 1 & 0 & 0 & 0 & 1 \\
0 & 1 & 1 & 0 & 1 & 0 & 1 & 0 & 1 & 1 & 0 & 0 \\ \hline
1 & 0 & 0 & 1 & 1 & 0 & 0 & 1 & 1 & 0 & 0 & 1 \\
1 & 1 & 0 & 0 & 1 & 0 & 0 & 0 & 0 & 1 & 1 & 1
\end{array}\right)
\end{array}.
$$

Returning to step CONTINUE, we see that the minimum residual degree in H_1 is 1 (recall that for the residual degree we consider only the rows in the physical range $[t+1, n-k-g]$; these are the rows between the two horizontal bars). The columns with this property are 4, 5, 9, 11, and 12. We therefore go to step EXTEND. Assume we select column 4. To bring the 1 in column 4 into the right position along the diagonal we swap column 4 with column 2. We are lucky; no row swaps are necessary. The parameters are now $t = 2$ and $g = 2$ and the resulting matrix H_2 is

$$
\begin{array}{c} \\ 2 \\ 1 \\ 3 \\ 4 \\ 5 \\ 6 \end{array}
\begin{array}{c}
\begin{array}{cc|cccccccccc} 1 & 4 & 3 & 2 & 5 & 6 & 7 & 8 & 9 & 10 & 11 & 12 \end{array} \\
\left(\begin{array}{cc|cccccccccc}
\mathbf{1} & 1 & 0 & 0 & 0 & 1 & 1 & 0 & 1 & 0 & 1 & 0 \\
0 & \mathbf{1} & 1 & 0 & 0 & 1 & 0 & 1 & 0 & 1 & 1 & 0 \\ \hline
0 & 0 & 1 & 1 & 0 & 1 & 1 & 1 & 0 & 0 & 0 & 1 \\
0 & 0 & 1 & 1 & 1 & 0 & 1 & 0 & 1 & 1 & 0 & 0 \\ \hline
1 & 1 & 0 & 0 & 1 & 0 & 0 & 1 & 1 & 0 & 0 & 1 \\
1 & 0 & 0 & 1 & 1 & 0 & 0 & 0 & 0 & 1 & 1 & 1
\end{array}\right)
\end{array}.
$$

The minimum residual degree in H_2 is again 1. The columns with this property are 5, 6, 8, 9, 10, and 12. Assume we select column 10. To get the 1 into the right position we swap columns 10 and 3 and then rows 3 and 4. The parameters after this step are $t = 3$ and $g = 2$ and the resulting matrix H_3 is

$$
\begin{array}{c} \\ 2 \\ 1 \\ 4 \\ 3 \\ 5 \\ 6 \end{array}
\begin{array}{c}
\begin{array}{ccc|ccccccccc} 1 & 4 & 10 & 2 & 5 & 6 & 7 & 8 & 9 & 3 & 11 & 12 \end{array} \\
\left(\begin{array}{ccc|ccccccccc}
\mathbf{1} & 1 & 0 & 0 & 0 & 1 & 1 & 0 & 1 & 0 & 1 & 0 \\
0 & \mathbf{1} & 1 & 0 & 0 & 1 & 0 & 1 & 0 & 1 & 1 & 0 \\
0 & 0 & \mathbf{1} & 1 & 1 & 0 & 1 & 0 & 1 & 1 & 0 & 0 \\ \hline
0 & 0 & 0 & 1 & 0 & 1 & 1 & 1 & 0 & 1 & 0 & 1 \\ \hline
1 & 1 & 0 & 0 & 1 & 0 & 0 & 1 & 1 & 0 & 0 & 1 \\
1 & 0 & 1 & 1 & 1 & 0 & 0 & 0 & 0 & 0 & 1 & 1
\end{array}\right)
\end{array}.
$$

We perform one more step. We see that the minimum residual degree is once more 1. The columns with this property are 2, 6, 7, 8, 3, and 12. The algorithm asks that we choose one of those columns with uniform probability. Assume that we choose column 3. Swapping columns 2 and 3 and increasing t by 1 gives us the matrix H_4:

$$
\begin{array}{c} \\ 2 \\ 1 \\ 4 \\ 3 \\ 5 \\ 6 \end{array}
\begin{array}{c}
\begin{array}{cccc|cccccccc} 1 & 4 & 10 & 3 & 5 & 6 & 7 & 8 & 9 & 2 & 11 & 12 \end{array} \\
\left(\begin{array}{cccc|cccccccc}
\mathbf{1} & 1 & 0 & 0 & 0 & 1 & 1 & 0 & 1 & 0 & 1 & 0 \\
0 & \mathbf{1} & 1 & 1 & 0 & 1 & 0 & 1 & 0 & 0 & 1 & 0 \\
0 & 0 & \mathbf{1} & 1 & 1 & 0 & 1 & 0 & 1 & 1 & 0 & 0 \\
0 & 0 & 0 & \mathbf{1} & 0 & 1 & 1 & 1 & 0 & 1 & 0 & 1 \\ \hline
1 & 1 & 0 & 0 & 1 & 0 & 0 & 1 & 1 & 0 & 0 & 1 \\
1 & 0 & 1 & 0 & 1 & 0 & 0 & 0 & 0 & 1 & 1 & 1
\end{array}\right)
\end{array}.
$$

Since $t + g = n - k = 6$ the upper triangulation procedure is complete and we see that the final gap is $g = 2$. ◇

If you sensed something familiar in the encoder construction it is because it is erasure decoding (in the sense of the peeling decoder) in disguise: the roles of check nodes and variable nodes are simply reversed. We refer to this as *dual erasure decoding*. This connection is the key to our analysis of the algorithm. To illustrate the connection let us look again at the example but this time from the Tanner graph perspective.

Figure A.13 shows the Tanner graph corresponding to H_0. Let us consider the

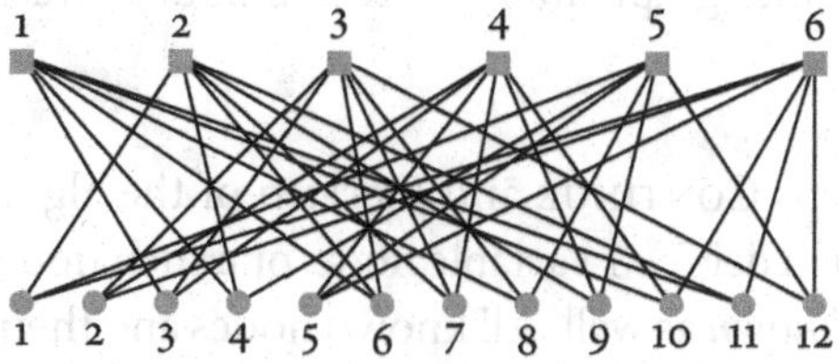

Figure A.13: Tanner graph corresponding to H_0.

first step, consisting of selecting variable (column) 1 in step CHOOSE. Let us represent that step by splitting that node into three (the degree of the node) degree 1 nodes as illustrated in Figure A.14: node 1 is split threefold and is now represented by nodes 1_a, 1_b, and 1_c. The check nodes attached to nodes 1_b and 1_c correspond to

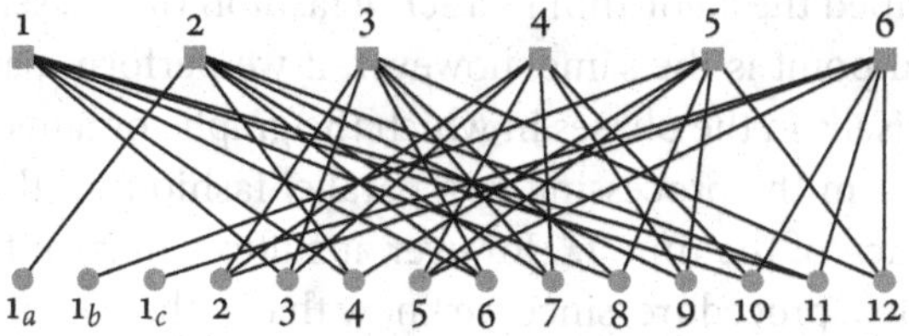

Figure A.14: Tanner graph after splitting of node 1.

the two rows which were placed at the bottom of the matrix (increasing the gap by 2). The check node attached to node 1_a corresponds to the row which was placed at the top of the matrix and the connected edge corresponds to the 1 which became the first diagonal element of T.

As t is increased by 1 and g is increased by 2, the nodes 1_a, 1_b, and 1_c and their neighboring check nodes, nodes 2, 5, and 6 are effectively removed from H. Note that this is equivalent to performing one step of dual erasure decoding on the graph in Figure A.14. More precisely, we assume that degree 1 variable nodes transmit *known* messages to their neighbors which in turn become *known*. The known nodes are removed from the graph. The remaining residual graph is shown in Figure A.15. This is the graph corresponding to the residual matrix, i.e., the submatrix of H_1 which lies between the two horizontal lines and is to the right of the vertical line (physical columns 2 through 12 and physical rows 2 through 4).

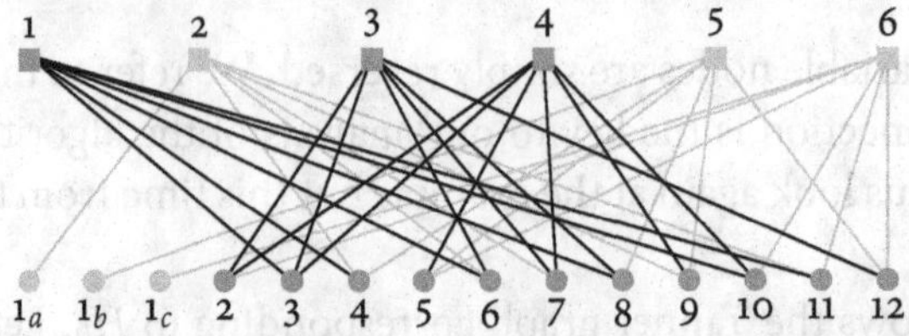

Figure A.15: Tanner graph after one round of dual erasure decoding.

We now repeat steps Continue and Extend of the algorithm. As long as the residual graph contains a degree 1 variable node, one such node becomes *known* and its neighbor becomes *known* as well. All known nodes and their connected edges are eliminated. In the next iteration of our example, variable node 4 becomes known and it is peeled off together with its neighboring check node 1 and all the edges that are connected to either of these two nodes. This process continues until there are no more degree 1 variable nodes in the residual graph. That may happen because the residual graph becomes empty or because the algorithm gets stuck in the maximum dual stopping set. When the latter happens, the algorithm returns to step Choose and the process reinitiates.

We have described the algorithm in a *serial* fashion (in the sense of the peeling decoder). The final point is the same, however, if we perform parallel erasure decoding. More precisely, in the phases in which the graph contains degree 1 variable nodes we can perform the processing in a parallel fashion. In that case it is more difficult to keep track of the time t. However, it suffices to keep track of the gap g throughout the whole procedure, since we know that at the end of the algorithm we have $t = n - k - g$. Thus, the algorithm can be analyzed from the perspective of dual erasure decoding.

§A.3. Linear Encoding Complexity

We are now ready to give a sufficient condition on a degree distribution pair (λ, ρ) to give rise to linear encoding complexity. We start with ensembles that contain degree 1 variable nodes.

Theorem A.16 (Ensembles with Linear Encoding Complexity). Let (λ, ρ) be a degree distribution pair satisfying $1 - z - \rho(1 - \lambda(z)) > 0$ for $z \in [0, 1)$ (this implies $\lambda_1 > 0$) and $\mathtt{r}_{\min} > 2$. Let $\mathtt{G}$ be chosen uniformly at random from the ensemble LDPC(n, λ, ρ). Then, the gap is zero with probability $1 - O(1/n)$.

Proof. Consider dual erasure decoding. The update equations follow directly from the standard density evolution equations by exchanging the roles of the variable and check nodes. Let z_ℓ denote the fraction of *known* check-to-variable messages at the

ℓ-th iteration. Then

$$z_{\ell+1} = 1 - \rho(1 - \lambda(z_\ell)), \qquad z_0 = 0. \tag{A.17}$$

Recall that for dual erasure decoding the roles of λ and ρ are exchanged. Consider therefore the "dual" ensemble LDPC $(n(1-r), \rho, \lambda)$.

By assumption we have $\lambda_1 > 0$. The upper triangulation process therefore starts by itself (without having to enter the CHOOSE step). Further, by assumption, $1 - z - \rho(1 - \lambda(z)) > 0$ for all $z \in [0, 1)$. We therefore know from the analysis of the peeling decoder in Section 3.19 that the process continues until the graph has shrunk to a size less than ηn (where η is an arbitrary strictly positive constant) with probability at least $1 - e^{-c\sqrt{n}}$. Here, c is some strictly positive constant.

Further, from our analysis of the weight distribution of LDPC ensembles in Section 3.24, in particular the arguments in the proof of Lemma 3.166, we know that if $\mathtt{r}_{\min} > 2$ then there exists a strictly positive constant η so that a randomly chosen element from this ensemble has *no* stopping sets of size less than ηn with probability converging to 1 at the speed $1/n$.

This means that typical elements from this ensemble have a linear erasure correcting radius. We conclude that if $\lambda_1 > 0$, then with probability converging to 1 the matrix can be completely upper triangulated, i.e., we conclude that typically $g = 0$. □

THEOREM A.18 (ENSEMBLES WITH LINEAR ENCODING COMPLEXITY). Let (λ, ρ) be a degree distribution pair satisfying $1 - z - \rho(1 - \lambda(z)) > 0$ for $z \in (0, 1)$ (here we allow equality at 0), $\mathtt{r}_{\min} > 2$, and the *strict* inequality $\lambda'(0)\rho'(1) > 1$. Let $\mathtt{G}$ be chosen uniformly at random from the ensemble LDPC (n, λ, ρ). Then, there exists a strictly positive number c such that for each $k \in \mathbb{N}$ the probability that $g \leq k$ is asymptotically (in the blocklength) lower bounded by $1 - (1 - c)^k$.

Proof. Our preceding discussion started with the assumption that $\lambda_1 > 0$. This condition guaranteed that the upper triangulation process starts. We know already from Chapter 3 (see Problem 3.6) that good irregular ensembles do not fulfill this condition. What happens if $\lambda_1 = 0$? In this case we see from the update equations that the upper triangulation process does not start by itself. But since by assumption $\lambda'(0)\rho'(1) > 1$, the fixed point $z = 0$ is unstable and a small "nudge" gets the process started. More precisely, consider the subgraph induced by the variable nodes of degree 2. Since by assumption we have $\lambda'(0)\rho'(1) > 1$, this subgraph contains a "giant" (linear-sized) connected component (see the discussion in Section C.5 and in particular Lemma C.38). Assume we choose one of those nodes of degree 2 in the first step CHOOSE. The process then continues until at least all variable nodes in this giant component have been upper triangulated. If we wait until a linear-sized portion of the graph has been processed then we can use the technique of density evolution to show that the process continues with high probability until the residual graph has reached size essentially zero. It follows that in such a case we have achieved an approximate upper triangulation with $g = 1$. Assume now that we proceed as stated in the algorithm and choose nodes of minimum residual degree (which in the beginning is degree 2) at random in step CHOOSE. Then at each such step we have a non-zero chance of choosing such a node of degree 2 which is part of this linear-sized giant component. Let this linear fraction be denoted by c. If we do, which happens with probability c, then we increase the gap by 1 and we are done with high probability. If not, then we also increase the gap by one but most likely the process stops after a finite number of steps. Since in the latter case we only touch a finite number of nodes and edges, a finite number of consecutive trials behave asymptotically like independent trials. □

In the preceding statement we can relax the condition $\mathtt{r}_{\min} > 2$ to $\lambda'(1)\rho'(0) < 1$. In this case we know from our analysis of the error floor in Chapter 3 that most codes only contain a small number of low-weight (sublinear-sized) stopping sets. Optimized degree distributions tend to fulfill the conditions of Theorem A.18.

EXAMPLE A.19 (ENSEMBLE WITH LINEAR ENCODING COMPLEXITY). Consider the degree distribution pair

$$\lambda(x) = 0.272536x + 0.237552x^2 + 0.070380x^3 + 0.419532x^9,$$
$$\rho(x) = 0.7x^6 + 0.3x^7.$$

The ensemble LDPC (n, λ, ρ) is of rate one-half and has a BP threshold of $\sigma^{\text{BP}} \approx$

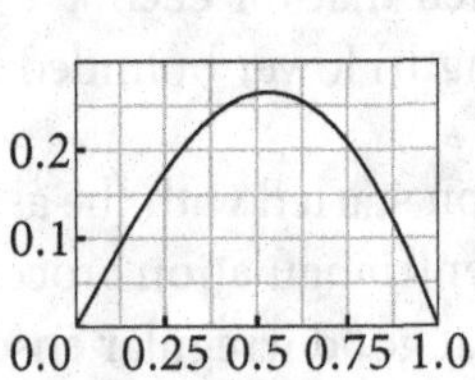

Figure A.20: $1 - z - \rho(1 - \lambda(z))$.

0.9564880 for transmission over the BAWGNC(σ). This is less than 0.2 dB away from capacity. Figure A.20 shows that the function $1 - z - \rho(1 - \lambda(z))$ is strictly positive for $z \in (0, 1)$. Further, $\lambda'(0)\rho'(1) = 1.71694 > 1$ and $\mathtt{r}_{\min} = 7 > 2$. Therefore, if we apply the greedy algorithm to a sufficiently long randomly chosen element from this ensemble we expect to see a small gap. Figure A.21 shows the result of applying the greedy algorithm to a randomly chosen element of length 1024. Indeed, the resulting gap is 1. ◇

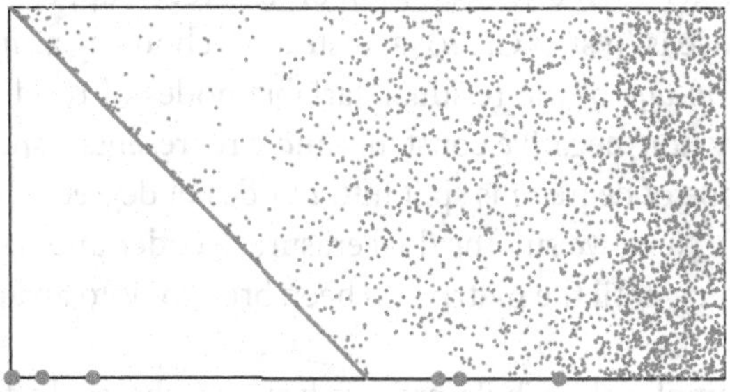

Figure A.21: Element chosen uniformly at random from LDPC $(1024, \lambda, \rho)$, with (λ, ρ) as described in Example A.19, after the application of the greedy algorithm. For the particular experiment we get $g = 1$. The non-zero elements in the last row (in the gap) are drawn larger to make them more visible.

Example A.22 (Gap of Finite-Length Experiment). Although optimized ensembles tend to fulfill the conditions of Theorem A.18, it is easy to construct ensembles which do not. A simple such case is the $(3, 6)$-regular ensemble. Figure A.23 shows the result of applying the greedy algorithm to an element chosen uniformly at random from LDPC $\left(2048, x^2, x^5\right)$. The result is $g = 39$. This corresponds to a relative gap of $39/2048 \approx 0.019$. This is quite close to the asymptotic value of 0.01709, which we compute in the next section. ◇

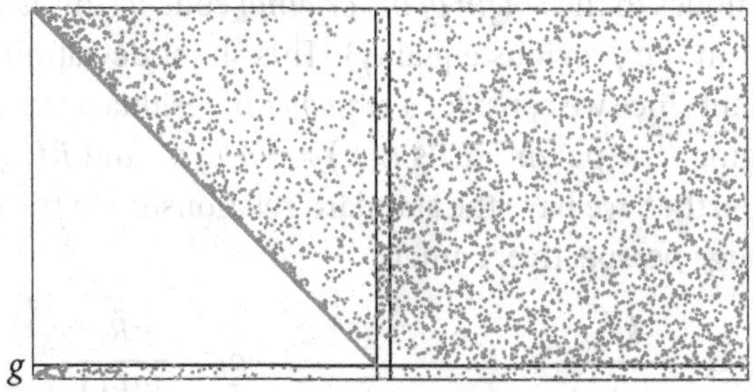

Figure A.23: Element chosen uniformly at random from LDPC $\left(2048, x^2, x^5\right)$ after the application of the greedy algorithm. The result is $g = 39$.

§A.4. Analysis of Asymptotic Gap

So far we have focused on cases where the asymptotic gap is zero. How can we determine the asymptotic gap in the general case?

Consider an easier-to-analyze variation of the greedy upper triangulation algorithm. The difference is in step Choose: instead of choosing a *single* variable node we choose a *fraction* $\omega_i \epsilon$ of the residual variable nodes of residual degree i. Here the ω_i are arbitrary non-negative constants and ϵ represents a small step size. Each chosen variable node of degree i is split into i nodes of degree 1 (with $i-1$ of them going into the gap). Next, we run the dual erasure decoder until it gets stuck and we iterate these two steps until the matrix has been brought into an approximate upper triangular form.

For typical examples the whole process has two phases. In the first phase the process proceeds in small steps and iterates between CHOOSE and EXTEND. If we let the step size in this phase tend to zero we can describe the behavior of the system by a differential equation. At the beginning of the second phase the residual degree distribution has been transformed to one which has (essentially) zero expected gap and the algorithm finishes in one giant step. We know from our previous discussion that such a giant step appears if the parameter $\lambda'(0)\rho'(1)$ takes on the critical value 1 (or exceeds it). Of course, there are degree distribution pairs that exhibit a more complicated behavior and in which we have several of the preceding two phases interlaced. But in the sequel we limit our discussion to the simplest case.

The analysis is most easily expressed from the node perspective of the residual graph. We use the notation $\tilde{\cdot}$ to indicate quantities associated to the residual graph. Let n denote the number of variable nodes in the original graph and let $\tilde{L}_k n$ be the number of variable nodes of degree k in the residual graph. Thus, initially $\tilde{L}_k = L_k$. Similarly, let $\tilde{R}_k n$ be the number of check nodes of degree k in the residual graph. Initially we have $\tilde{R}_k = (1-r)R_k$. Note that we normalized the numbers of both variable and check nodes to the *length of the original* code n (this is why we have the extra factor $(1-r)$ in the previous equality). This slight deviation from our usual notation is convenient since we consider a *sequence* of residual graphs whose lengths and rates are changing. The polynomials $\tilde{L}(x) = \sum_k \tilde{L}_k x^k$ and $\tilde{R}(x) = \sum_k \tilde{R}_k x^k$ are defined as before. Further, we have the standard relationship to the degree distribution pair from an edge perspective, namely

$$\tilde{\lambda}_k = \frac{k\tilde{L}_k}{\tilde{L}'(1)}, \qquad \tilde{\rho}_k = \frac{k\tilde{R}_k}{\tilde{R}'(1)}.$$

Initially, when the residual graph is the entire graph, we have $\tilde{\lambda} = \lambda$ and $\tilde{\rho} = \rho$.

Lemma A.24 (Asymptotic Gap of Greedy Algorithm). Assume we apply the greedy upper triangulation algorithm to an element G chosen uniformly at random from the ensemble LDPC (n, λ, ρ) where $\lambda'(0)\rho'(1) < 1$. Let $\{\omega_i\}_i$, $2 \le i \le 1_{\max}$, denote some fixed set of positive real numbers. Define

$$r(u) = \frac{e^{-u}\rho'(e^{-u})}{\rho(e^{-u})}.$$

Consider the solution to the following system of differential equations

$$\frac{d\tilde{L}_k}{du} = -\frac{\omega_k \tilde{L}_k(\tilde{L}'(1) - 2\tilde{L}_2 r)}{\sum_{j\ge 2} j\omega_j \tilde{L}_j} + r\big((k+1)\tilde{L}_{k+1} - k\tilde{L}_k\big), k \ge 2,$$

$$\frac{dg}{du} = \frac{(\sum_{j\ge 2}(j-1)\omega_j \tilde{L}_k)(\tilde{L}'(1) - 2\tilde{L}_2 r)}{\sum_{j\ge 2} j\omega_j \tilde{L}_j},$$

with the boundary condition $\tilde{L}_k(0) = L_k$. Let u^* be the smallest u where $\tilde{\lambda}'(0)r = 1$. If the resulting solution $(\tilde{\lambda}(u^*), \tilde{\rho}(u^*))$ satisfies the conditions of Theorem A.18, except that here $\tilde{\lambda}'(0)\tilde{\rho}'(1) = 1$, then the gap concentrates (up to order n) around the solution $g(u^*)$ of this system.

Proof. The basic idea of the proof is to take tiny steps in the greedy algorithm and take the limit as the step size goes to zero, arriving at the given differential equation. We are quite casual in our analysis. Strictly speaking we should proceed as follows. We are given a degree distribution pair (λ, ρ), a set of weights $\{\omega_i\}_i$, and a *small* step size ϵ. The algorithm proceeds in small phases. The first phase consists of choosing a random subset of the variables which are then split. The second phase consists of running the dual erasure decoder. These phases alternate until the whole matrix has been processed. Each phase can be analyzed separately by the *Wormald* method discussed in Section C.4 and we get the overall solution by pasting the individual solutions together. We consider right away the behavior of the system for the limit in which we let the step size approach zero.

Let us track the evolution of the residual variables at each of the steps. At the beginning of a cycle the number of degree 1 variable nodes in the residual graph is zero. Indeed, this is the reason why we perform a CHOOSE step. In the CHOOSE step each chosen variable node of degree k is split into k nodes of degree 1. As a result, the number of degree 1 variable nodes becomes $\sum_{k\ge 2} k\tilde{L}_k \omega_k \epsilon$. Further, for $k \ge 2$, $\tilde{L}_k$ is *decreased* to $\tilde{L}_k(1 - \omega_k \epsilon)$. Note that $n\tilde{L}'(1)$ is the number of edges in the graph and that this number is not altered in the CHOOSE step. Hence, $\tilde{\lambda}_1$ is linear in ϵ and

$$\frac{d\tilde{\lambda}_1}{d\epsilon} = \frac{\sum_k k\omega_k \tilde{L}_k}{\tilde{L}'(1)}. \tag{A.25}$$

Now consider running dual erasure decoding until a fixed point z of (A.17) is reached. Just before the CHOOSE step we had $z = 0$. Further, in the CHOOSE step $\tilde{\rho}$ stays unchanged and the coefficients of $\tilde{\lambda}$ change proportional to ϵ as we have just discussed.

Since the step size is very small we can expand the right-hand side of the fixed point equation $z = 1 - \tilde{\rho}(1 - \tilde{\lambda}(z))$ in a series keeping only the terms linear in ϵ and z to get

$$z = \tilde{\rho}'(1)\frac{\mathrm{d}\tilde{\lambda}_1}{\mathrm{d}\epsilon}\epsilon + \tilde{\rho}'(1)\tilde{\lambda}'(0)z + O(\epsilon^2). \tag{A.26}$$

From this we conclude that

$$\frac{\mathrm{d}z}{\mathrm{d}\epsilon} = \frac{\tilde{\rho}'(1)}{1 - \tilde{\lambda}'(0)\tilde{\rho}'(1)}\frac{\mathrm{d}\tilde{\lambda}_1}{\mathrm{d}\epsilon}. \tag{A.27}$$

At the fixed point the probability of an edge carrying a *known* message into a *variable* node is z. These edges are removed from the residual graph (see descriptions of Figures A.13–A.15). A variable node of degree k in the residual graph receives a single known message (and therefore its degree is decreased by 1) with probability $kz(\epsilon) + O(\epsilon^2)$ and more than 1 with probability $O(\epsilon^2)$. Therefore, for $k \geq 2$,

$$\frac{\mathrm{d}\tilde{L}_k}{\mathrm{d}\epsilon} = \underbrace{-\omega_k\tilde{L}_k}_{\text{CHOOSE}} + \underbrace{\big((k+1)\tilde{L}_{k+1} - k\tilde{L}_k\big)\frac{\mathrm{d}z}{\mathrm{d}\epsilon}}_{\text{EXTEND and CONTINUE}}.$$

The probability of carrying a *known* message into a *check* node is

$$\frac{\mathrm{d}\tilde{\lambda}_1}{\mathrm{d}\epsilon}\epsilon + \tilde{\lambda}'(0)z(\epsilon) + O(\epsilon^2) \overset{\text{(A.26)}}{=} \frac{z}{\tilde{\rho}'(1)} + O(\epsilon^2).$$

We conclude that a check node of degree k receives a known message with probability $k\tilde{R}_k\frac{z(\epsilon)}{\tilde{\rho}'(1)} + O(\epsilon^2)$ and such a check node is removed. Hence,

$$\frac{\mathrm{d}\tilde{R}_k}{\mathrm{d}\epsilon} = -\frac{k\tilde{R}_k}{\tilde{\rho}'(1)}\frac{\mathrm{d}z}{\mathrm{d}\epsilon}.$$

It is convenient to switch the independent variable to z. We get

$$\frac{\mathrm{d}\tilde{L}_k}{\mathrm{d}z} = -\omega_k\tilde{L}_k\frac{\mathrm{d}\epsilon}{\mathrm{d}z} + \big((k+1)\tilde{L}_{k+1} - k\tilde{L}_k\big), \tag{A.28}$$

$$\frac{\mathrm{d}\tilde{R}_k}{\mathrm{d}z} = -\frac{k\tilde{R}_k}{\tilde{\rho}'(1)}, \tag{A.29}$$

where

$$\frac{\mathrm{d}\epsilon}{\mathrm{d}z} = \frac{1 - \tilde{\lambda}'(0)\tilde{\rho}'(1)}{\tilde{\rho}'(1)}\frac{\tilde{L}'(1)}{\sum_k k\omega_k\tilde{L}_k} = \frac{(\tilde{L}'(1) - 2\tilde{L}_2 r)}{\sum_{j\geq 2} j\omega_j\tilde{L}_j}, \tag{A.30}$$

which we get by combining (A.25) and (A.27). The gap g is a derived variable; its rate of growth is given by

$$\frac{\mathrm{d}g}{\mathrm{d}z} = \sum_k (k-1)\omega_k \tilde{L}_k \frac{\mathrm{d}\epsilon}{\mathrm{d}z}.$$

The rate of growth of the gap g plus the rate of growth of the size of the triangular system t equals in magnitude the rate at which check nodes are removed. Hence

$$\frac{\mathrm{d}(g+t)}{\mathrm{d}z} = -\sum_k \frac{\mathrm{d}\tilde{R}_k}{\mathrm{d}z}.$$

Now, we can simplify the equations and solve for $\tilde{R}$ in closed form by expressing the equations in terms of a new parameter u defined by $\frac{\mathrm{d}u}{\mathrm{d}z} = \frac{1}{\tilde{\rho}'(1)}$, so that

$$\frac{\mathrm{d}\tilde{R}_k}{\mathrm{d}u} \overset{(A.29)}{=} -k\tilde{R}_k.$$

The solution to this differential equation is $\tilde{R}_k(u) = \mathrm{e}^{-ku}\tilde{R}_k(0)$. We can now write down $\tilde{\rho}'(1)$ in terms of $\tilde{R}$ if we recall that $\tilde{\rho}_j = (j\tilde{R}_j)/(\sum_k k\tilde{R}_k)$. This gives

$$\begin{aligned}\tilde{\rho}'(1)(u) &= \frac{\sum_k k(k-1)\mathrm{e}^{-ku}\tilde{R}_k(0)}{\sum_k k\mathrm{e}^{-ku}\tilde{R}_k(0)} = \frac{\sum_k (k-1)\mathrm{e}^{-ku}\rho_k}{\sum_k \mathrm{e}^{-ku}\rho_k} \\ &= \frac{\mathrm{e}^{-2u}\rho'(\mathrm{e}^{-u})}{\mathrm{e}^{-u}\rho(\mathrm{e}^{-u})} = r(u).\end{aligned}$$

We already have an explicit solution of $\tilde{R}_k$ in terms of u. It remains to rewrite the solution of $\tilde{L}_k$ given in (A.28) in terms of u. This can be done by multiplying both sides of (A.28) by $\frac{\mathrm{d}z}{\mathrm{d}u}$ and using (A.30) as well as the relationship $\frac{\mathrm{d}u}{\mathrm{d}z} = \frac{1}{\tilde{\rho}'(1)}$. This gives the equation stated in the theorem.

Once the point where $\tilde{\lambda}'(0)\tilde{\rho}'(1) = 1$ has been reached, a tiny Choose step gives us $\lambda_1 > 0$ and the conditions of Theorem A.16 are met. Thus the gap increases by only an arbitrarily small fraction thereafter. □

Example A.31 (Asymptotic Gap of $(3,6)$ Ensemble). For the regular $(3,6)$ ensemble the initial condition is $\tilde{L}_3(0) = 1$. To mimic the behavior of the greedy algorithm we choose $\omega_2 = \gamma$ and $\omega_3 = \gamma^2$ for a small positive γ and consider the solution as γ tends to zero. We have $r(u) = 5$. Figure A.32 shows the solution of the differential equation. As we can see, we have $u^* \approx 0.0247856$, i.e., for this u^* we have $\tilde{\lambda}'(0)r = 1$. Further, the resulting degree distribution at this point is

$$\tilde{\lambda}(x) = 0.2x + 0.8x^2, \qquad \tilde{\rho}(x) = x^5.$$

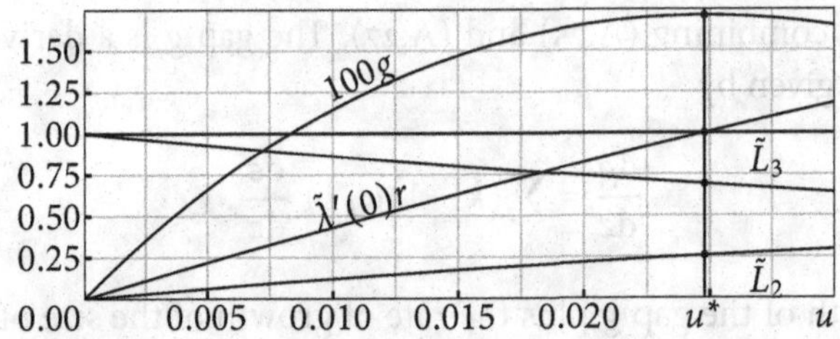

Figure A.32: Evolution of the differential equation for the $(3,6)$-regular ensemble. For $u^* \approx 0.0247856$ we have $\tilde{\lambda}'(0)r = 1$, $\tilde{L}_2 \approx 0.2585$, $\tilde{L}_3 \approx 0.6895$, and $g \approx 0.01709$.

Since for this residual degree distribution $x < \tilde{\rho}(1-\tilde{\lambda}(1-x))$ on $(0,1)$, the assumptions of Lemma A.24 are fulfilled. Therefore, the asymptotic gap is $g(u^*) \approx 0.01709$. ◇

Notes

The material presented in this chapter first appeared in the paper by Richardson and Urbanke [4]. This paper also contains the description and analysis of several variants of the greedy upper triangulation process. There are many concerns which bear on practical implementations of the preceding algorithm. For example, variable nodes which are selected in the Choose step end up as parity bits. Typically, nodes of degree 2 have the highest bit error rate. Thus, it is often preferable to use as many low-degree variable nodes for parity bits as possible. Therefore, it may be preferable to select nodes in the Choose step whose *original* degree is minimal.

MacKay, Wilson, and Davey [3] proposed to redefine the ensemble to consist only of codes that have a parity-check matrix in approximate upper triangular form. In a similar spirit, Freundlich, Burshtein, and Litsyn proposed an ensemble of irregular LDPC codes that have a linear encoding complexity and that asymptotically have the same performance as standard irregular LDPC ensembles [1]. Alternatively, Sipser and Spielman [5] as well as Luby, Mitzenmacher, Shokrollahi, Spielman, and Stemann [2] used a *cascaded* graph which leads to a linear time encoding complexity as well.

Problems

A.1 (Encoding Example). Consider the code of Example A.10. Compute the codeword that corresponds to $s = (0,1,0,0,0,0)$ according to the recipe described in Table A.9. ◇

A.2 (Asymptotic gap for $(3,5)$ Ensemble). Proceed as in Example A.31 and show that the asymptotic gap for the $(3,5)$-regular ensemble is approximately 0.0275113.

References

[1] S. Freundlich, D. Burshtein, and S. L. Litsyn, *Approximately lower triangular ensembles of LPDC, USA codes with linear encoding complexity*, in Proc. of the IEEE Int. Symposium on Inform. Theory, Seattle, WA, USA, July 2006, pp. 821–825. [456]

[2] M. Luby, M. Mitzenmacher, A. Shokrollahi, D. A. Spielman, and V. Stemann, *Practical loss-resilient codes*, in Proc. of the 29th annual ACM Symposium on Theory of Computing, 1997, pp. 150–159. [157, 262, 434, 456]

[3] D. J. C. MacKay, S. T. Wilson, and M. C. Davey, *Comparison of constructions of irregular Gallager codes*, in Proc. of the Allerton Conf. on Commun., Control, and Computing, Monticello, IL, USA, Sept. 1998, pp. 220–229. [456]

[4] T. Richardson and R. Urbanke, *Efficient encoding of low-density parity-check codes*, IEEE Trans. Inform. Theory, 47 (2001), pp. 638–656. [159, 456]

[5] M. Sipser and D. A. Spielman, *Expander codes*, IEEE Trans. Inform. Theory, 42 (1996), pp. 1710–1722. [434, 456]

Appendix B

EFFICIENT IMPLEMENTATION OF DENSITY EVOLUTION

Density evolution plays a fundamental role in the analysis of iterative systems; it is also a valuable tool in the design of such systems. Actual computation of density evolution for low-density parity-check (LDPC) codes requires an algorithmic implementation. Finding an efficient such implementation is a challenging problem. In this section we show how this can be done.

When implementing density evolution one may or may not assume symmetry of the densities. Working directly in the space of symmetric distributions yields the most efficient implementations. The interaction of the symmetry with the practical constraints of finite support typically leads to optimistic predictions of the threshold. On the other hand, sometimes one is interested specifically in consequences of non-symmetry. Moreover, by allowing densities to be non-symmetric one can compute density evolution for a message-passing decoder which corresponds to a quantized version of belief propagation. Since belief propagation is optimal, thresholds computed this way are lower bounds on the true belief propagation threshold. We will not assume strict symmetry in general, but we will assume that densities are "nearly" symmetric.

We use the notation $\star$ to denote standard convolution over $\mathbb{R}$, $\mathbb{Z}$, or $\mathbb{Z}/N\mathbb{Z}$ – the ring of integers modulo N. Variable-node domain convolution, which we have denoted by $\circledast$, is standard convolution but we shall use $\star$ to emphasize computational aspects.

In our discussions of density evolution we focus on *pairwise* convolutions of two densities. What we typically need in an implementation of density evolution, however, is the computation of $\lambda(\mathsf{a})$ for a given polynomial $\lambda(x)$ and a given density a (and the equivalent operation at the check-node side). Consider the case $\lambda(x) = \lambda_2 x + \lambda_3 x^2 + \lambda_6 x^5 + \lambda_{13} x^{12}$, where $\lambda_i > 0$ for each given i. What is the most efficient way of computing $\lambda(\mathsf{a})$ using only pairwise convolutions? We can, e.g., compute, in the stated order,

$$\mathsf{a}^{\star 2} = \mathsf{a} \star \mathsf{a}, \qquad \mathsf{a}^{\star 3} = \mathsf{a} \star \mathsf{a}^{\star 2}, \qquad \mathsf{a}^{\star 5} = \mathsf{a}^{\star 2} \star \mathsf{a}^{\star 3},$$
$$\mathsf{a}^{\star 6} = \mathsf{a}^{\star 3} \star \mathsf{a}^{\star 3}, \qquad \mathsf{a}^{\star 12} = \mathsf{a}^{\star 6} \star \mathsf{a}^{\star 6},$$

which requires five convolutions. If we are interested in computing efficiently a single, let's say, d-fold, convolution by means of pairwise convolutions, then the problem is identical to the well-studied problem of how to efficiently compute a given integer power of a number. It is known that the minimum number of required pairwise convolutions is at least $\lceil \log_2 d \rceil$ (with equality if d is a power of 2) and at most $\lfloor \log_2 d \rfloor + \nu(d) - 1$, where $\nu(d)$ denotes the number of 1 in the binary representation of d. The latter upper bound is achieved by the so-called *binary method* (see Problem B.1). The actual optimal value (and the corresponding sequence of pairwise convolutions) has been tabulated for numbers up to several hundred. For density evolution we are usually required to compute a set of powers, so our problem is a little more general. Even more generally, for multi-edge type graphs the same exponent realization problem arises but now over integer *vectors* rather than over integers. A viable (suboptimal) strategy consists of determining the required powers of each type (component of the vector), finding the optimal computation path for each type, and taking the union. Regardless of the decomposition, the computational efficiency ultimately rests on that of the pairwise convolutions. In the sequel we therefore focus on the efficient computation of pairwise convolutions.

§B.1. Quantization

Choosing a density representation suitable for computation is an important aspect of the implementation problem. When discussing EXIT charts (see Sections 3.14, 4.10, and 6.7) we already introduced the simplest approach: assume that all intermediate message densities are Gaussian. In this case, and under the symmetry assumption, the space of densities is parameterized by a single real parameter and density evolution collapses to a one-dimensional recursion.

We are interested in implementing generic density evolution, making minimal a priori assumptions on the form of the message densities. As a first practical step we restrict ourselves to densities that are supported on (finitely many) discrete points. Attempting to choose a discretization "optimally" is challenging and ambiguous. A practical solution which yields good results, and is often applied in the implementation of message-passing decoders, is to space the samples uniformly in the log-likelihood domain. More precisely, we consider L-densities that are supported on the set $\delta\,\mathbb{Z}$, where δ denotes the quantization step size. We will carry the constant δ throughout this chapter. We will also allow the densities to have point masses at $+\infty$. Symmetric densities can have point masses at $+\infty$ but not at $-\infty$. Moreover, the target of density evolution is usually the density $\Delta_{+\infty}$.

§B.2. Variable-Node Update via Fourier Transform

Since log-likelihood ratios add at variable nodes under belief propagation (BP) decoding, the convolution in the variable-node domain is relatively straightforward. In practice, however, we must restrict ourselves to finitely supported densities. This restriction creates a computational problem in that the restricted space of densities is not invariant under convolution. Thus, some approximation, projecting onto the finitely supported space, is required.

Let us consider the support set of the densities to be $\delta[-N,N]_{\mathbb{Z}} \cup \{+\infty\}$. Put differently, we consider densities with point masses on the points $-\delta N, -\delta(N-1), \ldots, \delta N$, together with $+\infty$. Given a density a supported on this set, let a_i denote the value of the density at the point $i\delta \in \delta[-N,N]_{\mathbb{Z}}$ and let a_∞ denote the magnitude of the point mass at $+\infty$. Thus, $\mathsf{a} = \sum_{i=-N}^{N} \mathsf{a}_i \Delta_{\delta i} + \mathsf{a}_\infty \Delta_{+\infty}$. Then

$$\mathsf{a} \circledast \mathsf{b} = \Big(\sum_{i=-N}^{N} \mathsf{a}_i \Delta_{\delta i}\Big) \star \Big(\sum_{i=-N}^{N} \mathsf{b}_i \Delta_{\delta i}\Big) + (\mathsf{a}_\infty + \mathsf{b}_\infty - \mathsf{a}_\infty \mathsf{b}_\infty)\Delta_{+\infty}.$$

In general, $(\sum_{i=-N}^{N} \mathsf{a}_i \Delta_{\delta i}) \star (\sum_{i=-N}^{N} \mathsf{b}_i \Delta_{\delta i})$ is supported on $\delta[-2N,2N]_{\mathbb{Z}}$. In the symmetric case (i.e., if $\mathsf{a}_i = \mathsf{a}_{-i} e^{\delta i}$), the total probability mass on $\delta[-2N,-(N+1)]_{\mathbb{Z}}$ is upper bounded by $e^{-(N+1)\delta}$ so, if $(N+1)\delta$ is sufficiently large, this probability is negligible. In a practical implementation one can choose either to neglect this mass or to gather it and place it at $-N\delta$.

The probability mass on $\delta[N+1,2N]_{\mathbb{Z}}$, on the other hand, is potentially large. In a practical implementation we gather this mass and place it either at $+\infty$ or at δN. The latter choice corresponds to placing the mass from $\delta[-2N,-(N+1)]_{\mathbb{Z}}$ at $-\delta N$. This is the choice one should make if it is desired to implement density evolution in such a way that, up to numerical precision, the algorithm corresponds to density evolution for a message-passing algorithm (a quantized version of BP in which message magnitudes are saturated at δN at variable nodes). In practice it is simpler and nearly equivalent to gather the mass and to place it at $+\infty$.

The calculation of $(\sum_{i=-N}^{N} \mathsf{a}_i \Delta_{\delta i}) \star (\sum_{i=-N}^{N} \mathsf{b}_i \Delta_{\delta i})$ can be efficiently performed using the FFT (fast Fourier Transform). But in order to move the mass from $\delta[N+1,2N]_{\mathbb{Z}}$ to δN or $+\infty$ we need to transform the result back into the real domain. This implies that each pairwise convolution requires three FFTs – two forward (one for each of the two densities) and one reverse. It would be preferable to work entirely in the Fourier domain, returning to the real domain only at the end of the update (after *all* powers have been computed). It turns out that one can exploit symmetry, or approximate symmetry, to circumvent the need to return to the real domain prematurely.

Let a be a symmetric density on $\delta\mathbb{Z} \cup \{+\infty\}$, $\mathsf{a} = \sum_i \mathsf{a}_i \Delta_{i\delta} + \mathsf{a}_\infty \Delta_{+\infty}$. For $i \in \mathbb{Z}$ define $\hat{\mathsf{a}}$ by $\hat{\mathsf{a}}_i = e^{-\frac{1}{2}i\delta}\mathsf{a}_i$. This makes $\hat{\mathsf{a}}$ even, $\hat{\mathsf{a}}_i = \hat{\mathsf{a}}_{-i}$. The densities $\hat{\mathsf{a}}$ (no longer probability densities) still convolve at variable nodes, i.e., $(\hat{\mathsf{a}} \star \hat{\mathsf{b}})_i = (\mathsf{a} \star \mathsf{b})_i e^{-\frac{1}{2}i\delta}$. Since the total mass of a is 1, the value of a_∞ can be recovered from $\hat{\mathsf{a}}$. Let $P_T(\hat{\mathsf{a}})$ denote the projection of $\hat{\mathsf{a}}$ onto functions supported on $\mathbb{Z}/T\mathbb{Z}$, the ring of integers modulo T. We defined it by

$$P_T(\hat{a})_i = \sum_{k\in\mathbb{Z}} \hat{a}_{i+Tk}.$$

Note that

$$P_T(\hat{a} \star \hat{b}) = P_T(\hat{a}) \star P_T(\hat{b}),$$

where the second convolution is over the integers modulo T. In practice we choose $T = 2^n$ so that $P_T(\hat{a}) \star P_T(\hat{b})$ can be efficiently computed using the FFT.

The key observation, and the motivation for scaling a by $e^{-\frac{1}{2}x}$, is that aliasing is very small and that when $N \ll T/2$ we can accurately recover the density a in $\delta[-N, N]_{\mathbb{Z}}$ from $P_T(\hat{a})$. Indeed,

$$\begin{aligned}
a_i \le e^{\frac{1}{2}\delta i} P_T(\hat{a})_i &= \sum_{k\in\mathbb{Z}} e^{\frac{1}{2}\delta i}\hat{a}_{i+Tk} = a_i + \sum_{k\in\mathbb{Z}\setminus 0} e^{-\frac{1}{2}\delta Tk} a_{i+Tk} \\
&= a_i + \sum_{k>0} e^{-\frac{1}{2}\delta Tk} a_{i+Tk} + \sum_{k>0} e^{\frac{1}{2}\delta Tk} a_{i-Tk} \\
\text{by symmetry} \quad &= a_i + \sum_{k>0} e^{-\frac{1}{2}\delta Tk} a_{i+Tk} + \sum_{k>0} e^{-\frac{1}{2}\delta(Tk-2i)} a_{Tk-i} \\
&\le a_i + \frac{e^{-\frac{1}{2}\delta T} + e^{-\frac{1}{2}\delta(T-2i)}}{1 - e^{-\frac{1}{2}\delta T}}.
\end{aligned}$$

Notice that only the second-to-last step invokes symmetry. For the bound to be good it is not necessary that strict symmetry holds; the actual requirement is that $e^{\frac{1}{2}\delta Tk} a_{i-Tk}$ be small for positive k and i in the desired range. The preceding bound shows that a_i can be accurately recovered from $P_T(\hat{a})_i$ for $i \in [-N, N]_{\mathbb{Z}}$ provided that $\frac{1}{2}\delta(T - 2N)$ is sufficiently large.

In summary, the method is the following. Given an incoming density a we first form $P_T(\hat{a})$ by zero padding. Next we perform the FFT to obtain the discrete Fourier transform of $P_T(\hat{a})$. Pairwise convolution is performed by pointwise multiplication of the Fourier transforms. If we need to compute higher powers of a density we can also do so directly. Only once all powers have been computed do we perform the inverse FFT to go back to the real domain. We then recover the density on $\delta[-N, N]_{\mathbb{Z}}$ directly (after scaling) and at $+\infty$ from the total sum.

§B.3. Check-Node Update via Table Method

The first method we present to perform check-node updates is conceptually simple and can be implemented using look-up tables. Suppose we have two L-densities a and b supported on $\delta[-N,N]_{\mathbb{Z}}$,

$$\mathsf{a} = \sum_{k=-N}^{N} \mathsf{a}_k \Delta_{k\delta}, \qquad \mathsf{b} = \sum_{k=-N}^{N} \mathsf{b}_k \Delta_{k\delta}.$$

For the moment we assume that there is no mass at $+\infty$. We want to approximate the result of the check-node domain convolution $\mathsf{a} \boxast \mathsf{b}$ with a density supported on $\delta[-N,N]_{\mathbb{Z}}$.

Consider a (degree 3) check-node update with incoming messages $i\delta$ and $j\delta$. Let us focus on $i, j \geq 0$, $i \neq j$. Under BP the outgoing message is

$$2\tanh^{-1}(\tanh(\frac{1}{2}i\delta)\tanh(\frac{1}{2}j\delta)).$$

Given incoming densities a and b, such a message has probability

$$\mathsf{a}_i\mathsf{b}_j + \mathsf{a}_j\mathsf{b}_i + \mathsf{a}_{-i}\mathsf{b}_{-j} + \mathsf{a}_{-j}\mathsf{b}_{-i} = \frac{1}{2}(\mathsf{a}_i^+\mathsf{b}_j^+ + \mathsf{a}_j^+\mathsf{b}_i^+) + \frac{1}{2}(\mathsf{a}_i^-\mathsf{b}_j^- + \mathsf{a}_j^-\mathsf{b}_i^-),$$

where we define a^+ and a^- supported on $\delta[0,N]_{\mathbb{Z}}$ by $\mathsf{a}_i^+ = \mathsf{a}_i + \mathsf{a}_{-i}$ ($\mathsf{a}_0^+ = \mathsf{a}_0$) and $\mathsf{a}_i^- = \mathsf{a}_i - \mathsf{a}_{-i}$ ($\mathsf{a}_0^- = 0$). In a similar manner, the probability of observing $-2\tanh^{-1}(\tanh(\frac{1}{2}i\delta)\tanh(\frac{1}{2}j\delta))$ is given by

$$\mathsf{a}_{-i}\mathsf{b}_j + \mathsf{a}_{-j}\mathsf{b}_i + \mathsf{a}_i\mathsf{b}_{-j} + \mathsf{a}_j\mathsf{b}_{-i} = \frac{1}{2}(\mathsf{a}_i^+\mathsf{b}_j^+ + \mathsf{a}_j^+\mathsf{b}_i^+) - \frac{1}{2}(\mathsf{a}_i^-\mathsf{b}_j^- + \mathsf{a}_j^-\mathsf{b}_i^-).$$

Since $2\tanh^{-1}(\tanh(\frac{1}{2}i\delta)\tanh(\frac{1}{2}j\delta))$ is typically not an element of $\delta[-N,N]_{\mathbb{Z}}$ we need to relocate its associated probability mass. Thus, for $i, j \geq 0$ let us define the quantizer map

$$Q(i,j) = \lfloor 2\tanh^{-1}(\tanh(\tfrac{1}{2}i\delta)\tanh(\tfrac{1}{2}j\delta))/\delta + \tfrac{1}{2}\rfloor.$$

We have $Q(i,j) \leq \min\{i,j\}$, so that $Q(i,j) \in [0,N]_{\mathbb{Z}}$ if $i, j \in [0,N]_{\mathbb{Z}}$. Now, define the *approximate* convolution of a and b as the density c supported on $\delta[-N,N]_{\mathbb{Z}}$ as follows. Set

$$\mathsf{c}_k^+ = \sum\{\mathsf{a}_i^+\mathsf{b}_j^+ \mid \{i,j\} : Q(i,j) = k\}, \quad \mathsf{c}_k^- = \sum\{\mathsf{a}_i^-\mathsf{b}_j^- \mid \{i,j\} : Q(i,j) = k\}.$$

Recover c by $\mathsf{c}_0 = \mathsf{c}_0^+$, $\mathsf{c}_i = \frac{1}{2}(\mathsf{c}_i^+ + \mathsf{c}_i^-)$ for $i > 0$, and $\mathsf{c}_i = \frac{1}{2}(\mathsf{c}_{-i}^+ - \mathsf{c}_{-i}^-)$ for $i < 0$.

A significant reduction in complexity can be extracted from the observation that $Q(i,j) = \min\{i,j\}$ in most cases and that the difference between these two quantities cannot be too large. More precisely, by Problem B.2 we have

$$\min\{i,j\} - Q(i,j) < \delta^{-1}\ln(2) - \frac{1}{2}. \tag{B.1}$$

By computing the cumulative distributions of a^+, a^-, b^+, and b^- we can compute the contribution to c_i^+ for all pairs $a_i^+ b_j^+$ and $b_i^- a_j^-$ for fixed i and all $j \geq i$ in about $2\lceil \delta^{-1} \ln(2) \rceil$ multiplications and subtractions. Letting $L = N\delta$ denote the maximum log-likelihood magnitude, we see that the number of operations is of order $(2\ln 2)L\delta^{-2}$. This is quite a bit better than the order $L^2\delta^{-2}$ operations required by the naive approach.

Including the point mass at ∞ is very simple: extend the definition of Q so that $Q(\infty, i) = Q(i, \infty) = i$.

We now summarize the method. Given the density a compute

$$a_i^+ = a_i + a_{-i}, \qquad a_i^- = a_i - a_{-i},$$

$$A_i^+ = a_\infty + \sum_{j=i}^{N} a_j^+, \qquad A_i^- = a_\infty + \sum_{j=i}^{N} a_j^-,$$

for $i \in [0, N]$ (with the special convention that $a_0^+ = a_0$ and $a_0^- = 0$). Set $A_{N+1}^+ = A_{N+1}^- = 0$. Compute the equivalent quantities for b.

Initially set $c^+ = c^- = 0$. Set $c_\infty^+ = c_\infty^- = a_\infty b_\infty$. Then, for $i \in [0, N]$ and $k \in [0, \lceil \delta^{-1} \ln(2) - \frac{1}{2} \rceil]$, perform

$$c_{i-k}^+ \mathrel{+}= a_i^+ (B_{T_Q(i,k)}^+ - B_{T_Q(i,k-1)}^+) + b_i^+ (A_{T_Q(i,k)}^+ - A_{T_Q(i,k-1)}^+),$$

$$c_{i-k}^- \mathrel{+}= a_i^- (B_{T_Q(i,k)}^- - B_{T_Q(i,k-1)}^-) - b_i^- (A_{T_Q(i,k)}^- - A_{T_Q(i,k-1)}^-),$$

where, for $k \geq 0$, $T_Q(i, k)$ is defined as

$$T_Q(i, k) = \min\{j \geq i : Q(i, j) \geq i - k\}$$

and $T_Q(i, -1) = N + 1$. It is efficient to store the values of T_Q in a table.

Once c^+ and c^- are computed we recover c. Note that it is not necessary to recover c until the end of the check-node update. In other words, one can remain in the Fourier domain over $\mathbb{F}_2$ until the end of the check-node update.

Since the convolution in the variable-node domain can be computed nearly exactly while, with the table approach, the check-node computation introduces quantization noise with each pairwise convolution, it is sensible to oversample for the check-node convolutions, requantizing only at the completion of the message density calculation. This of course increases complexity.

§B.4. Check-Node Update via Fourier Method

The second method we discuss to perform check-node updates operates entirely in the check-node Fourier domain. The convolution operations are *exact* in the sense that no quantization noise is introduced. At the end of the check-node update, however, we must approximate the L-density with one supported on $\delta[-N,N]_{\mathbb{Z}} \cup \{+\infty\}$. In this method we combine the quantization with the inverse Fourier transform to obtain the quantized approximation directly.

At check nodes BP reduces to addition over $\mathbb{F}_2 \times [0,\infty]$. To avoid introducing further notation, here and throughout this section, we will reinterpret the group operation in $\mathbb{F}_2$ as addition over $\{0,1\}$ (modulo 2) rather than as multiplication over $\{1,-1\}$.

Let a denote an L-density and let X be a random variable with density a. Let $(S,Y) = (\mathfrak{H}(X), \ln\coth(X/2))$ (as defined in (4.17) but with range of $\mathfrak{H}$ now $\{0,1\}$) be the corresponding random message represented in the G-domain. In this representation, messages add at check nodes under BP and, therefore, the densities convolve under density evolution. The Fourier transform in the G-domain can be written as (see Definition 4.52)

$$\mathcal{G}_{\mathsf{a}}(\mu,\nu) = \mathbb{E}[\mathrm{e}^{-\mu S-\nu Y}],$$

where $\mu \in \{0, i\pi\}$ and ν is in the right half of the complex plane.

We have seen on page 181 that it is possible to define a Fourier transform in the G-domain. This appears to give a direct and simple way to compute the evolution of the densities at the check nodes, namely pointwise multiplication of the Fourier transform. Computing the transform is problematic though because the random variable Y is not supported on uniformly spaced points when X is. Indeed, note that $\ln\frac{1+u}{1-u} = \ln(1+u) - \ln(1-u)$ has the expansion $2\sum_{i\geq 0}\frac{u^{2i+1}}{2i+1}$. The function $\ln\coth(x/2)$ therefore has the expansion

$$\text{(B.2)} \qquad \ln\coth(x/2) \overset{u=\mathrm{e}^{-x}}{=} \ln\frac{1+u}{1-u} = 2\sum_{i\geq 0}\frac{u^{2i+1}}{2i+1} \overset{u=\mathrm{e}^{-x}}{=} 2\sum_{i\geq 0}\frac{\mathrm{e}^{-(2i+1)x}}{2i+1}.$$

Assuming that we choose a uniform spacing $\delta[-N,N]_{\mathbb{Z}}$ of the samples in the L-domain, the support in the G-domain is not uniformly spaced; rather it is approximately exponentially spaced. We therefore cannot directly apply standard discrete Fourier methods. The implementation problem we address is how to perform the G domain Fourier transform of densities that are uniformly sampled in the L-domain.

With a little algebra we can represent the Fourier transform directly in the L-domain, where the sampling is uniform:

$$\mathcal{G}_{\mathsf{a}}(\mu,\nu) = \int_{-\infty}^{\infty} \mathrm{e}^{-\mu S(x)-\nu Y(x)}\mathsf{a}(x)\mathrm{d}x = \int_{-\infty}^{\infty} \mathrm{e}^{-\mu\,\mathfrak{H}(x)-\nu\ln\coth(x/2)}\mathsf{a}(x)\mathrm{d}x$$

$$\text{(B.3)} \qquad = \int_{0}^{\infty} \tanh^{\nu}(x/2)(\mathsf{a}(x) + \mathrm{e}^{\mu}\mathsf{a}(-x))\,\mathrm{d}x.$$

If a is symmetric then $\mathcal{G}_{\mathsf{a}}(0,\nu) = \mathcal{G}_{\mathsf{a}}(i\pi, \nu-1)$. We will not assume or exploit symmetry, although it is possible to do so.

Our method for check-node density evolution is the following. We compute $\mathcal{G}_{\mathrm{a}}(\mu, \nu)$ for a finite set of ν (and both values of μ). Then, check-node convolution consists of pointwise multiplication of these functionals. After performing the pointwise multiplication we find a density that, approximately, has the given transform at the given points.

At the beginning of the update incoming densities are supported on $\delta[-N, N]_{\mathbb{Z}} \cup \{+\infty\}$. At the end of the true density evolution update the underlying density is discrete but in general it is supported on arbitrary points in the set $[-N\delta, N\delta] \cup +\infty$. When we invert the transform at the end of the update, we seek an approximation on $\delta[-N, N]_{\mathbb{Z}} \cup \{+\infty\}$, so some quantization is inevitable. There are immediately two fundamental problems: how do we compute $\mathcal{G}_{\mathrm{a}}(\mu, \nu)$ efficiently for incoming densities and how do we compute the approximate inverse. Related to these two fundamental problems is the question of choosing appropriate values for ν.

The following result, whose proof is left as Problem B.3, hints at a convenient solution:

$$\lim_{\alpha \to \infty} \tanh^{e^{\alpha}}\left(\frac{x+\alpha}{2}\right) = e^{-2e^{-x}}. \tag{B.4}$$

Let us put ν in the form $\nu = e^{\alpha}(1 + i\omega)$. The preceding result says that if we fix ω then for large enough α (it need not be very large) we have

$$\tanh^{\nu}(x/2) \simeq e^{-2e^{-(x-\alpha)}(1+i\omega)}. \tag{B.5}$$

Thus, in this representation, the integration kernel $\tanh^{\nu}(x/2)$ effectively *translates* by α. Since log-likelihood samples are uniformly spaced, it appears that a δ-uniformly spaced α will be a good choice. Moreover, the computation of the transform almost takes the form of a convolution, suggesting a method for efficient calculation.

It is perhaps unfortunate that the approximation (B.5) is not exact since then the desired calculation would in fact be a convolution. With a certain trick, however, we can still convert the calculation to a convolution. A naive approach would be to approximate and simply replace $\tanh^{\nu}(x/2)$ with $e^{-2e^{-(x-\alpha)}(1+i\omega)}$ in (B.3). Although this is not our approach it is very closely related. The equations resulting from the substitution turn out to be the density evolution for an approximation to BP that we will call AppBP. This approximation corresponds to a practical decoding algorithm whose performance is very close to that of BP. To understand the solution for density evolution of BP it is simplest to first understand the density evolution for AppBP.

§B.4.1. Approximating Belief Propagation: AppBP

The AppBP algorithm is very similar to BP. The only difference is that it replaces the function $(\mathfrak{H}(x), \ln\coth(\frac{|x|}{2}))$ with $(\mathfrak{H}(x), 2e^{-|x|})$ as the quantities to be added at check nodes. In order for this to have a well-defined inverse we also saturate the sum of the second component (the terms corresponding to $2e^{-|x|}$) at 2. The approximation of $\ln \coth(\frac{|x|}{2})$ by $2e^{-|x|}$ comes directly from the expansion (B.2). Under this update rule, if $a_1, \dots, a_{d-1}$ are incoming L-messages to a degree d check node, then the outgoing L-message on the d-th edge has magnitude

$$\Big(-\ln \frac{\sum_{i=1}^{d-1} 2e^{-|a_i|}}{2}\Big)^+ = \Big(-\ln\big(\sum_{i=1}^{d-1} e^{-|a_i|}\big)\Big)^+$$

where $(x)^+$ equals x for $x \geq 0$ and is equal to 0 for $x \leq 0$. As we see by looking at the right-hand term of the preceding expression we could have used the function $e^{-|x|}$ instead of $2e^{-|x|}$, the final result is the same. We use $2e^{-|x|}$ in order to emphasize the relationship to true BP.

THEOREM B.6 (BP VERSUS APPBP). Let $a_1, \dots, a_{d-1}$ be positive reals. Then

$$\Big(-\ln\big(\frac{1}{2}\sum_{i=1}^{d-1} 2e^{-a_i}\big)\Big)^+ \leq 2\tanh^{-1}\Big(\prod_{i=1}^{d-1} \tanh\frac{a_i}{2}\Big).$$

Proof. Problem B.4. □

Thus, the preceding approximation has the effect of reducing the magnitudes of check-to-variable messages as compared to belief propagation.

Consider again the L-density a and let X be a random variable with density a. Define

$$(S, Y) = (\mathfrak{H}(X), 2e^{-|X|}).$$

Note that this random variable is supported in the G-domain, i.e., in $\mathbb{F}_2 \times [0, +\infty]$. Under check-node updates in AppBP, incoming messages are added. As in the BP case, the desired Fourier transform in the G-domain can be written as

$$\mathcal{G}_{\mathsf{a}}(\mu, \nu) = \mathbb{E}[e^{-\mu S - \nu Y}],$$

where $\mu \in \{0, i\pi\}$ and ν is in the right half of the complex plane. In the L-domain this takes the form

$$\mathcal{G}_{\mathsf{a}}(\mu, \nu) = \int_{-\infty}^{\infty} e^{-\mu S(x) - \nu Y(x)} \mathsf{a}(x) dx = \int_{-\infty}^{\infty} e^{-\mu \mathfrak{H}(x) - \nu 2e^{-|x|}} \mathsf{a}(x) dx$$

$$= \int_0^{\infty} e^{-\nu 2e^{-x}} (\mathsf{a}(x) + e^{-\mu}\mathsf{a}(-x))\, dx. \tag{B.7}$$

Since the sum of the incoming messages of the form $2e^{-|x|}$ can exceed 2 (so that the outgoing value, which is $-\ln(\cdot/2)$ of this sum, can be negative) it is convenient to extend the preceding definition to allow for this. Thus, for a density (not necessarily a probability density) f whose support might extend to the negative real axis and include a point mass at $+\infty$ let

$$\mathcal{H}_f(v) = \int_{-\infty}^{\infty} e^{-v2e^{-x}} f(x)\,dx\,. \tag{B.8}$$

Then

$$\mathcal{G}_a(0,v) = \mathcal{H}_{a^+}(v), \qquad \mathcal{G}_a(i\pi,v) = \mathcal{H}_{a^-}(v),$$

where $a^{\pm}(x) = a(x) \pm a(-x)$ for $x \geq 0$ and zero otherwise. Let us set $v = e^{\alpha}(1+i\omega)$ and fix ω. Abusing notation, we have

$$\mathcal{H}_f(\alpha) = \int_{-\infty}^{\infty} e^{-2(1+i\omega)e^{\alpha-x}} f(x)\,dx. \tag{B.9}$$

We observe that $\mathcal{H}_f$ is the convolution of $f(x)$ and

$$\Psi(x) = e^{-2(1+i\omega)e^{x}}.$$

If f is discretely supported, on $\delta\mathbb{Z}$ say, $\mathcal{H}_f(\alpha)$ for $\alpha \in \delta\mathbb{Z}+\delta'$ can be computed using the discrete Fourier transform to perform the convolution. Note that while $\lim_{x\to\infty}\Psi(x) = 0$, we have $\lim_{x\to-\infty}\Psi(x) = 1$. Because of this effective non-locality of Ψ, in practice we convolve with effectively locally supported

$$\psi(x) = \Psi(x) - \Psi(x+\delta)$$

rather than with $\Psi(x)$.

Let A_1 and A_2 be independent real random variables with probability densities g_1 and g_2, and define the random variable B by

$$e^{-B} = e^{-A_1} + e^{-A_2}.$$

This is how the magnitudes of the log-likelihood ratios are combined at check nodes under AppBP. Solving for the density of B is the key problem for implementing density evolution for AppBP. We have

$$\begin{aligned}\mathbb{P}\{B \leq x\} &= \mathbb{P}\{e^{-A_1} + e^{-A_2} \geq e^{-x}\} = \int \mathbb{P}\{e^{-A_1} \geq e^{-x} - e^{-y}\} g_2(y)dy \\ &= \int \mathbb{P}\{A_1 \leq -\ln(e^{-x} - e^{-y})\} g_2(y)dy.\end{aligned}$$

Taking the derivative with respect to x we see that the density of B is the "convolution"

$$(g_1 \tilde{\otimes} g_2)(x) = \int_x^{\infty} \frac{1}{1-e^{x-y}} g_1(x - \ln(1-e^{x-y}))\, g_2(y)\,dy\,. \tag{B.10}$$

Let us discuss some basic properties of $\tilde{\otimes}$. Let $g_1, \dots, g_d$ be non-negative densities supported on $\{-\infty\} \cup [z_1, z_2] \cup \{+\infty\}$ for some real z_1, z_2 and define

$$f = g_1 \tilde{\otimes} g_2 \tilde{\otimes} \cdots \tilde{\otimes} g_d.$$

Then

$$\int f = \prod_i \int g_i,$$

$$\Big(\int f - f(-\infty)\Big) = \prod_i \Big(\int g_i - g_i(-\infty)\Big),$$

$$f(+\infty) = \prod_i g_i(+\infty),$$

$$f \text{ is supported on } \{-\infty\} \cup [z_1 - \ln d, z_2] \cup \{+\infty\}.$$

The first three equations tell us how the total mass of the result, as well as the masses located at $\pm\infty$, can be computed from the corresponding quantities of the operands. We soon make use of these facts.

Consider a check node of degree d and let $\mathsf{a}_1, \dots, \mathsf{a}_{d-1}$ be incoming L-densities on $d-1$ of the edges. Define the densities

$$\tilde{\mathsf{a}}^+ = \mathsf{a}_1^+ \tilde{\otimes} \dots \tilde{\otimes} \mathsf{a}_{d-1}^+, \qquad \tilde{\mathsf{a}}^- = \mathsf{a}_1^- \tilde{\otimes} \dots \tilde{\otimes} \mathsf{a}_{d-1}^-.$$

Since the densities $\mathsf{a}_i^\pm$ are supported on $[0, +\infty]$, the densities $\tilde{\mathsf{a}}^\pm$ are supported on $[-\ln(d-1), +\infty]$. The densities $\mathsf{a}^\pm$ associated to the outgoing L-density a under AppBP are obtained from $\tilde{\mathsf{a}}^\pm$ by moving the mass on $[-\ln(d-1), 0]$ to $\{0\}$. This is the density evolution analog to the $(\cdot)^+$ operation in the check-node update under AppBP.

From our earlier analysis (see also Problem B.6) we know that in the Fourier domain we can compute the densities $\tilde{\mathsf{a}}^\pm$ as

$$\text{(B.11)} \qquad \mathcal{H}_{\tilde{\mathsf{a}}^+}(\alpha) = \prod_{i=1}^{d-1} \mathcal{H}_{\mathsf{a}_i^+}(\alpha), \qquad \mathcal{H}_{\tilde{\mathsf{a}}^-}(\alpha) = \prod_{i=1}^{d-1} \mathcal{H}_{\mathsf{a}_i^-}(\alpha).$$

Let us address the question of how to efficiently compute $\mathcal{H}_{\mathsf{a}_i^\pm}(\alpha)$ for $\alpha \in \delta\mathbb{Z} + \delta'$. Assume the parameters δ, δ', and ω are fixed. Recall that it is more convenient to first convolve with ψ and then later to recover from this the convolution with Ψ. Therefore, define

$$\psi_i = \psi(i\delta + \delta').$$

Assume incoming L-densities supported on $\delta[-N, N]_{\mathbb{Z}} \cup \{+\infty\}$, i.e., a_i is defined for $i \in [-N, N]_{\mathbb{Z}} \cup \{+\infty\}$, so that $\mathsf{a}_i^\pm$ is defined for $i \in [0, N]_{\mathbb{Z}} \cup \{+\infty\}$. To simplify notation, let us take a positive density g supported on $\delta[0,N]_{\mathbb{Z}} \cup \{+\infty\}$; we want to compute $\mathcal{H}_g$ using the FFT. First, we zero pad g to $\delta[-T',T]_{\mathbb{Z}} \cup \{+\infty\}$, setting $g_i = 0$ for $i \in [-T',T]_{\mathbb{Z}} \backslash [0,N]_{\mathbb{Z}}$. We choose $T - T' = 2^n - 1$ for some n to allow computation using the FFT. Next we convolve $\{g_i\}_{i=-T'}^{T}$ with $\{\psi_i\}$ (with support suitably truncated) using the FFT to obtain $\{h_i\}_{i=-T'}^{T}$. Note that g_∞ does not affect h_i for $i \in [-T',T]_{\mathbb{Z}}$. We should choose T' and T sufficiently large to avoid aliasing under the desired precision. Fortunately $\psi(x) \to 0$ like e^x for $x \to -\infty$ and like $\mathrm{e}^{-\mathrm{e}^x}$ for $x \to \infty$. Choose $N' > N$ so that $\psi(x) \simeq 0$ for $x \geq (N'-N)\delta$. Let us now recover the convolution with Ψ from the convolution with ψ. Call the result H. Set $H_{N'} = H_\infty = g_\infty$ and for $i \in [-T', N'-1]$ set $H_i = H_{i+1} + h_i$. As a result, we have, up to numerical precision

$$H_i = \mathcal{H}_g(i\delta + \delta')$$

for $i \in [-T',N']_{\mathbb{Z}} \cup \{+\infty\}$. Furthermore, for $i > N'$, we have $H_i = H_{N'} = g_\infty$ and we assume that the total mass of g is computed. These quantities are then multiplied pointwise, in accordance with (B.11). (The total masses multiply too.) The remaining question is how to recover $\tilde{\mathsf{a}}^\pm$ from (the total masses and) the products formed in (B.11).

To avoid notational clutter, let us assume that the densities $\tilde{\mathsf{a}}_i^\pm$ are supported on $\delta\mathbb{Z}$ (and not only on a finite set of points). If we knew that the densities $\tilde{\mathsf{a}}^\pm$ were supported on $\delta\mathbb{Z}$ then we could recover them from $\mathcal{H}_{\tilde{\mathsf{a}}^\pm}(\alpha)$, $\alpha \in \delta\mathbb{Z} + \delta'$, using a discrete inverse convolution. Unfortunately this is not the case and applying the discrete inverse convolution produces unusable results. Understanding this is key to understanding the method.

Let us consider the general inverse problem: assume that we have a positive density g supported on $(-\infty,\infty)$. We know the total mass $\int g$ and we are given

$$H_j = \int_{-\infty}^{\infty} \Psi(j\delta + \delta' - x) g(x) \mathrm{d}x, \quad j \in \mathbb{Z}.$$

We want to approximate g by a density of the form $\sum g_j \Delta_{j\delta}$, where the coefficients g_j, $j \in \mathbb{Z}$, are non-negative. First we form

$$h_j = H_j - H_{j+1} = \int_{-\infty}^{\infty} \psi(j\delta + \delta' - x) g(x) \mathrm{d}x.$$

Second, we want to obtain the sequence g_i as a convolution of the sequence h_j. (This ensures an efficient implementation using FFTs.) If we assume that $g(x)$ is supported on $\delta\mathbb{Z}$, i.e., if $g = \sum g_i \Delta_{i\delta}$, then the sequence $\{h_j\}$ is the discrete convolution of the sequence $\{\psi_i\}$ with the sequence $\{g_i\}$,

$$h_j = \int_{-\infty}^{\infty} \psi(j\delta + \delta' - x) \Big(\sum_i g_i \Delta_{i\delta}(x)\Big) \mathrm{d}x = \sum_i g_i \psi((j-i)\delta + \delta') = \sum_i g_i \psi_{j-i}.$$

Assuming invertibility of the convolution with $\{\psi_i\}$, we can recover g by performing the inverse convolution. Consider, though, what happens if the support of g is not restricted to $\delta\mathbb{Z}$. Let $\{\hat{\psi}_i\}$ denote the inverse convolution sequence (assuming it exists) to $\{\psi_i\}$, i.e.,

$$\sum_j \hat{\psi}_j \psi_{i-j} = \sum_j \hat{\psi}_{i-j}\psi_j = \mathbb{1}_{\{i=0\}}.$$

Then, assuming that we still perform the inverse convolution operation, we have defined g_i by

$$g_i = \int_{-\infty}^{\infty} \Big(\sum_j \hat{\psi}_{i-j}\psi(j\delta + \delta' - x)\Big) g(x) \mathrm{d}x. \tag{B.12}$$

Now, the function $\sum_j \hat{\psi}_{-j}\psi(j\delta + \delta' - x)$ restricted to $\delta\mathbb{Z}$ is equal to 1 at $x = 0$ and is equal to 0 elsewhere. If, however, $g(x)$ is not supported on $\delta\mathbb{Z}$ then we need to be concerned about $\sum_j \hat{\psi}_{-j}\psi(j\delta + \delta' - x)$ for x not in $\delta\mathbb{Z}$. This function is not real in general and the real part is not non-negative so that even for a positive density g (which is the case of interest) we may obtain a negative value for g_i; this is highly undesirable given the intended use of the inverse. In equation (B.12) the kernel $\sum_j \hat{\psi}_j\psi(x - j\delta - \delta')$ "accumulates" mass from g. More generally, we may *choose* a kernel $\mathfrak{K}$ defined by

$$\mathfrak{K}(x) = \operatorname{Re}\sum_j \beta_j \psi(x - j\delta - \delta'), \tag{B.13}$$

and obtain

$$g_i = \int_{-\infty}^{\infty} \mathfrak{K}(i\delta - x) g(x). \tag{B.14}$$

This approach is attractive because it leads to a linear inverse operation that can be easily computed using the FFT (convolve $\{\beta_i\}$ with $\{h_i\}$.) Other approaches, non-negative least squares, linear programming, etc., are possible but none of these approaches can yield the desired efficiency.

How shall we choose $\mathfrak{K}$? As we discussed, the choice $\beta_i = \hat{\psi}_i$ results in a $\mathfrak{K}$ which is the indicator function on $\delta\mathbb{Z}$. This, however, is not actually the most desirable property to require of $\mathfrak{K}$. Perhaps the ideal target for $\mathfrak{K}$ is the quantizer function $\mathbb{1}_{[\delta/2,\delta/2]}(x)$: in this case we have $g_i = \int_{(i-\frac{1}{2})\delta}^{(i+\frac{1}{2})\delta} g(x)\,\mathrm{d}x$ and if $g = \sum_i g_i \Delta_i$ then $g_i = g_i$. Unfortunately, this will not be achievable in general. The following are some desirable properties for $\mathfrak{K}$:

P1 $\mathfrak{K}(x) \geq 0$.

P2 $\sum_i \mathfrak{K}(x - i\delta) = 1$.

If these two conditions are met, then after the inversion we get a probability density (an L-density) and the resulting density evolution is the exact density evolution for some message-passing algorithm that approximates BP (see Problem B.8).

An approach we have found effective for constructing $\mathfrak{K}$ is to

$$\text{minimize: } \int_{-\infty}^{\infty} x^2 \mathfrak{K}(x) \mathrm{d}x,$$

under the constraints P1 and P2. From (B.13) we have

$$\sum_i \mathfrak{K}(x - i\delta) = \sum_j \operatorname{Re}(\beta_j) \tag{B.15}$$

(see Problem B.5) so P2 is a linear condition on $\{\beta_j\}$. The entire setup is a linear programming (LP) problem and that the ideal quantizing function $\mathbb{1}_{[\delta/2,\delta/2]}(x)$ would be the optimal solution if it were a feasible solution of the LP; unfortunately, it is not. Up to now we have not discussed the choice of ω or δ'. For a given δ one should select these parameters for the numerical properties inherent in the inverse problem. Choosing ω too small tends to make the inverse problem numerically unstable, especially for small δ, and choosing it too large tends to make it difficult to control $\mathfrak{K}$. A practical rule of thumb that we have found effective is to set $\omega \simeq 1/(2\delta)$.

Let us now focus on the inverse problem for the finitely supported case and describe the approach in a little more detail. Assume we are given

$$H_i = \mathcal{H}_g(i\delta)$$

for $i \in [-T', N']_{\mathbb{Z}}$ together with $\int g$ and that g is a positive density supported on $[-\delta T'', \delta N] \cup +\infty$ where, typically, $T'' < T'$ will be quite small. In the case of AppBP, $\delta T''$ will be approximately $-\ln(d-1)$ where d is the maximum check-node degree. We set $h_i = 0$ for $i = N', N'+1, \ldots, T$ and otherwise for $i \in [-T', T]$ define $h_i = H_{i+1} - H_i$. We compute

$$\{\tilde{g}_i\} = \mathfrak{R}(\{h_i\} \otimes \{\beta_i\})$$

using the FFT and set $\tilde{g}_\infty = H_{N'}$. Finally, we define $\{g_i\}, i \in [0, N]_{\mathbb{Z}} \cup \{+\infty\}$ by setting $g_i = \tilde{g}_i$ for $i \in [1, N-1]_{\mathbb{Z}} \cup \{+\infty\}$ and setting

$$g_N = \sum_{i=N}^{N+K} \tilde{g}_i, \qquad g_0 = \int g - (\sum_{i=1}^{N} g_i + g_\infty),$$

where K is chosen large enough so that we may assume $\mathfrak{K}(x) \simeq 0$ for $x > K$. Using this calculation (twice) we obtain $\mathsf{a}^{\pm}$ for an outgoing density a.

§B.4.2. Check-Node Update for Belief Propagation

We now return to the original problem of performing check-node updates for BP. We shall reuse the method developed for AppBP by applying a slight reparameterization. Recall that for BP the appropriate Fourier transform is given by

$$\mathcal{G}_{\mathrm{a}}(\mu,\nu) = \int_0^\infty \tanh^\nu(x/2)\,(\mathrm{a}(x) + \mathrm{e}^\mu\,\mathrm{a}(-x))\mathrm{d}x,$$

where a is an L-density, whereas for AppBP we have (for incoming densities)

$$\mathcal{G}_{\mathrm{a}}(\mu,\nu) = \int_0^\infty (\mathrm{e}^{-2\mathrm{e}^{-x}})^\nu\,(\mathrm{a}(x) + \mathrm{e}^\mu\,\mathrm{a}(-x))\,\mathrm{d}x.$$

For $x \geq 0$ define $\tilde{x}(x)$ implicitly by $\exp(-2\mathrm{e}^{-\tilde{x}}) = \tanh(|x|/2)$. This is equivalent to $2\mathrm{e}^{-\tilde{x}} = \ln\coth(|x|/2)$ so that, by definition, the quantities $2\mathrm{e}^{-\tilde{x}}$ are added at check nodes under BP. Now consider the attendant implications to the computation of density evolution. For an incoming L-density a_i we first form $\mathrm{a}_i^\pm$ and then make a change of variables from x to $\tilde{x}$. This amounts to re-sampling or requantizing the densities. The resulting densities are then $\tilde{\otimes}$ convolved. A reverse change of variables, performed by re-sampling, from $\tilde{x}$ to x then gives $\mathrm{a}^\pm$ for the outgoing density a. The re-sampling is done so as to preserve the total mass of the density.

Under the proposed change of variables we have

$$\tilde{x}(0) = -\infty, \qquad\qquad \tilde{x}(\delta) \simeq -\ln(-\frac{1}{2}\ln\frac{\delta}{2}).$$

Hence, the mass at $x = 0$ needs special handling, we will return to that. Ignoring the mass at $x = 0$ for the moment, the first step is to take incoming densities g over $\delta[1,N]_{\mathbb{Z}} \cup \{+\infty\}$ and redistribute the mass over $\delta[-S,N]_{\mathbb{Z}} \cup \{+\infty\}$ where $\delta S \simeq \tilde{x}(\delta)$ to form the re-sampled density $\tilde{g}$. For i sufficiently large we will have $\tilde{g}_i \simeq g_i$. Note that for small i the sampling resolution increases after the re-sampling so this step does not result in any significant loss of information about g. Once the incoming densities $\mathrm{a}_i^\pm$ have been thus re-sampled, approximating the change of variables, they are $\tilde{\otimes}$ convolved using the method described for AppBP. As a result one obtains

$$\mathcal{H}_{\tilde{\mathrm{a}}_i^\pm}(\alpha) = \mathcal{G}_{\mathrm{a}}(\omega,\nu)$$

with $\nu = \mathrm{e}^\alpha(1 + i\omega)$. After forming the products of the transforms of the incoming densities we obtain $\mathcal{H}_{\tilde{g}}(\alpha)$ for densities supported on $\delta[-S',N]\cup+\infty$, where $S' \geq S$ depends on the maximum check-node degree. Here $\tilde{g}$ represents $\tilde{\mathrm{a}}^\pm$ for some outgoing density a. We compute the approximate inverse exactly as in the AppBP case. The resulting discrete density $\tilde{g}$ is then re-sampled back to $\delta[0,N]_{\mathbb{Z}} \cup \{+\infty\}$ using the inverse change of variables. This yields $\mathrm{a}^\pm$ from which a is easily obtained.

δ	AWGN		BSC	
	BP	AppBP	BP	AppBP
1/2	0.5726	0.5733	0.5864	0.5868
1/4	0.5715	0.5720	0.5845	0.5846
1/8	0.5710	0.57172	0.5839	0.5843
1/16	0.5709	0.57167	0.5837	0.5842

Table B.16: Computed thresholds (expressed in channel capacity) for the (3,6) regular degree distribution.

parameter	value	parameter	value	parameter	value
δ	0.125	FFT(check)	512	c_f	9.18
δ'	0.0225	FFT(variable)	1024	N	311
ω	4.8	T'	80	N'	334

Table B.17: Parameters used in a particular implementation.

The problematic mass at $x = 0$ can be handled separately, treating it as a point mass at $\tilde{x} = -\infty$. Since we track the total mass, it can simply be ignored and we can recover the mass at $x = 0$ by the imposing correctness of the total mass.

Table B.16 shows thresholds computed using the method described herein for BP and AppBP decoding of the regular $(3, 6)$ graph structure for various values of δ. For $\delta = 1/2^k$ a $64 \cdot 2^k$ point FFT was used for computing the check-node update and twice as many points are used for the variable-node update.

§B.4.3. An Example and Some Final Tips

Table B.17 lists parameters used for one case associated to the results in Table B.16. The left-hand graph in Figure B.18 plots the quantization function $\mathfrak{K}$ obtained for this example. The right-hand graph shows $\mathfrak{K}$ on a log scale to indicate the achieved effective localization of the quantization function.

Some of the parameters in Table B.17 have been discussed and others refer to further details of the implementation beyond what has been discussed so far. In this section we describe some of the finer points of implementing this approach to density evolution.

The check-node domain transform values are maintained on δ-sample points $v = e^{\delta k}(1 + i\omega)$ for

$$k = -T', -T' + 1, \ldots, -1, 0, 1, \ldots, N, N + 1, \ldots, N'.$$

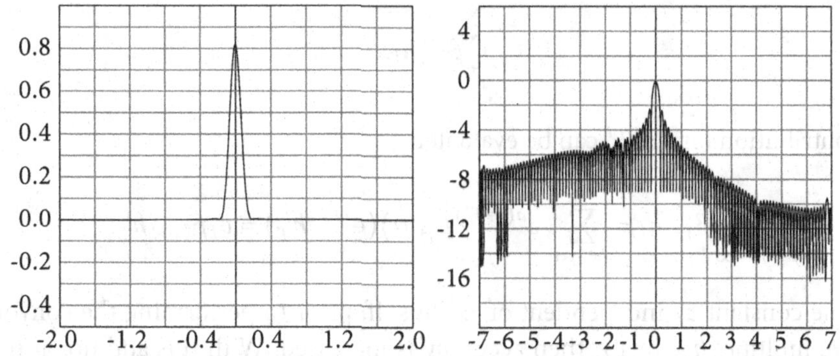

Figure B.18: Left: Example $\mathfrak{K}(x)$, $\delta = 0.125$. Right: Logarithm (base 10) $\mathfrak{K}(x)$.

The values of T' and $N' < T$ should be large enough to support accurate inversion. The choice of N' is relatively simple since probability mass moves to the left under check-node domain convolution and the forward transform kernel decays like e^{-2e^x}. We simply need to choose N' large enough so that $\mathrm{e}^{-2(1+i\omega)\mathrm{e}^{x-N\delta}} \simeq 0$ for $x \geq N'\delta$. The selection of T' is more involved.

When we perform the inversion to obtain the probability mass at $x = \delta$, we, approximately, recover mass for $x \in [\delta/2, 3\delta/2]$. Since for small values of x we have

$$\tilde{x}(x) = -\ln(-\frac{1}{2}\ln\tanh\frac{x}{2}) \simeq -\ln(-\frac{1}{2}\ln\frac{x}{2})$$

it follows that the leftmost point $\tilde{x}$ at which we want to compute the convolution $\int \mathfrak{K}(\tilde{x}-u)\tilde{a}(u)du$ is approximately $\tilde{x} = -\ln(-\frac{1}{2}\ln\frac{\delta}{4})$. For our example $\delta = 0.125$ this evaluates to -0.54 or about -4.4δ. Thus, in our example, we want to choose T' large enough so that the inverse convolution is accurate down to about -5δ. For incoming densities the leftmost location of positive mass is about $-\ln(-\frac{1}{2}\ln\delta/2)$ which is about $-0.33 \simeq -2.6\delta$. If the maximum degree is d then the leftmost location of positive mass in the transform to be inverted will be $-0.54 - \ln(d)$. For $d = 1000$ this is approximately -60δ. In our example we chose $T' = 80$. We use an additional trick to make the effective T' even larger.

The forward transform ψ decays like $C\mathrm{e}^x$, where C is a complex constant, for negative x of large magnitude. Thus, if T' is sufficiently large compared to S' then the transform value decays exponentially for for $i < -T'$. That is, for $i < -T'$ we have $h_i \simeq \mathrm{e}^{(i+T')\delta}h_{-T'}$. Now, it turns out that the constructed sequence $\{\beta_i\}$ decays exponentially to the right, i.e., like $C\mathrm{e}^{-ai}$ for some positive a. Consider the convolution of $\{\beta_i\}$ and h_i evaluated at $j \gg -T'$,

$$\sum_i \beta_{j-i} h_i.$$

The contribution for $i \leq T'$ can be evaluated

$$\sum_{i=-\infty}^{-T'} \beta_{j-i} h_i = \sum_{i=-\infty}^{-T'} (\mathrm{e}^{a(i+T')} \beta_{j+T'})(\mathrm{e}^{i+T'} h'_T) = c_f \beta_{j+T'} h'_T$$

for some constant c_f independent of j. Thus, if, prior to performing the convolution, we multiply $h_{T'}$ by c_f, then accuracy is increased. With this additional trick choosing $T' = 80$ is quite adequate.

Numerical experiments show that, even if we put all of the probability mass near $-T'\delta$, the inverse is still accurate to about 10^{-5}. In practice degrees of 1000 are never needed and, even if they were, the probability mass at this extreme values would actually be very small. So accuracy is ensured for arbitrary degrees.

Finding $\mathfrak{K}$

The linear programming problem used to define $\mathfrak{K}$ is in principle infinite and, once sampled, tends to be numerically unstable. In this section we give some suggestions for dealing with this problem.

Given a real function f and a discrete sequence c_i we call the function $\sum_i f(x - i\delta)c_i$ the pseudo-convolution of f with $\{c_i\}$. For convenience let us assume that $\delta' = 0$. Note that $\mathfrak{K}$ is a sum of pseudo-convolutions of $\psi_{\mathfrak{R}} = \mathfrak{R}(\psi)$ and $\psi_{\mathfrak{I}} = \mathfrak{I}(\psi)$ and that this space is closed under pseudo-convolution and linear combination. Thus, instead of using $\psi_{\mathfrak{R}}$ and $\psi_{\mathfrak{I}}$ directly as the basis for the LP, we can use any pseudo-convolutions whose shifts span the same space.

One way to reduce the difficulty of the LP is to use a pseudo-convolution that is already close to the optimal solution. Let γ be a positive constant and consider $\psi_{\mathfrak{R}} + i\gamma\psi_{\mathfrak{I}}$. Let $\{\psi_i^\gamma\}$ be this function sampled on $\delta\mathbb{Z}$ and let $\{\hat{\psi}_i^\gamma\}$ be the discrete inverse convolution of $\{\psi_i^\gamma\}$. Let $F_\gamma(x)$ be the pseudo-convolution of $\psi_{\mathfrak{R}} + i\gamma\psi_{\mathfrak{I}}$ and $\{\hat{\psi}_i^\gamma\}$. Both the real and imaginary parts of F_γ are pseudo-convolutions of $\psi_{\mathfrak{R}}$ and $\psi_{\mathfrak{I}}$. Restricted to $\delta\mathbb{Z}$, we have $\mathfrak{R}(F_\gamma) = \mathbb{1}_{\{x=0\}}$ and $\mathfrak{I}(F_\gamma) = 0$. Thus, any pseudo-convolution of $\mathfrak{I}(F_\gamma)$ added to $\mathfrak{R}(F_\gamma)$ will still be equal to $\mathbb{1}_{\{x=0\}}$ on $\delta\mathbb{Z}$. One can use this, for example, to find a candidate function that is equal to $\mathbb{1}_{\{x=0\}}$ on $\delta\mathbb{Z}$, and equal to 0 on $\delta\mathbb{Z} + \delta/2 \backslash \{-\delta/2, \delta/2\}$. Such a function tends to be well localized and is already similar to the optimal $\mathfrak{K}$. As another example, one can find a pseudo-convolution of $\mathfrak{I}(F_\gamma)$ that is equal to 1 at $x = \delta/2$, equal to -1 at $x = -\delta/2$, and equal to 0 on $\delta\mathbb{Z} \cup \delta\mathbb{Z} + \delta/2 \backslash \{-\delta/2, \delta/2\}$. The freedom to choose γ can be used to find a numerically stable choice for the preceding procedure.

Introducing such functions into the LP problem can significantly reduce the difficulty of solving it. Our example was found this way using $\gamma = 0.5$. By using multiple values for γ, further refinements are possible.

Notes

An extensive discussion of how to find the most efficient way to compute the integer power of a function can be found in the book by Knuth [2] in Section 4.6.3 (see also Exercise 32 on page 465).

The table method to compute the update at the check nodes is the approach taken by Chung, Forney, Richardson, and Urbanke [1]. It leads to a lower bound on the threshold since it corresponds to a quantized version of BP. The considerably faster method of density evolution which uses the Fourier transform to compute check-node updates (Section B.4) is due to Richardson and Jin [4].

A systematic study of the relationship of the countless forms of discrete cosine transforms to the discrete Fourier transforms, in particular the appropriate definition of convolution in each case, was performed by Martucci [3]. We also recommend the overview article by Strang [5].

Problems

B.1 (Efficient Computation via Pairwise Convolutions). Explain how to compute the d-fold convolution $\mathrm{a}^{\circledast d}$ with at most $\lfloor \log_2 d \rfloor + \nu(d) - 1$ pairwise convolutions, where $\nu(d)$ denotes the number of ones in the binary representation of d.

B.2 (Bound on $Q(i,j)$). Prove (B.1).

B.3 (Limit Form of Tanh). Prove (B.4).

B.4 (BP versus AppBP). Prove Theorem B.6.

B.5 (Linear Constraint). Prove (B.15).

B.6 (Efficient Computation of Density). Let A_1 and A_2 be independent random variables supported on $\delta[T, T]$, and let B be defined by

$$\mathrm{e}^{-B} = \mathrm{e}^{-A_1} + \mathrm{e}^{-A_2}.$$

How can one efficiently compute a discrete approximation of the distribution of B?

B.7 (B versus e^{-B}). Let B be a real random variable with density g. Let G denote the density of e^{-B}. Show that

$$\int_0^{\infty} \mathrm{e}^{-\nu x} G(x)\, dx = \int_{-\infty}^{\infty} \mathrm{e}^{-\nu \mathrm{e}^{-x}} g(x)\, dx.$$

B.8 (Desirable Properties of $\mathfrak{K}$). Explain how choosing $\mathfrak{K}$ to satisfy $P1$ and $P2$ ensures that the resulting density evolution (assuming numerical exactness) gives rise to a lower bound (in terms of code performance, e.g., thresholds) on the true BP density evolution.

References

[1] S.-Y. Chung, G. D. Forney, Jr., T. Richardson, and R. Urbanke, *On the design of low-density parity-check codes within 0.0045 dB of the Shannon limit*, IEEE Commun. Lett., 5 (2001), pp. 58–60. [264, 477]

[2] D. E. Knuth, *The Art of Computer Programming: Seminumerical Algorithms*, vol. 2, Addison-Wesley, Reading, MA, USA, 1981. [477]

[3] S. A. Martucci, *Symmetric convolution and the discrete sine and cosine transforms*, IEEE Trans. Sig. Processing, 42 (1994), pp. 1038–1051. [477]

[4] T. Richardson and H. Jin, *A new fast density evolution*, in Proc. of the IEEE Inform. Theory Workshop, Monte Video, Uruguay, Feb. 2006. pp. 183–187. [477]

[5] G. Strang, *The discrete cosine transform*, SIAM Rev., 41 (1999), pp. 135–147. [477]

Appendix C
CONCENTRATION INEQUALITIES

Asserting a specific property about an individual code is typically a hard task. To the contrary, it is often easy to show that *most* codes in a properly chosen ensemble possess this property. In the realm of classical coding theory an important such instance is the minimum distance of a code. For Elias's generator ensemble $\mathcal{G}$ a few lines suffice to show that with high probability an element chosen uniformly at random has a relative minimum distance of at least δ_{GV}, where δ_{GV} is the Gilbert-Varshamov distance discussed on page 7. But as mentioned on page 33, it is known that the corresponding decision problem – whether a given code has relative minimum distance at least δ_{GV} – is NP-complete.

We encounter a similar situation in the realm of message-passing decoding. The whole analysis rests on the investigation of ensembles of codes and, therefore, concentration theorems which assert that most codes in this ensemble behave close to the ensemble average are at the center of the theory. There is one big difference though which makes concentrations theorems invaluable for message-passing decoding, whereas in the classical setting they only play a marginal role. The main obstacle which we encounter in classical coding is that a random code (in $\mathcal{G}$) is unlikely to have an efficient decoding algorithm, and, therefore, a random code is unlikely to be of much practical value. In the message-passing world the choice of the ensemble (e.g., LDPC, turbo, multi-edge, etc.) guarantees that every element can be decoded with equal ease.

We discuss various techniques to show that individual instances of a random system are concentrated around the ensemble mean. We start with the rather pedestrian, yet quite useful, first and second moment method. As an application we show that the weight distribution of codes in $\mathcal{G}$ is concentrated. Next we discuss the various versions of Bernstein's inequality, leading to Chebyshev's inequality. We then venture into the realm of Martingales and discuss the Hoeffding-Azuma inequality. The basic concentration theorem of message-passing decoding, asserting that most codes perform closely to the ensemble average, is based on this technique. We continue our discussion with Wormald's theorem, which asserts that discrete stochastic processes whose expected behavior can be approximated by a differential equation show a concentration of the individual behavior around the ensemble average. We demonstrate this technique by giving a proof of Theorem 3.106 which characterizes the behavior of the peeling decoder. The Wormald technique is also used in Section A.4 to analyze the asymptotic gap of the greedy upper triangulation process.

The final technique which we discuss is useful in showing that the distribution of a sequence of random variables converges to the Poisson one. We use this technique to show that the number of minimal low-weight codewords of a random element converges to a Poisson random variable.

In general, it is fair to say that all concentration inequalities which have been proved in the setting of message-passing decoding so far are rather weak and that a much stronger concentration can be observed in practice. Therefore, to date, these concentration theorems serve mostly as a moral justification of the ensemble approach.

§C.1. First and Second Moment Method

Lemma C.1 (Markov Inequality). Let X be a non-negative random variable with finite mean. For $\alpha > 0$,

$$\mathbb{P}\{X \geq \alpha\} = \mathbb{E}[\mathbb{1}_{\{X \geq \alpha\}}] \leq \mathbb{E}[X/\alpha] = \mathbb{E}[X]/\alpha. \tag{C.2}$$

The Markov inequality represents the strongest bound we can assert assuming we only know $\mathbb{E}[X]$ and the fact that X is non-negative. As shown in Problem C.1, the basic idea of this inequality can be extended beyond recognition. A particularly important generalization is the following.

Lemma C.3 (Chebyshev's Inequality). Let X be a random variable with finite mean and variance. Then for $\alpha > 0$,

$$\mathbb{P}\{|X - \mathbb{E}[X]| \geq \alpha\} \leq \mathbb{E}[(X - \mathbb{E}[X])^2]/\alpha^2.$$

Proof. We have

$$\begin{aligned}
\mathbb{E}[(X - \mathbb{E}[X])^2] &= \int (x - \mathbb{E}[X])^2 p_X(x) \mathrm{d}x \\
&\geq \int_{|x - \mathbb{E}[X]| \geq \alpha} (x - \mathbb{E}[X])^2 p_X(x) \mathrm{d}x \\
&\geq \alpha^2 \, \mathbb{P}\{|X - \mathbb{E}[X]| \geq \alpha\}. \qquad \square
\end{aligned}$$

The following lemma is a straightforward application of both the Markov inequality and the Chebyshev inequality.

Lemma C.4 (Weight Distribution of Codes in the Elias Ensemble $\mathcal{G}$). Let $r \in (0, 1]$ and for $n \in \mathbb{N}$ so that $nr \in \mathbb{N}$ consider the sequence of ensembles $\mathcal{G}(n, k = rn)$ (see Definition 1.26). Let C be chosen uniformly at random from $\mathcal{G}(n, k = rn)$. Let $A(C, w)$ denote the number of codewords in C of weight w. Define $g(r, \omega)$, $\omega \in [0, \frac{1}{2}]$, $g(r, \omega) = h_2(\omega) + r - 1$. Given r, let $\delta_{\text{GV}}(r)$ be the unique root in $[0, \frac{1}{2}]$ of $g(r, \omega) = 0$. Then

(i) The rate of C is equal to r with probability at least $1 - 2^{-n(1-r)}$.

(ii) For $\omega \in (0, 1/2]$,

$$\mathbb{P}\Big\{\sum_{w=1}^{n\omega} A(C, w) \geq 1\Big\} \leq 2^{ng(r,\omega)}.$$

(iii) For $\omega \in (0, 1/2]$ so that $n\omega \in \mathbb{N}$ and $\alpha > 0$,

$$\mathbb{P}\{A(C, w)/\mathbb{E}[A(C, w)] \notin (1 - \alpha, 1 + \alpha)\} \leq \frac{1}{\alpha^2\, \mathbb{E}[A(C, w)]}.$$

Proof. From Problem 3.21 we know that the probability that a random binary $k \times n$, $1 \leq k \leq n$, generator matrix G has (full) rank k is equal to

$$\prod_{i=0}^{k-1}(1 - 2^{i-n}) \geq 1 - 2^{k-n} = 1 - 2^{-n(1-r)},$$

which proves statement (i). In Problem 1.17 you are asked to show that, for $w \geq 1$,

$$\mathbb{E}[A(C, w)] = \binom{n}{w}\frac{2^{nr} - 1}{2^n}, \tag{C.5}$$

$$\mathbb{E}[(A(C, w) - \mathbb{E}[A(C, w)])^2] = \mathbb{E}[A(C, w)] - \frac{2^{nr} - 1}{2^{2n}}\binom{n}{w}^2. \tag{C.6}$$

If $\omega \in (0, 1/2]$ then

$$\begin{aligned}
&\text{by (C.5)} && \mathbb{E}\Big[\sum_{w=1}^{n\omega} A(C, w)\Big] \leq \frac{2^{nr} - 1}{2^n}\sum_{w=0}^{n\omega}\binom{n}{w} \\
&\text{by (1.59), Problem 1.25} && \leq 2^{n(h_2(\omega)+r-1)} = 2^{ng(r,\omega)}.
\end{aligned}$$

Statement (ii) now follows by an application of the Markov inequality.

To see the final claim let $\omega \in (0, 1/2]$. Then we have

$$\begin{aligned}
&\mathbb{P}\{A(C, w)/\mathbb{E}[A(C, w)] \notin (1 - \alpha, 1 + \alpha)\} \\
&\qquad = \mathbb{P}\{|A(C, w) - \mathbb{E}[A(C, w)]| \geq \alpha\, \mathbb{E}[A(C, w)]\} \\
&\text{by (C.6), Lemma C.3} \qquad \leq \frac{\mathbb{E}[A(C, w)] - \frac{2^{nr}-1}{2^{2n}}\binom{n}{w}^2}{\alpha^2\, \mathbb{E}[A(C, w)]^2} \\
&\qquad \leq \frac{1}{\alpha^2\, \mathbb{E}[A(C, w)]}.
\end{aligned}$$

□

Note from (C.5) that $\lim_{n\to\infty}\frac{1}{n}\log_2 \mathbb{E}[A(C, n\omega)] = g(r,\omega)$ (for any increasing sequence of n so that $n\omega \in \mathbb{N}$). We know from Lemma C.4 that most random instances contain no non-zero codewords of weight less than $n\delta_{\text{GV}}(r)$ (and no codewords of weight exceeding $n(1-\delta_{\text{GV}}(r))$). Further, the number of codewords of weight w so that $n\delta_{\text{GV}}(r) \le w \le n(1-\delta_{\text{GV}}(r))$ concentrates on the average such number. This motivates the introduction of the *exponent* of the weight distribution:

$$G(r,\omega) = \begin{cases} g(r,\omega), & \delta_{\text{GV}} \le \omega \le 1-\delta_{\text{GV}}, \\ 0, & \text{otherwise.} \end{cases}$$

As an example, Figure C.7 shows $G(r = 1/2, \omega)$.

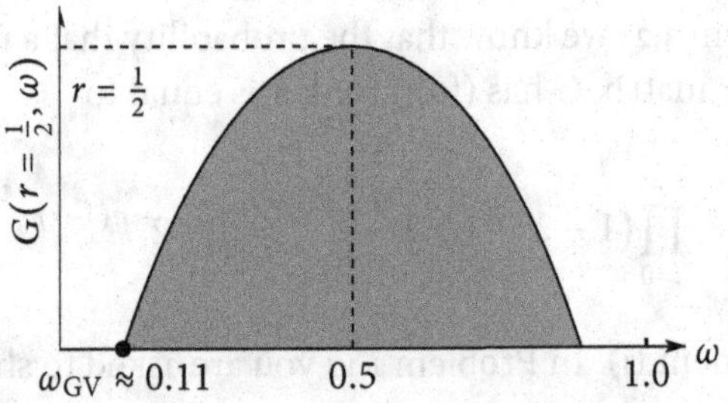

Figure C.7: Exponent $G(r = 1/2, \omega)$ of the weight distribution of typical elements of $\mathcal{G}(n, k = n/2)$ as a function of the normalized weight ω. For $w/n \in (\delta_{\text{GV}}, 1-\delta_{\text{GV}})$ the number of codewords of weight w in a typical element of $\mathcal{G}(n,k)$ is $2^{n(G(r,w/n)+o(1))}$.

Although we have not included an application of the *second moment method* it has already proved useful in the realm of message-passing decoding. We therefore state it for completeness.

Lemma C.8 (Second Moment Method). Assume that $\mathbb{E}[X]^2 \ge a$ and $\mathbb{E}[X^2] \le b$ for some non-negative real numbers a and b. Then

$$\mathbb{P}\{X=0\} \le \frac{\mathbb{E}[X^2]-\mathbb{E}[X]^2}{\mathbb{E}[X^2]} \le \frac{\mathbb{E}[X^2]-\mathbb{E}[X]^2}{\mathbb{E}[X]^2} \le \frac{b-a}{a}.$$

Proof. Recall that the Cauchy-Schwarz inequality applied to random variables Y and Z reads $\mathbb{E}[YZ]^2 \le \mathbb{E}[|YZ|]^2 \le \mathbb{E}[Y^2]\,\mathbb{E}[Z^2]$. Write the random variable X as $X = X\mathbb{1}_{\{X\neq 0\}}$ and apply the Cauchy-Schwarz inequality: $\mathbb{E}[X]^2 \le \mathbb{E}[X^2]\,\mathbb{P}\{X \neq 0\}$. This is equivalent to the claim. □

§C.2. Bernstein's Inequality

In some instances we want to prove that an event is *very* unlikely, i.e., that it decays *exponentially* (in the size of the underlying parameter). The standard approach is to apply (a variant of) Bernstein's inequality.

Assume we are given a sequence $X_1, X_2, \ldots, X_n$ of random variables and that we want to derive an upper bound on $\mathbb{P}\{\sum_{i=1}^n X_i \geq \alpha\}$, $\alpha > 0$. If the random variables are non-negative and have finite mean, then the Markov inequality gives

$$\mathbb{P}\Big\{\sum_{i=1}^n X_i \geq \alpha\Big\} \leq \sum_{i=1}^n \mathbb{E}[X_i]/\alpha.$$

This bound decreases only linearly in α. The basic trick introduced by Bernstein reads

$$\text{if } s>0 \qquad \mathbb{P}\Big\{\sum_{i=1}^n X_i \geq \alpha\Big\} = \mathbb{P}\Big\{s\sum_{i=1}^n X_i \geq s\alpha\Big\} = \mathbb{P}\big\{e^{s\sum_{i=1}^n X_i} \geq e^{s\alpha}\big\}$$

$$\text{by Lemma C.1} \qquad \leq \mathbb{E}\big[e^{s\sum_{i=1}^n X_i}\big]e^{-s\alpha}.$$

An upper bound on $\mathbb{E}\big[e^{s\sum_{i=1}^n X_i}\big]$ therefore gives us the desired upper bound on $\mathbb{P}\{\sum_{i=1}^n X_i \geq \alpha\}$. In a similar manner we get

$$\text{if } s>0 \qquad \mathbb{P}\Big\{\sum_{i=1}^n X_i \leq \alpha\Big\} \leq \mathbb{E}\big[e^{-s\sum_{i=1}^n X_i}\big]e^{s\alpha}.$$

Example C.9 (Tail Bounds for Binary Random Variables). Consider the case of binary iid random variables $X_1, \ldots, X_n$ with $\mathbb{P}\{X_i = 1\} = p = 1 - \mathbb{P}\{X_i = 0\}$. Then

$$\mathbb{E}\big[e^{s\sum_{i=1}^n X_i}\big]e^{-s\alpha} = \Big(\mathbb{E}\big[e^{sX_1}\big]e^{-s\alpha/n}\Big)^n = \big((1-p+pe^s)e^{-s\alpha/n}\big)^n.$$

Let α be of the form $\alpha = (1+\delta)np$. We are free to choose s, $s > 0$. Some calculus reveals that the optimal choice is $s = \ln\big(\frac{\bar{p}(1+\delta)}{1-p(1+\delta)}\big)$. With this choice we get the bound

$$\mathbb{P}\Big\{\sum_{i=1}^n X_i \geq np(1+\delta)\Big\} \leq \frac{1}{(1+\delta)^n}\Big(\frac{\bar{p}(1+\delta)}{1-p(1+\delta)}\Big)^{n(1-p(1+\delta))}.$$

If we use the bound $1 - p + pe^s = 1 + p(e^s - 1) \overset{s>0}{<} e^{p(e^s-1)}$ and then choose $s = \ln(1+\delta)$, we get the weaker but more convenient form,

$$\mathbb{P}\Big\{\sum_{i=1}^n X_i \geq np(1+\delta)\Big\} \leq \Big(\frac{e^\delta}{(1+\delta)^{1+\delta}}\Big)^{np}.$$

In the same manner we get

$$\text{(C.10)} \quad \mathbb{P}\Big\{\sum_{i=1}^n X_i \leq np(1-\delta)\Big\} \leq \Big(\frac{\bar{p}}{\bar{p}\bar{\delta}}\Big(\frac{1-p\bar{\delta}}{\bar{p}\bar{\delta}}\Big)^{p\bar{\delta}}\Big)^n \leq \Big(\frac{e^{-\delta}}{(1-\delta)^{1-\delta}}\Big)^{np} \leq e^{-n\delta^2 p/2}.$$

◇

The technique extends easily to the case where the variables X_i are no longer identically distributed, and the same method can handle the case of (weakly) *dependent* variables as shown in Problem C.2.

§C.3. Martingales

We get to the heart of this chapter – concentration inequalities for Martingales.

§C.3.1. Basic Definitions

The following is probably the simplest definition of a Martingale.

Definition C.11 (Martingale: Special Case). A sequence $X_0, X_1, \dots$ of random variables is called a *Martingale* if for every $i \geq 1$ we have

$$\mathbb{E}[X_i \,|\, X_0, X_1, \dots, X_{i-1}] = X_{i-1}.$$

The prototypical albeit fairly unrealistic example of a Martingale sequence is the following: assume a gambler is playing in a fair casino, i.e., the expected gain from each bet is zero. Let X_i, $i \geq 1$, denote the wealth of the gambler after the i-th bet and let X_0 represents the initial wealth. At the i-th bet the gambler may wager any amount up to X_{i-1} and the amount he wagers can be any function of $X_0, \dots, X_{i-1}$. Because the bet is assumed to be fair, the expected change in wealth at the i-th bet is zero and, therefore, $\mathbb{E}[X_i \,|\, X_0, X_1, \dots, X_{i-1}] = X_{i-1}$, i.e., $X_0, X_1, \dots, X_n$ forms a Martingale.

The preceding definition includes Martingales where the "history" can be summarized by the values of the past outcomes. The following is a more general definition which allows for more complex histories. Recall first that a *σ-field* $(\Omega, \mathcal{F})$ consists of a probability space Ω and a collection of subsets of Ω, called $\mathcal{F}$ such that (i) $\varnothing \in \mathcal{F}$, (ii) $\mathcal{E} \in \mathcal{F} \Rightarrow \mathcal{E}^c \in \mathcal{F}$, and (iii) any countable union of elements of $\mathcal{F}$ is again an element of $\mathcal{F}$. We say that a random variable X is measurable with respect to $\mathcal{F}$ if for any $x \in \mathbb{R}$, $\{\omega \in \Omega : X(\omega) \leq x\} \in \mathcal{F}$. For later reference we state the following two useful properties of conditional expectation.

Fact C.12 (Basic Properties of Conditional Expectation). Let $\mathcal{F}_0 \subseteq \mathcal{F}_1 \subseteq \cdots$ be a *filter* with respect to a given probability space Ω. This means that each $\mathcal{F}_i$ forms a σ-field $(\Omega, \mathcal{F}_i)$ with respect to the same probability space Ω and that the $\mathcal{F}_i$ are ordered by refinement in the sense that each subset of Ω contained in $\mathcal{F}_i$ is also a subset of $\mathcal{F}_j$ for $j \geq i$. Further, $\mathcal{F}_0 = \{\varnothing, \Omega\}$. Let X be a random variable in the given probability space and let Y be $\mathcal{F}_i$-measurable, where $i \geq 0$. Then for $j \geq 0$

$$\mathbb{E}\left[\mathbb{E}[X \,|\, \mathcal{F}_i] \,|\, \mathcal{F}_j\right] = \mathbb{E}[X \,|\, \mathcal{F}_{\min\{i,j\}}], \tag{C.13}$$

and

$$\mathbb{E}[XY \,|\, \mathcal{F}_i] = \mathbb{E}[X \,|\, \mathcal{F}_i] Y. \tag{C.14}$$

We are now ready to state our general definition.

Definition C.15 (Martingale). Let $\mathcal{F}_0 \subseteq \mathcal{F}_1 \subseteq \cdots$ be a filter with respect to a given probability space. Let $X_0, X_1, \dots$ denote a sequence of random variables on this probability space such that X_i is $\mathcal{F}_i$-measurable. Then we say that $X_0, X_1, \dots$ forms a Martingale with respect to the filter $\mathcal{F}_0 \subseteq \mathcal{F}_1 \subseteq \cdots$ if $\mathbb{E}[X_i \,|\, \mathcal{F}_{i-1}] = X_{i-1}$.

Discussion: Note that if $X_0, X_1, \dots$ forms a Martingale with respect to the filter $\mathcal{F}_0 \subseteq \mathcal{F}_1 \subseteq \cdots$ then X_i is measurable with respect to $\mathcal{F}_j$ for $j \geq i$. This is true since by definition of conditional expectation $\mathbb{E}[X_{i+1} \,|\, \mathcal{F}_i]$ is a $\mathcal{F}_i$-measurable random variable. But since $\mathbb{E}[X_{i+1} \,|\, \mathcal{F}_i] = X_i$ we conclude that X_i is $\mathcal{F}_i$-measurable and so it is also $\mathcal{F}_j$-measurable, for $j \geq i$.

We are concerned with a particular type of Martingale which is commonly known as Doob's Martingale.

Definition C.16 (Doob's Martingale). Let $\mathcal{F}_0 \subseteq \mathcal{F}_1 \subseteq \cdots$ be a filter with respect to a given probability space. Let X be a random variable on this probability space. Then the sequence of random variables $X_0, X_1, \dots$, where $X_i = \mathbb{E}[X \,|\, \mathcal{F}_i]$, is a *Doob's Martingale*.

To see that $X_0, X_1, \dots$ indeed forms a Martingale use (C.13) to conclude that

$$\mathbb{E}[X_i \,|\, \mathcal{F}_{i-1}] = \mathbb{E}[\mathbb{E}[X \,|\, \mathcal{F}_i] \,|\, \mathcal{F}_{i-1}] = \mathbb{E}[X \,|\, \mathcal{F}_{i-1}] = X_{i-1}.$$

In our context the σ-fields $(\Omega, \mathcal{F}_i)$ are generated by revealing more and more about a random experiment.

§C.3.2. Hoeffding-Azuma Inequality

Theorem C.17 (Hoeffding-Azuma Inequality). Let $X_0, X_1, \dots$ be a Martingale with respect to the filter $\mathcal{F}_0 \subseteq \mathcal{F}_1 \subseteq \cdots$ such that for each $i \geq 1$,

$$|X_i - X_{i-1}| \leq \gamma_i, \gamma_i \in [0, \infty).$$

Then, for all $n \geq 1$ and any $\alpha > 0$,

$$\mathbb{P}\{|X_n - X_0| \geq \alpha\sqrt{n}\} \leq 2e^{-\frac{\alpha^2 n}{2\sum_{i=1}^{n} \gamma_i^2}}.$$

Proof. We restrict ourselves to the case $\gamma_i = 1$. The general case is left as Problem C.3. Consider the function

$$f(x) = e^{\gamma x}, \tag{C.18}$$

where $\gamma \geq 0$. Define the line $c(x)$ which intersects $f(x)$ in the two points $(-1, e^{-\gamma})$ and $(+1, e^{+\gamma})$. Explicitly,

$$c(x) = \frac{1}{2}[(e^{\gamma} + e^{-\gamma}) + x(e^{\gamma} - e^{-\gamma})] = \cosh(\gamma) + \sinh(\gamma)x.$$

Since $f(x)$ is convex-∪ we have

$$f(x) \leq c(x), \quad x \in [-1, 1]. \tag{C.19}$$

We also need the inequality

$$\cosh(x) \leq e^{x^2/2}, \quad x \in \mathbb{R}. \tag{C.20}$$

This is most easily seen by noting that

$$\cosh(x) = \sum_{i \geq 0} \frac{x^{2i}}{(2i)!} \text{ and } e^{x^2/2} = \sum_{i \geq 0} \frac{x^{2i}}{2^i i!}.$$

Since the terms in both series are non-negative for $x \in \mathbb{R}$, the claim follows by observing that for $i \in \mathbb{N}$, $2^i i! \leq (2i)!$.

Let $Y_i = X_i - X_{i-1}$, $i \geq 1$. The condition $|X_i - X_{i-1}| \leq 1$ becomes

$$|Y_i| \leq 1. \tag{C.21}$$

Further, we have

$$\mathbb{E}[Y_i \,|\, \mathcal{F}_{i-1}] = \mathbb{E}[X_i - X_{i-1} \,|\, \mathcal{F}_{i-1}] = \mathbb{E}[X_i \,|\, \mathcal{F}_{i-1}] - X_{i-1} = 0. \tag{C.22}$$

Therefore, we get with $f(x)$ as defined in (C.18)

by (C.19) and (C.21)	$\mathbb{E}[f(Y_i) \,\|\, \mathcal{F}_{i-1}] \leq \mathbb{E}[c(Y_i) \,\|\, \mathcal{F}_{i-1}]$
due to linearity	$= c(\mathbb{E}[Y_i \,\|\, \mathcal{F}_{i-1}])$
by (C.22)	$= c(0) = \cosh(y)$
by (C.20)	$\leq e^{y^2/2}$.

Note that $X_n - X_0 = \sum_{i=1}^n Y_i$ and that $\mathbb{E}[\cdot]$ is equivalent to $\mathbb{E}[\cdot \,|\, \mathcal{F}_0]$. Therefore,

$$\mathbb{E}[f(X_n - X_0)] = \mathbb{E}\Big[\prod_{i=1}^n f(Y_i)\Big]$$

by (C.13)

$$= \mathbb{E}\Big[\mathbb{E}\Big[\Big(\prod_{i=1}^n f(Y_i)\Big) \,\Big|\, \mathcal{F}_{n-1}\Big]\Big]$$

by (C.14)

$$= \mathbb{E}\Big[\Big(\prod_{i=1}^{n-1} f(Y_i)\Big) \mathbb{E}\Big[f(Y_n) \,\Big|\, \mathcal{F}_{n-1}\Big]\Big]$$

$$\leq \mathbb{E}\Big[\prod_{i=1}^{n-1} f(Y_i)\Big] e^{y^2/2} \leq e^{ny^2/2}.$$

It follows that

$$\mathbb{P}[X_n - X_0 \geq \alpha\sqrt{n}] = \mathbb{P}[f(X_n - X_0) \geq f(\alpha\sqrt{n})]$$

by Lemma C.1
$$\leq \frac{\mathbb{E}[f(X_n - X_0)]}{f(\alpha\sqrt{n})} = e^{\gamma^2 n/2} e^{-\gamma\alpha\sqrt{n}} = e^{-\alpha^2/2},$$

where the last step follows if we choose $\gamma = \alpha/\sqrt{n}$. In the same manner we can prove $\mathbb{P}[X_n - X_0 \leq -\alpha\sqrt{n}] \leq e^{-\alpha^2/2}$, which together with the previous result proves the claim. □

As a primary application of the Hoeffding-Azuma inequality we give the proof of the main concentration theorem of message-passing decoding. To be concrete, we prove that a particular performance measure, namely the error probability, concentrates around its expected value, but almost verbatim the same proof can be used to show the concentration of a variety of other measures.

Proof of Theorems 3.30 and 4.92. To simplify the notation we focus on the $(\mathtt{l}, \mathtt{r})$-regular case. The general case can be dealt with in essentially the same way. The basic idea of the proof is to form a Doob's Martingale by revealing more and more about the object of interest. In our case, it is natural to order the $n\mathtt{l}$ edges on the variable-node side in some arbitrary but fixed way and then to reveal one by one for each edge the destination socket on the check-node side. We say that an edge is being *exposed*. We then reveal in the following n steps the channel observations. At the beginning of the exposure process we deal with the average (over all graphs and channel realizations). At the end of the $n(\mathtt{l} + 1)$ exposure steps we have revealed the exact graph which is used and the given channel realization. Assuming that we can show that at each exposure step the quantity of interest does not change by too much irrespective of the revealed information, we can use the Hoeffding-Azuma inequality to link the average case to specific instances.

Let Z denote the number of incorrect variable-to-check node messages among all $n\mathtt{l}$ variable-to-check node messages passed in the ℓ-th iteration for a $(\mathtt{G}, Y) \in \Omega$, where $\mathtt{G}$ is a graph in the ensemble LDPC $(n, x^{\mathtt{l}-1}, x^{\mathtt{r}-1})$, Y is the channel observation, and Ω is the probability space. Let $=_i$, $0 \leq i \leq (\mathtt{l} + 1)n$, be a sequence of equivalence relations on Ω ordered by refinement, i.e., $(\mathtt{G}', Y') =_i (\mathtt{G}'', Y'')$ implies $(\mathtt{G}', Y') =_{i-1} (\mathtt{G}'', Y'')$. These equivalence classes are defined by partial equalities.

Suppose we expose the $\mathtt{l}n$ edges of the graph one at a time, i.e., at step $i \in [n\mathtt{l}]$ we expose the check-node socket $\pi(i)$ which is connected to the i-th variable-node socket, and, similarly, in the following n steps we expose the n received values, one at a time. We have $(\mathtt{G}', Y') =_i (\mathtt{G}'', Y'')$ if and only if the information revealed in the first i steps is the same for both pairs. Let h denote the number of steps in this exposure procedure, $h = (\mathtt{l} + 1)n$.

Define $Z_0, Z_1, \ldots, Z_h$ by

$$Z_i(\mathtt{G}, Y) = \mathbb{E}[Z(\mathtt{G}', Y') \mid (\mathtt{G}', Y') =_i (\mathtt{G}, Y)].$$

By construction $Z_0, Z_1, \ldots, Z_h$ forms a *Doob's Martingale*. The crucial step in the proof is to find a (tight) bound

$$|Z_i(\mathtt{G}, Y) - Z_{i-1}(\mathtt{G}, Y)| \leq \gamma_i, i \in [h], \tag{C.23}$$

for some suitable constants γ_i which may depend on $\mathtt{l}$, $\mathtt{r}$ and ℓ, but preferably not on n.

We first prove (C.23) for $i \in [n\mathtt{l}]$, i.e., for the steps where we expose the edges. Recall that $\pi(i) = j$ means that the i-th variable-node socket is connected to the j-th check-node socket. Denote by $\mathcal{G}(\mathtt{G}, i)$ the subset of graphs in LDPC $(n, x^{\mathtt{l}-1}, x^{\mathtt{r}-1})$ such that the first i edges are equal to the edges in $\mathtt{G}$, i.e., $\mathcal{G}(\mathtt{G}, i) = \{\mathtt{G}' : (\mathtt{G}', Y) =_i (\mathtt{G}, Y)\}$. Let $\mathcal{G}_j(\mathtt{G}, i)$ be the subset of $\mathcal{G}(\mathtt{G}, i)$ consisting of those graphs for which $\pi(i+1) = j$. Thus, $\mathcal{G}(\mathtt{G}, i) = \cup_j \mathcal{G}_j(\mathtt{G}, i)$.

We have

$$\begin{aligned} Z_{i-1}(\mathtt{G}, Y) &= \mathbb{E}[Z(\mathtt{G}', Y') \mid (\mathtt{G}', Y') =_{i-1} (\mathtt{G}, Y)] = \mathbb{E}[Z(\mathtt{G}', Y') \mid \mathtt{G}' \in \mathcal{G}(\mathtt{G}, i-1)] \\ &= \sum_{j \in [n\mathtt{l}]} \mathbb{E}[Z(\mathtt{G}', Y') \mid \mathtt{G}' \in \mathcal{G}_j(\mathtt{G}, i-1)] \cdot \\ &\qquad \Pr\{\mathtt{G}' \in \mathcal{G}_j(\mathtt{G}, i-1) \mid \mathtt{G}' \in \mathcal{G}(\mathtt{G}, i-1)\}. \end{aligned} \tag{C.24}$$

We claim that if j and k are such that $\mathbb{P}\{\mathtt{G}' \in \mathcal{G}_j(\mathtt{G}, i-1) \mid \mathtt{G}' \in \mathcal{G}(\mathtt{G}, i-1)\} \neq 0$ and $\mathbb{P}\{\mathtt{G}' \in \mathcal{G}_k(\mathtt{G}, i-1) \mid \mathtt{G}' \in \mathcal{G}(\mathtt{G}, i-1)\} \neq 0$ then

$$\left|\mathbb{E}[Z(\mathtt{G}', Y') \mid \mathtt{G}' \in \mathcal{G}_j(\mathtt{G}, i-1)] - \mathbb{E}[Z(\mathtt{G}', Y') \mid \mathtt{G}' \in \mathcal{G}_k(\mathtt{G}, i-1)]\right| < 8(\mathtt{l}\mathtt{r})^{\ell}. \tag{C.25}$$

To prove this claim define a map $\phi_{j,k} : \mathcal{G}_j(\mathtt{G}, i-1) \to \mathcal{G}_k(\mathtt{G}, i-1)$ as follows. Let π be the permutation defining the edge assignment for a given graph $\mathtt{H} \in \mathcal{G}_j(\mathtt{G}, i-1)$ and let $k' = \pi^{-1}(k)$. Define a permutation π' by $\pi' = \pi$ except that $\pi'(i) = k$ and $\pi'(k') = j$. Let $\mathtt{H}'$ denote the resulting graph. Note that $\mathtt{H}' \in \mathcal{G}_k(\mathtt{G}, i-1)$. By definition, $\mathtt{H}' = \phi_{j,k}(\mathtt{H})$. The construction is shown in Figure C.27. Clearly, $\phi_{j,k}$ is a bijection and, since every (edge-labeled) graph in the ensemble has uniform probability, such a bijection preserves probabilities. We claim that for a fixed $\mathtt{H}$ and Y

$$|Z(\mathtt{H}, Y) - Z(\phi_{j,k}(\mathtt{H}), Y)| \leq 8(\mathtt{l}\mathtt{r})^{\ell}. \tag{C.26}$$

To see this, note that the message along a given edge sent in iteration ℓ is only

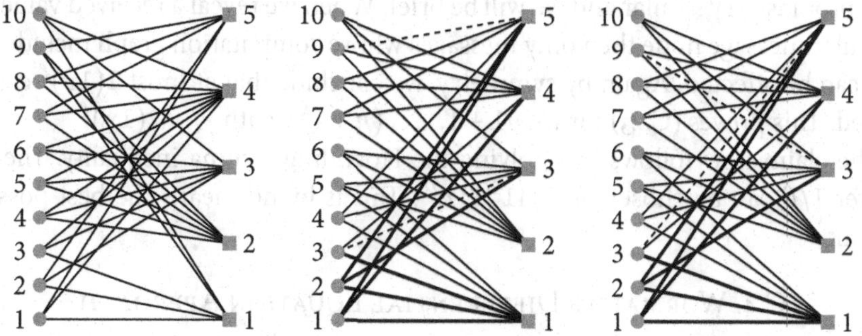

Figure C.27: Left: Graph G from the ensemble LDPC $(10, x^2, x^5)$; Middle: Graph H from the ensemble $\mathcal{G}_7(\mathtt{G}, 7)$ (note that the labels of the sockets are not shown – these labels should be inferred from the order of the connections in the middle figure); the first 7 edges that H has in common with G are drawn in bold; Right: the associated graph $\phi_{7,30}(\mathtt{H})$. The two dashed lines correspond to the two edges whose endpoints are switched.

a function of the associated computation graph, call it $\vec{\mathcal{C}}_\ell$. Therefore, a message is only affected by an exchange of the endpoints of two edges if one (or both) of the two edges is (are) elements of $\vec{\mathcal{C}}_\ell$. Note that $\vec{\mathcal{C}}_\ell$ contains at most $2(\mathtt{lr})^\ell$ distinct edges and, by symmetry, an edge can be part of at most $2(\mathtt{lr})^\ell$ such computation graphs. It follows that at most $8(\mathtt{lr})^\ell$ computation graphs can be affected by the exchange of the endpoints of two edges, which proves claim (C.26).

Since $\phi_{j,k}$ is a bijection and preserves probability, it follows that

$$\mathbb{E}[Z(\mathtt{G}', Y') \,|\, \mathtt{G}' \in \mathcal{G}_k(\mathtt{G}, i-1)] = \mathbb{E}[Z(\phi_{j,k}(\mathtt{G}'), Y') \,|\, \mathtt{G}' \in \mathcal{G}_j(\mathtt{G}, i-1)].$$

By (C.26) any pair $Z(\mathtt{H}, Y)$ and $Z(\phi_{j,k}(\mathtt{H}), Y)$ has difference bounded by $8(\mathtt{lr})^\ell$ and, since for any random variable $|\mathbb{E}[W]| \leq \mathbb{E}[|W|]$, claim (C.25) follows.

By definition, $Z_i(\mathtt{G}, Y)$ is equal to $\mathbb{E}[Z(\mathtt{G}', Y') \,|\, \mathtt{G}' \in \mathcal{G}_j(\mathtt{G}, i-1)]$ for some $j \in [n\mathtt{l}] \backslash \{\pi(l) : l < i\}$. Hencc,

$$\begin{aligned}
&|Z_i(\mathtt{G}, Y) - Z_{i-1}(\mathtt{G}, Y)| \\
&\leq \max_j |\mathbb{E}[Z(\mathtt{G}', Y') \,|\, \mathtt{G}' \in \mathcal{G}_j(\mathtt{G}, i-1)] - Z_{i-1}(\mathtt{G}, Y)| \\
&\leq \max_{j,k} |\mathbb{E}[Z(\mathtt{G}', Y') \,|\, \mathtt{G}' \in \mathcal{G}_j(\mathtt{G}, i-1)] - \mathbb{E}[Z(\mathtt{G}', Y') \,|\, \mathtt{G}' \in \mathcal{G}_k(\mathtt{G}, i-1)]| \\
&\leq 8(\mathtt{lr})^\ell,
\end{aligned}$$

where we have used the representation of $Z_i(\mathtt{G}, Y)$ given in (C.24). This proves (C.23) for $i \in [n\mathtt{l}]$ with $\gamma_i = 8(\mathtt{lr})^\ell$.

It remains to show that the inequality is also fulfilled for the last n steps. The idea of the proof is very similar and we will be brief. When we reveal a received value at a particular message node then only messages whose computation graph include this node can be affected. Again by symmetry, we conclude that at most $2(\mathtt{lr})^{\ell}$ can be affected. This proves (C.23) for $i \in n\mathtt{l}+1, \dots, (n+1)\mathtt{l}$ with $\gamma_i = 2(\mathtt{lr})^{\ell}$.

The claim now follows by applying the Hoeffding-Azuma inequality. The parameter $1/\beta$ can be chosen as $544\mathtt{l}^{2\ell-1}\mathtt{r}^{2\ell}$. This is by no means the best possible constant. □

§C.4. Wormald's Differential Equation Approach

Consider the peeling decoder for the binary erasure channel (BEC) as described in Section 3.19: the decoder "peels off" one degree 1 check node at a time. In this manner we observe a sequence of residual graphs and an associated sequence of degree distributions. We are interested in describing the "typical" evolution of this sequence of degree distributions as a function of "time" in the limit of large blocklengths. The Wormald method is the perfect tool to accomplish that.

Consider a function $f : \mathbb{R}^d \to \mathbb{R}$, $d \geq 1$. We say that f is *Lipschitz continuous* if there exists a constant L such that for any pair $x, y \in \mathbb{R}^d$

$$|f(x) - f(y)| \leq L\|x - y\|_1 = L \sum_{i=1}^{d} |x_i - y_i|.$$

We formulate the Wormald method in the setting of the evolution of a Markov process. More general versions are available but we do not need them. Consider the discrete-time Markov process Z with *state space* $\{0, \dots, k\}^d$, $k, d \in \mathbb{N}$. More precisely, if Ω denotes the probability space then Z is a sequence $Z(t = 0), Z(t = 1), \dots$ of random variables, $Z(t) : \Omega \to \{0, \dots, k\}^d$. To every $\omega \in \Omega$ corresponds a *realization* $z(t = 0, \omega), z(t = 1, \omega), \dots$. In our setting, each (residual) graph is characterized in terms of a degree distribution pair. We therefore choose the random variable $Z(t)$ at step t to be the (un-normalized) degree distribution pair $(\Lambda^{(t)}, P^{(t)})$. Why does the sequence $Z(t = 0), Z(t = 1), \dots$ form a Markov process? Associate to each (un-normalized) degree distribution pair $(\Lambda^{(t)}, P^{(t)})$ the corresponding ensemble LDPC$\left(\Lambda^{(t)}, P^{(t)}\right)$. The degree distribution pair $(\Lambda^{(t)}, P^{(t)})$ completely specifies the probability distribution of the degree distribution pair $\left(\Lambda^{(t+1)}, P^{(t+1)}\right)$. Even more, given that the degree distribution is $(\Lambda^{(t)}, P^{(t)})$, every (labeled) graph in LDPC$\left(\Lambda^{(t)}, P^{(t)}\right)$ has the same probability of being the residual graph at time t in the decoding process.

In the Wormald method we do not look at a single Markov process but at a sequence of such processes, denoting them by $\{Z^{(m)}\}_{m \geq 1}$. In our setting, this corresponds to looking at ensembles of (increasing) length n. In particular, it is convenient to choose $m = nL'(1)$, i.e., m is equal to the number of edges in the graph. The Wormald method consists of showing that, for increasing m, with high probability the random variables $Z^{(m)}(t=0), Z^{(m)}(t=1), \ldots$ stay close to an "average" path and that this average path is equal to the solution of a differential equation. In the sequel, $Z_i^{(m)}(t)$ denotes the i-th component (out of d) of $Z^{(m)}(t)$. Also, given a subset $D \subseteq \mathbb{R}^{d+1}$, we define the *stopping time* $T_D(Z_1^{(m)}, \ldots, Z_d^{(m)})$ to be the minimum t so that $(t/m, Z_1^{(m)}/m, \ldots, Z_d^{(m)}/m) \notin D$.

THEOREM C.28 (WORMALD METHOD). Let $\{Z^{(m)}\}_{m \geq 1}$ denote a sequence of Markov processes, where the m-th process has state space $\{0, \ldots, \lfloor m\alpha \rfloor\}^d$, for some $\alpha > 0$. Let D be some open connected bounded set containing the closure of

$$\{(0, z_1, \ldots, z_d) : \mathbb{P}\{Z_i^{(m)}(t=0)/m = z_i, 1 \leq i \leq d\} > 0 \text{ for some } m\}. \tag{C.29}$$

For $1 \leq i \leq d$, let $f_i : \mathbb{R}^{d+1} \to \mathbb{R}$ such that the following four conditions are fulfilled.

(i) (Boundedness) There exists a constant c such that for all $m \geq 1$ and $1 \leq i \leq d$,

$$\left|Z_i^{(m)}(t+1) - Z_i^{(m)}(t)\right| \leq c,$$

for all $0 \leq t < T_D$.

(ii) (Trend) For all $m \geq 1$ and $1 \leq i \leq d$,

$$\mathbb{E}[Z_i^{(m)}(t+1) - Z_i^{(m)}(t) \,|\, Z^{(m)}(t)] = f_i\Big(\frac{t}{m}, \frac{Z_1^{(m)}(t)}{m}, \ldots, \frac{Z_d^{(m)}(t)}{m}\Big) + O\Big(\frac{1}{m}\Big),$$

uniformly for all $0 \leq t < T_D$.

(iii) (Lipschitz) Each function f_i, $1 \leq i \leq d$, is Lipschitz continuous on the intersection of D with the half-space $\{(t, z_1, \ldots, z_d) : t \geq 0\}$.

(iv) (Initial Concentration) For $1 \leq i \leq d$,

$$\mathbb{P}\{|Z_i^{(m)}(t=0)/m - \mathbb{E}[Z_i^{(m)}(t=0)/m]| \geq m^{-\frac{1}{6}}\} \leq O\left(m^{\frac{1}{6}} e^{-\frac{\sqrt{m}}{c^3}}\right).$$

Then the following is true.

(i) For $(0, \hat{z}_1, \ldots, \hat{z}_d) \in D$ the system of differential equations

$$\frac{dz_i}{d\tau} = f_i(\tau, z_1, \ldots, z_d), \quad 1 \leq i \leq d,$$

has a unique solution in D for $z_i : \mathbb{R} \to \mathbb{R}$ passing through $z_i(0) = \hat{z}_i$. This solution extends to points arbitrarily close to the boundary of D.

(ii) There exists a strictly positive constant a such that

$$\mathbb{P}\{|Z_i^{(m)}(t)/m - z_i(t/m)| \geq am^{-\frac{1}{6}}\} \leq O\left(m^{\frac{1}{6}} e^{-\frac{\sqrt{m}}{c^3}}\right),$$

for $0 \leq t \leq n\tau_{\max}$ and for each $1 \leq i \leq d$. Here, $z_i(\tau)$ is the solution in (i) with $\hat{z}_i = \mathbb{E}[Z_i^{(m)}(t=0)]/m$ and $\tau_{\max} = \tau_{\max}(m)$ is the supremum of those τ to which the solution can be extended before reaching within L_1-distance b of the boundary of D, where b is a strictly positive constant.

Our main application of the Wormald technique is to the analysis of the peeling decoder. In particular we are now in the position to prove Theorem 3.106. A second application is to the analysis of an efficient encoder discussed in Appendix A. The Wormald technique is also the basis for the finite-length scaling result discussed in Section 3.23. In this case the corresponding differential equation not only tracks the mean of the process but also the second moment.

Proof of Theorem 3.106. Consider the evolution of the peeling decoder for a specific element $\mathtt{G}$ from LDPC (n, L, R). Time t starts at zero. Just before time zero, each variable node together with its connected edges is removed independently from all other choices with probability $1 - \epsilon$. These are the nodes that have been transmitted successfully. Time increases by one for each subsequent variable node which is removed by the decoder.

Choose a code $\mathtt{G}$ uniformly at random from LDPC (n, L, R). Without loss of generality we can assume that the all-zero codeword was transmitted. Let m denote the total number of edges in the graph, i.e., $m = nL'(1)$. At time t, let $V_i^{(m)}(t)$ denote the number of edges in the residual graph which are connected to *variable* nodes of degree i, where $2 \leq i \leq \mathtt{l}_{\max}$. This random variable depends on the choice of the graph $\mathtt{G}$, the channel realization, and the random choices made by the decoder. In the same manner, let $C_i^{(m)}(t)$ denote the number of edges in the residual graph which are connected to *check* nodes of degree i, where $1 \leq i \leq \mathtt{r}_{\max}$. To keep the density of sub- and superscripts at a manageable level we drop in some of the subsequent equations the superscript m which indicates the size of the system we are considering. For example, we write $V_i(t)$ instead of $V_i^{(m)}(t)$. Note that $(V(t), C(t))$ is the pair of un-normalized degree distributions from an edge perspective and the sequence of such random variables forms a Markov process.

Consider one decoding step. We claim that as long as $C_1(t) > 0$

(C.30)

$$\mathbb{E}[V_i(t+1) - V_i(t) \mid V(t), C(t)] = -\frac{iV_i(t)}{\sum_j V_j(t)}, \quad i \geq 2,$$

$$\mathbb{E}[C_1(t+1) - C_1(t) \mid V(t), C(t)] = (C_2(t) - C_1(t))\frac{(\sum_j (j-1)V_j(t))}{(\sum_j V_j(t))^2} - 1 + O(1/m), \tag{C.31}$$

$$\mathbb{E}[C_i(t+1) - C_i(t) \mid V(t), C(t)] = (C_{i+1}(t) - C_i(t))\frac{i(\sum_j (j-1)V_j(t))}{(\sum_j V_j(t))^2} + O(1/m), \quad i \geq 2. \tag{C.32}$$

Consider (C.30). This equation says that the expected decrease in the number of edges which are connected to variable nodes of degree i equals $iV_i(t)/\sum_j V_j(t)$: assume that there is at least one degree 1 check node left so that the decoding process continues. The decoder randomly picks such a degree 1 check node. Its outgoing edge is connected to one of the $\sum_j V_j(t)$ remaining sockets uniformly at random. The fraction of edges which are connected to degree i nodes is $V_i(t)/\sum_j V_j(t)$, and this equals the probability that a variable node of degree i is removed. If a variable node of degree i is removed then exactly i edges which are connected to a variable node of degree i are removed.

The interpretation for the check-node side as described by (C.31) and (C.32) is similar. We always remove the one edge which is connected to the chosen degree 1 check node. This explains the -1 on the right-hand side of (C.31). In addition we remove in expectation $\sum_j jV_j(t)/\sum_j V_j(t) - 1 = \sum_j (j-1)V_j(t)/\sum_j V_j(t)$ other edges: each such edge we remove is connected to the sockets on the right in a uniform manner. If a removed edge is connected to a check node of degree $i+1$, which happens with probability $C_{i+1}(t)/\sum_j C_j(t) = C_{i+1}(t)/\sum_j V_j(t)$, then the residual degree of this node changes from $i+1$ to i. This means that there are $i+1$ fewer edges of degree $i+1$ and i more of degree i. The extra term $O(1/m)$ comes from the fact that we delete *several* edges at each step but that in the preceding derivation we assumed that the degree distribution was constant during one step.

Assume that the given instance evolves according to this expected value and that the decoding process has not stopped. Introduce the scaled time $\tau = t/m$. If we let m tend to infinity then, motivated by Theorem C.28, we are lead to consider the set of first-order differential equations

$$\frac{\mathrm{d}v_i(\tau)}{\mathrm{d}\tau} = -\frac{iv_i(\tau)}{\sum_j v_j(\tau)}, \qquad i \geq 2,$$

$$\frac{\mathrm{d}w_1(\tau)}{\mathrm{d}\tau} = (w_2(\tau) - w_1(\tau))\frac{(\sum_j (j-1)v_j(\tau))}{(\sum_j v_j(\tau))^2} - 1,$$

$$\frac{\mathrm{d}w_i(\tau)}{\mathrm{d}\tau} = (w_{i+1}(\tau) - w_i(\tau))\frac{i(\sum_j (j-1)v_j(\tau))}{(\sum_j v_j(\tau))^2}, \qquad i \geq 2.$$

To establish the relationship, first recall that we assumed that the process evolves like its expected value so that we can drop the expectation. Then write with $\Delta t = 1$

$$\frac{V_i(t+\Delta t) - V_i(t)}{\Delta t} = \frac{V_i(m\tau + m\mathrm{d}\tau) - V_i(m\tau)}{m\mathrm{d}\tau} \approx \frac{\mathrm{d}v_i(\tau)}{\mathrm{d}\tau},$$

where we introduced the function $v_i(\tau) \approx V_i(m\tau)/m$. The same line of argument can be used to establish the relationship for the check-node side.

Two tasks remain. We have to solve the set of differential equations and we have to verify that the conditions of Theorem C.28 are fulfilled so that we can assert that typical instances evolve like the solution of the differential equation.

We start by solving the system of differential equations. Define $e(\tau) = \sum_j v_j(\tau) = \sum_j w_j(\tau)$. Note that $e(\tau)$ represents the fraction (out of m) of edges remaining in the residual graph at time τ. If we assume that $e(\tau)$ is a known function then the solution of $v_i(\tau)$ can be written as

$$v_i(\tau) = \epsilon\lambda_i \mathrm{e}^{-i\int_0^\tau \mathrm{d}s/e(s)}.$$

Let us verify this. Clearly, for $\tau = 0$, i.e., at the beginning of the process right after the received variable nodes and all connected edges have been removed, this gives us the correct result, namely $v_i(\tau = 0) = \epsilon\lambda_i$. The general solution is verified by checking that it fulfills the differential equation.

The "time" τ has a direct operational significance – it measures (normalized by m) the number of decoding steps taken. To write down the solution in its final form it is more convenient to consider an alternative "time." Define $y(\tau) = \mathrm{e}^{-\int_0^\tau \mathrm{d}s/e(s)}$. Then we can write the solution as $v_i(y) = \epsilon\lambda_i y^i$. Recall that $v_i(y)$ is equal to the fraction (with respect to m) of edges in the residual graph which are connected to variable nodes of residual degree i. Let us compute the fraction (with respect to the original number n) of nodes in the residual graph of degree i. Always i edges of degree i form one node of degree i. Therefore the fraction of variable nodes of degree i in the residual graph, call it $\tilde{L}_i$, is equal to

$$\tilde{L}_i = L'(1)\epsilon\lambda_i y^i/i = \epsilon\frac{\lambda_i/i}{\sum_j \lambda_j/j} y^i = \epsilon L_i y^i,$$

which confirms (3.107). It remains to find the evolution of the check-node degree distribution. We claim that

$$w_1(y) = \epsilon\lambda(y)[y - 1 + \rho(1 - \epsilon\lambda(y))],$$
$$w_i(y) = \sum_j \rho_j \binom{j-1}{i-1} (\epsilon\lambda(y))^i (1 - \epsilon\lambda(y))^{j-i}, \quad i \geq 2.$$

Let us check that this solution fulfills the correct initial condition. For $y = 1$ ($\tau = 0$) the preceding equations specialize to

$$w_i(1) = \sum_j \rho_j \binom{j-1}{i-1} \epsilon^i (1-\epsilon)^{j-i}, \quad i \geq 1.$$

This is indeed correct. An edge which just before time $t = 0$ has degree j stays in the graph with probability ϵ and has degree i afterward with probability $\binom{j-1}{i-1}\epsilon^{i-1}(1-\epsilon)^{j-i}$. This gives the preceding expression. That the stated equations not only fulfill the initial conditions but indeed solve the differential equations needs some extra work which we skip.

Let us convert this solution into the node perspective. Let $\tilde{R}_i$ denote the fraction (with respect to the original size $n(1-r) = m/R'(1)$) of check nodes in the residual graph. Since i edges of degree i form one node of degree i and since the fraction of edges is normalized with respect to m, we get for $i \geq 2$

$$\begin{aligned}
\tilde{R}_i &= R'(1) \sum_j \rho_j / i \binom{j-1}{i-1} (\epsilon\lambda(y))^i (1-\epsilon\lambda(y))^{j-i} \\
&= R'(1) \sum_{j \geq 2} \rho_j / j \binom{j}{i} (\epsilon\lambda(y))^i (1-\epsilon\lambda(y))^{j-i} \\
&= \sum_{j \geq 2} R_j \binom{j}{i} (\epsilon\lambda(y))^i (1-\epsilon\lambda(y))^{j-i}.
\end{aligned}$$

This confirms (3.109). For $i = 1$ we get

$$\tilde{R}_1 = R'(1)\epsilon\lambda(y)[y - 1 + \rho(1-\epsilon\lambda(y))],$$

as stated in (3.108).

We now use Theorem C.28 to justify the preceding steps. Let us check that all conditions are fulfilled.

To simplify our notation let us assume that *all* components L_i, $i = 2, \ldots, \mathtt{l}_{\max}$, are strictly positive. The general case where some components are zero can be handled in a similar way by restricting the region D to only the non-zero components. Without loss of generality we assume that $R_{\mathtt{r}_{\max}}$ is strictly positive. Let D be $(-\eta, 1) \times \mathcal{V} \times \mathcal{C}$, where $\mathcal{V} = (0,1)^{\mathtt{l}_{\max}-1} \smallsetminus \{(v_2, \ldots, v_{\mathtt{l}_{\max}}) : \sum v_i \leq \eta\}$, $\mathcal{C} = (0,1)^{\mathtt{r}_{\max}} \smallsetminus \{(c_1, \ldots, c_{\mathtt{r}_{\max}}) : \sum c_i \leq \eta\}$. Here, η is strictly positive but arbitrarily small. As required, D is an open connected bounded set.

Consider the initial condition. As discussed earlier, on the variable side the expected normalized number of nodes of degree i at the beginning of the process equals ϵL_i, $i = 2, \ldots, \mathtt{l}_{\max}$. Since we assumed that all components are strictly positive, this point is in the interior of $\mathcal{V}$ for η sufficiently small. In a similar manner, we can check that the vector whose components consists of the expected normalized number of check nodes is in the interior of C for η sufficiently small. We conclude that the *expected* initial degree distribution is in the interior of D. Further, using the tools discussed in this appendix, one can show that the probability that for a random instance a given component deviates by more than $m^{-\frac{1}{6}}$ from this expected normalized value is $O\left(e^{-m^{\frac{2}{3}}\kappa}\right)$, where κ is a strictly positive constant. Therefore, let us limit our focus to those instances whose initial degree distribution deviates from the expected one by an L_1-norm of at most $m^{-\frac{1}{6}}$. For all other instances we assume that their (normalized) trajectories are more than the allowed $am^{-\frac{1}{6}}$ away from the expected value and so we count those instances as errors on the right. Since $m^{\frac{2}{3}} \geq \sqrt{m}$ these instances do not increase our error bound. In other words, for this subset of instances condition (C.29) as well as condition (iv) are fulfilled.

We claim that

$$|V_i(t+1) - V_i(t)| \leq \mathtt{l}_{\max}, \quad |C_j(t+1) - C_j(t)| \leq \mathtt{l}_{\max}\mathtt{r}_{\max}.$$

This is true since if a variable of degree i is removed then this means that the number of edges which have variable-degree i is decreased by i. Further, for each edge we removed the number of check-degree j edges can be decreased/increased by at most $\mathtt{r}_{\max}$. We see that condition (i) is fulfilled.

Next, define the functions

$$g_i(t, v_2, \dots, v_{\mathtt{l}_{\max}}, c_1, \dots, c_{\mathtt{r}_{\max}}) = -\frac{iv_i}{\sum_j v_j}, \quad 2 \leq i \leq \mathtt{l}_{\max},$$

$$h_1(t, v_2, \dots, v_{\mathtt{l}_{\max}}, c_1, \dots, c_{\mathtt{r}_{\max}}) = (c_2 - c_1)\frac{(\sum_j (j-1)v_j)}{(\sum_j v_j)^2} - 1,$$

$$h_i(t, v_2, \dots, v_{\mathtt{l}_{\max}}, c_1, \dots, c_{\mathtt{r}_{\max}}) = (c_{j+1} - c_j)\frac{i(\sum_j (j-1)v_j)}{(\sum_j v_j)^2}, \quad 2 \leq j \leq \mathtt{r}_{\max} - 1,$$

$$h_{\mathtt{r}_{\max}}(t, v_2, \dots, v_{\mathtt{l}_{\max}}, c_1, \dots, c_{\mathtt{r}_{\max}}) = -c_{\mathtt{r}_{\max}}\frac{i(\sum_j (j-1)v_j)}{(\sum_j v_j)^2}.$$

We see from (C.30), (C.31), and (C.32) that condition (ii) is fulfilled as well.

Finally, some calculus verifies that the Lipschitz condition (iii) is fulfilled.

Consider first the case $\epsilon < \epsilon^{\text{BP}}$, where ϵ^{BP} is the threshold computed according to density evolution. We see from (3.107)–(3.109) that the solution of the differential equation stays bounded away from the boundary of D from the start of the decoding process until the number of nodes in the residual graph has reached size less than ηm, where η is an arbitrary but strictly positive constant. We conclude that with probability at least $1 - O(m^{1/6} e^{-\frac{\sqrt{m}}{(1_{\max} r_{\max})^3}})$ the normalized degree distribution of the sequence of residual graphs of a specific instance deviates from the expected degree distribution by at most $O(m^{-1/6})$ uniformly from the start of the process until the total number of nodes in the residual graph has reached size ηm. In words, below the parameter ϵ^{BP} random instance are decoded successfully (at least up to an arbitrarily small fraction of the size of the graph) under the peeling decoder with probability approaching 1 as m tends to infinity.

Second consider the case $\epsilon > \epsilon^{\text{BP}}$. Now we see that the solution of the differential equation stays inside D until "time" y so that $y - 1 + \rho(1 - \epsilon\lambda(y)) = 0$, i.e., at the time when the number of degree 1 check nodes has decreased to zero. At this point the degree distribution of the residual graph concentrates around the one given by (3.107)–(3.109). We conclude that if $\epsilon > \epsilon^{\text{BP}}$ then with probability at least $1 - O(m^{1/6} e^{-\frac{\sqrt{m}}{(1_{\max} r_{\max})^3}})$ the normalized degree distribution of the sequence of residual graphs of a specific instance deviates from the expected degree distribution by at most $O(m^{-1/6})$ uniformly from the start of the process until the decoder gets stuck.

□

§C.5. Convergence to Poisson Distribution

So far we have discussed tools which are helpful in establishing the concentration of a random variable around its mean. Let us now see a standard method of proving that a sequence of random variables converges in distribution to a Poisson. A proof of the following theorem can be found, e.g., in Bollobás [4, p. 26].

Theorem C.33 (Convergence to Poisson Distribution). Let μ_i, $i = 1, \dots, m$, $m \in \mathbb{N}$, be non-negative numbers. For each $n \in \mathbb{N}$, let $(X_1(n), \dots, X_m(n))$ be a vector of non-negative integer-valued random variables defined on the same space. If for all $(r_1, \dots, r_m) \in \mathbb{N}^m$

$$\text{(C.34)} \qquad \lim_{n\to\infty} \mathbb{E}\Big[(X_1(n))_{r_1} \dots (X_m(n))_{r_m}\Big] = \prod_{i=1}^{m} \mu_i^{r_i},$$

where $(x)_r = x(x-1)\dots(x-r+1)$, then the random vector $(X_1(n), \dots, X_m(n))$ converges in distribution to a vector of independent Poisson random variables with mean $(\mu_1, \dots, \mu_m)$.

When dealing with sums of indicator random variables, the computation of the moments is made easier by the following observation.

FACT C.35. If $x = \sum_i v_i$, where the $v_i \in \{0,1\}$, then for any r

$$(x)_r = x(x-1)\dots(x-r+1) = \sum_{(i_1,\dots,i_r)} v_{i_1}\dots v_{i_r},$$

where the summation is taken over all *ordered* sets of *distinct* indices.

We can apply Theorem C.33 to show that the number of minimal codewords of small (constant) weight contained in a randomly chosen (suitable low-density parity-check (LDPC)) code converges to a Poisson random variable. This is done in the proof of Lemma 6.49 in the setting of turbo ensembles. We can proceed in the same fashion to prove the equivalent statement for LDPC (n, λ, ρ) ensembles. Let us give here however an alternative approach which highlights a useful connection between LDPC ensembles and standard random graphs.

Define the ensemble $G(n, R)$ of graphs, where $R(x)$ is a degree distribution from the node perspective so that $nR'(1) \in 2\mathbb{N}$. More precisely, consider a graph on $nR(1)$ variables, out of which nR_i have degree i. To sample from $G(n, R)$ proceed as follows. Label the $nR'(1)$ sockets of the $nR(1)$ nodes in some fixed but arbitrary way with the set of integers $[nR'(1)]$. Pair up the $nR'(1)$ sockets uniformly at random. The resulting graph might contain self-loops and multiple edges. A cycle in such a graph is defined in the usual way as a closed path. A cycle of length 1 corresponds to a self-loop and a cycle of length 2 corresponds to a pair of parallel edges. We call a cycle *primitive* if it cannot be decomposed into smaller cycles.

LEMMA C.36 (CYCLE DISTRIBUTION IN $G(n, R)$ – [4, THEOREM 2.16]). Let Γ be chosen uniformly at random from the ensemble $G(n, R)$. For a fixed $m \in \mathbb{N}$, let $(C_1, \dots, C_m)$ denote the random vector whose i-th component equals the number of primitive cycles of Γ of length i. Then with $\mu = R''(1)/R'(1)$ and $\mu_i = \mu^i/(2i)$,

$$\lim_{n\to\infty} p_{C_1,\dots,C_m}(c_1,\dots,c_m) = \prod_{i=1}^{m} \frac{\mu_i^{c_i} e^{-\mu_i}}{c_i!}.$$

Consider the bipartite (Tanner) graph associated with an LDPC code. Since the graph is bipartite, a cycle of such a graph always has even length. As before, we call a cycle *primitive* if it cannot be decomposed into smaller cycles.

LEMMA C.37 (CYCLE DISTRIBUTION IN LDPC (n, L, R)). Let G be chosen uniformly at random from the ensemble LDPC $(n, \lambda, \rho) \triangleq$ LDPC (n, L, R) and let G_2 denote the associated bipartite graph induced by degree 2 variable nodes. For a fixed $m \in \mathbb{N}$, let $(C_1, \dots, C_m)$, denote the random vector whose i-th component equals the number of primitive cycles of length $2i$ of G_2. Then with $\mu = \lambda'(0)\rho'(1)$ and $\mu_i = \mu^i/(2i)$,

$$\lim_{n\to\infty} p_{C_1,\dots,C_m}(c_1,\dots,c_m) = \prod_{i=1}^{m} \frac{\mu_i^{c_i} e^{-\mu_i}}{c_i!}.$$

Proof. Consider an element G chosen uniformly at random from LDPC (n, L, R). Let G_2 denote the bipartite graph induced by its degree 2 variable nodes. This residual graph has nL_2 variable nodes, all of degree 2. Further, the number of check nodes of degree i concentrates on the coefficient in front of x^i of $n(1-r)R(1-\lambda'(0)+\lambda'(0)x)$, where r is the rate of the original code. This is true since edges in the original graph stay edges in the residual graph with probability $\lambda'(0)$. A node of degree d therefore gives rise to the binomial distribution of degrees of the form $\sum_i \binom{d}{i}(\lambda'(0)x)^i(1-\lambda'(0))^{d-i} = (1-\lambda'(0)+\lambda'(0)x)^d$, which explains the preceding expression. (The "degree distribution" $n(1-r)R(1-\lambda'(0)+\lambda'(0)x)$ is non-standard in that it contains zero-degree nodes.)

Compare LDPC $\big(nL_2x^2, n(1-r)R(1-\lambda'(0)(1-x))\big)$ with $G(n(1-r), R(1-\lambda'(0)(1-x)))$. We claim that each element $G \in$ LDPC can be mapped into an element of G and that there are exactly $\binom{n}{nL_2}(nL_2)!2^{nL_2}$ such graphs G that map into each element of G. This map is accomplished by associating to each variable node in G one edge in the graph Γ, connecting the corresponding two check-node sockets (the factor 2^{nL_2} is due to the freedom in choosing the labels of the 2 sockets of each degree-2 variable node).

A cycle of length i in Γ corresponds to a cycle of length $2i$ in G and vice versa. The claim is therefore a direct consequence of Lemma C.36. It remains to verify the value of μ. By Lemma C.36 we have

$$\mu = \frac{R''(1-\lambda'(0)(1-x))}{R'(1-\lambda'(0)(1-x))}\Big|_{x=1} = \frac{(\lambda'(0))^2\sum_i R_i i(i-1)}{\lambda'(0)\sum_i R_i i} = \rho'(1)\lambda'(0). \qquad \square$$

Discussion: You might wonder why we only considered the cycles of G_2. As discussed in Theorem D.32, in the limit of large blocklengths with probability one the cycles of fixed length involve only degree 2 variable nodes. Therefore, for large blocklengths the cycles (of fixed length) of G are those of G_2. Further, these cycles all give rise to codewords. Some thought shows that a cycle which involves only degree 2 variable nodes and which is *primitive*, i.e., which cannot be decomposed into smaller cycles, corresponds to a *minimal* codeword, i.e., a codeword which is not the sum of two codewords each of which has support set contained in the support set of the original word. It follows that in the limit of large blocklengths the number of minimal low-weight codewords follows a Poisson distribution.

This connection between the graphs which represent LDPC codes and the ensemble G is not only useful to prove that the number of minimal low-weight codewords follows a Poisson distribution. It also gives rise to a simple proof that the stability condition applies to MAP decoding as well. This is discussed in Lemma 3.67 for the BEC. To this end we need the following lemma.

Lemma C.38 (Giant Component and Cycles in $G(n, R)$ – [14, Theorem 1]). Let Γ be chosen uniformly at random from the ensemble $G(n, R)$ and define $\mu = R''(1)/R'(1)$. If $\mu > 1$ then almost surely Γ contains a unique giant component (a component with a positive fraction of nodes), the number of primitive cycles in Γ grows at least linearly in n, and the fraction of nodes that lie on cycles is strictly positive.

Notes

A good account of the many applications of the first and second moment method (Section C.1) can be found in the book by Alon and Spencer [1]. Very readable introductions to Martingales (Section C.3) are contained in the books by Motwani and Raghavan [15] as well as Mitzenmacher and Upfal [13]. The method of using the Hoeffding-Azuma inequality to prove concentration theorems was introduced into the computer science literature by Shamir and Spencer [16]. In this paper the authors showed the tight concentration of the chromatic number around its mean (see Problem C.5). The inequality itself is due to Hoeffding [6] and Azuma [3]. The proof of Theorem C.17 follows the one given by Alon and Spencer [1]. The particular type of Martingale which results if we reveal more and more about a process is called *Doob's Martingale Process* [15, p. 90]. The migration of the technique into coding theory was initiated in the work of Luby, Mitzenmacher, Shokrollahi, and Spielman, [9, 11]. They analyzed LDPC codes in an essentially combinatorial setting: transmission over the binary symmetric channel (BSC) or the BEC with hard decision message-passing decoding. The technique was then extended by Richardson and Urbanke to encompass general message-passing decoding algorithms, including belief propagation, and general BMS channels.

One of the first applications of analyzing stochastic processes in terms of a differential equation which characterizes the average behavior and proving a concentration around this mean is due to Kurtz in the area of population dynamics [8]. The method was first applied in the computer science field by Karp and Sipser [7]. This method was extended and applied to a large set of problems in computer science by Wormald [18, 19]. The first application in the setting to message-passing systems is due to Luby, Mitzenmacher, Shokrollahi, and Spielman [9, 10]. The extension of the method to include the evolution of the second moment as presented in Section 3.23 is due to Amraoui, Montanari, Richardson, and Urbanke [2].

Theorem C.33, Fact C.35, as well as Lemma C.36 can be found in the book of Bollobás [4]. The result concerning the cycle distribution in LDPC ensembles appeared in the paper of Di, Richardson, and Urbanke [5].

The article by McDiarmid [12] is an excellent review of methods to prove concentration.

Problems

C.1 (Generalized Chebyshev Inequality). Assume that $\phi(x)$ is a strictly positive and increasing function on $(0,\infty)$, $\phi(u) = \phi(-u)$, and that X is a random variable such that $\mathbb{E}[\phi(X)] < \infty$. Prove that for any $\lambda > 0$

$$\mathbb{P}\{|X| \geq \lambda\} \leq \mathbb{E}[\phi(X)]/\phi(\lambda).$$

C.2 (Tail Bounds for Weakly Dependent Random Variables). Assume that the random variables X_i take integral values in the range $[0,d]$ and that

$$\mathbb{E}[X_i \,|\, X_1,\dots,X_{i-1}] \geq p.$$

We want to find a bound for $\mathbb{P}\{\sum_{i=1}^n X_i \leq np(1-\delta)\}$. Start with

$$\mathbb{E}[\mathrm{e}^{-s\sum_{i=1}^n X_i}] = \prod_{i=1}^n \mathbb{E}[\mathrm{e}^{-sX_i} \,|\, X_1,\dots,X_{i-1}].$$

Show that

$$\mathbb{E}[\mathrm{e}^{-s\sum_{i=1}^n X_i}] \leq (1 - \frac{p}{d} + \frac{p}{d}\mathrm{e}^{-sd})^n.$$

Optimizing over the choice of s this gives rise to the bound

$$\mathbb{P}\{\sum_{i=1}^n X_i \leq np(1-\delta)\} \leq \left(\frac{d-p}{d-p\bar{\delta}}\left(\frac{d-p\bar{\delta}}{(d-p)\bar{\delta}}\right)^{\frac{p\bar{\delta}}{d}}\right)^n \leq \mathrm{e}^{-n\delta^2\frac{p}{2d}}.$$

C.3 (Hoeffding-Azuma: General Case). Prove the Hoeffding-Azuma inequality for general constants γ_i.

C.4 (Expansion of Random Bipartite Graphs – Spielman [17]). Let $\mathtt{G}$ be chosen uniformly at random from the ensemble $\text{LDPC}(n, x^{\mathtt{l}-1}, x^{\mathtt{r}-1})$. Show that for all $0 < \alpha < 1$ and $\delta > 0$, with probability at least $\mathrm{e}^{-\delta n}$, all sets of αn variables in $\mathtt{G}$ have at least

$$n\Big(\mathtt{l}/\mathtt{r}(1-(1-\alpha)^{\mathtt{r}}) - \sqrt{2\mathtt{l}\alpha h_2(\alpha)} - \delta\sqrt{\frac{\mathtt{l}\alpha}{2h_2(\alpha)}}\Big)$$

neighbors.

Hint: First look at the expected number of neighbors. Then expose one neighbor at a time to bound the probability that the number of neighbors deviates significantly from this mean.

C.5 (Concentration of the Chromatic Number – Spencer and Shamir [16]). Consider a graph G on n vertices. The *chromatic number* of a graph G, denoted by $\chi(\mathtt{G})$, is the smallest number of colors needed to color all vertices so that no two vertices which are joined by an edge have the same color. Consider the standard ensemble of *random graphs* on n vertices with parameter p: to sample from this ensemble, pick n vertices and connect each of the $\binom{n}{2}$ ordered pairs of vertices independently from all other connections with probability p. Show that for this ensemble

$$\mathbb{P}[|\chi(\mathtt{G}) - \mathbb{E}[\chi(\mathtt{G})]| > \lambda\sqrt{n-1}] \leq 2e^{-\lambda^2/2}.$$

References

[1] N. Alon, J. Spencer, and P. Erdös, *The Probabilistic Method*, Wiley, New York, NY, USA, 1992. [500]

[2] A. Amraoui, A. Montanari, T. Richardson, and R. Urbanke, *Finite-length scaling for iteratively decoded LDPC ensembles*, in Proc. of the Allerton Conf. on Commun., Control, and Computing, Monticello, IL, USA, Oct. 2003. [159, 266, 500]

[3] K. Azuma, *Weighted sums of certain dependent random variables*, Tohoku Math. J., 19 (1967), pp. 357–367. [500]

[4] B. Bollobás, *Random Graphs*, Cambridge Univ. Press, 2001. [497, 498, 500]

[5] C. Di, T. Richardson, and R. Urbanke, *Weight distribution of low-density parity-check codes*, IEEE Trans. Inform. Theory, 52 (2006), pp. 4839–4855. [159, 500]

[6] W. Hoeffding, *Probability inequalitites for sums of bounded random variables*, J. Am. Stat. Assoc., 58 (1963), pp. 13–30. [500]

[7] R. M. Karp and M. Sipser, *Maximum matchings in sparse random graphs*, in Proc. of the Twenty-Second Annual IEEE Symposium on Foundations of Computing, 1981, pp. 364–375. [500]

[8] T. Kurtz, *Approximation of Population Processes*, vol. 36, CBMS-NSF Regional Conference Series in Applied Mathematicas (SIAM), Philadelphia, PA, USA, 1981. [500]

[9] M. Luby, M. Mitzenmacher, A. Shokrollahi, and D. A. Spielman, *Analysis of low density codes and improved designs using irregular graphs*, in Proc. of the 30th Annual ACM Symposium on Theory of Computing, 1998, pp. 249–258. [157, 262, 500]

[10] ——, *Efficient erasure correcting codes*, IEEE Trans. Inform. Theory, 47 (2001), pp. 569–584. [157, 262, 500]

[11] ——, *Improved low-density parity-check codes using irregular graphs*, IEEE Trans. Inform. Theory, 47 (2001), pp. 585–598. [157, 262, 500]

[12] C. McDiarmid, *Concentration*, in Probabilistic Methods for Algorithmic Discrete Mathematics, M. Habib, C. McDiarmid, J. Ramirez-Alfonsin, and B. Reed, eds., Springer Verlag, New York, NY, USA, 1998, pp. 195–248. [500]

[13] M. MITZENMACHER AND E. UPFAL, *Probability and Computing*, Cambridge Univ. Press, Cambridge, MA, USA, 2005. [500]

[14] M. MOLLOY AND B. REED, *A critical point for random graphs with a given degree sequence*, Random Structures and Algorithms, 6 (1995), pp. 161–180. [500]

[15] R. MOTWANI AND P. RAGHAVAN, *Randomized Algorithms*, Cambridge Univ. Press, Cambridge, MA, USA, 1995. [500]

[16] A. SHAMIR AND J. SPENCER, *Sharp concentration of the chromatic number on random graphs $G_{n,p}$*, Combinatorica, 7 (1987), pp. 121–129. [500, 502]

[17] D. A. SPIELMAN, *Computationally Efficient Error-Correcting Codes and Holographic Proofs*, PhD thesis, MIT, June 1995. [501]

[18] N. C. WORMALD, *Differential equations for random processes and random graphs*, Ann. Appl. Probab., 5 (1995), pp. 1217–1235. [500]

[19] ——, *The differential equation method for random graph processes and greedy algorithms.* Notes on lectures given at the Summer School on Randomized Algorithms in Antonin, Poland, Published in: Karonski and Proemel, eds., Lectures on Approximation and Randomized Algorithms PWN, Warsaw, 1999, pp. 73-155, 1999. [500]

Appendix D
FORMAL POWER SUMS

§D.1. Definition

A *formal power sum* $F(D)$ over a field $\mathbb{F}$ is a sum of the form

$$F(D) = \sum_{n\geq 0} F_n D^n, \ F_n \in \mathbb{F}.$$

Such objects appear frequently in this book: in the context of the finite-length analysis for the BEC, during the study of convolutional and turbo codes, or as generating functions of weight distributions. We collect the basic facts about formal power sums that we have used throughout the text.

§D.2. Basic Properties

In the sequel we assume that the field $\mathbb{F}$ is fixed and we usually omit any reference to it. The two most important examples for us are $\mathbb{F} = \mathbb{F}_2$ and $\mathbb{F} = \mathbb{R}$. Given two formal power sums $F(D) = \sum_{n\geq 0} F_n D^n$ and $G(D) = \sum_{n\geq 0} G_n D^n$, we *define* their *addition* by

$$F(D) + G(D) = \sum_{n\geq 0} (F_n + G_n) D^n.$$

In a similar way we *define* their *multiplication* by

$$F(D) \cdot G(D) = \sum_{n\geq 0} \Big(\sum_{i=0}^{n} G_i F_{n-i} \Big) D^n,$$

which is the rule familiar from polynomial multiplication. Note that this is well defined – to compute the n-th coefficient of the product we only need to perform a *finite* number of operations. Therefore, we do not encounter issues of convergence. Also note that multiplication is *commutative.*

Is is possible to define *division*? Recall that over the reals we say that y is the *multiplicative inverse* of x, $x \neq 0$, which we write as $y = 1/x$, if $xy = 1$. Dividing by x is then the same as multiplying by y. We proceed along the same lines for formal power sums. Consider the formal power sum $F(D)$. We want to find the formal power sum $G(D)$, if it exists, such that $H(D) = F(D)G(D) = 1$. We say that $G(D)$ is the multiplicative inverse of $F(D)$ and dividing by $F(D)$ corresponds to multiplying with $G(D)$. Using the preceding multiplication rule, we get the following set of equations:

$$1 = H_0 = F_0 G_0,$$

$$0 = H_n = \sum_{i=0}^{n} G_i F_{n-i}, \quad n \geq 1.$$

This set of equations has a solution, and this solution is unique, if and only if $F_0 \neq 0$. In this case we get

$$G_0 = \frac{1}{F_0},$$

$$G_n = -\frac{1}{F_0} \sum_{i=0}^{n-1} G_i F_{n-i}, \quad n \geq 1.$$

Since the evaluation of each coefficient F_n only involves a finite number of algebraic operations and only makes use of the values of F_i, $0 \leq i < n$, this gives rise to a well defined formal power sum. In summary, a formal power sum $F(D)$ has a multiplicative inverse if and only if $F_0 \neq 0$. We write this multiplicative inverse as $1/F(D)$.

Example D.1 (Inverse of $1 - D$ over $\mathbb{R}$). Let $\mathbb{F} = \mathbb{R}$ and consider the example $F(D) = 1 - D$. Since $F_0 \neq 0$, $1/(1 - D)$ exists. We get

$$G_0 = \frac{1}{F_0} = 1, G_1 = -\frac{1}{F_0} G_0 F_1 = 1, G_2 = -\frac{1}{F_0}(G_0 F_2 + G_1 F_1) = G_1 = 1,$$

and, in general,

$$G_n = -\frac{1}{F_0} \sum_{i=0}^{n-1} G_i F_{n-i} = -\frac{1}{F_0} G_{n-1} F_1 = G_{n-1} = 1.$$

Therefore, $G(D) = \sum_{n=0}^{\infty} D^n$. ◇

The resemblance to the identity $\sum_{i=0}^{\infty} x^i = 1/(1 - x)$, which is familiar from analysis and valid for $|x| < 1$, is not a coincidence. In general, as a rule of thumb any identity which is valid for Taylor series and which can be meaningfully interpreted in the realm of formal power sums is still valid if considered as an identity of formal power sums. Further basic properties of formal power sums are developed in Problem D.1.

§D.3. Summation of Subsequences

Consider a formal power sum over $\mathbb{R}$,

$$F(D) = \sum_{n \geq 0} F_n D^n,$$

so that the corresponding real sum $\sum_{n\geq 0} F_n$ converges. Assume we have succeeded in finding a "closed-form solution" $F(D)$. Then we have the compact representation of the sum $\sum_{n\geq 0} F_n$ as $F(D)|_{D=1}$. What can we say about the related power sum

$$F_q(D) = \sum_{n\geq 0: n/q\in\mathbb{N}} F_n D^n,$$

and the related sum $\sum_{n\geq 0: n/q\in\mathbb{N}} F_n$, where q is a natural number? In words, we would like to find the sum of all q-th terms. We claim that

$$F_q(D) = \frac{1}{q}\sum_{i=0}^{q-1} F\left(D e^{i\frac{2\pi j}{q}}\right), \qquad F_q(1) = \frac{1}{q}\sum_{i=0}^{q-1} F\left(e^{i\frac{2\pi j}{q}}\right).$$

This is true since

$$\frac{1}{q}\sum_{i=0}^{q-1} F\left(D e^{i\frac{2\pi j}{q}}\right) = \sum_{n\geq 0} F_n\left(\frac{1}{q}\sum_{i=0}^{q} e^{ni\frac{2\pi j}{q}}\right)D^n = \sum_{n\geq 0: n/q\in\mathbb{N}} F_n D^n.$$

Example D.2 (Probability of Check Node Satisfaction). Assume we have $\mathtt{r}$ binary iid random variables X_n, $n = 1,\dots,\mathtt{r}$, with $\mathbb{P}\{X_n = 1\} = p$. We want to determine $\mathbb{P}\{\sum_{n=1}^{\mathtt{r}} X_n = 0\}$. This can be interpreted as the probability that a check node is fulfilled, assuming that each of its attached variable nodes is in error with probability p and assuming that all these errors are independent. We have

$$\mathbb{P}\Big\{\sum_{n=1}^{\mathtt{r}} X_n = 0\Big\} = \sum_{n:n \text{ even}} \binom{\mathtt{r}}{n} p^n (1-p)^{\mathtt{r}-n} D^n\Big|_{D=1}.$$

Since we know that $\sum_n \binom{\mathtt{r}}{n} p^n(1-p)^{\mathtt{r}-n}D^n = (pD+1-p)^{\mathtt{r}}$, the solution is

$$\begin{aligned}\mathbb{P}\Big\{\sum_{n=1}^{\mathtt{r}} X_n = 0\Big\} &= \frac{1}{2}\left((pe^{\frac{2\pi j}{2}0}+1-p)^{\mathtt{r}} + (pe^{\frac{2\pi j}{2}1}+1-p)^{\mathtt{r}}\right)\\ &= \frac{1+(1-2p)^{\mathtt{r}}}{2}.\end{aligned}$$

◇

§D.4. Coefficient Growth of Powers of Polynomials

Generating functions are an indispensable tool in enumerative combinatorics. Often the exact growth of a combinatorial object is complicated but the asymptotic behavior of this growth is quite simple. This is, e.g., the case for the weight distribution problem. As a motivation, let us start with the following simple example. Assume we want to compute

$$N_n(k) = \operatorname{coef}\{F(D)^n, D^k\},$$

where $F(D)$ is a function analytic around zero, which has a power-series expansion with non-negative coefficients, and $n, k \in \mathbb{N}$. Let x be any strictly positive number. Then $N_n(k)x^k \leq \sum_i N_n(i)x^i = (\sum_{i\geq 0} F_i x^i)^n = F(x)^n$, since all coefficients are non-negative. It follows that

$$N_n(k) \leq F(x)^n/x^k. \tag{D.3}$$

Therefore, $N_n(k) \leq \inf_{x>0} F(x)^n/x^k$. It may come as a surprise that such a simple bound is essentially tight.

Example D.4 (Bound on Binomials). If we take $F(x) = 1 + x$, then a little bit of calculus shows that $x = k/(n-k)$ is the optimum choice. It is a valid choice since it is positive if $k \in [n]$. With this choice we get $\binom{n}{k} \leq 2^{nh_2(k/n)}$, a result we saw in slightly strengthened form in Problem 1.25 (but see Problem D.5). ◇

The preceding argument can easily be extended to polynomials in several variables and we skip the details. In some cases we are not only interested in an upper bound but we are asking for a "good" approximation of the actual value. The *Hayman method* is a powerful tool in this respect. Although the basic idea of the Hayman method is straightforward (recast the problem in the form of an integral and use the Laplace method to estimate this integral), the details are cumbersome. Let us introduce the idea by revisiting our simple example. This hopefully clarifies the basic concept. We want to compute

$$N_n(n\omega) = \text{coef}\{F(D)^n, D^{n\omega}\},$$

where $F(D)$ is a polynomial with non-negative coefficients, and $\omega \in \mathbb{R}$ and $n \in \mathbb{N}$ are such that $n\omega \in \mathbb{N}$. Without loss of generality we can assume that $F_0 = 1$, since common factors of D can be taken out and we can rescale the polynomial if needed. By the residue theorem we have

$$N_n(n\omega) = \frac{1}{2\pi j}\oint \frac{F(z)^n}{z^{n\omega+1}}\mathrm{d}z,$$

where we can choose the integration path to be a circle of radius x, $x > 0$, centered around the origin. We get a trivial upper bound if we replace the integrand with the maximum of $|F(z)|^n$. Since the length of the contour is $2\pi x$, this gives

$$N_n(n\omega) \leq \frac{2\pi x}{2\pi}\frac{|F(x)|^n}{x^{n\omega+1}} = \frac{F(x)^n}{x^{n\omega}},$$

where we used the fact that $|F(x e^{j\theta})|$ takes on a maximal value for $\theta = 0$ since all coefficients of $F(D)$ are non-negative. We can optimize (minimize) this bound by taking the optimal value of x. The condition for this optimum reads $a(x_\omega) = \omega$, where $a(x) = \frac{x\partial F}{F\partial x}$. To summarize, we have

$$N_n(n\omega) \leq \frac{F(x_\omega)^n}{x_\omega^{n\omega}}, \quad \text{where} \quad a(x_\omega) = \omega, \; x_\omega \in \mathbb{R}^+.$$

This is the same bound which we derived previously using only the positivity of the coefficients F_i. But from the new point of view we can see how the bound can be improved. Note that the integrand is much smaller than its maximal value $F(x)$ for most of the contour – unless the polynomial has a special structure, the only way to "line up" (over $\mathbb{C}$) the individual vectors $F_i x^i e^{j\theta i}$ is by choosing $\theta = 0$. Assume for the moment that for the given polynomial the integral is dominated by a small interval around $\theta = 0$. Expand $\ln(F(xe^{j\theta})/F(x))$ around $\theta = 0$. We get $\ln(F(xe^{j\theta})/F(x)) = ja(x)\theta - \frac{1}{2}xa'(x)\theta^2 + O(\theta^3)$. We therefore have

$$\begin{aligned} N_n(n\omega) &= \frac{1}{2\pi j}\oint \frac{F(z)^n}{z^{n\omega+1}}\mathrm{d}z \overset{z=xe^{j\theta}}{=} \frac{1}{2\pi}\int_{-\pi}^{\pi}\frac{F(xe^{j\theta})^n}{x^{n\omega}e^{jn\omega\theta}}\mathrm{d}\theta \\ &= \frac{1}{2\pi}\int_{-\pi}^{\pi}\frac{F(x)^n e^{jna(x)\theta} e^{-\frac{1}{2}xa'(x)\theta^2} e^{nO(\theta^3)}}{x^{n\omega}e^{jn\omega\theta}}\mathrm{d}\theta. \end{aligned}$$

If, as preceding, we choose $x = x_\omega$ then this reduces to

$$N_n(n\omega) = \frac{1}{2\pi}\int_{-\pi}^{\pi}\frac{F(x_\omega)^n e^{-\frac{1}{2}x_\omega a'(x_\omega)\theta^2} e^{nO(\theta^3)}}{x_\omega^{n\omega}}\mathrm{d}\theta.$$

Assume now that $a'(x_\omega) > 0$. In this case the integral is up to the error term the integral over a Gaussian kernel with variance $1/(x_\omega a'(x_\omega))$. One can check that the influence of the error term vanishes as n tends to infinity. If we compute the integral over the Gaussian kernel ignoring the error term we see that the value of the integral converges to $\frac{1}{\sqrt{2\pi x_\omega a(x_\omega)}}\frac{F(x_\omega)^n}{x_\omega^{n\omega}}$. This is the Hayman approximation. In fact, more is true. Under the same conditions we also have the following *local expansion*:

$$N_n(n\omega + \Delta w) = N_n(n\omega)x_\omega^{-\Delta w}e^{-\frac{(\Delta w)^2}{2nb(x_\omega)}}(1 + o(1)),$$

where $b(x) = xa'(x)$ and where the expansion is valid for $\Delta w \in o(\sqrt{n}\log(n))$.

Let us summarize what is needed for the Hayman approximation to be valid. First, we need that $F(xe^{j\theta})$ takes on its maximum only at $\theta = 0$. It is easy to see how this condition can fail: if the powers of $F(x)$ with non-zero coefficients are all divisible by k then $F(xe^{j\theta})$ takes on a maximum at all angles which are multiples of $\frac{2\pi}{k}$. Consider as example $F(x) = 1 + x^2$. This function takes on a maximum value at both $\theta = 0$ and $\theta = \pi$. Conversely, as we confirm soon in a more general setting, if $\gcd\{k : F_k > 0\} = 1$, then $F(x\mathrm{e}^{j\theta})$ has a unique maximum which is located at $\theta = 0$. Next, we need that $a(x) = \omega$ has a strictly positive solution for all $\omega \in (0, d)$, where d is the degree of $F(x)$. We claim that this is fulfilled if $F_0 > 0$ and if $F(x)$ contains at least one other non-zero term: by direct inspection of $a(x) = \frac{\sum_i iF_i x^i}{\sum_i F_i x^i}$ we see that in this case $a(0) = 0$. Further, we then have $\lim_{x\to\infty} a(x) = d$. The claim now follows by the mean-value theorem. But more is true. As discussed in Problem D.6, $a'(x) > 0$ for $x > 0$, so that $a(x)$ is strictly increasing. There is hence a unique such positive solution. Finally, because of $a'(x) > 0$ the integrand is indeed well approximated by a Gaussian.

In the one-dimensional case we say that a polynomial is *Hayman* admissible if the two conditions $F_0 > 0$ and $\gcd\{k : F_k > 0\} = 1$ are fulfilled. What happens if $F_0 > 0$ but $\gcd\{k : F_k > 0\}$ is strictly larger than 1? As we discussed earlier, $F(x_\omega \mathrm{e}^{2\pi j\theta})$ has in this case $\gcd\{k : F_k > 0\}$ maxima. But the asymptotic expansion is essentially still valid. We just have to multiply it by $\gcd\{k : F_k > 0\}$ to account for the contribution to the integral of each of these maxima. The following definition extends to, and formalizes these notions for, higher dimensions.

Definition D.5 (Hayman Admissibility of Multivariate Functions). Consider a function $F(\underline{x})$ in m variables. We say that $F(\underline{x})$ is *Hayman admissible* if the following two conditions are satisfied.

For each $\underline{x} \in \mathbb{R}^m$ with strictly positive components, there exists a constant c such that for all $\underline{\theta} \in [-\pi, \pi)^m \setminus [-n^{-\frac{2}{5}}, n^{-\frac{2}{5}}]^m$

$$|F(\underline{x}\mathrm{e}^{j\underline{\theta}})/F(\underline{x})|^n \leq c n^{-\frac{1}{5}}. \tag{D.6}$$

For each $\underline{x} \in \mathbb{R}^m$ with strictly positive components

$$\ln\big(F(\underline{x}\mathrm{e}^{j\underline{\theta}})/F(\underline{x})\big) = j\underline{a}(\underline{x})\underline{\theta}^T - \frac{1}{2}\underline{\theta}\,\underline{B}(\underline{x})\underline{\theta}^T + O(\|\underline{\theta}\|_1^3), \tag{D.7}$$

where the error term is uniform over all $\underline{\theta} \in \big[-\delta, \delta\big)^m$, for some $\delta > 0$.

Remark: In the earlier definition, the choice of the exponent $-\frac{2}{5}$ is arbitrary and is purely done for convenience.

The following lemma simplifies the task of determining if a multivariate *polynomial* is Hayman admissible.

Lemma D.8 (Hayman Admissibility of Multivariate Polynomials). Let $F(\underline{D})$ be a multivariate polynomial in m variables with non-negative coefficients and at least two strictly positive terms. Associate to $F(\underline{D})$ the following integral matrix $\underline{A}$:

for each pair of distinct coefficients $(\underline{k}, \underline{l})$ with $F_{\underline{k}}F_{\underline{l}} > 0$, $\underline{A}$ contains the difference vector $\underline{k} - \underline{l}$ as a row. If the image of the map $[0,1)^m \setminus \{\underline{0}\} : \underline{x} \mapsto \underline{x}\,\underline{A}^T$ contains no element with all integral components then $F(\underline{D})$ is Hayman admissible.

Proof. To simplify the notation in the proof, let us agree that c denotes a suitably chosen strictly positive constant (only depending on F and $\underline{x}$), *not always the same.* We prove the claims in the order stated. Some calculus shows that

$$\text{(D.9)}\qquad \left|F(\underline{x}e^{j\underline{\theta}})/F(\underline{x})\right|^2 = 1 - \sum_{\underline{k} \neq \underline{l}} F_{\underline{k}}F_{\underline{l}}\underline{x}^{\underline{k}+\underline{l}}\left(1 - \cos((\underline{k} - \underline{l})\underline{\theta}^T)\right)/|F(\underline{x})|^2,$$

where the sum ranges over all ordered pairs $(\underline{k}, \underline{l})$ with $\underline{k} \neq \underline{l}$. We claim that for a fixed $\underline{x} \in \mathbb{R}^m$, for all $\underline{\theta} \in [0, 2\pi)^m$

$$\text{(D.10)}\qquad |F(\underline{x})|^2 \geq \sum_{\underline{k} \neq \underline{l}} F_{\underline{k}}F_{\underline{l}}\underline{x}^{\underline{k}+\underline{l}}(1 - \cos((\underline{k} - \underline{l})\underline{\theta}^T) \geq c\,\underline{\theta}\,\underline{A}^T\underline{A}\,\underline{\theta}^T \geq 0.$$

Let us postpone the proof of this claim for a moment and see how it can be used to verify condition (D.6). We first claim that $\underline{A}$ has full column rank. This is true since if $\underline{A}$ did not have full column rank then there would exist a *non-zero* vector $\underline{z}$ so that $\underline{A}\,\underline{z}^T = \underline{0}^T$. Take $\underline{x} = \underline{z} \pmod 1$. Prescaling $\underline{z}$ if necessary, we can assume that $\underline{x}$ is not the all-zero vector. But since $\underline{A}$ is integral, $\underline{A}\,\underline{x}^T$, and so also $\underline{x}\,\underline{A}^T$, would have all integral components, a contradiction. Since $\underline{A}$ has full column rank, $\underline{A}^T\underline{A}$ is strictly positive definite so that

$$\text{(D.11)}\qquad \underline{\theta}\,\underline{A}^T\underline{A}\,\underline{\theta}^T \geq c\,\|\underline{\theta}\|_2^2.$$

Now if $\underline{\theta} \in [-\pi, \pi)^m \setminus [-n^{-\frac{2}{5}}, n^{-\frac{2}{5}}]^m$ then

$$\text{(D.12)}\qquad \|\underline{\theta}\|_2^2 \geq n^{-\frac{4}{5}}.$$

Therefore,

$$\begin{aligned}
\text{by (D.9)}\quad \left|F(\underline{x}e^{j\underline{\theta}})/F(\underline{x})\right|^n &= \Big(1 - \sum_{\underline{k} \neq \underline{l}} F_{\underline{k}}F_{\underline{l}}\underline{x}^{\underline{k}+\underline{l}}\left(1 - \cos((\underline{k} - \underline{l})\underline{\theta}^T)\right)/|F(\underline{x})|^2\Big)^{n/2} \\
\text{by (D.10)}\quad &\leq \left(1 - c\underline{\theta}\,\underline{A}^T\underline{A}\,\underline{\theta}^T/|F(\underline{x})|^2\right)^{n/2} \\
\text{by (D.11) and (D.12)}\quad &\leq \left(1 - cn^{-\frac{4}{5}}/|F(\underline{x})|^2\right)^{n/2} \\
&\overset{\text{(i)}}{=} \left(1 - cn^{-\frac{4}{5}}\right)^{n/2} \leq e^{-cn^{\frac{1}{5}}} \leq cn^{-\frac{1}{5}}.
\end{aligned}$$

Step (i) follows since by assumption $\underline{x}$ has strictly positive components so that $F(\underline{x})$ is strictly positive. This shows that $F(D)$ fulfills condition (D.6) assuming we can prove (D.10).

Let us now see how this can be accomplished. Consider

$$\min\{F_{\underline{k}}F_{\underline{l}}\underline{x}^{\underline{k}+\underline{l}} : F_{\underline{k}}F_{\underline{l}} > 0, \underline{k} \neq \underline{l}\}. \tag{D.13}$$

Since by assumption $F(\underline{D})$ has at least two strictly positive terms the set in (D.13) is non-empty. Since further $F(\underline{D})$ has only a finite number of non-zero coefficients, all of them are positive, and each component of $\underline{x}$ is strictly positive, the minimum is well defined and strictly positive. We continue:

$$
\begin{aligned}
& \sum_{\underline{k}\neq\underline{l}} F_{\underline{k}}F_{\underline{l}}\underline{x}^{\underline{k}+\underline{l}}\left(1-\cos((\underline{k}-\underline{l})\underline{\theta}^T)\right) \\
\text{from (D.13)} \quad & \geq c\sum_{\underline{k}\neq\underline{l}}\left(1-\cos((\underline{k}-\underline{l})\underline{\theta}^T)\right) \\
\text{since } \cos(z)\leq 1-z^2/2+|z|^3/6 \quad & \geq \frac{c}{2}\sum_{\underline{k}\neq\underline{l}}\left|(\underline{k}-\underline{l})\underline{\theta}^T\right|^2 - \frac{c}{6}\sum_{\underline{k}\neq\underline{l}}\left|(\underline{k}-\underline{l})\underline{\theta}^T\right|^3.
\end{aligned}
$$

Choose δ to be equal to the inverse of $\frac{2}{3}\max\left\{\|\underline{k}-\underline{l}\|_1 : F_{\underline{k}}F_{\underline{l}} > 0\right\}$. Then if $\underline{\theta} \in [0,\delta)^m$ it follows that $\sum_{\underline{k}\neq\underline{l}}\left|(\underline{k}-\underline{l})\underline{\theta}^T\right|^3 \leq \frac{3}{2}\sum_{\underline{k}\neq\underline{l}}\left|(\underline{k}-\underline{l})\underline{\theta}^T\right|^2$. If we insert this bound into the last paragraph then this proves the desired bound (D.10) for the region $\underline{\theta} \in [0,\delta)^m$.

By assumption, the image of the map $[0,1)^m \smallsetminus \{\underline{0}\} : \underline{x} \mapsto \underline{x}\underline{A}^T$ does not contain an element with all integral components. This is equivalent to the condition that

$$\underline{A}\,\underline{x}^T = \underline{0}^T \pmod 1$$

has no non-zero solutions for $\underline{x} \in [0,1)^m$. If we multiply this equation by 2π this is in turn equivalent to the statement that

$$\underline{A}\,\underline{\theta}^T = \underline{0}^T \pmod{2\pi}$$

has no non-zero solution for $\underline{\theta} \in [0,2\pi)^m$. Since the domain $[0,2\pi]^m \smallsetminus (0,\delta)^m$ is compact, it follows by a continuity argument that in the range $\underline{\theta} \in [0,2\pi]^m \smallsetminus (0,\delta)^m$

$$\sum_{\underline{k}\neq\underline{l}} F_{\underline{k}}F_{\underline{l}}\underline{x}^{\underline{k}+\underline{l}}(1-\cos((\underline{k}-\underline{l})\underline{\theta}^T))$$

is lower bounded by a strictly positive constant. This in turn shows that (D.10) is also true for $\underline{\theta} \in [0,2\pi)^m \smallsetminus (0,\delta)^m$. This concludes the proof of condition (D.6).

It remains to prove that condition (D.7) holds as well. This is easier. If $\underline{x}$ has strictly positive components then $\ln\left(F(\underline{x}e^{j\underline{\theta}})/F(\underline{x})\right)$, seen as a function of θ, is analytic in $[-\pi,\pi)^m$. It therefore has a Taylor-series representation. If we expand the function up to second-order terms then the error term can be bounded by $\|\underline{\theta}\|_1^3$ times a constant, where the constant is equal to the supremum of all partial derivatives of third order in the domain $[-\pi,\pi)^m$. This supremum is finite which shows that the expansion is uniform. □

Lemma D.14 (Hayman Method: Multivariate Polynomials). Let $F(\underline{D})$ be a multivariate polynomial in m variables with non-negative coefficients. Let $\underline{\omega} \in \mathbb{R}^m$ and $n \in \mathbb{N}$ so that $n\underline{\omega} \in \mathbb{N}^m$. Define $N_n(n\underline{\omega}) = \text{coef}\{F(\underline{D})^n, \underline{D}^{n\underline{\omega}}\}$ and $\underline{a}(\underline{x}) = ((x_i \partial F)/(F \partial x_i))_{i=1}^m$. Further, let the $m \times m$ symmetric matrix $\underline{B}$ have components $\underline{B}_{i,j} = x_j \partial \underline{a}_i / \partial x_j$. Let $\underline{x}_\omega$ be a real strictly positive solution to $\underline{a}(\underline{x}) = \underline{\omega}$. If $F(\underline{D})$ is *Hayman admissible* then $\underline{B}$ is strictly positive definite and

$$N_n(n\underline{\omega}) = \frac{F\left(\underline{x}_\omega\right)^n}{\sqrt{(2\pi n)^m |\underline{B}(\underline{x}_\omega)|}\left(\underline{x}_\omega\right)^{n\underline{\omega}}}(1+o(1)). \tag{D.15}$$

Also, $N_n(n\underline{\omega}+\underline{\Delta w})$ can be approximated in terms of $N_n(n\underline{\omega})$. This approximation is called the *local limit theorem* of N_n around $n\underline{\omega}$. Explicitly, if $\|\underline{\Delta w}\|_1 = o(\sqrt{n \log n})$, then

$$N_n(n\underline{\omega} + \underline{\Delta w}) = N_n(n\underline{\omega})\left(\underline{x}_\omega\right)^{-\underline{\Delta w}} \mathrm{e}^{-\frac{1}{2n}\underline{\Delta w}\,\underline{B}^{-1}(\underline{x}_\omega)\underline{\Delta w}^T}(1+o(1)).$$

Example D.16 (Binomial Coefficient). Consider the binomial coefficient $\binom{n}{k}$. We have $\binom{n}{k} = \text{coef}\{F(D)^n, D^k\}$ with $F(D) = 1 + D$. Since $F_0 = 1 > 0$ and $\gcd\{k : F_k > 0\} = 1$, $F(D)$ is Hayman admissible. We have $a(x) = x/(1+x) = \kappa$, so that $x_\kappa = \kappa/(1-\kappa)$. Further, $\underline{B}$ consists of a single term, call it $b(x) = x/(1+x)^2$. We get $b(x_\kappa) = \kappa(1-\kappa)$. Therefore, for $\kappa \in (0,1)$ and $\Delta = o(\sqrt{n \log(n)})$,

$$\begin{aligned}\binom{n}{n\kappa} &= \frac{F(\kappa/(1-\kappa))^n}{\sqrt{2\pi n\kappa(1-\kappa)}(\kappa/(1-\kappa))^{n\kappa}}(1+o(1)),\\ &= \frac{2^{nh_2(\kappa)}}{\sqrt{2\pi n\kappa(1-\kappa)}}(1+o(1)),\\ \binom{n}{n\kappa+\Delta} &= \binom{n}{n\kappa}(\kappa/(1-\kappa))^{-\Delta}\mathrm{e}^{-\frac{\Delta^2}{2n\kappa(1-\kappa)}}(1+o(1)),\end{aligned}$$

where $h_2(x) = -x\log_2(x) - (1-x)\log_2(1-x)$. ◇

Although we have proved the preceding statement for $\kappa \in (0,1)$ and $k = \kappa n$, i.e., for the case where k scales linearly with n, it actually remains true with $\kappa = k/n$ as long as both k and $n-k$ tend to infinity. For a slightly less straightforward application consider the weight distribution of the ensemble LDPC $(n, x^{\mathtt{l}-1}, x^{\mathtt{r}-1})$.

Lemma D.17 (Average Weight Distribution of Regular Ensembles). Consider the ensemble LDPC $(n, x^{\mathtt{l}-1}, x^{\mathtt{r}-1})$ with $\mathtt{l}, \mathtt{r} \geq 2$. For $\mathtt{G}$ chosen uniformly random from the ensemble let $A_{\text{cw/ss}}(\mathtt{G}, w)$ denote the number of codewords/stopping sets of *weight* w in the code defined by $\mathtt{G}$, respectively. Then

$$\mathbb{E}[A_{\text{cw/ss}}(\mathtt{G}, w)] = \binom{n}{w} \frac{\operatorname{coef}\{F_{\text{cw/ss}}(D)^{n\frac{\mathtt{l}}{\mathtt{r}}}, D^{w\mathtt{l}}\}}{\binom{n\mathtt{l}}{w\mathtt{l}}},$$

where $F_{\text{cw}}(D) = \sum_i \binom{\mathtt{r}}{2i} D^{2i}$ and $F_{\text{ss}}(D) = \sum_{i \neq 1} \binom{\mathtt{r}}{i} D^i$. Let x_ω be the unique real solution to $a(x) = \omega \mathtt{r}$, where

$$a_{\text{cw}}(x) = (\sum_i \binom{\mathtt{r}}{2i} 2ix^{2i}) / (\sum_i \binom{\mathtt{r}}{2i} x^{2i}), \quad a_{\text{ss}}(x) = (\sum_{i \neq 1} \binom{\mathtt{r}}{i} ix^i) / (\sum_{i \neq 1} \binom{\mathtt{r}}{i} x^i),$$

and define $b_{\text{cw/ss}}(x) = x a'_{\text{cw/ss}}(x)$. Let $n\omega\mathtt{l} \in \mathbb{N}$, $\Delta w = o(\sqrt{n \log(n)})$, and $\Delta w\mathtt{l} \in \mathbb{N}$. If $n\omega\mathtt{l}$ is odd then $\mathbb{E}[A_{\text{cw}}(\mathtt{G}, n\omega)] = 0$. If we are looking at the weight distribution of codewords for even $n\omega\mathtt{l}$ or if we are looking at the weight distribution of stopping sets, then

$$\mathbb{E}[A_{\text{cw/ss}}(\mathtt{G}, n\omega)] = \frac{\mu_{\text{cw/ss}} \sqrt{\mathtt{r}}\, 2^{-(\mathtt{l}-1) n h_2(\omega)} F_{\text{cw/ss}}(x_\omega)^{n\frac{\mathtt{l}}{\mathtt{r}}}}{\sqrt{2\pi n b(x_\omega)}\, x_\omega^{n\omega\mathtt{l}}} (1 + o(1)),$$

$$\frac{\mathbb{E}[A_{\text{cw/ss}}(\mathtt{G}, n\omega + \Delta w)]}{\mathbb{E}[A_{\text{cw/ss}}(\mathtt{G}, n\omega)]} = \frac{w^{(\mathtt{l}-1)\Delta w}}{x_\omega^{\mathtt{l}\Delta w} (n - w)^{(\mathtt{l}-1)\Delta w}} \mathrm{e}^{-\frac{(\Delta w)^2}{2} \left(\frac{\mathtt{l}\mathtt{r}}{n b(x_\omega)} - \frac{\mathtt{l}-1}{w(1-w/n)} \right)} (1 + o(1)),$$

where $\mu_{\text{cw}} = 2$, $\mu_{\text{ss}} = 1$ if $\mathtt{r} > 2$, and $\mu_{\text{ss}} = 2$ if $\mathtt{r} = 2$.

Proof. If $\mathtt{r} = 2$ then codewords and stopping sets are identical. Therefore, consider first the weight distribution of stopping sets assuming that $\mathtt{r} > 2$. In this case $F_{\text{ss}}(D)$ is Hayman admissible since $F_0 = 1 > 0$ and $\gcd\{k : F_k > 0\} = 1$ so that the statement is a straightforward application of the Hayman method (to the term $\operatorname{coef}\{F_{\text{ss}}(D)^{n\frac{\mathtt{l}}{\mathtt{r}}}, D^{w\mathtt{l}}\}$ as well as to the Binomial coefficients).

We cannot apply the Hayman method directly to the weight distribution of codewords since F_{cw} has only even powers and so $\gcd\{k : F_k > 0\} = 2$. But we know already from our previous discussion that this is corrected easily by multiplying the expansion by a factor of 2. Alternatively, the case of codewords can be treated by rewriting the problem in the form

$$\mathbb{E}[A_{\text{cw}}(\mathtt{G}, w)] = \binom{n}{w} \frac{\operatorname{coef}\{F(\sqrt{D})^{n\frac{\mathtt{l}}{\mathtt{r}}}, D^{\frac{1}{2}w\mathtt{l}}\}}{\binom{n\mathtt{l}}{w\mathtt{l}}},$$

keeping in mind that $\frac{1}{2}w\mathtt{l}$ must be an integer. Now we can apply the Hayman method since $F_{\text{cw}}(\sqrt{D})$ is Hayman admissible. □

Although we have stated Lemma D.14 in *explicit* form, it is often more convenient to use the result in *implicit* form. If we want, e.g., to plot the growth rate for the weight distribution in Lemma D.17 over the whole range, rather than solving for x_ω for each choice of ω, we let x run over all elements of $\mathbb{R}^+$ and the corresponding value of ω is then given by $\omega = a(x)/\mathrm{r}$.

It remains to prove Lemma D.14. Let $\underline{\omega}$ and $n \in \mathbb{N}$ be such that $n\underline{\omega} \in \mathbb{N}^n$. Let $\underline{x}_\omega$ be a strictly positive solution of $\underline{a}(\underline{x}) = \underline{\omega}$. First note that if $F(D)$ is Hayman admissible then $\underline{B}$ is strictly positive definite: by explicit computations we see that for $\underline{y} \in \mathbb{R}^m$

$$\underline{y}\underline{B}(\underline{x})\underline{y}^T = \sum_{\underline{k},\underline{l}} \Big(\sum_{i=1}^{m} (\underline{k}_i - \underline{l}_i) y_i\Big)^2 F_{\underline{k}} F_{\underline{l}} \underline{x}^{\underline{k}+\underline{l}}.$$

We want to prove that the right-hand side is strictly positive for non-zero $\underline{y}$. If $\underline{x}$ is strictly positive then this is the same as proving that $\sum_{\underline{k},\underline{l}}\big(\sum_{i=1}^m (\underline{k}_i - \underline{l}_i) y_i\big)^2 F_{\underline{k}} F_{\underline{l}}$ is strictly positive. If $F(D)$ is Hayman admissible then we have seen that the matrix $\underline{A}$ whose rows are formed by the differences $\underline{k} - \underline{l}$ has full column rank so that there cannot exist a vector $\underline{y}$ which is orthogonal to all the rows of $\underline{A}$. We conclude that in the preceding sum there must be at least one strictly positive term, which shows that $\underline{B}$ is strictly positive definite.

Note that both the asymptotic expansion as well as the local limit theorem is proved if we can show that for $\|\underline{\Delta w}\|_1 = o(\sqrt{n \log n})$

$$\left| n^{m/2} \frac{N_n(n\underline{\omega} + \underline{\Delta w}) \left(\underline{x}_\omega\right)^{n\underline{\omega}+\underline{\Delta w}}}{F(\underline{x}_\omega)^n} - \frac{1}{\sqrt{(2\pi)^m |B(\underline{x}_\omega)|}} e^{-\frac{1}{2n}\underline{\Delta w}\,\underline{B}(\underline{x}_\omega)\underline{\Delta w}^T} \right| \tag{D.18}$$

is of order $O\left(n^{-1/10}\right)$. To see that this is true consider the right-hand side of the preceding expression. If $\|\underline{\Delta w}\|_1 = o(\sqrt{n \log n})$ then $\underline{\Delta w}\,\underline{B}(\underline{x}_\omega)\underline{\Delta w}^T/n = o(\log n) \le \Theta(\log n^{1/20})$, let's say. It follows that the right-hand side is at least of order $\Theta(n^{-1/20})$. The claim now follows since $\Theta(n^{-1/20}) > O\left(n^{-1/10}\right)$. Note that

$$F(\underline{x})^n = \sum_{\underline{k} \in \mathbb{N}^m} N_n(\underline{k}) \underline{x}^{\underline{k}}. \tag{D.19}$$

We continue:

$$\frac{N_n(n\underline{\omega} + \underline{\Delta w}) \underline{x}_\omega^{n\underline{\omega}+\underline{\Delta w}}}{F(\underline{x}_\omega)^n} \tag{D.20}$$

using $\frac{1}{2\pi}\int_{[-\pi,\pi)} e^{jk\theta} d\theta = \delta(k)$

$$= \sum_{\underline{k}} \frac{N_n(\underline{k}) \underline{x}_{\underline{\omega}}^{\underline{k}}}{F(\underline{x}_{\underline{\omega}})^n} \left(\frac{1}{(2\pi)^m} \int_{[-\pi,\pi)^m} e^{-j(n\underline{\omega}+\Delta w-\underline{k})\underline{\nu}^T} d\underline{\nu} \right)$$

$$= \frac{1}{(2\pi)^m} \int_{[-\pi,\pi)^m} \frac{\sum_{\underline{k}} N_n(\underline{k}) \underline{x}_{\underline{\omega}}^{\underline{k}} e^{j\underline{\nu}\,\underline{k}^T}}{F(\underline{x}_{\underline{\omega}})^n} e^{-j(n\underline{\omega}+\Delta w)\,\underline{\nu}^T} d\underline{\nu}$$

by (D.19)

$$= \frac{1}{(2\pi)^m} \int_{[-\pi,\pi)^m} \frac{F(\underline{x}_{\underline{\omega}} e^{j\underline{\nu}})^n}{F(\underline{x}_{\underline{\omega}})^n} e^{-j(n\underline{\omega}+\Delta w)\,\underline{\nu}^T} d\underline{\nu}.$$

Therefore, (D.18) can be written equivalently as

(D.21)
$$\Bigg| n^{m/2} \frac{1}{(2\pi)^m} \int_{[-\pi,\pi)^m} \frac{F(\underline{x}_{\underline{\omega}} e^{j\underline{\nu}})^n}{F(\underline{x}_{\underline{\omega}})^n} e^{-j(n\underline{\omega}+\Delta\omega)\,\underline{\nu}^T} d\underline{\nu} - \frac{1}{\sqrt{(2\pi)^m |B(\underline{x}_{\underline{\omega}})|}} e^{-\frac{1}{2n}\Delta w\, B(\underline{x}_{\underline{\omega}}) \Delta w^T} \Bigg| = O\left(n^{-10}\right).$$

Let us divide the integration region $[-\pi,\pi)^m$ into the part $\Gamma = [-n^{-\frac{2}{5}}, n^{-\frac{2}{5}})^m$ and its complement $[-\pi,\pi)^m \setminus \Gamma$. By assumption F is Hayman admissible. Therefore, the contribution to the integral from the region $[-\pi,\pi)^m \setminus \Gamma$ is negligible, since

$$\left| \frac{1}{(2\pi)^m} \int_{[-\pi,\pi)^m \setminus \Gamma} \frac{F(\underline{x}_{\underline{\omega}} e^{j\underline{\nu}})^n}{F(\underline{x}_{\underline{\omega}})^n} e^{-j(n\underline{\omega}+\Delta\omega)\,\underline{\nu}^T} d\underline{\nu} \right|$$

by (D.6)

$$\leq \frac{1}{(2\pi)^m} \int_{[-\pi,\pi)^m \setminus \Gamma} \left| \frac{F(\underline{x}_{\underline{\omega}} e^{j\underline{\nu}})}{F(\underline{x}_{\underline{\omega}})} \right|^n d\underline{\nu} = O(n^{-1/5}).$$

On the other hand, for the integral over the region Γ, we have

$$\frac{1}{(2\pi)^m} \int_{\Gamma} \frac{F(\underline{x}_{\underline{\omega}} e^{j\underline{\nu}})^n}{F(\underline{x}_{\underline{\omega}})^n} e^{-j(n\underline{\omega}+\Delta w)\,\underline{\nu}^T} d\underline{\nu}$$

by (D.7)

$$= \frac{1}{(2\pi)^m} \int_{\Gamma} e^{jn\underline{a}(\underline{x}_{\underline{\omega}})\underline{\nu}^T - \frac{n}{2}\underline{\nu} B(\underline{x}_{\underline{\omega}})\underline{\nu}^T + O\left(n^{-\frac{1}{5}}\right)} e^{-j(n\underline{\omega}+\Delta w)\,\underline{\nu}^T} d\underline{\nu}$$

$\underline{a}(\underline{x}_{\omega}) = \underline{\omega}$

$$= \frac{1}{(2\pi)^m} \int_{\Gamma} e^{-\frac{n}{2}\underline{\nu} B(\underline{x}_{\underline{\omega}})\underline{\nu}^T + O\left(n^{-\frac{1}{5}}\right)} e^{-j\Delta w\,\underline{\nu}^T} d\underline{\nu}$$

$$= \frac{1}{(2\pi)^m} \int_{\Gamma} e^{-\frac{n}{2}\underline{\nu} B(\underline{x}_{\underline{\omega}})\underline{\nu}^T} e^{-j\Delta w\,\underline{\nu}^T} d\underline{\nu} \left(1 + O(n^{-\frac{1}{5}})\right)$$

$$
\begin{aligned}
\underline{y} = \sqrt{n}\underline{v} \\
&= n^{-\frac{m}{2}} \frac{1}{(2\pi)^m} \int_{[-n^{\frac{1}{10}}, n^{\frac{1}{10}})^m} e^{-\frac{1}{2}\underline{y}\underline{B}(\underline{x}_\omega)\underline{y}^T} e^{-j\Delta w/\sqrt{n}\,\underline{y}^T} d\underline{y}(1 + O(n^{-\frac{1}{5}})) \\
&= n^{-\frac{m}{2}} \frac{1}{(2\pi)^m} \int_{\mathbb{R}^m} e^{-\frac{1}{2}\underline{y}\underline{B}(\underline{x}_\omega)\underline{y}^T} e^{j\Delta w/\sqrt{n}\,\underline{y}^T} d\underline{y}(1 + O(n^{-\frac{1}{5}})) + O(n^{-1/10} n^{-m/2}) \\
&= n^{-\frac{m}{2}} \sqrt{\frac{(2\pi)^m}{|\underline{B}(\underline{x}_\omega)|}} e^{-\frac{1}{2n}\Delta w \underline{B}^{-1}(\underline{x}_\omega)\Delta w^T} (1 + O(n^{-\frac{1}{5}})) + O(n^{-1/10} n^{-m/2}).
\end{aligned}
$$

A few remarks are in order. In the last step we have used the fact that the Fourier transform of a Gaussian is again a Gaussian. To justify the second-to-last step we claim that, for any $m \times m$ real positive-definite matrix $\underline{B}^{-1}$ and any $K \in \mathbb{R}^+$,

$$
\text{(D.22)} \qquad \left| \int_{\mathbb{R}^m \setminus (-K,K)^m} e^{-\frac{\underline{s}\underline{B}^{-1}\underline{s}^T}{2}} e^{-j\underline{u}\,\underline{s}^T} d\underline{s} \right| = O(1/K).
$$

To see this last claim write $\underline{B}^{-1} = \underline{U}\,\underline{\Lambda}\,\underline{U}^T$ where $\underline{U}$ is a rotation matrix and $\underline{\Lambda}$ is a diagonal matrix with positive entries along the diagonal, call them λ_i. Let $\lambda_{\min} = \{\lambda_i\}_i$, $\lambda_{\min} > 0$. Then we have

$$
\begin{aligned}
\underline{t} = \underline{s}\,\underline{U} \qquad \left| \int_{\mathbb{R}^m \setminus (-K,K)^m} e^{-\frac{\underline{s}\underline{B}^{-1}\underline{s}^T}{2}} e^{-j\underline{u}\,\underline{s}^T} d\underline{s} \right| &= \int_{\underline{t}\,\underline{U}^T \in \mathbb{R}^m \setminus (-K,K)^m} e^{-\frac{\sum_{i=1}^m \lambda_i t_i^2}{2}} d\underline{t} \\
&\leq \int_{\|\underline{t}\|_2 \geq K} e^{-\frac{\lambda_{\min} \sum_{i=1}^m t_i^2}{2}} d\underline{t} = O(1/K),
\end{aligned}
$$

where the estimate in the last step is quite crude but sufficient for our purpose. □

As a first serious application of the Hayman method, let us furnish the proof of Lemma 3.22 that gives a sufficient condition under which the actual rate of a randomly chosen element of an ensemble is close to the design rate.

Proof of Lemma 3.22. For any element G we have $r(\mathrm{G}) \geq r(\lambda, \rho)$. If it is true that the expected value of the rate (more precisely, the logarithm of the expected number of codewords divided by the length) is close to the design rate, then we can use the Markov inequality to show that most codes have a rate close to the design rate.

First, we ask where the growth rate of the weight distribution has its maximum, i.e., which codeword "type" dominates. It is natural to assume that for "most" codes the dominant type is the one with roughly half the components equal to 1. Let us check when this is true. We could use the expression of the growth rate as stated in Lemma 3.158. It is slightly more convenient though to start anew and derive a more compact representation. We know that the expected number of codewords involving e edges is given by

$$\mathbb{E}[A_{\text{cw}}(\mathtt{G}, e)] = \frac{\text{coef}\left\{\prod_i (1+y^i)^{nL_i} \prod_j q_j(z)^{n\frac{L'(1)}{R'(1)}R_j}, y^e z^e\right\}}{\binom{nL'(1)}{e}},$$

where $q_j(z) = ((1+z)^j + (1-z)^j)/2$. Let n tend to infinity and define $\epsilon = e/(nL'(1)$. We are interested in the exponent of the growth rate. Recall from our previous discussion that the very simple upper bound which we introduced in the beginning of Section D.4 is exponentially tight. If we use this bound on the numerator and the standard estimate of the binomial in terms of the binary entropy function, then we get that, for a fixed ϵ, the exponent $\lim_{n\to\infty} \frac{1}{n} \log_2(\mathbb{E}[A_{\text{cw}}(\mathtt{G}, \epsilon n L'(1))])$ is given by the infimum with respect to $y, z > 0$ of

$$\text{(D.23)} \quad \sum_i L_i \log_2(1+y^i) - L'(1)\epsilon \log_2 y + \frac{L'(1)}{R'(1)} \sum_j R_j \log_2 q_j(z) \\ - L'(1)\epsilon \log_2 z - L'(1) h(\epsilon).$$

We want to determine the exponent corresponding to the expected number of codewords, i.e., $\lim_{n\to\infty} \frac{1}{n} \log_2(\mathbb{E}[A_{\text{cw}}(\mathtt{G})])$, where $A_{\text{cw}}(\mathtt{G}) = \sum_e A_{\text{cw}}(\mathtt{G}, e)$. Since the number of "types" (numbers e) grows only linearly this exponent is equal to the supremum of (D.23) over all $0 \le \epsilon \le 1$. In summary, the sought-after exponent is given by a stationary point of the function stated in (D.23) with respect to y, z, and ϵ. Take the derivative with respect to ϵ. This gives $\epsilon = yz/(1+yz)$. If we substitute this expression for ϵ into (D.23), subtract the design rate $r(L, R)$, and rearrange the terms we get (3.23). Next, if we take the derivative with respect to y and solve for z we get (3.24). In summary, $\Psi(y)$ is a function so that

$$\log_2 \mathbb{E}[A_{\text{cw}}(\mathtt{G})] = n\{r(L, R) + \sup_{y \in [0,\infty)} \Psi(y) + \omega_n\},$$

where $\omega_n = o(1)$. In particular, by explicit computation we see that $\Psi(y = 1) = 0$. A closer look shows that $y = 1$ corresponds to the exponent of codewords of weight $n/2$. This is most easily seen by considering

$$\text{coef}\left\{\prod_i (1+y^i)^{nL_i}, y^{n\alpha}\right\} \le \frac{\prod_i (1+y_\alpha^i)^{nL_i}}{y_\alpha^{n\alpha}},$$

where we know that the bound is exponentially tight if we choose the optimum y_α. By explicit calculation we see that $a(1) = \frac{\sum_i L_i i}{2}$, i.e, $y = 1$ corresponds to $\alpha = \frac{\sum_i L_i i}{2}$.

This corresponds to the case in which we take take half of each variable-node type, so that the global weight is one-half.

By an application of (the Markov Inequality) (C.1) we have for any $\xi > 0$

$$\mathbb{P}\{r(\mathtt{G}) \geq r(\lambda, \rho) + \xi\} = \mathbb{P}\{A_{\text{cw}}(\mathtt{G}) \geq 2^{n(\xi-\omega_n)}\mathbb{E}[A_{\text{cw}}(\mathtt{G})]\} \leq e^{-n\xi\ln(2)/2}.$$

□

Proof of Lemma 3.27. We start by showing that the dominant codeword type has relative weight one-half. For the $(\mathtt{l}, \mathtt{r})$-regular case the function $\Psi(y)$ introduced in Lemma 3.22 simplifies to

$$\Psi(y) = \log\Big(\frac{1}{2}(1+y^{\mathtt{l}})^{\mathtt{l}-1}((1+y^{\mathtt{l}-1})^{\mathtt{r}} + (1-y^{\mathtt{l}-1})^{\mathtt{r}})^{\frac{\mathtt{l}}{\mathtt{r}}}\Big).$$

Define $x = y^{\mathtt{l}-1}$. Then the condition $\Psi(y) \leq 0$, with strict inequality except for $y = 1$, is equivalent to $f(x, \mathtt{r}) \leq g(x, \mathtt{l})$, with strict inequality except for $x = 1$, where $f(x, \mathtt{r}) = ((1+x)^{\mathtt{r}} + (1-x)^{\mathtt{r}})^{\frac{1}{\mathtt{r}}}$ and $g(x, \mathtt{l}) = 2^{\frac{1}{\mathtt{l}}}(1+x^{\frac{\mathtt{l}}{\mathtt{l}-1}})^{\frac{\mathtt{l}-1}{\mathtt{l}}}$. We start by showing that for $\mathtt{r} \geq 2$ and $x \geq 0$, $f(x, \mathtt{r}) \leq g(x, \mathtt{r})$. To see this, consider the equivalent statement $2\sum_i \binom{\mathtt{r}}{2i}x^{2i} = f(x, \mathtt{r})^{\mathtt{r}} \leq g(x, \mathtt{r})^{\mathtt{r}} = 2\sum_j \binom{\mathtt{r}-1}{j}x^{\frac{\mathtt{r}}{\mathtt{r}-1}j}$. For $\mathtt{r} = 2$ a direct check shows that the two sides are equal and the same is true for $x = 0$. Consider therefore the case $\mathtt{r} \geq 3$ and $x > 0$. First cancel the factor 2 from both sides. Next note that both series start with the term 1 and if $\mathtt{r}$ is even then the last term on both sides is $x^{\mathtt{r}}$. For each remaining term $\binom{\mathtt{r}}{2i}x^{2i}$, $2 \leq 2i < \mathtt{r}$, on the left there are exactly two terms on the right of the form $\binom{\mathtt{r}-1}{2i-1}x^{\frac{(2i-1)\mathtt{r}}{\mathtt{r}-1}} + \binom{\mathtt{r}-1}{2i}x^{\frac{2i\mathtt{r}}{\mathtt{r}-1}}$. Now note that for $x > 0$, x^{α} is a convex-∪ function in α for $\alpha > 0$ and that $\binom{\mathtt{r}-1}{2i-1} + \binom{\mathtt{r}-1}{2i} = \binom{\mathtt{r}}{2i}$. Therefore by Jensen's inequality (1.61),

$$\frac{\binom{\mathtt{r}-1}{2i-1}}{\binom{\mathtt{r}}{2i}}x^{\frac{(2i-1)\mathtt{r}}{\mathtt{r}-1}} + \frac{\binom{\mathtt{r}-1}{2i}}{\binom{\mathtt{r}}{2i}}x^{\frac{2i\mathtt{r}}{\mathtt{r}-1}} \geq \Big(x^{(\binom{\mathtt{r}-1}{2i-1}\frac{(2i-1)\mathtt{r}}{\mathtt{r}-1} + \binom{\mathtt{r}-1}{2i}\frac{2i\mathtt{r}}{\mathtt{r}-1})/\binom{\mathtt{r}}{2i}}\Big) = x^{2i}.$$

Now where we know that $f(x, \mathtt{r}) \leq g(x, \mathtt{r})$ for $\mathtt{r} \geq 2$ and $x \geq 0$, the proof is complete if we can show that $g(x, \mathtt{l})$ is a decreasing function in $\mathtt{l}$ and that it is strictly decreasing except for $x = 1$: we write $f(x, \mathtt{r}) \leq g(x, \mathtt{r}) \overset{\mathtt{l} \leq \mathtt{r}}{\leq} g(x, \mathtt{l})$, where the last inequality is strict for $x \neq 1$. It is the task of Problem D.7 to show that $g(x, \mathtt{l})$ is indeed decreasing in $\mathtt{l}$.

Let us now use the local Hayman expansion for codewords of relative weight close to one-half to determine the total number of codewords. In order to avoid technicalities assume that n is divisible by 4, so that $n/2$ is even. From Lemma D.17 we have

$$\mathbb{E}[A_{\text{cw}}(\mathtt{G}, n\omega)] = \frac{2\sqrt{\mathtt{r}}2^{-(\mathtt{l}-1)nh_2(\omega)}F_{\text{cw}}(x_{\omega})^{n\frac{\mathtt{l}}{\mathtt{r}}}}{\sqrt{2\pi n b(x_{\omega})}x_{\omega}^{n\omega\mathtt{l}}}(1 + o(1)),$$

$$\frac{\mathbb{E}[A_{\text{cw}}(\mathtt{G}, n\omega + \Delta w)]}{\mathbb{E}[A_{\text{cw}}(\mathtt{G}, n\omega)]} = \left(\frac{w}{n-w}\right)^{(\mathtt{l}-1)\Delta w} \frac{\mathrm{e}^{-(\Delta w)^2(\frac{\mathtt{l}\mathtt{r}}{2nb(x_\omega)} - \frac{\mathtt{l}-1}{2w(1-w/n)})}}{x_\omega^{\mathtt{l}\Delta w}}(1+o(1)),$$

where in the local expansion it is understood that $\mathtt{l}\Delta w$ must be even. Consider the expansion around $w = n/2$. Explicit computations show that in this case $x_\omega = 1$, $b(x_\omega) = b(1) = \frac{\mathtt{r}}{4}$, $F_{\text{cw}}(x_\omega) = F_{\text{cw}}(1) = 2^{\mathtt{r}-1}$. If we insert these values into the preceding two expressions and set $w = n/2$ we get

$$\mathbb{E}[A_{\text{cw}}(\mathtt{G}, n/2)] = \frac{2\sqrt{\mathtt{r}}2^{-(\mathtt{l}-1)n}2^{n\frac{\mathtt{l}}{\mathtt{r}}(\mathtt{r}-1)}}{\sqrt{2\pi n\mathtt{r}/4}}(1+o(1)) = \frac{2 2^{n(1-\frac{\mathtt{l}}{\mathtt{r}})}}{\sqrt{2\pi n/4}}(1+o(1)),$$

$$\frac{\mathbb{E}[A_{\text{cw}}(\mathtt{G}, n\omega + \Delta w)]}{\mathbb{E}[A_{\text{cw}}(\mathtt{G}, n\omega)]} = \mathrm{e}^{-(\Delta w)^2(\frac{2\mathtt{l}}{n} - \frac{2(\mathtt{l}-1)}{n})}(1+o(1)) = \mathrm{e}^{-(\Delta w)^2\frac{2}{n}}(1+o(1)).$$

This gives us for $\mathtt{l}\Delta w$ even

$$\mathbb{E}[A_{\text{cw}}(\mathtt{G}, n/2 + \Delta w)] = \frac{2 2^{n(1-\frac{\mathtt{l}}{\mathtt{r}})}}{\sqrt{2\pi n/4}}\mathrm{e}^{-(\Delta w)^2\frac{2}{n}}(1+o(1)).$$

If $\mathtt{l}$ is even then Δw ranges over all integers and by summing we see that

$$\mathbb{E}[\sum_{w=0}^{n} A_{\text{cw}}(\mathtt{G}, w)] = 2 2^{n(1-\frac{\mathtt{l}}{\mathtt{r}})}(1+o(1)).$$

If $\mathtt{l}$ is odd then we must only sum over all even integers and we get

$$\mathbb{E}[\sum_{w=0}^{n} A_{\text{cw}}(\mathtt{G}, w)] = 2^{n(1-\frac{\mathtt{l}}{\mathtt{r}})}(1+o(1)).$$

In summary, we have

$$\mathbb{E}[\sum_{w=0}^{n} A_{\text{cw}}(\mathtt{G}, w)] = 2^{n(1-\frac{\mathtt{l}}{\mathtt{r}})+\nu}(1+o(1)), \tag{D.24}$$

where $\nu = 1$ if $\mathtt{l}$ is even and $\nu = 0$ otherwise.

To finish the proof note that if $\mathtt{l}$ is odd then the rate is always at least equal to the design rate $r(\lambda, \rho) = (1 - \mathtt{l}/\mathtt{r})$, whereas if $\mathtt{l}$ is even then the rate is always at least $(1 - \mathtt{l}/\mathtt{r}) + 1/n$. This is true since the corresponding parity-check matrix has always at least one dependent row (over $\mathbb{F}_2$ the sum of all rows is equal to zero). Look at the case of odd $\mathtt{l}$. The even case is handled in the same manner. Note that $r(\mathtt{G}) - r(\lambda, \rho)$ is a non-negative random variable. We have

$$\mathbb{P}\{r(\mathtt{G}) = r(\lambda, \rho)\} = 1 - \mathbb{P}\{nr(\mathtt{G}) - nr(\lambda, \rho) \geq 1\}$$

$$
\begin{aligned}
\text{Markov inequality (C.1)} \quad & \geq 1 - \mathbb{E}[nr(\mathtt{G}) - nr(\lambda, \rho)] \\
& = 1 - \mathbb{E}\Big[\log_2 \frac{\sum_w A(G,w)}{2^{n(1-\frac{1}{\mathtt{r}})}}\Big] \\
\text{Jensen's inequality (1.61)} \quad & \geq 1 - \log_2 \mathbb{E}\Big[\frac{\sum_w A(G,w)}{2^{n(1-\frac{1}{\mathtt{r}})}}\Big] \\
\text{by (D.24)} \quad & = 1 - \log_2(1 + o_n(1)) \\
& = 1 + o_n(1).
\end{aligned}
$$

□

The following lemma is useful when proving that projections of the global code C onto the local code $C(\mathtt{T})$, where $\mathtt{T}$ is the computation tree of a variable node, are proper with high probability (see Lemma 3.47 and Problem 3.8).

Lemma D.25 (Weight Distribution of Perturbed Regular Ensemble). Consider a $(\mathtt{l}, \mathtt{r})$-regular degree distribution pair. Let $a, b, c \in \mathbb{N}$ be fixed and let n tend to infinity. Define

$$
p(x) = \frac{(1+x)^{\mathtt{r}} + (1-x)^{\mathtt{r}}}{2}, \qquad q(x) = \frac{(1+x)^{\mathtt{r}-1} + (1-x)^{\mathtt{r}-1}}{2},
$$

and

$$
N(n,a,b,c) = \sum_w \frac{\binom{n-a}{w}}{\binom{(n-a)\mathtt{l}}{w\mathtt{l}}} \mathrm{coef}\Big(p(x)^{n\frac{\mathtt{l}}{\mathtt{r}} - b\mathtt{l}} q(x)^{c\mathtt{l}}, x^{w\mathtt{l}}\Big).
$$

Then

$$
N(n,a,b,c) = N(n,0,0,0) 2^{a(\mathtt{l}-1) - b\mathtt{l}(\mathtt{r}-1) + c\mathtt{l}(\mathtt{r}-2)} (1 + o(1)).
$$

Proof. Proceeding as in Lemma D.17, we have for $\Delta = o(\sqrt{m \ln(m)})$ and $m\mathtt{r}$ even

$$
\text{(D.26)} \qquad \mathrm{coef}\{p(x)^m, x^{m\mathtt{r}/2+\Delta}\} = 2^{m(\mathtt{r}-1)} \frac{1}{\sqrt{\pi m \mathtt{r}/2}} e^{-\frac{2\Delta^2}{m\mathtt{r}}} (1 + o(1)).
$$

Therefore, for a fixed $k \in \mathbb{N}$,

$$
\begin{aligned}
\mathrm{coef}\{p(x)^m x^{2k}, x^{m\mathtt{r}/2+\Delta}\} &= \mathrm{coef}\{p(x)^m, x^{m\mathtt{r}/2-2k+\Delta}\} \\
&= 2^{m(\mathtt{r}-1)} \frac{1}{\sqrt{\pi m \mathtt{r}/2}} e^{-\frac{2(-2k+\Delta)^2}{m\mathtt{r}}} (1 + o(1)) \\
&= 2^{m(\mathtt{r}-1)} \frac{1}{\sqrt{\pi m \mathtt{r}/2}} e^{-\frac{2\Delta^2}{m\mathtt{r}}} (1 + o(1))
\end{aligned}
$$

$$=\text{coef}\{p(x)^m, x^{m\mathtt{r}/2+\Delta}\}.$$

Note that $q(x)$ is a polynomial with only even terms and that it has bounded degree (at most $\mathtt{r}-1$). Therefore,

$$\text{(D.27)}\qquad \text{coef}\{p(x)^m q(x)^k, x^{m\mathtt{r}/2+\Delta}\}$$

$$= \sum_{i=0}^{(\mathtt{r}-1)k/2} \text{coef}\{q(y), y^{2i}\}\text{coef}\{p(x)^m x^{2i}, x^{m\mathtt{r}/2+\Delta}\}$$

$$= \sum_{i=0}^{(\mathtt{r}-1)k/2} \text{coef}\{q(y), y^{2i}\}\text{coef}\{p(x)^m, x^{m\mathtt{r}/2+\Delta}\}(1+o(1))$$

$$=q(1)^k\text{coef}\{p(x)^m, x^{m\mathtt{r}/2+\Delta}\}(1+o(1))$$

$$\text{(D.28)}\qquad =2^{(\mathtt{r}-2)k}\text{coef}\{p(x)^m, x^{m\mathtt{r}/2+\Delta}\}(1+o(1)).$$

From the Hayman expansion (D.26) we also conclude that if $k\mathtt{r}$ is even then

$$\text{coef}\{p(x)^{m-k}, x^{m\mathtt{r}/2+\Delta}\} =\text{coef}\{p(x)^{m-k}, x^{(m-k)\mathtt{r}/2+(k\mathtt{r}/2+\Delta)}\}$$

$$=2^{(m-k)(\mathtt{r}-1)}\frac{1}{\sqrt{\pi(m-k)\mathtt{r}/2}}e^{-\frac{2(k\mathtt{r}/2+\Delta)^2}{(m-k)\mathtt{r}}}$$

$$\text{(D.29)}\qquad =2^{-k(\mathtt{r}-1)}\text{coef}\{p(x)^m, x^{m\mathtt{r}/2+\Delta}\}(1+o(1)).$$

As a final ingredient note that

$$\frac{\binom{n-a}{w}}{\binom{(n-a)\mathtt{l}}{w\mathtt{l}}} = \frac{\binom{n}{w}}{\binom{n\mathtt{l}}{w\mathtt{l}}}(1-w/n)^{-a(\mathtt{l}-1)}(1+O(n^{-2})).$$

If $w = n/2+\Delta$ this gives

$$\frac{\binom{n-a}{n/2+\Delta}}{\binom{(n-a)\mathtt{l}}{(n/2+\Delta)\mathtt{l}}} = \frac{\binom{n}{n/2+\Delta}}{\binom{n\mathtt{l}}{(n/2+\Delta)\mathtt{l}}}(\frac{1}{2}-\Delta/n)^{-a(\mathtt{l}-1)}(1+O(n^{-2}))$$

$$\text{(D.30)}\qquad =2^{a(\mathtt{l}-1)}\frac{\binom{n}{n/2+\Delta}}{\binom{n\mathtt{l}}{(n/2+\Delta)\mathtt{l}}}(1+O(n^{-1})).$$

Let us now combine all these estimates. We get

$$\sum_w \frac{\binom{n-a}{w}}{\binom{(n-a)\mathtt{l}}{w\mathtt{l}}}\text{coef}\{p(x)^{n\frac{\mathtt{l}}{\mathtt{r}}-b\mathtt{l}}q(x)^{c\mathtt{l}}, x^{w\mathtt{l}}\}$$

$$= \sum_{\Delta} \frac{\binom{n-a}{n/2+\Delta}}{\binom{(n-a)\mathtt{l}}{(n/2+\Delta)\mathtt{l}}} \text{coef}\{p(x)^{n\frac{\mathtt{l}}{\mathtt{r}}-b\mathtt{l}} q(x)^{c\mathtt{l}}, x^{n\mathtt{l}/2+\Delta\mathtt{l}}\}$$

$$\overset{(D.30)}{=} 2^{a(\mathtt{l}-1)} \sum_{\Delta} \frac{\binom{n}{n/2+\Delta}}{\binom{n\mathtt{l}}{(n/2+\Delta)\mathtt{l}}} \text{coef}\{p(x)^{n\frac{\mathtt{l}}{\mathtt{r}}-b\mathtt{l}} q(x)^{c\mathtt{l}}, x^{n\mathtt{l}/2\Delta\mathtt{l}}\}(1+o(1))$$

$$\overset{(D.28)}{=} 2^{a(\mathtt{l}-1)+c\mathtt{l}(\mathtt{r}-2)} \sum_{\Delta} \frac{\binom{n}{n/2+\Delta}}{\binom{n\mathtt{l}}{(n/2+\Delta)\mathtt{l}}} \text{coef}\{p(x)^{n\frac{\mathtt{l}}{\mathtt{r}}-b\mathtt{l}}, x^{n\mathtt{l}/2+\Delta\mathtt{l}}\}(1+o(1))$$

$$\overset{(D.29)}{=} 2^{a(\mathtt{l}-1)-b\mathtt{l}(\mathtt{r}-1)+c\mathtt{l}(\mathtt{r}-2)} \sum_{\Delta} \frac{\binom{n}{n/2+\Delta}}{\binom{n\mathtt{l}}{(n/2+\Delta)\mathtt{l}}} \text{coef}\{p(x)^{n\frac{\mathtt{l}}{\mathtt{r}}}, x^{n\mathtt{l}/2+\Delta\mathtt{l}}\}(1+o(1))$$

$$\overset{\text{Lem. 3.27}}{=} 2^{a(\mathtt{l}-1)-b\mathtt{l}(\mathtt{r}-1)+c\mathtt{l}(\mathtt{r}-2)} N(n,0,0,0)(1+o(1)). \qquad \square$$

Proof of Lemma 3.47. We only give the proof for the case $\ell = 0$. The general proof proceeds in a similar fashion and is relegated to Problem 3.8.

Consider the ensemble LDPC $(n, \mathtt{1}, \mathtt{r})$. Let $A_i^{\alpha}(\mathtt{G})$ denote the number of codewords of the code $\mathtt{G}$ which have an α at position i, $i \in [n]$, and $\alpha \in \{0,1\}$. Note that position i of the code is proper if and only if $A_i^1(\mathtt{G}) > 0$. We know from Problem 1.5 that in this case $A_i^1(\mathtt{G}) = A_i^0(\mathtt{G})$.

Without loss of generality we can assume that $i = 1$ since the definition of the ensemble is invariant under permutations of the components. Define r implicitly by $2^{nr} = 2^{n\frac{\mathtt{l}}{\mathtt{r}}-1}(1 + \mathbb{1}_{\{\mathtt{l} \text{ is even}\}})$. We know from Lemma D.25 that

$$\mathbb{E}[A(\mathtt{G})] = 2^{nr}(1+o(1)).$$

Note that $A(\mathtt{G})$ can only take on the values 2^j, $j \geq nr$. Let $a_j = \mathbb{P}\{A(\mathtt{G}) = 2^j\}$, $j \geq nr$. Then we have $1 + o(1) = \sum_{j\geq 0} a_{nr+j} 2^j = 1 + \sum_{j\geq 1} a_{nr+j}(2^j - 1)$. This implies that

$$\sum_{j>nr} a_j 2^j = o(1). \tag{D.31}$$

Consider $A_1^0(\mathtt{G})$. This means that we set bit 1 to zero. With probability $1 - O(1/n)$ the $\mathtt{l}$ edges emanating from this variable connect to distinct check nodes. We can therefore restrict ourselves to this case. To simplify notation we omit the conditioning in the following. Setting bit 1 to zero is equivalent to removing this bit from the graph together with its outgoing edges. Computing $\mathbb{E}[A_1^0(\mathtt{G})]$ is hence equivalent to computing the average rate of a code which has length $n-1$, where all variable nodes have degree $\mathtt{l}$, $n\frac{\mathtt{l}}{\mathtt{r}} - \mathtt{l}$ check nodes have degree $\mathtt{r}$, and $\mathtt{l}$ check nodes have degree $\mathtt{r} - 1$. Using Lemma D.25 with $a = b = c = 1$, we see that $\mathbb{E}[A_1^0(\mathtt{G})] = 2^{nr(\mathtt{l},\mathtt{r})-1}$. This is not surprising. It is natural that in expectation such a code (which has one degree of freedom less) has half the codewords.

Now note that $A_1^0(\mathtt{G}) \geq A_1^1(\mathtt{G})$ so that

$$2^{nr}(1+o(1)) = \mathbb{E}[A(\mathtt{G})] = \mathbb{E}[A_1^0(\mathtt{G})] + \mathbb{E}[A_1^1(\mathtt{G})] \leq 2\,\mathbb{E}[A_1^0(\mathtt{G})] = 2^{nr}(1+o(1)).$$

Since by (D.31) the dominant contribution to the expectation of $A(\mathtt{G})$ stems from the case where $A(\mathtt{G}) = 2^{nr}$, we conclude from the last sequence of equalities that $\mathbb{P}\{A_1^1(\mathtt{G}) \neq A_1^0(\mathtt{G})\} = o(1)$, as claimed. □

Theorem D.32 (More Than You Want to Know about the Weight Distribution). Consider the ensemble LDPC $(n, \lambda, \rho) \triangleq$ LDPC (n, L, R) and let $G_{\text{cw/ss}}(\omega)$ denote the growth rate of the weight distribution of codewords/stopping sets. Define the *critical* parameter $\omega^{\text{crit}}_{\text{cw/ss}} = \inf\{0 < \omega \leq 1/2 : G_{\text{cw/ss}}(\omega) > 0\}$. Let $A_{\text{cw/ss}}(\mathtt{G}, w)$ denote the number of codewords/stopping sets of weight w of a code $\mathtt{G}$ and let $\hat{A}_{\text{cw/ss}}(\mathtt{G}, w)$ denote the corresponding number of minimal codewords/stopping sets. Further, let $\mathtt{l}_{\min}/\mathtt{l}_{\max}$ denote the minimum/maximum variable-node degree, respectively, and let $\mathtt{l}_{\text{e}}$ be the minimum *even* variable-node degree. Set $L_{\mathtt{l}_{\text{e}}} = 0$ if no such degree exists. Finally, let $\mu = \lambda'(0)\rho'(1)$, $\mu_i = \frac{\mu^i}{2i}$, $i \in \mathbb{N}$, and $\kappa = \frac{R''(1)}{2L'(1)R'(1)}$.

Then

$$\text{(D.33)} \qquad \mathbb{E}[A_{\text{cw/ss}}(\mathtt{G}, w)] \leq \mathrm{e}^{O(\sqrt{\frac{w^3}{n}})} \frac{\left((4\kappa)^{\mathtt{l}_{\min}/2} L_{\mathtt{l}_{\min}}\right)^w (w\mathtt{l}_{\min}/2)!}{n^{w(\mathtt{l}_{\min}/2-1)} w!},$$

$$\text{(D.34)} \qquad \lim_{n\to\infty} \frac{\mathbb{E}[A_{\text{cw/ss}}(\mathtt{G}, w \in 2\mathbb{N})]}{n^{w(1-\mathtt{l}_{\min}/2)}} = \kappa^{w\mathtt{l}_{\min}/2} L^w_{\mathtt{l}_{\min}} \frac{(w\mathtt{l}_{\min})!}{w!(w\mathtt{l}_{\min}/2)!},$$

$$\text{(D.35)} \qquad \lim_{n\to\infty} \frac{\mathbb{E}[A_{\text{cw}}(\mathtt{G}, w \in (2\mathbb{N}+1))]}{n^{w-((w-1)\mathtt{l}_{\min}+\mathtt{l}_{\text{e}})/2}} = \kappa^{((w-1)\mathtt{l}_{\min}+\mathtt{l}_{\text{e}})/2} L^{w-1}_{\mathtt{l}_{\min}} L_{\mathtt{l}_{\text{e}}} \frac{((w-1)\mathtt{l}_{\min}+\mathtt{l}_{\text{e}})!}{w!(((w-1)\mathtt{l}_{\min}+\mathtt{l}_{\text{e}})/2)!},$$

$$\text{(D.36)} \qquad \lim_{n\to\infty} \frac{\mathbb{E}[A_{\text{ss}}(\mathtt{G}, w \in (2\mathbb{N}+1))]}{n^{3/2+w(1-\mathtt{l}_{\min}/2)}} = \kappa^{(w\mathtt{l}_{\min}-3)/2} L^w_{\mathtt{l}_{\min}} \frac{\bar{r}R'''(1)(w\mathtt{l}_{\min}-3)!}{6((w\mathtt{l}_{\min}-3)/2)!}.$$

For $\mathtt{l}_{\min} = 2$ and $W \in \mathbb{N}$,

$$\text{(D.37)} \qquad \mathbb{E}[A_{\text{cw/ss}}(\mathtt{G}, w)] \leq \mu^w \mathrm{e}^{O(\sqrt{\frac{w^3}{n}})},$$

$$\text{(D.38)} \qquad \lim_{n\to\infty} \mathbb{E}[A_{\text{cw/ss}}(\mathtt{G}, w)] = \mu^w \binom{2w}{w} 4^{-w}, \quad \lim_{n\to\infty} \mathbb{E}[\hat{A}_{\text{cw/ss}}(\mathtt{G}, w)] = \frac{\mu^w}{2w},$$

$$\text{(D.39)} \qquad \lim_{n\to\infty} \mathbb{P}\Big\{\hat{A}_{\text{cw/ss}}(\mathtt{G}, 1) = a_1, \dots, \hat{A}_{\text{cw/ss}}(\mathtt{G}, W) = a_W\Big\} = \prod_{w=1}^{W} \frac{\mu_i^{a_i} \mathrm{e}^{-\mu_i}}{a_i!}.$$

If $0 < \mu < 1$, then for $0 < \omega < \omega^{\text{crit}}_{\text{cw/ss}}$

$$\lim_{n\to\infty} \mathbb{P}\Big\{\sum_{w=1}^{n\omega} A_{\text{cw/ss}}(\mathtt{G}, w) = 0\Big\} = \lim_{n\to\infty} \mathbb{P}\Big\{\sum_{w=1}^{n\omega} \hat{A}_{\text{cw/ss}}(\mathtt{G}, w) = 0\Big\} = \sqrt{1-\mu}, \tag{D.40}$$

and, more generally, for any fixed $W \in \mathbb{N}$

$$\lim_{n\to\infty} \mathbb{P}\Big\{\sum_{w=W}^{n\omega} \hat{A}_{\text{cw/ss}}(\mathtt{G}, w) = 0\Big\} = e^{-\sum_{i=W}^{\infty} \mu_i}. \tag{D.41}$$

If $\mathtt{l}_{\min} \geq 3$ ($\mu = 0$), then for any $0 < \omega < \omega^{\text{crit}}_{\text{cw/ss}}$ and $W \in \mathbb{N}$,

$$\mathbb{P}\Big\{\sum_{w=W}^{n\omega} A_{\text{cw/ss}}(\mathtt{G}, w) > 0\Big\} \leq O(n^{-W(\mathtt{l}_{\min}-1)+\lfloor \frac{W\mathtt{l}_{\min}}{2} \rfloor}). \tag{D.42}$$

Discussion: Let us dissect the many statements which are contained in this theorem. We start with the case $\mathtt{l}_{\min} = 2$ ($\mu > 0$). From (D.37) we see that the expected number of codewords/stopping sets of size w is essentially upper bounded by μ^w. More precisely, if we fix the weight w and increase the blocklength then this number converges to a definite limit for both regular as well as minimal codewords/stopping sets as stated in (D.38). Equation (D.39) in addition asserts that the *distribution* of minimal codewords/stopping sets converges to a Poisson distribution. As long as $\mu < 1$, the expected number of "small" codewords/stopping sets (smaller than $n\omega^{\text{crit}}_{\text{cw/ss}}$) stays finite and the probability that a code chosen uniformly at random contains no small non-zero codewords/stopping sets converges to the non-zero limit $\sqrt{1-\mu}$ (see (D.40)). In other words, a fraction $\sqrt{1-\mu}$ of all codes has linear minimum distance. According to (D.41), this fraction can be made arbitrarily close to one by looking only at words of size at least W, where W is chosen sufficiently large.

The case $\mathtt{l}_{\min} \geq 3$ behaves in a fundamentally different way. In this case the expected number of non-zero codewords/stopping sets of weight strictly less than $\omega^{\text{crit}}_{\text{cw/ss}} n$ converges to zero as n tends to infinity. This convergence is quite rapid. For fixed w, words of weight w tend to zero at a rate $n^{-w(\mathtt{l}_{\min}-1)+\lfloor \frac{w\mathtt{l}_{\min}}{2} \rfloor}$ as stated in (D.42). Therefore, almost all codes in such an ensemble have linear minimum distance.

Proof of Theorem D.32. We are looking for a good upper bound on the expected weight distribution stated in Lemma 3.157. According to Problem D.13, it is relatively straightforward to derive a bound of the form (D.37) with an additional factor

$\mathtt{l}\mathtt{l}\mathtt{l}_{\max}w$. It takes more effort if we want to eliminate this additional factor. We start with $\alpha(w,e)$. For any $y > 0$ we have

$$\sum_e \alpha(w,e)y^e = \operatorname{coef}\{\prod_i (1+xy^i)^{nL_i}, x^w\} = \sum_{(w_1,\dots,w_{\mathtt{l}_{\max}}):\sum w_i = w} \prod_i \binom{nL_i}{w_i} y^{iw_i}$$
$$\overset{(1.57)}{\leq} \binom{n}{w} \Big/ \binom{n}{w} \sum_{(w_1,\dots,w_{\mathtt{l}_{\max}}):\sum w_i = w} \frac{n^w}{w_1!\dots w_{\mathtt{l}_{\max}}!} \prod_i (L_i y^i)^{w_i}$$
$$\overset{(1.57)}{\leq} e^{w^2/n}\binom{n}{w} \sum_{(w_1,\dots,w_{\mathtt{l}_{\max}}):\sum w_i = w} \frac{w!}{w_1!\dots w_{\mathtt{l}_{\max}}!} \prod_i (L_i y^i)^{w_i}$$
$$= e^{w^2/n}\binom{n}{w} L(y)^w. \tag{D.43}$$

Therefore, if we find an upper bound for $\beta_{\mathrm{cw/ss}}(e)$ of the form y^e for some y then we can insert this bound into the preceding summation to derive an upper bound on the expected weight distribution. We claim that we have the uniform (in e) bound

$$\beta_{\mathrm{cw/ss}}(e) \leq e^{O(\sqrt{\frac{e^3}{n}})} \int_0^\infty e^{-s}(4s\kappa/n)^{e/2} ds. \tag{D.44}$$

Let us postpone the proof of (D.44) for a moment. We continue:

$$\mathbb{E}[A_{\mathrm{cw/ss}}(\mathtt{G},w)] = \sum_e \alpha(w,e)\beta_{\mathrm{cw/ss}}(e)$$
$$\overset{(D.44)}{\leq} \sum_e \alpha(w,e) e^{O(\sqrt{\frac{e^3}{n}})} \int_0^\infty e^{-s}(4s\kappa/n)^{e/2} ds.$$

Since the only non-zero terms in the sum are those with $\mathtt{l}_{\min}w \leq e \leq \mathtt{l}_{\max}w$, we can rewrite the error bound $e^{O(\sqrt{\frac{e^3}{n}})}$ as $e^{O(\sqrt{\frac{w^3}{n}})}$. If we exchange the order of summation and integration (which is justified since all terms are non-negative) and use (D.43) we get

$$\mathbb{E}[A_{\mathrm{cw/ss}}(\mathtt{G},w)] \leq e^{O(\sqrt{\frac{w^3}{n}})} \binom{n}{w} \int_0^\infty e^{-s} L(\sqrt{4s\kappa/n})^w ds.$$

Define $\tilde{L}(x) = L(x)/(L_{\mathtt{l}_{\min}} x^{\mathtt{l}_{\min}})$, so that the last expression is equal to

$$e^{O(\sqrt{\frac{w^3}{n}})} \binom{n}{w} L_{\mathtt{l}_{\min}}^w (4\kappa/n)^{\mathtt{l}_{\min}w/2} \int_0^\infty e^{-s} s^{\mathtt{l}_{\min}w/2} \tilde{L}(\sqrt{4s\kappa/n})^w ds.$$

Using (1.54) we have[1]

$$\int_0^\infty e^{-s} s^{\mathtt{l}_{\min}w/2} \tilde{L}(\sqrt{4s\kappa/n})^w ds = \sum_i \operatorname{coef}\{\tilde{L}(z)^w, z^i\}(4\kappa/n)^{i/2}((\mathtt{l}_{\min}w+i)/2)!$$

[1]With some abuse of notation we continue to write $m!$ even if m is not an integer as proxy for $\Gamma(m+1)$.

$$\overset{\text{(i)}}{\leq} \sum_i \text{coef}\{\tilde{L}(z)^w, z^i\}(2\kappa \mathtt{l}_{\max} w/n)^{i/2}(\mathtt{l}_{\min} w/2)!$$
$$= \tilde{L}(\sqrt{2\kappa \mathtt{l}_{\max} w/n})^w (\mathtt{l}_{\min} w/2)!\,.$$

In step (i) we used the bound $((\mathtt{l}_{\min} w + i)/2)! \leq ((\mathtt{l}_{\min} w + i)/2)^{i/2}(\mathtt{l}_{\min} w/2)!$. Because of the definition of $\tilde{L}$ we know that $((\mathtt{l}_{\min} w + i)/2) \leq \mathtt{l}_{\max} w/2$. The contribution of $\tilde{L}(\sqrt{2\kappa \mathtt{l}_{\max} w/n})^w$ can be upper bounded by $e^{O(\sqrt{\frac{w^3}{n}})}$. Collecting our results we get the upper bound (D.33). Further, for $\mathtt{l}_{\min} = 2$, and taking into account that $4\kappa L_2 = \mu$, we get (D.37).

Let us now justify the upper bound (D.44) which we have used to prove (D.33). Let $p_{\text{cw}}(z) = \prod_i \left((1+z)^i/2 + (1-z)^i/2\right)^{R_i}$ and $p_{\text{ss}}(z) = \prod_i ((1+z)^i - iz)^{R_i}$. We can treat both cases together. We therefore simply write $p(z)$. Consider the polynomial $q(z) = \prod_i (1 + \binom{i}{2} z^2)^{R_i}$. First assume that e is even. Using the results of Problems D.11 and D.12, we see that with $\gamma = \bar{r}R''(1)$

$$\text{coef}\left\{p(z)^{n\bar{r}}, z^e\right\} \leq \text{coef}\left\{q(z)^{n\bar{r}}, z^e\right\} e^{O(\sqrt{\frac{e^3}{n}})} \leq \frac{(\gamma n/2)^{e/2}}{(e/2)!} e^{O(\sqrt{\frac{e^3}{n}})}.$$

We claim that the bound stays valid if e is odd. In this case $e \geq 3$. We then have

$$\text{coef}\left\{p(z)^{n\bar{r}}, z^e\right\} \overset{\text{(i)}}{\leq} n\bar{r}2^{\mathtt{r}_{\max}} \text{coef}\left\{p(z)^{n\bar{r}}, z^{e-3}\right\}$$
$$\leq n\bar{r}2^{\mathtt{r}_{\max}} \text{coef}\left\{q(z)^{n\bar{r}}, z^{e-3}\right\} e^{O(\sqrt{\frac{e^3}{n}})}$$
$$\overset{\text{(ii)}}{\leq} n\bar{r}2^{\mathtt{r}_{\max}} \frac{(\gamma n/2)^{(e-3)/2}}{((e-3)/2)!} e^{O(\sqrt{\frac{e^3}{n}})} \leq \frac{(\gamma n/2)^{e/2}}{(e/2)!} e^{O(\sqrt{\frac{e^3}{n}})}.$$

In step (i) we first choose a check node to which we connect an odd number of times (≥ 3). At least one such node must exist if e is odd. In step (ii) we then use the previously derived bound for even e.

We continue:

$$\beta(e) \overset{(1.57)}{\leq} \frac{(\gamma n/2)^{e/2} e! e^{O(\sqrt{\frac{e^3}{n}})}}{L'(1)^e n^e (e/2)!}$$
$$= \sqrt{\kappa/n}^{\,e} \binom{e}{e/2} (e/2)! e^{O(\sqrt{\frac{e^3}{n}})}$$
$$\overset{\binom{e}{e/2} \leq 2^e}{\leq} \sqrt{4\kappa/n}^{\,e} (e/2)! e^{O(\sqrt{\frac{e^3}{n}})}$$
$$\overset{(1.54)}{=} \sqrt{4\kappa/n}^{\,e} e^{O(\sqrt{\frac{e^3}{n}})} \int_0^\infty e^{-s} s^{e/2} ds.$$

Statement (D.37) is a special case of (D.33) (just set $\mathtt{l}_{\min} = 2$). We can refine our bounds. If we fix e and let n tend to infinity we get

$$\beta_{\text{cw}}(e) = (\kappa/n)^{e/2} \frac{e!}{(e/2)!} \mathbb{1}_{\{e \in 2\mathbb{N}\}} (1 + O(1/n)),$$

$$\beta_{\text{ss}}(e) = \begin{cases} \beta_{\text{cw}}(e), & \text{if } e \text{ is even,} \\ \frac{(1-r)R'''(1)}{6n^{3/2}} \beta_{\text{cw}}(e-3)(1+O(1/n)), & \text{if } e \text{ is odd.} \end{cases}$$

The statements (D.34), (D.35), (D.36), and (D.38) follow by combining these estimates.

Although we have not explicitly mentioned it, the proof of (D.33) actually shows that the only words of "small" weight which appear in large graphs are those that only involve variable nodes of degree 2. This follows from the fact that the contribution of $\tilde{L}(\sqrt{2\kappa \mathtt{l}_{\max} w/n})^w$ is small and that the main contribution of $L(x)$ comes from the lowest-order term.

Statement (D.39), i.e., that the distribution of the number of minimal codewords/stopping sets converges to a Poisson distribution, is shown in Lemma C.37.

To deduce from (D.39) statement (D.40) we proceed as follows. If E and D are two events and $\mathbb{P}\{D\} \geq 1 - \delta$ then

$$\mathbb{P}\{E\} - \delta \leq \mathbb{P}\{E, D\} \leq \mathbb{P}\{E\}.$$

Choose as E the event $\sum_{w=1}^{W} \hat{A}(\mathtt{G}, w) = 0$ and as D the event $\sum_{w=W+1}^{n\omega} \hat{A}(\mathtt{G}, w) = 0$, where $\omega < \omega^{\text{crit}}_{\text{cw/ss}}$. Then

$$\mathbb{P}\{D\} \geq 1 - \mathbb{P}\{\sum_{w=W+1}^{n\omega} A(\mathtt{G}, w) \geq 1\}$$

by Markov inequality

$$\geq 1 - \mathbb{E}[\sum_{w=W+1}^{n\omega} A(\mathtt{G}, w)].$$

Pick a strictly positive but small $\tilde{\omega}$. Then

by (D.37)

$$1 - \mathbb{E}[\sum_{w=W+1}^{n\tilde{\omega}} A(\mathtt{G}, w)] \geq 1 - \sum_{w=W+1}^{n\tilde{\omega}} \mu^w e^{O(\sqrt{w^3/n})} \geq 1 - \frac{\mu^{W+1}}{1-\mu} e^{O(\sqrt{w^3/n})}.$$

In the last step we have made use of the fact that $e^{O(\sqrt{w^3/n})}$ is uniform in w, that $0 < \mu < 1$, and that $\tilde{\omega}$ is chosen sufficiently small. Finally, the contribution of $\mathbb{E}[\sum_{w=\tilde{\omega}n}^{n\omega} A(\mathtt{G}, w)]$ tends to zero as n increases by definition of $\omega^{\text{crit}}_{\text{cw/ss}}$. We see that by an appropriate choice of W we can make $\mathbb{P}\{D\}$ arbitrarily small. It follows that $\mathbb{P}\{\sum_{w=1}^{n\omega} \hat{A}(\mathtt{G}, w) = 0\}$ behaves like $\mathbb{P}\{\sum_{w=1}^{W} \hat{A}(\mathtt{G}, w) = 0\}$, for W large. But by (D.39) this probability converges for increasing W to (D.40). □

§D.5. Unimodality

Definition D.45 (Unimodality and Log-Concavity). We say that a sequence $\{F_i\}_{i=0}^n$ of real numbers is *unimodal* if for some $0 \leq k \leq l \leq n$

$$F_0 \leq F_1 \leq \cdots \leq F_k = \cdots = F_l \geq F_{l+1} \geq \cdots F_n.$$

We say that a sequence $\{F_i\}_{i=0}^n$ of positive real numbers is *log-concave* if for $1 \leq i < n$

$$\frac{1}{2}(\ln F_{i-1} + \ln F_{i+1}) \leq \ln F_i.$$

Fact D.46 (Log-Concavity Implies Unimodality). Let $\{F_i\}_{i=0}^n$ be a log-concave sequence of positive numbers. Then $\{F_i\}_{i=0}^n$ is unimodal.

Lemma D.47 (Unimodality of Generating Functions). Let $F(D) = F_0 + F_1 D + \cdots + F_n D^n$ be a polynomial all of whose roots are negative. Then $\{F_i\}_{i=0}^n$ is log-concave.

Example D.48 (Binomial Coefficient Sequence). Consider the sequence

$$\{F_i\}_{i=0}^n = \left\{\binom{n}{i}\alpha^i\right\}_{i=0}^n,$$

where $\alpha > 0$. Then $F(\alpha) = \sum_{i=0}^n F_i \alpha^i = (1+\alpha)^n$, which has a root of multiplicity n at $D = -\frac{1}{\alpha} < 0$. Therefore $\{\binom{n}{i}\alpha^i\}_{i=0}^n$ is log-concave and a fortiori unimodal. Further, the maximum is taken on at $i = \frac{\alpha}{1+\alpha}n$. ◇

Notes

A highly recommended source for further reading and the inspiration for these brief notes is Wilf's book *generatorfunctionology* [18]. Some of the many other useful references in this respect are Graham, Knuth, and Patashnik [10], Sedgewick and Flajolet [15], Egorychev [7], and Stanley [16, 17].

The simple but extremely useful inequality (D.3) was applied by Zyablov and Pinsker [19] to the analysis of the flipping algorithm. Burshtein and Miller applied the inequality in [3] to the weight and stopping set enumeration problem and pointed out that the inequality is asymptotically tight. The inequality was also put to good use by Orlitsky, Viswanathan, and Zhang [11].

The *Hayman method* can be seen as a refinement of inequality (D.3). It is a special case of the so-called *saddle-point method* which is due to Riemann and Debye [5, 13]. A very lucid introduction into this method was written by de Brujin [4]. Gardy and Solé discussed the saddle-point method in the context of asymptotic coding theory

[9]. We stated the Hayman method for the special case of determining the coefficient growth of powers of polynomials. This particular case was studied by Bender and Richmond [1] as well as Gardy [8]. The proof we give is essentially the one appearing in [1] presented as in the paper by Rathi [12]. The Hayman method was used for the determination of the weight distribution by Di [6].

The most important application of the Hayman method in this book is the determination of the average weight distribution of ensembles as described in Chapter 3 but there are several other important potential applications. To mention just one, in Section 6.9.1 we discuss how the weight distribution of convolutional codes can be computed in terms of powers of suitable matrices. Although we have not stated such a result in the main text, it was shown by Bender, Richmond, and Williamson [2] that central and local limit theorems can be derived in this case as well. It was shown by Sason, Telatar, and Urbanke [14] how to compute the asymptotic growth rate in this case.

For references on the weight distribution problem and its historical development see the Notes at the end of Chapter 3.

Problems

D.1 (Further Properties of Formal Power Sums). In this example we investigate some further basic properties of formal power sums. Consider the set of all formal power sums with coefficients in the field $\mathbb{F}$. We have seen that we can endow the set of such power sums with some algebraic structure. In particular, we can add, subtract, multiply, and we can even divide by some elements. The set of all formal power sums over $\mathbb{F}$ is usually denoted by $\mathbb{F}[[D]]$ and is called the *ring* of formal power sums. As a general rule, in $\mathbb{F}[[D]]$ all those operations are meaningful which require for the determination of each coefficient of the output only a *finite number* of (algebraic) operations. We saw how addition, multiplication, and the determination of the multiplicative inverse (if it exists) of a given element all fall in this category. We now discuss some more operations which we can perform on elements of $\mathbb{F}[[D]]$, we see how we can use formal powers sums as generating functions to solve problems in the area of enumerative combinatorics, and we investigate the relationship between formal power sums and Taylor series.

1. Assume we are given a sequence $\{F_i\}_{i\geq 0}$. We then say that $F(D) = \sum_{i=0}^{\infty} F_i D^i$ is the *generating function* of $\{F_i\}_{i\geq 0}$ and we write $F(D) \leftrightarrow \{F_i\}_{i\geq 0}$.

   a) If $F(D) \leftrightarrow \{F_i\}_{i\geq 0}$, then what are the generating functions of $\{F_{i+1}\}_{i\geq 0}$ and of $\{F_{i+2}\}_{i\geq 0}$?

   b) Define the *derivative* of a formal power sum $F(D) = \sum_{i=0}^{\infty} F_i D^i$ to be $F'(D) = \sum_{i=0}^{\infty} iF_i D^{i-1}$. If $F(D) \leftrightarrow \{F_i\}_{i\geq 0}$, then what is the formal power sum corresponding to $\{iF_i\}_{i\geq 0}$?

2. Let $f(z)$ be a function such that for some region of convergence $f(z)$ has the Taylor-series expansion $f(z) = \sum_{i=0}^{\infty} f_i z^i$. If $F(D) = \sum_{i=0}^{\infty} f_i D^i$ we then also write with some abuse of notation $F(D) = f(D)$. Using this notation, what would you write for $F_1(D) = \sum_{i=0}^{\infty} \frac{D^i}{i!}$ and for $F_2(D) = \sum_{i=0}^{\infty} (-1)^i \frac{D^{2i}}{(2i)!}$? How about $F_1'(D)$ and $F_2'(D)$? Any comments?

3. Let $\mathbb{F} = \mathbb{R}$ and consider the recurrence $a_{i+2} = a_{i+1} + a_i$, $(i \geq 0; a_0 = a_1 = 1)$. Define the formal power sum $a(D) = \sum_{i=0}^{\infty} a_i D^i$ and use it to solve this recursion. You hopefully got an answer of the form $a(D) = \frac{p(D)}{q(D)}$, where $p(D), q(D) \in \mathbb{F}[D]$. Find now a_0, a_1, a_2, and a_3 by *formally* finding the first four coefficients of the resulting power sum. Note: Do not use the recursion for that; start with $a(D)$ and use only algebraic operations. Now use a *partial fraction expansion* to write $a(D)$ as a sum of rational terms each of which has only one pole. Can this procedure be again defined in a purely formal way, i.e., only using algebraic operations but making no use of any analytic properties? Finally, use this partial fraction expansion to give an expression of the coefficients a_i in a more explicit form.

4. Let $\mathbb{F} = \mathbb{R}$ and consider the recurrence $(i+1)a_{i+1} = 3a_i$, $(i \geq 0; a_0 = 1)$. Define the formal power sum $a(D) = \sum_{i=0}^{\infty} a_i D^i$ and use it to solve this recursion.

5. Let $F(D), G(D) \in \mathbb{F}[[D]]$. We are interested in *compositions* of formal power sums. Can you find a meaningful definition for the expression $G(F(D))$? Does such an expression always make sense? Using your findings: Does $e^{e^D - 1}$ have a well defined formal power series? How about e^{e^D}?

D.2 (Inverse, Composition, and Derivative). Define the two formal power sums $F(D) = \sum_{i=0}^{\infty} \frac{1}{i!} D^i$ and $G(D) = -\sum_{i=1}^{\infty} \frac{(-1)^i}{i} D^i$, where all coefficients are over $\mathbb{R}$.

1. Do $\frac{1}{F(D)}$ and $F(G(D))$ exist? If so, determine their first three coefficients.

2. Show that $F'(D) = F(D)$ and $G'(D) = \frac{1}{1+D}$, where all operations are interpreted formally. [Recall that the formal derivative of a formal power sum $z(D) = \sum_{i=0}^{\infty} z_i D^i$ is equal to $\sum_{i=0}^{\infty} i z_i D^{i-1} = \sum_{i=0}^{\infty} (i+1) z_{i+1} D^i$.]

3. Find functions $f(D)$ and $g(D)$ such that $F(D)$ and $G(D)$ are their respective Taylor series around zero.

Hint: You might recognize the functions $f(D)$ and $g(D)$ from their respective Taylor series $\sum_{i=0}^{\infty} \frac{1}{i!} D^i$ and $-\sum_{i=1}^{\infty} \frac{(-1)^i}{i} D^i$ directly. If not, observe from earlier that $f(D)$ and $g(D)$ fulfill the equations $f'(D) = f(D)$ and $g'(D) = \frac{1}{1+D}$.

4. Use the preceding functions to write down $\frac{1}{F(D)}$ and $F(G(D))$ explicitly as formal power sums.

D.3 (Summation of Shifted Subsequences). As in Section D.3 consider a formal power sum over $\mathbb{R}$,

$$F(D) = \sum_{n \geq 0} F_n D^n,$$

so that the corresponding real sum $\sum_{n \geq 0} F_n$ converges. Express

$$F_{q,k}(D) = \sum_{n \geq 0 : (n+k)/q \in \mathbb{N}} F_n D^n,$$

and the related sum $\sum_{n \geq 0 : (n+k)/q \in \mathbb{N}} F_n$, where q is a natural number, in a compact way.

D.4 (Efficient Computation of Power of Polynomial – Wilf [18, p. 26]). Let $p(x)$ be a polynomial with $p_0 = 1$ and assume that we want to compute the first few coefficients of $p(x)^n$. If $q(x) = p(x)^n$, this means that we want to compute the first few coefficients of $q(x)$. Take the derivative of both sides of the equation $q(x) = p(x)^n$. The resulting equation reads $np'(x)q(x) = p(x)q'(x)$. Now use your mastery of formal power sums to conclude that the coefficient q_i can be computed recursively via $iq_i = \sum_{j=1}^{i} (j(n+1) - i) p_j q_{i-j}$. Apply this procedure to compute, for generic n and $\mathtt{r}$, the first three coefficients of $q(x)$ when $p(x) = (1+x)^{\mathtt{r}} - \mathtt{r}x$.

D.5 (Bound on Sum of Binomials – The Return). Prove the estimate $\sum_{k=0}^{m} \binom{n}{k} \leq 2^{n h_2(m/n)}$, $0 \leq m \leq n/2$, using the technique of Example D.4.

D.6 ($a(x)$ Is Increasing). Let $F(x)$ denote a polynomial with non-negative coefficients, $F_0 > 0$, and at least one further non-zero term. Define $a(x) = \frac{xF'(x)}{Fx}$. Show that $a'(x) > 0$ for $x > 0$.

Hint: Show, equivalently, that $xa'(x) > 0$ for $x > 0$.

D.7 (Rate versus Design Rate for Regular Ensembles). Show that $g(x, \mathtt{l})$ introduced in the proof of Lemma 3.27 on page 519 is a decreasing function in $\mathtt{l}$ and that it is strictly decreasing except for $x = 1$.

D.8 (Reduction of Trinomial to Binomial). Let $\alpha, \beta \geq 0$, with a strict inequality for α. Justify each of the following steps:

$$
\begin{aligned}
\text{coef}\{(1+\alpha x+\beta x^2)^n, x^k\} &\overset{\text{(i)}}{=} \sum_{j\leq k} \binom{n}{j}\binom{j}{k-j}\alpha^{2j-k}\beta^{k-j} \\
&\overset{\text{(ii)}}{\leq} \sum_{j\leq k} \binom{n}{j} \frac{j^{k-j}\alpha^{2j-k}\beta^{k-j}}{(k-j)!} \\
&\overset{\text{(iii)}}{\leq} \binom{n}{k} \sum_{j\leq k} \frac{\binom{n}{j}}{\binom{n}{k}} \frac{k^{k-j}\alpha^{2j-k}\beta^{k-j}}{(k-j)!} \\
&\overset{\text{(iv)}}{\leq} \binom{n}{k} \sum_{j\leq k} \frac{n^j k!}{j! n^k} e^{\frac{k^2}{n}} \frac{k^{k-j}\alpha^{2j-k}\beta^{k-j}}{(k-j)!} \\
&\overset{\text{(v)}}{=} \binom{n}{k} e^{\frac{k^2}{n}} \sum_j \binom{k}{j} \alpha^k \Big(\frac{k\beta}{n\alpha^2}\Big)^{k-j} \\
&\overset{\text{(vi)}}{=} \binom{n}{k} e^{\frac{k^2}{n}} \alpha^k \Big(1+\frac{k\beta}{n\alpha^2}\Big)^k \\
&\overset{\text{(vii)}}{\leq} \binom{n}{k} \alpha^k e^{\frac{k^2}{n}(1+\frac{\beta}{\alpha^2})} \\
&\overset{\text{(viii)}}{=} \text{coef}\{(1+\alpha x)^n, x^k\} e^{\frac{k^2}{n}(1+\frac{\beta}{\alpha^2})}.
\end{aligned}
$$

D.9 (Reduction of Polynomial with Four Terms to Binomial). Let $\alpha, \beta, \gamma \geq 0$, with strict inequality for α. Justify the following steps.

$$
\begin{aligned}
\text{coef}\{(1+\alpha x+\beta x^2+\gamma x^3)^n, x^k\} &\overset{\text{(i)}}{=} \sum_j \binom{n}{j}\gamma^j \text{coef}\{(1+\alpha x+\beta x^2)^{n-j}, x^{k-3j}\} \\
&\overset{\text{(ii)}}{\leq} \sum_j \binom{n}{j}\gamma^j \text{coef}\{(1+\alpha x+\beta x^2)^{n}, x^{k-3j}\} \\
&\overset{\text{(iii)}}{\leq} \sum_j \binom{n}{j}\gamma^j \text{coef}\{(1+\alpha x)^{n}, x^{k-3j}\} e^{\frac{k^2}{n}(1+\frac{\beta}{\alpha^2})} \\
&\overset{\text{(iv)}}{=} \binom{n}{k}\alpha^k e^{\frac{k^2}{n}(1+\frac{\beta}{\alpha^2})} \sum_j \frac{\binom{n}{j}\binom{n}{k-3j}}{\binom{n}{k}} \gamma^j \alpha^{-3j} \\
&\overset{\text{(v)}}{\leq} \binom{n}{k}\alpha^k e^{\frac{k^2}{n}(1+\frac{\beta}{\alpha^2})} e^{\frac{k^2}{n}} \sum_{0\leq j\leq k/3} \frac{n^j k! n^{k-3j}}{j! n^k (k-3j)!} \gamma^j \alpha^{-3j} \\
&\overset{\text{(vi)}}{\leq} \binom{n}{k}\alpha^k e^{\frac{k^2}{n}(2+\frac{\beta}{\alpha^2})} \sum_j \binom{k}{j} \Big(\frac{k^2\gamma}{n^2\alpha^3}\Big)^j
\end{aligned}
$$

$$\stackrel{\text{(vii)}}{=} \binom{n}{k}\alpha^k e^{\frac{k^2}{n}(2+\frac{\beta}{\alpha^2})}\left(1+\frac{k^2\gamma}{n^2\alpha^3}\right)^k$$

$$\stackrel{\text{(viii)}}{\leq} \binom{n}{k}\alpha^k e^{\frac{k^2}{n}(2+\frac{\beta}{\alpha^2})+\frac{k^3\gamma}{n^2\alpha^3}}$$

$$\stackrel{\text{(ix)}}{=} \text{coef}\{(1+\alpha x)^n, x^k\} e^{\frac{k^2}{n}(2+\frac{\beta}{\alpha^2})+\frac{k^3\gamma}{n^2\alpha^3}}.$$

D.10 (Reduction of General Polynomial to Binomial). Consider a polynomial $p(x) = 1 + \sum_{i\geq 1} p_i x^i$ with $p_1 > 0$ and $p_i \geq 0$, $i \geq 2$. Show that

$$\text{coef}\{(1+\sum_{i\geq 1} p_i x^i)^n, x^k\} \leq \text{coef}\{(1+p_1 x+\beta x^2+\gamma x^3)^n, x^k\}$$

$$\leq \text{coef}\{(1+p_1 x)^n, x^k\} e^{\frac{k^2}{n}(2+\frac{\beta}{p_1^2})+\frac{k^3\gamma}{n^2 p_1^3}},$$

where $\beta = p_2 + \sum_{i\geq 4} p_i \lfloor \frac{i}{2} \rfloor$ and $\gamma = p_3 + \sum_{i\geq 5: i \text{odd}} p_i$.

D.11 (Reduction of Codeword Polynomial to Binomial). Consider $p(x) = 1 + \sum_{i\geq 1}\binom{\mathtt{r}}{2i} x^{2i}$. Prove that

$$\text{coef}\{p(x)^n, x^k\} \leq \text{coef}\{(1+\binom{\mathtt{r}}{2}x^2)^n, x^k\} e^{\frac{k^2}{4n}(2+\frac{\beta}{\alpha^2})+\frac{k^3\gamma}{8n^2\alpha^3}},$$

where $\alpha = \binom{\mathtt{r}}{2}$, $\beta = \binom{\mathtt{r}}{4} + \sum_{i\geq 4}\binom{\mathtt{r}}{2i}\lfloor \frac{i}{2} \rfloor$, and $\gamma = \binom{\mathtt{r}}{6} + \sum_{i\geq 5: i \text{odd}}\binom{\mathtt{r}}{2i}$. ◇

D.12 (Reduction of Stopping Set Polynomial to Binomial). Consider $p(x) = 1 + \sum_{i\geq 2}\binom{\mathtt{r}}{i}x^i$. Show that for k even

$$\text{coef}\{p(x)^n, x^k\} \leq \text{coef}\{(1+\binom{\mathtt{r}}{2}x^2)^n, x^k\} e^{O(\sqrt{\frac{k^3}{n}})}.$$

D.13 (Bound on Weight Distribution). Consider the pair $(\lambda, \rho) \triangleq (L, R)$ with $\lambda'(0) > 0$ and no check nodes of degree 1. Show that

$$\mathbb{E}[A_{\text{cw}}(\mathtt{G}, w)] \leq \mathbb{E}[A_{\text{ss}}(\mathtt{G}, w)] = 111_{\max} w\left(\lambda'(0)\rho'(1)\right)^w e^{O(\sqrt{\frac{w^3}{n}})},$$

using the bound

$$\mathbb{E}[A_{\text{ss}}(\mathtt{G}, w)] \leq \min_{x,y,z>0} \sum_{e\geq 2w} \frac{p(x,y,z)^n}{x^w y^e z^e} \frac{1}{\binom{nL'(1)}{e}},$$

where $p(x,y,z) = \left(\prod_j (1+xy^j)^{L_j}\right)\left(\prod_i \left((1+z)^i - iz\right)^{R_i \bar{r}}\right)$, r is the design rate of the code, and $\bar{r} = 1 - r$.

Hint: Choose $x = \frac{w}{nL(y)}$, $y = \sqrt{\frac{1}{\kappa}\frac{w}{n}}$, where $\kappa = \frac{1}{2}\frac{L'(1)}{\rho'(1)1_{\max}}$, and $z = \sqrt{\frac{e}{2n\bar{r}\sum_{j\geq 2} R_j \binom{j}{2}}}$.

References

[1] E. A. Bender and L. B. Richmond, *Central and local limit theorems applied to asymptotic enumeration II: Multivariate generating functions*, J. Combin. Theory, A 34 (1983), pp. 255–265. [530]

[2] E. A. Bender, L. B. Richmond, and S. G. Williamson, *Central and local limit theorems applied to asymptotic enumeration III: Matrix recursions*, J. Combin. Theory, A 35 (1983), pp. 263–278. [367, 530]

[3] D. Burshtein and G. Miller, *Asymptotic enumeration methods for analyzing LDPC codes*, IEEE Trans. Inform. Theory, 50 (2004), pp. 1115–1131. [159, 529]

[4] N. G. de Brujin, *Asymptotic Methods in Analysis*, North Holland, Amsterdam, 3 ed., 1981. Reprinted by Dover, New York. [529]

[5] P. Debye, *Näherungsformeln für die Zylinder Funktionen für grosse Werte des Arguments und unbeschränkt veränderliche Werte des Index*, Math. Ann., (1909), pp. 535–558. [529]

[6] C. Di, *Asymptotic and Finite-Length Analysis of Low-Density Parity-Check Codes*, PhD thesis, EPFL, Lausanne, Switzerland, 2004. Number 3072. [160, 530]

[7] G. P. Egorychev, *Integral Representation and the Computation of Combinatorial Sums*, American Mathematical Society, 1984. [529]

[8] D. Gardy, *Some results on the asymptotic behavior of coefficients of large powers of functions*, Discrete Math., 139 (1995), pp. 189–217. [530]

[9] D. Gardy and P. Solé, *Saddle point techniques in asymptotic coding theory*. Technical report, CS-91-29, Apr. 1991. [530]

[10] R. L. Graham, D. E. Knuth, and O. Patashnik, *Concrete mathematics: A foundation for computer science*, Addison-Wesley, Reading, MA, USA, 1989. [529]

[11] A. Orlitsky, K. Viswanathan, and J. Zhang, *Stopping set distribution of LDPC code ensembles*, IEEE Trans. Inform. Theory, 51 (2004), pp. 929–953. [529]

[12] V. Rathi, *On the asymptotic weight and stopping set distribution of regular LDPC ensembles*, IEEE Trans. Inform. Theory, 52 (2006), pp. 4212–4218. [160, 530]

[13] B. Riemann, *Riemann's Gesammelte Mathematische Werke und Wissenschaftlicher Nachlass, hrsg. unter Mitwirkung von R. Dedekind und H. Weber*, Teubner, Leipzig, Germany, 2. Auflage ed., 1892. [529]

[14] I. Sason, E. Telatar, and R. Urbanke, *On the asymptotic input-output weight distributions and thresholds of convolutional and turbo-like encoders*, IEEE Trans. Inform. Theory, 48 (2002), pp. 3052–3061. [367, 530]

[15] R. Sedgewick and P. Flajolet, *An Introduction to the Analysis of Algorithms*, Addison-Wesley, Reading, MA, USA, 1996. [529]

[16] R. P. Stanley, *Enumerative Combinatorics*, vol. 1 of Cambridge Studies in Advanced Math., Cambridge Univ. Press, 1999. [529]

[17] ——, *Enumerative Combinatorics*, vol. 2 of Cambridge Studies in Advanced Math., Cambridge Univ. Press, 1999. [529]

[18] H. S. Wilf, *Generatingfunctionology*, Academic Press, 2 ed., 1994. [418, 529, 532]

[19] V. Zyablov and M. Pinsker, *Estimation of the error-correction complexity of Gallager low-density codes*, Problemy Peredachi Informatsii, 11 (1975), pp. 23–36. [261, 434, 529]

Appendix E

CONVEXITY, DEGRADATION, AND STABILITY

We collect here a couple of lengthy proofs.

Proof of Theorem 4.74. The proof proceeds (i) $\Rightarrow$ (ii), (ii) $\Rightarrow$ (iii), and (iii) $\Rightarrow$ (i).

The implication (i) $\Rightarrow$ (ii) is a consequence of (Jensen's inequality) (1.61). Let $p_{Y|X}$ and $p_{Z|X}$ denote the two BMS channels corresponding to $\mathfrak{A}(x)$ and $\mathfrak{B}(x)$, respectively. Assume that the output of both channels is in D-representation. Property (i) states that $X \to Y \to Z$, i.e., that $p_{Z|X}$ is degraded with respect to $p_{Y|X}$. As both $p_{Y|X}$ and $p_{Z|X}$ are symmetric, Lemma 4.73 assures us that there exists a *symmetric* channel $p_{Z|Y} : [-1,1] \to [-1,1]$ which effects this degradation. Let $p_{X,Y,Z}(x,y,z)$ denote the joint density. As always we assume that the input X to the BMS channel is uniform. Using this fact and the relationship $X \to Y \to Z$,

$$\begin{aligned}
p_{X,Y,Z}(1,y,z) &= \frac{1}{2} p_{Y|X}(y\,|\,1) p_{Z|Y}(z\,|\,y), \\
p_{X,Y,Z}(-1,y,z) &= \frac{1}{2} p_{Y|X}(y\,|\,-1) p_{Z|Y}(z\,|\,y) = \frac{1}{2} p_{Y|X}(-y\,|\,1) p_{Z|Y}(z\,|\,y) \\
&= \frac{1}{2} p_{Y|X}(y\,|\,1) \frac{1-y}{1+y} p_{Z|Y}(z\,|\,y),
\end{aligned}$$

where in the last step we have used the symmetry relation (4.13). We conclude that

$$\begin{aligned}
p_{Y,Z}(y,z) &= \sum_{x\in\{\pm 1\}} p_{X,Y,Z}(x,y,z) = \frac{1}{2} p_{Y|X}(y\,|\,1) \frac{2}{1+y} p_{Z|Y}(z\,|\,y) \\
&= \frac{2}{1-y} p_{X,Y,Z}(-1,y,z). \qquad \text{(E.1)}
\end{aligned}$$

Recall that the output Z of the channel $p_{Z|Y}$ is in D-representation. From Problem 4.59 we know that if the output of the channel is z then $(1-|z|)/2$ equals the error probability. If we assume that the output takes on the non-negative value z then the error probability can also be expressed as $p_{X|Z}(-1\,|\,z)$. We conclude that for $z \geq 0$, $(1-z)/2 = p_{X|Z}(-1\,|\,z)$. Therefore, for $z \geq 0$

$$\frac{1-z}{2} = p_{X|Z}(-1\,|\,z) = \frac{p_{X,Z}(-1,z)}{p_Z(z)} = \frac{\int_{-1}^{1} p(-1,y,z)\mathrm{d}y}{p_Z(z)}$$

$$\overset{(\text{E.1})}{=} \frac{\int_{-1}^{1}\left(\frac{1-y}{2}\right)p(y,z)\mathrm{d}y}{p_Z(z)} = \frac{1-\mathbb{E}[Y\,|\,z]}{2}.$$

We conclude that $\mathbb{E}[Y\,|\,z] = z$ for $z \geq 0$. The equivalent argument for $z \leq 0$ shows that $\mathbb{E}[Y\,|\,z] = z$ is valid also for $z \leq 0$, and hence for all z. This implies $\mathbb{E}[|Y|\,|\,z] \geq \mathbb{E}[Y\,|\,z] = z$. Further, since X has a uniform prior and since the channels $p_{Y\,|\,X}$ as well as $p_{Z\,|\,Y}$ are both symmetric it follows that $p_{Y\,|\,Z}(y\,|\,z) = p_{Y\,|\,Z}(-y\,|\,-z)$. Therefore, $\mathbb{E}[|Y|\,|\,z] = \mathbb{E}[|Y|\,|\,-z]$ so that we have $\mathbb{E}[|Y|\,|\,z] \geq |z|$. We write

$$f(|z|) \geq f(\mathbb{E}[|Y|\,|\,z]) \geq \mathbb{E}[f(|Y|)\,|\,z],$$

where the first step follows since $f(\cdot)$ is non-increasing and the second step is a consequence of (Jensen's inequality) (1.61) since $f(\cdot)$ is convex-$\cap$. Taking expectations with respect to Z on both sides we get

$$\mathbb{E}[f(|Z|)] \geq \mathbb{E}[f(|Y|)],$$

which is (ii).

Consider the implication (ii) $\Rightarrow$ (iii). Define $h(x,z) = (1-z) - (x-z)\mathbb{1}_{\{x \geq z\}}$. Fix $z \in [0,1]$ and note that $h(x,z)$ is non-increasing and convex-$\cap$ (as a function of x). Integrating by parts we get

$$\begin{aligned}\int_0^1 h(x,z)\mathrm{d}|\mathfrak{A}|(x) &= \int_0^1 ((1-z)-(x-z)\mathbb{1}_{\{x\geq z\}})\mathrm{d}|\mathfrak{A}|(x) \\ &= ((1-z)-(x-z)\mathbb{1}_{\{x\geq z\}})|\mathfrak{A}|(x)\,|_0^1 + \int_z^1 |\mathfrak{A}|(x)\mathrm{d}x \\ &= \int_z^1 |\mathfrak{A}|(x)\mathrm{d}x.\end{aligned}$$

By assumption we know that $\int_0^1 h(x,z)\mathrm{d}|\mathfrak{A}|(x) \leq \int_0^1 h(x,z)\mathrm{d}|\mathfrak{B}|(x)$. By the preceding derivation this implies $\int_z^1 |\mathfrak{A}|(x)\mathrm{d}x \leq \int_z^1 |\mathfrak{B}|(x)\mathrm{d}x$, as claimed.

Let us close the circle by proving that (iii) $\Rightarrow$ (i). This is the most difficult part of the proof. The proof uses a new symmetric channel representation. This representation takes the form $x \to (S,Z)$, where $S \in \{-1,+1\}$ and Z is uniform on $[0,1]$ and carries reliability information. Such a representation is naturally related to the $|D|$-distribution: let $x \xrightarrow{|\mathfrak{A}|} (S,Y)$ and assume that $|\mathfrak{A}|$ is smooth and monotonically increasing on $[0,1]$ with $|\mathfrak{A}|(0) = 0$ and $|\mathfrak{A}|(1) = 1$. In this case define $Z = |\mathfrak{A}|(Y)$. In the general case where $|\mathfrak{A}|$ has jumps, given Y, choose Z to be uniformly distributed on the interval $[|\mathfrak{A}|(Y^-), |\mathfrak{A}|(Y)]$ (define $|\mathfrak{A}|^{-1}(0^-) = 0$). It follows that Z is uniformly distributed on $[0,1]$. In this representation channel symmetry takes the form

$$p_{S\,|\,Z,X}(1\,|\,z,1) = p_{S\,|\,Z,X}(-1\,|\,z,-1).$$

We claim that if $|\mathfrak{A}|$ is continuous at the point $|\mathfrak{A}|^{-1}(z)$ then

$$p_{S|Z,X}(1\,|\,z,1) - p_{S|Z,X}(-1\,|\,z,1) = |\mathfrak{A}|^{-1}(z).$$

To see this, note that the left-hand side of the preceding equation is equal to one minus twice the error probability, assuming that the Z-reliability takes on the value z. Similarly, in the $|D|$-representation, the value y also represents one minus twice the error probability incurred assuming the $|D|$-reliability equals y. The claim now follows since by definition of the map $y = |\mathfrak{A}|^{-1}(z)$. To cover the general case (including points of discontinuity) set

$$|\mathfrak{A}|^{-1}(z) = \min\{y : |\mathfrak{A}|(y) \geq z\}.$$

Since $|\mathfrak{A}|$ is right continuous the minimum exists and $|\mathfrak{A}|^{-1}$ is right continuous.

Before proceeding to the general case let us consider a simple special case. Assume we have $|\mathfrak{A}| \leq |\mathfrak{B}|$, or, equivalently $|\mathfrak{A}|^{-1} \geq |\mathfrak{B}|^{-1}$. Assuming transmission over $\mathfrak{A}$, we have

$$p_{S|Z,X}(1\,|\,z,1) - p_{S|Z,X}(-1\,|\,z,1) = |\mathfrak{A}|^{-1}(z) \geq |\mathfrak{B}|^{-1}(z).$$

In this case we can degrade $(S, Z) \rightarrow (S', Z)$ by passing S through a BSC with cross-over probability depending on Z. The goal is to have

$$p_{S'|Z,X}(1\,|\,z,1) - p_{S'|Z,X}(-1\,|\,z,1) = |\mathfrak{B}|^{-1}(z).$$

We claim that the required cross-over probability is

$$\frac{1}{2}\left(1 - \frac{|\mathfrak{B}|^{-1}(z)}{|\mathfrak{A}|^{-1}(z)}\right).$$

To see where this map is coming from consider the simplest case: two BSCs, one with parameter ϵ and one with parameter ϵ', with $\epsilon < \epsilon'$. In this case the two $|D|$-distributions are step functions, one at $1 - 2\epsilon$ and the second one at $1 - 2\epsilon'$. Compute the cross-over probability according to the preceding formula for a point $z \in (0, 1]$: according to our definition we have for any such point $|\mathfrak{A}|^{-1}(z) = 1 - 2\epsilon$ and $|\mathfrak{B}|^{-1}(z) = 1 - 2\epsilon'$, so that the cross-over probability of the degrading channel has parameter $\Delta\epsilon = \frac{1}{2}(1 - \frac{1-2\epsilon'}{1-2\epsilon}) = \frac{\epsilon'-\epsilon}{1-2\epsilon}$. But we have already seen in Example 4.71 that this is the correct parameter, i.e., that the concatenation of the BSC(ϵ) with the BSC$(\Delta\epsilon)$ gives the BSC(ϵ'). The resulting channel is equivalent to $\mathfrak{B}$ (see Problem 4.30).

The general case is solved by reducing it to the simple case. The remainder of the proof is quite technical and is intended only for those that believe nothing

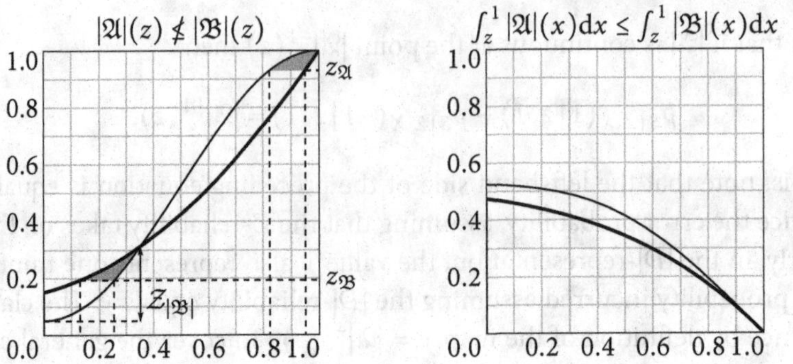

Figure E.2: Left: Two $|D|$-distributions $|\mathfrak{A}|$ (thick line) and $|\mathfrak{B}|$ (thin line). Right: Since $\int_z^1 |\mathfrak{A}|(x)\mathrm{d}x \leq \int_z^1 |\mathfrak{B}|(x)\mathrm{d}x$ we know that $|\mathfrak{A}| \rightarrow |\mathfrak{B}|$.

unless they have checked every step themselves. To help convey the ideas of the proof we use an example. Figure E.2 shows a plot of $|\mathfrak{A}|$ and $|\mathfrak{B}|$ satisfying (iii) but not satisfying $|\mathfrak{A}| \leq |\mathfrak{B}|$. Starting from the channel $|\mathfrak{A}|$ we construct a channel q, $(S, Z) \xrightarrow{q} (S, Z')$ such that Z' is uniform on $[0,1]$ and

$$p_{S\,|\,Z',X}(1\,|\,z',1) - p_{S\,|\,Z',X}(-1\,|\,z',1) \geq |\mathfrak{B}|^{-1}(z').$$

From there it can be degraded to $\mathfrak{B}$ as in the simple case. Let us define the set of "troublesome" points z (since these are the ones that prevent us from using the previous method), $\mathcal{Z}_{|\mathfrak{B}|}$,

$$\mathcal{Z}_{|\mathfrak{B}|} = \{z : |\mathfrak{A}|^{-1}(z) < |\mathfrak{B}|^{-1}(z)\}.$$

Since $|\mathfrak{A}|^{-1}(z) - |\mathfrak{B}|^{-1}(z)$ is right continuous it follows that $\mathcal{Z}_{|\mathfrak{B}|}$ is right open, i.e., if $z \in \mathcal{Z}_{|\mathfrak{B}|}$ then $[z, z+\epsilon) \subset \mathcal{Z}_{|\mathfrak{B}|}$ for some $\epsilon > 0$. The basic idea of the proof is to find for each $z_{\mathfrak{B}} \in \mathcal{Z}_{|\mathfrak{B}|}$ a unique companion point $z_{\mathfrak{A}}$, see Figure E.2, such that

$$|\mathfrak{A}|^{-1}(z_{\mathfrak{A}}) \geq |\mathfrak{B}|^{-1}(z_{\mathfrak{A}}) \geq |\mathfrak{B}|^{-1}(z_{\mathfrak{B}}) > |\mathfrak{A}|^{-1}(z_{\mathfrak{B}}). \tag{E.3}$$

The degrading channel q appropriately mixes the two z-values. For the point $z_{\mathfrak{B}}$ the error rate under $\mathfrak{A}$ is too high, while for $z_{\mathfrak{A}}$ it is lower than that for $z_{\mathfrak{B}}$ and lower than we require at $z_{\mathfrak{A}}$. We work our way down $\mathcal{Z}_{|\mathfrak{B}|}$ and exchange mass between $z_{\mathfrak{A}}$ and $z_{\mathfrak{B}}$ so that both

$$p_{S\,|\,Z,X}(1\,|\,z_{\mathfrak{B}},1) - p_{S\,|\,Z,X}(-1\,|\,z_{\mathfrak{B}},1) = |\mathfrak{B}|^{-1}(z_{\mathfrak{B}}), \quad \text{and} \tag{E.4}$$

$$p_{S\,|\,Z,X}(1\,|\,z_{\mathfrak{A}},1) - p_{S\,|\,Z,X}(-1\,|\,z_{\mathfrak{A}},1) = |\mathfrak{B}|^{-1}(z_{\mathfrak{A}}) \tag{E.5}$$

hold. Figure E.2 illustrates a particular $z_{\mathfrak{B}}$ with its matching $z_{\mathfrak{A}}$. The condition (iii) ensures that we will not run out of mass associated with $z_{\mathfrak{A}}$ prior to covering all of $\mathcal{Z}_{|\mathfrak{B}|}$. For each pair $(z_{\mathfrak{B}}, z_{\mathfrak{A}})$ we define ϵ_1 and ϵ_2 (depending on $(z_{\mathfrak{B}}, z_{\mathfrak{A}})$) so that

$$q_{Z'|Z}(z_{\mathfrak{A}} \,|\, z_{\mathfrak{B}}) = \epsilon_1, \qquad q_{Z'|Z}(z_{\mathfrak{B}} \,|\, z_{\mathfrak{B}}) = 1 - \epsilon_1,$$
$$q_{Z'|Z}(z_{\mathfrak{A}} \,|\, z_{\mathfrak{A}}) = 1 - \epsilon_2, \qquad q_{Z'|Z}(z_{\mathfrak{B}} \,|\, z_{\mathfrak{A}}) = \epsilon_2,$$

see Figure E.6. We will define $z_{\mathfrak{A}}(z_{\mathfrak{B}})$, ϵ_1, and ϵ_2 presently.

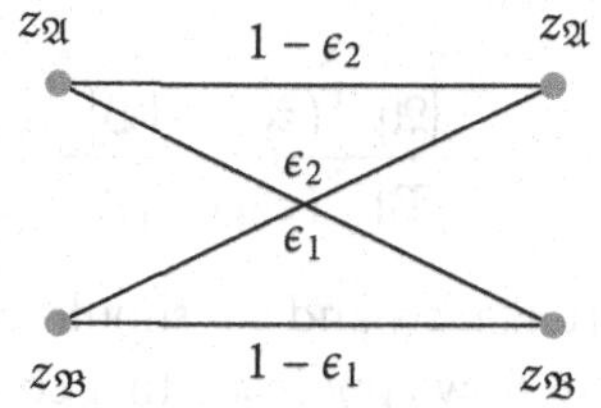

Figure E.6: Definition of q on $(z_{\mathfrak{B}}, z_{\mathfrak{A}})$ pair.

Let us first consider what constraints are placed on ϵ_1 and ϵ_2 so that Z' is uniformly distributed. To keep our exposition simple, let us assume that $z_{\mathfrak{A}}$ and $z_{\mathfrak{B}}$ are smoothly parameterized by a parameter α. Consider an infinitesimal $\mathrm{d}z_{\mathfrak{B}}$ and $\mathrm{d}z_{\mathfrak{A}}$ that are matched; i.e., the two intervals correspond to $\mathrm{d}\alpha$. Then the probability mass of Z' in the interval $\mathrm{d}z_{\mathfrak{A}}$ is $(1-\epsilon_2)\mathrm{d}z_{\mathfrak{A}} + \epsilon_1 \mathrm{d}z_{\mathfrak{B}}$ and the probability mass of Z' in the interval $\mathrm{d}z_{\mathfrak{B}}$ is $(1-\epsilon_1)\mathrm{d}z_{\mathfrak{B}} + \epsilon_2 \mathrm{d}z_{\mathfrak{A}}$. For Z' to be uniform therefore requires $\epsilon_2 \mathrm{d}z_{\mathfrak{A}} = \epsilon_1 \mathrm{d}z_{\mathfrak{B}}$ or

(E.7) $$\frac{\epsilon_2}{\epsilon_1} = \frac{\mathrm{d}z_{\mathfrak{A}}}{\mathrm{d}z_{\mathfrak{B}}}.$$

We also need to ensure that at the output of the channel q we have the desired reliability values as described by (E.4) and (E.5). Consider first the top output. It is the result of mixing a "mass" $\mathrm{d}z_{\mathfrak{A}}(1-\epsilon_2)$ with reliability $|\mathfrak{A}|^{-1}(z_{\mathfrak{A}})$ with a "mass" $\mathrm{d}z_{\mathfrak{B}}\epsilon_1$ corresponding to the reliability $|\mathfrak{A}|^{-1}(z_{\mathfrak{B}})$. By our previous discussion the total mass $\mathrm{d}z_{\mathfrak{A}}(1-\epsilon_2) + \mathrm{d}z_{\mathfrak{B}}\epsilon_1$ equals $\mathrm{d}z_{\mathfrak{A}}$. The two relative masses are therefore

$$\frac{\mathrm{d}z_{\mathfrak{A}}(1-\epsilon_2)}{\mathrm{d}z_{\mathfrak{A}}(1-\epsilon_2) + \mathrm{d}z_{\mathfrak{B}}\epsilon_1} = 1 - \epsilon_2, \qquad \frac{\mathrm{d}z_{\mathfrak{B}}\epsilon_1}{\mathrm{d}z_{\mathfrak{A}}(1-\epsilon_2) + \mathrm{d}z_{\mathfrak{B}}\epsilon_1} = \epsilon_2,$$

respectively. In a similar manner, the relative masses for the bottom output are ϵ_1 (for the branch from the top) and $1-\epsilon_1$, respectively. According to (E.4) and (E.5) we should have

(E.8) $$\epsilon_1|\mathfrak{A}|^{-1}(z_{\mathfrak{A}}) + (1-\epsilon_1)|\mathfrak{A}|^{-1}(z_{\mathfrak{B}}) = |\mathfrak{B}|^{-1}(z_{\mathfrak{B}}), \text{ and}$$

$$(1-\epsilon_2)|\mathfrak{A}|^{-1}(z_{\mathfrak{A}}) + \epsilon_2|\mathfrak{A}|^{-1}(z_{\mathfrak{B}}) = |\mathfrak{B}|^{-1}(z_{\mathfrak{A}}). \tag{E.9}$$

These equations can be solved uniquely for ϵ_1 and ϵ_2 and we take this solution as our definition:

$$\epsilon_1 = \frac{|\mathfrak{B}|^{-1}(z_{\mathfrak{B}}) - |\mathfrak{A}|^{-1}(z_{\mathfrak{B}})}{|\mathfrak{A}|^{-1}(z_{\mathfrak{A}}) - |\mathfrak{A}|^{-1}(z_{\mathfrak{B}})}, \qquad \epsilon_2 = \frac{|\mathfrak{A}|^{-1}(z_{\mathfrak{A}}) - |\mathfrak{B}|^{-1}(z_{\mathfrak{A}})}{|\mathfrak{A}|^{-1}(z_{\mathfrak{A}}) - |\mathfrak{A}|^{-1}(z_{\mathfrak{B}})}.$$

Both are well defined; i.e., both lie in $[0,1]$, if we can establish (E.3). From (E.7) and the previous equations we see that

$$\frac{dz_{\mathfrak{A}}}{dz_{\mathfrak{B}}} = \frac{\epsilon_2}{\epsilon_1} = \frac{|\mathfrak{A}|^{-1}(z_{\mathfrak{A}}) - |\mathfrak{B}|^{-1}(z_{\mathfrak{A}})}{|\mathfrak{B}|^{-1}(z_{\mathfrak{B}}) - |\mathfrak{A}|^{-1}(z_{\mathfrak{B}})}.$$

Thus, in the example in Figure E.2, $z_{\mathfrak{A}}$ and $z_{\mathfrak{B}}$ should be paired up so that the two shaded areas are equal. We can now appreciate that condition (iii) ensures that we will not run out of mass associated with $z_{\mathfrak{A}}$ prior to covering all of $\mathcal{Z}_{|\mathfrak{B}|}$. That a mapping from $z_{\mathfrak{A}}$ to $z_{\mathfrak{B}}$ which fulfills (E.3) is possible also in the general case is hopefully sufficiently clear that we can skip the tedious technical details.

The definition of q is now complete. Given (s,z) we check if $z = z_{\mathfrak{B}} \in \mathcal{Z}_{|\mathfrak{B}|}$. If so then we set $z' = z_{\mathfrak{B}}$ with probability $1-\epsilon_1$ and set $z' = z_{\mathfrak{A}}(z_{\mathfrak{B}})$ with probability ϵ_1. If $z = z_{\mathfrak{A}} \in \mathcal{Z}_{|\mathfrak{A}|}$ then we set $z' = z_{\mathfrak{A}}$ with probability $1-\epsilon_2$ and set $z' = z_{\mathfrak{B}}(z_{\mathfrak{A}})$ with probability ϵ_2. Finally, if $z \notin \mathcal{Z}_{|\mathfrak{A}|} \cup \mathcal{Z}_{|\mathfrak{B}|}$ then we set $z' = z$.

For all $z' \in \mathcal{Z}_{|\mathfrak{B}|} \cup \mathcal{Z}_{|\mathfrak{A}|}$ we have argued that

$$p_{S\,|\,Z',X}(1\,|\,z',1) - p_{S\,|\,Z',X}(-1\,|\,z',1) = |\mathfrak{B}|^{-1}(z').$$

If $z' \notin \mathcal{Z}_{|\mathfrak{B}|} \cup \mathcal{Z}_{|\mathfrak{A}|}$ then

$$\begin{aligned} &p_{S\,|\,Z',X}(1\,|\,z',1) - p_{S\,|\,Z',X}(-1\,|\,z',1) \\ &= p_{S\,|\,Z,X}(1\,|\,z',1) - p_{S\,|\,Z,X}(-1\,|\,z',1) \geq |\mathfrak{B}|^{-1}(z'). \end{aligned}$$ □

Proof of Lemma 4.75. Let us assume that the sequence of densities are $|D|$-densities $|\mathfrak{a}|_i$ and let $|\mathfrak{A}|_i$ be the associated cumulative distributions. By Lemma 4.25 there exists a limit $|\mathfrak{A}|_\infty$ for some subsequence. This is our candidate for the limit. By Theorem 4.74 (iii) the sequences $\int_z^1 |\mathfrak{A}|_i(x)\,dx$ are monotonic (in i) and hence converge for all $z \in [0,1]$. Clearly, the limit must be $\int_z^1 |\mathfrak{A}|_\infty(x)\,dx$.

It remains to show that the convergence of the integrals implies the convergence of the distributions themselves. Let $z \in (0,1)$ be a point of continuity of $|\mathfrak{A}|_\infty$. Then, for small $\epsilon > 0$,

$$|\mathfrak{A}|_\infty(z)\epsilon + o(\epsilon) = \int_{z-\epsilon}^{z} |\mathfrak{A}|_\infty(x)\,dx = \lim_{i\to\infty} \int_{z-\epsilon}^{z} |\mathfrak{A}|_i(x)\,dx$$

$$= \liminf_{i\to\infty} \int_{z-\epsilon}^{z} |\mathfrak{A}|_i(x)\, \mathrm{d}x \le \liminf_{i\to\infty} |\mathfrak{A}|_i(z)\epsilon,$$

where the last step follows since $|\mathfrak{A}|_i$ is non-decreasing. Letting ϵ tend to zero we obtain $|\mathfrak{A}|_\infty(z) \le \liminf_{i\to\infty} |\mathfrak{A}|_i(z)$. Similarly,

$$|\mathfrak{A}|_\infty(z)\epsilon + o(\epsilon) = \int_{z}^{z+\epsilon} |\mathfrak{A}|_\infty(x)\, \mathrm{d}x = \lim_{i\to\infty} \int_{z}^{z+\epsilon} |\mathfrak{A}|_i(x)\, \mathrm{d}x$$
$$= \limsup_{i\to\infty} \int_{z}^{z+\epsilon} |\mathfrak{A}|_i(x)\, \mathrm{d}x \ge \limsup_{i\to\infty} |\mathfrak{A}|_i(z)\epsilon.$$

Hence $|\mathfrak{A}|_\infty(z) \ge \limsup_{i\to\infty} |\mathfrak{A}|_i(z)$, and so $|\mathfrak{A}|_\infty(z) = \lim_{i\to\infty} |\mathfrak{A}|_i(z)$. □

Proof of Theorem 4.76. From Theorem 4.74 we know that if f is a non-increasing convex-∩ function then the implication holds.

To prove the inverse implication assume first that f is increasing. This means that there exist two points $x_1 < x_2$, $x_1, x_2 \in [0, 1]$, such that $f(x_1) < f(x_2)$. Let $|\mathfrak{A}|$ correspond to the BSC$((1 - x_2)/2)$ and let $|\mathfrak{B}|$ correspond to the BSC$((1 - x_1)/2)$. We know from Example 4.71 that the family $\{\text{BSC}(\epsilon)\}$ is ordered by degradation. Since $(1 - x_2)/2 < (1 - x_1)/2$, this proves that $\mathfrak{A} \prec \mathfrak{B}$. From page 183 we know that $|\mathfrak{B}|(x) = \mathsf{H}_{x_1}(x)$ and $|\mathfrak{A}|(x) = \mathsf{H}_{x_2}(x)$. Therefore,

$$\int_0^1 f(x)\, \mathrm{d}|\mathfrak{A}|(x) = f(x_2) > f(x_1) = \int_0^1 f(x)\, \mathrm{d}|\mathfrak{B}|(x).$$

On the other hand, if f is not convex-∩ then there exists a probability distribution $|\mathfrak{A}|$ such that

$$\int_0^1 f(x)\, \mathrm{d}|\mathfrak{A}|(x) > f\Big(\int_0^1 x\, \mathrm{d}|\mathfrak{A}|(x)\Big).$$

Consider the sequence $X \xrightarrow{|\mathfrak{A}|} (S, Y) \xrightarrow{q} (S)$, i.e., the channel q makes a hard decision on the received bit. The overall channel is therefore equivalent to a BSC. Let $|\mathfrak{B}|$ be this channel. If we receive the value y then the associated error probability is $(1 - |y|)/2$ (see Problem 4.59). Define $\alpha = \int_0^1 x\, \mathrm{d}|\mathfrak{A}|(x)$, so that $|\mathfrak{B}|$ corresponds to the BSC$((1 - \alpha)/2)$. Clearly $\mathfrak{A} \prec \mathfrak{B}$ and yet

$$\int_0^1 f(x)\, \mathrm{d}|\mathfrak{A}|(x) > f(\alpha) = \int_0^1 f(x)\, \mathrm{d}|\mathfrak{B}|(x).$$ □

Proof of Lemma 4.66. The right-hand side of (4.67) follows from the left bound in (4.65) and Lemma 4.63.

Let us show the left-hand side of (4.67) for $k = 1$. The basic idea is to express the probability of error in the (variable-node) Fourier domain. Recall from Definition 4.53 that for a symmetric L-density a we have

$$\mathfrak{E}(\mathsf{a}) = \frac{1}{2}\int \mathrm{e}^{-|x/2|}(\mathrm{e}^{-x/2}\mathsf{a}(x))\mathrm{d}x.$$

We view this as the inner product of the functions $\mathrm{e}^{-|x/2|}$ and $\mathrm{e}^{-x/2}\mathsf{a}(x)$ and apply the Parseval theorem. Let us denote the Fourier transform (restricted to the imaginary axis) of $\mathrm{e}^{-x/2}\mathsf{a}(x)$ by $\hat{\mathsf{a}}(f)$. Note that $\hat{\mathsf{a}}(f)$ is $\mathcal{F}_{\mathsf{a}}(\frac{1}{2} + 2\pi j f)$. Due to the symmetry of a, $\mathrm{e}^{-x/2}\mathsf{a}(x)$ is even and $\hat{\mathsf{a}}(f)$ is therefore real valued and even in f. The Fourier transform of $\mathrm{e}^{-|x/2|}$ is $\int \mathrm{e}^{-|x/2|}\mathrm{e}^{-2\pi j f x}\mathrm{d}x = \frac{4}{1+4(2\pi f)^2}$. Parseval's theorem gives us

$$\mathfrak{E}(\mathsf{a}) = \frac{1}{2}\int_{-\infty}^{\infty} \frac{4}{1+4(2\pi f)^2}\hat{\mathsf{a}}(f)\,\mathrm{d}f = \int_{0}^{\infty} \frac{4}{1+4(2\pi f)^2}\hat{\mathsf{a}}(f)\,\mathrm{d}f. \tag{E.10}$$

We want to bound $\hat{\mathsf{a}}(f)$ from below:

$$\begin{aligned}\hat{\mathsf{a}}(f) &= \int_{-\infty}^{\infty} \mathrm{e}^{-x/2}\mathsf{a}(x)\mathrm{e}^{-2\pi j f x}\,\mathrm{d}x = \int_{-\infty}^{\infty} \mathrm{e}^{-x/2}\mathsf{a}(x)\cos(2\pi f x)\,\mathrm{d}x \\ &\overset{\cos(x)\geq 1-x^2/2}{\geq} \int_{-\infty}^{\infty} \mathrm{e}^{-x/2}\mathsf{a}(x)(1-\frac{1}{2}(2\pi f x)^2)\,\mathrm{d}x \\ &= \mathfrak{B}(\mathsf{a}) - \int_{0}^{\infty} \mathrm{e}^{-x/2}\mathsf{a}(x)(2\pi f x)^2\,\mathrm{d}x \geq \mathfrak{B}(\mathsf{a}) - 16\mathrm{e}^{-2}(2\pi f)^2,\end{aligned}$$

where, for the last step, we used the inequality $\mathrm{e}^{-x/2}x^2 \leq 16\mathrm{e}^{-2}$, $x \geq 0$, and the fact that $\int_0^\infty \mathsf{a}(x)\mathrm{d}x \leq 1$. Recall that $\mathcal{F}_{\mathsf{a}^{\circledast d}} = \mathcal{F}_{\mathsf{a}}^d$ so that the Fourier transform of $\mathrm{e}^{-x/2}\mathsf{a}^{\circledast d}(x)$ is $\hat{\mathsf{a}}^d(f)$. If d is an even integer then $\hat{\mathsf{a}}^d(f)$ is positive. We then have

$$\hat{\mathsf{a}}^d(f) \geq \max\{0, \mathfrak{B}^d(\mathsf{a})(1-\frac{16\mathrm{e}^{-2}}{\mathfrak{B}(\mathsf{a})}(2\pi f)^2)^d\} \geq \mathfrak{B}^d(\mathsf{a})\Big(1-d\frac{16\mathrm{e}^{-2}}{\mathfrak{B}(\mathsf{a})}(2\pi f)^2\Big)^+.$$

Plugging this into (E.10) we get

$$\begin{aligned}\mathfrak{E}(\mathsf{a}^{\circledast d}) &= \int_0^\infty \frac{4}{1+4(2\pi f)^2}\hat{\mathsf{a}}^d(f)\,\mathrm{d}f \\ &\geq \int_0^\infty \frac{4}{1+4(2\pi f)^2}\mathfrak{B}^d(\mathsf{a})\Big(1-d\frac{16\mathrm{e}^{-2}}{\mathfrak{B}(\mathsf{a})}(2\pi f)^2\Big)^+\mathrm{d}f \\ &\overset{\frac{4\pi f}{\mathrm{e}}=u}{=} \frac{\mathrm{e}}{4\pi}\int_0^\infty \frac{4}{1+\mathrm{e}^2u^2}\mathfrak{B}^d(\mathsf{a})\Big(1-\frac{4d}{\mathfrak{B}(\mathsf{a})}u^2\Big)^+\mathrm{d}u \\ &\geq \frac{1}{1+\mathrm{e}^2\frac{\mathfrak{B}(\mathsf{a})}{4d}}\frac{\mathrm{e}}{\pi}\mathfrak{B}^d(\mathsf{a})\int_0^{\sqrt{\frac{\mathfrak{B}(\mathsf{a})}{4d}}}\Big(1-\frac{4d}{\mathfrak{B}(\mathsf{a})}u^2\Big)\mathrm{d}u\end{aligned}$$

$$= \frac{1}{1+e^{2\frac{\mathfrak{B}(\mathsf{a})}{4d}}} \frac{e}{\pi} \mathfrak{B}^d(\mathsf{a})\Big(\frac{2}{3}\sqrt{\frac{\mathfrak{B}(\mathsf{a})}{4d}}\Big) = \frac{2}{3\pi} \frac{(e^{2\frac{\mathfrak{B}(\mathsf{a})}{4d}})^{\frac{1}{2}}}{1+e^{2\frac{\mathfrak{B}(\mathsf{a})}{4d}}} \mathfrak{B}^d(\mathsf{a}).$$

To handle odd d we note that $\mathfrak{B}(\mathsf{a}) \leq 1$ and that $\mathfrak{E}(\mathsf{a}^{\circledast d})$ is non-decreasing in d and write

$$\mathfrak{E}(\mathsf{a}^{\circledast d}) \geq \frac{2}{3\pi} \frac{(e^{2\frac{\mathfrak{B}(\mathsf{a})}{4d}})^{\frac{1}{2}}}{1+e^{2\frac{\mathfrak{B}(\mathsf{a})}{4d}}} \mathfrak{B}^{d+1}(\mathsf{a}),$$

which then holds in general.

The proof for $k > 1$ follows essentially the same line, substituting $\mathfrak{B}_{\min}$ for $\mathfrak{B}(\mathsf{a}_i)$ at the appropriate places. □

Second Proof of Sufficiency in Theorem 4.125. On page 234 we gave a proof of Theorem 4.125. The proof uses the notion of extremes of information combining. Here is a second proof based on entirely different ideas.

The second proof uses the idea of minimal codewords of a tree. Consider the density evolution equations $\mathsf{a}_\ell = \mathsf{a}_{\text{BMSC}} \circledast \lambda(\rho(\mathsf{a}_{\ell-1}))$, $\ell \geq 1$, but initialized with $\mathsf{a}_0 = \mathsf{b}$, where b is a symmetric L-density. The claim is that if $\mathfrak{B}(\mathsf{a}_{\text{BMSC}})\lambda'(0)\rho'(1) < 1$, then there exists a strictly positive constant ξ so that if $\mathfrak{E}(\mathsf{b}) \leq \xi$ then $\mathfrak{E}(\mathsf{a}_\ell)$ converges to zero.

Consider the tree channel $(\vec{\mathcal{T}}_\ell, \mathsf{a}_{\text{BMSC}}, \mathsf{b})$. The message density emitted at the root node is a_ℓ. Let T_ℓ denote an element of the tree ensemble. Recall that on T_ℓ, the BP decoder is equivalent to a MAP decoder of the root node, i.e., it is optimal. Since we want to show that $\mathfrak{E}(\mathsf{a}_\ell)$ converges to zero and since the BP decoder on a tree is optimal it suffices if we can show that some (well-chosen suboptimal) decoder has a probability of error which converges to zero.

Since a_{BMSC} and b are symmetric we can assume that the all-one codeword was transmitted. Recall from Definition 3.45 the notion of a tree code. Without loss of generality we can assume that the channel outputs log-likelihood ratios. Let L_{v} denote the log-likelihood ratio associated to variable v, where in the sequel v goes over all variables contained in T_ℓ. Some thought shows that the message relating to the root node which the BP decoder computes can be expressed as

$$\ln \frac{\sum_{c \in C^+(\mathrm{T}_\ell)} \exp(\frac{1}{2}\sum_{\mathrm{v}\in \mathrm{T}_\ell} c_{\mathrm{v}} L_{\mathrm{v}})}{\sum_{c \in C^-(\mathrm{T}_\ell)} \exp(\frac{1}{2}\sum_{\mathrm{v}\in \mathrm{T}_\ell} c_{\mathrm{v}} L_{\mathrm{v}})}.$$

Therefore, the BP decoder makes a mistake if the numerator inside the log is smaller than the denominator. We upper bound this error probability by considering a suboptimal (in terms of bit error probability) decoder which compares the likelihood of the most likely codeword in $C^+(\mathtt{T}_\ell)$ with the likelihood of the most likely word in $C^-(\mathtt{T}_\ell)$ and extracts the value of the root node accordingly. We get a further upper bound if we compare the likelihood of the all-one word with the likelihood of the most likely word in $C^-(\mathtt{T}_\ell)$. We conclude that

$$\begin{aligned}\mathfrak{E}(\mathsf{a}_\ell) &\leq \mathbb{P}\{\sum_{\mathtt{v}\in\mathtt{T}_\ell} L_\mathtt{v} < \max_{c\in C^-(\mathtt{T}_\ell)} \sum_{\mathtt{v}\in\mathtt{T}_\ell} c_\mathtt{v} L_\mathtt{v}\} + \frac{1}{2}\mathbb{P}\{\sum_{\mathtt{v}\in\mathtt{T}_\ell} L_\mathtt{v} = \max_{c\in C^-(\mathtt{T}_\ell)} \sum_{\mathtt{v}\in\mathtt{T}_\ell} c_\mathtt{v} L_\mathtt{v}\} \\ &= \mathbb{P}\{\sum_{\mathtt{v}\in\mathtt{T}_\ell} L_\mathtt{v} < \max_{c\in C^-_{\min}(\mathtt{T}_\ell)} \sum_{\mathtt{v}\in\mathtt{T}_\ell} c_\mathtt{v} L_\mathtt{v}\} + \frac{1}{2}\mathbb{P}\{\sum_{\mathtt{v}\in\mathtt{T}_\ell} L_\mathtt{v} = \max_{c\in C^-_{\min}(\mathtt{T}_\ell)} \sum_{\mathtt{v}\in\mathtt{T}_\ell} c_\mathtt{v} L_\mathtt{v}\} \\ &= \mathbb{P}\{\min_{c\in C^-_{\min}(\mathtt{T}_\ell)} \sum_{\mathtt{v}:c_\mathtt{v}=-1} L_\mathtt{v} < 0\} + \frac{1}{2}\mathbb{P}\{\min_{c\in C^-_{\min}(\mathtt{T}_\ell)} \sum_{\mathtt{v}:c_\mathtt{v}=-1} L_\mathtt{v} = 0\}.\end{aligned}$$

The second-to-last step needs a justification. Assume that all most likely codewords are elements of $C^-(\mathtt{T}_\ell)$. Let c denote one of them. Either $c \in C^-_{\min}(\mathtt{T}_\ell)$, i.e., it is minimal *itself*, or it *contains* a minimal codeword. In either case, denote this minimal codeword by $c_{\min}$. Define $\tilde{c} = c \cdot c_{\min}$ so that $c = c_{\min} \cdot \tilde{c}$, where the multiplication is component-wise. Note that $\tilde{c} \in C^+(\mathtt{T}_\ell)$ and that $c_{\min}$ and $\tilde{c}$ do not have any positions in common where they take on the value -1. We claim that

$$\sum_{\mathtt{v}\in\mathtt{T}_\ell} L_\mathtt{v} < \sum_{\mathtt{v}\in\mathtt{T}_\ell} c_{\min,\mathtt{v}} L_\mathtt{v}.$$

In words, if all most likely codewords are elements of $C^-(\mathtt{T}_\ell)$ then there exists a minimal codeword which is strictly more likely than the all-one codeword. This is seen as follows. Since by assumption all most likely codewords are elements of $C^-(\mathtt{T}_\ell)$, we know that c is strictly more likely than $\tilde{c}$, which is an element of $C^+(\mathtt{T}_\ell)$. Therefore,

$$\sum_{\mathtt{v}\in\mathtt{T}_\ell} c_\mathtt{v} L_\mathtt{v} = \sum_{\mathtt{v}\in\mathtt{T}_\ell} c_{\min,\mathtt{v}} \tilde{c}_\mathtt{v} L_\mathtt{v} > \sum_{\mathtt{v}\in\mathtt{T}_\ell} \tilde{c}_\mathtt{v} L_\mathtt{v}.$$

We conclude that $\sum_{\mathtt{v}\in\mathtt{T}_\ell}(c_{\min,\mathtt{v}} - 1)\tilde{c}_\mathtt{v} L_\mathtt{v} > 0$. Since $c_{\min}$ and $\tilde{c}$ do not share a position in which both of them take on the value -1 it follows that

$$\sum_{\mathtt{v}\in\mathtt{T}_\ell : c_{\min,\mathtt{v}}=-1} c_{\min,\mathtt{v}} L_\mathtt{v} = \frac{1}{2}\sum_{\mathtt{v}\in\mathtt{T}_\ell}(c_{\min,\mathtt{v}} - 1)\tilde{c}_\mathtt{v} L_\mathtt{v} > 0.$$

The claim now follows from

$$\begin{aligned}\sum_{\mathtt{v}\in\mathtt{T}_\ell} c_{\min,\mathtt{v}} L_\mathtt{v} &= \sum_{\mathtt{v}\in\mathtt{T}_\ell : c_{\min,\mathtt{v}}=+1} c_{\min,\mathtt{v}} L_\mathtt{v} + \sum_{\mathtt{v}\in\mathtt{T}_\ell : c_{\min,\mathtt{v}}=-1} c_{\min,\mathtt{v}} L_\mathtt{v} \\ &> \sum_{\mathtt{v}\in\mathtt{T}_\ell : c_{\min,\mathtt{v}}=+1} c_{\min,\mathtt{v}} L_\mathtt{v} - \sum_{\mathtt{v}\in\mathtt{T}_\ell : c_{\min,\mathtt{v}}=-1} c_{\min,\mathtt{v}} L_\mathtt{v} = \sum_{\mathtt{v}\in\mathtt{T}_\ell} L_\mathtt{v}.\end{aligned}$$

In the event that the set of most likely codewords contains both elements from $C^-(\mathtt{T}_\ell)$ as well as $C^+(\mathtt{T}_\ell)$ then essentially the same argument shows that the inequality stays valid but might no longer be strict.

Consider the random variable

$$X_\ell = \min_{c \in C^-_{\min}(\mathtt{T}_\ell)} \sum_{\mathtt{v} \in \mathtt{T}_\ell : c_\mathtt{v} = -1} L_\mathtt{v},$$

where the randomness resides both in the choice of the tree $\mathtt{T}_\ell$ as well as the realization of the channel represented by the log-likelihood ratios $\{L_\mathtt{v}\}_{\mathtt{v} \in \mathtt{T}_\ell}$. Denote its distribution by B_ℓ, $\mathsf{B}_\ell(x) = \mathbb{P}\{X_\ell \leq x\}$, and its density by b_ℓ. Let us derive a recursion for B_ℓ. We claim that we have

$$\mathsf{b}_{\ell+1} = \mathsf{a}_{\text{BMSC}} \circledast \lambda(\rho'(1 - \mathsf{B}_\ell)\mathsf{b}_\ell), \quad \mathsf{b}_0 = \mathsf{b}.$$

This can be seen as follows. Consider a variable node v and its subtree and assume that v has associated value -1. We then want to compute the sum of the log-likelihoods associated to v itself and the minimum of the sum of log-likelihoods computed over all of its minimal subtrees. Assume for the moment that v has degree $\mathtt{l}$ and that each of its connected child check nodes have degree $\mathtt{r}$. For each of its child check nodes the incoming cumulative distribution is $1 - (1 - \mathsf{B}_{\ell-1})^{\mathtt{r}-1}$, since it is the minimum of $\mathtt{r} - 1$ incoming independent random variables. If we average over the edge degree distribution and take the derivative to get the density we see that the density entering the variable node v along a particular edge is $\rho'(1 - \mathsf{B}_\ell)$. At the variable node itself we add $\mathtt{l} - 1$ such independent samples and add a sample form the received distribution. This gives the stated formula.

Since this recursion is hard to handle we introduce the following simpler one,

$$\tilde{\mathsf{b}}_{\ell+1} = \mathsf{a}_{\text{BMSC}} \circledast \lambda(\rho'(1)\tilde{\mathsf{b}}_\ell), \quad \tilde{\mathsf{b}}_0 = \mathsf{b}.$$

Since $\tilde{\mathsf{b}}_0 = \mathsf{b}_0 = \mathsf{b}$, it is not hard to check that $\tilde{\mathsf{b}}_\ell \geq \mathsf{b}_\ell$ pointwise for each $\ell \geq 0$. This follows from the fact that λ and ρ are polynomials with non-negative coefficients. Note that $\tilde{\mathsf{b}}_\ell$ is in general *not* a probability density (it is non-negative but does not in general integrate to 1).

We now show that if $\mathfrak{B}(\mathsf{a}_{\text{BMSC}})\lambda'(0)\rho'(1) < 1$ then there exists a strictly positive constant η such that if $\mathfrak{B}(\mathsf{b}) < \eta$ then $\lim_{\ell \to \infty} \mathfrak{B}(\tilde{\mathsf{b}}_\ell) \to 0$. Let us first see how this proves the claim. The right-hand side of inequality (4.65) states that $\mathfrak{B}(\mathsf{b}) \leq 2\sqrt{\mathfrak{E}(\mathsf{b})(1 - \mathfrak{E}(\mathsf{b}))}$. Thus, there exists $\xi > 0$ such that $\mathfrak{E}(\mathsf{b}) < \xi$ implies $\mathfrak{B}(\mathsf{b}) < \eta$. By our claim this implies that $\mathfrak{B}(\tilde{\mathsf{b}}_\ell) \to 0$ as $\ell \to \infty$. The proof now follows from the sequence of inequalities

$$\mathfrak{B}(\tilde{\mathsf{b}}_\ell) \geq \mathfrak{B}(\mathsf{b}_\ell) \geq \mathfrak{E}(\mathsf{b}_\ell) \geq \mathfrak{E}(\mathsf{a}_\ell),$$

where in the second step we use the left-hand side of inequality (4.65).

The proof of the claim proceeds by induction. Define

$$\xi_\ell = \mathfrak{B}(\mathsf{a}_{\text{BMSC}})(\lambda_2 + (1-\lambda_2)\rho'(1)\,\mathfrak{B}(\tilde{\mathsf{b}}_\ell))\rho'(1).$$

Note that since $\mathfrak{B}(\mathsf{a}_{\text{BMSC}})\lambda'(0)\rho'(1) < 1$ there exists a strictly positive η such that $\xi_0 = \mathfrak{B}(\mathsf{a}_{\text{BMSC}})(\lambda_2 + (1-\lambda_2)\rho'(1)\eta)\rho'(1) < 1$. We claim that $\mathfrak{B}(\tilde{\mathsf{b}}_\ell) \leq \xi_0^\ell\,\mathfrak{B}(\mathsf{b})$. This not only shows that $\mathfrak{B}(\tilde{\mathsf{b}}_\ell)$ converges to 0 but it shows that the convergence is exponential in ℓ.

Let us first show that for $\ell \geq 0$, (i) $\xi_\ell < 1$, and (ii) $\rho'(1)\,\mathfrak{B}(\tilde{\mathsf{b}}_\ell) < 1$. We have already seen that the first condition is fulfilled for $\ell = 0$ and we can redefine η if necessary so that also the second condition is fulfilled at the beginning. This serves as our anchor.

Assume that (i) and (ii) hold for some $\ell \geq 0$. We will show that they then also hold for $\ell + 1$. Since $\lambda(1) = 1$ and (ii) holds we conclude that

$$\lambda(\rho'(1)\,\mathfrak{B}(\tilde{\mathsf{b}}_\ell)) \leq (\lambda_2 + (1-\lambda_2)\rho'(1)\,\mathfrak{B}(\tilde{\mathsf{b}}_\ell))\rho'(1)\,\mathfrak{B}(\tilde{\mathsf{b}}_\ell).$$

From Lemma 4.63 we know that the Bhattacharyya operator is multiplicative at the variable-node side. Combined with the previous inequality, and recalling the definition of ξ_ℓ, this gives

$$\mathfrak{B}(\tilde{\mathsf{b}}_{\ell+1}) \leq (\lambda_2 + (1-\lambda_2)\rho'(1)\,\mathfrak{B}(\tilde{\mathsf{b}}_\ell))\rho'(1)\,\mathfrak{B}(\tilde{\mathsf{b}}_\ell)\,\mathfrak{B}(\mathsf{a}_{\text{BMSC}}) = \xi_\ell\,\mathfrak{B}(\tilde{\mathsf{b}}_\ell).$$

Therefore, $\mathfrak{B}(\tilde{\mathsf{b}}_{\ell+1}) \leq (\prod_{i=0}^{\ell} \xi_i)\,\mathfrak{B}(\mathsf{b}) \leq \mathfrak{B}(\mathsf{b})$. Now recall that by induction hypothesis $\xi_i < 1$ for $0 \leq i \leq \ell$. The previous inequality then shows that both (i) and (ii) also hold for $\ell + 1$.

Further, it follows that ξ_ℓ is non-increasing so that we have the bound $\mathfrak{B}(\tilde{\mathsf{b}}_\ell) \leq (\prod_{i=0}^{\ell-1} \xi_i)\,\mathfrak{B}(\mathsf{b}) \leq \xi_0^\ell\,\mathfrak{B}(\mathsf{b})$, as claimed. □

Proof of Lemma 4.165. Since the derivatives in (4.166) are known to exist almost everywhere, the lemma is in fact equivalent to saying that, for any $\mathtt{h}_i' \geq \mathtt{h}_i$,

$$H(X_i \,|\, Y_i(\mathtt{h}_i'), Y_{\sim i}) - H(X_i \,|\, Y_i(\mathtt{h}_i), Y_{\sim i}) \leq H(X_i \,|\, Y_i(\mathtt{h}_i'), \Phi_i) - H(X_i \,|\, Y_i(\mathtt{h}_i), \Phi_i).$$

By definition the family $\{\text{BMSC}(\mathtt{h}_i)\}$ is ordered by degradation: $X_i \to Y_i(\mathtt{h}_i) \to Y_i(\mathtt{h}_i')$. Also, $\Phi_i = \phi_i(Y_{\sim i})$, i.e., Φ_i is a function of $Y_{\sim i}$. Finally, since the channel is memoryless, we know that $(Y_i(\mathtt{h}_i), Y_i(\mathtt{h}_i')) \to X_i \to (Y_{\sim i}, \Phi_i)$. The thesis is therefore a consequence of Lemma E.11 by making the substitutions $X_i \mapsto X$, $Y_i(\mathtt{h}_i) \mapsto Y$, $Y_i(\mathtt{h}_i') \mapsto Y'$, $Y_{\sim i} \mapsto Z$, $\Phi_i \mapsto Z'$. □

Lemma E.11. Consider the random variables X, Y, Y', Z, Z'. Assume that $X \to Y \to Y'$, $X \to Z \to Z'$, and $(Y, Y') \to X \to (Z, Z')$. Then

$$H(X \mid Y', Z) - H(X \mid Y, Z) \leq H(X \mid Y', Z') - H(X \mid Y, Z'). \tag{E.12}$$

Proof. We verify (E.12) by proving the following sequence of steps:

$$\begin{aligned} H(X \mid Y', Z) - H(X \mid Y, Z) &\overset{\text{(i)}}{\leq} H(X \mid Y', Z, Z') - H(X \mid Y, Y', Z, Z') \\ &\overset{\text{(ii)}}{\leq} H(X \mid Y', Z') - H(X \mid Y, Y', Z') \\ &\overset{\text{(iii)}}{=} H(X \mid Y', Z') - H(X \mid Y, Z'). \end{aligned}$$

Consider claim (i). Since conditioning can only decrease entropy it suffices to show that $H(X \mid Y', Z) = H(X \mid Y', Z, Z')$. Equivalently, we can show that $I(X; Y', Z) = I(X; Y', Z, Z')$. In turn this is equivalent to $I(X; Z \mid Y') = I(X; Z, Z' \mid Y')$. Now note that $I(X; Z, Z' \mid Y') = I(X, Y'; Z, Z') - I(Y'; Z, Z')$. First consider $I(X, Y'; Z, Z')$. Since $(Y, Y') \to X \to (Z, Z')$ we have $I(X, Y'; Z, Z') = I(X; Z, Z')$. Further, since $X \to Z \to Z'$ it follows that $I(X; Z, Z') = I(X; Z)$, so that $I(X, Y'; Z, Z') = I(X; Z)$. Next consider $I(Y'; Z, Z')$. Note that $I(Y, Y'; Z, Z' \mid X) = 0$ since $(Y, Y') \to X \to (Z, Z')$. Expand $I(Y, Y'; Z, Z' \mid X)$ to get

$$0 = I(Y, Y'; Z, Z' \mid X) = I(Y, Y'; Z \mid X) + I(Y'; Z' \mid X, Z) + I(Y; Z' \mid Y', X, Z).$$

Since mutual information is non-negative, each part of this expansion must be zero. This shows that $I(Y'; Z' \mid X, Z) = 0$. Using $X \to Z \to Z'$, we see that $I(Y'; Z' \mid X, Z) = I(X, Y'; Z' \mid Z) = I(Y'; Z' \mid Z) + I(X; Z' \mid Y', Z)$. This shows that $I(Y'; Z' \mid Z) = 0$. It follows that $I(Y'; Z, Z') = I(Y'; Z) + I(Y'; Z' \mid Z) = I(Y'; Z)$. Combining the two statements we see that $I(X; Z, Z' \mid Y') = I(X; Z) - I(Y'; Z)$. Clearly, $I(X; Z \mid Y') \leq I(X; Z, Z' \mid Y')$. On the other hand, $I(X; Z \mid Y') = I(X, Y'; Z) - I(Y'; Z) \geq I(X; Z) - I(Y'; Z) = I(X; Z, Z' \mid Y')$. This shows that we have in fact equality.

Step (iii) requires us to prove that $H(X \mid Y, Z') = H(X \mid Y, Y', Z')$. By symmetry (of the assumptions and the statement) the proof is identical to the previous one if we exchange the roles of (Y, Y') and (Z, Z').

It remains to prove step (ii). It can alternatively be written as $I(Y; X \mid Y', Z, Z') \leq I(Y; X \mid Y', Z')$. Expand $I(Y; X, Z \mid Y', Z')$ both ways to get

$$\begin{aligned} I(Y; X, Z \mid Y', Z') &= I(Y; X \mid Y', Z') + I(Y; Z \mid Y', X, Z') \\ &= I(Y; Z \mid Y', Z') + I(Y; X \mid Y', Z, Z'). \end{aligned}$$

If we can show that $I(Y; Z \mid Y', X, Z') = 0$ then our claim follows from the previous equality by noting that mutual information is non-negative, so that in particular

$I(Y;Z\,|\,Y',Z') \geq 0$. We proceed similarly as for step (i). Expand $I(Y,Y';Z,Z'\,|\,X)$ to get

$$0 = I(Y,Y';Z,Z'\,|\,X) = I(Y,Y';Z'\,|\,X) + I(Y';Z\,|\,X,Z') + I(Y;Z\,|\,Y',X,Z').$$

We conclude that $I(Y;Z\,|\,Y',X,Z') = 0$. □

AUTHORS

INDEX

Printed in the United States
By Bookmasters